FUEL CELL CATALYSIS

WILEY SERIES ON ELECTROCATALYSIS AND ELECTROCHEMISTRY

Andrzej Wieckowski, Series Editor

Fuel Cell Catalysis: A Surface Science Approach, Edited by Marc T. M. Koper

Electrochemistry of Functional Supramolecular Systems, Margherita Venturi, Paola Ceroni, and Alberto Credi

Catalysis in Electrochemistry: From Fundamentals to Strategies for Fuel Cell Development, Elizabeth Santos and Wolfgang Schmickler

Fuel Cell Science: Theory, Fundamentals, and Biocatalysis, Andrzej Wieckowski and Jens Norskov

FUEL CELL CATALYSIS
A SURFACE SCIENCE APPROACH

Edited by

Marc T. M. Koper

The Wiley Series on Electrocatalysis and Electrochemistry

A JOHN WILEY & SONS, INC., PUBLICATION

Copyright © 2009 by John Wiley & Sons, Inc. All rights reserved

Published by John Wiley & Sons, Inc., Hoboken, New Jersey
Published simultaneously in Canada

No part of this publication may be reproduced, stored in a retrieval system, or transmitted in any form or by any means, electronic, mechanical, photocopying, recording, scanning, or otherwise, except as permitted under Section 107 or 108 of the 1976 United States Copyright Act, without either the prior written permission of the Publisher, or authorization through payment of the appropriate per-copy fee to the Copyright Clearance Center, Inc., 222 Rosewood Drive, Danvers, MA 01923, (978) 750-8400, fax (978) 750-4470, or on the web at www.copyright.com. Requests to the Publisher for permission should be addressed to the Permissions Department, John Wiley & Sons, Inc., 111 River Street, Hoboken, NJ 07030, (201) 748-6011, fax (201) 748-6008, or online at http://www.wiley.com/go/permission.

Limit of Liability/Disclaimer of Warranty: While the publisher and author have used their best efforts in preparing this book, they make no representations or warranties with respect to the accuracy or completeness of the contents of this book and specifically disclaim any implied warranties of merchantability or fitness for a particular purpose. No warranty may be created or extended by sales representatives or written sales materials. The advice and strategies contained herein may not be suitable for your situation. You should consult with a professional where appropriate. Neither the publisher nor author shall be liable for any loss of profit or any other commercial damages, including but not limited to special, incidental, consequential, or other damages.

For general information on our other products and services or for technical support, please contact our Customer Care Department within the United States at (800) 762-2974, outside the United States at (317) 572-3993 or fax (317) 572-4002.

Wiley also publishes its books in variety of electronic formats. Some content that appears in print may not be available in electronic formats. For more information about Wiley products, visit our web site at www.wiley.com.

Library of Congress Cataloging-in-Publication Data:

Fuel cell catalysis : a surface science approach / edited by Marc Koper.
    p. cm.—(The Wiley series on electrocatalysis and electrochemistry)
  Includes index.
  ISBN 978-0-470-13116-9 (cloth)
 1. Electrocatalysis. 2. Fuel cells. 3. Solid-liquid interfaces. I. Koper, Marc.
  QD569.F84 2009
  621.31′2429—dc22

2009004193

Cover image courtesy of Matthew Neurock

Printed in the United States of America
10 9 8 7 6 5 4 3 2 1

CONTENTS

Preface to the Wiley Series on Electrocatalysis and Electrochemistry ... vii

Preface ... ix

List of Contributors ... xi

1. **Electrocatalysis of Oxygen Reduction in Polymer Electrolyte Fuel Cells: A Brief History and a Critical Examination of Present Theory and Diagnostics** ... 1
   Shimshon Gottesfeld

2. **Electrochemical Electron Transfer: From Marcus Theory to Electrocatalysis** ... 31
   E. Santos and W. Schmickler

3. **Electrocatalysis and Catalyst Screening from Density Functional Theory Calculations** ... 57
   J. Rossmeisl, J. Greeley, and G.S. Karlberg

4. **First-Principles Simulation of the Active Sites and Reaction Environment in Electrocatalysis** ... 93
   Michael J. Janik, Sally A. Wasileski, Christopher D. Taylor, and Matthew Neurock

5. **Ab Initio Atomistic Thermodynamics for Fuel Cell Catalysis** ... 129
   Timo Jacob

6. **Mechanisms of the Oxidation of Carbon Monoxide and Small Organic Molecules at Metal Electrodes** ... 159
   Marc T.M. Koper, Stanley C.S. Lai, and Enrique Herrero

7. **Clues for the Molecular-Level Understanding of Electrocatalysis on Single-Crystal Platinum Surfaces Modified by *p*-Block Adatoms** ... 209
   V. Climent, N. García-Aráez, and J.M. Feliu

8. **Electrochemistry at Well-Characterized Bimetallic Surfaces** ... 245
   Vojislav R. Stamenkovic and Nenad M. Markovic

9. **Recent Developments in the Electrocatalysis of the O_2 Reduction Reaction** — 271
   Ye Xu, Minhua Shao, Manos Mavrikakis, and Radoslav R. Adzic

10. **Electrocatalysis at Platinum and Bimetallic Alloys** — 317
    Masahiro Watanabe and Hiroyuki Uchida

11. **Electrocatalysis for the Direct Alcohol Fuel Cell** — 343
    J.-M. Leger, C. Coutanceau, and C. Lamy

12. **Broadband Sum Frequency Generation Studies of Surface Intermediates Involved in Fuel Cell Electrocatalysis** — 375
    G.Q. Lu, A. Lagutchev, T. Takeshita, R.L. Behrens, Dana D. Dlott, and A. Wieckowski

13. **Methanol, Formaldehyde, and Formic Acid Adsorption/Oxidation on a Carbon-Supported Pt Nanoparticle Fuel Cell Catalyst: A Comparative Quantitative DEMS Study** — 411
    Z. Jusys and R.J. Behm

14. **The Effect of Structurally Well-Defined Pt Modification on the Electrochemical and Electrocatalytic Properties of Ru(0001) Electrodes** — 465
    H.E. Hoster and R.J. Behm

15. **Size Effects in Electrocatalysis of Fuel Cell Reactions on Supported Metal Nanoparticles** — 507
    Frederic Maillard, Sergey Pronkin, and Elena R. Savinova

16. **Support and Particle Size Effects in Electrocatalysis** — 567
    Brian E. Hayden and Jens-Peter Suchsland

17. **Electrocatalysis for Fuel Cells at Enzyme-Modified Electrodes** — 593
    K.A. Vincent, S.C. Barton, G.W. Canters, and H.A. Heering

18. **Metalloporphyrin Catalysts of Oxygen Reduction** — 637
    Roman Boulatov

Index — 695

PREFACE to the Wiley Series on Electrocatalysis and Electrochemistry

This series covers recent advances in electrocatalysis and electrochemistry and depicts prospects for their contribution into the present and future of the industrial world. It illustrates the transition of electrochemical sciences from a solid chapter of physical electrochemistry (covering mainly electron transfer reactions, concepts of electrode potentials and structure of the electrical double layer) to the field in which electrochemical reactivity is shown as a unique chapter of heterogeneous catalysis, is supported by high-level theory, connects to other areas of science, and includes focus on electrode surface structure, reaction environment, and interfacial spectroscopy.

The scope of this series ranges from electrocatalysis (practice, theory, relevance to fuel cell science and technology) to electrochemical charge transfer reactions, biocatalysis, and photoelectrochemistry. While individual volumes may look quite diverse, the series promises updated and overall synergistic reports on insights to further the understanding of properties of electrified solid/liquid systems. Readers of the series will also find strong reference to theoretical approaches for predicting electrocatalytic reactivity by such high-level theories as DFT. Beyond the theoretical perspective, further vehicles for growth are the sound experimental background and demonstration of significance of such topics as energy storage, syntheses of catalytic materials via rational design, nanometer-scale technologies, prospects in electrosynthesis, new instrumentation, surface modifications in basic research on charge transfer, and related interfacial reactivity. In this context, readers will notice that new methods that are being developed for a specific field may be readily adapted for application in others.

Electrochemistry has benefited from numerous monographs and review articles due to its unique character and significance in the practical world (including electroanalysis). Electrocatalysis has also been the subject of individual reviews and compilations. The Wiley Series on Electrocatalysis and Electrochemistry is dedicated to be complementary with respect to the current activity by focusing each volume on a specific topic of choice. The chapters also demonstrate electrochemistry's connections to other areas of chemistry and physics, such as biochemistry, chemical engineering, quantum mechanics, chemical physics, surface science, and to biology, and illustrate the wide range of literature that each topic contains. While the title of each volume informs of the specific focus chosen by the volume editors and chapter authors, the integral outcome offers a broad-based analysis of the total development of the field. The progress of the series will provide a global definition of what electrocatalysis and electrochemistry are concerned with now and how they evolve with time.

The purpose is manifold, mainly to provide a modern reference for graduate instruction and for active researchers in the two disciplines, as well as to document that electrocatalysis and electrochemistry are dynamic fields that expand rapidly and likewise rapidly change in their scientific profiles.

Creation of each volume required the editor's involvement, vision, enthusiasm, and time. The Series Editor thanks all Volume Editors who graciously accepted his invitations. Special thanks are for Ms. Anita Lekhwani, the Series Acquisition Editor, who extended the invitation to the Series Editor and is a wonderful help in the Series assembling process.

ANDRZEJ WIECKOWSKI

Series Editor

PREFACE

The collection of chapters in this first volume of the "Electrocatalysis and Electrochemistry" series originates from a workshop organized by Andrzej Wieckowski, Jens Nørskov, and myself at the Lorentz Center of Leiden University, Leiden, The Netherlands, from October 16–20, 2006. The "Lorentz Workshop" on "Fuel Cell Catalysis: A Surface Science Approach" (http://www.lorentzcenter.nl/lc/web/2006/220/info.php3?wsid=220) brought together some 70 electrochemists and fuel cell scientists from Europe, North America and Asia to discuss and identify major research themes in the "surface science approach" to catalytic reactions at the solid–liquid interface relevant to fuel cells. Emphasis was on a molecular-level description of catalysis for low-temperature polymer-electrolyte membrane fuel cells, both hydrogen–oxygen fuel cells and direct alcohol fuel cells, based on well-defined systems probed with state-of-the-art experimental and theoretical tools. The Workshop in Leiden served as a continuation of two earlier workshops. The first one, organized in Denmark in July 2003 by Wieckowski and Nørskov, was entitled "Theory and Surface Measurements of Fuel Cell Catalysis." That workshop was one of the first scientific meetings that brought together experimentalists and theoreticians to discuss molecular-level understanding of fuel cell reactions, and clearly stimulated and identified some important research questions that are still reflected in the current scientific literature. A second one-day workshop was organized by Wieckowski in May 2005, as a satellite meeting to the Electrochemical Society Meeting in Quebec City, Canada, and also provided a discussion forum dedicated the interplay between theory and experiment in fuel cell catalysis. The idea for the Leiden Fuel Cell Catalysis workshop was born in Quebec City, and finally materialized at the unique and special facilities of the Lorentz Center at Leiden University.

The 18 chapters in this book were written by leading experts in the field, and are to a large extent based on the presentations given by the authors at the Lorentz Workshop. The first chapter by Shimshon Gottesfeld, who was actually the last speaker at the Workshop, gives an historical account of fuel cell catalysis, but also includes the author's reflections on some of issues discussed at the meeting, in particular in relation to the oxygen reduction reaction. The rest of the book is divided into five parts. Chapters 2–5 constitute the "Theory Part" of the book, discussing the classical, quantum-mechanical, and statistical-mechanical aspects of the modeling of electrocatalytic reactions, from Marcus theory and its various extensions to large-scale first-principles quantum-chemical simulations based on Density Functional Theory. Chapters 6–11 give an experimental account of the suggested mechanisms of some of the most important fuel cell reactions, including oxygen reduction, carbon monoxide oxidation, and alcohol oxidation. Emphasis in these chapters is on well-defined and well-characterized monometallic and bimetallic electrode surfaces, probed by both electrochemical and spectroscopic techniques. Chapters 12–14 are primarily devoted to the application of a specific high-level technique or strategy, which have a strong surface-science character, to interrogate some of the pertinent issues in fuel cell catalysis. This includes broadband sum frequency generation as an extremely powerful spectroscopic technique, the detailed quantitative application of Differential Electrochemical Mass Spectrometry, and the combination of UHV preparation techniques with voltammetry allowing for an atomic-level control and insight into electrocatalytic reactions. The next two chapters, comprising the fourth part of the book, are dedicated to the properties and electrocatalytic activity of supported metal nanoparticles. Size effects in electrocatalysis are central to these chapters and were intensely discussed during the meeting. Finally, the fifth and last part of the book consists of two chapters devoted to a bio-inspired approach to fuel cell catalysis, based on natural enzymes or smaller molecular catalysts inspired by enzymes, and their immobilization on electrode surfaces. These modified electrodes offer a unique level of molecular control and unique catalytic properties, and may ultimately serve as leads or as a source of inspiration for cheaper fuel cell catalysts containing a smaller amount of precious metal.

I would like to acknowledge the financial support from the Lorentz Center, Leiden University, the Netherlands Organization for Scientific Research (NWO), the Delft Institute for Sustainable Energy (DIDE), the Royal Netherlands Academy for Arts and Sciences, the International Society of Electrochemistry, Toyota, Johnson Matthey, the Energy Research Centre of the Netherlands (ECN), Umicore, and CMR Fuel Cells for their sponsorship of the Workshop, Martje Kruk and Gerda Filippo of the Lorentz Center for their friendly help and efficient support, and my co-organizers Andrzej Wieckowski and Jens Nørskov for helping me setting up the program. I would also like to thank all the authors of the chapters in this book for their efforts and contributions, which I believe together form an excellent account of the state-of-the-art of our molecular understanding of fuel cell catalytic reactions. Finally, I would like to express my gratitude to Anita Lekhwani, Senior Acquisitions Editor at John Wiley & Sons, and to Andrzej Wieckowski, the Series Editor, for the pleasant collaboration and for their continuous enthusiasm and patience in the preparation of this book.

MARC T. M. KOPER

Leiden, August 2008

Note: Several figures in this work that appear in black and white throughout this print version may be viewed in color free of charge on ftp://ftp.wiley.com/public/sci_tech_med/advanced_biomaterials

LIST OF CONTRIBUTORS

Barton S.C., Department of Chemical Engineering and Materials Science, Michigan State University, East Lansing, MI 48824, USA

Behm R.J., Institute of Surface Chemistry and Catalysis, Ulm University, D-89069 Ulm, Germany

Behrens R.L., Department of Chemistry, University of Illinois at Urbana-Champaign, Urbana, IL 61801, USA

Brian E. Hayden, School of Chemistry, University of Southampton, Southampton SO17 1BJ, UK

Canters G.W., Leiden Institute of Chemistry, Leiden University, Einsteinweg 55, 2333 CC Leiden, The Netherlands

Christopher D. Taylor, Materials Science and Technology Division, Los Alamos National Laboratory, Los Alamos, NM 87545, USA

Climent V., Instituto de Electroquímica, Universidad de Alicante, Ap. 99, E-03080 Alicante, Spain

Coutanceau C., Laboratoire de Catalyse en Chimie Organique, UMR 6503 CNRS, Equipe Electrocatalyse, Université de Poitiers, 40 Avenue du Recteur Pineau, F-86022 Poitiers, France

Dana D. Dlott, Department of Chemistry, University of Illinois at Urbana-Champaign, Urbana, IL 61801, USA

Elena R. Savinova, LMSPC UMR-7515, ECPM, Louis Pasteur University, 67087 Strasbourg, France

Enrique Herrero, Instituto de Electroquimica, Universidad de Alicante, Apartado 99, E-03080, Alicante, Spain

Feliu J.M., Instituto de Electroquímica, Universidad de Alicante, Ap. 99, E-03080 Alicante, Spain

Frederic Maillard, Laboratoire d'Electrochimie et de Physico-chimie des Matériaux et des Interfaces, UMR 5631 CNRS/INPG/UJF, 1130 rue de la piscine, BP75-38402 Saint Martin d'Hères, France

GarcíA-Ará Ez N., Instituto de Electroquímica, Universidad de Alicante, Ap. 99, E-03080 Alicante, Spain

Greeley J., Center for Nanoscale Materials, Argonne National Laboratory, Argonne, IL 60439, USA

Heering H.A., Kavli Institute of NanoScience, Delft University of Technology, Lorentzweg 1, 2628, CJ Delft,The Netherlands; and Leiden Institute of Chemistry

Hiroyuki Uchida, Clean Energy Research Center, University of Yamanashi, Takeda 4, Kofu 400-8510, Japan

Hoster H.E., Institute of Surface Chemistry and Catalysis, Ulm University, D-89069 Ulm, Germany

Jens-Peter Suchsland, School of Chemistry, University of Southampton, Southampton SO17 1BJ, UK

Jusys, Z., Institute of Surface Chemistry and Catalysis, Ulm University, D-89069 Ulm, Germany

Karlberg G.S., Center for Atomic-Scale Materials Design, NanoDTU, Department of Physics, Technical University of Denmark, DK-2800 Kongens Lyngby, Denmark

Lagutchev A., Department of Chemistry, University of Illinois at Urbana-Champaign, Urbana, IL 61801, USA

Lamy C., Laboratoire de Catalyse en Chimie Organique, UMR 6503 CNRS, Equipe Electrocatalyse, Université de Poitiers, 40 Avenue du Recteur Pineau, F-86022 Poitiers, France

Leger J.-M., Laboratoire de Catalyse en Chimie Organique, UMR 6503 CNRS, Equipe Electrocatalyse, Université de Poitiers, 40 Avenue du Recteur Pineau, F-86022 Poitiers, France

Lu G.Q., Department of Chemistry, University of Illinois at Urbana-Champaign, Urbana, IL 61801, USA

Manos Mavrikakis, Department of Chemical and Biological Engineering, University of Wisconsin-Madison, Madison, WI 53705, USA

Marc T.M. Koper Leiden Institute of Chemistry, Leiden University, PO Box 9502, 2300 RA Leiden, The Netherlands

Masahiro Watanabe, Clean Energy Research Center, University of Yamanashi, Takeda 4, Kofu 400-8510, Japan

Matthew Neurock, Departments of Chemical Engineering and Chemistry, University of Virginia, Charlottesville, VA 19904, USA

Michael J. Janik, Department of Chemical Engineering, Pennsylvania State University, University Park, PA 16802, USA

Minhua Shao, Chemistry Department, Brookhaven National Laboratory, Upton, NY 11973, USA

Nenad M. Markovic, Argonne National Laboratory, Materials Science Division, University of Chicago, Argonne, IL 60439, USA

Radoslav R. Adzic, Chemistry Department, Brookhaven National Laboratory, Upton, NY 11973, USA

Roman Boulatov, Department of Chemistry, University of Illinois, 600 South Mathews Avenue, Urbana, IL 61801, USA

Rossmeisl J., Center for Atomic-Scale Materials Design, NanoDTU, Department of Physics, Technical University of Denmark, DK-2800 Kongens Lyngby, Denmark

Sally A. Wasileski, Department of Chemistry, University of North Carolina at Asheville, Asheville, NC 28804, USA

Santos E., Faculdad de Matemática, Astronomía y Física, Universidad Nacional de Cordoba, Cordoba, Argentina, and Zentrum für Sonnenenergie und Wasserstoff-Forschung Helmholtzstrasse 8 D-89081 Ulm, Germany

Schmickler W., Institute of Theoretical Chemistry, University of Ulm, D-89069 Ulm, Germany

Sergey Pronkin, LMSPC UMR-7515, ECPM, Louis Pasteur University, 67087 Strasbourg, France

Shimshon Gottesfeld, CTO, Cell ERA Technologies, Caesarea, Israel and President, Fuel Cell Consulting, LLC Niskayuna, NY, USA

Stanley C.S. Lai, Leiden Institute of Chemistry, Leiden University, PO Box 9502, 2300 RA Leiden, The Netherlands

Takeshita T., Department of Chemistry, University of Illinois at Urbana-Champaign, Urbana, IL 61801, USA

Timo Jacob, Fritz-Haber-Institut der Max-Planck-Gesellschaft, Faradayweg 4–6, Berlin, D-14915 Germany

Vincent K.A., Inorganic Chemistry Laboratory, University of Oxford, South Parks Road, Oxford OX1 3QR, UK

Vojislav R. Stamenkovic, Argonne National Laboratory, Materials Science Division, University of Chicago, Argonne, IL 60439, USA

Wieckowski A., Department of Chemistry, University of Illinois at Urbana-Champaign, Urbana, IL 61801, USA

Ye Xu, Center for Nanophase Materials Sciences and Chemical Sciences Division, Oak Ridge National Laboratory, Oak Ridge, TN 37831, USA

CHAPTER 1

Electrocatalysis of Oxygen Reduction in Polymer Electrolyte Fuel Cells: A Brief History and a Critical Examination of Present Theory and Diagnostics

SHIMSHON GOTTESFELD

Cellera Technologies, Caesarea, Israel

1.1 INTRODUCTION

The title that I chose for my talk in Leiden was "Electrocatalysis—a scan over 20 years + one great meeting." Although the superlative used in the name of the talk was chosen well before the meeting itself, it will be seen from this chapter that the meeting indeed provided significant further food for thought, at least in the case of this participant, regarding the fundamentals of fuel cell electrocatalysis. Of particular interest has been the dialogue between the recent advances in the theory of electrocatalysis and the corresponding experimental data collected to date, that should serve as a basis for solid mechanistic conclusions and define a road map for the future. This chapter will first describe, in line with its original assignment, key milestones in the development of electrocatalysts for low temperature fuel cells, leading next to a discussion of the state of the art in the field. Both the historical comments and, particularly, the latter part of the chapter, devoted to a critical examination of the state of the art, are focused on the oxygen reduction reaction (ORR). The ORR remains the most serious challenge to both fuel cell experimentalists and theorists, as is evident from the major contribution that the air cathode loss makes at present to the overall voltage loss of a polymer electrolyte fuel cell (PEFC), or any other low temperature fuel cell. This is despite the significant experimental and, particularly, theoretical advances in air cathode electrocatalysis in PEFCs that have been made since the year 2000.

Fuel Cell Catalysis. Edited by Marc T. M. Koper
Copyright © 2009 John Wiley & Sons, Inc.

Specific aspects examined here include insights and conclusions derived from the most recently performed density functional theory (DFT) calculations, which have been based on a comprehensive model of the electrochemical interface, and the strong disagreements (which seem to defy all recent theoretical efforts) that remain regarding proper interpretation of experimental ORR results and proper identification of the ORR mechanism in a PEFC cathode employing Pt catalysts.

1.2 KEY MILESTONES ON THE WAY TO THE PRESENT STATE OF THE ART OF FUEL CELL ELECTROCATALYSIS

When surveying the central milestones in the development of electrocatalysis for low temperature fuel cells operating in acidic environments, the following, listed in chronological order, seem to be the most outstanding:

- Establishment of carbon-supported Pt catalysts as a means to achieve higher and more stable dispersion of the precious metal electrocatalyst on an electronically conducting support [Petrow and Allen, 1977].
- Implementation of Pt/C catalysts in PEFC technology using recast Nafion® as a proton conducting and bonding agent [Raistrick, 1986; Wilson and Gottesfeld, 1992].
- Optimization of the catalyst layer composition and thickness in PEFCs for maximum catalyst utilization in operation on air and on impure hydrogen feed streams [Wilson, 1993; Springer et al., 1993].
- Advancing from carbon-supported Pt to carbon-supported Pt alloy catalysts, to enhance the performance per milligram Pt by three- to four-fold [Mukerjee and Srinivasan, 1993; Mukerjee et al., 1995].
- Moving on from preparation of homogenous Pt alloy particles to tailoring of core-and-shell alloy particles, targeting (i) further lowering of the mass of precious metal per unit power output and (ii) further boost of catalytic activity per square centimeter of catalyst area [Zhang et al., 2004].

Two parallel efforts common to all of the above critical steps and milestones are (i) maximizing of catalyst dispersion and enhancing electrochemical utilization of the overall surface area of the catalyst incorporated in the fuel cell electrode and (ii) further fine tuning of the electronic and, consequently, surface chemistry properties of Pt catalysts by alloying, typically with electropositive metals such as Co or Ni, to achieve higher activity per unit surface area of the optimized alloy catalysts. The former part of the effort led to established methods of preparation of carbon-supported Pt alloy particles in the diameter range of 2–5 nm, i.e., of overall catalyst surface area between 600 and 1500 cm^2 Pt (or Pt alloy) per square centimeter of electrode cross-sectional area achieved at a mass loading of only 1 mg Pt (or Pt alloy) per square centimeter of cross-sectional area. Such high "electrode surface amplification factors" are enabled by carefully selected mild reduction processes that generate, typically from chloroplatinic acid, a large number of nanometer-size Pt metal nuclei on the carbon support, while slowing down the rate of excessive Pt crystallite growth. Alloying of the Pt nanoparticles is typically achieved by reaction of the carbon-supported Pt crystallites with added oxide of the electropositive metal, e.g., Co or Ni, and results in some increase in size of the catalyst particles, typically from around 2 nm in unalloyed form to 4–5 nm in the alloyed form. These high electrode surface area amplification factors, of the order of 10^2-10^3, have clearly been key for achieving the specific activity of hydrogen/air PEFC electrodes, enabling lowering of the loading required in fuel cell stacks to only 0.2 g Pt catalyst per kilowatt of power generated [Gasteiger, 2005].

Particle size effect: Since the total surface area of the dispersed catalyst available at some given mass loading (in milligrams per geometric square centimeter, mg/cm^2 geo.), is inversely proportional to the particle diameter, further reduction of the Pt particle size to less than 2 nm has been pursued by several groups, with associated efforts to modify the carbon support so as to reduce inter-Pt particle distance (see Chapter 10). However, this route to further increase of the Pt catalyst surface area has encountered difficulties, interpreted by most researchers by the lower intrinsic activity obtainable per cm^2 Pt at such ultrasmall particle sizes. This lower catalyst surface activity reported for such small particle sizes, particularly in the oxygen reduction process, has been understood to be the result of a larger fraction of Pt atoms located in edge site, rather than in terraces of catalytically preferred Pt crystal surfaces, or in steps that could enable multisite interaction. Strong evidence for the benefit of trading off ultrahigh surface area for a "smoother" catalyst surface morphology has been provided by the unique development at 3M Company of electrocatalyst layers based on sputter coating by Pt of an array of micrometer-long inert dendrites that are subsequently embedded into the ionomeric membrane [Debe et al., 2003]. Activities per cm^2 Pt obtained with this structure of a supported catalyst are about five times higher than those recorded for 2 nm size Pt particles supported on carbon, fully compensating for the similar loss in Pt surface area per unit mass compared with the case of 2 nm supported Pt particles. It should be noted here that a dissenting opinion in this regard has been expressed through the years by Watanabe and co-workers, namely that the loss of activity with drop in Pt particle size has only to do with exacerbated, localized mass transport limitations within the structure of the catalyst layer, caused by low interparticle distance (see Chapter 10). Whereas this dissenting opinion suggests that further optimization of catalyst layer composition and structure could enable the use of Pt particles of diameters smaller than 2 nm, i.e., enable higher surface area per unit mass of Pt, most researchers in the field have accepted the conclusion that reducing particle size to below 2–3 nm does not enhance the activity per unit mass of Pt, because of intrinsic surface catalysis reasons, apparently to do with the surface atomic structure of such very small Pt particles.

Catalyst layer architecture: As a consequence of the diminishing returns from ever higher dispersion, the effort to increase the active catalyst surface area per unit mass of Pt has centered in recent years primarily on optimization of catalyst *layer* properties, aiming to maximize "catalyst utilization" in fuel cell electrodes based on Pt catalyst particle sizes of 2–5 nm. High catalyst utilization is conditioned on access to the largest possible percentage of the total catalyst surface area embedded in a catalyst

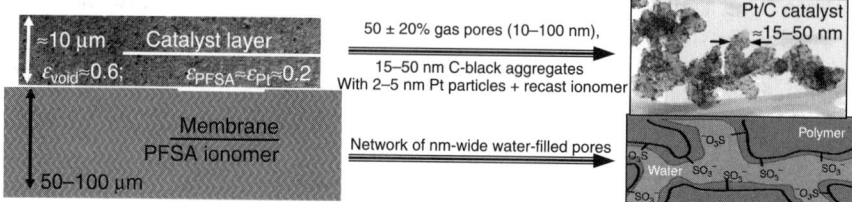

Figure 1.1 The nature of a composite, Pt/C/recast ionomer layer with a structure that enables high electronic and gas mobilities as well as sufficient proton mobility [Gasteiger, 2005].

layer by the three participants in the electrochemical process—gaseous reactant, protons, and electrons—all at the rates called for by the demand current density. Fulfilling the latter condition requires a composite catalyst layer structure in which the electron mobility, the proton mobility, and the effective gas diffusivity across the thickness dimension of the catalyst layer are all sufficient, at the demand current, to access the maximum fraction of the catalyst particles dispersed uniformly in the (5–20 μm thick) catalyst layer. The type of structure satisfying high catalyst utilization in catalyst layers of PEFCs is shown in Fig. 1.1.

The figure shows catalyst layer porosity at both micrometer and nanometer scales. The larger pores form as a result of the highly structured nature of the carbon support, that prevents closer packing of the submicrometer carbon particles, facilitating gas transport through the catalyst layer. It is easy to see that the carbon structure obtained will support electron percolation with the roughly 30% volume fraction occupied by the carbon particles. However, achieving higher proton mobility is usually a bigger challenge. As can be seen from Fig. 1.1, the proton-conducting component of the catalyst layer, typically applied as a solution of the ionomer which recasts around carbon particles [Gasteiger, 2005], needs to reach a volume fraction that would satisfy the ionic conductivity demand, but, at the same time, would leave the network of micropores sufficiently open to gas transport. The specific conductivity of the ionomer is several orders of magnitude smaller than that of the carbon and, as a result, the performance of the membrane/electrode assembly will typically be limited by the effective *protonic* conductivity within the catalyst layer, thereby diminishing the effective overpotential at catalyst particles located away from the surface of the ionomeric membrane [Springer et al., 1993]. A reverse situation, in which catalyst particles away from the gas diffusion layer are the least well utilized, will apply when the gas permeability through the catalyst layer becomes the main transport limiting factor.

The most recent improvements in Pt catalyst utilization U by optimization of catalyst layer composition and structure have led to catalyst utilizations as high as 80%, or more, determined as the ratio between measured ORR current per geometric square centimeter of electrode area and the current expected from the total measured Pt surface area per geometric square centimeter of the electrode, i.e.,

$$U = \frac{J_{ORR}(E, T)}{A_{Pt}^* J_{ORR}^*(E, T)} \quad (1.1)$$

where $J^*_{ORR}(E, T)$ is the ORR current per cm^2 Pt at cathode potential E, determined by an independent measurement at a Pt electrode of known surface area (and well-defined mass transport characteristics) and A^*_{Pt} is the number of cm^2 Pt per geometric cm^2 of the fuel cell cathode, determined from the hydrogen adsorption charge or the CO desorption charge as measured by voltammetry. The ability to achieve catalyst utilizations as high as 80% in PEFC cathodes (even more easily achieved on the anode side) by optimizing the composition and structure of the catalyst layer, means that this degree of freedom in enhancing further the net electrocatalytic activity has been exhausted to a significant degree. This said, some remaining opportunity for further enhancing the rate of the cathode process through better catalyst utilization has been argued very recently, based on (i) higher utilizations determined from Reaction (1.1) demonstrated for types of carbon support material of higher surface area [Toyota R&D Center, 2007] and (ii) the argument that A^*_{Pt}, the "electrochemically active" surface area determined by voltammetry, may not include the area of Pt catalyst particles dislodged from the carbon support during catalyst layer preparation. This argument implies that the performance of the air cathode in the fuel cell can likely be further improved, at some given temperature, by a factor of about 2 with further perfection of the carbon/PFSA ionomer composite catalyst layer composition and mode of fabrication. Such a performance gain is not insignificant, but, at the same time, leaves ample room for further, larger reductions of air electrode losses.

Pt demand per kW versus PEFC performance and cost targets: The presently achieved PEFC air cathode (initial) performance with Pt/C catalysts is basically defined, as explained above, by a Pt cathode catalyst surface area of 600–1500 cm^2/mg Pt and by the intrinsic ORR catalytic activity at the (bulk) Pt/ionomer interface at some given temperature and ionomer hydration level, $J_{ORR}(V_{cath}, T_{cell}, RH)$/cm^2 Pt. At the lower end of catalyst dispersion specified above, the intrinsic ORR activity per cm^2 Pt is maintained similar to that of the bulk metal. Reading the quality of catalytic activity obtained state-of-the-art dispersed Pt/C catalyst is done in somewhat different ways from a PEFC technology perspective and from an ORR electrocatalysis science perspective. From a technology implementation perspective, the question is: "How close is this PEFC technology based on carbon-supported Pt catalyst to market-entry targets of PEFC stack performance and cost?" From the perspective of electrocatalysis fundamentals, the question is: "With the cathode remaining the largest source of PEFC voltage loss, by far, what can be further done to improve performance by moving away from the Pt/C ORR catalyst to another catalyst of higher intrinsic electrocatalytic activity?" Starting from the technology perspective, in a PEFC operating at 80 °C, the electrocatalytic activity derived from state-of-the-art catalyst dispersion and catalyst utilization described above, the PEFC cathode activity measured at $V_{cath} = 0.90$ V is 0.11 A per mg Pt and increases 10-fold at $V_{cath} = 0.84$ V [Gasteiger, 2005]. This would translate to nearly 1 W per mg Pt at $V_{cath} = 0.84$ V, enabling one to achieve with 1 mg Pt per square centimeter of electrode geometric area an areal power density of 1 W/cm^2, translating to a bulk power density of the cell of around 1 W/cm^3, i.e., a respectable power density of 1 kW per liter of the stack at a cell voltage as high as 0.8 V. However, the Pt catalyst mass required for operation at such level of power density and at a conversion efficiency higher than 60%, i.e., 1 mg Pt/W, or 1 g Pt/kW, translates at the present market price of Pt metal to $50/kW. Such a cost of the catalyst component of a power source, may possibly enable technology implementation for some stationary or portable power applications, where the cost of the relevant incumbent power technology is in the hundreds and thousands of dollars per kilowatt, respectively. The same catalyst cost is prohibitive, however, in transportation applications, where the overall cost per kilowatt of the incumbent technology, i.e., the internal combustion engine, is similar to, or lower than the cost of just the Pt catalyst required per kilowatt of fuel cell stack at the present state of PEFC electrode technology and the present market price of Pt. Consequently, the greatest drive during the last few years to lower the catalyst loading in a PEFC stack to 0.1–0.2 g/kW, i.e., 5–10 times lower than is possible with Pt/C catalysts, has come from programs and teams pursuing transportation applications of PEFCs.

Pt alloy catalysts: While driven technologically by the demanding cap on Pt usage per unit power generated, the development of Pt alloy catalysts of higher ORR activity has been driven scientifically by the pursuit of the fundamental physical parameter(s) that determine the electrocatalytic activity of metals and metal alloys in the ORR. As this aspect of electrocatalysis research is covered extensively in this book, mention will be made here only of recent central achievements in (i) the actual introduction of Pt alloy/C catalysts into PEFC cathodes, demonstrating for such PEFCs an advantage in cell power per mg Pt of about fourfold compared with the non-alloyed Pt/C catalyst, and (ii) the coupled theoretical and experimental research efforts since the year 2000, leading to an understanding of the origin of the higher ORR activity observed following proper alloying of Pt and, indeed, to forecasts on possible further improvements in ORR catalysis.

The specific Pt alloy that has been introduced more than any other as a carbon-supported cathode catalyst in PEFCs has been PtCo. It belongs to the first group of catalysts considered as possible ORR activity enhancers in PEFCs, a group that also included Pt_3Cr and Pt_3Ni [Mukerjee and Srinivasan, 1993; Mukerjee et al., 1995]. In fact, this choice was inspired by the similar ORR activity enhancement observed for this group of alloys in earlier development of cathode catalysts for the phosphoric acid fuel cell (PAFC). Incorporation of these alloys, starting more than 20 years ago, in cathodes of commercial PAFC power units fabricated by UTC (Connecticut, US), was based on the superior activity of the alloy per unit mass of Pt compared with the unalloyed Pt/C catalyst. When the first measurements of the changes in electronic properties of Pt on alloying with Co, Ni, or Cr were reported [Mukerjee and Srinivasan, 1993; Mukerjee et al., 1995], it was still not clear why an increase should be observed in Pt d-band vacancy following alloying and why that would necessarily lead to higher ORR activity compared with unalloyed Pt. What helped here was electrochemical surface characterization by cyclic voltammetry, which revealed a shift of the onset of OH or O electrosorption to higher anodic potential as a result of alloying, suggesting a lowered tendency of the metal alloy surface to chemisorb O or OH by water discharge, as compared with unalloyed Pt. Such lowered affinity for oxygen would be difficult to rationalize for a surface containing, e.g., Co atoms in addition to Pt atoms; however, experimental evidence then appeared of spontaneous formation on such PtCo alloy particles of a single-atom-thick shell of

Pt atoms around a PtCo alloy core. With such a core-and-shell structure, the lowered drive of the Pt shell atoms to bond to surface oxygen atoms could be now understood as the result of the involvement of d-electrons of shell Pt atoms in bonding to Co atoms in the adjacent atomic layer underneath the shell, resulting in lower availability of d-electrons in the surface Pt atoms for formation of a surface bond to an OH group or an O atom. This observed correlation between higher ORR activity and lowered surface affinity to chemisorbed O or OH revealed that the surface affinity to oxygen of unalloyed Pt surfaces is apparently somewhat too high to achieve optimized cathode catalytic activity, and, consequently, a slight lowering of the catalyst surface affinity to oxygen through alloying enhances the rate of the ORR.

Interpretation of this observed correlation between a lowered affinity of the metal surface to oxygen and a higher rate of ORR measured at a Pt shell over a Pt-alloy core has also been at the center of recent theoretical work, based primarily on DFT calculations of electronic properties and surface bond strengths for a variety of expected ORR intermediates at metal and metal alloy catalysts. The second part of this chapter contains a discussion of these valuable contributions and of outstanding issues in tying together this recent theoretical work and ORR experimental data.

Core-and-shell-type electrocatalyst particles have most recently defined the frontier in ORR electrocatalysis research. Not only can this structure enable fine tuning of the electronic properties of surface Pt atoms, it could also allow placing of all the Pt atoms only on the outer surface of the catalyst particle, i.e., where the catalytic process takes place, using non-precious metal atoms in the particle core. By having all Pt atoms located on the outer surface (shell) of a core-and-shell catalyst particle, the mass of Pt required to generate some given Pt catalyst surface area would drop by a factor of 5–10 compared with catalyst particles built exclusively of Pt atoms. Consequently, the cost of the catalyst per unit power generated could drop by a similar factor of 5–10 if the particle core were made of non-precious metal(s). Several recent demonstrations of this approach have included a Pt shell over a Ru core as a CO-tolerant anode catalyst and a Pt shell on a Pd core for the ORR process, both made by Adzic et al. [Zhang et al., 2004 and references therein]. The most recent work in this area is covered in Chapters 8 and 9.

A remaining great challenge in the introduction of such atomic-level tailored nanoparticles of electrocatalysts is maintenance of the stability of the preferred surface atomic structure under fuel cell operation conditions. Encouraging results in this regard for Pt shell/Ru core anode catalysts tested for 1000 hours were facilitated by the reducing environment in the anode. The most researched cathode alloy catalyst, carbon-supported Pt shell/Pt_3Co core, was first reported to be even more stable than unalloyed carbon-supported Pt, but more recently a performance decay pattern reported by Johnson Matthey [Thompset, 2007] showed accelerated decay following the first 1000 hours, which was explained by loss of near-surface Co atoms, leaving behind a faulty structure of surface Pt atoms. Basically, under fuel cell cathode operating conditions, it will be highly nontrivial to achieve the operational stability required with shell-and-core catalyst particles employing highly electropositive atoms such as Co, Fe, or Ni, because of the strong tendency of such atoms to leave the metal alloy crystal and form metal ions.

1.3 ELECTROCATALYSIS OF OXYGEN REDUCTION IN THE FUEL CELL CATHODE: NEW INSIGHTS AND NEW QUESTIONS

Significant advances have been made since the year 2000 in the theoretical analysis of the ORR at Pt metal electrocatalysts. These recent theoretical advances have been made using DFT calculations considering model systems of good resemblance to the actual electrochemical interface. Models of the interface used in such recent calculations include a significant ensemble of metal atoms (Fig. 1.2) [Nørskov et al., 2004] (rather than a single metal atom or a pair of metal atoms as was done in pioneering contributions made somewhat earlier [Anderson et al., 2000]), as well as the aqueous molecular environment adjacent to the metal catalyst surface. Last, but not least, the effects of variations in the electrode–electrolyte potential difference on interfacial thermodynamics and dynamics have also been included in these models [Nørskov et al., 2004; Panchenko et al., 2004; Desai and Neurock, 2003].

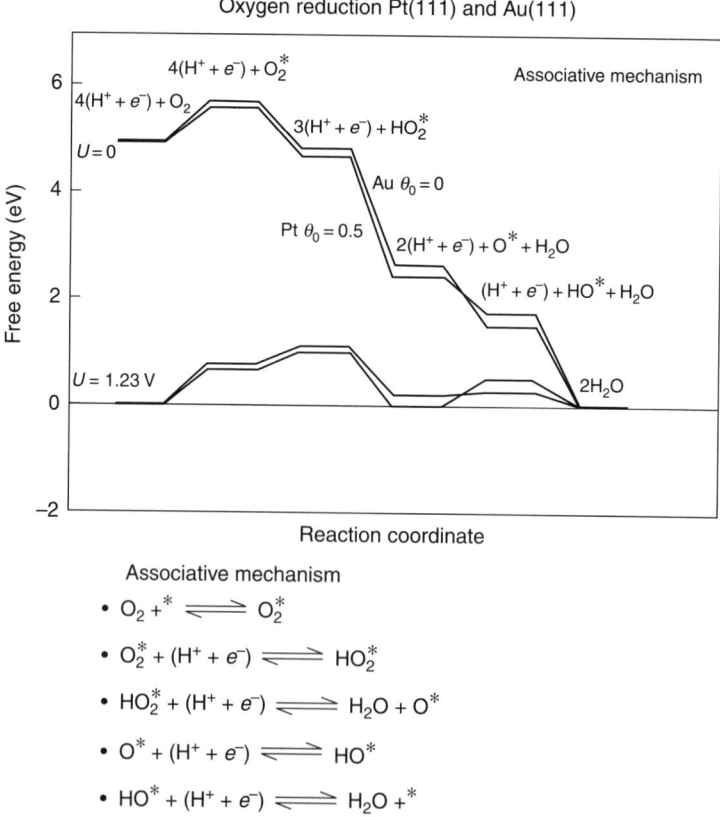

Figure 1.2 Free energy map for the "associative mechanism" of ORR at Pt metal and the steps considered in this mechanism [Nørskov et al., 2004].

1.3 ELECTROCATALYSIS OF OXYGEN REDUCTION IN THE FUEL CELL CATHODE

With the four-electron ORR process in the fuel cell cathode well recognized as the principal challenge at both the theoretical and experimental/technical levels, it is interesting to examine the effects of the most recent theoretical developments on the fundamental understanding of ORR electrocatalysis. Such examination would naturally focus on the nature and quality of the links with experimental work, as reflected by:

- Predictive abilities regarding future promising directions in pursuit of more active ORR electrocatalysts;
- Elucidation of the actual values of key diagnostic parameters that have been used for a long time by ORR experimentalists, including primarily the measured "Tafel slope" and the reaction order with regard to O_2, and their predictive power regarding the ORR mechanism.

Starting with predictive abilities of ORR activity, although DFT calculations have provided a much more accurate map of the free energy of formation and free energy of conversion of the possible intermediates in the ORR in acid electrolytes [Nørskov et al., 2004], the tool used for prediction of the ORR rate at the surface of some metal or alloy has remained very simple when considering the complexity of the ORR process. The M—OH and/or M—O bond strength, proposed as a yardstick for the expected ORR rate at the surface of a metal M, seems to explain the tendencies observed in ORR measured rates at various metal surfaces. This dependence is characterized by a "volcano"-shaped plot of ORR rate versus M—O or M—OH bond, presented many years ago, albeit using less accurate values for ORR rates. The metal surfaces at the top of the "volcano" correspond to the optimized M—Ox bond strength enabling the highest ORR activity, with the metals on the ascending branch of the plot having bonds to surface oxygen that are too strong to allow further activity as ORR intermediate, and metals on the descending branch having affinity to surface oxygen which is insufficient to activate the dioxygen molecule. While intuitively compelling, a challenge to such a simple yardstick is to provide a sound justification for the predictive power it seems to have for a complex, multistep process like the ORR (a question in this spirit was raised by W. Schmickler in a discussion during the Leiden meeting).

Another challenge stands in the way of attempts to evaluate new theoretical insights into the mechanism of the ORR in light of Tafel slopes and reaction orders reported for the Pt/hydrous poly{PFSA} interface: as late as 2006, results of measurements of these experimental parameters reported from different sources have been in strong apparent disagreement. In the following discussion, an explanation is offered for the origin of such apparent disagreements between various reported Tafel slopes and reaction orders for the ORR at Pt. These disagreements can be reconciled through a proper description of the actual catalyst surface in a PEFC cathode employing a Pt catalyst. The same, fuller description of the cathode catalyst surface in an operating fuel cell also sheds a different light on the possible origin of the power of the M—Ox bond strength to predict rates of ORR at different metal catalyst surfaces.

Mechanistic conclusions from DFT calculations: The recent DFT calculations aided in the clarification of the relative probabilities of various proposed multistep ORR routes from reactant O_2 to product H_2O. This has been enabled by the accurate calculation of formation energies for relevant intermediates in any mechanism considered, where the interactions of an intermediate with both the Pt metal surface and the adjacent molecular layers of water are both considered. For example, the probability that the first step in the ORR process is direct, dissociative chemisorption of O_2 to form two O_{ads} species on the Pt metal surface appears to be low, not only in light of the low rates of this process at the Pt/oxygen gas interface at such low temperatures, but also in light of the deep energy sink associated with chemisorption of an oxygen atom on Pt [Nørskov et al., 2004]. Accordingly, there seems to be an advantage for the ORR route starting with a single electron and proton transfer to O_2 to generate the HOO_{ads} intermediate and thereby facilitate subsequent breaking of the O—O bond, which in HOO_{ads} has an order significantly lower than 2 [Panchenko et al., 2004]. OH_{ads}, rather than O_{ads}, seems the likely next intermediate on the way to the final H_2O product, as can be concluded, again, from the deep energy sink associated with O_{ads} on Pt. The recent DFT calculations further support the O_2-to-OOH_{ads}-to-OH_{ads} route by showing that OOH is indeed an intermediate of significant bonding energy to Pt, at least on some Pt crystal surfaces [Panchenko et al., 2004]. Another important contribution of the theoretical calculations has been in providing an explanation for the higher measured ORR activity of Pt-alloy catalysts versus unalloyed Pt. In accordance with the premise that the M—Ox bond strength is an effective yardstick for predicting relative ORR rates at metal/electrolyte interfaces, the key to the higher ORR activity observed following proper alloying of Pt seems to be the calculated increase in *d*-band vacancy which causes reduction in the affinity of Pt surface sites to chemisorbed OH or O species.

Paradoxically, all these significant recent contributions to the theory of the ORR, together with most recent experimental efforts to characterize the ORR at a fuel cell cathode catalyst, have *not* led at all to a consensus on either the mechanism of the ORR at Pt catalysts in acid electrolytes or even on how to properly determine this mechanism with available experimental tools. To elucidate the present mismatch of central pieces in the ORR puzzle, one can start from the identification of the slow step in the ORR sequence. With the O_2-to-HOO_{ads}-to-HO_{ads} route appearing from recent DFT calculations to be the likely mechanism for the ORR at a Pt metal catalyst surface in acid electrolyte, the first electron and proton transfer to dioxygen, according to the reaction

$$O_2 + H^+ + e^- \longrightarrow OOH_{ads} \qquad (1.2)$$

becomes the likely slow step that determines the overall interfacial ORR rate at a Pt catalyst. One reason for this tentative assignment is the demanding requirement of forming a significant bond to the Pt catalyst surface before O—O dissociation can take place—a requirement that can apparently be fulfilled on Pt surface sites of specific atomic geometries, available on some Pt crystal surfaces [Panchenko et al., 2004]. The second reason for assigning (1.2) as the slow step is the results of past experimental determinations of the reaction order with respect to O_2 and the Tafel slope for the ORR at Pt in acid electrolytes. A reaction order of (or slightly under) 1.0 and a Tafel slope near 120 mV/decade, measured in the potential range where the Pt surface is practically free of chemisorbed oxygen, are both in agreement with the Reaction (1.2) as the slow step in the ORR sequence. Such measured Tafel slope is expected for a first, one-electron-transfer process like (1.2), when α, the "symmetry factor" in the electrochemical rate equation, has a value of 0.5. Such a value of α means that the change in activation energy in the Reaction (1.2) per some increase in overpotential is 50% of the corresponding change in the reaction free energy, i.e., $\{\delta(\Delta G^{\#})/\delta V\}/\{\delta(\Delta G)/\delta V\} = 0.5$. There is no a priori reason to expect this specific value for α; however, a value of 0.5 for what is basically a "Brønsted factor" that describes $\delta(\Delta G^{\#})$ as some constant fraction of $\delta(\Delta G)$ is commonly found in electrochemical reactions. Consequently, the reported Tafel slope of 120 mV/decade of current density seems to be in accordance with the first, one-electron-transfer step being the slow step in the ORR sequence, as does the first order in oxygen partial pressure documented for the ORR at the Pt/hydrated ionomer interface [Parthasarathy et al., 1992a, b].

It would seem that to test the predictive power of recent theoretical work with regard to the ORR mechanism at Pt in acid electrolytes, the Tafel slope predicted by theory should be compared with a reported value of 120 mV/decade measured for ORR at Pt metal, whereas the reaction order predicted by theory be compared with the reported value of, or slightly below 1.0 [Parthasarathy et al., 1992a, b]. However, testing of the recent theoretical predictions against these two experimentally measured parameters has just become complicated by recent reports suggesting strong disagreement between Tafel slopes and reaction orders reported previously from measurements of ORR kinetics at rotating disk or other, ionomer-filmed bulk Pt electrodes and the slope and reaction order measured for cathodes in an operating PEFC. In one such recent experimental report [Neyerlin et al., 2006], the conclusion offered was that, for a fuel cell cathode employing a Pt/C catalyst, the full description of ORR kinetics in the potential range relevant to fuel cell operation should be based on a constant Tafel slope of 60 mV/decade of current density and on a reaction order with respect to oxygen of 0.5. Such reports of a measured constant slope of 60 mV/decade for ORR at the temperatures of an operating PEFC and in the potential range relevant to fuel cell operation have been identified by theorists [Nørskov et al., 2004] as experimental support for a conjecture that the rate-determining step in the ORR may be a single electron and proton transfer to an OH_{ads} or O_{ads} surface intermediate. According to that conjecture, this step is likely to be rate-determining in light of the significant energy sink from which these intermediates need to be lifted to complete the four-electron process [Nørskov et al., 2004]. The latter authors did not address, however, one consequence expected of a rate-determining step located lower in the sequence of one-electron steps in the ORR process, namely that the reaction order with respect to O_2 should then be 0.5, rather than 1.0 consistently reported from ORR investigations at model, Pt/electrolyte interfaces, including the Pt/hydrated Nafion interface [Parthasarathy et al., 1992a, b]. As if to strengthen the growing question mark regarding widely accepted values of key ORR kinetic parameters, subsequent to the 2004 ORR theoretical paper [Nørskov et al., 2004] came the 2006 ORR experimental paper [Neyerlin et al., 2006] and provided apparent support for an ORR reaction order of 0.5 with respect to O_2 from measurements done on Pt catalyst in the PEFC cathode. This combination of reports from experimental ORR studies [Neyerlin et al., 2006] and from a theoretical study of the ORR in acid electrolytes [Nørskov et al., 2004], thus seemed to provide significant evidence against the assignment of the Reaction (1.2) as the rate-determining step in the ORR.

The apparent challenges presented by these recent readings of experimental ORR results, winning early support from a DFT study, have left some important questions to answer:

- What is behind the apparent disagreements between Tafel slopes and reaction orders reported from recent investigations of the ORR at PEFC cathode catalysts and the slopes and reaction orders measured earlier for model systems of low Pt surface area? Is the ORR process at a dispersed Pt catalyst possibly different in nature from the ORR process at low-surface-area Pt?
- Can an ORR mechanism at Pt metal in an acid electrolyte with the Reaction (1.2) as the first and rate-limiting step be defended in light of the recently reported apparent Tafel slope and reaction order for ORR in the PEFC cathode?
- With the remaining question marks regarding the mechanism of the ORR and the nature of the slow step in the ORR, does the viability of the simple yardstick (M—OH or M—O bond strength), offered for predicting the rate of the complex ORR process [Nørskov et al., 2004] remain unscathed? Or does its predictive power possibly depend strongly on the rate-limiting step being reduction of surface oxygen intermediates [Nørskov et al., 2004], rather than the Reaction (1.2)?

One point that should be raised before trying to answer the above questions is the actual significance of reconciliation between new information and understanding at the atomic and molecular level, as provided by DFT calculations, and "old-fashioned" macroscopic parameters such as the Tafel slope and the reaction order. Arguments have been made, including at the Leiden meeting, to the effect that such macroscopic parameters should not be considered legitimate diagnostic tools in the age of advanced calculations of interfacial electronic structures. The experience of the present author in critically reviewing very recent ORR work, both theoretical and experimental, has led to a renewed conviction that those "old-fashioned" parameters have significant value and should, in fact, be considered carefully in theoretical work done at the atomic/molecular level.

1.4 DOCUMENTED INFORMATION ON THE ACTUAL NATURE OF THE SURFACE OF A Pt CATALYST IN THE CATHODE OF AN OPERATING FUEL CELL

Proper answers to the above questions can be provided once it is realized that the surface of the catalyst in a fuel cell cathode using a Pt or Pt/C catalyst cannot be

1.4 DOCUMENTED INFORMATION ON ACTUAL NATURE OF SURFACE OF Pt CATALYST

described for the purposes of either theory or diagnostics as "Pt metal." While Pt surface coverage by chemisorbed oxygen species at cathode potentials higher than 0.75 V has been recognized, at least to a degree, in ORR studies, in most examinations of ORR kinetics at "Pt," it has been considered mostly as an afterthought, rather than the major factor it really is in determining the ORR rate and, in fact, in defining the documented dependence of the ORR rate on the M—Ox bond strength. As ORR takes place at "Pt," parallel interfacial processes that need highlighting in ORR diagnostics are driven by the substantial reactivity of the Pt surface with water and with dioxygen. Such processes cause potential- and time-dependent modification of the interfacial composition and structure of an ORR cathode. A recent study by Paik et al. [2004] highlights the significant reactivity of a Pt metal surface in a PEFC cathode under fuel cell cathode operation conditions, bringing about slow build-up of Pt surface OH_{ads} and/or O_{ads}, above and beyond that measured under ordinary cyclic voltammetry conditions in an inert atmosphere. The coverage of a Pt catalyst surface by chemisorbed oxygen species under a set, constant cathode potential can be estimated from the charge passed during a cathodic potential scan following some timed exposure of the catalyst surface to an oxidizing gaseous atmosphere as shown in Fig. 1.3. As amply documented, Pt is covered by a chemisorbed oxygen species through the potential domain relevant for a fuel cell cathode, even in the complete absence of dioxygen, and this chemisorbed species which is formed by water discharge is quite irreversibly reduced. As seen in Fig. 1.3, the surface oxygen coverage at some cathode potential could actually grow substantially further when dioxygen is introduced into the system. The extra growth in oxide coverage following introduction of O_2 is most substantial at 0.85 V and significantly lower at 0.95 V. This suggests that the extra chemisorbed oxygen species probably forms on Pt metal by a mechanism different than the one operative in the absence of dioxygen. The extra "chemical" surface oxidation process is apparently inoperative when the surface population of metal sites is too low, as would be the case when the applied potential is 0.95 V or above.

The Pt surface electro-oxidation process observed in the absence of dioxygen to form chemisorbed OH from water is driven by the potential difference at the Pt/electrolyte interface, according to the reaction

$$H_2O_{ads} \longrightarrow OH_{ads} + H^+ + e^- \tag{1.3}$$

whereas the additional surface oxygen collected on adding dioxygen at an electrode potential around 0.85 V, is likely driven by O_2 reduction at surface metal sites, according to the reaction

$$\text{(a) } \tfrac{1}{2}O_2 + 2H^+ + 2e^- \longrightarrow H_2O$$
$$+$$
$$\text{(b) } 2H_2O_{ads} \longrightarrow 2OH_{ads} + 2H^+ + 2e^-$$
$$\overline{}$$
$$\tfrac{1}{2}O_2 + H_2O_{ads} \longrightarrow 2OH_{ads} \tag{1.4}$$

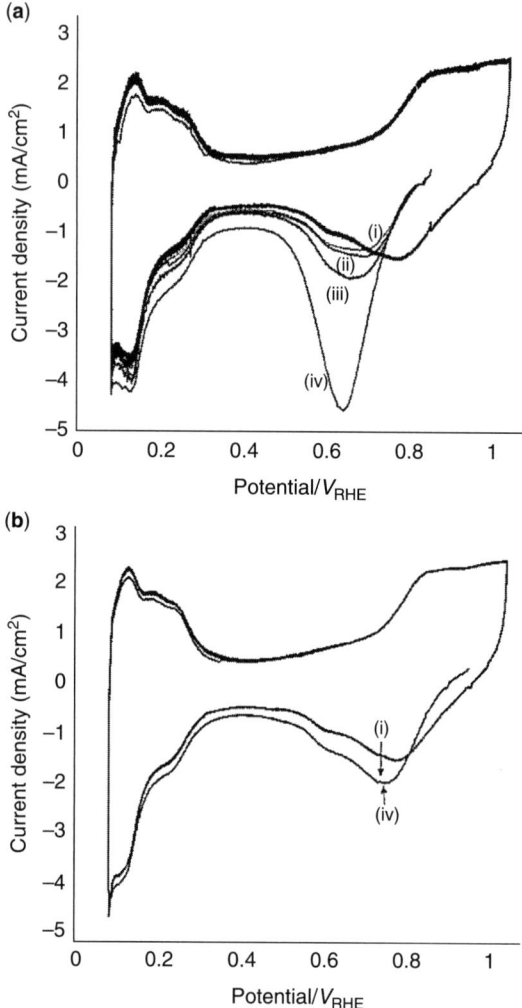

Figure 1.3 Pt catalyst coverage by chemisorbed oxygen species (for Pt/C in a PEFC cathode at 25 °C): (a) 0.85 V, 3 h of exposure to N_2, 4% O_2 in N_2 and air, and 30 min exposure to O_2; (b) 0.95 V, 3 h exposure to N_2 and 30 min exposure to O_2. From Paik et al. [2004].

Considering the standard potential of 1.23 V associated with Reaction (1.4a), it can be understood why a dioxygen molecule can be more reactive in the formation of OH_{ads} by water oxidation (1.4) than the purely anodic discharge of water driven by a potential of 0.85 V in an oxygen-free atmosphere. Interestingly, dissociative chemisorption of O_2 from the gas phase at Pt-group metal surfaces is reported to be strongly accelerated in the presence of water vapor [Weaver, 2002], suggesting a surface oxidation process involving dioxygen and surface water molecules, identical or similar to the Reaction (1.4). As Fig. 1.3 suggests, Reaction (1.4) apparently takes place at cathode potentials over 0.75 V but under 0.95 V, such that coverage by the OH_{ads} formed by the Reaction (1.3) is still incomplete, enabling ORR activity (1.4a) at some non-oxidized (water-covered) Pt metal sites.

The lesson to be taken from this report by Paik et al. [2004] is that a Pt catalyst in contact with a hydrous electrolyte is so active in forming chemisorbed oxygen at temperatures and potentials relevant to an operating PEFC, that the description of the cathode catalyst surface as "Pt," implying "Pt metal," is seriously flawed. Indeed, that a Reaction (1.4) actually takes place at a Pt catalyst surface, "exposes," Pt to be less noble than usually considered (although it remains a precious metal nevertheless ...). Such a surface oxidation process, taking place on exposure to O_2 and water and driven by electronically shorted ORR cathode site and metal anode site, is ordinarily associated with surface oxidation (and corrosion) of the less noble metals.

Turning next to the effect of these catalyst surface oxidation processes on the rate of ORR at "Pt", it has been well established that the chemisorbed oxygen species is an inhibitor of the ORR at a "Pt" catalyst surface, as would be expected for any process that requires Pt *metal* sites to proceed. This has been repeatedly demonstrated experimentally and should be readily understood from theoretical considerations, including the notion of optimized M—Ox bond strength as a yardstick for the rate of ORR where M is a *metal* surface site. One direct experimental demonstration of this inhibiting effect is the observed continuous PEFC performance reduction on extended operation time scales, shown to be caused by the continuous build-up of chemisorbed oxygen [Eickes et al., 2006], likely formed by both Reactions (1.3) and (1.4). The chemisorbed oxygen formed practically instantaneously on a Pt surface by Reaction (1.3), for example during a potential sweep at several millivolts per second into the "Pt oxide region," is itself well documented to be an ORR inhibitor. This is readily seen from higher ORR currents measured in the fuel-cell-relevant potential domain (>0.75 V) during the anodic half-cycle of a triangular potential scan (1.2 V– 0.4 V–1.2 V) [Eickes et al., 2006]. During the cathodic half-cycle, the Pt surface is still covered at $V > 0.75$ V by (irreversibly reduced) chemisorbed oxygen species formed by the Reaction (1.3) at the high end potential (1.2 V), whereas the anodic half-cycle starts from a low potential of 0.4 V where all the chemisorbed oxygen has been reductively removed, and, consequently, the coverage by this inhibiting species is quite small when the fuel-cell-relevant potential domain is traversed, resulting in higher observed ORR currents.

With these documented experimental findings and the clear understanding of the criticality of *metal* sites free of chemisorbed oxygen for ORR catalysis, it remains to be explained why any analysis of the ORR process at Pt catalysts would consider a purely metallic Pt catalyst surface in contact with surface water molecules as the relevant interface. In some cases, this could have been a choice made consciously, aiming to reduce the complexity of the system addressed by leaving full consideration of the actual state of the Pt catalyst surface for later. However, one other factor could have been insufficient knowledge of earlier literature, covering significant efforts to characterize and study the formation of surface and subsurface oxygen on Pt as function of Pt electrode potential, temperature, and time. The latter work was done to a large degree by Conway and co-workers in the 1970s and 1980s. Conway et al. reported that the long-term potentiostatic build-up of surface oxide charge on Pt (in an inert atmosphere) was proportional to the logarithm of the time, and suggested the schematic description shown in Fig. 1.4 for processes at the atomic scale on and in the Pt surface, when holding the Pt sample at a constant positive potential for extended periods of time. The figure is just a schematic; however, it can serve as a basis for consideration of two important factors. One is identification of the origin of the irreversible reduction of chemisorbed oxygen on Pt (and the Q_{oxide} versus log (time) rate law documented) as a "place exchange" process between chemisorbed oxygen atoms (or OH_{ads}) and surface Pt atoms. This type of process can be understood as an early step in the formation of a three-dimensional oxide phase from the very initial state of oxygen chemisorption on the metal surface. According to Conway et al. [1990], the place exchange process described schematically in Fig. 1.4 results in the conversion of an initial sub-monoatomic layer of chemisorbed oxygen on top of a Pt metal surface into a structure of mixed layers of Pt and oxygen atoms (or OH groups). In line with this reasoning, the outer surface of a "Pt" catalyst in a PEFC cathode, particularly following long-term exposure to high potentials at elevated temperatures, could in fact be a mixed Pt/oxygen atomic layer.

Figure 1.4 also highlights the possibility of a significant range of Pt—OH bond strengths, considering on one end of the spectrum the OH groups well surrounded by metal atoms and, at the other end, OH groups on top of Pt surface atoms, likely corresponding to those "last formed" surface OH groups that have not undergone further reorganization into a more stable sub-monolayer lattice of adsorbed intermediates, not to mention insertion under the metal skin. The latter type of OH species was identified

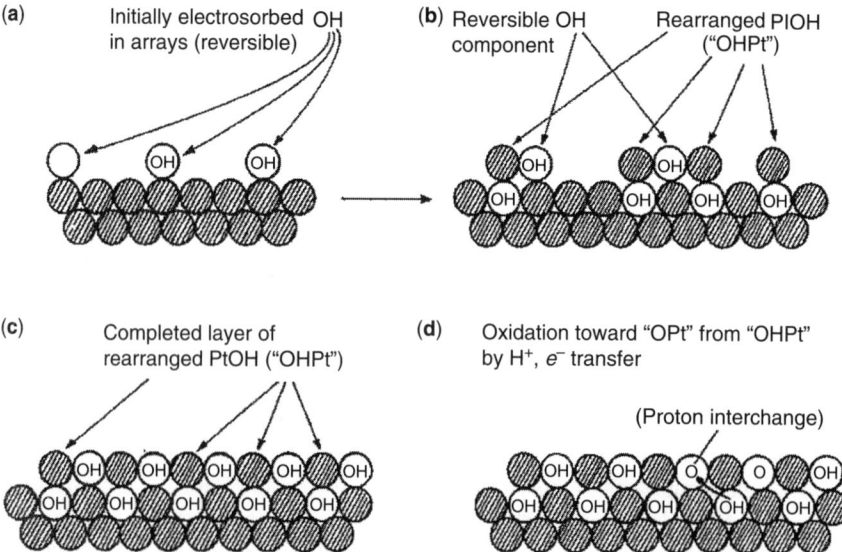

Figure 1.4 Proposed steps in the chemisorption of OH on/in Pt, starting with arrays of OH groups over the uppermost metal atom layer, increasing the coordination number of the adsorbed OH by place exchange, and next generating a mixed, metal/oxygen overlayer while further oxidizing to form O atoms. From Conway et al. [1990].

[Conway et al., 1990] as "reversible OH groups," implying a fast OH_{ads} formation/reduction cycle—a property not shared by the majority of surface oxygen on Pt, which requires a significant cathodic overpotential to be electroreduced. Such surface oxygen groups (the majority) have either been anodically deposited directly on surface sites providing higher coordination (e.g., steps) or have achieved a state of lower energy by surface migration and/or place exchange. The presence of a much more "reversible component of OH_{ads}" was concluded by Conway and co-workers from measurements of frequency-dependent pseudocapacitance at Pt at potentials above 0.75 V [Conway et al., 1990], and Pt surface coverage by this "reversible" type of OH_{ads} was determined to be near 10% of the total surface oxygen species. This insight, provided back in the 1980s, reveals that chemisorbed oxygen, or chemisorbed OH on Pt, can exhibit a wide range of bond strengths to metal surface atoms, with only a small population of (likely "newly formed"), surface OH groups being associated with a relatively high rate of reduction, i.e., having the characteristics expected of an active surface intermediate in a faradaic process.

A distribution of adsorption energies of chemically identical surface species formed on a metal electrocatalyst, with a large fraction playing "spectator" and/or "site blocking" roles and only a small fraction being active intermediates in a faradaic process, is not a situation unique to chemisorbed oxygen species and the ORR process at Pt. A recent analysis of the hydrogen oxidation process on Pt reveals a similar situation, as described by Wang et al. [2006].

1.5 IMPACTS OF THE ACTUAL STATE OF THE CATALYST SURFACE ON THE RATE OF THE ORR AT "Pt"

To evaluate the effects of the actual nature of a "Pt" catalyst surface on measured ORR characteristics, one has to abandon the prevailing assumption that measured changes of the ORR rate with potential, or with oxygen pressure, can be properly analyzed based on a conceptual model of a catalyst surface of invariable composition, ordinarily assumed to be Pt metal. To evaluate the Tafel slope expected for the ORR at "Pt" in the potential range 0.90–0.75 V, the variation with potential of the surface coverage by blocking chemisorbed oxygen species must be considered, in addition to the "classical" effect of enhancement of the rate of the ORR at Pt metal sites with increased cathode overpotential [Uribe et al., 1992; Wang et al., 2004]. In an early treatment of the expected combined effects of cathodic polarization in (i) removing blocking surface oxide species and (ii) enhancing the rate of ORR at "free" Pt metal sites available on the surface, Uribe et al. [1992] used the following assumptions:

- The ORR at Pt metal sites has a rate dependence on potential described by a Tafel slope of 120 mV/decade (as determined for Pt model systems at high cathodic overpotential, where Pt is practically surface oxide-free [Parthasarathy et al., 1992a, b]).
- The effect of the surface oxide species on the rate of the ORR is explained by metal site blocking, and can be described mathematically by including a $1 - \theta_{ox}$ term in the pre-exponential factor of the rate expression.

- Assuming that the electrode has not been left at potentials over 0.75 V for hours, the variation of θ_{ox} with electrode potential can be derived from the voltammetric charge associated with bringing the Pt electrode up from a chemisorbed oxygen-free state to some given potential above 0.75 V. A further assumption required here is on the number of electrons passed per surface oxygen species deposited, and this number was assumed by Uribe et al. [1992] to be 1, corresponding to an OH_{ads} intermediate formed on a single Pt surface site by the Reaction (1.3).

The rate expression considering the blocking of active sites by surface oxide can then be written as

$$J_{ORR} = J_0 (1 - \theta_{ox}) \exp\left(\frac{E^o - E}{b_{int}}\right) \quad (1.5)$$

where E is the cathode potential, E^o is the redox potential for the O_2/H_2O couple under the relevant operation conditions, θ_{ox} is the coverage by chemisorbed oxygen at potential E, and b_{int} is the "intrinsic" value of the ORR Tafel slope that would be measured at a Pt metal surface free of any blocking surface species. The variation of ORR current with Pt cathode overpotential to be expected from (1.5) is given by

$$\frac{d(\log J_{ORR})}{d(E^o - E)} = \frac{1}{b_{int}} + \frac{1}{1 - \theta_{ox}} \frac{d\theta_{ox}}{dE} \quad (1.6)$$

where the dependence of θ_{ox} on cathode potential E appears as an additional source of ORR rate variation with potential. The qualitative result of the dependence of surface oxygen coverage on E_{cath} can be seen from (1.6): the *apparent* Tafel slope for ORR is expected to be smaller than the intrinsic slope b_{int}, i.e., to be smaller than 120 mV per decade of current density near ambient conditions. Furthermore, when b_{int} is constant through the relevant potential range, the *apparent* Tafel slope will not be constant necessarily, as can be realized from the nature of the second term on the right-hand side of (1.6).

It is not difficult to understand the physical picture behind the apparent low Tafel slope expected for the ORR according to (1.5). The two terms on the right-hand side of (1.6) correspond to two contributions to ORR rate enhancement with increase in cathode overpotential: (a) lowering of the activation energy for the ORR process at Pt metal surface sites and (b) generation of more active (metal) surface sites by reducing the coverage θ_{ox} of the ORR-blocking chemisorbed oxygen species. It should be also easy to understand why interpretation of an *apparent* low Tafel slope, resulting from two different contributions to ORR rate enhancement with cathode overpotential, cannot be presented and further analyzed as if it were the *intrinsic* Tafel slope for the process at the metal surface. The suggestion by Neyerlin et al. [2006] that a Tafel slope of 60 mV/decade describes well the *ORR kinetics* at a Pt electrode in the fuel-cell-relevant potential range, and is therefore the value to use in fuel cell cathode diagnostics, is not defensible. A measured, practically constant slope of 60 mV per

1.5 IMPACTS OF THE ACTUAL STATE OF THE CATALYST SURFACE

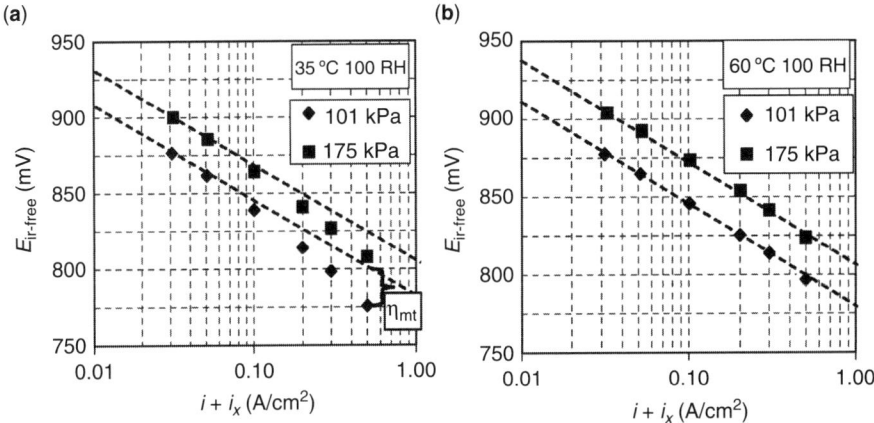

Figure 1.5 The slope of E_{cath} versus $\log J_{orr}$ through the fuel-cell-relevant potential range has an apparently constant value near RT/F (measured current density, here designated i, is corrected for hydrogen crossover current, designated i_x, and the measured cell voltage is ir-corrected to provide the cathode potential E) [Neyerlin et al., 2006].

decade of current density has indeed been experimentally established for the process in the PEFC cathode between 0.90 and 0.75 V, as shown in Fig. 1.5 and can obviously be used as some "pragmatic parameter" to describe the PEFC cathode behavior. However, to link it with *ORR kinetics* at the Pt/hydrated ionomer interface, an analysis along the lines of the expressions (1.5) and (1.6) is crucial. For one thing, if the apparent Tafel slope is still to provide, as it should, some information on the nature of the slow step in the ORR process, the *intrinsic* value b_{int} has to be extracted first from the combined two effects of cathode overpotential as described in (1.6).

To provide quantitative support for the interpretation of the observed Tafel slope in terms of the expression (1.6), one needs to show quantitatively that the experimentally observed behavior, i.e., the *apparent* slope of about 60 mV/decade in the cathode potential range 0.90–0.75 V, can indeed be the result of an intrinsic slope b_{int} of 120 mV/decade for ORR at a Pt metal surface, modified for the actual, Pt/Pt-OH$_{ads}$ mixed surface by the documented variations of θ_{ox} with E. This was indeed shown by Uribe et al. [1992] using a θ_{ox} versus E dependence based on reported voltammetric data [Conway et al., 1990] and an assumption that one electron is passed per surface site in forming an ORR-blocking, chemisorbed oxygen species. The result of that treatment, shown in Fig. 1.6, clearly demonstrates that the observation of a 60 mV/decade Tafel slope for the PEFC cathode between 0.90 and 0.75 V can be effectively explained by the two combined effects of a change in cathode overpotential as in Equation (1.6). Figure 1.6 also shows that, according to Equation (1.6), the apparent and intrinsic Tafel slopes become identical at high cathode overpotentials, where the coverage by chemisorbed oxygen is negligible and where the second term on the right-hand side of (1.6) goes to zero. Such a rise of the Tafel slope, from near 60 to near 120 mV/decade with increasing overpotential,

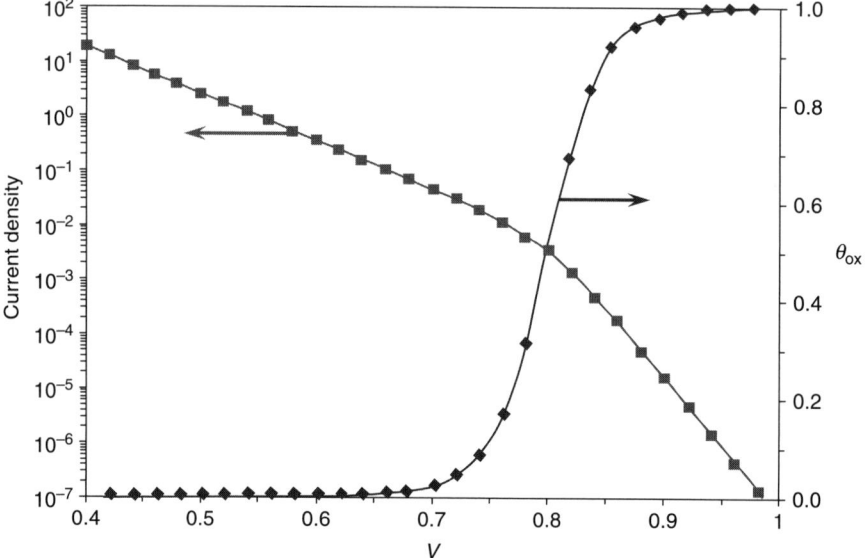

Figure 1.6 The variation of log J_{ORR} with V_{cath} expected based on (1.5), with the coverage by surface oxygen derived from voltammetry [Uribe et al., 1992].

has indeed been observed near ambient conditions in a number of investigations at relevant model Pt/electrolyte interfaces (see, e.g., Parthasarathy et al., 1992a, b).

In this regard, the skepticism expressed by Neyerlin et al. [2006] about the accuracy of mass transport corrections in evaluating ORR kinetic currents, and hence about the Tafel slope reported for the ORR at Pt at cathode potentials below 0.75 V, may be justified for high surface area electrodes, where the rate of the surface process is significantly "amplified" compared with the rate of mass transport processes. However, this criticism does not hold to the same degree for model systems of much lower Pt surface area per geometric electrode area. Consequently, the 120 mV/decade slope consistently reported for mass-transport-corrected ORR currents measured near room temperature at potentials under 0.75 V at RDEs, can be considered as established data that, with high probability, fully applies to the dispersed Pt metal catalyst in the PEFC. The fuel-cell-relevant potential region (see Fig. 1.5), falls fully within one branch of the complete behavior described by Fig. 1.6, and consequently this low, practically unchanged, value of the Tafel slope measured throughout this limited potential range (see Fig. 1.5) does not challenge the general analysis based on the expression (1.6).

Is there a way to verify experimentally not only that the measured low Tafel slope of 60 mV/decade in the fuel-cell-relevant potential range is fully explainable by an "intrinsic" slope of 120 mV/decade for ORR at Pt metal, but also that this intrinsic Tafel slope can be experimentally verified within the fuel-cell-relevant potential range? In fact, an elegant response to this challenge has recently been provided by impedance spectra measured for the ORR process at Pt/C in a PEFC cathode

[Makharia et al., 2005]. These spectra display a major loop in the Z'' versus Z' plot that cuts the Z' axis at some frequency in the range 0.1–1 Hz, followed by an inductive loop that cuts the Z' axis again at a much lower frequency. This frequency response of the interfacial faradaic process likely reflects variations of ORR current in response to a cyclic potential perturbation, originating from two effects of the potential on ORR rate, which are well resolved by their different response times. A relevant expression describing this behavior is likely of the form

$$\frac{dJ}{dE}(E, \omega) = \left(\frac{dJ}{dE}\right)_{\theta_{ox}} (E, \omega) + \left(\frac{dJ}{d\theta_{ox}}\right)_E \frac{d\theta_{ox}}{dE}(E, \omega) \qquad (1.7)$$

This expression suggests that the dependence of the ORR interfacial impedance on potential at a Pt surface of "frozen coverage," corresponding to the first term on the right-hand side, could possibly be measured at perturbation frequencies sufficiently high that the variation of oxide coverage with potential (appearing in the second term) is too slow to follow. The additional contribution of $d\theta_{ox}/dE$ to the overall impedance measured under DC conditions can then be derived from the additional changes of the measured impedance at lower frequencies, i.e., when θ_{ox} follows the slow potential perturbation. Quite strikingly, the variation of the ORR impedance under frozen coverage conditions, derived from variation with E_{cath} of the diameter of the large loop in Fig. 1.7, corresponded to a Tafel slope close to 120 mV/decade, whereas the apparent Tafel slope under DC conditions, corresponding to the value of Z' at the lowest end of the frequency spectrum, was found close to 60 mV (Mark Mathias, personal communication, 2007).

The other challenging ORR finding reported recently and mentioned above was of measured apparent reaction order with respect to oxygen of near 0.5 [Neyerlin et al., 2006]. Added to the suggestion that the measured Tafel slope of 60 mV/decade is representing "ORR kinetics" at Pt fuel cell catalysts above 0.75 V, the report of a reaction order with respect to O_2 of 0.5 seemed to support the possibility of the slow step in the ORR belonging further down the ORR sequence, as actually suggested in Nørskov et al. [2004]. Such a slow step could possibly be the reduction of OH_{ads} according to

$$OH_{ads} + H^+ + e^- \longrightarrow H_2O \qquad (1.8)$$

Having explained above why the *apparent* low Tafel slope is, in fact, hiding an *intrinsic* Tafel slope no smaller than 120 mV/decade, it remains to be seen if the reported reaction order of 0.5 [Neyerlin et al., 2006] can be understood along similar lines. For the purpose of this consideration, it is important to note the experimental conditions employed in Neyerlin et al. [2006] to study the ORR rate dependence on P_{O_2}. Unlike previous studies at relevant Pt/ionomer interfaces, which were performed near room temperature and employed oxygen partial pressures no higher than 1 atm [Parthasarathy et al., 1992a, b], the most recent study by Neyerlin et al. [2006] used elevated temperatures typical of an operating PEFC and oxygen pressures between 1 and 4 atm. The likely result of such experimental conditions is that an increase in the oxygen pressure again has more than the single effect on the rate of the ORR

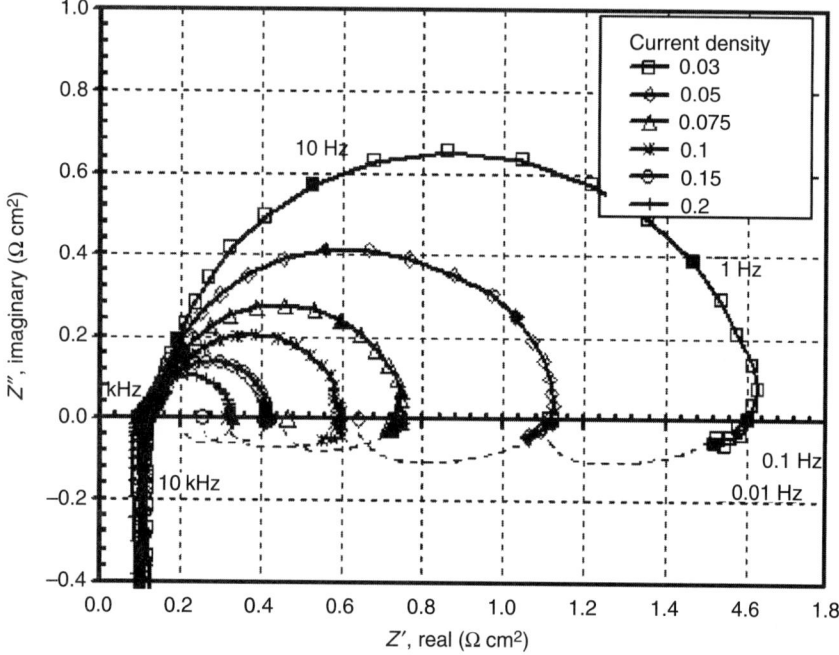

Figure 1.7 ORR impedance plots for a Pt electrode at several cathode DC currents, showing a low frequency branch corresponding to slow increase of the rate (slow lowering of the faradaic impedance) following cathodic perturbation [Makharia et al., 2005].

that is ordinarily considered. This time, the two effects of (i) an enhanced ORR rate at surface metal sites and (ii) changes in catalyst surface composition with increasing P_{O_2} [Paik et al., 2004] will be opposing each other. For a dependence of J_{ORR} on P_{O_2} at Pt metal at a constant cathode overpotential η described by

$$J_{ORR} = J_0 P_{O_2}^\gamma (1 - \theta_{ox}) \tag{1.9a}$$

where J_0 corresponds to the standard state $P_{O_2} = 1$ atm, the resulting *apparent* reaction order will be

$$\frac{d(\log J_{ORR})}{d(\log P_{O_2})} = \gamma - P_{O_2} \left[\frac{1}{1-\theta_{ox}}\right]_\eta \left[\frac{d\theta_{ox}}{dP_{O_2}}\right]_\eta \tag{1.9b}$$

Equation (1.9b) shows that, when the surface coverage by blocking chemisorbed oxygen increases with P_{O_2}, the apparent reaction order with respect to O_2 is expected to be smaller than the actual intrinsic reaction order with respect to dioxygen, γ, for ORR at a Pt metal catalyst. An increase in surface coverage by chemisorbed oxygen with enhanced P_{O_2} will likely take place to a larger extent under elevated temperatures and with oxygen pressures of several atmospheres, as used in Neyerlin et al. [2006], possibly leading to an apparent reaction order significantly smaller than reported, for example, in Parthasarathy et al. [1992a, b], under much milder conditions with respect to Pt surface oxidation.

1.6 A COMPREHENSIVE EXPRESSION FOR J_{ORR} WITH CONSIDERATION OF THE REAL NATURE OF THE CATHODE CATALYST SURFACE

The ORR rate expression used in most previous ORR studies—theoretical as well as experimental—has not included proper consideration of the actual, potential-dependent catalyst surface composition under relevant PEFC cathode operating conditions. From the discussion above, inclusion of the effect of chemisorbed oxygen species in terms of *site blocking*, i.e., in terms of lowering of the number of active metal sites available for the ORR, can be seen to be essential for the proper description of ORR in a PEFC cathode. This effect of site blocking belongs in the pre-exponential factor of the rate expression, which includes, in general, a frequency factor that is multiplied by the probability of the surface reaction to generate a fruitful encounter of the reactant molecule with the surface containing the catalytically active sites. Accordingly, an expression that considers the site blocking effect under discussion here will have the form

$$J_{ORR}(E) = k N_{total} P_{O_2}^{\gamma} (1 - \theta_{ox}) \exp\left(-\frac{\Delta H_{act}^*}{RT}\right) \exp\left(-\frac{E - E^o_{O_2/H_2O}}{b_{int}}\right) \quad (1.10)$$

where k is a frequency factor, N_{total} is the total number of active sites, and ΔH_{act}^* is the activation energy for the (slow step in the) ORR process at $E^o_{O_2/H_2O}$ with respect to an RHE. In Reaction (1.10), the $1 - \theta_{ox}$ term is included in the pre-exponential factor to describe the effect of site blocking on the probability of fruitful encounters, as has been done previously in Uribe et al. [1992] and Wang et al. [2004]. But the effect of the electrode potential in (1.10) is still confined to the exponential factor, i.e., to the effect of cathode potential on the ORR activation energy at metal sites. In reality, as discussed above, θ_{ox} is a function of E_{cath}, and this dependence can be described more explicitly.

As is well documented, formation of chemisorbed oxygen species on a Pt surface at $V > 0.75$ V occurs in an inert atmosphere on Pt in contact with an aqueous, or hydrous polymer electrolyte, by "anodic discharge" of water molecules to form OH_{ads} on metal sites, according to the Reaction (1.3). It is this chemisorbed oxygen species, derived from "water discharge," that will be considered in the following discussion. Significantly, the Reaction (1.3) is associated with a redox potential $E^o_{Pt(H_2O)/Pt\text{-}OH_{ads}}$ which is quite different from the redox potential for the faradaic ORR process, $E^o_{O_2/H_2O}$. In fact, $E^o_{Pt(H_2O)/Pt\text{-}OH_{ads}}$ was determined from recent DFT calculations of formation energies for various oxygen species adsorbed on a Pt metal surface. The reported result was $E^o_{Pt(H_2O)/Pt\text{-}OH_{ads}} = 0.80$ V with respect to an RHE, with $\theta_{OH} = \frac{1}{3}$ selected to define the standard state [Nørskov et al., 2004]. This calculated value of

$E^o_{Pt(H_2O)/Pt-OH_{ads}}$ agrees well with voltammetric measurements of the potential corresponding to similar partial coverage by OH_{ads} on a Pt surface in contact with aqueous electrolytes free of strongly adsorbed anions. It can be readily seen that it is the value of the cathode potential with respect to $E^o_{Pt(H_2O)/Pt-OH_{ads}}$ (not with respect to $E^o_{O_2/H_2O}$) that determines the value of $1 - \theta_{OH}$ in (1.10). Consequently, to "ignite" the ORR current at Pt, a key requirement is seen to be a pre-exponential factor in (1.10) significantly larger than zero, which requires, in turn that θ_{ox} be lowered significantly under 1 by bringing E_{cath} close to $E^o_{Pt(H_2O)/Pt-OH_{ads}}$, i.e., down from $E^o_{O_2/H_2O}$ by almost 400 mV.

Can this demand for a significant number of metal active sites be further quantified by a general expression in terms of cathode potential demand? The answer is, in principle, yes, although the dependence of the relative populations of metal surface sites and oxidized surface sites on cathode potential could depend on $E - E^o_{Pt(H_2O)/Pt-OH_{ads}}$ in a somewhat different way, depending on the degree to which the free energy of oxygen chemisorption is coverage-dependent. In the simplest case of noninteracting adsorbed species, where the relative populations are determined by a Nernst-type relationship with one electron assumed to be required for conversion of a metal site to an oxide-covered site (and vice versa), the expression for the surface population ratio will be

$$\frac{\theta_{OH}}{1 - \theta_{OH}} = \exp\left\{\frac{F}{RT}\left(E - E^o_{Pt(H_2O)/Pt-OH_{ads}}\right)\right\} \quad (1.11)$$

A more spread-out dependence of OH_{ads} on $E - E^o_{Pt(H_2O)/Pt-OH_{ads}}$ will occur when interaction energies between chemisorbed species become significant; however, it is always the potential difference $E - E^o_{Pt(H_2O)/Pt-OH_{ads}}$ that determines the availability of active metal sites, reflected by the $1 - \theta_{OH}$ term in the pre-exponential factor. The pre-exponential factor can consequently be expressed directly as a function of $E - E^o_{Pt(H_2O)/Pt-OH_{ads}}$ by replacing the $1 - \theta_{OH}$ term in (1.10) by a term derived from (1.11), to yield

$$J_{ORR}(E) = kP^\gamma_{O_2} N_{total} \left(\frac{1}{Z+1}\right) \exp\left\{-\frac{\Delta H^*_{act}}{RT}\right\} \exp\left\{-\frac{E - E^o_{O_2/H_2O}}{b_{int}}\right\} \quad (1.12)$$

where

$$Z = \exp\left\{\frac{F}{RT}\left(E - E^o_{Pt(H_2O)/Pt-OH_{ads}}\right)\right\} \quad (1.13)$$

Equation (1.12) describes an ORR rate dependence on potential that derives from *two different redox potentials*, one affecting the exponential part of the expression and the other affecting the pre-exponential part. The term depending on $E - E^o_{O_2/H_2O}$ reflects the lowering of the activation energy at an active metal site by an increase in cathode overpotential, whereas the term depending on $(E - E^o_{Pt(H_2O)/Pt-OH_{ads}})$ describes the fraction of active metal sites, $(1/Z + 1)$, at some value of E. Equation (1.12) suggests that it is the value of $E - E^o_{Pt(H_2O)/Pt-OH_{ads}}$ that could be the key for the "ignition" of the ORR current on a gradual increase of the cathode overpotential. The simple-case expression used here for Z yields a ratio of number of metal sites to number of chemisorbed oxygen covered sites of $1:100$ at $E - E^o_{Pt(H_2O)/Pt-OH_{ads}} = 0.12\,V$, increasing to $1:10$ at $E - E^o_{Pt(H_2O)/Pt-OH_{ads}} = 0.06\,V$. That "ignition" of the ORR process is indeed observed about $0.1\,V$ positive of $E^o_{Pt(H_2O)/Pt-OH_{ads}}$ clearly confirms the criticality of cathodic generation of active metal sites and, consequently, of proper representation of site availability by the $(1/1+Z)$ term in (1.12).

1.7 POSSIBLE DESCRIPTION OF ORR AT A Pt/Pt-Ox CATALYST SURFACE AS A REDOX-MEDIATED PROCESS

The notion of a surface redox system that determines an "ignition potential" for a faradaic process, including for the ORR, is well recognized in cases where such a surface redox system is added onto a metal or a carbon substrate. For example, to activate the ORR on various high surface area carbon structures, Co and/or Fe ion centers are attached, using appropriate ligation, which affects the standard potential of the redox couple and assists in surface bonding. Redox mediation in cases of such ORR catalysts can be conceptually divided into three steps:

1. A significant steady-state population of reduced surface sites is generated by the cathodic overpotential, as set by the fuel cell load.
2. The reduced sites donate electrons to the reactant oxygen molecule and to intermediates formed, this electron transfer being coupled with bond breaking and making involved in the ORR process.
3. Electrons donated to the surface oxygen species are instantaneously replenished from the ohmic contact to maintain a steady state population of electron-filled (reduced) surface states determined by E_{cath}.

A description of the above sequence in terms of equations for the electrochemical steps takes the form exemplified by the following specific case:

(a) $4Co^{+3}_{surface} + 4e \longrightarrow 4Co^{+2}_{surface}$

(b) $4Co^{+2}_{surface} + O_2 + 4H^+ \longrightarrow 4Co^{+3}_{surface} + 2H_2O$ (1.14)

(c) $4Co^{+3}_{surface} + 4e \longrightarrow 4Co^{+2}_{surface}$ [followed by repeat of step (a)]

In such cases of redox mediation, it seems clear why step (a) would maintain a steady state population of reduced sites according to the value of $E - E^o_{surface\ redox\ couple}$, whereas the rate of step (b) will be enhanced with $E - E^o_{O_2/H_2O}$, the cathode overpotential driving the faradaic four-electron reduction of oxygen. And it is also clear why approaching $E^o_{surface\ redox\ couple}$ could be the prerequisite for "igniting" the ORR process, as the proximity to $E^o_{\text{surface redox couple}}$ determines the population of active, reduced sites, according to

$$N_{\text{red}} = N_{\text{total}} \frac{1}{Z+1} \tag{1.15}$$

where Z is a function of $E - E^o_{\text{surface redox couple}}$.

Along the same lines, the population of the active metal sites on a Pt/Pt–Ox catalyst surface can be seen to be well described by (1.15), with the parameter Z being given as $Z = \exp\{(F/RT)(E - E^o_{\text{Pt-OH}_2/\text{Pt-OH}})\}$, and with the ORR process proceeding along conceptual steps similar to those in the process (1.14), according to

(a) $4\text{Pt-OH}_{\text{surface}} + 4\text{H}^+ + 4e \longrightarrow 4\text{Pt}_{\text{surface}} + 2\text{H}_2\text{O}$

(active reduced site generation)

(b) $4\text{Pt}_{\text{surface}} + \text{O}_2 + 4e + 4\text{H}^+ \longrightarrow 4\text{Pt-OH}_{\text{surface}}$ (1.16)

(faradaic reaction of O_2 at/with the reduced site followed by repeat of (A), and so on)

This analogy to a surface redox mediated process is significant. In a way very similar to the reaction sequence (1.14), the standard potential of the redox surface system Pt(H$_2$O)/Pt-OH$_{\text{ads}}$ (0.80 V with respect to RHE) determines the active (reduced) site population at any cathode potential E, and consequently is the critical parameter in determining the "ignition potential" for the ORR process.

The critical role of the M/M—OH redox system in determining the population of the surface active metal sites is, with high probability, the actual reason for the strong predictive power of the M—Ox bond strength with regard to the relative rates of ORR at different "metal" surfaces. In fact, a better presentation of the "volcano plot" would be obtained by using, for the ordinate of the plot the value $(1/Z + 1)\exp(-\Delta H^*_{\text{act}}/RT)$, at a constant cathode potential of interest for fuel cell technology, e.g., 0.85 V, where Z for a metal M is given by

$$Z = \exp\left[\frac{F}{RT}\left(E - E^o_{\text{M(H}_2\text{O)/M-OH}_{\text{ads}}}\right)\right] \tag{1.17}$$

For the ascending branch of the volcano plot, the term $(1/Z + 1)$ could serve by itself as an effective ORR activity predictor, whereas, for the descending branch, $(1/Z + 1)$ becomes close to unity at 0.85 V, and the exponential factor $\exp(-\Delta H^*_{\text{act}}/RT)$, then determines the ORR rate based on the residual interaction of dioxygen with the (excessively) noble metal catalyst surface.

The expression $(1/Z + 1)]\exp[-\Delta H^*_{\text{act}}/RT]$ at 0.85 V, better reflects the reality of a partially oxidized Pt surface and the critical effect of active site availability on the rate of the ORR. Effects of site availability were not considered in the calculations in Nørskov et al. [2004] of ORR "activity" for various metals. The expression used to calculate "activity" defined the ordinate parameter in the ORR volcano plots presented. This parameter was defined in Nørskov et al. [2004] as $kT \min_i \log(k_i/k_o)$, where $k_i = \nu_i \exp(-E_a^i/kT)$, with ν_i a frequency factor and E_a^i the activation energy of the slower step, i, in the ORR sequence at the metal surface. Nørskov et al. [2004], in fact, explain that such a calculation of "activity" yields a maximum value for the rate expected when the maximum possible surface sites required for the rate-limiting process are all available. The latter means that, for the case of the Reaction (1.2) being the slow ORR step, all surface sites are oxide-free metal sites—an unrealistic assumption even in the case of Pt, not to mention more electropositive metals. *It is the site availability effect that dominates the actual activity*, rather than the activation energy of the slow step in the faradaic process.

So why then do volcano plots in Nørskov et al. [2004] generate reasonable fits to experiment while completely disregarding the major factor of metal site availability? This is because of the assumption made there that the reduction of surface oxygen is the slow step *in the faradaic ORR process*. The result is the appearance of the potential associated with the M/M–OH couple in the activation energy of the assumed slow step of the ORR process occurring at metal sites. In reality, however, the significance of the potential of the M/M–OH couple is in determining the active (metal) site availability near 0.9 V, not in determining the rate of the faradaic ORR process at metal sites. The latter sites are, in fact, completely unavailable at the relevant potentials on all metals less noble than Pt appearing on the ascending branch of the ORR volcano plots.

Using $(1/Z + 1)$ as a gauge of ORR activity therefore also addresses a critical remark frequently raised regarding the applicability of volcano plots based on metal–oxygen bond strengths to those metals that are fully covered by a surface oxide in the relevant cathode potential range, i.e., metals that cannot actually form new M–Ox bonds when interacting with dioxygen. Once the value of the pre-exponential factor $(1/Z + 1)$ appears as part of or as the complete ORR activity yardstick, metals strongly covered by surface oxide at $E_{cath} > 0.75$ V are clearly predicted to exhibit negligible ORR activity, as a direct result of their excessive coverage by surface oxide.

1.8 SOME CONCLUSIONS

A description of ORR at "Pt" that disregards the potential-dependent blocking of active metal surface sites is fundamentally unsatisfactory, because it neglects a central physical/chemical feature of the real interfacial system. Furthermore, such descriptions can lead to numerical values and mechanistic conclusions in *apparent* conflict with understandings of the ORR mechanism that have been established and widely accepted to date. Indeed, such conflict may, at least in some cases, be the result of a "pragmatic approach," targeting system parameterization while knowingly detaching the result from any physical or mechanistic meaning. Unfortunately, it does not take much for such reported *apparent* parameters to be considered relevant to real interfacial kinetics and to be of mechanistic significance, leaving uncertainty regarding the key reported ORR kinetic parameters and proposed ORR mechanisms.

Once the key role of the overpotential with respect to $E^0_{Pt(H_2O)/Pt-OH_{ads}}$ in securing active surface sites has been realized, new light is shed on the Mn—O or M—OH bond energy as a yardstick for the activity of a metal or metal alloy surface in the ORR. Because the generation of active metal sites is likely the important key to achieving practical ORR rates at high, relevant cathode potentials, it can be understood why an optimized M—Ox bond strength may indeed become an effective yardstick, almost irrespective of the details of the mechanism of the ORR at an exposed active metal site.

The new analysis of the ORR offered here helps to reconcile experimental findings, particularly those reported recently for ORR at dispersed Pt catalysts in the PEFC cathodes [Neyerlin et al., 2006], and the likely molecular level mechanism of the ORR at Pt metal. The apparent Tafel slope of 60 mV/decade through the fuel-cell-relevant potential domain can be fully explained by an intrinsic "classical" Tafel slope of 120 mV/decade for ORR at Pt metal, modified by the added effect of the cathode overpotential in generating more active metal surface sites (see Fig. 1.6). True, a Tafel slope cannot by itself serve to establish the overall mechanism of a complex process such as the ORR. However, it is certainly important, on the other hand, to examine why the prevalent assignment of the first electron and proton transfer step as rate-determining in the ORR is in apparent conflict with results of measurements on a fuel cell cathode catalyst [Neyerlin et al., 2006]. The elucidation that the value of the *intrinsic* Tafel slope for the ORR on Pt metal is, in fact, not different than 120 mV/decade at the dispersed fuel cell Pt cathode catalyst [see Equation (1.6) and Fig. 1.6], whereas the apparent smaller reaction order of around 0.5 could be the result of enhanced Pt surface oxidation (i.e., enhanced ORR catalyst deactivation) with increase in oxygen pressure at elevated temperatures [see Equation (1.9b)], restores confidence in the assignment of the Reaction (1.2) as rate-determining in the ORR at a Pt cathode catalyst. The analysis used here also brings home the full impact of the complex dependence of the ORR process on the nature of the *actual* Pt catalyst surface in the fuel-cell-relevant potential range, this catalyst surface having a composition and structure that vary with potential, temperature, and time, resulting in a behavior that can be explained only when such surface modification effects are considered in the rate expressions.

ACKNOWLEDGMENTS

Thanks are due to the organizers of the Leiden meeting for a very pleasant and stimulating event.

The idea that the cathode potential with respect to $E^o_{Pt(H_2O)/Pt\text{-}OH_{ads}}$ determines the value of the pre-exponential factor in the ORR rate expression was inspired by a comment by Andy Gewirth (Urbana) in his talk in Leiden, pointing to the value of Pourbaix diagrams for understanding ORR electrocatalysis. Indeed, the information on these ORR-mediating and facilitating M/M–OH surface redox systems is to be found in Pourbaix's Atlas.

Extensive discussions with Hubert Gasteiger of the GM Fuel Cell Project (and now with ACTA) and with Jan Rossmeisl of Lyngby, Denmark, are gratefully acknowledged. It would be fair to add that these discussions did not resolve key disagreements; however, they were certainly conducted in and with good spirit.

REFERENCES

Anderson AB, Albu TV. 2000. Catalytic effect of platinum on oxygen reduction. An *ab initio* model including electrode potential dependence. J Electrochem Soc 147: 4229–4238.

Conway BE, Barnett B, Angerstein-Kozlowska H, Tilak BV. 1990. A surface-electrochemical basis for the direct logarithmic growth law for initial stages of extension of anodic oxide films formed at noble metals. J Chem Phys 93: 8361–8373.

Debe MK. 2003. In: Vielstich W, Lamm A, Gasteiger HA, eds. Handbook of Fuel Cells. Volume 3. New York: Wiley, Chapter 45.

Desai SK, Neurock M. 2003. First-principles study of the role of solvent in the dissociation of water over a Pt-Ru alloy. Phys Rev B 68: 075420.

Eickes C, Piela P, Davey J, Zelenay P. 2006. Recoverable cathode performance loss in direct methanol fuel cells. J Electrochem Soc 153: A171–A178.

Gasteiger HA. 2005. In: Proceedings of the International Fuel Cell Workshop, Kofu, Japan; September.

Makharia R, Mathias MF, Baker DR. 2005. Measurement of catalyst layer electrolyte resistance in PEFCs using electrochemical impedance spectroscopy. J Electrochem Soc 152: A970–A977.

Mukerjee S, Srinivasan S. 1993. Enhanced electrocatalysis of oxygen reduction on platinum alloys in proton exchange membrane fuel cells. J Electroanal Chem 357: 201–224.

Mukerjee S, Srinivasan S, Soriaga MP, McBreen J. 1995. Role of structural and electronic properties of Pt and Pt alloys on electrocatalysis of oxygen reduction. J Electrochem Soc 142: 1409–1422.

Neyerlin KC, Gu W, Jorne J, Gasteiger HA. 2006. Determination of catalyst unique parameters for the oxygen reduction reaction in a PEMFC. J Electrochem Soc 153: A1955–A1963.

Nørskov JK, Rossmeisl J, et al. 2004. Origin of the overpotential for oxygen reduction at a fuel-cell cathode. J Phys Chem B 108: 17886.

Paik CH, Jarvi TD, O'Grady WE. 2004. Extent of PEMFC cathode surface oxidation by oxygen and water measured by CV. Electrochem Solid State Lett 7: A82–A84.

Panchenko A, Koper MTM, et al. 2004. *Ab initio* calculations of intermediates of oxygen reduction on low-index platinum surfaces. J Electrochem Soc 151: A2016–A2027.

Parthasarathy A, Srinivasan S, Appleby AJ, et al. 1992a. Temperature dependence of the electrode kinetics of oxygen reduction at the platinum/Nafion® interface—A microelectrode investigation. J Electrochem Soc 139: 2530–2537.

Parthasarathy A, Srinivasan S, Appleby AJ, et al. 1992b. Pressure dependence of the oxygen reduction reaction at the platinum microelectrode/Nafion interface: Electrode kinetics and mass transport. J Electrochem Soc 139: 2856–2862.

Petrow HG, Allen RJ. 1977. US Patent No. 4,044,193; August.

Raistrick ID. 1986. In: Van Zee JW, White RE, Kinoshita K, et al., eds. Diaphragms, Separators and Ion Exchange Membranes. Pennington: The Electrochemical Society Softbound Proceedings Series. PV 86-13. p. 172.

Springer TE, Wilson MS, Gottesfeld S. 1993. Modeling and experimental diagnostics in polymer electrolyte fuel cells. J Electrochem Soc 140: 3513–3526.

Thompset D, 2007. ECS Fall Meeting, Washington, DC.

Toyota R&D Center 2007. ECS Meeting Report, Washington, DC; October.

Uribe FA, Wilson MS, Gottesfeld S. 1992. Electrochem Soc Proc 92-11: 494.

Wang JX, Markovic NM, Adzic RR. 2004. Kinetic analysis of oxygen reduction on Pt(111) in acid solutions: Intrinsic kinetic parameters and anion adsorption effects. J Phys Chem B 108: 4127–4133.

Wang JX, Springer TE, Adzic RR. 2006. Dual-pathway kinetic equation for the hydrogen oxidation reaction on Pt electrodes. J Electrochem Soc 153: A1732–A1740.

Weaver MJ. 2002. Surface-enhanced Raman spectroscopy as a versatile *in situ* probe of chemisorption in catalytic electrochemical and gaseous environments. J Raman Spectrosc 33: 309.

Wilson MS, Gottesfeld S. 1992. High performance catalyzed membranes of ultra-low Pt loadings for polymer electrolyte fuel cells. J Electrochem Soc 139: L28–L30.

Wilson MS. 1993. US Pat Nos. 5,211,984 and 5,234,777.

Zhang J, Mo Y, Vukmirovic MB, Klie R, Sasaki K, Adzic RR. 2004. Platinum monolayer electrocatalysts for O_2 reduction: Pt monolayer on Pd(111) and on carbon-supported Pd nanoparticles. J Phys Chem B 108: 10955–10964.

CHAPTER 2

Electrochemical Electron Transfer: From Marcus Theory to Electrocatalysis

E. SANTOS

Faculdad de Matemática, Astronomía y Física, IFFaMAF-CONICET, Universidad Nacional de Córdoba, Córdoba, Argentina, and Department of Theoretical Chemistry, Ulm University D-89069 Ulm, Germany

W. SCHMICKLER

Department of Theoretical Chemistry, Ulm University, D-89069 Ulm, Germany

2.1 INTRODUCTION

Early attempts at a theory of electrochemical reactions were focused on the hydrogen evolution reaction. In retrospect, these attempts were premature, because catalytic reactions, of which hydrogen evolution is a prime example, cannot be understood without detailed knowledge of the electronic structure and the interactions of the electrode and the reactant, information that can only be obtained by quantum chemical calculations requiring high speed computers. Therefore, these valiant efforts were unsuccessful, even though some qualitative aspects, such as the role of tunneling events investigated by Gurney [1931, 1936], or the role of the adsorption of intermediates [Parsons, 1958; Gerischer, 1956; Trasatti, 1972], were elucidated. It has been claimed, with some credibility, that this preoccupation with hydrogen evolution delayed the development of theoretical electrochemistry by at least a decade [Bockris and Khan, 1993].

The first successful theories of electrochemical reactions, proposed by Marcus [1956] and Hush [1958], dealt with outer-sphere electron transfer, and focused on the role of solvent reorganization, while electronic interactions and catalysis played a minor, merely formal, role. This remained largely true for the quantum mechanical version of that theory developed by Levich [1970], and even for the first theories describing electron transfer involving the breaking of a bond [Savéant, 1993; German and Dogonadze, 1974; German and Kuznetsov, 1994]. Generally, it was assumed that the electronic interactions between the reactant and the electrode are

Fuel Cell Catalysis. Edited by Marc T. M. Koper
Copyright © 2009 John Wiley & Sons, Inc.

in the weakly adiabatic range, a concept that we will discuss below: sufficiently strong to ensure adiabaticity, but too weak to affect the energy of activation. Only very recently has the role of the reactant–metal interaction, in particular the effect of metal d-bands, been explored [Santos and Schmickler, 2007a, b, c].

In this chapter, we will review electrochemical electron transfer theory on metal electrodes, starting from the theories of Marcus [1956] and Hush [1958] and ending with the catalysis of bond-breaking reactions. On this route, we will explore the relation to ion transfer reactions, and also cover the earlier models for noncatalytic bond breaking. Obviously, this will be a tour de force, and many interesting side-issues will be left unexplored. However, we hope that the unifying view that we present, based on a framework of model Hamiltonians, will clarify the various aspects of this most important class of electrochemical reactions.

Figure 2.1 (Plate 2.1) shows a classification of the processes that we consider; they all involve interaction of the reactants both with the solvent and with the metal electrode. In simple outer sphere electron transfer, the reactant is separated from the electrode by at least one layer of solvent; hence, the interaction with the metal is comparatively weak. This is the realm of the classical theories of Marcus [1956], Hush [1958], Levich [1970], and German and Dogonadze [1974]. Outer sphere transfer can also involve the breaking of a bond (Fig. 2.1b), although the reactant is not in direct contact with the metal. In inner sphere processes (Fig. 2.1c, d) the reactant is in contact with the electrode; depending on the electronic structure of the system, the electronic interaction can be weak or strong. Naturally, catalysis involves a strong

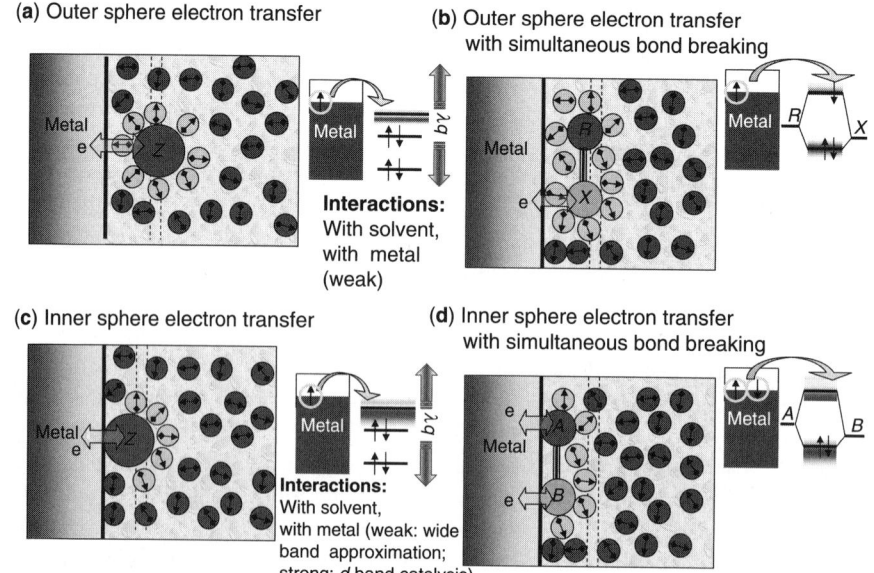

Figure 2.1 Classification of electrochemical electron transfer reaction on metal electrodes. (See color insert.)

interaction. For geometric reasons, the interaction with the solvent is somewhat weaker for inner sphere reactions, than for outer sphere ones. Besides the reactions shown in Fig. 2.1, we shall also briefly discuss ion transfer, in which an ion passes from the outer to the inner sphere position on the surface.

2.2 OUTER SPHERE ELECTRON TRANSFER

The theory of outer sphere electron transfer was first developed for homogeneous reactions in polar solvents. A simple electron exchange between two partners in close proximity was expected to be extremely fast, since the electron can tunnel rapidly from one molecule to the other. Nevertheless, such electron exchange is comparatively slow, and requires an energy of activation. The reason is that the electron transfer is accompanied by a reorganization of the solvent: the initial and the final states of the reactants carry different charges, and therefore their solvation, which is governed by the interaction of the molecular charge with the dipole moment of the solvent, changes during the reaction. Electron transfer is therefore preceded by a fluctuation of the solvation to an intermediate state, and this requires an energy of activation. The key concept of the theories of simple electron transfer is therefore the *energy of reorganization* of the solvent, which determines the energy required to reach this intermediate state; its exact definition will be given below.

These ideas can be applied to electrochemical reactions, treating the electrode as one of the reacting partners. There is, however, an important difference: electrodes are electronic conductors and do not posses discrete electronic levels but electronic bands. In particular, metal electrodes, to which we restrict our subsequent treatment, have a wide band of states near the Fermi level. Thus, a model Hamiltonian for electron transfer must contains terms for an electronic level on the reactant, a band of states on the metal, and interaction terms. It can be conveniently written in second quantized form, as was first proposed by one of the authors [Schmickler, 1986]:

$$H_\mathrm{e} = \varepsilon_a n_a + \sum_k \varepsilon_k n_k + \sum_k \left(V_k c_k^+ c_a + V_k^* c_a^+ c_k\right) \qquad (2.1)$$

Here, n denotes a number operator, c^+ a creation operator, c an annihilation operator, and ε an energy. The first term with the label a describes the reactant, the second term describes the metal electrons, which are labeled by their quasi-momentum k, and the last term accounts for electron exchange between the reactant and the metal; V_k is the corresponding matrix element. This part of the Hamiltonian is similar to that of the Anderson–Newns model [Anderson, 1961; Newns, 1969], but without spin. The neglect of spin is common in theories of outer sphere reactions, and is justified by the comparatively weak electronic interaction, which ensures that only one electron is transferred at a time. We shall consider spin when we treat catalytic reactions.

As discussed above, a crucial aspect is the interaction of the reactant with the solvent. In a quantum theory, the solvent can be represented as a bath of harmonic oscillators, which interact linearly with the reactant. The corresponding Hamiltonian is written in the form

$$H_{\text{osc}} = \frac{1}{2}\sum_{\nu} \hbar\omega_\nu(p_\nu^2 + q_\nu^2) - n_a \sum_{\nu} \hbar\omega_\nu g_\nu q_\nu \qquad (2.2)$$

where ν labels the oscillator modes, which have frequencies ω_ν, dimensionless momenta p_ν, and coordinates q_ν; g_ν denote the coupling constants. In many cases, it is not only the solvent that is reorganized during electron transfer, but also inner sphere modes of the reactants. In a first approximation, these can also be included in the harmonic oscillator bath. However, often the frequencies of such modes also change during the reaction; for a systematic treatment of this effect, we refer to the literature [Schmickler, 1976; Schmickler and Koper, 1999]. To a good approximation, the solvent coordinates can be considered as classical, and we shall do this here. Inner sphere modes are typically in the quantum range of frequencies. Such quantum effect are an interesting topic in their own right [Schmickler, 1996; Kuznetsov, 1995], but they cannot be considered here. We shall assume all harmonic oscillator modes to be classical. The coupling of the electron transfer to the solvent can then be characterized by a single energy of reorganization defined as $\lambda = \sum_\nu \frac{1}{2}\hbar\omega_\nu g_\nu^2$.

The total Hamiltonian is the sum of the two terms: $H = H_e + H_{\text{osc}}$. The way in which the rate constant is obtained from this Hamiltonian depends on whether the reaction is adiabatic or nonadiabatic, concepts that are explained in Fig. 2.2, which shows a simplified, one-dimensional potential energy surface for the reaction. In the absence of an electronic interaction between the reactant and the metal (i.e., all $V_k = 0$), there are two parabolic surfaces: one for the initial state labeled A, and one for the final state B. In the presence of an electronic interaction, the two surfaces split at their intersection point. When a thermal fluctuation takes the system to the intersection, electron transfer can occur; in this case, the system follows the path

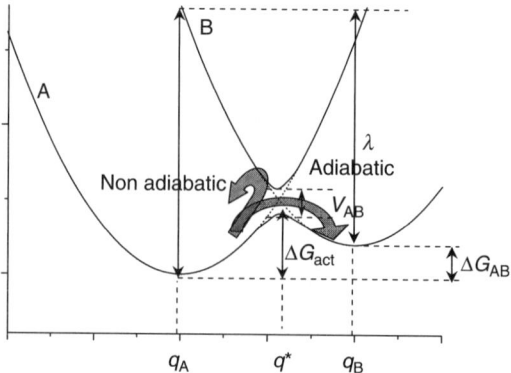

Figure 2.2 Adiabatic and non-adiabatic electron transfer (schematic). The splitting at the intersection has been exaggerated.

$q_A \to q^* \to q_B$. If the interaction is sufficiently strong, electron transfer takes place every time the system reaches the intersection q^*; in this case, the reaction is adiabatic. On the other hand, if the interaction is weak, electron transfer will seldom occur, and the reaction is said to be nonadiabatic. In the latter case, the last two terms in the electronic Hamiltonian (2.1) can be considered as small, and first-order perturbation theory can be used to obtain the rate. However, if the interaction is strong, the adiabatic energy surface can be calculated by Green function techniques [Davison and Sulston, 2006], and the rate obtained from Kramers [1940] theory.

In first-order perturbation theory, the rate can be calculated in a straightforward manner. The rate for a transition from a metal state k to the reactant orbital a is, after thermal averaging [Schmickler, 1996],

$$v(k \to a) = \frac{|V_k|^2}{h} \sqrt{\frac{\pi}{\lambda k_B T}} \exp\left[-\frac{(\varepsilon_a - \varepsilon_k)^2}{4\lambda k_B T}\right] \tag{2.3}$$

where k_B is Boltzmann's constant; a corresponding expression holds for the reverse process. In order to obtain the total rate constant k_{red} for the reduction of an oxidized reactant one first has to fix the energy scale. For this purpose, the Fermi level of the metal is usually taken as the energy zero. The energy scale in the solution is then obtained by noting that at equilibrium the forward and backward rates to the Fermi level must be equal. This gives $\varepsilon_a = -\lambda$ at equilibrium. Application of an overpotential η shifts the energy scales of the metal and the reactant by an amount $e_0\eta$. The coupling constants are usually taken as constant: $V_k = V = $ const. The total rate is then obtained by integrating over all occupied states ε_k. This results in

$$k_{red} = \frac{|V|^2}{h} \sqrt{\frac{\pi}{\lambda k_B T}} \int d\varepsilon \, \rho_m(\varepsilon) f(\varepsilon) \exp\left[-\frac{(\lambda + \varepsilon - e_0\eta)^2}{4\lambda k_B T}\right] \tag{2.4}$$

Here, $f(\varepsilon)$ denotes the Fermi–Dirac distribution, and the integral over ε_k has been converted into an integral over the energy ε by means of the metal density of states $\rho_m(\varepsilon)$. The rate at equilibrium has an energy of activation $\frac{1}{4}\lambda$ and is thus solely determined by the energy of reorganization. The interaction between the reactant and the metal enters only into the pre-exponential factor, and the rate is proportional to the square of the corresponding coupling constant, as is common for theories based on first-order perturbation. On bare metals, the electronic interaction is typically too strong for first-order perturbation theory to be valid. Therefore, (2.4) is usually applied to electrodes covered by an insulating film such as oxides or organic layers.

When the electronic interaction is sufficiently strong, first-order perturbation theory no longer appplies. The reaction proceeds adiabatically, and the system is in electronic equilibrium for all solvent configurations. In order to calculate adiabatic potential energy surfaces, the solvent coordinates q_ν are replaced by a single effective coordinate q; this is permissible as long as all modes are classical. Further, it is convenient to normalize q by the transformation: $\tilde{q} = q/g$; the transformed variable has a simple interpretation: a solvent configuration $\tilde{q}$ would be in equilibrium with a reactant of charge $-\tilde{q}$. Thus, $\tilde{q}$ plays the role of a reaction coordinate in the sense of Kramers [1940] theory; all other modes form part of the heat bath and give rise to friction effects, which will be discussed below. For simplicity, we shall drop the tilde henceforth, and assume that q is normalized.

The energy of the system can be calculated as a function of q by Green function techniques [Davison and Sulston, 2006]. For this purpose, it is convenient to gather the terms containing n_a in the Hamiltonian and introduce an electronic energy that depends on the solvent configuration through

$$\tilde{\varepsilon}_a(q) = \varepsilon_a - 2\lambda q \tag{2.5}$$

The density of states of the reactant is then given by [Schmickler, 1986]

$$\rho(\varepsilon) = \frac{1}{\pi} \frac{\Delta(\varepsilon)}{[\varepsilon - \tilde{\varepsilon}_a(q) - \Lambda(\varepsilon)]^2 + \Delta(\varepsilon)^2} \tag{2.6}$$

where the so-called chemisorption functions [Davison and Sulston, 2006] are

$$\Delta(\varepsilon) = \pi \sum_k |V_{ak}|^2 \delta(\varepsilon - \varepsilon_k), \qquad \Lambda(\varepsilon) = \mathcal{P} \sum_k \frac{|V_{ak}|^2}{\varepsilon - \varepsilon_k} \tag{2.7}$$

$\mathcal{P}$ denotes the principal value.

In outer sphere electron transfer, the reactant is not adsorbed; therefore, the interaction with the metal is not as strong as with the catalytic reactions discussed below. Hence, the details of the metal band structure are not important, and the coupling $\Delta(\varepsilon)$ can be taken as constant. This is the so-called *wide band approximation*, because it corresponds to the interaction with a wide, structureless band on the metal. In this approximation, the function $\Lambda(\varepsilon)$ vanishes, and the reactant's density of states takes the form of a Lorentzian. The situation is illustrated in Fig. 2.3.

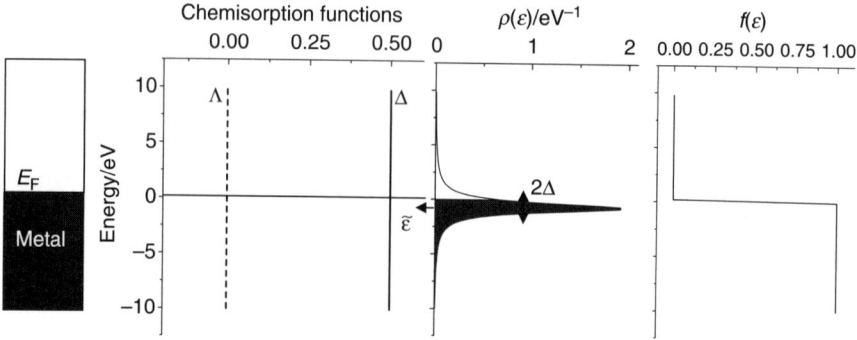

Figure 2.3 The wide band approximation.

2.2 OUTER SPHERE ELECTRON TRANSFER

The electronic contribution to the energy is obtained by integrating over all occupied states. To a good approximation, the Fermi–Dirac distribution can be replaced by a step function, and the integral can be performed up to the Fermi level:

$$E_{\text{elec}} = \int \varepsilon \rho(\varepsilon) f(\varepsilon) \, d\varepsilon \approx \int_{-\infty}^{E_F} \varepsilon \rho(\varepsilon) \, d\varepsilon \qquad (2.8)$$

This integral diverges—a consequence of the wide band approximation—however, this poses no problem. The relevant quantities are the differences in energy between various states. It is natural to take the initial state as a reference. This gives [Schmickler, 1986]

$$E(q) = \tilde{\varepsilon}_a(q)\langle n_a(q)\rangle + \frac{1}{2}\hbar\omega q^2 + \frac{\Delta}{2\pi} \ln \frac{\tilde{\varepsilon}_a(q)^2 + \Delta^2}{\tilde{\varepsilon}_a(0)^2 + \Delta^2} \qquad (2.9)$$

The first term is just the electronic energy of the reactant, including the interaction with the solvent; the second is the energy of the solvent. The sum of these two terms is the adiabatic part of the potential energy in the original Marcus and Hush theory. The last part is the correction due to the finite width of the density of states; the denominator in the logarithmic term is often left out, because it just adds a constant.

Figure 2.4 shows several potential energy curves for various values of the interaction strength Δ. With increasing interaction strength Δ, the energy of activation is reduced. This effect, which corresponds to a catalytic action of the electrode, becomes noticeable when Δ is of the same order of magnitude as the reorganization energy λ.

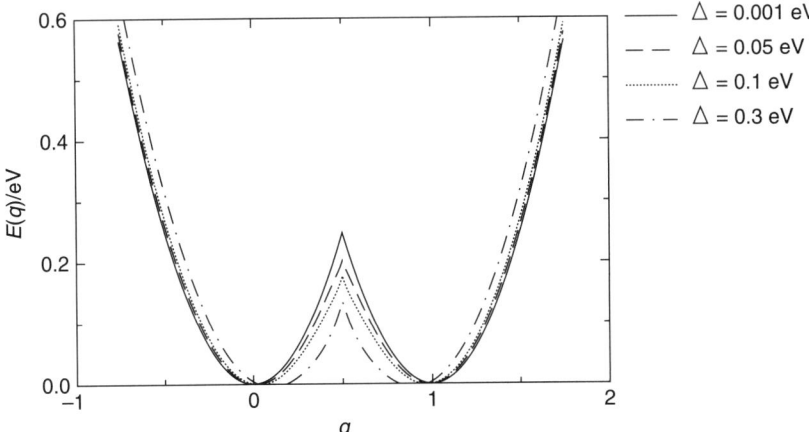

Figure 2.4 Adiabatic potential energy $E(q)$ for various values of Δ at equilibrium. The solvent coordinate q has been normalized so that the initial state corresponds to $q = 0$ and the final state to $q = 1$. The reorganization energy was taken as $\lambda = 1$ eV.

From these potential energy curves, the reaction rate can be calculated with the aid of Kramers theory. In the limit of a high solvent friction γ, the rate is given by Kramers [1940] and Zusman [1980]

$$k = \gamma \exp\left(-\frac{E_{\text{act}}}{k_B T}\right) \qquad (2.10)$$

where the energy of activation is obtained by finding the saddle point from (2.9). In contrast to (2.4), this expression does not contain an integral over ε; this has already be performed in the calculation of the energy from the density of states.

Thus, for weak electronic interactions, the pre-exponential factor is proportional to the strength of the electron interaction; in contrast, in the adiabatic limit, it is determined by the solvent friction. These two cases are bridged by a region with a mixed regime. A typical plot of the dependence of the rate on the interaction strength Δ is shown in Fig. 2.5. For small interactions, the rate is proportional to Δ; with increasing Δ, it reaches a plateau, where it becomes almost constant. This is the so-called weakly adiabatic region, in which the Marcus and Hush theories are valid. On further increase, the energy of activation is lowered, and the rate increases again.

In typical outer sphere electron transfer on metal electrodes, Δ is in the weakly adiabatic region and thus sufficiently large to ensure adiabaticity, but too small to lead to a noticeable reduction of the activation energy. In this case, the rate is determined by solvent reorganization, and is independent of the nature of the metal [Iwasita et al., 1985; Santos et al., 1986].

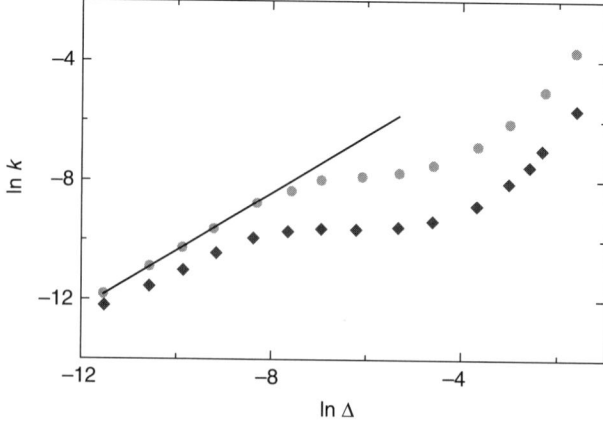

Figure 2.5 Electron transfer rate as a function of the electronic interaction Δ. The full line is the prediction of first-order perturbation theory. The upper points correspond to a solvent with a low friction; the lower points to a high friction. The data have been taken from Schmickler and Mohr [2002].

2.3 SIMPLE ION TRANSFER REACTIONS

In a simple electron transfer reaction, the reactant is situated in front of the electrode, and the electron is transferred when there is a favorable solvent fluctuation. In contrast, during ion transfer, the reactant itself moves from the bulk of the solution to the double layer, and then becomes adsorbed on, or incorporated into, the electrode. Despite these differences, ion transfer can be described by essentially the same formalism [Schmickler, 1995], but the interactions both with the solvent and with the metal depend on the position of the ion. In addition, the electronic level on the reactant depends on the local electric potential in the double layer, which also varies with the distance. These complications make it difficult to perform quantitative calculations.

As a reacting ion moves from the bulk of the solution towards the electrode surfaces, it loses a part of its solvation sheath, and therefore its energy increases. Quantitatively, this effect is described by the potential of mean force (PMF), which is the average potential energy that the ion experiences at a given position. We focus here on the PMF due to the interactions with the solvent, which is the average energy of interaction of the ions with the solvent at a given position. Good estimates can be obtained by molecular dynamics simulations; for the simple case of an iodide ion I^-, the PMF is shown in Fig. 2.6 [Pecina et al., 1995]. In the bulk of the solution, the PMF is constant; in the plot, it has been set to zero. When the ion starts to penetrate the first layer of water in contact with the electrode, its energy rises, until it is in contact with the electrode. The total rise is about 1.5 eV; the hydration energy of the ion in the bulk is about -3 eV, so it loses roughly half of its solvation energy when it moves to the surface.

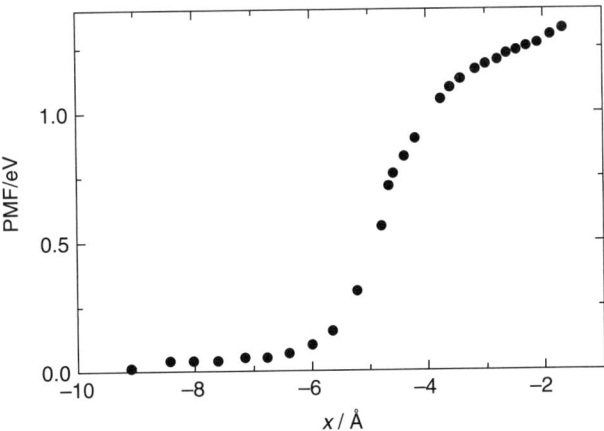

Figure 2.6 Potential of mean force of an iodide ion in front of a Pt(111) electrode due to its interaction with the solvent. The electrode is situated at $x = 0$; the value in the bulk has been set to zero. The data have been taken from Pecina et al. [1995].

The strength of the chemical interaction with the metal, can be expressed by the quantity Δ, which decays roughly exponentially with distance; typically, the decay lengths are of the order of 1 Å [Koper and van Santen, 1999]. Combining the results of quantum chemical calculation with the potential of mean force, the potential energy surface of the ion as a function of the separation and the solvent coordinate can be constructed at the potential of zero charge (PZC) [Schmickler, 1995]—for other electrode potentials, one would need the distribution of the electric potential in the double layer, which is not well known. For the case of the adsorption of an iodide ion, such a plot is shown in Fig. 2.7. At the PZC, adsorption of the ion is favored owing to a strong chemical interaction, but to get from the bulk towards the surface, the ion has to overcome a barrier, whose height is mainly determined by the partial desolvation. In the adsorbed state, the ion is partially discharged, so the minimum is situated near $q = -0.5$. Far from the electrode, the surface shows a high lying valley centered near $q = 0$ corresponding to the uncharged atom. Obviously, at the PZC, the atom is in an energetically unfavorable state.

Similar surfaces have been constructed for other reactions, but this example suffices to illustrate the important point that ion transfer, just like electron transfer, is governed by the interplay of the interactions with the solvent and the electrode.

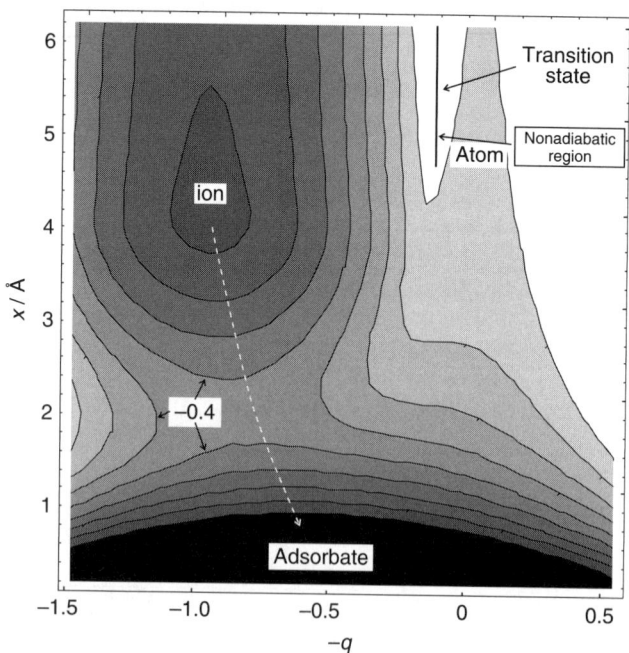

Figure 2.7 Potential energy surface for the adsorption of an iodide ion on Pt(111) [Schmickler, 1995] as a function of the solvent coordinate q and the distance x from the electrode.

2.4 PROTON TRANSFER

Just like the electron, the proton is an elementary particle, which in an aqueous solution never exists just by itself. It associates with water, forming species such as the Eigen H_3O^+ [Eigen, 1963] or the Zundel $H_5O_2^+$ [Zundel and Metzger, 1968] ion, and exchanges rapidly with the hydrogen atoms of the water molecules. Its hydration energy is extremely high, of the order of 12 eV, and so is the ionization energy of the hydrogen atom (about 13.6 eV). Thus, the ionization of a hydrogen atom to form a proton requires much energy, but the gain in solvation energy is of the same order of magnitude. Because of its quantum nature, and its rapid exchange with water, the transport of a proton towards an electrode surface cannot be viewed as the motion of an individual particle, but must occur by a Grotthus-type mechanism. It thus differs essentially from the transfer of a simple ion, which we have discussed above. In particular, it cannot involve a barrier amounting to half of its solvation energy, since this would be far too high to allow the proton to reach the surface.

Besides these generalities, little is known about proton transfer towards an electrode surface. Based on classical molecular dynamics, it has been suggested that the rate-determining step is the orientation of the H_3O^+ with one proton towards the surface [Pecina and Schmickler, 1998]; this would be in line with proton transport in bulk water, where the proton transfer itself occurs without a barrier, once the participating molecules have a suitable orientation. This is also supported by a recent quantum chemical study of hydrogen evolution on a Pt(111) surface [Skulason et al., 2007], in which the barrier for proton transfer to the surface was found to be lower than 0.15 eV. This extensive study used a highly idealized model for the solution—a bilayer of water with a few protons added—and it is not clear how this simplification affects the result. However, a fully quantum chemical model must necessarily limit the number of particles, and this study is probably among the best that one can do at present.

Of course, proton transfer can also occur between two reactants in the solution. As such, it is not an electrochemical reaction, unless it is combined with an electron exchange with the electrode. Such a combined electron–proton transfer can be represented by the scheme of squares shown in Fig. 2.8. Both electron and proton transfer

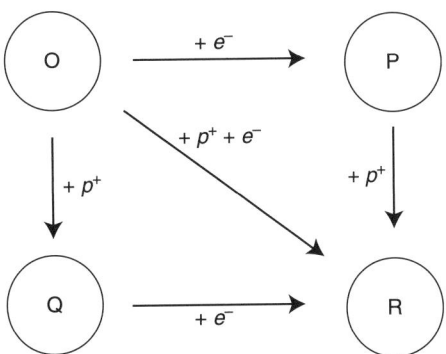

Figure 2.8 Reaction scheme for combined electron and proton transfer.

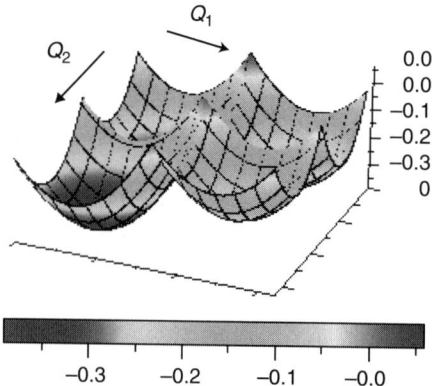

Figure 2.9 Model potential energy surface for combined electron and proton transfer. Q_1 is the solvent coordinate for electron transfer and Q_2 that for proton transfer. (See color insert.)

involve a reorganization of the solvent, and the reaction coordinate can be identified with the corresponding solvent coordinates. This idea has been explored in a recent model [Grimminger et al., 2005] that allows the construction of model potential surfaces. An example is shown in Fig. 2.9 (Plate 2.2); the four minima correspond to the four corners of the scheme of squares. Experimental results are usually discussed in terms of a consecutive mechanism, in which proton and electron transfer occur as separate steps; Fig. 2.9 corresponds to this case. However, in some cases, a concerted mechanism, in which both particles are transferred simultaneously, may predominate; Fig. 2.10 (Plate 2.3) shows a potential energy surface for this case. A concerted process would generally involve less reorganization of the solvent, since both the initial and final states have the same charge. In addition, the Coulomb attraction between electron and proton transfer would favor a simultaneous exchange. Note that only the reaction steps involving electron transfer depend on the electrode potential, since the proton is not transferred to the electrode. Therefore, the dominant mechanism may well change with the potential.

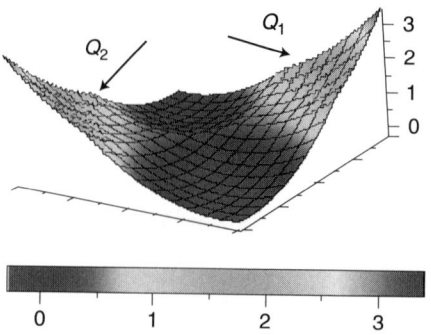

Figure 2.10 Potential energy surface for combined electron and proton transfer. (See color insert.)

2.5 SIMPLE BOND-BREAKING ELECTRON TRANSFER

The breaking of a chemical bond requires additional activation energy. In the simplest case, the reaction proceeds according to

$$R-X + e^- \longrightarrow R^\bullet + X^- \qquad (2.11)$$

and neither the reactant nor the products are adsorbed. A quantitative theory can be based on an extension of the Hamiltonian for simple electron transfer. Koper and Voth [1998] proposed to add terms describing the interaction between R and X in the molecule, and R and X^- after the reaction, as a function of the separation r. The corresponding expression in the Hamiltonian can be written in the form

$$H_{bb} = (1 - n_a)H_i(r) + n_a H_f(r) \qquad (2.12)$$

and thus consists of a term that acts before and one that acts after electron transfer.

Explicit forms for the potential energy in the terms H_i and H_f have been proposed by Savéant [1993], who has developed a semiclassical version, along the lines of the Marcus theory, and applied it successfully to several reactions. In his model, the potential curve for the reactants is a Morse curve, and that for the products is the repulsive branch of a Morse curve:

$$H_i = -\frac{1}{2m\hbar^2}\frac{d^2}{dr^2} + D(e^{-2r/l} - 2e^{-r/l}) \qquad (2.13)$$

$$H_f = -\frac{1}{2m\hbar^2}\frac{d^2}{dr^2} + De^{-2r/l} \qquad (2.14)$$

where m is the effective mass along the bond, $-D$ is the binding energy, and l is the decay length of the Morse potential. The corresponding potential energy surfaces now depend on the solvent coordinate q and the bond distance r; a typical example is shown in Fig. 2.11. The initial state, the intact molecule, corresponds to the minimum centered at $q = 0$, $r = 0$. In the final state, the bond has been broken, and the energy surface shows a trough centered at $q = 1$. The two regions are separated by an energy barrier, whose height is determined by both the energy of reorganization and the binding energy.

In the case where the bond coordinate can be treated as classical, and when the electronic interaction Δ is much smaller than the solvent reorganization, the energy of activation can be calculated explicitly in Savéant's [1993] model:

$$E_{act} = \frac{(\lambda + D - e_0\eta)^2}{4(\lambda + D)} \qquad (2.15)$$

Thus, the energy of reorganization λ from Marcus theory is replaced by the sum $\lambda + D$, and the activation energy is significantly enhanced.

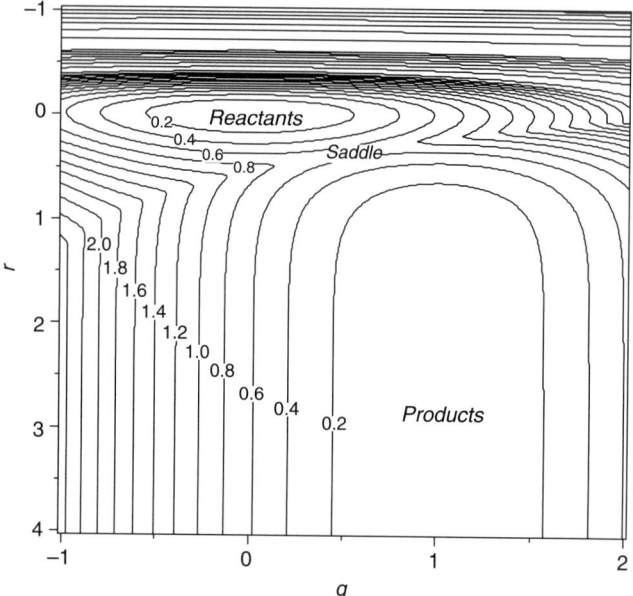

Figure 2.11 Potential energy surface for a simple bond-breaking reaction in Savéant's model [Savéant, 1993].

2.6 INTERACTION OF A REACTANT ORBITAL WITH A NARROW BAND

In the electron transfer theories discussed so far, the metal has been treated as a structureless donor or acceptor of electrons—its electronic structure has not been considered. Mathematically, this view is expressed in the wide band approximation, in which Δ is considered as independent of the electronic energy ε. For the sp-metals, which near the Fermi level have just a wide, structureless band composed of s- and p-states, this approximation is justified. However, these metals are generally bad catalysts; for example, the hydrogen oxidation reaction proceeds very slowly on all sp-metals, but rapidly on transition metals such as platinum and palladium [Trasatti, 1977]. Therefore, a theory of electrocatalysis must abandon the wide band approximation, and take account of the details of the electronic structure of the metal near the Fermi level [Santos and Schmickler, 2007a, b, c; Santos and Schmickler, 2006].

Before considering our model for electrocatalysis, it is instructive to investigate the interaction of a single reactant orbital with a model metal containing a wide sp-band and a narrow d-band. For this purpose, it is convenient to use the model of a semi-elliptic band [Newns, 1969], for which several important quantities can be calculated explicitly. A single such metal band has the form

$$\rho_m(\varepsilon) = \left[1 - \left(\frac{\varepsilon - \varepsilon_c}{w}\right)^2\right]^{1/2} \theta[w^2 - (\varepsilon - \varepsilon_c)^2] \qquad (2.16)$$

2.6 INTERACTION OF A REACTANT ORBITAL WITH A NARROW BAND

where w denotes the width of the band and ε_c its center. We consider two such bands: a wide sp-band centered at the Fermi level for simplicity and a narrow d-band of arbitrary width w_d and center ε_c, and assume that each band couples to a reactant level with a certain coupling constant: Δ_{sp} for the sp-band and Δ_d for the d-band. Thus, we obtain for the chemisorption function Δ,

$$\Delta(\varepsilon) = \Delta_{sp}\left[1 - \left(\frac{\varepsilon}{w_0}\right)^2\right]^{1/2} \theta(w_0^2 - \varepsilon^2)$$

$$+ \Delta_d\left[1 - \left(\frac{\varepsilon - \varepsilon_c}{w_d}\right)^2\right]^{1/2} \theta[w_d^2 - (\varepsilon - \varepsilon_c)^2] \quad (2.17)$$

For the explicit calculations presented below, we have chosen a width $w_0 = 10\,\text{eV}$ for the sp-band, and a coupling strength $\Delta_{sp} = 0.2\,\text{eV}$, and have varied the parameters for the d-band. The level shift $\Lambda(\varepsilon)$ is obtained from the second part of (2.7). The resulting functions are illustrated in Fig. 2.12.

Figure 2.13 shows the effect that a d-band has on the density of states of a level positioned at its center. When the interaction is weak, the level is just broadened. However, when the interaction is strong, the level is split into a bonding orbital (with respect to the metal) and an antibonding orbital, which lie below and above the d-band, respectively. The same effects can be observed when the level lies originally above or below the d-band (Fig. 2.14). For a weak interaction, the level just

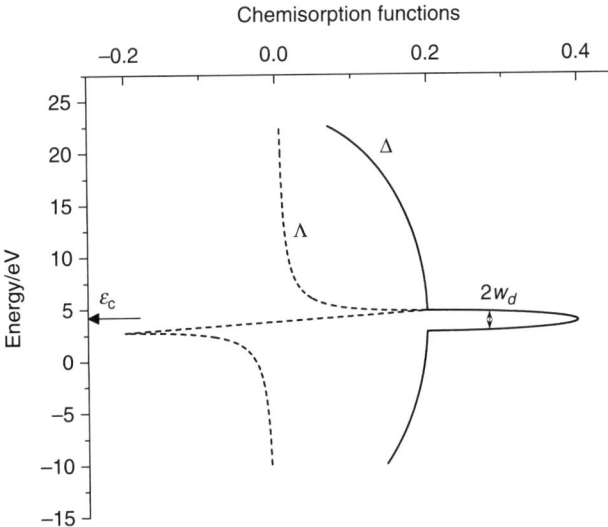

Figure 2.12 Chemisorption functions for the model metal band structure.

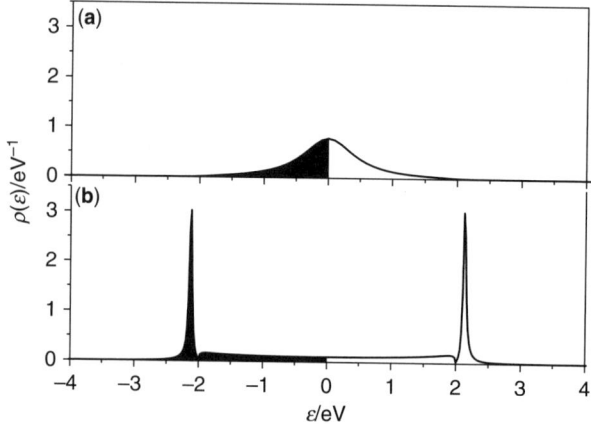

Figure 2.13 Density of states of a level positioned at the center of a d-band for the cases of a strong and a weak interaction. The d-band is centered at $\varepsilon = 0$ and has a half-width of 2 eV. (a) $\Delta_d = 0.3$ eV; (b) $\Delta_d = 3$ eV.

acquires a little density of states at the energy of the d-band; for a strong interaction, the level is again split into an antibonding orbital, which has been shifted upwards, and a bonding orbital, which has been shifted downwards. However, the two levels do not have equal weight. For a more detailed discussion, we refer to the paper by [Newns, 1969]. This splitting of an orbital by a strong interaction with a narrow d-band is of central importance for electrocatalysis, as we shall see below.

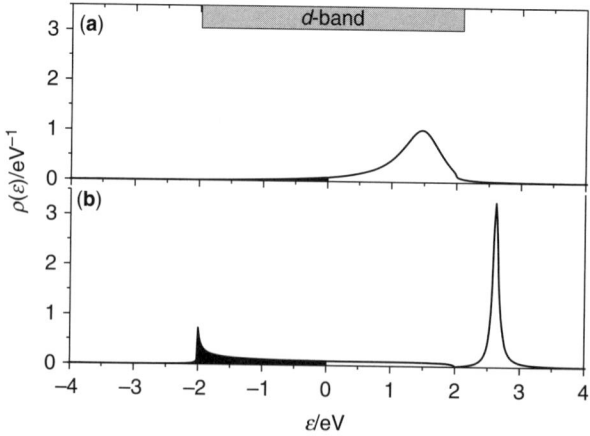

Figure 2.14 Density of states of a level originally positioned at $\varepsilon = 1.2$ eV above a d-band, for the cases of a strong and a weak interaction. The position of the d-band is as in Fig. 2.13. (a) $\Delta_d = 0.3$ eV; (b) $\Delta_d = 3$ eV.

2.7 SIMPLE INNER SPHERE ELECTRON TRANSFER

As demonstrated in Section 2.2, the energy of activation of simple electron transfer reactions is determined by the energy of reorganization of the solvent, which is typically about 0.5–1 eV. Thus, these reactions are typically much faster than bond-breaking reactions, and do not require catalysis by a d-band. However, before considering the catalysis of bond breaking in detail, it is instructive to apply the ideas of the preceding section to simple electron transfer, and see what effects the abandonment of the wide band approximation has.

In inner sphere electron transfer, the interaction is so strong that the adiabatic theory applies. Equations (2.6)–(2.8) remain valid, but, because of the more complicated forms of the chemisorption functions $\Delta(\varepsilon)$ and $\Lambda(\varepsilon)$, the electronic energy can no longer be calculated analytically. However, numerical integration is easy, and potential energy curves analogous to those shown in Fig. 2.4 can be calculated with ease. A few examples are shown in Fig. 2.15. A coupling with a localized d-band can greatly diminish the energy of activation—in extreme cases, it may even disappear [Santos and Schmickler, 2007a, b, c]. Also, the coupling with a metal band not centered at the Fermi level leads to a lack of symmetry between the forward and backward reactions. The curves shown in Fig. 2.15 refer to equilibrium, even though the energies of the initial states are somewhat lower than those for the final state. Equilibrium is determined by the situation in the bulk, where the electronic interactions are absent and where, for the energy $\varepsilon_a = -\lambda$ chosen (see Section 2.2), the initial and final states have the same energy.

In the activated state, the valence orbital passes the Fermi level of the metal, and the electron transfer occurs. A d-band situated near the Fermi level will induce a strong broadening of the reactant's density of states, or even to a splitting, as shown in

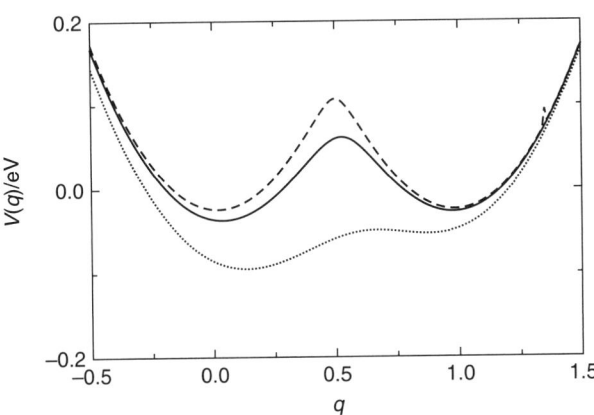

Figure 2.15 Potential energy curves for a simple electron transfer reaction at equilibrium. The system parameters are $\lambda = 0.75$ eV, $w_0 = 10$ eV, $\Delta_{sp} = 0.05$ eV, $w_d = 1$ eV, and $\varepsilon_c = -0.5$ eV. Dashed line, $\Delta_d = 0$; full line, $\Delta_d = 0.1$ eV; dotted line, $\Delta_d = 0.5$ eV.

the previous section. This entails a reduction of the electronic energy. We will discuss this effect in greater detail for the catalysis of bond breaking, where it is much more pronounced.

2.8 ELECTROCATALYSIS OF BOND BREAKING

In electrocatalysis, the reactants are in contact with the electrode, and electronic interactions are strong. Therefore, the one-electron approximation is no longer justified: at least two spin states on a valence orbital must be considered. Further, the form of the bond Hamiltonian (2.12) is not satisfactory, since it simply switches between two electronic states. This approach becomes impractical with two spin states in one orbital; also, it has an ad hoc nature, which is not satisfactory.

In order to obtain a better model for the molecular bond, [Santos et al., 2006] employed the extended Hückel, or tight binding, theory. For the breaking of the bond in a diatomic molecule according to the schemes

$$A-B + 2e^- \longrightarrow A^- + B^- \quad \text{or} \quad A-B \longrightarrow A^+ + B^+ + 2e^- \qquad (2.18)$$

we consider one atomic orbital for each atom, which can take up two electrons with opposite spin. We shall later specialize to the case of a homonuclear molecule, but it is convenient to start with the general case. In the tight binding theory, the Hamiltonian of the isolated molecule is

$$H_{\text{mol}} = \sum_\sigma \left(\varepsilon_a n_{a,\sigma} + \varepsilon_b n_{b,\sigma} + V_{ab} c^+_{a\sigma} c_{b\sigma} + V^*_{ab} c^+_{b\sigma} c_{a\sigma} \right)$$
$$+ U_a n_{a,\sigma} n_{a,-\sigma} + U_b n_{b,\sigma} n_{b,-\sigma} \qquad (2.19)$$

Here the indices a and b stand for the valence orbitals on the two atoms; as before, n is a number operator, c^+ and c are creation and annihilation operators, and σ is the spin index. The third and fourth terms in the parentheses effect electron exchange and are responsible for the bonding between the two atoms, while the last two terms stand for the Coulomb repulsion between electrons of opposite spin on the same orbital. As is common in tight binding theory, we assume that the two orbitals a and b are orthogonal; we shall correct for this neglect of overlap later. The coupling V_{ab} can be taken as real; we set $V_{ab} = \beta \leq 0$.

The Hamiltonian for the metal is the same as before, but the metal states now couple to two atomic orbitals:

$$H_{\text{met}} = \sum_{k,\sigma} [\varepsilon_k n_{k,\sigma} + (V_{a,k} c^+_{k,\sigma} c_{a,\sigma} + V_{b,k} c^+_{k,\sigma} c_{b,\sigma} + \text{h.c.})] \qquad (2.20)$$

where h.c. stands for the hermitian conjugate. The phonon bath interacts with the charges on both ions. The region of interest is immediately after the bond breaking, when the ions are close; therefore, we can assume that the interaction is proportional to the charge on both ions. Using normalized coordinates, the corresponding terms are

$$H_{sol} = \lambda\left[q^2 + p^2 + 2(Z-n)q\right], \quad \text{where } n = \sum_\sigma (n_{a,\sigma} + n_{b,\sigma}) \quad (2.21)$$

Z is the charge number of the molecule when the orbitals a and b are empty.

By using the Hartree–Fock approximation, the Coulomb repulsion terms in the molecular Hamiltonian H_{mol} can be reduced to one-electron terms, and the density of states can be calculated. The isolated molecule shows the expected splitting into a bonding and an antibonding orbital. In the homonuclear case, which we consider from here on, the energy separation between the two orbitals is determined by the interaction β and by the Coulomb repulsion $U = U_a = U_b$. The interaction β depends on the separation r between the two atoms, usually exponentially: $\beta = -A \exp(-r/l)$, $A > 0$. If we re-introduce the neglected orbital overlap and use the Wolfsberg–Helmholtz approximation, the potential energy of the isolated molecule assumes a Morse shape, similar to Savéant's model. The details are given in [Santos et al., 2006, 2008].

From the given Hamiltonian, adiabatic potential energy surfaces for the reaction can be calculated numerically [Santos and Schmickler 2007a, b, c; Santos and Schmickler 2006]; they depend on the solvent coordinate q and the bond distance r, measured with respect to its equilibrium value. A typical example is shown in Fig. 2.16a (Plate 2.4) it refers to a reduction reaction at the equilibrium potential in the absence of a d-band ($\Delta_d = 0$). The stable molecule correspond to the valley centered at $q = 0$, $r = 0$, and the two separated ions correspond to the trough seen for larger r and centered at $q = 2$. The two regions are separated by an activation barrier, which the system has to overcome.

In order to understand the nature of electrocatalysis, it is instructive to consider the evolution of the density of states of the reactant (Fig. 2.17). In the initial state, it shows a filled bonding orbital below the Fermi level and an empty antibonding orbital above. In the final state, the bond has been broken, the difference between bonding and antibonding orbitals has vanished, and the system has just one doubly degenerate orbital below the Fermi level, filled with four electrons. The crucial stage in the reaction occurs when, owing to a thermal fluctuation, the antibonding orbital passes through the Fermi level. For a reaction at equilibrium, the saddle point corresponds to the situation where the antibonding orbital is about half filled. In the absence of d-band catalysis, the antibonding orbital is somewhat broadened by the interactions with the sp-band. This broadening reduces the energy of activation somewhat, because the electronic energy of the system is obtained by integrating $\varepsilon\rho(\varepsilon)$ up to the Fermi level.

Figure 2.16b shows a potential energy surface in the presence of a band situated near the Fermi level and coupling strongly to the reactant. The energy of activation is significantly reduced. This catalysis can be readily understood by examining the density of states (Fig. 2.18). At the saddle point, the antibonding orbital of the molecule is split into two parts: a bonding and an antibonding part with respect to the metal d-band. This is the splitting mechanism for strong interactions that we discussed

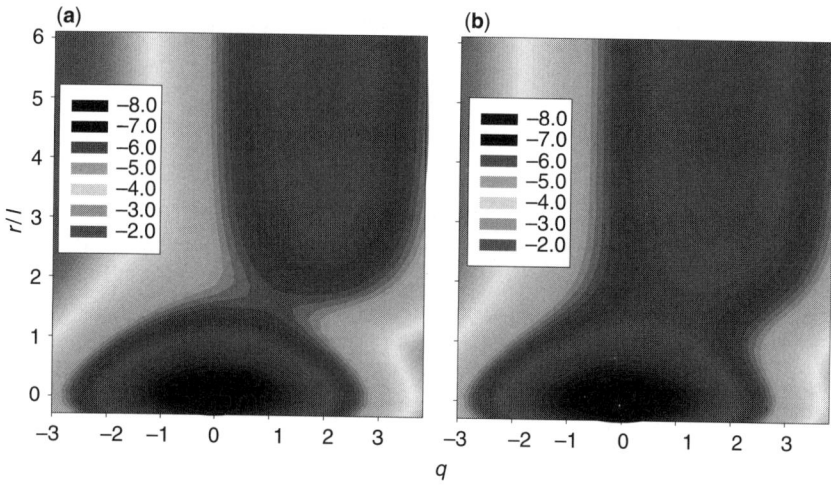

Figure 2.16 Potential energy surfaces for the reduction of a diatomic molecule: (a) without a d-band; (b) in the presence of a d-band. The parameters are $\Delta^0_{sp} = 0.1$ eV, $\lambda = 0.5$ eV $A = 4$ eV; in (b), $\Delta^0_d = 1.0$ eV, $\varepsilon_c = -0.5$ eV, and $w_d = 1$ eV. (See color insert.)

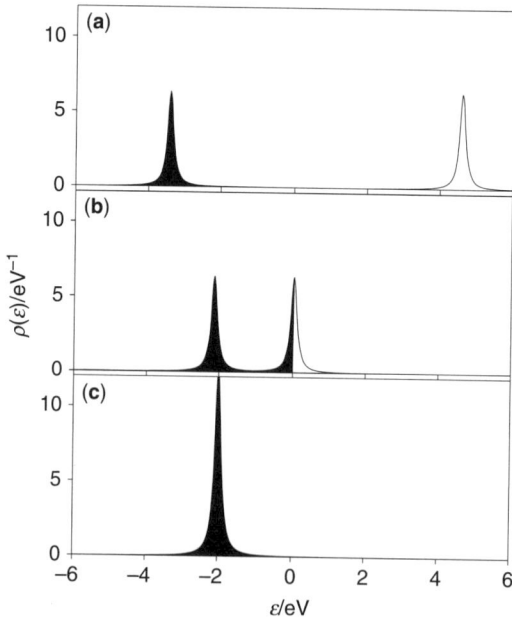

Figure 2.17 Evolution of the molecular density of states in the absence of a d-band, corresponding to Fig. 2.16a. (a) Initial state, stable molecule. (b) Activated state. (c) Final state, two anions.

2.8 ELECTROCATALYSIS OF BOND BREAKING

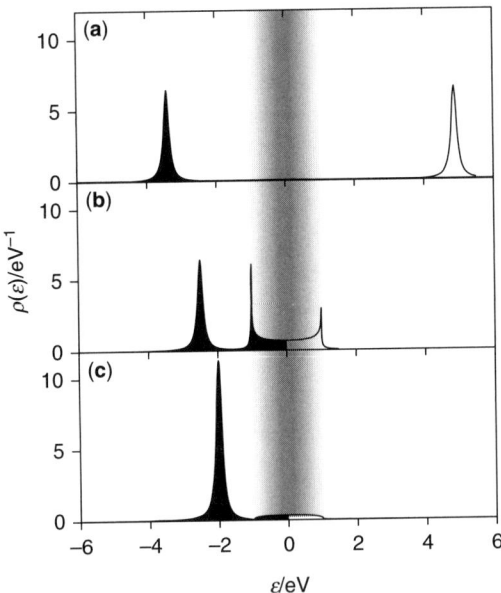

Figure 2.18 Evolution of the molecular density of states in the presence of a d-band, corresponding to Fig. 2.16b. (a) Initial state, stable molecule. (b) Activated state. (c) Final state, two anions. The shading indicates the position of the d-band.

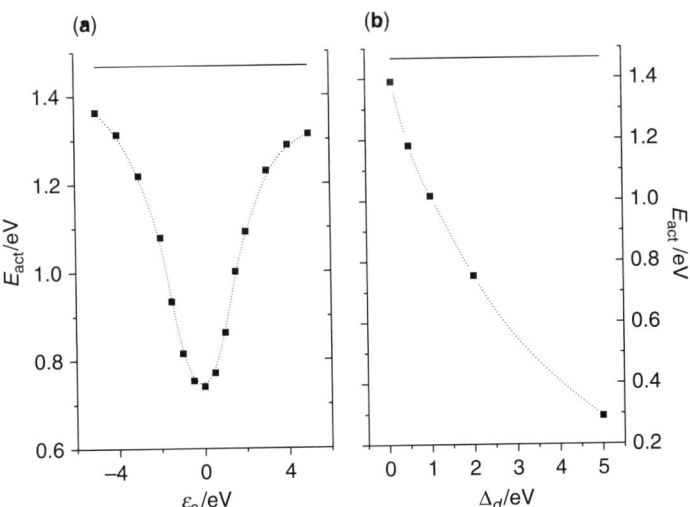

Figure 2.19 The energy of activation of the reduction reaction for various interactions with the d-band: (a) as a function of the position of the d-band center; (b) as a function of the coupling strength. The parameters are $\Delta_{sp}^0 = 0.1$ eV, $w_d = 1$ eV, $\lambda = 0.5$ eV, and $\mathbf{A} = 4$ eV; in (a), $\Delta_d^0 = 2.0$ eV; in (b), $\varepsilon_c = -0.5$ eV. The horizontal line indicates the value in the absence of the d-band.

in the previous section. Only the bonding part is filled, and, since it lies well below the Fermi level, the electronic energy of the transition state is significantly reduced.

A natural question is: for a given width of a d-band and a fixed interaction strength, what is the optimum position of the center of the band for catalysis? As Fig. 2.19a shows, this is near the Fermi level—a fact that is easy to understand in terms of the splitting mechanism. Conversely, for a fixed position of the band, an increase in the coupling strength can reduce the activation energy significantly. In the model calculation presented here, the activation energy is reduced by more than 1 eV; thus, a process that at sp-metals is practically too slow to be observable can become quite fast at transition metals. Of course, this is exactly what is observed for the hydrogen oxidation reaction: at sp-metals such as mercury and lead it is extremely slow, while at platinum, palladium, and several other transition metals it is very fast. Indeed, a first application of our model to the hydrogen oxidation on a series of metals shows a good correlation with experimental results [Santos and Schmickler, 2007a, b, c; Santos and Schmickler, 2008].

2.9 CONCLUSIONS

Electrochemical electron transfer reactions take place at the interface between an electrode and a solution, and therefore they are governed by the interaction of the reactant both with the solvent and with the electrode. To a good approximation, the interaction of ions with the solvent can be treated by classical statistical mechanics, while the chemical interaction between a reactant and a metal electrode must be calculated by advanced quantum chemical methods, which have only become available during the last decade. Therefore, it is logical that the theory of electron transfer at first focused on the solvent reorganization, and that the electrode was simply considered as a reservoir of electrons. This is the essence of the theories of outer sphere electron transfer reactions originated by Marcus and Hush. Since in this class of reactions the reactants are not adsorbed, they interact relatively weakly with the metal. Indeed, most reactions of this type seem to fall into the class of weakly adiabatic interactions, which are sufficiently strong to ensure adiabaticity, but too weak to affect the energy of activation. Therefore, these theories are quite successful for outer sphere reactions, and for a long time they formed the basis of our understanding of electrochemical electron transfer.

Reactions of practical interest involve the breaking or formation of chemical bonds, which require extra energy. The theory of Savéant and its subsequent developments are an ingenious extension of the Marcus–Hush type of theory to the breaking of a simple bond. The binding energy enters into the energy of activation, but the interaction with the metal electrode is still assumed to be weak, i.e., the reactants are not adsorbed. In this sense, the reaction is not catalyzed by the electronic interaction with the metal.

In a simple ion transfer reaction, the distance of the reactant to the surface changes, and it becomes quite strong when it is actually in contact with the metal. Thus, a full description requires a good treatment of the interaction both with the solvent and with the metal. Nevertheless, the energy of activation is mainly determined by the partial loss of the solvation shell. Unfortunately, reliable calculations for these reaction can presently only be made for the potential of zero charge, since the distribution of the double-layer potential is not known sufficiently well.

Many important reactions, in particular those that occur in fuel cells, take place right on the electrode surface, and require catalysis for technological applications. Obviously, the concepts familiar from outer sphere reactions no longer apply; the electronic structure of the electrode and the interaction with the reactant are of prime importance. In this respect, electrochemical reactions resemble catalytic reactions in the gas phase. However, a decisive difference is that, during the course of an electrochemical reaction, the electrode exchanges electrons with the reactant. Either the initial or the final states, involves ions, which are stabilized by the solvent. This also implies that in the activated state either the bonding or the antibonding orbital passes through the Fermi level of the electrode. This complicated scenario makes it difficult, if not impossible, to calculate the course of an electrochemical reaction by standard quantum chemical methods. The merit of the model for electrocatalysis proposed by us is, we believe, that it provides a framework describing the whole course of bond-breaking electron transfer, which explains how a catalyst acts. In applications to real systems, the key parameters of that model—interaction strengths, densities of states, and energies of reorganization—will have to be obtained from quantum chemical methods and from simulations. The power of these methods, together with the speed of computers, has increased tremendously during the last decades, so that quantitative calculations of catalytic activity now seem to be in reach. We may hope that ultimately we will be able to predict what should be a good catalyst for a particular reaction.

ACKNOWLEDGMENTS

Financial support by the Deutsche Forschungsgemeinschaft (Schm 344/34-1, Sa 1770/1-1), the European Union (NENA project), the Deutsche Akademische Austauschdienst, and CONICET (Argentina) is gratefully acknowledged.

REFERENCES

Anderson PW. 1961. Localized magnetic states in metals. Phys Rev 124: 41–53.

Bockris JO'M, Khan SUM. 1993. Surface Electrochemistry. New York: Plenum.

Davison SG, Sulston KW. 2006. Green Function Theory of Chemisorption. London: Springer Verlag.

Eigen M. 1963. Protonenübertragung, Säure-Base-Katalyse und enzymatische Hydrolyse. Teil I: Elementarvorgänge. Angewandte Chemie, 75: 489–508.

Gerischer H. 1956. Über den Zusammenhang zwischen dem Mechanismus der elektrolytischen Wasserstoffabscheidung und der Adsorptionsenergie des atomaren Wasserstoffs an verschiedenen Metallen. Z Phys Chem NF 8: 137–153.

German ED, Dogonadze RR. 1974. The kinetics of nucleophilic substitution processes in the alkyl halides. Part A—theory. Int J Chem Kin 6: 457–466, and references therein.

German ED, Kuznetsov AM. 1994. Quantum mechanical theory of dissociative electron transfer in polar solvents. J Phys Chem 98: 6120–6127.

Grimminger J, Bartenschlager S, Schmickler W. 2005. A model for combined electron and proton transfer in electrochemical systems. Chem Phys Lett 416: 316–320.

Gurney RW. 1936. Ions in Solution. Proc Roy Soc A. Cambridge University Press. 134 (1931): 137.

Hush NS. 1958. Adiabatic rate processes at electrodes. I. Energy-charge relationships. J Chem Phys 28: 962–971.

Iwasita T, Schmickler W, Schultze JW. 1985. The influence of the metal on the kinetics of outer sphere redox reactions. Ber Bunsenges Phys Chem 89: 138–140.

Koper MTM, Voth GA. 1998. A theory for adiabatic bond breaking electron transfer reactions at metal electrodes. Chem Phys Lett 282: 100–106.

Koper MTM, van Santen RA. 1999. Interaction of halogens with Hg, Ag and Pt surfaces: A density functional study. Surf Sci 422: 118–131.

Kramers HA. 1940. Brownian motion in a field of force and the diffusional model of chemical reactions. Physica 7: 284–304.

Kuznetsov AN. 1995. Charge Transfer in Physics, Chemistry and Biology. Gordon & Breach.

Levich VG. 1970. Kinetics of reactions with charge transfer. In: Eyring H, Henderson D, Jost W, editors. Physical Chemistry, an Advanced Treatise. Vol. Xb. New York: Academic Press.

Marcus RA. 1956. On the theory of oxidation-reduction reactions involving electron transfer. I. J Chem Phys 24: 966–979.

Newns DM. 1969. Self consistent model of hydrogen chemisorption. Phys Rev 178: 1123–1135.

Parsons R. 1958. The rate of electrolytic hydrogen evolution and the heat of adsorption of hydrogen. Trans Farad Soc 94: 1053–1063.

Pecina O, Schmickler W, Spohr E. 1995. On the mechanism of electrochemical ion transfer reactions. J Electroanal Chem 394: 29–34.

Pecina O, Schmickler W. 1998. On the dynamics of electrochemical ion-transfer reactions. J Electroanal Chem 450: 303–311.

Santos E, Iwasita T, Vielstich W. 1986. On the use of the coulostatic method for the investigation of fast redox systems. Electrochim Acta 31: 431–437.

Santos E, Koper MTM, Schmickler W. 2006. A model for bond-breaking electron transfer at metal electrodes. Chem Phys Lett 419: 421–425.

Santos E, Schmickler W. 2006. d-Band catalysis in electrochemistry. Chem Phys Chem 7: 2282–2285.

Santos E, Schmickler W. 2007a. Catalyzed bond-breaking electron transfer: Effect of the separation of the reactant from the electrode. J Electroanal Chem 607: 101–106.

Santos E, Schmickler W. 2007b. Electrocatalysis of hydrogen oxidation—Theoretical foundations. Angew Chem Int Ed 46: 8262–8265.

Santos E, Schmickler W. 2007c. Fundamental aspects of electrocatalysis. Chem Phys 332: 39–48.

Santos E, Koper MTM, Schmickler W. 2008. Bond-breaking electron transfer of diatomic reactants at metal electrodes. Chem Phys 344: 195–201.

Santos E, Schmickler W. 2008. Electronic interactions decreasing the activation barrier for the hydrogen electro-oxidation reaction. Electrochim Acta 53: 6149–6156.

Savéant JM. 1993. Electron transfer, bond breaking, and bond formation. Acc Chem Res 26: 455–461.

Schmickler W. 1976. The effect of quantum vibrations on electrochemical outer sphere redox reactions. Electrochim Acta 21: 161–168.

Schmickler W. 1986. A theory of adiabatic electron transfer reactions. J Electroanal Chem 204: 31–43.

Schmickler W. 1995. A unified model for electrochemical electron and ion transfer reactions. Chem Phys Lett 152–160.

Schmickler W. 1996. Interfacial Electrochemistry. New York: Oxford University Press.

Schmickler W, Koper MTM. 1999. Adiabatic electrochemical electron-transfer reactions involving frequency changes of inner-sphere modes. Electrochem Comm 1: 402–405.

Schmickler W, Mohr J. 2002. The rate of electrochemical electron-transfer reactions. J Chem Phys 117: 2867–2872.

Skulason E, Karlberg GS, Rossmeisl J, Bligaard T, Greeley J, Jonsson H, Nørskov JK. 2007. Phys Chem Chem Phys 25: 3241.

Trasatti S. 1972. Work function, electronegativity, and electrochemical behaviour of metals: III. Electrolytic hydrogen evolution in acid solutions. J Electroanal Chem 39: 163–184.

Trasatti S. 1977. Electrocatalysis of hydrogen evolution. Adv Electrochem Electrochem Eng 10: 213.

Zundel G, Metzger H. 1968. Zeitschrift Physikal. Chemie NF 58: 225.

Zusman LD. 1980. Outer-sphere electron transfer in polar solvents. Chem Phys 49: 295–304.

CHAPTER 3

Electrocatalysis and Catalyst Screening from Density Functional Theory Calculations

J. ROSSMEISL

Center for Atomic-Scale Materials Design, NanoDTU, Department of Physics, Technical University of Denmark, DK-2800 Kongens Lyngby, Denmark

J. GREELEY

Center for Nanoscale Materials, Argonne National Laboratory, Argonne, IL 60439, USA

G.S. KARLBERG

Center for Atomic-Scale Materials Design, NanoDTU, Department of Physics, Technical University of Denmark, DK-2800 Kongens Lyngby, Denmark

3.1 INTRODUCTION

A realistic atomistic description of electrochemical reactions taking place at the water–metal interface is a very demanding task from a computational point of view. To begin with, it is necessary to describe the adsorption of atoms and molecules on the surface. This part of the task is analogous to the description of chemical processes in heterogeneous catalysis, for which quite detailed calculations based mainly on density function theory (DFT) are beginning to appear [Behler et al., 2005; Bucko et al., 2005; Hensen et al., 2005; Honkala et al., 2005; Kieken et al., 2005; Xu et al., 2005; Zellner et al., 2005]. In addition, the description of electrochemical reactions at the water–solid interface presents a number of new challenges; solvation effects due to the presence of water, effects of the applied bias, and effects due to counter-ions. Finally, nonadiabatic effects in connection with electron transfer may be important. Recently a number of different approaches have been proposed to handle various aspects of these challenges [Filhol and Neurock, 2006; Koper, 2005; Nørskov et al., 2004; Roques et al., 2005].

Whereas electrochemistry is a very rich and many-facetted science, we will focus here on systems with metal electrodes in contact with acid or alkaline solutions.

Fuel Cell Catalysis. Edited by Marc T. M. Koper
Copyright © 2009 John Wiley & Sons, Inc.

In spite of the importance of having an accurate description of the real electrochemical environment for obtaining absolute values, it seems that for these systems many trends and relative features can be obtained within a somewhat simpler framework. To make use of the wide range of theoretical tools and models developed within the fields of surface science and heterogeneous catalysis, we will concentrate on the effect of the surface and the electronic structure of the catalyst material. Importantly, we will extend the analysis by introducing a simple technique to account for the electrode potential. Hence, the aim of this chapter is to link the successful theoretical surface science framework with the complicated electrochemical environment in a model simple enough to allow for the development of both trends and general conclusions.

Once we have developed our basic model and shown how it may be used to establish trends in electrochemical reactivity, we will take the further step of applying it to the identification of new bimetallic electrocatalysts. We will introduce simple procedures to rapidly screen bimetallic alloys for promising electrocatalytic properties, and we will demonstrate the importance of including estimates of the alloys' stability in the screening procedure. Finally, we will give examples of successful application of this method to specific problems in the area of electrocatalyst development.

3.2 THEORETICAL STANDARD HYDROGEN ELECTRODE

Ab initio atomic simulations are computationally demanding; present day computers and theoretical methods allow simulations at the quantum mechanical level of hundreds of atoms. Since an electrochemical cell contains an astronomical number of atoms, however, simplifications are essential. It is therefore obvious that it is necessary to study the half-cell reactions one by one. This, in turn, implies that a reference electrode with a known fixed potential is needed. For this purpose, a theoretical counterpart to the standard hydrogen electrode (SHE) has been established [Nørskov et al., 2004]. We will describe this model in some detail below.

Consider the following reaction:

$$\tfrac{1}{2}H_2(g) \longleftrightarrow H^* \longleftrightarrow H^+(aq) + e^- \tag{3.1}$$

where H^* is hydrogen bound on the surface. Ignoring the H^*, we get

$$\tfrac{1}{2}H_2(g) \longleftrightarrow H^+(aq) + e^- \tag{3.2}$$

At the conditions used to define the standard hydrogen electrode potential, $T = 300$K, $p_{H_2} = 1$ bar, and pH $= 0$, the reaction free energy of any of these reactions, ΔG, is zero. Hence, the free energies of $\tfrac{1}{2}H_2(g)$ and $H^+(aq) + e^-$ are equal. This defines $U = 0$ V. At another potential, the chemical potential of the electrons e^- is changed by a factor of $-eU$ with respect to H_2 in the gas phase. Taking this into account, the free energy of H adsorbed on the surface can be related to the free energy of H_2

in the gas phase at a given potential U via:

$$\Delta G(U) = \Delta G_0 + eU \qquad (3.3)$$

where ΔG_0 is the free energy of reaction for

$$\tfrac{1}{2}H_2 \longleftrightarrow H^* \qquad (3.4)$$

at standard conditions. This free energy can be directly calculated using DFT and standard molecular tables via

$$\Delta G_0 = \Delta E + \Delta \text{ZPE} - T\Delta S \qquad (3.5)$$

Here ΔE and ΔZPE are the differential adsorption energy for H^* and the difference in zero point energy for Reaction (3.4) as given by DFT. ΔS is the difference in entropy. At a pH different from zero, the entropy for the H^+ ion will change. We can correct expression (3.5) for this effect by adding $\Delta G(\text{pH}) = -k_B T \ln[H^+]$.

Equation (3.3) gives the potential dependence of the reaction free energy of Reaction (3.2). Since this reaction equilibrium defines the standard hydrogen electrode potential, we now have a direct link between quite simple DFT calculations and the electrode potential. In a similar way, we can now calculate potential-dependent reaction free energies for other reactions, such as $O^* + H^+ + e^- \rightarrow OH^*$ or $OH^* + H^+ + e^- \rightarrow H_2O$.

At this point, it is appropriate to describe in more detail how to calculate the relevant reaction energies using DFT. Taking the electrochemical adsorption of hydrogen as an example, the calculation of the adsorption energy of H^* should actually be done not at ultrahigh vacuum (UHV) conditions, but at realistic electrochemical conditions. This means that we need not only the platinum electrode and the hydrogen atoms present, but also the electrolyte. To simplify matters, neglecting the effect of anion, we could divide the effect of the electrolyte on the hydrogen adsorption energy into two major effects, not necessarily decoupled from each other: the effect of the liquid surrounding (water) and the effect of electric field (the electric double layer).

The influence of water can be included by adding water molecules to the DFT calculation. Whereas the interaction with water will be discussed in more detail later, in short, the water interaction will be most important for adsorbates that easily form hydrogen bonds, react with water, or form strong ionic bonds to the surface. For other adsorbates, such as H^*, the effect of water is negligible [Jerkiewicz, 1998; Roudgar and Gross, 2005].

The dependence on the electrical field can be approximated either analytically, e.g., with a dipole field interaction, or by simulations including an external field. In either case, the electric field would give rise to a correction term, $\Delta G^{\text{field}}(U)$, to be added to Equation (3.3). As will also be discussed later, this correction will in most cases be small.

60 ELECTROCATALYSIS AND CATALYST SCREENING

In practice, ΔG_0 will also depend on the coverage of the adsorbate, θ. Since this effect has direct implications for cyclic voltammetry, it will be discussed in more detail in Section 3.3.1 on theoretical cyclic voltammograms.

To summarize the model described above, we highlight the close connection this theoretical standard electrode provides between surface science and electrochemistry. Neglecting the effect of the electrolyte, the left and central parts of Reaction (3.1) represent a typical surface reaction, where the change in energy is the dissociative adsorption energy. Such reactions have been studied extensively within surface science, both experimentally and using DFT. The central and right part of Reaction (3.1), on the other hand, represent a typical electrochemical reaction. Hence, by introducing this theoretical counterpart to the standard hydrogen electrode, we have provided a close link between surface science and electrochemistry.

3.3 CYCLIC VOLTAMMOGRAMS AND POURBAIX DIAGRAMS

3.3.1 Theoretical Cyclic Voltammograms

Cyclic voltammetry is perhaps the most important and widely used technique within the field of analytical electrochemistry. With a theoretical standard hydrogen electrode at hand, one of the first interesting and challenging applications may be to try to use it to make theoretical cyclic voltammograms (CVs). In following, we set out to do this by attempting to calculate the CV for hydrogen adsorption on two different facets of platinum: the (111) and the (100) facets.

Cyclic voltammetry entails the measurement of current flowing through an electrode during a linear sweep of its potential versus a known electrochemical reference, such as the standard hydrogen electrode. In the following, we will construct a theoretical CV that only concerns adsorption and desorption of hydrogen. The process we study is thus the reaction of protons from the aqueous solution with electrons in the electrode, i.e., from the last to the second-to-last state in Reaction (3.1). This reaction is generally believed to be fast, a conclusion supported both by experiments [Markovic and Ross, 2002] and calculations [Skúlason et al., 2007]. In the case that the reaction is close to equilibrium at each electrode potential, the free energy for the reaction must be zero:

$$\Delta G(U, \theta) = 0 \tag{3.6}$$

Combining Equations (3.6) with (3.3), one obtains

$$U(\theta) = -\Delta G_0(\theta)/e \tag{3.7}$$

This relation, analogous to the Nernst equation at pH = 0, implies that there is a relation between the coverage of hydrogen at the surface and the potential. Based on this observation, the starting point for the derivation of a theoretical cyclic voltammogram is the derivative of U with respect to time:

$$\frac{dU}{dt} = \frac{dU}{d\theta}\frac{d\theta}{dQ}\frac{dQ}{dt} \qquad (3.8)$$

Here, dU/dt is the linear sweep rate of the experiment and dQ/dt is the measured current. With a transfer of one electron per adsorbed H, the relation between the charge transferred per area Q, and the coverage θ is $Q = Q_{tot}\theta$, where Q_{tot} is e times the density of Pt atoms in the surface layer. Introducing $\pm K$ for the sweep rate and $i(t)$ for the current, we obtain

$$i(t) = \pm K Q_{tot} \frac{1}{dU/d\theta} = \pm K Q_{tot} \frac{d\theta}{dU} \qquad (3.9)$$

Equations (3.7) and (3.9) imply that we need only the hydrogen coverage dependence of the reaction free energy of Reaction (3.4) as input in the theoretical CV calculation. This reaction energy should be calculated for H adsorbed at the platinum electrode under realistic conditions for electrochemical measurements, i.e., in the presence of both the electrolyte and the electric field. To assess the importance of these effects, we could first look at the effect of water and electric field on the reaction energy. As was mentioned previously, and as can be seen from Fig. 3.1a, it turns out that the presence of both water and electric field has a very minor effect on the reaction energy of Reaction (3.4). Hence, we can simply do the calculations in a normal surface science setup, i.e., with vacuum above the metal surface. In Fig. 3.1a and b, the reaction energy for Reaction (3.4) is shown for Pt(111) and Pt(100) versus the number of nearest neighbors and versus coverage, respectively.

Before inserting the coverage dependence shown in Fig. 3.1 into the formulas above, we must ensure that we have taken all temperature effects into account. Whereas the reaction free energy, as defined in Equation (3.5), already contains an entropy term, when considering the macroscopic coverage θ, we will also have to include differential configurational entropy. In principle, this can be done in two ways. We can use the differential configurational entropy of noninteracting particles,

$$\Delta S_{conf} = k_B \ln \frac{1-\theta}{\theta} \qquad (3.10)$$

in which case we obtain the mean field, or Frumkin, isotherm. Alternatively, we can model the system by use of a lattice gas model and solve for the free energy as a function of coverage using, for instance, Metropolis Monte Carlo. In the latter case, the configurational entropy of interacting particles is taken into account to the limits of the accuracy of the lattice gas model. We will not go into all details here (see [Karlberg et al., 2007a]), but rather go directly to the resulting CVs shown in Fig. 3.2. The agreement with experiments, such as the CVs reported by [Markovic et al., 1997], is good, a result that is in line with a similar investigation

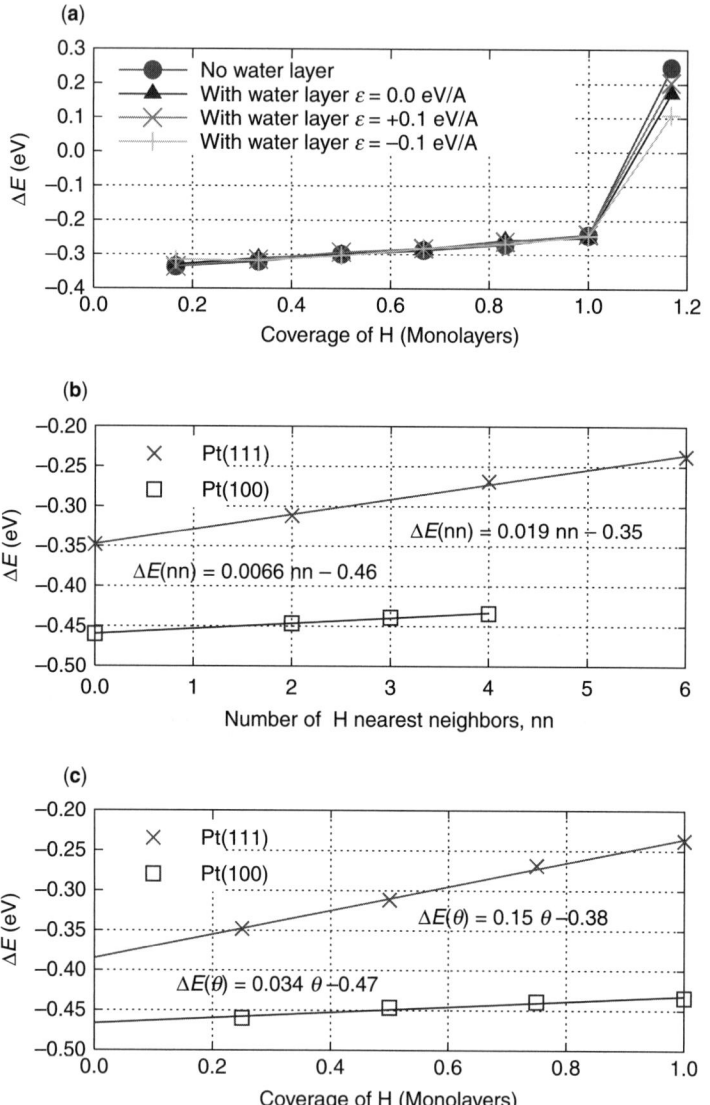

Figure 3.1 Reaction energy ΔE for Reaction (3.4) as calculated using DFT. (a) Effect of water and electric field. (b) ΔE versus number of H nearest neighbors on the surface. (c) ΔE versus H coverage. For more details see [Karlberg et al., 2007a].

based on experimental input for the hydrogen adsorption and interaction energies [Koper and Lukkien, 2000]. The derivation of the cyclic voltammogram directly from first principles calculations and standard molecular tables, however, clearly demonstrates the applicability of the electrochemical model introduced in [Nørskov et al., 2004].

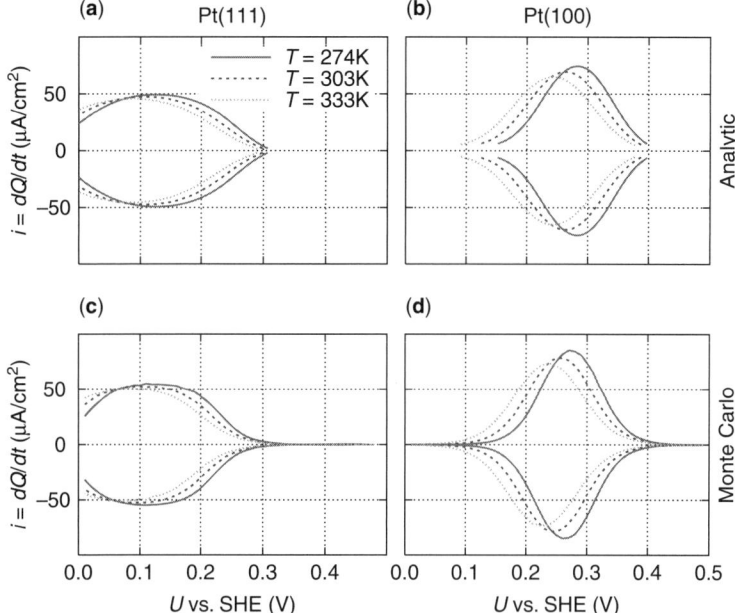

Figure 3.2 Cyclic voltammograms for H adsorption on Pt(111) and Pt(100). Two different methods have been applied. In (a) and (b), the H particles were assumed not to interact in the expression for the configurational entropy. In (c) and (d), the more elaborate model involving Metropolis Monte Carlo was applied. As can be seen, for these homogenous surfaces, the simple method suffices. The figure is adopted from [Karlberg et al., 2007a], where the full details of the calculations can also be found.

3.3.2 Electrochemical Phase Diagram

Having established the theoretical reference electrode, we are now able to estimate electrochemical phase diagrams. The starting point of the electrochemical phase diagram is the relative stability of the different species on the surface. If we consider a metal electrode in an acid electrolyte, we obtain

$$H^+ + e^- \longleftrightarrow H^*$$
$$\Delta G^{H^*}(U) = \Delta G_0^{H^*} + eU \tag{3.11}$$

$$H_2O \longleftrightarrow H^+ + e^- + OH^*$$
$$\Delta G^{OH^*}(U) = \Delta G_0^{OH^*} - eU \tag{3.12}$$

$$H_2O \longleftrightarrow 2H^+ + 2e^- + O^*$$
$$\Delta G^{O^*}(U) = \Delta G_0^{O^*} - 2eU \tag{3.13}$$

ELECTROCATALYSIS AND CATALYST SCREENING

TABLE 3.1 Calculated Adsorption Free Energies (in eV) for OH, O, and H on Ni, Pt, and Au(111) Surfaces at Standard Conditions ($T = 298K$, pH $= 0$, $U = 0$ V)

	$OH^* + e^- + H^+$	$O^* + 2(e^- + H^+)$	$H^* - (e^- + H^+)$
Pt:			
ΔE	1.05	1.57	-0.33
$\Delta E_{w.\ water}$	0.46	1.54	-0.33
$\Delta G_{w.\ water}$	0.81	1.59	-0.09
Ni:			
ΔE	0.13	0.34	-0.51
$\Delta G_{w.\ water}$	-0.11	0.36	-0.27
Au:			
ΔE	1.49	2.75	0.21
$\Delta G_{w.\ water}$	1.25	2.77	0.45

The binding energy determines the potential at which the surface is reduced by binding hydrogen, the potential window where the surface is free of any adsorbates, and the potential at which the surface starts being oxidized by OH or oxygen. These potentials are material-dependent: gold is free of absorbates in a large potential window, but nickel, on the other hand, goes directly from being reduced to oxidized, and Pt is somewhere in between these two extremes. See Table 3.1 and Fig. 3.3.

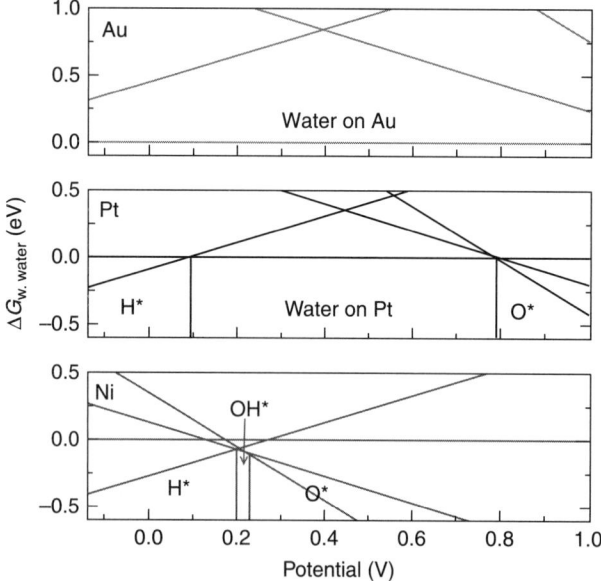

Figure 3.3 Phase diagram showing the free energy for different surface structures for water at pH $= 0$ in contact with Au(111), Pt(111), and Ni(111). The figure is based on the free energy values in Table 3.1. All free energies are shown relative to those of the clean surface with the water bilayer.

3.3 CYCLIC VOLTAMMOGRAMS AND POURBAIX DIAGRAMS

As can be guessed from the linear dependence on the electrode potential U in the relations above, the thermochemical approach for handling the potential and pH described above is in fact totally analogous to the method used in the classic Pourbaix diagram. Figure 3.3 corresponds directly to reading of the redox potentials in a Pourbaix diagram at the pH = 0 line. The main difference is that the ab initio binding energies can be obtained for all intermediates, i.e., also the species for reactions and adsorbates not stable in the electrochemical environment. Whereas Pourbaix diagrams were originally made for bulk transitions, DFT can provide the numbers related to the surface adsorption, which are actually more relevant for electrocatalytic reactions. In Fig. 3.4, we compare the first oxidation of the surface obtained with ab initio DFT with the classic Pourbaix bulk values. As can be seen, the results are very similar; the main difference is a downward shift in potential for the surface oxidation compared with the bulk oxidation, which is what could be expected. The computational Pourbaix diagrams can include all other surfaces and dissolution potentials as well; Fig. 3.4 is just a simple example of what is possible with this technique.

These results show that the electrochemical measurements can, via ab initio simulations, be linked to phenomena at the atomic level, such as structural and electronic effects and, in this case, binding energies on the surfaces.

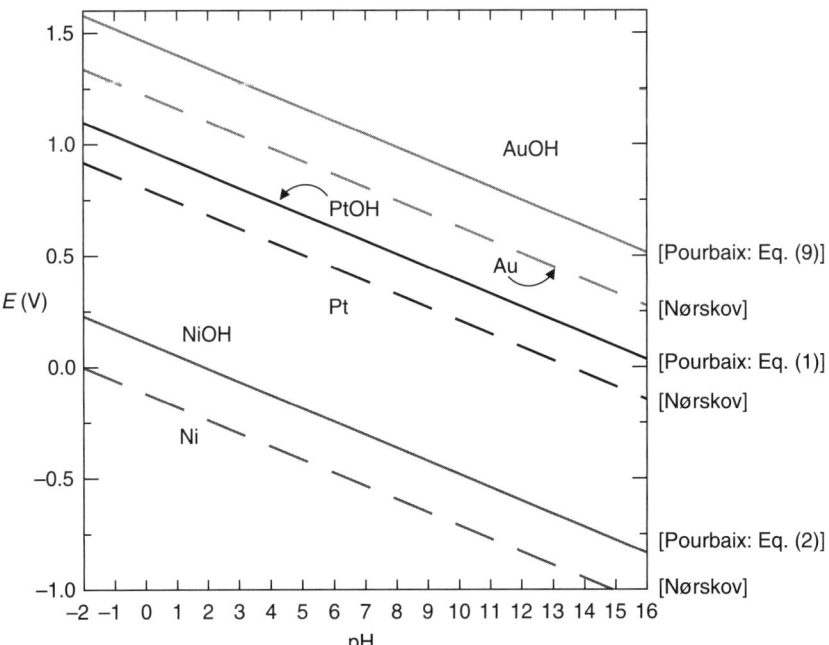

Figure 3.4 Pourbaix diagram for the first oxidation step of Ni, Pt, and Au for the (111) surface (dashed lines) [Nørskov et al., 2004] and the bulk (solid lines) [Pourbaix, 1966]. The oxidation reactions corresponding to the solid lines are indicated to the right in the figure.

3.4 CATALYTIC ACTIVITY: THE OXYGEN REDUCTION REACTION

3.4.1 Reaction Mechanism

As with the phase diagrams and Pourbaix diagrams, the theoretical standard hydrogen electrode also allows us to calculate the relative energies of intermediates in electrochemical reactions. As an example, we investigate the oxygen reduction reaction (ORR). We look at the four proton and electron transfer elementary steps:

$$O_2 + * + (H^+ + e^-) \longrightarrow HO_2^* \tag{3.14}$$

$$HO_2^* + (H^+ + e^-) \longrightarrow H_2O + O^* \tag{3.15}$$

$$O^* + (H^+ + e^-) \longrightarrow HO^* \tag{3.16}$$

$$HO^* + (H^+ + e^-) \longrightarrow H_2O + * \tag{3.17}$$

This analysis cannot tell whether or not atomic oxygen is formed on the surface during the ORR. One could imagine alternative reaction pathways through H_2O_2, but O^* is a more stable intermediate, and since our analysis will show that either the formation of HO_2 or the formation of water is potential-determining, the conclusions will not change if the reaction goes through a hydrogen peroxide intermediate.

3.4.2 Potential Energy Surface

Based on the theoretical electrochemistry method outlined above in combination with DFT calculations, the potential energy of the intermediates can be obtained at a given potential, (Fig. 3.5). Since all steps involve exactly one proton and electron transfer, the height of the different steps scales directly with the potential. To calculate the potential energy landscape at the equilibrium potential, the levels are moved down by $n \times 1.23$ eV, where n is the number of the electrons at the given state (the horizontal axis in Fig. 3.5).

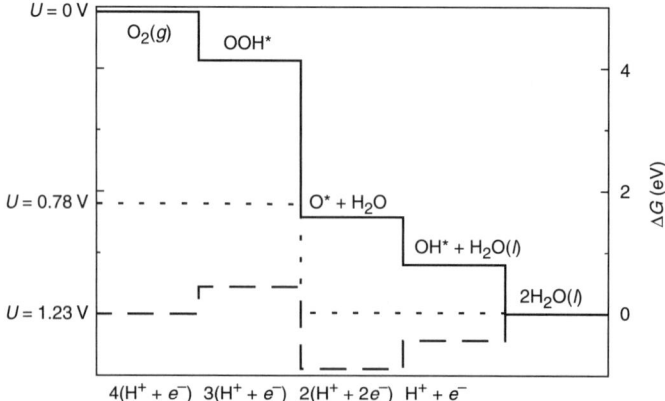

Figure 3.5 The free energy of the intermediates along ORR on Pt(111) at three potentials: $U = 0$ V with respect to a standard hydrogen electrode, the equilibrium potential $U = 1.23$ V, and the highest potential where all steps are downhill in free energy, $U = U_{ORR}^{Max} = 0.78$ V.

At the equilibrium potential, some steps are uphill in free energy, meaning that the reaction on the surface is slow. A perfect catalyst in this analysis would be characterized by a flat potential energy landscape at the equilibrium potential, i.e., by all steps having the same height at zero potential. Whereas no such catalyst has yet been found, we can define the highest potential at which all steps are just downhill in free energy, U_{ORR}^{Max}. Below U_{ORR}^{Max}, we would say that the reaction starts to be transport-limited. At potentials above U_{ORR}^{Max}, the catalytic reactions on the surface are limiting. The higher U_{ORR}^{Max} is the better the catalyst, and, as mentioned above, for the perfect material, U_{ORR}^{Max} corresponds exactly to the equilibrium potential.

This analysis only deals with the change in free energy for each reaction step. To do better, we would have to make detailed calculations of the barrier for proton and electron transfer between the surface and the electrolyte. At present, this is too computationally demanding. Another challenge associated with that approach is how to connect the output of such calculations with the standard hydrogen electrode in a rigorous manner. The reason why the simple no-barrier analysis is still valid is linked to the fact that the transition state often scales with the change in free energy of the reaction; this is the so-called Brønsted–Evans–Polanyi relation found throughout surface catalysis [Nørskov et al., 2002].

In Fig. 3.5, we consider the reaction free energy diagram for Pt(111) at different potentials. As can be seen, some steps have larger free energy changes than others, meaning that Pt is not a perfect catalyst for the ORR. Atomic oxygen, O*, and hydroxyl, OH*, seem to be too strongly bound to the Pt surface, whereas peroxide, OOH*, is too weakly bound. This can also be seen looking at the free energy surface at a realistic potential of 0.9 V (Fig. 3.6). In this graph, two reactions are uphill: the formation of OOH* and the reaction of OH*.

It is important to realize that the binding energies of O*, OH*, and OOH* most likely cannot be varied independently by changing the catalyst. When looking deeper into the

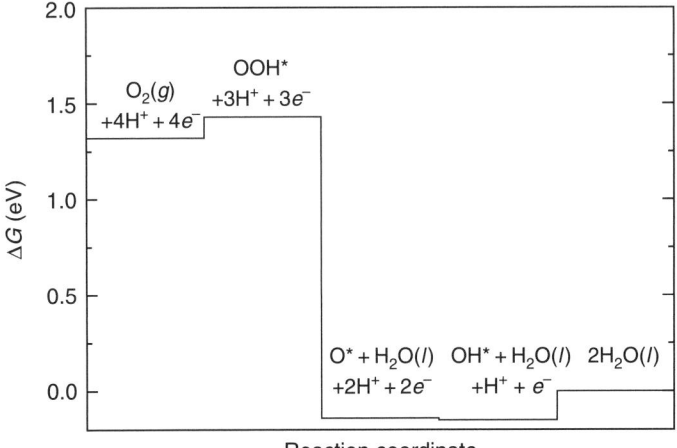

Figure 3.6 The free energy surface at $U = 0.9$ V; the two uphill reactions, $O_2(g) \rightarrow OOH^*$ and $OH^* \rightarrow H_2O(l)$, are seen directly.

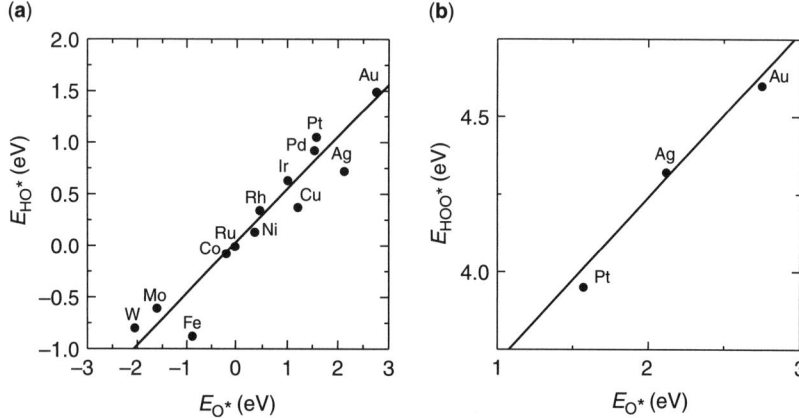

Figure 3.7 (a) Adsorption energy of HO* as function of the adsorption energy of O*, both on the terrace. The best linear fit is $E_{HO^*} = 0.50 E_{O^*} + 0.05$ eV. (b) Adsorption energy of HOO* as function of the adsorption energy of O*, both on the terrace. The best linear fit is $E_{HOO^*} = 0.53 E_{O^*} + 3.18$ eV.

correlations between binding energies on transition metal surfaces, it turns out that the binding energies are, in fact, linearly dependent [Abild-Pedersen et al., 2007]. A material with a strong binding of OH* can be expected to have a strong binding of O* and OOH* as well. In Fig. 3.7, the binding of OH* is shown as function of the binding of O*, and a similar relationship is shown for OOH. Since what we need is a catalyst where all steps have equal free energy barriers, when looking at transition metal electrodes, we can already at this stage tell that there will be intrinsic limitations.

When put into an appropriate model [Nørskov et al., 2004], the binding energy correlations directly define a limit to U_{ORR}^{Max} on the metals obeying the linear relations shown in Fig. 3.7. Since all intermediates are dependent on E_O, it is possible to plot the heights of all the steps ΔG_{1-4} as functions of E_O at zero potential. The step with the smallest free energy change will define U_{ORR}^{Max} (Fig. 3.8):

$$U_{ORR}^{Max} = \text{Max}[-\Delta G_1(\Delta E_O)/e, -\Delta G_2(\Delta E_O)/e, -\Delta G_3(\Delta E_O)/e, -\Delta G_4(\Delta E_O)/e]$$

(3.18)

The first proton transfer, ΔG_1, to oxygen forming OOH* defines the limit to U_{ORR}^{Max} for weakly binding metals such as Au, whereas ΔG_4, the proton transfer to OH forming water, defines the limit to U_{URR}^{Max} for the strongly binding metals. Notice that Pt is found close to the top of the "volcano," meaning that, among the pure metal (111) surfaces, Pt is the best catalyst for the ORR.

This illustrates the Sabatier principle: a good catalyst is a material with an optimal trade-off between being reactive (strong binding of intermediates) and not being poisoned by reaction products (weak binding). Obviously, this principle also holds for electrocatalysts, and, using the linear relations between the binding of

3.4 CATALYTIC ACTIVITY: THE OXYGEN REDUCTION REACTION

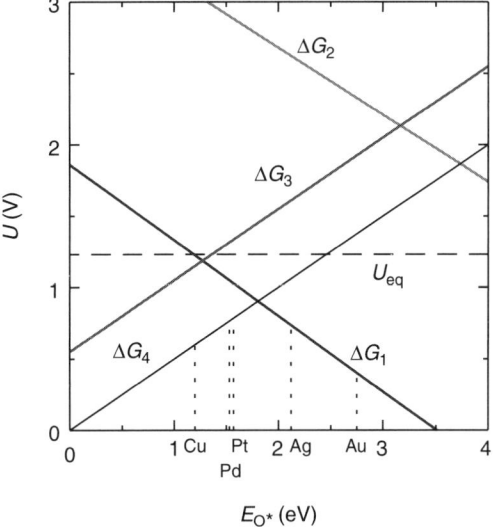

Figure 3.8 Plots of the heights of the steps in Fig. 3.5 divided by the electron charge in order to get a potential. Via the relations in Fig. 3.7, these steps are all functions of E_{O^*}. On the vertical axis is the highest potential at which a step in the ORR is downhill in free energy, depicted as a function of the binding of O^*. The step that first becomes uphill in free energy defines U_{ORR}^{Max} (ΔE_O)U. Steps 1 and 4 (lines labeled ΔG_1 and ΔG_4, respectively) define the lower "volcano" and thereby U_{ORR}^{Max} (ΔE_O). Pt is the pure metal closest to the top.

different intermediates, it is possible to obtain a *quantitative* version of the Sabatier principle, stating at which value of the "descriptor," in this case E_O, the best tradeoff is obtained. We note that the binding of OH^* would have served as an equally valid descriptor.

Whereas U_{ORR}^{Max} is relatively easy to determine from the calculated binding energies, it is not easy to measure experimentally, since the measured potentials are always related to a specific current. Therefore, in order to compare directly with experiment, we have to calculate polarization curves, i.e., the current. The link between U_{ORR}^{Max} and the current is the Tafel equation,

$$j_c = j_0 \exp(\alpha \eta e / k_B T) \tag{3.19}$$

where j_c is the cathode current density, j_0 is the exchange current, η is the overpotential, and α is the transfer coefficient. We will here only consider $\alpha = 1$, but the model could be constructed for other values of α. The overpotential is written as $\eta \equiv U_0 - U$ in order to rewrite the Tafel equation as

$$j_c = j_0 \exp\left[\frac{\alpha(eU_0 - eU)}{k_B T}\right] = j_{\text{limit}} \exp\left[\frac{\alpha(U_{ORR}^{Max} - eU)}{k_B T}\right] \tag{3.20}$$

where j_limit can be written as

$$j_\text{limit} = j_0 \exp\left[\frac{\alpha(eU_0 - U_\text{ORR}^\text{Max})}{k_\text{B}T}\right] \quad (3.21)$$

The physical meaning of the term j_limit is the current density achieved if all surface reactions are exothermic, i.e., the highest possible turnover frequency per site. In previous work on hydrogen evolution, we found that $j_\text{limit} \approx 200$ sites^{-1} s^{-1} or, in terms of surface area, 96 mA/cm^2 for Pt(111), fitting experimental data well [Nørskov et al., 2004, 2005]. Since the lattice parameters of various metals change only by a few percent relative to pure Pt, the number of sites per square centimeter is fairly constant, and j_limit can be effectively considered material-independent, unlike exchange current density. By incorporating j_limit into the analysis, we can thus rewrite the Tafel equation (3.19) in such a manner that all material dependence is concentrated in only one computable parameter, U_ORR^Max.

3.4.3 Improved ORR on Pt$_3$X Alloys

Since Pt is so close to the top of the volcano curve for the ORR, it can be difficult to find a better catalyst that does not contain Pt. A more fruitful approach is to modify Pt slightly by introducing a subsurface layer of another material. The electronic structure of the Pt at the surface is then changed so that oxygen binding becomes a little weaker. In general, the subsurface atoms act by changing the density of states, especially the position of the d-band center relative to the Fermi level [Hammer and Nørskov, 1995; Kitchin et al., 2004]. Similar effects can be obtained by compressing or stretching the lattice constant of Pt [Mavrikakis et al., 2000].

In principle, it could be very difficult or impossible to experimentally produce such surfaces. Fortunately, some Pt alloys spontaneously form Pt skins. For this reason, Pt skins on Pt$_3$X alloys are of significant interest. The Markovic group, in particular, has made well-controlled and well-characterized experiments for the ORR on Pt skins on Pt$_3$X [Stamenkovic et al., 2006, 2007a, b].

We have performed calculations on this type of alloys. The atomic setup used is a stochiomeric Pt$_3$X slab where the surface X atoms are swapped with Pt subsurface atoms, leaving a pure Pt skin on top of a subsurface layer enriched with 50% X atoms. The density of states is affected both by the X atoms in the subsurface layer and by the changed lattice constant of the alloy compared with Pt. However, both effects are included when looking at the position of the d-band center. A downward shift of the d-band results in weaker adsorbate binding, owing to increased occupation of the antibonding orbital [Hammer and Nørskov, 2000]. The model predicts the U_ORR^Max for the Pt$_3$X alloys to be a bit higher than that of pure Pt. Note that a small change in U_ORR^Max can lead to a significant change in the current. Experiments on nanoparticles [Stamenkovic et al., 2006] showed Pt$_3$Co to be the most active, while the theory indicated that Pt$_3$Ni should be closest to the top of the volcano. One possible reason for this modest discrepancy is that the model only includes (111) surfaces, whereas the nanoparticles have many different facets. If, for some reason, Pt$_3$Co and Pt$_3$Ni

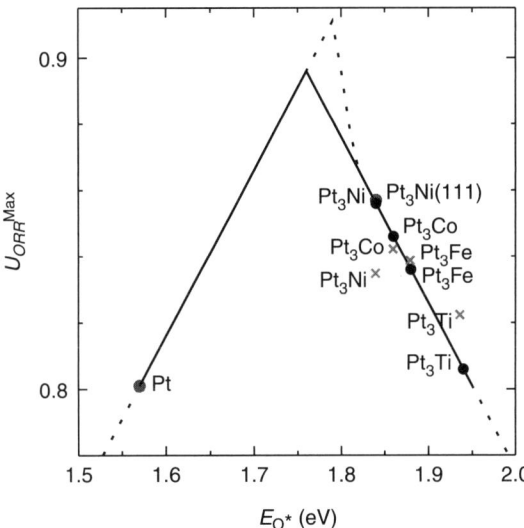

Figure 3.9 The relative activity of Pt and Pt skins on Pt$_3$X alloys as a function of the adsorption energy of O*.

particles do not have the same shape, this could be part of the reason. This explanation is supported by experiments on Pt$_3$Ni single-crystal (111) surfaces that show an activity exactly where the model predicts it to be (Fig. 3.9). We note that we cannot, in general, expect the model to predict experiments this well, and the perfect agreement is most likely fortuitous. Future single-crystal experiments on some of the other alloys could give a more complete picture of how good the correlation is between the model and the experiments.

Applying the Tafel equation with U_{ORR}^{Max}, we obtain the polarization curves for Pt and Pt$_3$Ni (Fig. 3.10). The experimental polarization curves fall off at the transport limiting current; since the model only deals with the surface catalysis, this part of the polarization curve is not included in the theoretical curves. Looking at the low current limit, the model actually predicts the relative activity semiquantitatively. We call it semiquantitative since the absolute value for the prefactor on Pt is really a fitting parameter.

3.4.4 Polarization Curves for OH/Water Adsorption

The potential-determining structure for the ORR on Pt is OH*. It is therefore interesting to compare, in more detail, models of OH adlayer structures with experimental results. Experiments of the OH coverage as function of the potential on Pt(111) and Pt$_3$Ni(111) are found in [Stamenkovic et al., 2007a]. Our model translates a free energy to a potential, which means that the coverage as function of the binding energy—or vice versa—has to be found.

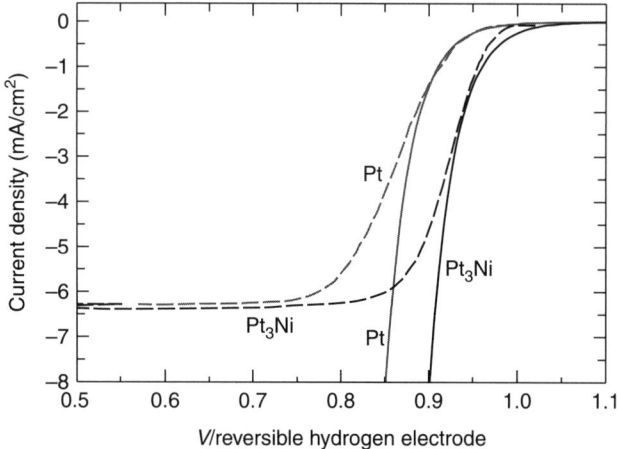

Figure 3.10 Calculated current density based on Equation (3.5) (solid curves) and experimental current density from [Stamenkovic et al., 2007a] (dashed curves).

In the water/OH* layer, there is a perfect one-to-one ratio between water and OH*, giving a coverage of $\frac{1}{3}$ monolayer (ML) of OH*. Going beyond $\frac{1}{3}$ ML of OH, we have to replace a water—OH bond with an OH—OH bond. In the following, we estimate the OH—OH interaction energy at a coverage of $\frac{1}{3}$ ML of OH, without the presence of water, and compare this energy with the corresponding result for the OH/water layer (Fig. 3.11). Since the step is high, about 0.3 V, it is reasonable to assume that the coverage never will be larger than $\frac{1}{3}$ ML at potentials around $U = 0.9$ V. Although there is no consensus in the literature regarding the coverage of OH during water splitting, we note that our maximum coverage of $\frac{1}{3}$ ML is in good agreement with the maximum coverage observed by [Stamenkovic et al., 2007a].

We include the configurational entropy of noninteracting particles, $\Delta S = k_B \ln[(1 - \theta_{OH})/\theta_{OH}]$, for $0 < \theta_{OH} < \frac{1}{3}$. This expression assumes that OH molecules do not interact, which we expect to be a good assumption, since all OH* only have water as nearest neighbors as long as the coverage is smaller than $\frac{1}{3}$ ML [Karlberg and Wahnstrom, 2005]. We can therefore write the potential and coverage dependence of the reaction free energy of Reaction (3.17) as

$$\Delta G(\theta_{OH}, U) = \Delta G_{OH} - k_B \ln\left(\frac{1 - \theta_{OH}}{\theta_{OH}}\right) - eU \quad (3.22)$$

Here ΔG_{OH} is calculated for the standard condition of $\frac{1}{3}$ ML OH and 1/3 ML H$_2$O. Assuming that Reaction (3.18) is in equilibrium for all potentials [$\Delta G(U, \theta_{OH}) = 0$], and, furthermore, that the excess barrier for water splitting is small, this leads to the following expression for the coverage:

$$\theta_{OH} = \frac{1}{3} \frac{1}{1 + \exp[(\Delta G_{OH} - eU)/k_B T]} \quad (3.23)$$

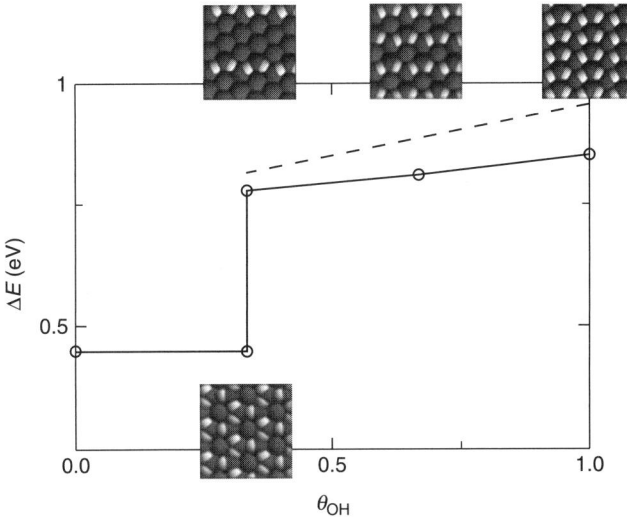

Figure 3.11 Estimation of the binding enthalpy of OH* as function of the coverage of OH*. For coverages higher than $\frac{1}{3}$ ML, the water—OH interactions are locally replaced with OH—OH interactions. The energy of the latter interaction is estimated with a zigzag OH pattern on the surface. It can be seen that the OH—OH interaction is much smaller than the water—OH interaction.

In the following, we will take $\Delta G_{OH} = 0.80$ eV and 0.93 eV for Pt and P$_3$Ni, respectively, as was reported in [Stamenkovic et al., 2007a]. The resulting potential/coverage curves are shown in Fig. 3.12, together with the experimentally obtained curves. We conclude that the method seems to predict many of the features also seen in well-controlled experiments even at a semiquantitative level.

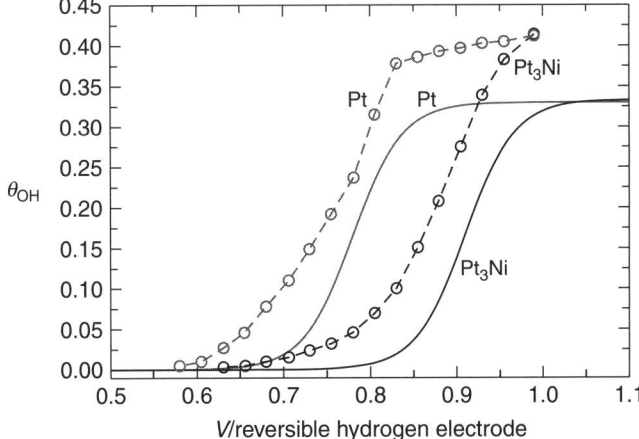

Figure 3.12 Coverage of OH* as function of the potential for Pt and Pt$_3$Ni. Experimental values from [Stamenkovic et al., 2007a] are represented by points connected with dashed lines.

3.5 MODEL DETAILS

The results shown above build on a thorough investigation of water interaction and effects related to the electrical field in the double layer near the surface. While these details are important when doing the simulations, they are less important for the understanding of the results presented above. We have therefore chosen to present them here at the end of the section.

3.5.1 Water Interaction

As mentioned above, interaction with water may affect the adsorption energy, especially for species that form hydrogen bonds. The most accurate way of including the effect of water is to explicitly add water molecules into the simulations. At the temperatures and pressures relevant for an electrochemical experiment, the water-containing electrolyte will be liquid. However, since in this context we are mainly interested in the effect of water on adsorption energies and not so much the actual structure of liquid water itself, we can probably simplify the problem.

Instead of adding liquid water, we add a water bilayer. The water bilayer on Pt(111) has been studied extensively in surface science [Thiel and Madey, 1987; Henderson, 2002; Ogasawara et al., 2002]. The water tends to form an ice-like layer structure oriented in hexagonal planes parallel to the surface, with a coverage roughly corresponding to two-thirds of a monolayer (Fig. 3.13). The vertical hydrogen bonds in the water layer can either point the hydrogen towards or away from the surface. At zero potential, the two structures are very close in energy. However, the relative energy of the two types of water bilayers can change, depending both on the strength of the electric field and on which other adsorbates are present [Rossmeisl et al., 2006].

From the analysis described above, we now know that a very important molecule that may be adsorbed together with water is OH. Also, this system has been studied quite extensively within surface science [Thiel and Madey, 1987; Bedurftig et al., 1999; Clay et al., 2004; Karlberg and Wahnstrom, 2005]. It appears that a mixed water—OH system forms a hexagonal structure much like the water structure discussed above (see Fig. 3.13c, d). Both from DFT calculations and UHV experiments, the most stable structure appears to be that where every other molecule is water and every other OH. This is interesting, since it coincides with the electrochemical observation, discussed above, where the maximum OH coverage was measured to be about one-third of a monolayer [Stamenkovic et al., 2007a].

With only small variations, water interaction appears to be constant for the different metals. From Fig. 3.14, which is based on data from [Karlberg, 2006], we can see that the interaction with water shifts the OH* binding energy down by an approximately constant amount. Another important point to note is the linear scaling relation between the binding of O* and the binding of OH*. Interestingly, it appears that the correlation becomes better in the presence of water. The reason for this improvement is that in the water/OH layer, OH always binds on top of a surface metal atom. Without water, OH can sit on top, as a bridge, or in a hollow, depending on the metal, which results in some scatter in the linear relation in Fig. 3.14. This is an example where trends give

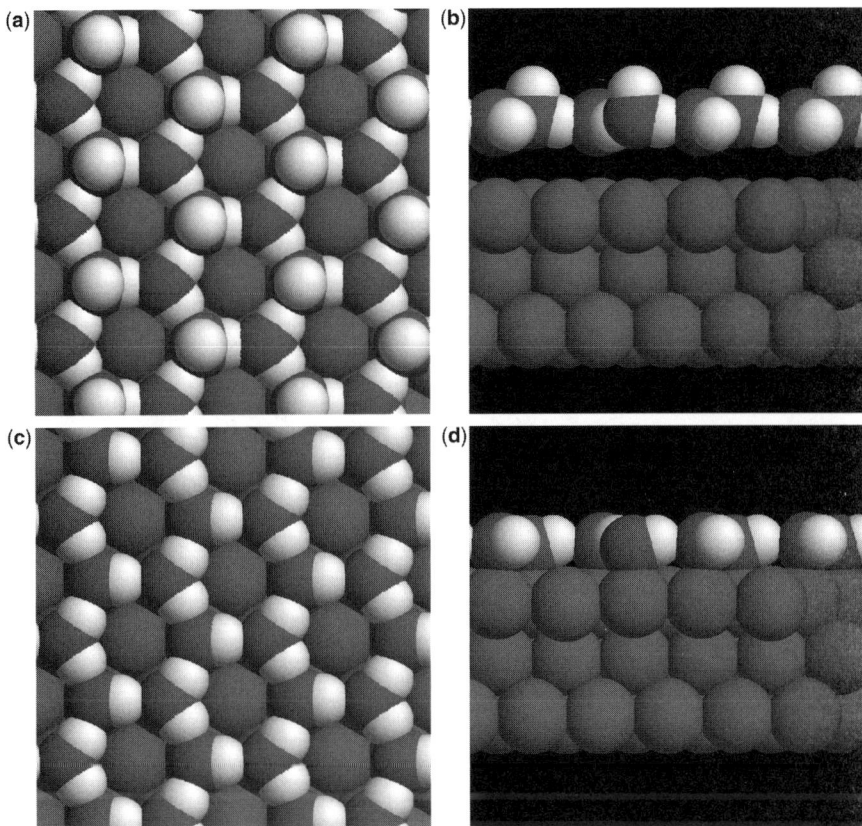

Figure 3.13 A pure H_2O (a, b) and a mixed H_2O/OH (c, d) hexagonal structure on Pt(111), top view and side view.

better estimates than individual points on the viewgraph, since in this case the solvent reduces the scatter.

The advantages of the simple approach outlined above are the limited number of water molecules needed in the simulation and the well-defined water structure. The major drawback is that, owing to the periodicity, this water structure fits best on a (111) or (111)-like surface, e.g., (211). There are at least two other approximations to model the water interaction. One is to include a large number of water molecules and apply molecular dynamics to determine a structure for the water and include this water arrangement in the simulations [Filhol and Neurock, 2006]. The drawbacks of this approach are the computational time required and the results' sensitivity to the water structure.

Another, more semiempirical, method is to assume that the only effect of the catalyst is to change the binding to the surface; entropy and solvation effects are taken to be the same as for the solvated spices. This assumes that the hydrogen bonds to, e.g., OH^- are the same as the hydrogen bonding to OH^* [Roques and Anderson, 2004]. We

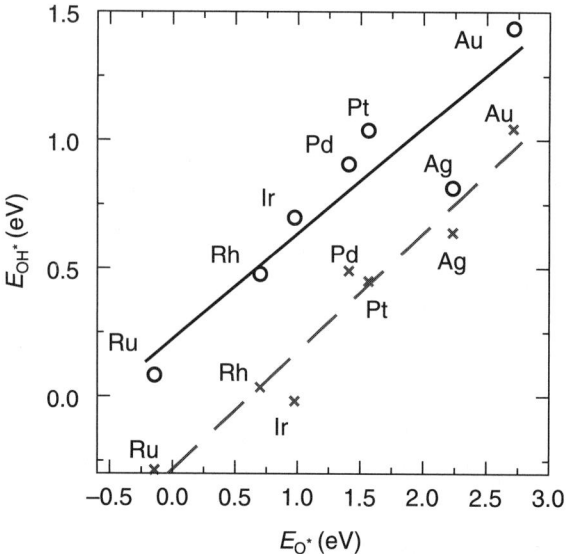

Figure 3.14 The full line and circles show the linear relation between E_{OH^*} and E_{O^*} for OH* adsorbed alone on the surface in the most stable site. The dashed line and crosses represent the same relationship, but calculated for OH in the OH/H$_2$O layer. The downward shift is related to hydrogen bonding between water and OH. Another important effect is that the deviation from the line is much smaller in the OH/H$_2$O layer. The correlation coefficient is improved from $R^2 = 0.92$ to 0.97 by including the water. This is partly because in the OH/H$_2$O layer all metals binds OH on top. The figure is based on data from [Karlberg, 2006].

note that this most likely is a good approximation. The results obtained with the water bilayer compare reasonably well with the numbers obtained by this approach, meaning that the bilayer is a robust model for water interaction.

3.5.2 Electric Field Effects

Apart from the water interaction, another interesting approximation in the model outlined above regards the electric field. In the electrochemical model described in this chapter, the effect of the potential is taken into account by shifting the free energy of states containing $(H^+ + e^-)$ by $-eU$. However, since the electrode will be at a different electric potential than the bulk of the conducting electrolyte, there will be a potential drop over the electrolyte in the electrolyte region closest to the electrode. This, in turn, means that there will be an electric field in this region. Whereas the effect of this electric field on the binding energy of the relevant adsorbates can be estimated to be small based on simple calculations [Nørskov et al., 2004], it could still be interesting to look a bit more into this approximation.

In Fig. 3.15, the calculated shift in free energy of adsorption due to the electric field is shown for various intermediates of interest for the ORR on Pt(111). The effect of the electric field was taken into account explicitly by adding an external electric field to the

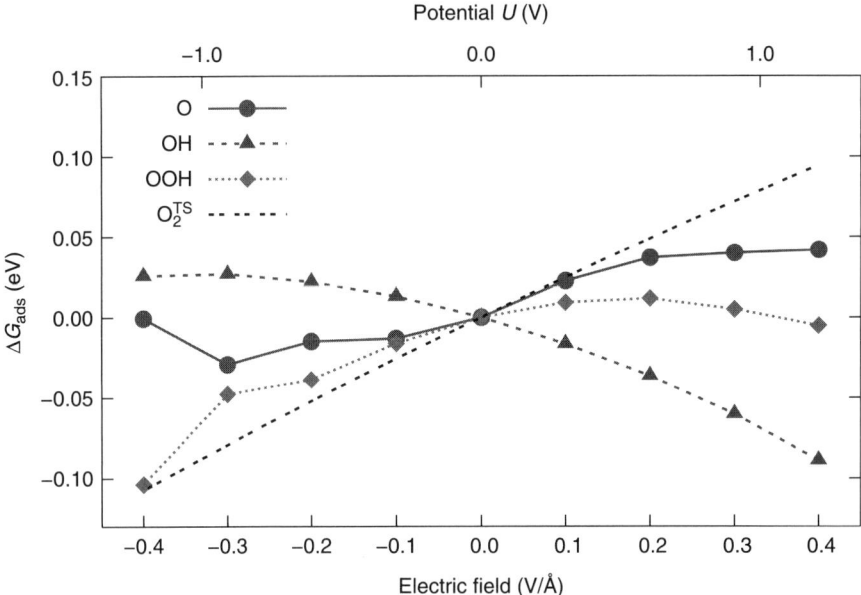

Figure 3.15 The change in free adsorption energy as a function of the electric field. At the top of the figure, we estimate the corresponding potential change by assuming that the potential drops over a Helmholtz layer of thickness 3 Å.

DFT calculations (for details, see [Karlberg et al., 2007b]). However, as can be seen from Fig. 3.15, the change in free energy is small, of the order of 0.1 eV, for most species. Owing to cancellation, the effect is even smaller when the electric field-related energy shifts are inserted into the Sabatier model. Hence, also according to this more detailed analysis, neglecting the effect of the electric field on the free energy of adsorption appears to be a good approximation.

3.6 INTRODUCTION TO COMPUTATIONAL ELECTROCATALYST SCREENING

Now that we have developed a fundamental, surface-science-based, first-principles approach to modeling electrocatalytic processes, we have the fundamental tools needed to begin to address a problem of significant importance in electrocatalyst research: that of new electrocatalyst discovery. We have already shown how careful application of fundamentally derived insights has led to the discovery of improved electrocatalysts for the ORR (see above). However, although extremely important, this approach has been limited to considering only a few (five or so) alloys with structures very similar to pure Pt (i.e., Pt skins); in effect, the approach successfully made small perturbations to the structure of a good catalyst (pure Pt) to obtain an improved catalyst (a Pt skin). To have a chance of finding catalytic materials with fundamentally different structures and properties from known catalysts, however, a different type of search strategy must be developed. For this purpose, we propose a *computational, combinatorial* search strategy. Combinatorial screening of catalysts using *experimental* techniques is a well-established technique in the catalysis industry [Corma et al., 2005; Hagemeyer et al., 2001; Liu et al., 2003; Saalfrank and Maier, 2004; Yaccato et al., 2004]. To date, however, only limited efforts have been made to use *computational* techniques to screen for improved catalysts [Andersson et al., 2006; Besenbacher et al., 1998; Faglioni and Goddard, 2005; Gong et al., 2003; Greeley and Mavrikakis, 2004, 2005; Greeley et al., 2002; Hammer et al., 1996; Jacobsen et al., 2001; Kitchin et al., 2004; Linic et al., 2004; Muller et al., 2003; Pallassana and Neurock, 2000], and no successful computational, combinatorial searches have been performed to find improved electrocatalysts. Given the expense associated with experimental combinatorial electrocatalyst searches, it would seem to be well worth the effort to develop corresponding computational techniques; these techniques could be used to identify a limited number of highly promising candidate electrocatalysts that could, in turn, be tested experimentally. By reducing the number of electrocatalysts to be experimentally tested, considerable speedup of catalyst discovery could be realized. In addition, by pointing to particular catalyst *structures* that might not be intuitively obvious, computational screening techniques might identify novel structures for known catalysts; these new structures, in turn, could exhibit improved catalytic properties.

3.6.1 The Hydrogen Evolution Reaction

To illustrate the power and flexibility of computational, combinatorial electrocatalyst searches, we first present a general search paradigm, and we apply this paradigm to the hydrogen evolution reaction (HER). The HER is an attractive electrochemical reaction with which to illustrate the power of computational search. There is significant technological interest in the reaction, because of the important role that it plays in electrodeposition and corrosion of metals in acids, in storage of energy via H_2 production, and as the microscopic reverse of the hydrogen oxidation reaction in low temperature fuel cells [Hamann et al., 1998; Jacobson et al., 2005]. These applications have motivated a significant amount of research on the HER, from the development of fundamental reactivity theories to the investigation of biomimetic catalysts for the reaction [Bockris et al., 1998; Conway and Bockris, 1957; Gerischer, 1958; Harinipriya and Sangaranarayanan, 2002; Hinnemann et al., 2005; Krishtalik, 1966; Medvedev, 2004; Parsons, 1958; Tard et al., 2005; Trasatti, 1972, 1977, 1995]. An additional benefit of choosing the HER to illustrate the potential of theoretical search techniques is that the reaction is quite tractable from a computational point of view; indeed, it has recently been shown that a single adsorption energy can accurately predict trends in HER activity [Greeley et al., 2006b; Nørskov et al., 2005] (see also the discussion below).

3.6.2 General Computational Screening Procedure

Although there are many procedures that could, in principle, be used to apply computational techniques to catalyst discovery processes, we have found that the following simple strategy is effective. The scheme permits the estimation of the activity and stability of a large number of transition metal alloys with reasonable computational effort.

1. Determine the smallest possible set of catalytic descriptors (binding energies, activation barriers, pre-exponential factors, etc.) that is needed to determine trends in the catalytic activity of the reaction of interest. This determination can be made, for example, by finding the smallest set of descriptors that correlates well with empirical activity data. Alternatively, microkinetic models [Cortright and Dumesic, 2001] or Sabatier analyses [Bligaard et al., 2004] can be used to develop rate expressions as functions of selected groups of descriptors. The latter approach is exactly the technique that was used to developed activity-based descriptions of the ORR earlier in this chapter.
2. Determine the values of the descriptors from step 1 that yield optimal catalytic activity. This determination can, again, be made empirically, via microkinetic modeling, or via Sabatier analysis.
3. Select a suitable pool of transition metal alloys and, using DFT techniques, evaluate the values of the catalytic descriptors on these alloys. A more approximate, but still very useful, alternative is to estimate the descriptors by some form of linear interpolation [Andersson et al., 2006; Jacobsen et al., 2001].
4. Using the results of steps 2 and 3, find the alloys with the highest predicted activity.
5. Using results from the DFT calculations, combined with databases of segregation energies, estimate the stability of the alloys in working reaction environments.
6. Test the best candidates experimentally.

3.6.3 Electrochemical Activity Model for the HER

The development of our HER model uses many of the basic principles developed for the ORR [Nørskov et al., 2004, 2005], and only a brief overview of the approach will be repeated here. The focus is on predicting trends in the HER exchange current density. As in the case of the ORR, the free energy of surface intermediates is calculated, including the effects of solvation (found to be negligible for hydrogen adsorption energies) and electrode potential (we note, however, that such potential effects are by definition absent for exchange current determinations). These free energies are then used to evaluate rates from simple mechanistic models; below, we consider the well-known Tafel–Volmer mechanism:

$$H^+ + e^- + * \longrightarrow H^*$$
$$H^* \longrightarrow \tfrac{1}{2} H_2(g) + *$$

A single catalytic descriptor, the free energy of hydrogen adsorption ΔG_H, turns out to be sufficient to predict trends in the exchange current density. For very exothermic hydrogen adsorption ($\Delta G_H < 0$), the coverage of adsorbed hydrogen will be high, and will, in essence, poison the hydrogen evolution process. An expression for the exchange current density in this regime is derived in [Nørskov et al., 2005]:

$$i_0 = -ek_0 \frac{1}{1 + \exp(-\Delta G_H/k_B T)} \quad (3.24)$$

where the pre-exponential factor k_0 is the only unknown parameter in the model. As described above, it has been fitted to experimental data for the exchange current density for the elemental metals [Nørskov et al., 2005]. For very endothermic adsorption ($\Delta G_H > 0$), the model predicts that a high barrier for H* formation from solvated protons will lead to low exchange current densities. A rate expression in this regime is also derived [Nørskov et al., 2005]:

$$i_0 = -ek_0 \frac{\exp(-\Delta G_H/k_B T)}{1 + \exp(-\Delta G_H/k_B T)} \quad (3.25)$$

For ($\Delta G_H \approx 0$), these two regimes are approximately balanced, and a maximum in the exchange current density is predicted. This is the optimal value of the descriptor, and it immediately suggests that a reasonable goal for a computational, combinatorial electrocatalyst search is to find alloys with ΔG_H values close to zero.

3.6.4 Validation of the Model

The model described above is only the first step in a comprehensive description of HER kinetics. Important extensions would be to consider the effect of alternative mechanisms (e.g., the Heyrovsky–Volmer mechanism), the detailed effects of changing hydrogen coverage, and the effects of non-unity transfer coefficients (Brønsted–Evans–Polanyi coefficients) on the calculated rates. We have, indeed, developed a substantially more detailed model that incorporates all of these features [Skúlason et al., 2007]. However, while a detailed model of this sort provides significant physical and chemical insight into the details of the HER, it does not change the fundamental conclusion that a maximum in activity is found for materials with ΔG_H values of zero. Hence, the detailed model is, in fact, too detailed to be truly useful for screening of large numbers of alloys—a simplified model that captures the same activity trends is more appropriate.

The simplified model does, in fact, do an excellent job of predicting trends in HER activities and exchange current densities. In Fig. 3.16, we show a comparison of the model predictions with experimental data for a variety of pure, polycrystalline metals, for selected single-crystal metals, and for single-crystal Pd overlayers supported on the close-packed surfaces of a variety of metal substrates [Greeley et al., 2006b]. Clearly, both the theoretical models and the experimental data show rate maxima at $\Delta G_H \approx 0$, and the quantitative agreement between the experimental and theoretical rates is reasonable. Given that we are primarily interested in predicting trends in activities over the various metals and alloys, the model appears to have quite satisfactory accuracy.

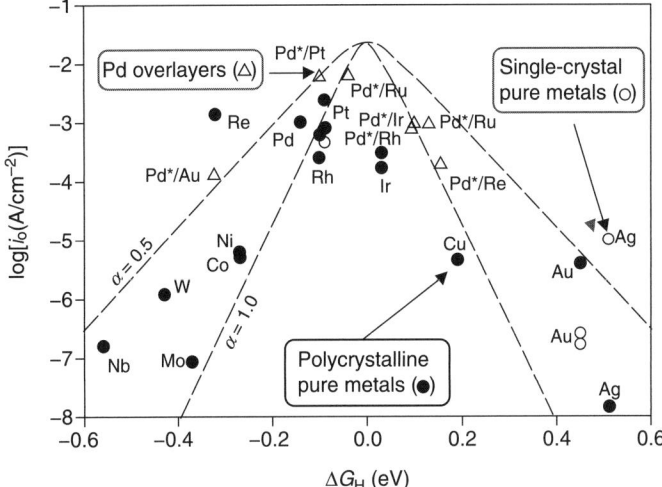

Figure 3.16 Volcano plot for the hydrogen evolution reaction (HER) for various pure metals and metal overlayers. Values are calculated at 1 bar of H_2 (298K) and at a surface hydrogen coverage of either 0.25 or 0.33 ML. The two curved lines correspond to the model (3.24), (3.25); transfer coefficients (not included in the indicated equations) of 0.5 and 1.0, respectively, have also been added to the model predictions in the figure. The current values for specific metals are taken from experimental data on polycrystalline pure metals, single-crystal pure metals, and single-crystal Pd overlayers on various substrates. Adapted from [Greeley et al., 2006a]; see this reference for more details.

3.6.5 Surface Alloys

Having established a descriptor and a reliable computational procedure for determining HER activity trends, we are in a position to begin our computational catalyst search. To start the search, we create a pool of alloys that we will later analyze for stability and activity properties. For our present purposes, we selected a pool of binary transition metal *surface alloys*. These alloys are composed of a homogeneous, pure substrate element (also called the "host"), with an alloying element (the "solute") incorporated into the surface layer of the host. Solute coverages of $\frac{1}{3}$, $\frac{2}{3}$, and 1 ML are considered for each solute/host pair, and, in all cases, the lattice constant of the alloy is naturally constrained to adopt the lattice constant of the pure host element (additional computational details are given in [Greeley and Nørskov, 2007]). In total, binary combinations of 16 elements (Fe, Co, Ni, Cu, Ru, Rh, Pd, Ag, Ir, Pt, Au, Re, Cd, As, Sb, and Bi) are considered; these metals are chosen because they are not thermodynamically favored to form bulk oxides in water at zero applied potential with respect to a standard hydrogen electrode. In total, 736 such surface alloys are considered.

3.6.6 Results of Combinatorial DFT Calculations

Using periodic DFT calculations [Greeley and Nørskov, 2007], we calculate the value of the HER descriptor ΔG_H on all of the surface alloys of interest. The results are

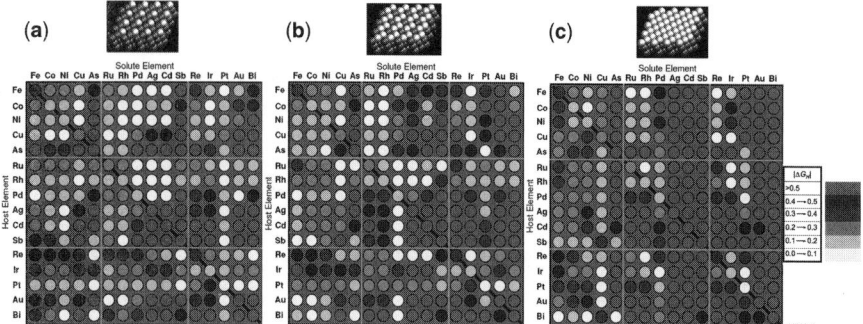

Figure 3.17 Computational high throughput screening for 736 pure metals and surface alloys. The rows indicate the identity of the pure metal substrates, and the columns indicate the identity of the solute embedded in the surface layer of the substrate. The solute coverage is (a) $\frac{1}{3}$ ML, (b) $\frac{2}{3}$ ML, and (c) 1 ML, and the adsorbed hydrogen coverage is also $\frac{1}{3}$ ML. The diagonals of the plots correspond to the hydrogen adsorption free energy on the pure metals. Adapted from [Greeley et al., 2006a]; see this reference for more details. (See color insert.)

summarized in Fig. 3.17 (Plate 3.1). In this figure, each circle corresponds to a particular surface alloy, and the color indicates the approximate value of ΔG_H; a yellow shading corresponds to $\Delta G_H \approx 0$. Although trends analyses are not the primary goal of the present chapter, it is nonetheless interesting to point out a few features of hydrogen adsorption that are revealed by Fig. 3.17. In general, at low solute metal coverages (Fig. 3.17a, b), alloys of late-group transition metals (and semimetals) have ΔG_H values that are highly positive [endothermic adsorption of hydrogen from $H_2(g)$]. The reverse is true for alloys of early-group metals. On average, the free energy goes from positive to negative as one moves from the lower right to the upper left of each periodic section of the figure. Intermediate metals and alloys (e.g., those found in the center and on the lower left–upper right diagonals of each periodic section) tend to have ΔG_H values closer to zero. These trends are consistent with the d-band model proposed by Hammer and Nørskov [Hammer and Nørskov, 1995; Ruban et al., 1997], mentioned above. That theory suggests that the decrease in the d-band center from left to right across the periodic table implies weaker adsorbate binding (i.e., a more positive value of the adsorption free energy); such a weakening is clearly consistent with the present results.

It is also interesting to consider the trends revealed within given columns or rows of Fig. 3.17a, b. Alloys containing Pt, Pd, or Rh (either as host or as solute) tend to have ΔG_H values close to zero; this behavior is consistent with the fact that these metals form good hydrogen evolution catalysts in their pure states [Nørskov et al., 2005], and the effect of the alloying elements is thus to introduce modest perturbations to the properties of these active metals. Some alloys, in contrast, have ΔG_H values that differ considerably from the adsorption free energies of the component pure metals; Ag on Ni (Fig. 3.17a) is a typical example. The binding in these cases can largely be understood as a simple interpolative combination of the ΔG_H values on the corresponding pure elements.

Finally, it is interesting to note that the ΔG_H trends change considerably for pure overlayers (Fig. 3.17c) compared with the cases with low solute coverage. For pure

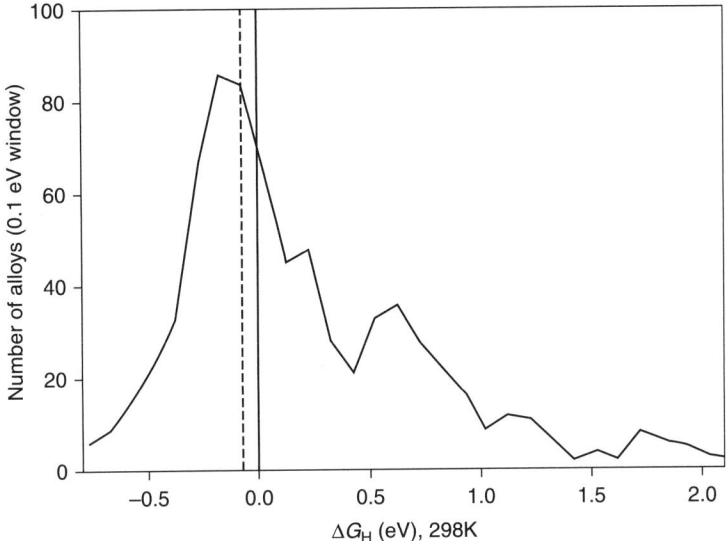

Figure 3.18 Spectrum of free energies of hydrogen adsorption, ΔG_H, on binary surface alloys at $T = 298K$. The vertical axis shows the number of elements with free energies within 0.1 eV windows (0.0–0.1 eV, 0.1–0.2 eV, etc.). The solid vertical line indicates $\Delta G_H = 0$. The dashed vertical line gives the hydrogen free energy adsorption for pure Pt. All free energies are referenced to gas phase H_2. Adapted from [Greeley and Nørskov, 2007]; see this reference for more details.

overlayers, there are relatively few alloys with free energies close to zero. This result may be due, in part, to the significant strain effects that are often associated with such overlayers. These effects lead to significant shifts in the d-band centers [Mavrikakis et al., 1998], thus resulting in extremes of ΔG_H.

For screening purposes, the most important result to emerge from the data in Fig. 3.17 is that there is a very large number of surface alloys with ΔG_H values roughly equal to zero (and hence a large number of such alloys with near-optimal values of the HER descriptor). This fact can be seen clearly in Fig. 3.18; the distribution of alloys with particular values of ΔG_H is peaked near $\Delta G_H = 0$.

3.6.7 Stability Considerations

The large number of binary transition metal surface alloys that are found to have near-optimal values of the HER descriptor suggests, at first glance, that a very large number of such alloys might have promising HER activities. However, such a large number immediately raises the question of whether or not it is realistic to expect that our simple screening procedure could have identified so many novel HER catalysts. In fact, this is quite unlikely to be the case, since the screening so far has not accounted for the *stability* of the surface alloys in the actual reactive environments. In acidic environments, such stability considerations could be particularly significant.

To probe the stability properties of the surface alloys for the HER, we perform a series of stability tests on all of the surface alloys that are predicted to have high HER activity. First, we estimate the free energy change associated with surface segregation events; such events can cause surface solute atoms to segregate into the bulk. Second, we determine the free energy change associated with intrasurface transformations such as island formation and surface dealloying. Third, we evaluate the free energy of oxygen adsorption, beginning with splitting of liquid water; facile oxygen adsorption can lead to surface poisoning and/or oxide formation. Finally, we estimate the likelihood that the surface alloys of interest will corrode in acidic environments (pH = 0). For the last test, we simply take the free energies of dissolution as reported in the electrochemical series [Lide, 1996]. These stability tests are summarized schematically in Fig. 3.19.

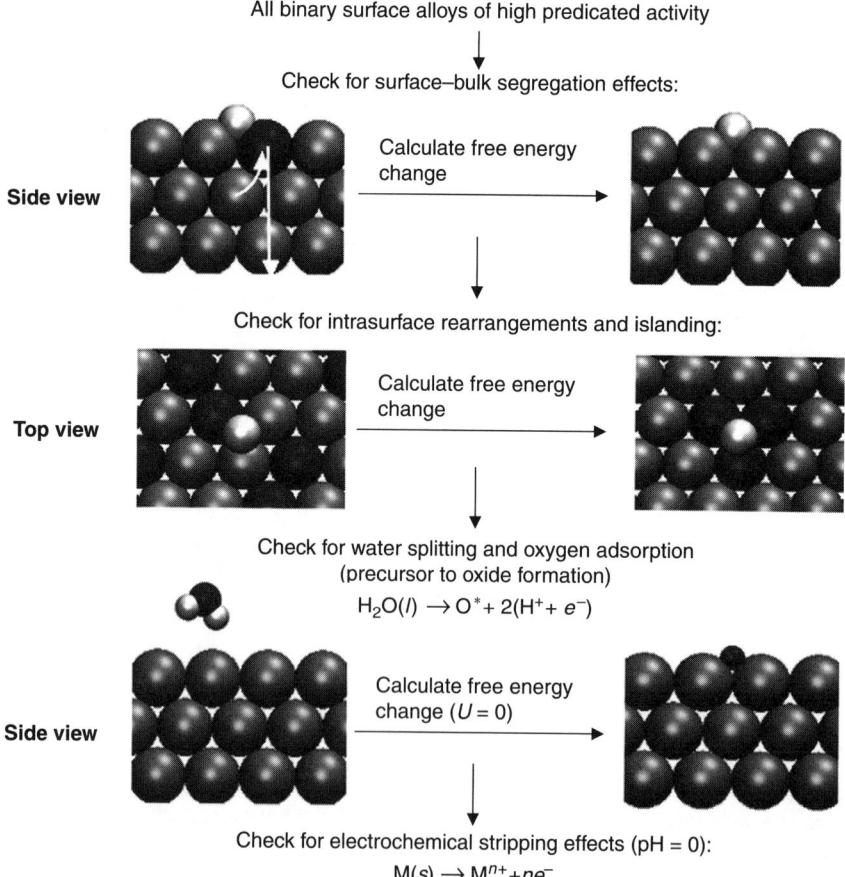

Figure 3.19 Schematic representation of surface alloy stability tests. White spheres denote adsorbed hydrogen, black spheres denote solute metal atoms, and gray spheres denote host metal atoms. Adapted from [Greeley and Nørskov, 2007]; see this reference for more details.

The combination of activity and stability considerations described above identifies approximately 30 surface alloys that could be robust catalysts for the HER [Greeley and Nørskov, 2007]. To further refine this list of candidates, we perform a *Pareto-optimal analysis* of the aforementioned activity and stability features. This analysis is implemented by plotting the most negative of the free energy transformation values determined for each alloy (i.e., the most pessimistic of the stability criteria for each alloy described above, which might vary from alloy to alloy) against the absolute magnitude of ΔG_H. In essence, we are simply putting the stability and activity criteria for the various alloys on a single plot. The stability considerations immediately eliminate a large number of alloys from consideration; although many alloys have high predicted HER activity, only a small fraction are predicted to be both active and stable in acidic HER environments (Fig. 3.20).

Using the raw data in Fig. 3.20, we can identify the Pareto-optimal set for the HER activity/stability criteria. This set represents the best possible compromise between activity and stability criteria for the surface alloys that we have considered; the alloys in the set are, thus, logical choices for further consideration. The presence of pure Pt on the Pareto-optimal set is, in effect, a "sanity check" for our computational screening procedure. Pt is well known to be the most active and stable pure metal for the HER in acidic conditions. The alloys seen on the Pareto-optimal set include RhRe and BiPt.

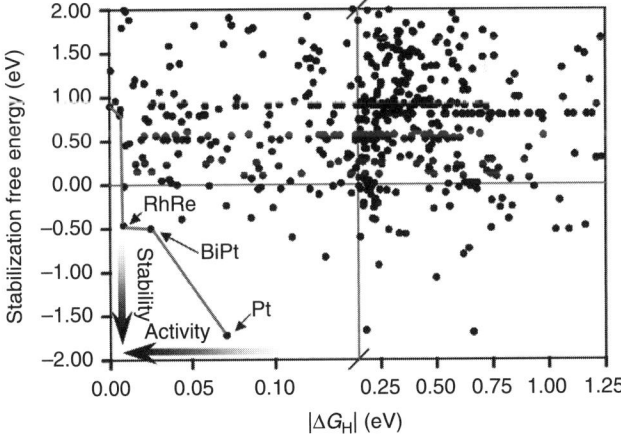

Figure 3.20 Pareto-optimal plot of stability and activity of surface alloys for the hydrogen evolution reaction (HER). The stabilization free energy can be thought of as a free energy of formation for the surface alloys; the stability of the alloys with respect to various reconstructive/deactivating processes (including surface segregation, island formation, water splitting/oxygen adsorption, and metal dissolution) is evaluated for each alloy, and the most pessimistic such energy (i.e., the energy that would give the maximum probability that the alloy would destabilize) is plotted. The Pareto-optimal line indicates the best possible compromise between activity and stability, but, given the simplicity of our model, other alloys could certainly be worth considering for use as HER catalysts; the alloys in the lower left quadrant, in particular, are promising. For the labeled points, a single element indicates a pure metal. For the bimetallic alloys, the solute is listed first; the solute coverages are $\frac{1}{3}$ ML (BiPt) and 1 ML (RhRe). Adapted from [Greeley et al., 2006a]; see this reference for more details.

RhRe, while intriguing, would likely be of too high a cost to be of practical interest. BiPt, in contrast, is worth further consideration. From a practical point of view, Bi is a very cheap metal, suggesting that some cost savings might be realized by using BiPt (as opposed to pure Pt) as an HER catalyst. More importantly, however, BiPt presents unusual properties from a fundamental scientific point of view. The novelty here is related to the stark contrast between its two constituent elements; pure Pt exhibits high HER activity, while pure Bi is not active at all. A surface alloy formed from these two elements, however, yields a material predicted by the calculations to have an activity comparable to, or even better than, pure Pt. This counterintuitive result suggest that BiPt is worthy of further experimental study and validation.

3.6.8 Experimental Results

Given that a BiPt surface alloy, identified by our combinatorial, computational search procedure, appears to be a novel and interesting HER candidate, we set out to synthesize and test it experimentally. A three-step approach was used: (i) an initial Pt film was electrodeposited onto an inert support, (ii) a sub-monolayer of Bi was spontaneously deposited onto the film by Bi-UPD, forming adsorbed Bi adatoms, and (iii) the Pt–Bi precursor film was annealed to form the surface alloy (see [Greeley et al., 2006a] for details). Current–voltage measurements of the Pt film, the Bi adatoms (denoted Pt–Bi_{ir}), and the annealed surface alloy are plotted in Fig. 3.21. Immediately after Bi adatom deposition, the measured activity of the Pt–Bi_{ir} sample is considerably less than that of the initial Pt film—the Bi_{ir} blocks Pt sites and poisons the surface for hydrogen evolution [Markovic and Ross, 2002]. However, after the

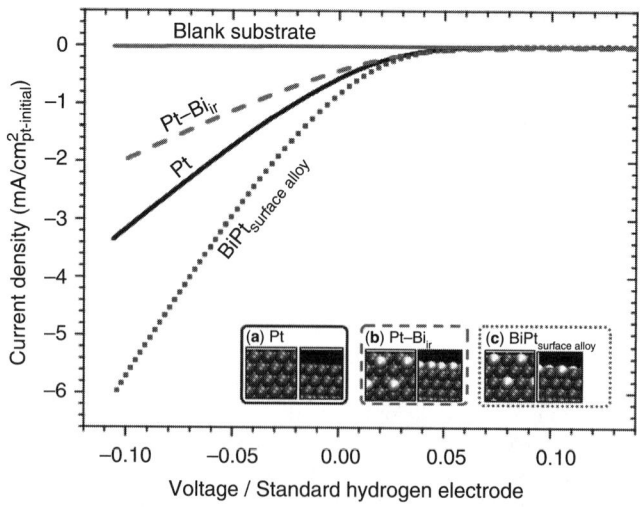

Figure 3.21 Hydrogen evolution after each stage of BiPt surface alloy synthesis. (a) Pt film after deposition and anneal; (b) immediately after Bi-UPD; (c) after second anneal to form the BiPt surface alloy. Adapted from [Greeley et al., 2006a]; see this reference for more details.

Pt–Bi$_{ir}$ precursor has been annealed to 500 °C and a surface alloy has formed, the activity is found to be superior to that of pure Pt (Fig. 3.21).

The experimental results are in complete agreement with the predictions of our computational screening approach; the annealed BiPt sample shows enhanced HER activity compared with pure Pt. As mentioned above, this result is rather counterintuitive, given that Bi itself is a notoriously poor electrocatalyst for the HER [Trasatti, 1972]. Hence, it appears that our computational, combinatorial screening procedure is capable of identifying improved catalysts for electrochemical reactions that are not immediately apparent from simple intuitive arguments.

The above results demonstrate that computational screening is promising technique for use in electrocatalyst searches. The screening procedure can be viewed as a general, systematic, DFT-based method of incorporating both activity and stability criteria into the search for new metal alloy electrocatalysts. By suggesting plausible candidates for further experimental study, the method can, ultimately, result in faster and less expensive discovery of new catalysts for electrochemical processes.

3.7 CONCLUSION

The goal in this chapter has been to show that it is possible to perform simulations relevant to electrochemistry-based ab initio surface calculations, without including all known physical effects. Focusing on trends and differences rather than absolute values, the approach in some cases yields not only qualitative results, but also (semi)-quantitative predictions.

With this approach, one is able to link electrochemical measurements, e.g., cyclic voltammograms, directly to atomic-level events taking place on the electrode surface. It is thereby possible to gain new insight into the electrochemical reactions and reaction "descriptors." These descriptors, in turn, form the basis for computational search and fast-screening algorithms for promising new electrocatalyst materials. The study of PtBi surface alloys is an example where insight at the atomic level has lead to experiments and to an improved catalyst at the macroscopic level. This type of knowledge-based computational screening may, in the future, become an important supplement to current experimental methods in catalyst development.

Development of our models and screening methods is ongoing, and, in order to further develop and benchmark this method, we will rely on well-defined and well-characterized experiments, varying one parameter at the time. Our hope is that our simulations, in turn, will provide new insight and thereby give inspiration to new experiments.

ACKNOWLEDGMENTS

The Center for Atomic-Scale Materials Design is supported by the Lundbeck Foundation. The Center for Nanoscale Materials/Argonne National Laboratory is supported by the US Department of Energy, Office of Science, Office of Basic Energy Sciences, under Contract DE-AC02-06CH11357.

REFERENCES

Abild-Pedersen F, Greeley J, Studt F, Moses PG, Rossmeisl J, Munter T, Bligaard T, Nørskov JK. 2007. Scaling properties of adsorption energies for hydrogen-containing molecules on transition-metal surfaces. Phys Rev Lett 99: 016105.

Andersson MP, Bligaard T, Kustov A, Larsen KE, Greeley J, Johannessen T, Christensen CH, Nørskov JK. 2006. Towards computational screening in heterogeneous catalysis: Pareto-optimal methanation catalysts. J Catal 239: 501–506.

Bedurftig K, Volkening S, Wang Y, Wintterlin J, Jacobi K, Ertl G. 1999. Vibrational and structural properties of OH adsorbed on Pt(111). J Chem Phys 111: 11147–11152.

Behler J, Delley B, Lorenz S, Reuter K, Scheffler M. 2005. Dissociation of O_2 at Al(111): The role of spin selection rules. Phys Rev Lett 94: 036104.

Besenbacher F, Chorkendorff I, Clausen BS, Hammer B, Molenbroek AM, Nørskov JK, Stensgaard I. 1998. Design of a surface alloy catalyst for steam reforming. Science 279: 1913–1915.

Bligaard T, Nørskov JK, Dahl S, Matthiesen J, Chistensen CH, Sehested J. 2004. The Brønsted–Evans–Polanyi relation and the volcano curve in heterogeneous catalysis. J Catal 224: 206–217.

Bockris JO'M, Reddy AKN, Gamboa-Aldeco M. 1998. Modern Electrochemistry 2A: Fundamentals of Electrodics. 2nd ed. New York: Plenum.

Bucko T, Hafner J, Benco L. 2005. Adsorption and vibrational spectroscopy of CO on mordenite: ab-initio density-functional study. J Phys Chem B 109: 7345–7357.

Clay C, Haq S, Hodgson A. 2004. Hydrogen bonding in mixed OH + H_2O overlayers on Pt(111). Phys Rev Lett 92: 046102.

Conway BE, Bockris JO'M. 1957. Electrolytic hydrogen evolution kinetics and its relation to the electronic and adsorptive properties of the metal. J Chem Phys 26: 532–541.

Corma A, Serra JM, Serna P, Moliner M. 2005. Integrating the high-throughput characterization into combinatorial heterogeneous catalysis: unsupervised construction of quantitative structure/property relationship models. J Catal 232: 335–341.

Cortright RD, Dumesic JA. 2001. Kinetics of heterogeneous catalytic reactions: analysis of reaction schemes. Adv Catal 46: 161–264.

Faglioni F, Goddard WA. 2005. Energetics of hydrogen coverage on group VIII transition metal surfaces and a kinetic model for adsorption/desorption. J Chem Phys 122: 014704.

Filhol JS, Neurock M. 2006. Elucidation of the electrochemical activation of water over Pd by first principles. Angew Chem Int Ed 45: 402–406.

Gerischer H. 1958. Mechanism of electrolytic discharge of hydrogen and adsorption energy of atomic hydrogen. Bull Soc Chim Belg 67: 506.

Gong X-Q, Liu Z-P, Raval R, Hu P. 2003. A systematic study of CO oxidation on metals and metal oxides: Density Functional Theory calculations. J Am Chem Soc 126: 8–9.

Greeley J, Mavrikakis M. 2004. Alloy catalysts designed from first-principles. Nature Mater 3: 810–815.

Greeley J, Mavrikakis M. 2005. Surface and subsurface hydrogen: adsorption properties on transition metals and near-surface alloys. J Phys Chem B 109: 3460–3471.

Greeley J, Nørskov JK. 2007. Large-scale density functional theory-based screening of alloys for hydrogen evolution. Surf Sci 601: 1590.

Greeley J, Nørskov JK, Mavrikakis M. 2002. Electronic structure and catalysis on metal surfaces. Ann Rev Phys Chem 53: 319–348.

Greeley J, Jaramillo T, Bonde J, Chorkendorff I, Nørskov JK. 2006a. Combinatorial high-throughput screening of electrocatalytic materials for hydrogen evolution. Nature Mater 5: 909–913.

Greeley J, Kibler L, El-Aziz AM, Kolb DM, Nørskov JK. 2006b. Hydrogen evolution over bimetallic systems: understanding the trends. Chem Phys Chem 7: 1032–1035.

Hagemeyer A, Jandeleit B, Liu YM, Poojary DM, Turner HW, Volpe AF, Weinberg WH. 2001. Applications of combinatorial methods in catalysis. Appl Catal A: Gen 221: 23–43.

Hamann CH, Hamnett A, Vielstich W. 1998. Electrochemistry. Weinheim: Wiley-VCH.

Hammer B, Nørskov JK. 1995. Electronic factors determining the reactivity of metal surfaces. Surf Sci 343: 211–220.

Hammer B, Nørskov JK. 2000. Theoretical surface science and catalysis—calculations and concepts. Adv Catal 45: 71–129.

Hammer B, Morikawa Y, Nørskov JK. 1996. CO chemisorption at metal surfaces and overlayers. Phys Rev Lett 76: 2141–2144.

Harinipriya S, Sangaranarayanan MV. 2002. Hydrogen evolution reaction on electrodes: Influence of work function, dipolar adsorption, and desolvation energies. J Phys Chem B 106: 8681–8688.

Henderson MA. 2002. Interaction of water with solid surfaces: fundamental aspects revisited. Surf Sci Rep 46: 5–308.

Hensen EJM, Zhu Q, van Santen RA. 2005. Selective oxidation of benzene to phenol with nitrous oxide over MFI zeolites. 2. On the effect of the iron and aluminum content and the preparation route. J Catal 233: 136–146.

Hinnemann B, Moses PG, Bonde J, Jorgensen KP, Nielsen JH, Horch S, Chorkendorff I, Nørskov JK. 2005. Biomimetic hydrogen evolution: MoS_2 nanoparticles as catalyst for hydrogen evolution. J Am Chem Soc 127: 5308–5309.

Honkala K, Hellman A, Remediakis IN, Logadottir A, Carlsson A, Dahl S, Christensen CH, Nørskov JK. 2005. Ammonia synthesis from first-principles calculations. Science 307: 555–558.

Jacobsen CJH, Dahl S, Clausen BS, Bahn S, Logadottir A, Nørskov JK. 2001. Catalyst design by interpolation in the periodic table: bimetallic ammonia synthesis catalysts. J Am Chem Soc 123: 8404–8405.

Jacobson MZ, Colella WG, Golden DM. 2005. Cleaning the air and improving health with hydrogen fuel-cell vehicles. Science 308: 1901–1905.

Jerkiewicz G. 1998. Hydrogen sorption at/in electrodes. Prog Surf Sci 57: 137–186.

Karlberg GS. 2006. Adsorption trends for water, hydroxyl, oxygen, and hydrogen on transition-metal and platinum-skin surfaces. Phys Rev B 74: 153414.

Karlberg GS, Wahnstrom G. 2005. An interaction model for $OH + H_2O$ mixed and pure H_2O overlayers adsorbed on Pt(111). J Chem Phys 122: 195705.

Karlberg GS, Jaramillo TF, Skúlason E, Rossmeisl J, Bligaard T, Nørskov JK. 2007a. Cyclic voltammograms for H on Pt(111) and Pt(100) from first principles. Phys Rev Lett 99: 126101.

Karlberg GS, Rossmeisl J, Nørskov JK. 2007b. Estimations of electric field effects on the oxygen reduction reaction based on the density functional theory. Phys Chem Chem Phys 37: 5158–5161.

Kieken LD, Neurock M, Mei DH. 2005. Screening by kinetic Monte Carlo simulation of Pt–Au(100) surfaces for the steady-state decomposition of nitric oxide in excess dioxygen. J Phys Chem B 109: 2234–2244.

Kitchin JR, Nørskov JK, Barteau MA, Chen JG. 2004. Modification of the surface electronic and chemical properties of Pt(111) by subsurface $3d$ transition metals. J Chem Phys 120: 10240–10246.

Koper MTM. 2005. Combining experiment and theory for understanding electrocatalysis. J Electroanal Chem 574: 375–386.

Koper MTM, Lukkien JJ. 2000. Modeling the butterfly: the voltammetry of $(\sqrt{3} \times \sqrt{3})R30°$ and p(2 × 2) overlayers on (111) electrodes. J Electroanal Chem 485: 161–165.

Krishtalik LI. 1966. On the influence of hydrogenation of the cathode metal upon the overvoltage of hydrogen. Electrokhimiya 2: 616.

Lide DR (Ed.). 1996. Handbook of Chemistry and Physics, 76th ed. New York: CRC Press.

Linic S, Jankowiak J, Barteau MA. 2004. Selectivity driven design of bimetallic ethylene epoxidation catalysts from first principles. J Catal 224: 489–493.

Liu YM, Cong PJ, Doolen RD, Guan SH, Markov V, Woo L, Zeyss S, Dingerdissen U. 2003. Discovery from combinatorial heterogeneous catalysis—a new class of catalyst for ethane oxidative dehydrogenation at low temperatures. Appl Catal A: Gen 254: 59–66.

Markovic NM, Ross PN. 2002. Surface science studies of model fuel cell electrocatalysts. Surf Sci Rep 45: 121–229.

Markovic NM, Grgur BN, Ross PN. 1997. Temperature-dependent hydrogen electrochemistry on platinum low-index single-crystal surfaces in acid solutions. J Phys Chem B 101: 5405–5413.

Mavrikakis M, Hammer B, Nørskov JK. 1998. Effect of strain on the reactivity of metal surfaces. Phys Rev Lett 81: 2819.

Mavrikakis M, Stoltze P, Nørskov JK. 2000. Making gold less noble. Catal Lett 64: 101.

Medvedev IG. 2004. To a theory of electrocatalysis for the hydrogen evolution reaction: The hydrogen chemisorption energy on the transition metal alloys within the Anderson–Newns model. Russ J Electrochem 40: 1123–1131.

Muller RP, Philipp DM, Goddard WA. 2003. Quantum mechanical–rapid prototyping applied to methane activation. Top Catal 23: 81–98.

Nørskov JK, Bligaard T, Logadottir A, Bahn S, Hansen LB, Bollinger M, Bengaard H, Hammer B, Sljivancanin Z, Mavrikakis M, Xu Y, Dahl S, Jacobsen CJH. 2002. Universality in heterogeneous catalysis. J Catal 209: 275.

Nørskov JK, Rossmeisl J, Logadottir A, Lindqvist L, Kitchin JR, Bligaard T, Jonsson H. 2004. Origin of the overpotential for oxygen reduction at a fuel-cell cathode. J Phys Chem B 108: 17886–17892.

Nørskov JK, Bligaard T, Logadottir A, Kitchin JR, Chen JG, Pandelov S, Stimming U. 2005. Trends in the exchange current for hydrogen evolution. J Electrochem Soc 152: J23–J26.

Ogasawara H, Brena B, Nordlund D, Nyberg M, Pelmenschikov A, Pettersson LGM, Nilsson A. 2002. Structure and bonding of water on Pt(111). Phys Rev Lett 89: 276102.

Pallassana V, Neurock M. 2000. Electronic factors governing ethylene hydrogenation and dehydrogenation activity of pseudomorphic Pd-ML/Re(0001), Pd-ML/Ru(0001), Pd(111), and Pd-ML/Au(111) surfaces. J Catal 191: 301–317.

Parsons R. 1958. The rate of electrolytic hydrogen evolution and the heat of adsorption of hydrogen. Trans Faraday Soc 54: 1053–1063.

Pourbaix M. 1966. Atlas of Electrochemical Equilibria. Oxford: Pergamon Press.

Roques RM, Anderson AB. 2004. Theory for the potential shift for OH_{ads} formation on the Pt skin on $Pt_3Cr(111)$ in acid. J Electrochem Soc 151: E85–E91.

Roques J, Anderson AB, Murthi VS, Mukerjee S. 2005. Potential shift for OH_{ads} formation on the Pt skin on $Pt_3Co(111)$ electrodes in acid. J Electrochem Soc 151: E85–E91.

Rossmeisl J, Norskov JK, Taylor CD, Janik MJ, Neurock M. 2006. Calculated phase diagrams for the electrochemical oxidation and reduction of water over Pt(111). J Phys Chem B 110: 21833–21839.

Roudgar A, Gross A. 2005. Hydrogen adsorption energies on bimetallic systems at the solid-vacuum overlayer and the solid–liquid interface. Surf Sci 597: 42–50.

Ruban A, Hammer B, Stoltze P, Skriver HL, Nørskov JK. 1997. Surface electronic structure and reactivity of transition and noble metals. J Mol Catal A: Chem 115: 421.

Saalfrank JW, Maier WF. 2004. Directed evolution of noble-metal-free catalysts for the oxidation of CO at room temperature. Angew Chem Int Ed 43: 2028–2031.

Skúlason E, Karlberg GS, Rossmeisl J, Bligaard T, Greeley J, Jónsson H, Nørskov JK. 2007. Density functional theory calculations for the hydrogen evolution reaction in an electrochemical double layer on the Pt(111) electrode. Phys Chem Chem Phys 9: 3241–3250.

Stamenkovic V, Mun BS, Mayrhofer KJJ, Ross PN, Markovic NM, Rossmeisl J, Greeley J, Nørskov JK. 2006. Changing the activity of electrocatalysts for oxygen reduction by tuning the surface electronic structure. Angew Chem Int Ed 45: 2897.

Stamenkovic V, Fowler B, Mun BS, Wang G, Ross PN, Lucas CA, Markovic NM. 2007a. Improved oxygen reduction activity on $Pt_3Ni(111)$ via increased surface site availability. Science 315: 493–497.

Stamenkovic VR, Mun BS, Arenz M, Mayrhofer KJJ, Lucas CA, Wang GF, Ross PN, Markovic NM. 2007b. Trends in electrocatalysis on extended and nanoscale Pt-bimetallic alloy surfaces. Nature Mater 6: 241–247.

Tard C, Liu XM, Ibrahim SK, Bruschi M, Gioia LD, Davies SC, Yang X, Wang LS, Sawers G, Pickett CJ. 2005. Synthesis of the H-cluster framework of iron-only hydrogenase. Nature 433: 610–613.

Thiel PA, Madey TE. 1987. The interaction of water with solid surfaces—fundamental aspects. Surf Sci Rep 7: 21–385.

Trasatti S. 1972. Work function, electronegativity, and electrochemical behaviour of metals. III. Electrolytic hydrogen evolution in acid solutions. Electroanal Chem Interfacial Electrochem 39: 163–184.

Trasatti S. 1977. The work function in electrochemistry. Adv Electrochem Electrochem Eng 10: 213–321.

Trasatti S. 1995. Surface science and electrochemistry: Concepts and problems. Surf Sci 335: 1–9.

Xu Y, Greeley J, Mavrikakis M. 2005. Effect of subsurface oxygen on the reactivity of the Ag(111) surface. J Am Chem Soc 127: 12823–12827.

Yaccato K, Hagemeyer A, Lesik A, Volpe A, Weinberg WH. 2004. High throughput screening of low temperature SCR and SCD de-NO_x catalysts in scanning mass spectrometer. Top Catal 30/31: 127–132.

Zellner MB, Goda AM, Skoplyak O, Barteau MA, Chen JG. 2005. Trends in the adsorption and decomposition of hydrogen and ethylene on monolayer metal films: A combined DFT and experimental study. Surf Sci 583: 281–296.

CHAPTER 4

First-Principles Simulation of the Active Sites and Reaction Environment in Electrocatalysis

MICHAEL J. JANIK

Department of Chemical Engineering, Pennsylvania State University, University Park, PA 16802, USA

SALLY A. WASILESKI

Department of Chemistry, University of North Carolina at Asheville, Asheville, NC 28804, USA

CHRISTOPHER D. TAYLOR

Materials Science and Technology Division, Los Alamos National Laboratory, Los Alamos, NM 87545, USA

MATTHEW NEUROCK

Departments of Chemical Engineering and Chemistry, University of Virginia, Charlottesville, VA 19904, USA

4.1 INTRODUCTION

Electrocatalysis is controlled by the complex interplay between the active catalytic sites and the influence of their complex environment at the electrified aqueous/metal interface. The most active electrocatalytic materials exquisitely integrate the atomic assembly of the active metal sites responsible for the elementary bond making and breaking steps, together with the carbon support to carry out efficient electron transfer, and polymer electrolyte and water to facilitate proton transfer, thus establishing an optimal three-phase interface. Understanding the elementary catalytic processes along with the atomic scale features that control them, however, is obscured by the complexity of this three-phase interface and the dynamic changes that occur to it under operating conditions. A simple schematic of the interface is shown in Fig. 4.1 in order to highlight some of the complexity. Many of the same issues that plague our ability to understand traditional vapor phase heterogeneous catalysts are also present

---

Fuel Cell Catalysis. Edited by Marc T. M. Koper
Copyright © 2009 John Wiley & Sons, Inc.

Figure 4.1 Schematic of the atomic structure of the active three-phase interface between the metal particle that catalyzes the reaction, the carbon support necessary to conduct electrons, and the polymer electrolyte and solution necessary to conduct protons for electrocatalytic systems.

here, including the influence of metal particle size and morphology effects, the role of metal defect sites (i.e., corner, edge, and kink sites), the influence of the support, alloying effects, and the influence of actual operating conditions that control the actual surface coverages. In addition, the unique electrocatalytic reaction environment of the "buried interface" presents a number of new issues. These include the influences of the following factors on the catalytic processes that occur at the three-phase interface:

- solution or humidity
- the electrochemical potential
- polymer electrolyte
- proton transfer
- electron transfer

While these features create a number of challenges in establishing fundamental insights for experiment as well as for theory and simulation, they also represent important opportunities, as they are the features by which one can control the design of new materials.

In this chapter, we will focus on some of the recent developments in understanding the influence of solution and electrochemical conditions over model single-crystal surfaces. Specifically, we will review work applying electronic structure methods to probe electrocatalytic mechanisms occurring at this complex interface.

4.2 THE ELECTRIFIED AQUEOUS METAL INTERFACE

While experiment and theory have made tremendous advances over the past few decades in elucidating the molecular processes and transformations that occur over ideal single-crystal surfaces, the application to aqueous phase catalytic systems has been quite limited owing to the challenges associated with following the structure and dynamics of the solution phase over metal substrates. Even in the case of a submersed ideal single-crystal surface, there are a number of important issues that have obscured our ability to elucidate the important surface intermediates and follow the elementary physicochemical surface processes. The ability to spectroscopically isolate and resolve reaction intermediates at the aqueous/metal interface has made it difficult to experimentally establish the surface chemistry. In addition, theoretical advances and CPU limitations have restricted ab initio efforts to very small and idealized model systems.

A number of significant experimental advances in spectroscopy, however, have occurred over the past decade that make it possible to interrogate and resolve surface reaction intermediates at complex solution/metal interfaces. This includes the development and application of broadband sum frequency generation (BB-SFG), attenuated total reflection (ATR), and surface enhanced infrared spectroscopy (SEIRS), which allow for the vibrational signatures of adsorbed intermediates to be separated from the background solution or solvent. In addition to these advances in spectroscopy, the continued increase in computational power coupled with novel algorithm developments have enabled the simulation of much larger and complex systems, including model aqueous/metal interfaces and, more generally, liquid/solid interfaces. It should be carefully noted, though, that these simulations have only recently become possible, and thus the size of the systems, as well as the ability to simulate their dynamics, is still limiting. Despite these issues, theory and spectroscopy have become invaluable partners in elucidating the chemistry at these complex interfaces.

As might be expected, the results from both theory and experiment suggest that the solution is more than a simple spectator, and can participate in the surface physicochemical processes in a number of important ways [Cao et al., 2005]. It is well established from physical organic chemistry that the presence of a protic or polar solvent can act to stabilize charged intermediates and transition states. Most C—H, O—H, C—O, and C—C bond breaking processes that occur at the vapor/metal interface are carried out homolytically, whereas, in the presence of aqueous media, the heterolytic pathways tend to become more prevalent. Aqueous systems also present the opportunity for rapid proton transfer through the solution phase, which opens up other options in terms of reaction and diffusion.

The presence of an (applied) potential at the aqueous/metal interface can, in addition, result in significant differences in the reaction thermodynamics, mechanisms, and structural topologies compared with those found in the absence of a potential. Modeling the potential has been a challenge, since most of today's ab initio methods treat chemical systems in a canonical form whereby the number of electrons are held constant, rather than in the grand canonical form whereby the potential is held constant. Recent advances have been made by mimicking the electrochemical model systems that employ a potentiostat whereby one can dial in a potential of interest to examine the resulting chemistry.

In this chapter, we present a brief background account of recent progress in the application of ab initio methods to elucidating the fundamental physicochemical processes that take place at the electrified aqueous metal interface. Our focus is predominantly on model single-crystal surfaces at lower surface coverages.

The fundamental understanding of the electronic and chemical interactions of molecules adsorbed to and reacting on extended surfaces has been significantly enhanced by quantum chemical studies employing metal slabs as surface models. For example, periodic density functional theory (DFT) models of surface–adsorbate interactions have been shown to provide considerable insight into a variety of chemical phenomena, such as elucidating reaction mechanisms for a variety of heterogeneous catalytic reactions [Honkala et al., 2005; van Santen and Neurock, 2006; Zhang and Hu, 2000] and the vibrational behavior of adsorbates [Greeley and Mavrikakis, 2004; Loffreda et al., 1999]. More recently, such methods have also been applied to more complex solvated surface–adsorbate systems (by including explicit solvent molecules surrounding the adsorbate and co-adsorbed on the surface) and more complex electrocatalytic systems that include both solvent and influences from the electrochemical potential [Cao et al., 2005; Halley et al., 1998; Janik and Neurock, 2006; Taylor et al., 2006b; Toney et al., 1994] The virtue of quantum chemical analyses of these more complex systems is the ability to elucidate the specific influence of how solvent or applied potential perturbs the chemistry occurring at such surfaces from both thermodynamic (e.g., calculations of reaction energies) and kinetic (e.g., calculations of activation barriers) standpoints. While we try to provide various references to work by others in the field of theoretical electrocatalysis, many of the examples discussed are taken from our own work.

4.3 ELECTRONIC STRUCTURE METHODS AND MODELS FOR THE ELECTROCATALYTIC INTERFACE

4.3.1 Periodic Density Functional Theory

The results discussed here were determined using plane wave DFT calculations within the generalized gradient approximation using the Vienna Ab initio Simulation Program (VASP) [Kresse and Hafner, 1993; Kresse and Furthmuller, 1996a, b]. The catalytic gas/solid interface was modeled using a standard supercell approach where 3–5 metal layers are chosen to mimic the metal substrate and a vacuum region of greater than 10 Å above the surface is used to represent the gas phase. The metal atoms are arranged so as to model a specific single-crystal surface. The unit cell was subsequently repeated in two dimensions along specific lattice vectors parallel to the surface plane and in a third dimension perpendicular to the surface to create parallel slabs. A schematic diagram of a typical supercell structure for a (111) surface is depicted in Fig. 4.2a, where the repeated unit cell is indicated by dashed lines. For ultrahigh vacuum (UHV) or vapor phase systems, the adsorbate is bound to one side of the slab, as shown in Fig. 4.2a. The interatomic distances of the

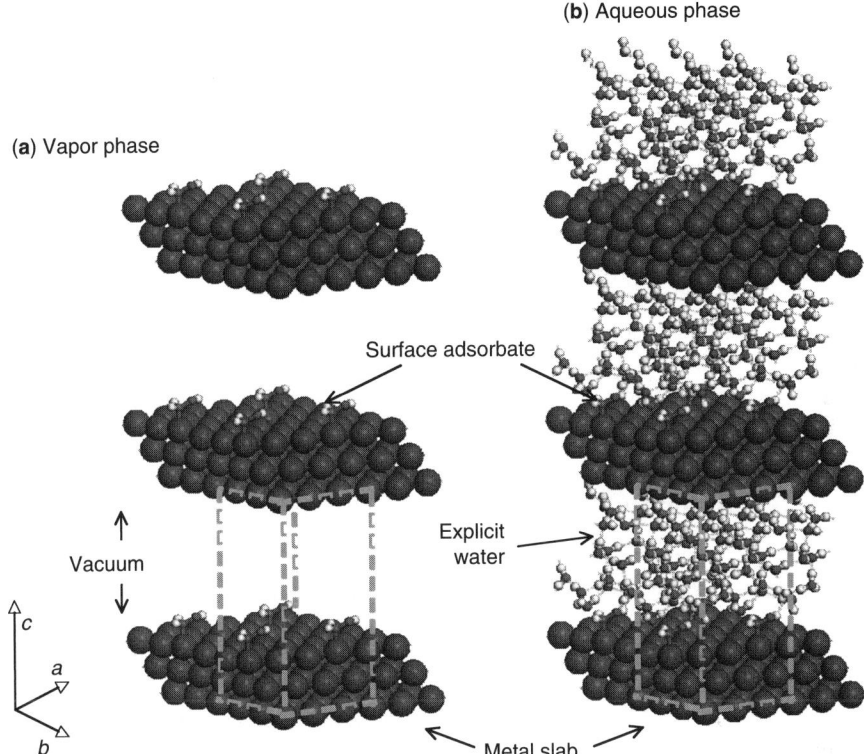

Figure 4.2 Graphical representation of the supercell structure, with a single (3 × 3) unit cell indicated by dashed lines that is repeated along lattice vectors a, b, and c, as indicated. (a) and (b) are vapor phase and aqueous phase models of the reaction environment, respectively, for an adsorbed CH_2OH intermediate with a surface coverage of $\frac{1}{9}$.

bottom layer of metal atoms are typically fixed at the experimental or DFT-optimized values for the bulk fcc metal. The reader is referred to the original literature cited in each example system for the specifics as to the generalized gradient approximation exchange-correlation functionals, Brillouin zone sampling, plane-wave cutoff energy, and pseudopotentials employed.

4.3.2 Aqueous Systems

The solution phase is modeled explicitly by the sequential addition of solution molecules in order to completely fill the vacuum region that separates repeated metal slabs (Fig. 4.2a) up to the known density of the solution. The inclusion of explicit solvent molecules allow us to directly follow the influence of specific intermolecular interactions (e.g., hydrogen bonding in aqueous systems or electron polarization of the metal surface) that influence the binding energies of different intermediates and the reaction energies and activation barriers for specific elementary steps.

For the aqueous solvated systems discussed in this chapter, the vacuum layer between slabs is filled with water molecules oriented in a hexagonal ice-like three-dimensional hydrogen-bonded structure, graphically depicted in the schematic diagram in Fig. 4.2b. The aqueous structures for each metal (111) surface are adapted from an original structure determined by ab initio molecular dynamics simulations for eight layers of water over a Pd(111) interface [Filhol and Neurock, 2006]. The water molecules closest to the metal surface form a bilayer structure as a network of hexagonal rings that match the registry of the metal surface, similar to experimental observations [Henderson, 2002; Michaelides et al., 2003a]. In modeling a solvated adsorbate, one H_2O molecule in the unit cell is replaced by the adsorbate of interest, and the entire system (metal surface, adsorbate, and water molecules) is re-optimized to account for variations in solvent–adsorbate and solvent–solvent hydrogen bonding and changes in the electronic structure of the metal surface. Electronic and geometric optimizations are held to the same convergence standards as for vapor phase calculations.

4.3.3 Simulating the Electrochemical Interface

In an electrochemical system, the surface–adsorbate and surface–solvent interactions are influenced by the resulting surface potential within the double-layer region. The potential ϕ at the interface induces a charge polarization in the metal (either negative or positive, owing to an excess or deficit of electron density at the metal surface); the magnitude of the excess charge depends on the difference between ϕ and the potential of zero charge ϕ_{pzc} of the metal. If ϕ is more negative than ϕ_{pzc}, the excess surface charge is negative. Ions, solute, and solvent molecules interact with the electron density and chemisorb or physisorb to the electrode surface. This adsorption occurs in the inner layer region, as shown in Fig. 4.3. At the outer edge of the inner layer is the outer Helmholtz plane (OHP), which is defined as the plane of closest approach of solvated counter-ions with charge opposite to the excess surface charge. The sum of charge on the counter-ions at the OHP is equal and opposite to the surface charge polarization, generating a capacitor for which the potential drops approximately linearly, forming an electric field with a typical magnitude of 10^8 V/Å. The magnitude of the field depends on ϕ and the thickness of the inner layer, x_{il}, which is governed by the radii of the counter-ions, adsorbates, and solvent molecules (and is typically a few Å).

The ability to simulate actual electrochemical systems from first principles is hindered by two predominant challenges. The first involves our inability to fully capture the anode, cathode, and electron and proton transfer channels within a single model system. The actual electrochemical system occurs at constant potential, which is dictated by the chemistry that occurs at the anode and cathode. The anode and cathode are connected electrically via macroscopic distances set by the conductor and supporting electrolyte solution. Ab initio simulations of such macroscopic length scales are currently not possible. The system, however, can be greatly simplified by modeling the anode and cathode separately as artificially charged half-cells, which is analogous to the electrode potential control set by a potentiostat in cyclic voltammetry experiments. The second major challenge involves maintaining a constant potential

Figure 4.3 Schematic diagram of the electrochemical metal–aqueous interface, with the electrode, inner layer, diffuse layer, outer Helmholtz plane (OHP), and inner-layer thickness x_{il} indicated.

in the half-cell model system while considering chemical transformations that alter the electronic structure. Current ab initio quantum mechanical methods are canonical, based on constant electron systems. Electrochemical half-cells, however, are based on constant chemical potential (instantaneously maintained by non-faradaic current from the potentiostat); this grand canonical ensemble is not feasible with current electronic structure methods. Despite such challenges, a variety and hierarchy of computational approaches to modeling the surface potential (and potential drop across the inner layer) have been developed [Anderson, 1981; Dominguez-Ariza et al., 2004; Halley et al., 1985; Head-Gordon and Tully, 1993].

The recent interest in the exploration of electrocatalytic phenomena from first principles can be traced to the early cluster calculations of Anderson [1990] and Anderson and Debnath [1983]. These studies considered the interaction of adsorbates with model metal clusters and related the potential to the electronegativity determined as the average of the ionization potential and electron affinity—quantities that are easily obtained from molecular orbital calculations. In some iterations of this model, changes in reaction chemistry induced by the electrochemical environment were modeled by shifting the energetics and/or the d-electron band of the electrode by a constant term given by neU, where n is the number of electrons and U is the potential [Anderson and Ray, 1982]. These calculations demonstrated the potential-dependent variation of reaction and adsorbate energies; however, the use of small metallic clusters and a limited solvation model have hindered its application. More recently, Anderson introduced the local reaction center model, which allows the calculation of potential-dependent activation barriers and methods to begin to introduce electrolyte [Anderson et al., 2005].

Approaches developed to maintain the integrity of the electrode structure (a semi-infinite planar electrode) while integrating electrochemical ideas were later introduced by Nørskov and co-workers [Nørskov et al., 2004], in parallel with the concepts developed by Anderson [Sidik and Anderson, 2006; Vayner et al., 2007]. These approaches also consider the variation in reaction energies to be linear in the potential; however, the improved description of the electrode leads to more reliable results. Furthermore, an implicit reference potential was cleverly introduced via the equivalence between the energy of H_2 and the energies of $2H^+ + 2e^-$ implied by the use of the normal hydrogen reference electrode. These models remain limited, however, as charges are not directly applied to the slab, and hence exploration of the thermodynamics of slab charging and the kinetics of electron transfer cannot readily be investigated.

A framework for probing these latter effects was laid out in a theoretical discussion by Alavi and Lozovoi [Lozovoi et al., 2001]. This framework depended on the concept of a reference electrode, against which the potential of a charged slab could be measured. Filhol and Neurock built upon this model by developing a more ambitious model for studying *aqueous* electrochemical interfaces with periodic DFT [Filhol and Neurock, 2006; Taylor et al., 2006c]. The resulting technique allows the simulation of electrochemical interfaces in a way that parallels the potentiostatic control of experimental electrochemical systems. The methodology, referred to as the *double-reference method*, employs two internal potential references within the calculation. The first is a vacuum reference that relates the interfacial potential to the vacuum scale. The second is an aqueous reference that ties the vacuum reference to a tunable potential that is varied by the addition or removal of electrons from the unit cell (i.e., system charging). The theoretical underpinning of the double-reference method is discussed briefly below. A more complete discussion of the approach and the theory behind it is given in Taylor et al. [2006b]. An alternative DFT-based, surface slab representation of charged surfaces by Otani and Sugino [2006] that removes the periodic boundary condition (PBC) in the surface normal direction to allow for calculations of charged surfaces is an encouraging approach for the consideration of electrochemical systems. Otani and Sugino couple the results from conventional periodic plane-wave electronic structure calculations for surfaces with a Poisson solver. This offers considerable flexibility in the treatment of electrochemical systems, although its implementation is nontrivial. Using the methods that we have developed to describe electrochemical systems within a PBC model, much of the physics proposed by Otani and Sugino is captured, without resorting to some of the continuum assumptions in their model (such as that water close to the interface has a dielectric constant of 78).

4.3.3.1 The Vacuum Reference
The first reference in the double-reference method enables the surface potential of the metal slab to be related to the vacuum scale. This relationship is determined by calculating the workfunction of the model metal/water/adsorbate interface, including a few layers of water molecules. The workfunction, $-\phi_{\text{Fermi}}$, is then used to calibrate the system Fermi level to an electrochemical reference electrode. It is convenient to choose the normal hydrogen electrode (NHE), as it has been experimentally and theoretically determined that the NHE potential is -4.8 V with respect to the free electron in a vacuum [Wagner, 1993]. We therefore apply the relationship

$$\phi_{\text{NHE}} = -4.8\,\text{V} - \phi_{\text{Fermi}} \tag{4.1}$$

to determine the electrochemical potential of our interfacial model.

4.3.3.2 The Aqueous Reference
The vacuum potential of the electron provides a convenient, absolute reference point for an uncharged system. In the presence of an electric field, however, the vacuum level is no longer meaningful, as the field permeates through free space, thus preventing the assignment of a unique free electron potential. In electrochemical systems, the electric field generated by a charged electrode is rapidly screened by the water molecules and ions in solution. Hence, at a point in the solution phase some ångströms from the surface (typically estimated at 5–10 Å), the electrochemical potential will again reach a constant level [Lozovoi et al., 2001; Sanchez et al., 2004; Taylor et al., 2006b]. Accordingly, we introduce a second reference point for monitoring the potential within the aqueous region of the unit cell. The electrochemical potential of this point is calibrated to the value $\phi_W(0)$ using the vacuum reference introduced above. As the system is subsequently charged with q electrons, the Fermi level $\phi_{\text{Fermi}}(q)$ of the model electrode/electrolyte system is measured against this reference point $\phi_W(q)$. The electrochemical potential of the charged system is therefore given by the relation

$$\phi_{\text{NHE}}(q) = -4.8 - [\phi_{\text{Fermi}}(q) - \phi_W(q) + \phi_W(0)] \tag{4.2}$$

4.3.3.3 Establishing the Potential-Dependent Free Energy
Thermodynamically, a variation of the number of electrons available to a system is equivalent to operating within the grand canonical ensemble (for a full discussion, see Lozovoi et al., 2001). Hence the free energy is a function not only of the electronic structure and field of the nuclei, but also of the electrochemical potential. Furthermore, the energy of the periodic slab must be corrected for interactions with the homogeneous background charge that is applied to restore the system to charge neutrality [a correction of $E_{\text{slab-bg}}(q)$]. This latter term is derived in Taylor et al. [2006b] and can be reduced to the following:

$$E_{\text{slab-bg}}(Q) = -\int_0^Q \Delta\phi_{\text{shift}}(q)\,dq \tag{4.3}$$

where $\Delta\phi_{shift}(q)$ is the change in potential resulting from the charging process up to the system charge Q. Accordingly, the grand canonical free energy is given by the equation:

$$E_{free} = E_{DFT} + \int_0^Q \Delta\phi_{shift}(q)\,dq - Q\phi_{Fermi} \tag{4.4}$$

The latter term arises owing to the loss or gain of Q electrons from a reservoir at potential ϕ_{Fermi}. The free energy E_{free} can now be used to compare reaction energies and activation barriers for electrocatalytic reactions *as a function of the electrochemical potential*. This is demonstrated by the schematic diagram in Fig. 4.4 for the generic reaction A $\rightarrow$ B, where E_{free} is plotted versus potential [calculated from (4.4)] and referenced to vacuum or the NHE. Such energies are shown as the squares and circles for reactant and product systems, respectively, versus calculated potential in Fig. 4.4. The encircled points correspond to the two different systems (A and B) calculated at the same external charge. However, owing to the variation in surface–adsorbate and surface–H_2O interactions in the two systems, A and B at the same charge do not correspond to the same surface potential. To calculate the reaction energy at a given potential, we must instead compare the vertical energy difference between the two best-fit curves for A and B, indicated by the length of the vertical arrow in Fig. 4.4. Therefore, the double-reference method enables the determination of potential-dependent aqueous phase reaction energies or activation barriers (for the latter case, curve B is the energy of the transition state).

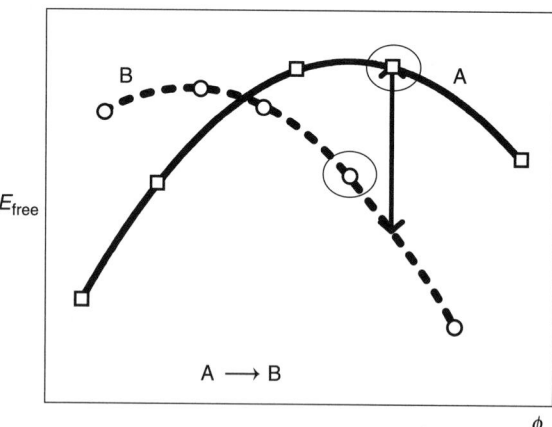

Figure 4.4 Schematic diagram of the free energy calculated from (4.4), E_{free}, versus potential ϕ for the generic electrocatalytic reaction A $\rightarrow$ B. Points indicated by squares and circles are for specific external charges (various q) for the systems A and B, respectively. Solid and dashed lines indicate the best-fit curves for the free energy versus potential relationship for systems A and B, respectively.

4.4 APPLICATIONS OF DENSITY FUNCTIONAL THEORY TO ELECTROCHEMICAL SYSTEMS

Water molecules, water clusters, water layers, or aqueous solution can influence the kinetics and thermodynamics governing catalytic reactions, even to the extent of completely altering the preferred reaction path. These mechanistic changes are due to the ability of water to stabilize charged or highly polarized adsorbed intermediates and transition-state species, which induces structural, electronic, and energetic variations in surface adsorption, diffusion, reaction, and desorption processes. We present various example studies here in which interfacial water significantly influences the reaction energetics at the electrode–aqueous electrolyte interface.

4.4.1 Structure of Water in the Double-Layer Region

The majority of fuel-cell-relevant reactions occur in or near the limits of the *double-layer potential region* for the pure electrolyte. In this region, charging of the electrode results in polarization of the electrode–electrolyte interface with negligible rates of redox reactions involving the electrolyte. Before examining electrocatalytic reaction mechanisms under these conditions, we first examine the structure of water at the electrode interface and the effects of potential on this structure. Studies of the structure of monolayer and multilayer adsorbed water at the metal interface have contributed substantially to our understanding of the aqueous interface. The adsorption of a single water molecule to Group VIII and IB metal surfaces is generally quite weak, with H_2O bound to atop sites on the metal surface with a tilted configuration and metal—O bond distances of around 2.1–2.3 Å [Michaelides et al., 2003b]. Adsorption energies calculated by DFT for vapor phase H_2O (low-coverage) bound to Group VIII and IB metal surfaces are listed in Table 4.1. The vapor phase binding energies, E_b (vapor phase), become stronger from right to left across the periodic table, and especially from the Group IB to VIII metals. At higher coverage, adsorbed water can form monolayers, bilayers, three-dimensional water clusters, and overlayers in

TABLE 4.1 Metal-Dependent Vapor Phase Binding Energy for H_2O Adsorption to Group VIII and IB (111) Metal Surfaces[a]

	E_b (vapor phase) (eV)
Cu(111)	−0.24
Ru(0001)	−0.38
Rh(111)	−0.42
Pd(111)	−0.33
Ag(111)	−0.18
Pt(111)	−0.35
Au(111)	−0.13

[a]Vapor phase (low-coverage) binding energies from [Michaelides et al., 2003b].

crystalline or amorphous structures [Henderson, 2002]. On most close-packed surfaces, H$_2$O adsorption forms a bilayer structure of hexagonal rings that aligns to the registry of the metal surface [Henderson, 2002; Michaelides et al., 2003a].

Surprisingly, molecular dynamics and Monte Carlo simulations have overwhelmingly shown that the structure of the metal–water interface at ambient temperatures and full solvation bears a striking resemblance to the metal–water interfaces described above under UHV conditions (for a review, see Taylor and Neurock [2005]). Hydrogen bonding creates time-fluctuating, metastable water networks at the discontinuity between the phases, and simulations have shown that these have a permanence of several picoseconds [Izvekov et al., 2001; Raghavan et al., 1991]. A number of these simulations use generic potentials in which the identity of the metal is omitted from the actual simulation (see, e.g., Christou et al. [1981]). More recent ab initio molecular dynamics simulations, however, paint a similar picture [Izvekov et al., 2001; Taylor, 2009b; Vassilev et al., 2005]. Figure 4.5 shows the results of ab initio density functional molecular dynamics simulations of the Cu(111)–H$_2$O interface, providing a clear indication of local structure, including the presence of bilayers extending up to about 1 nm from the interface, with a permanence of at least several picoseconds [Taylor, 2009b].

Accordingly, Neurock and co-workers have developed models for the electrochemical interface that retain this concept of hexagonal structure over close-packed metal surfaces [Filhol and Neurock, 2006; Taylor et al., 2006c]. With the use of a screening charge as described in Section 4.3, the sensitivity of the structural parameters of water with respect to the electrochemical environment were explored [Taylor et al., 2006a]. The predominant effect stems from the polar nature of the water molecule, in which the water molecules are observed to rotate as a function of the applied potential.

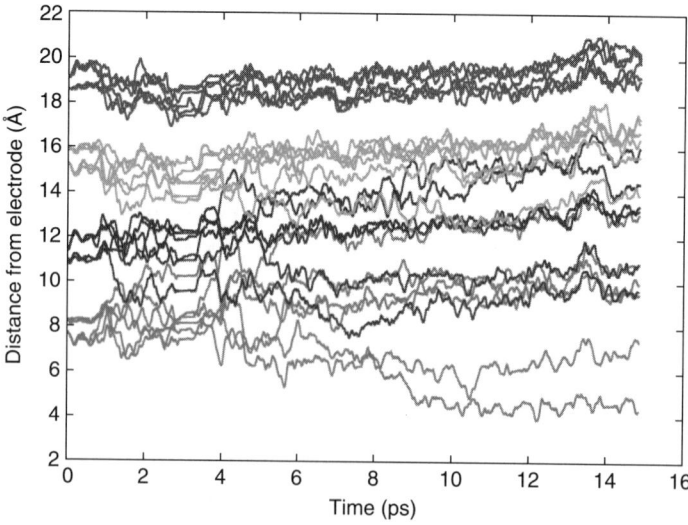

Figure 4.5 The trajectory of water molecules during an ab initio simulation of the electrode/electrolyte interface. The disentanglement and then reorganization of water molecules into bilayers can be clearly seen as time progresses [Taylor, 2009b].

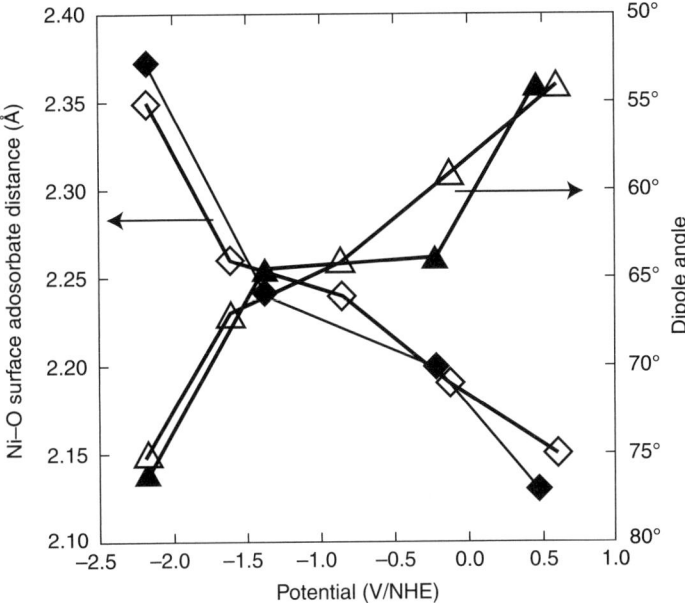

Figure 4.6 Geometric effects at the electrochemical Ni(111)/H$_2$O interface for adsorbed H$_2$O. Open symbols are for a five-layer slab and filled symbols for a three-layer slab; hence, slab size is shown to not influence greatly the outcome of these calculations [Taylor et al., 2006a].

Furthermore, the changes in the binding energy that result upon changes in the potential can result in dissociation and/or substitution by other solution-phase species. These last two topics will be explored in further detail later in this chapter.

As an example of the geometric changes that occur at the electrochemical interface, we review here the results of studies performed on Ni(111) in which the water–metal bond length (M—O) and the water dipole inclination angle are plotted with respect to the applied potential (Fig. 4.6) [Taylor et al., 2006a]. Here, the calculations have been controlled such that water dissociation is restricted in order to separate the topological effects for water from the chemical effects. We see that H$_2$O forms a closer bond to the Ni(111) surface as the surface charge is made more positive. Furthermore, the water dipole rotates such that the hydrogen atoms are directed away from the surface at more positive potentials. Similar effects have been observed for other metals using the same technique.

4.4.2 Activation of Water Outside the Double-Layer Region

Outside of the double-layer region, water itself may be oxidized or reduced, leaving stable hydride, hydroxyl, or oxide layers on the electrode surface. These species may adsorb strongly and block sites from participating in electrocatalysis, as for example, hydroxyl species present at the polymer electrolyte membrane fuel cell

(PEMFC)/direct methanol fuel cell (DMFC) cathode limit the available sites for reduction of molecular oxygen. Alternatively, at the anode of a PEMFC or DMFC, the oxidation of water is necessary to produce hydroxyl or oxygen species that participate in oxidation of strongly bound carbon monoxide species. Taylor and co-workers [Taylor et al., 2007b] have recently reported on a systematic study that examined the potential dependence of water redox reactions over a series of different metal electrode surfaces. For comparison purposes, we will start with a brief discussion of electronic structure studies of water activity with consideration of UHV model systems.

The coverage of water influences the mechanism for water dissociation. At low coverage, a single H_2O molecule binds atop a single metal atom and dissociates homolytically to form adsorbed OH and H species:

$$H_2O_{(ad)} \longrightarrow OH_{(ad)} + H_{(ad)} \quad (4.5)$$

The mechanism involves a metal atom insertion into the O—H bond, thus resulting in the formation of an adsorbed metal—OH species (at the same or similar binding site) and a new metal—H bond. This is a classic bond activation process, which involves a significant stretch of the O—H bond in order to lower the antibonding σ^*_{OH} orbital to enable it to accept electron density from the metal. The reaction has been calculated by DFT to be endothermic by +90 kJ/mol over Pt(111) surfaces with an activation barrier of +130 kJ/mol [Desai et al., 2003b].

Higher water coverages and the presence of solution both act to lower the barriers to activate water. The intermolecular interactions that result from hydrogen bonding with other water molecules stabilize the activated HO—H complex over the entire dissociation reaction coordinate. For metals with high workfunctions, the aqueous phase can enable heterolytic water activation

$$H_2O_{(ad)} \longrightarrow OH^-_{(ad)} + H^+_{(aq)} \quad (4.6)$$

in which hydrogen bonding within the aqueous phase stabilizes the formation of a solvated proton as well as the $OH^-_{(ad)}$ surface intermediate. In this mechanism, the hydroxide remains adsorbed to the metal surface at the same binding site (as before), and the high workfunction of the metal surface enables it to favorably accept the anionic hydroxide charge. For Pt(111), the heterolytic pathway has been calculated to be endothermic by +75 kJ/mol, with an activation barrier of +90 kJ/mol [Desai et al., 2003b]. The heterolytic barrier is 40 kJ/mol lower than the homolytic barrier. The heterolytic barrier is significantly lower because the aqueous phase significantly stabilizes the charged transition state by enabling O—H cleavage without the formation of a new metal—H bond. Instead, the hydrogen-bonded water network directly accepts and shuttles the proton into solution, thus forming $H_3O^+_{(aq)}$ or $H_5O_2^{+(aq)}$ species; the transition state is stabilized by the formation of the $HO^-_{(ad)}$—H^+—$OH_{2(aq)}$ state, without significant perturbation of H_2O over the metal surface. It is well established that polar solvents can significantly stabilize reactions involving charged transition states. From the chemistry discussed here, the water can be considered as a co-catalyst, and thus changes the mechanism.

In addition to enhancing surface reactions, water can also facilitate surface transport processes. First-principles ab initio molecular dynamics simulations of the aqueous/metal interface for Rh(111) [Vassilev et al., 2002] and PtRu(0001) alloy [Desai et al., 2003b] surfaces showed that the aqueous interface enhanced the *apparent* transport or diffusion of OH intermediates across the metal surface. Adsorbed OH and H_2O molecules engage in fast proton transfer, such that OH appears to *diffuse* across the surface. The oxygen atoms, however, remained fixed at the same positions, and it is only the proton that transfers. Transport occurs via the symmetric reaction

$$OH_{(ad)[site1]} + H_2O_{(ad)[site2]} \longrightarrow H_2O_{(ad)[site1]} + OH_{(ad)[site2]} \qquad (4.7)$$

where site 1 and site 2 are neighboring adsorption sites on the metal surface. In Reaction (4.7), proton transfer in one direction appears as OH diffusion in the opposite direction in a Grotthus-type mechanism.

These studies on the aqueous/metal interface provided a wealth of information on the influence of water that had not been fully understood, and they have been directly connected with recent spectroscopic characterizations of surface intermediates and elementary reactions at the aqueous/metal interface. In addition, they provided the basic foundation by which we began to explore chemical reactivity at the electrified aqueous/metal interface. By adopting the double-reference method described above, we found that, under anodic potentials, there is an increase in the demand of the metal for electron density—a demand that was manifested by increasingly strong metal–oxygen bonding at the metal–water interface. This additional bonding, however, weakens the bonding between the oxygen atom and its covalently attached hydrogen atoms, to the point at which dissociation may occur, which is apparent in the polarization study of water bound to the copper electrode in Fig. 4.7. Electron density diminishes on the copper atom, correlating with reduced metal–oxygen bond lengths and water dissociation (indicated by the points marked "b" and "c" in Fig. 4.7). The aqueous environment provides a "sea" of acceptors for the detached protons, and a rich set of pathways for the distribution of the excess charge produced in solution [Desai et al., 2003b]. The metal conducts the excess electron away through the external circuit, or to a cathodic site elsewhere on the surface. The potential at which this occurs controls the thermodynamic stability of this reaction, in addition to the environmental effects (i.e., electrical fields) that will hinder or stabilize the products in solution.

The aforementioned DFT model has been used to investigate the thermodynamic stability of water over a number of other metals [Filhol and Neurock, 2006; Rossmeisl et al., 2006; Taylor and Neurock, 2005; Taylor et al., 2006a–c, 2007a–c]. Detailed DFT analysis of the interplay between hydrogen bonding to solution, surface charging, and electric fields has shown that the metal–adsorbate bonding is influenced by a shift in the *d*-band of the metal, which is accompanied by a shift of the local orbitals of the adsorbate resulting from a change in the local electrochemical potential [Taylor et al., 2007a]. This effect leads to a change in the total binding energy of the adsorbate to the electrode. The magnitude of this effect upon the dissociation of water, however, has been shown to be rather small compared with the larger thermodynamic shifts induced in the energy of the conducting electron when the potential is

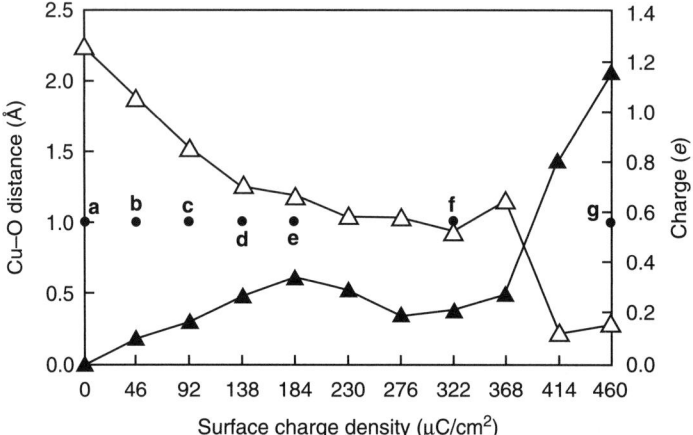

Figure 4.7 The correlation of metal atom charge with the metal–oxygen bond length in the water adsorption as a function of the total system charging in a representative copper–water electrode/electrolyte interface. Open triangles: height of the adsorbed oxygen above the surface plane (either as H_2O, OH, or O). Filled triangles: integrated charge density within a sphere of radius 1.45 Å about the coordinated copper atom directly coordinated to the dissociating water molecule. Points **a–g** correspond to the following: **a–b**, water migration from atop to bridge; **b–c**, dissociation to form OH and H^+; **c–d**, migration of OH to 3-fold coordination; **d–e**, dissociation of OH to O + H^+; **e–f**, gradual formation of oxide; **f–g**, oxide formation [Taylor, 2009b].

moved up or down [Rossmeisl et al., 2006]. Hence, plots of the dissociation reaction energy are linear in the potential with only slight deviations due to these bonding effects. This effect is demonstrated by the energy versus potential correlation in Fig. 4.8 for water dissociation over Pt(111), in which the deviation of the curve from linearity is shown by the variance of the solid lines (full model) from the dashed lines (linear term only model). Inclusion of the hydrogen chemical potential (i.e., the pH) can be incorporated easily into the thermodynamic model using the Nernst equation. The production of OH thus occurs at more positive potentials when the pH is lowered, whereas the production of H from H_2O occurs at more negative potentials when the pH is raised. An extensive survey of DFT predictions for electrochemical water dissociation has been made for a series of transition and noble metals, with good qualitative and semiquantitative agreement for the known double-layer regions for these metals [Taylor et al., 2007b]. It is important to be aware of the state and reactivity of water on the surface when exploring other chemistry that occurs in the presence of an aqueous environment (i.e., fuel cell activity). This synergism is discussed more fully in the following sections.

The results found here indicate that although the changes in potential can markedly influence oxidation reaction energy through altering the energy of the electron product, differences in the interactions of the reactant (water) and product (hydroxyl) species with the interfacial electric field appear to have only a small effect on the

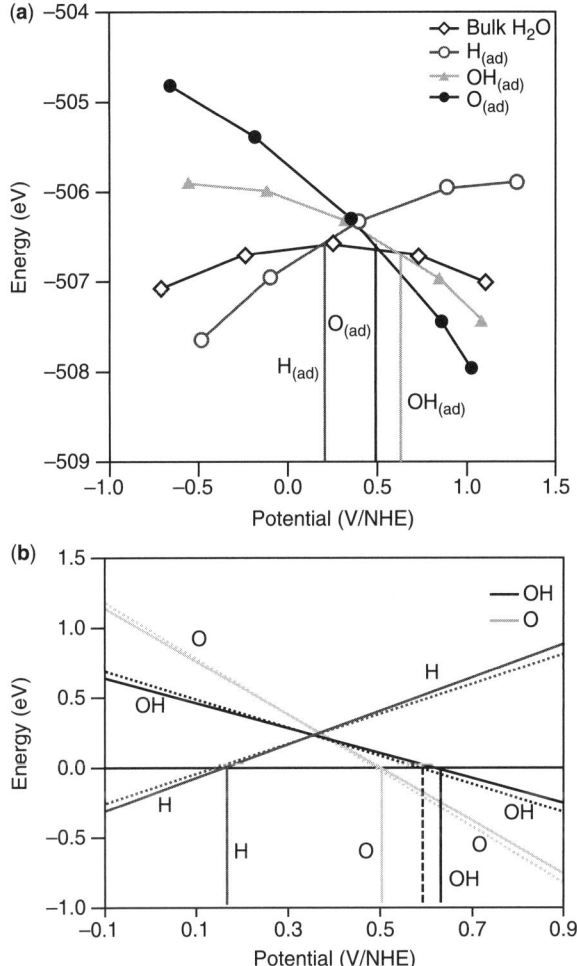

Figure 4.8 Potential-dependent reaction energies for water dissociation to form OH, O, and H over Pt(111). (a) Energy curves based on the full charge model: the nonlinearity of these plots expresses the capacitance of the interface. (b) Differences of the curves indicate the reaction energies. The nonlinear terms cancel almost completely. The dashed lines indicate predictions made from the linear model, whereas the solid lines are predictions made from a full solvation/charge-based model [Rossmeisl et al., 2006].

reaction energy. This holds only when the adsorbed reactant and product states have similar dipole moments, similar solvation structures, and similar effects on the interfacial capacitance. This similarity does not hold, however, when comparing, for example, the adsorption of molecular oxygen compared with water at the electrified interface. This competitive adsorption process is discussed in the following section.

4.4.3 Competitive Adsorption at the Aqueous, Electrified Metal Interface

The previous section addressed how ab initio methods can be used to examine the structure and state of water on the electrode surface as a function of electrochemical potential. The structure and binding interactions of water can affect the binding energies of reactant or product species at the electrocatalytic interface. The use of DFT to calculate the binding energies of species to metal surfaces is well established as a standard step in modeling heterogeneous catalytic systems. Binding energies can be used to determine the coverage of a single adsorbate as a function of gas-phase pressure and temperature, examine the competitive adsorption between gas phase species, compare various solid surfaces for the affinity to adsorption, or predict coverages of intermediate/product species developed during a catalytic reaction. The calculation of adsorption energies at the aqueous, electrified surface is less straightforward, as an adsorption process includes the displacement of another molecule from the surface and an exchange with a molecule in solution. While ab initio methods have been used to calculate relative binding energies of adsorbates on various metals and generate predictions of electrocatalytic activity [Greeley and Nørskov, 2005; Zhang et al., 2005a, b], they have not been commonly used to examine the competitive adsorption between the various species present in an electrocatalytic environment. Most simulations do not consider the energetics of supplying the reactant to the surface or the desorption of product species from within the inner layer when examining electrocatalytic pathways. However, the relative binding preference of water, electrolyte ions, or reactant/product species may directly affect the rates of electrocatalytic processes. Furthermore, the relative binding energies of these species will be a function of electrode potential, and thereby may affect the potential dependence of electrocatalytic rates and reaction paths. For these reasons, we felt that it was important to develop a theory of substitution chemistry at the electrode/electrolyte interface. In this section, we illustrate how the double-reference method may be used to consider the relative binding energies of adsorbates at the aqueous, electrified interface.

Let us reconsider the situation in which water on the surface is replaced with an OH species. This situation is analogous to the water dissociation reaction, discussed previously. We will use this as a model system to demonstrate how the methods used in examining water dissociation can be re-applied to consider competitive adsorption at the electrocatalytic interface. At a given pH, some concentration of hydroxyl ions is present in solution. The adsorption of a hydroxyl ion from solution to the surface requires moving the ion from the diffuse layer (outside the interfacial field region) to the surface, and results in the transfer of the negative charge (or partial negative charge) to the electrode. Thus, the water dissociation event could be just as easily cast as a H_2O/OH substitution event. In the aqueous environment, therefore, we assume that, rather than finding a bare metal site to adsorb to, the ion must displace a water molecule previously adsorbed, returning the water molecule to the diffuse layer (although the assumption of a one-for-one exchange is not obvious for any given adsorbate). The hydroxyl adsorption event can now be written as:

$$OH^-_{(aq)} + H_2O^* \longrightarrow H_2O_{(aq)} + OH^* + e^- \quad (4.8)$$

4.4 DENSITY FUNCTIONAL THEORY TO ELECTROCHEMICAL SYSTEMS

A "displacement free energy" ΔG_{disp} can then be defined as

$$\Delta G_{\text{disp}}(\text{pH}, U) = G_{\text{OH}_{(\text{ad})}}(U) + eU + G_{\text{H}_2\text{O}_{(\text{aq})}} - G_{\text{H}_2\text{O}_{(\text{ad})}}(U) - G_{\text{OH}^-_{(\text{aq})}}(\text{pH}) \quad (4.9)$$

where the free energy of the hydroxyl ion in solution, $G_{\text{OH}^-_{(\text{aq})}}$, is a function of the solution pH, and the free energies of the adsorbed hydroxyl and water species, $G_{\text{OH}_{(\text{ad})}}(U)$ and $G_{\text{H}_2\text{O}_{(\text{ad})}}$, are functions of the electrode potential. This displacement free energy is a thermodynamic descriptor for ion adsorption at the electrochemical interface.

Reaction (4.8) may be restated as the sum of two reactions:

$$\text{H}_2\text{O}^* \longrightarrow \text{OH}^* + \text{H}^+ + e^- \quad (4.10)$$

$$\text{OH}^-_{(\text{aq})} + \text{H}^+_{(\text{aq})} \longrightarrow \text{H}_2\text{O}_{(\text{aq})} \quad (4.11)$$

As Reaction (4.11) can be assumed to be at equilibrium, the free energy change of reaction is zero, and the values of ΔG_{disp} (Reaction 4.8), and of the free energy of the water oxidation reaction (4.10) are equivalent. This is, of course, necessary, because the Born–Haber paths from reaction to product must be thermodynamically equivalent (water dissociation, or a solution phase exchange, coming from the same reactant and leading to the same product states are therefore the same). For a general displacement reaction, however, the similarities to an interfacial electrochemical reaction are not as clear. This exercise serves, therefore, to illustrate that the double-reference method can be used to consider the potential dependence of water displacement at the surface by other solution phase species.

As is evident for water/hydroxyl exchange, the calculation of relative binding energies at the electrode surface is complicated by the need to consider the effects of the electrified interface on binding characteristics as well as the requirement that solution phase free energies of the various species be determined. For illustrative purposes, a rough approximation to the solution phase free energies can be generated by simply calculating the energy of a solute molecule in a periodic unit cell with a locally optimized solvation structure. In the future, ensemble averages could be taken from ab initio molecular dynamical simulations. The static approach with a locally optimized solvation shell is illustrated in Fig. 4.9 for the displacement of an adsorbed water molecule with an adsorbed oxygen molecule on the Pt(111) surface. The free energy of the solution phase species may be improved by combining the calculation of the gas phase free energy with experimental solvation free energies, or by more rigorous calculation of solvation free energies [Pliego and Riveros, 2000]. Although the energy value of the solution phase species affects the absolute value of the displacement energy, it does not alter the trend of the displacement energy with electrode potential, because the solution phase species are taken to reside beyond the region of potential drop at the interface (beyond the OHP). In other words, the energy of the solution phase species serves as a reference energy, which is independent of the electrochemical potential. A negative substitution energy implies an energy state for the adsorbed ion that is lower than its solution state (and hence thermodynamically favors adsorption), whereas a positive substitution energy implies a higher energy state for adsorption.

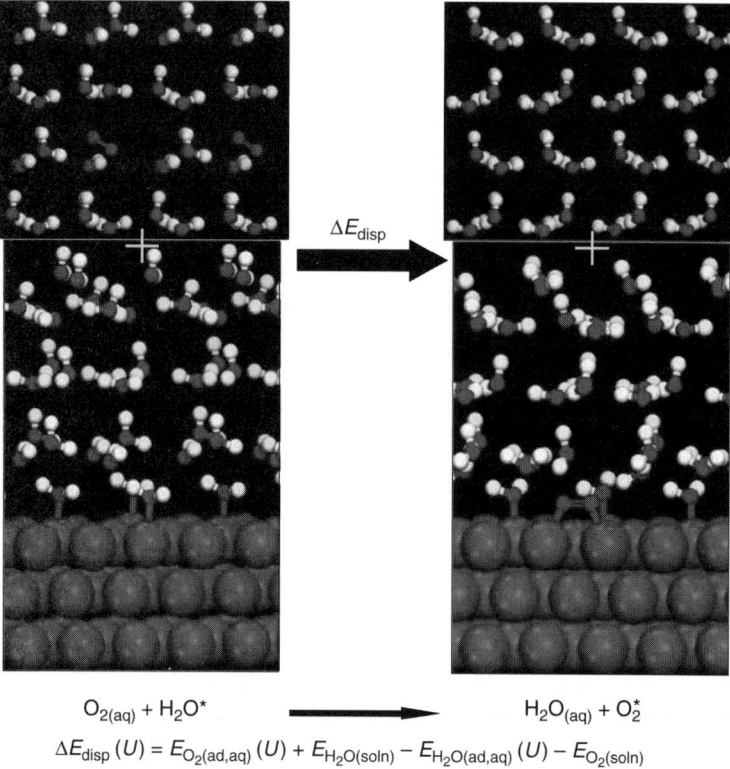

$$O_{2(aq)} + H_2O^* \longrightarrow H_2O_{(aq)} + O_2^*$$

$$\Delta E_{disp}(U) = E_{O_2(ad,aq)}(U) + E_{H_2O(soln)} - E_{H_2O(ad,aq)}(U) - E_{O_2(soln)}$$

Figure 4.9 Illustration of the structures considered for adsorbing an oxygen molecule from solution to the electrode surface, displacing a water molecule from the surface to solution.

The energy associated with the adsorption of O_2 at a Pt(111) electrode surface is relevant to the oxygen reduction reaction (ORR) occurring at the cathode of a PEMFC. Alloying a second metal with Pt can increase the rate of the ORR by weakening the binding of oxygen-containing species and limiting hydroxyl poisoning of the surface. However, in sacrificing strong oxygen binding, the ability of molecular oxygen to compete with water molecules for metal surface sites may be decreased as well. The relative binding preferences of molecular oxygen and water may be potential-dependent. The adsorption of water occurs primarily through the donation of electron density from the $1b_1$ lone pair of electrons on the oxygen atom into the unoccupied d_{z^2} orbital on the metal [Henderson, 2002], and binding energies may be expected to increase in strength as the electrode potential is increased (with respect to an NHE), or the workfunction of the metal is increased. Molecular oxygen, on the other hand, binds through both donation from molecular oxygen into the metal and back-donation from the metal into the $2\pi^*$ orbital on oxygen. Oxygen picks up electron density from the metal through back-donation, resulting in an elongation of the O—O bond when adsorbed to the surface. Both cluster [Hyman and Medlin, 2005; Panchenko et al., 2004] and periodic [Hyman and Medlin, 2005] DFT models of the Pt(111) surface indicate that lower electric fields (corresponding to more negative potentials on the NHE scale) cause stronger binding of O_2 through increased donation of electron density to the $2\pi^*$ orbital. The importance of back-donation in oxygen binding may lead to a noticeable difference in binding energy relative to water as the electrode potential is varied.

Figure 4.10 plots the potential-dependent displacement energy associated with replacing an adsorbed water molecule with an oxygen molecule on the Pt(111) surface. There is a 1.5 eV preference to bind O_2 compared with H_2O at 0 V/NHE. This value exceeds the UHV binding energy preference (about 0.25 eV) and is the result of the difference in solvation energy between O_2 and H_2O. However, as the potential is increased, water binding becomes relatively stronger, until, at a potential of 1.36 V/NHE, it becomes favorable to adsorb water rather than molecular oxygen. This potential, of course, is well past that of interest for the ORR, and well past the potential at which hydroxyl species would replace water molecules at the surface, changing the species with which oxygen must compete. In addition, the results here describe only the overall thermodynamic preference. In an aqueous medium, there may be an activation barrier for oxygen to actually displace water from the surface. The adsorption of O_2 includes electron transfer, to a varying extent depending on electrode potential, for which Hartnig and Koper suggest, based on molecular dynamics simulations, that solvent reorganization can have a dominant contribution to the overall rate [Hartnig and Koper, 2002].

The potential of relevance for competitive adsorption of water may shift upon alloying Pt. The calculated ratio of water to oxygen binding energies at the UHV

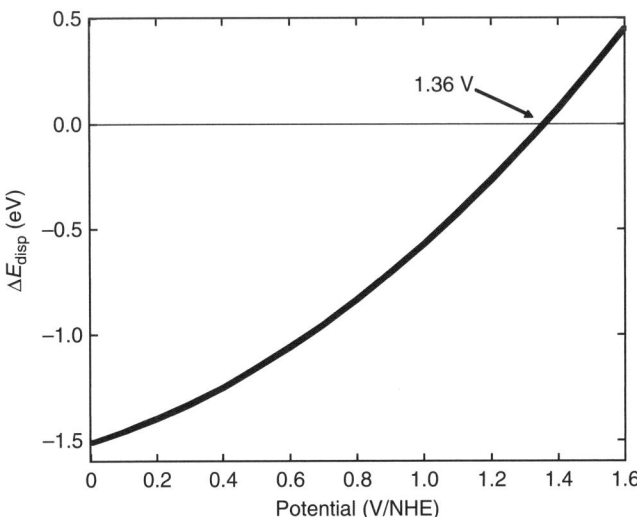

Figure 4.10 Energy of displacing water with molecular oxygen at the Pt(111) surface as a function of electrode potential. At potentials below 1.36 V, it is favorable to adsorb molecular oxygen, while at potentials above 1.36 V, water is more stable on the surface.

interface for the $Pt/Pt_2Co(111)$ overlayer model was found to be 0.54, which is greater than the value of 0.44 calculated for $Pt(111)$ (M.J. Janik and M. Neurock, unpublished work, 2007). These results show qualitatively that DFT calculations, together with the double-reference method to consider the potential-dependent energy of surface-bound species, can begin to examine the competition between species for surface sites in an electrochemical half-cell. This competition becomes more complex in an actual fuel cell environment, where, at the cathode for example, water, oxygen, and electrolyte anionic groups may compete for surface adsorption sites, and this competition will depend on the local concentrations and structures available to these components.

4.4.4 Elementary Reaction Thermodynamics at the Aqueous, Electrified Interface: Methanol Dehydrogenation

The presence of solution at a metal surface, as has been discussed, can significantly influence the pathways and energetics of a variety of catalytic reactions, especially electrocatalytic reactions that have the additional complexity of electrode potential. We describe here how the presence of a solution and an electrochemical potential influence the reaction pathways and the reaction mechanism for methanol dehydrogenation over ideal single-crystal surfaces.

The dehydrogenation of adsorbed methanol initially proceeds via one of two possible paths. The first involves the activation of the C—H bond to form an adsorbed hydroxymethyl ($CH_2OH_{(ad)}$) intermediate. The second path proceeds via the activation of the O—H bond of methanol to form an adsorbed methoxy ($CH_3O_{(ad)}$) intermediate. The resulting hydrogen atom can either be adsorbed to the metal surface or solvated by the aqueous phase. Periodic DFT investigations [Desai et al., 2002; Greeley and Mavrikakis, 2002, 2004] of vapor phase methanol dehydrogenation (in the absence of co-adsorbed water) over $Pt(111)$ show that the initial dehydrogenation step involves the homolytic activation of the C—H bond to form hydroxymethyl and hydride surface intermediates, with an overall reaction energy of -16 kJ/mol. The activation of the O—H bond of adsorbed methanol results in the formation of a methoxy and hydride surface intermediate with an overall reaction energy of $+64$ kJ/mol (80 kJ/mol less favorable). The energetics for the subsequent $CH_2OH_{(ad)}$ dehydrogenation steps indicate that the methanol dehydrogenation proceeds via the following reaction path:

$$CH_3OH_{(ad)} \longrightarrow CH_2OH_{(ad)} + H_{(ad)} \longrightarrow CHOH_{(ad)} + 2H_{(ad)}$$
$$\longrightarrow COH_{(ad)} + 3H_{(ad)} \longrightarrow CO_{(ad)} + 4H_{(ad)} \quad (4.12)$$

These results refer to dehydrogenation of methanol on the ideal close-packed $Pt(111)$ terrace. It is well established that step edges can be present on Pt surfaces and influence the surface chemistry. The resulting paths at the coordinatively unsaturated step sites lead to significantly enhanced binding energies. We showed previously that the overall reaction energy for the methanol to form methoxy in the vapor phase at a model $Pt(211)$ step site now becomes exothermic, with reaction energy that is 78 kJ/mol more exothermic than at the (111) terrace. The reaction of methanol to hydroxymethyl, on the other hand, was also found to be more exothermic, but was only improved by 26 kJ/mol. Experimental studies in UHV also show that defects on the (111) surface play an important role in methanol decomposition, and favor O—H cleavage through the methoxy intermediate [Ehlers et al., 1985]. O—H cleavage is also enhanced at (111) steps in electrocatalytic systems, favoring formation of formic acid [Housmans et al., 2006].

DFT investigations of methanol dehydrogenation over Pt(111) for the aqueous phase system show the influence of water on the methanol dehydrogenation mechanism. Okamoto and co-workers [Okamoto et al., 2003] investigated homolytic aqueous phase methanol dehydrogenation. A direct comparison of the homolytic and heterolytic paths at the aqueous/Pt(111) interface, however, showed that the heterolytic cleavage of the initial methanol C—H dehydrogenation step was preferred over homolytic cleavage by over 43 kJ/mol [Cao et al., 2005]. The differences between aqueous phase and vapor phase mechanisms parallels those discussed above for water activation, where the H^+ product is stabilized by the aqueous media and the high workfunction of the Pt(111) surface allows for the enhanced bonding with the charged $CH_xO^-_{(ad)}$ species. In addition, hydrogen bonding with the aqueous phase stabilizes both the adsorbed methanol ($CH_3OH_{(ad)}$) and the dehydrogenated species ($CH_xO_{(ad)}$), further influencing the energetics from the vapor phase.

Overall, the aqueous phase dehydrogenation mechanism [Cao et al., 2005] was found to follow a similar ordering of C—H and O—H cleavage steps to the vapor phase case [Desai et al., 2002; Greeley and Mavrikakis, 2002, 2004]. The calculated reaction energies for the aqueous phase and vapor phase systems are shown in Fig. 4.11 (Plate 4.1) (the vapor phase values are in parentheses), along with the optimum geometries for the adsorbed aqueous phase intermediates (showing the hydrogen bonds between H_2O and CH_xO species). The preferred aqueous phase pathway for heterolytic dehydrogenation over the ideal Pt(111) surface follows the order

$$CH_3OH_{(ad)} \longrightarrow CH_2OH_{(ad)} + H^+_{(aq)} \longrightarrow CHOH_{(ad)} + 2H^+_{(aq)}$$
$$\longrightarrow CHO_{(ad)} + 3H^+_{(aq)} \longrightarrow CO + 4H^+_{(aq)} \qquad (4.13)$$

However, the activation of the C—H bond of $CHOH_{(ad)}$ to form COH followed by O—H cleavage to $CO_{(ad)}$ is also a viable pathway (Fig. 4.11).

The results here clearly demonstrate some of the important differences between reactions in the vapor phase and those in the aqueous phase. Water solvates the ions that form and thus enhances the heterolytic bond activation processes. This leads to more significant stabilization of the charged transition and product states over the neutral reactant state. The changes that result in the overall energies and the activation barriers of particular elementary steps can also act to alter the reaction selectivity and change the mechanism.

The potential that develops in an electrochemical system such as a fuel cell can also act to significantly influence the energies, kinetics, pathways, and reaction mechanisms. The double-reference potential DFT method [Cao et al., 2005] described earlier was used to follow the influence of an external surface potential on the reaction

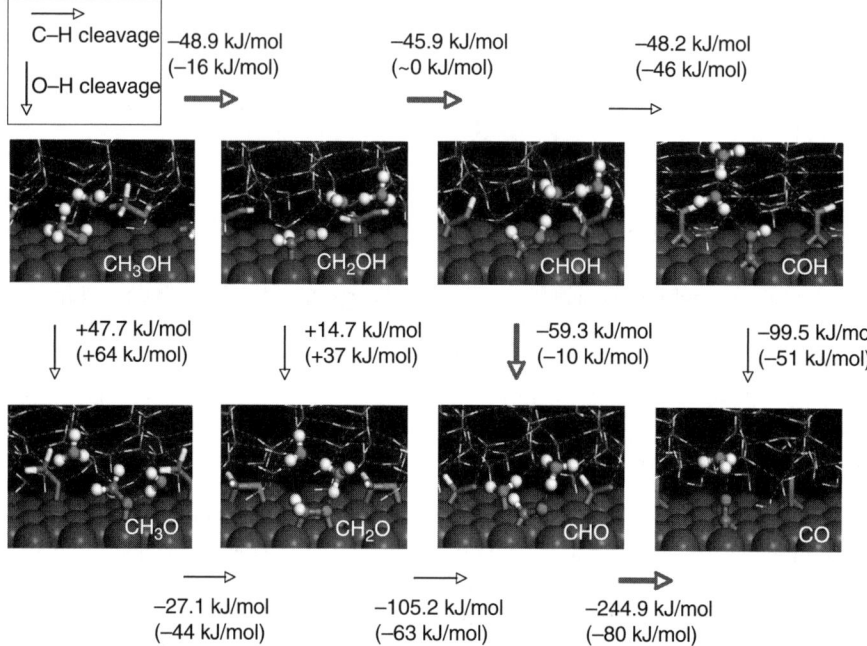

Figure 4.11 Optimized structures of CH$_x$O species, as indicated, over aqueous-solvated Pt(111) as determined by DFT in Cao et al. [2005]. Horizontal and vertical arrows indicate C—H and O—H cleavage steps, respectively. Reaction energies are included for the aqueous phase [Cao et al., 2005] and the vapor phase (in parentheses) [Desai et al., 2002]. The thermodynamically preferred aqueous phase pathway is indicated by bold arrows (in blue). (See color insert.)

energies and the reaction mechanisms for methanol decomposition. Plotted in the upper portions of each panel in Fig. 4.12 are the free energies for the various methanol dehydrogenation intermediates over the aqueous phase Pt(111) surface for a range of external potentials (tuned by the application of an external charge q). The filled symbols in Fig. 4.12 are for the $q = 0$ system; the open symbols are for the $q \neq 0$ systems. The reaction energies for different elementary steps are plotted in the lower portion of each panel as a function of potential. The reaction energies were determined by subtracting the free energy versus potential curves shown in the upper panels: $\Delta E_{\text{free}} = E_{\text{free(products)}} - E_{\text{free(reactants)}}$. Similar to the aqueous phase mechanism, the initial dehydrogenation step shows that C—H cleavage is heterolytic and thermodynamically favored over heterolytic O—H cleavage in the potential range of -0.5 to $+1.0$ V with respect to an NHE. However, the free energy of the methoxy species has a greater dependence on the potential than does the hydroxymethyl species, indicating that O—H cleavage becomes a competing path at higher potentials. This behavior is in agreement with experimental findings by Cao and co-workers [Cao et al., 2005], which show that methanol dehydrogenation involves a dual-path mechanism on Pt(111) at potentials greater than $+0.35$ V with respect to an NHE. Following the

4.4 DENSITY FUNCTIONAL THEORY TO ELECTROCHEMICAL SYSTEMS **117**

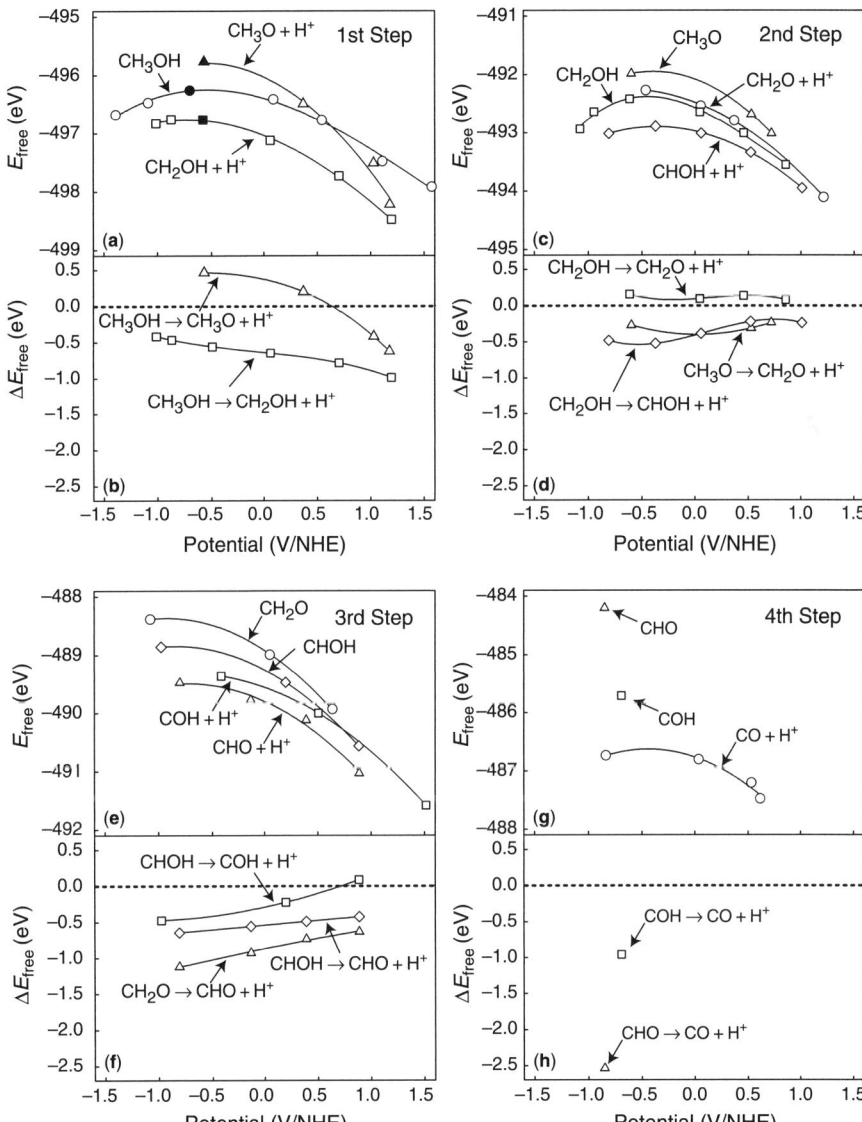

Figure 4.12 The upper plots (a, c, e, g) show the free energies (calculated by (4.4) from DFT) versus the estimated potential for reactants and products involved in the first, second, third, and fourth consecutive methanol dehydrogenation steps, as indicated, over Pt(111) from Cao et al. [2005]. Filled symbols in (a) refer to the energy and potential for the system at $q = 0$. The lower plots (b, d, f, h) show the corresponding reaction energies for the first, second, third, and fourth consecutive methanol dehydrogenation steps, as indicated.

most preferred pathway, methanol dehydrogenation involves the sequence

$$\text{CH}_3\text{OH}_{(ad)} \longrightarrow \text{CH}_2\text{OH}_{(ad)} + \text{H}^+_{(aq)} \longrightarrow \text{CHOH}_{(ad)} + 2\text{H}^+_{(aq)}$$
$$\longrightarrow \text{CHO}_{(ad)} + 3\text{H}^+_{(aq)} \longrightarrow \text{CO} + 4\text{H}^+_{(aq)} \qquad (4.14)$$

which is identical to the most favorable path determined in the aqueous phase (i.e, at a constant charge of $q = 0$ and not at constant potential). At more anodic potentials, however, a parallel path involving the initial O—H cleavage to form methoxy followed by C—H cleavage to form formaldehyde ($\text{CH}_2\text{O}_{(ad)}$) can also become favorable. The potential that results at the electrified aqueous/metal interface in this system has a strong influence on the reaction chemistry as well as the reaction mechanism. This again stresses the point that the extrinsic environment, which here includes the presence of a solution and an applied potential, can significantly influence the catalytic properties of the system.

4.4.5 Potential Dependence of Elementary Chemical Reactions in Electrocatalysis: Coupling of Carbon Monoxide and Hydroxyl Species

In the previous section, we described how the double-reference method could be used to follow the influence of water and potential on elucidating the elementary pathways and their thermodynamic reaction energies. The simulation of electrocatalytic kinetics is difficult owing to difficulty in appropriately tracking electron transfer kinetics over complex metal surfaces. The double-reference method was recently used to determine the elementary reaction activation barriers at the electrified interface as a function of potential for the oxidation of carbon monoxide over Pt(111) [Janik and Neurock, 2006] and Pt/Ru alloy [Janik et al., 2007] surfaces. The electro-oxidation of carbon monoxide from the Pt or alloy surface requires both the activation of water to form oxygen or hydroxyl species and the subsequent coupling of these species with adsorbed carbon monoxide to form the oxidized product (CO_2 or COOH, which is rapidly oxidized to CO_2). In this section, we concentrate on the coupling reaction between CO and OH, as both experimental studies [Gasteiger et al., 1994] and DFT calculations [Janik and Neurock, 2006; Liu et al., 2003] have identified OH as the likely oxidizing species. In addition, we have already discussed the activation of water over Pt in Section 3.2. Here, we compare the calculated activation barriers for the oxidation of CO for UHV, micro-solvated, and double-reference (solvated, electrified interface) DFT model systems.

DFT estimates in the literature report barriers for the coupling of CO and OH at UHV conditions over the Pt(111) surface of 0.89 eV at low coverage ($\frac{1}{9}$ ML) [Desai and Neurock, 2003a] and 0.58 eV [Shubina et al., 2004] and 0.44 eV for higher $\frac{1}{4}$ ML coverages [Gong et al., 2003]. Desai and Neurock examined the CO + OH coupling step at the aqueous–metal interface by filling the vacuum region of the unit cell with explicit water molecules to simulate an aqueous solution. They found that the barrier for CO oxidation at $\frac{1}{9}$ ML was reduced from 0.89 eV at UHV conditions to 0.62 eV in solution. The presence of water allowed for a direct heterolytic pathway in which the (OC)O—H bond was activated to form a CO_2^- and an H^+, which was transferred through the local water molecules to form an H_3O^+ or $H_5O_2^+$ intermediate as the CO—O bond is created. We reported that, in the solvated cell, the transition states for CO + OH coupling to COOH or with simultaneous O—H dissociation over Pt are equivalent [Janik and Neurock, 2006]. The barrier to CO + OH coupling is reduced with respect to the UHV environment, owing to hydrogen bond stabilization of the transition state (Fig. 4.13). The hydrogen bond distance between the hydroxyl species and a nearby water molecule was reduced from 2.03 Å in the initial state to 1.80 Å at the transition state, providing significant solvation stabilization to lower the barrier with respect to the UHV environment.

Although the coupling of CO and OH is not a redox reaction, the changes in electrode potential, and therefore the interfacial electric field, electrode Fermi level, and altered water structure can alter the activation barrier. The calculation of potential dependent barriers with the double-reference method requires the location of transition states with varying system charges. Transition states are located using the climbing image nudged elastic band method (CI-NEB) [Henkelman and Jonsson, 2000; Henkelman et al., 2000; Mills et al., 1995], which optimizes a series of images between initial and final states of an elementary reaction with the constraint of remaining spaced along the reaction path and with a single image forced to maximize its energy along the reaction path. This

Figure 4.13 The transition state for CO + OH coupling to form COOH on the Pt(111) surface. The transition state is stabilized in the aqueous model system through hydrogen bonding between the hydroxyl species and a nearby water molecule [Janik and Neurock, 2006].

method inherently requires that the number of electrons in the system remain constant along the reaction path, and, therefore, within a double-reference model system, the electrode potential may change along the reaction coordinate. However, for a nonredox chemical reaction, the potential change between reactants and products is small, and may be assumed negligible such that the constant charge activation barrier can be assigned to the potential of the reactant state.

The barrier to CO + OH coupling was found to be potential-dependent. As the surface of the electrode becomes increasingly positively charged, the activation barrier to coupling decreases from 0.55 to 0.30 eV [Janik and Neurock, 2006]. Although the complexity of the double-reference model system complicates assigning this trend to a specific interaction, changes in the solvation structure about the transition state suggests that the more positive surface charge facilitates greater solvation of the transition state. With increasing positive charge, the hydrogen bond distance between the hydroxyl species and a nearby water molecule decreases from 1.91 Å at the transition state at −0.53 V to 1.64 Å at 1.58 V. The potential dependence of the activation barrier appears to be a consequence of the dependence of the interfacial water structure on the electric field. The increasing rate of CO oxidation appears to be the result of both a more favorable reaction free energy to activate water and the decreasing coupling barrier.

4.4.6 Ion Transport through the Double Layer: Reduction of O_2^*

For elementary electrochemical steps, water plays an intrinsic role in the transport of ions across the electrochemical double layer. In Section 4.3.4, we discussed in detail the thermodynamics involved in the oxidation of methanol and generation of protons and electrons. The resulting proton product that formed in the activation of methanol remained in the model system, bound to an adjacent water molecule, forming a hydronium ion. The completion of the electrochemical step requires the transport of the ion across the electrochemical double layer, which, together with electron conduction across the electrochemical cell, maintains the constant potential drop across the interface. In the case of methanol oxidation, these processes are relatively rapid with respect to the breaking of C—H and O—H bonds. However, in the case of oxygen reduction, both experiment [Damjanovic and Brusic, 1967; Damjanovic and Sepa, 1990; Grgur et al., 1997] and theory [Anderson and Albu, 1999, 2000; Anderson et al., 2005] have led to speculation that the initial reduction of adsorbed molecular oxygen is the rate-limiting step. This reduction process is used as an initial example of the application of the double-reference method to examine the kinetics of elementary redox reactions. It should be noted that our work does not address whether electron transfer during the adsorption of O_2 itself should be referred to as the first reduction step; rather, we adopt the terminology of Anderson in referring to the electron transfer coinciding with the formation of OOH* as the first reduction step. Adsorption of O_2 occurs, with a net transfer of electron density from the electrode surface to the adsorbate, and this extent of electron transfer, as discussed above and by others [Hyman and Medlin, 2005; Panchenko et al., 2004], varies with the electrode potential.

Using the periodic DFT double-reference model, the reduction of adsorbed O_2 (O_2^*) was found to occur by two sequential steps:

$$H_{bulk}^+ + e_{reference}^- + O_2^* \longrightarrow H_{near}^+ + e^- + O_2^* \quad (4.15)$$

$$H_{near}^+ + O_2^* + e^- \longrightarrow OOH^* \quad (4.16)$$

where H_{bulk}^+ signifies a proton in the bulk electrolyte and H_{near}^+ refers to a proton in the model unit cell, but as far from the adsorbed O_2 as possible within the model system. Reaction (4.15) represents the diffusion of the proton from the bulk to the near-surface region, defined for our purposes as being the initial proton position within the DFT model unit cell. Within the double-reference model system, the completion of this diffusion process coincides with the addition of an electron to the model system, however, whether the electron resides in the electrode, on adsorbed O_2^*, or associated with the nearby proton is observed rather than prescribed. In the initial reactant state of Reaction (4.15), the energy of the proton and electron are included using a bulk reference rather than being included directly in the unit cell. This is done by invoking the equality of the free energy of a proton and electron with a gas phase hydrogen molecule at NHE conditions [Liu et al., 2003; Nørskov et al., 2004]:

$$G_{H^+,e}(U) = G_{H_2} + eU \quad (4.17)$$

where U represents the potential relative to the NHE. In the product species of Reactions (4.15) and (4.16) ("$H_{near}^+ + e^- + O_2^*$" and OOH^*), the proton and electron are included in the DFT model system and the double-reference method is used to explicitly consider whether the electron shifts the electrode potential or reduces the reaction center at a given electrode potential.

Figure 4.14 illustrates the structures considered and the potential dependence of the free energies of each species. Both of the steps involved in reducing adsorbed O_2 were found to endothermic at potentials of interest for the ORR, in line with the expectation that transport of the proton closer to the positively charged surface is successively more endothermic with increasing proximity. What is surprising, however, from this plot is the relative slopes of the two reaction steps. The slope of an elementary reaction free energy with potential is equal to the number of electrons transferred in that step. Clearly, the first step of moving the proton near to the adsorbed O_2 has a reaction free energy that increases with increasing potential, and, in fact, the slope of this reaction is exactly 1. This indicates that electron transfer has occurred prior to forming the OO—H bond. A detailed analysis of charge corroborates that reduction has occurred and that an additional electron resides in the adsorbate layer near to the surface as compared with the initial state. The additional electron cannot be directly assigned to the oxygen molecule, as the electron density increases across various local water molecules and the added H atom as well, suggesting that electron transfer to the adsorbed O_2 occurs together with a polarization to stabilize the positive charge on the hydronium ion. This result can only be attributed to the specific model system used in this study (i.e., the water structure, O_2 coverage, proton position, and

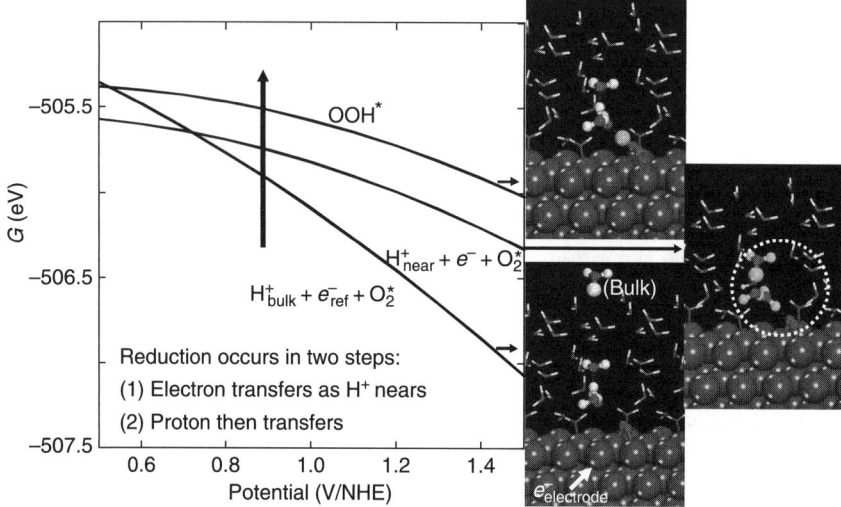

Figure 4.14 Free energy as a function of electrode potential for the species considered in the reduction of O_2^* to form OOH^*. In the atomistic diagrams, purple coloring is used to illustrate where the electron involved in reduction resides in each species. (See color insert.)

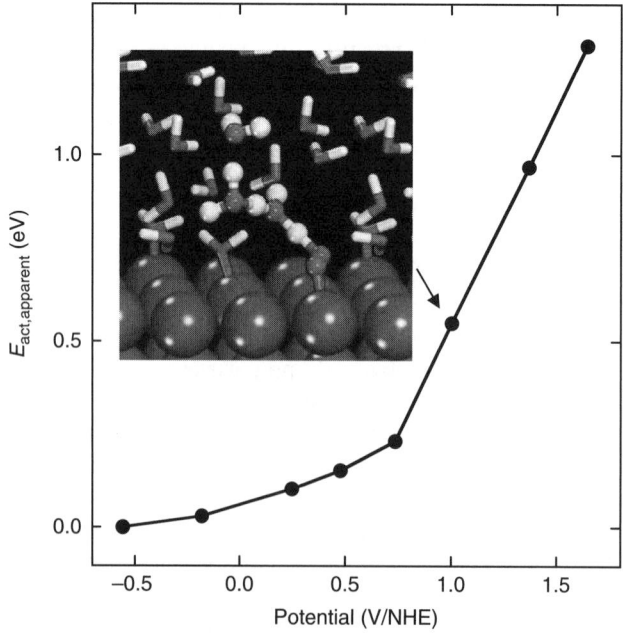

Figure 4.15 Overall barrier for O_2^* reduction to OOH^* over the Pt(111) surface as a function of electrode potential. The inset shows the transition state for reduction at 1 V/NHE.

concentration), however, it demonstrates that the electrolyte structure a few water layers from the adsorbate will affect the energetics associated with electron transfer, and cautious generalization could be made to other electrochemical reactions. In fact, this concept is not too far from that of OHP electron transfer reactions, such as the Fe^{2+}/Fe^{3+} redox event that occurs a few ångströms away from the electrode [Smith and Halley, 1994].

Subsequent diffusion of the proton through a Grotthius mechanism to form OOH^* was found to be an activated process. The barrier was calculated as a function of potential using the climbing nudges elastic band method [Henkelman and Jonsson, 2000; Henkelman et al., 2000; Mills et al., 1995] and added to the reaction energy of step (4.15) (i.e., referenced to the initial state with the proton and electron in bulk reference states) to give the overall barrier for the reduction of O_2^*. The potential-dependent reaction barrier and the transition state structure at 1.0 V/NHE are illustrated in Fig. 4.15. The calculated barrier at 0.8 V/NHE of 0.34 eV agrees well with that measured by Damjanovic and Sepa (0.26 eV) [Damjanovic and Sepa, 1990]. Although further work examining the dependence of this barrier on the water/electrolyte structure is planned, this initial agreement with experiment illustrates the ability of this method to examine ion transport and electron transfer at the aqueous, electrified interface.

4.5 SUMMARY AND CONCLUSIONS

The chemistry at the electrified aqueous/metal interface is quite fascinating, as its structure, properties, and dynamics can significantly influence reaction energetics, dictate the kinetics that control catalytic selectivity, and open up novel reaction pathways and mechanisms.

The presence of water on a metal substrate acts to stabilize both charged and polarized surface intermediates, thus favoring heterolytic bond activation processes. Water enhances reactions that result in charged or polarized transition states through classic charge stabilization. More interestingly, there are a number of reactions where water can act directly as a co-catalyst in facilitating proton transfer. This is seen in the heterolytic activation of water, CO oxidation, and methanol dehydrogenation. In a similar manner, water can help to "catalyze" diffusion by providing proton shuttling pathways.

The structure of water that forms in the double layer is controlled by an optimal balance between the interactions of adsorbed water and the metal surface and hydrogen bonding between co-adsorbed water molecules at the interface, as well as adsorbed water and water in solution. The interfacial water structure undergoes significant changes as a result of changes in the potential. Higher potentials show stronger affinities for the oxygen, lead to the migration of OH_x intermediates to higher-fold coordination sites, and ultimately aid in O—H bond activation.

Changes in the potential can significantly polarize the water structure at the aqueous/metal interface and thus alter the overall reaction energies. This is particularly true for systems in which there is a significant change in the dipole between the reactants and product states. The change in dipole leads to significant differences in the reactants' and products' responses to changes in the applied potential or electric field. The changes in applied potential for these systems are rather nonlinear, as the interfacial capacitance plays an important role. For reactions in which the reactant and product states have similar dipoles, however, the polarization of both states is similar and therefore tends to cancel out. The potential effects for these systems can then safely be decoupled from the reaction energies. The reaction energies for one-electron oxidation or reduction processes are then simply linear with respect to the potential, as capacitance differences do not alter relative energies.

The common framework for describing adsorption in the gas phase can break down when modeling adsorption at an aqueous/metal interface. While there is a gain in energy as solute molecules approach the surface, the water molecules at the surface must be displaced from the surface in order for the solute to adsorb. The displacement will depend upon the potential as well as upon both enthalpic and entropic considerations. We have shown here that water can only displace O_2 from the Pt(111) surface at higher potentials that tend to be outside the typical operating ranges of a PEMFC cathode. This analysis only describes the overall equilibrium behavior. Water may begin to compete at lower potentials if the system is kinetically limited. Sulfate anions in solution, on the other hand, have been shown to displace oxygen and water from the surface at potentials greater than 0.9 V, and thus act to inhibit the kinetics of different surface reactions.

The elementary reaction energies and thermodynamics for methanol dehydrogenation have been shown to be significantly influenced by electrode potential. The oxidation pathways become much more favorable at higher potentials. The relative barriers of O—H to C—H bond activation decrease with increasing potential, which decreases the overall selectivity to CO and CO_2 and increases the yield of formaldehyde. This is consistent with experimental studies. The oxidation of CO intermediates appears to occur via adsorbed hydroxyl intermediates. The hydroxyl intermediates are more weakly held to the surface than atomic oxygen, and thus have significantly lower barriers for the oxidation of CO.

Lastly, we have shown that transport of ions across the double layer is facilitated by water via proton transfer and that the barrier for the reduction of O_2 is controlled by electron transfer that occurs as the proton moves close to the adsorbed O_2 to form a reactive center. Electron transfer appears to occur before the actual formation of the OO—H bond.

REFERENCES

Anderson AB. 1981. Reactions and structures of water on clean and oxygen covered Pt(111) and Fe(100). Surf Sci 105: 159–176.

Anderson AB. 1990. The influence of electrochemical potential on chemistry at electrode surfaces modeled by MO theory. J Electroanal Chem 280: 37–48.

Anderson AB, Albu TV. 1999. Ab initio determination of reversible potentials and activation energies for outer-sphere oxygen reduction to water and the reverse oxidation reaction. J Am Chem Soc 121: 11855–11863.

Anderson AB, Albu TV. 2000. Catalytic effect of platinum on oxygen reduction: An ab initio model including electrode potential dependence. J Electrochem Soc 147: 4229–4238.

Anderson AB, Debnath NC. 1983. Mechanism of iron dissolution and passivation in an aqueous environment: active and transition ranges. J Am Chem Soc 105: 18–22.

Anderson AB, Ray NK. 1982. Structures and reactions of H_3O^+, H_2O, and OH on an Fe electrode. J Phys Chem 86: 488–494.

Anderson AB, Cai Y, Sidik RA, Kang DB. 2005. Advancements in the local reaction center electron transfer theory and the transition state structure in the first step of oxygen reduction over platinum. J Electroanal Chem 580: 17–22.

Cao D, Lu G-Q, Wieckowski A, Wasileski SA, Neurock M. 2005. Mechanisms of methanol decomposition on platinum: A combined experimental and an initio approach. J Phys Chem B 109: 11622–11633.

Christou NI, Whitehouse JS, Nicholson D, Parsonage NG. 1981. A Monte Carlo study of fluid water in contact with structureless walls. J Chem Soc Faraday Symp 16: 139–149.

Damjanovic A, Brusic V. 1967. Electrode kinetics of oxygen reduction on oxide-free platinum electrodes. Electrochim Acta 12: 615–628.

Damjanovic A, Sepa DB. 1990. An analysis of the pH dependence of enthalpies and Gibbs energies of activation for O_2 reduction at Pt electrodes in acid solutions. Electrochim Acta 35: 1157–1162.

Desai S, Neurock M. 2003a. A first principles analysis of CO oxidation over Pt and $Pt_{66.7\%}Ru_{33.3\%}$ (111) surfaces. Electrochim Acta 48: 3759–3773.

Desai SK, Neurock M. 2003b. First-principles study of the role of solvent in the dissociation of water over a Pt–Ru alloy. Phys Rev B 68: 075420.

Desai SK, Neurock M, Kourtakis K. 2002. A periodic density functional theory study of the dehydrogenation of methanol over Pt(111). J Phys Chem B 106: 2559–2568.

Dominguez-Ariza D, Hartnig C, Sousa C, Illas F. 2004. Combining molecular dynamics and ab initio quantum-chemistry to describe electron transfer reactions in electrochemical environments. J Chem Phys 121: 1066–1073.

Ehlers DH, Spitzer A, Luth H. 1985. The adsorption of methanol on Pt(111), an Ir reflection and Uv photoemission-study. Surf Sci 160: 57–69.

Filhol JS, Neurock M. 2006. Elucidation of the electrochemical activation of water over Pd by first principles. Angew Chem Int Ed 45: 402–406.

Gasteiger HA, Markovic N, Ross PN Jr, Cairns EJ. 1994. CO electrooxidation on well-characterized Pt–Ru alloys. J Phys Chem 98: 617–625.

Gong X, Hu P, Raval R. 2003. The catalytic role of water in CO oxidation. J Chem Phys 119: 6324–6334.

Greeley J, Mavrikakis M. 2002. A first-principles study of methanol decomposition on Pt(111). J Am Chem Soc 124: 7193–7201.

Greeley J, Mavrikakis M. 2004. Competitive paths for methanol decomposition on Pt(111). J Am Chem Soc 126: 3910–3919.

Greeley J, Nørskov JK. 2005. A general scheme for the estimation of oxygen binding energies on binary transition metal surface alloys. Surf Sci 592: 104–111.

Grgur BN, Markovic NM, Ross PN. 1997. Temperature-dependent oxygen electrochemistry on platinum low-index single crystal surfaces in acid solutions. Can J Chem 75: 1465–1471.

Halley JW, Johnson B, Price D, Schwalm M. 1985. Quantum theory of the double-layer – A model of the electrode-electrolyte interface. Phys Rev B 31: 7695–7709.

Halley JW, Mazzolo A, Zhou Y, Price D. 1998. First-principles simulations of the electrode vertical bar electrolyte interface. J Electroanal Chem 450: 273–280.

Hartnig C, Koper MTM. 2002. Molecular dynamics simulation of the first electron transfer step in the oxygen reduction reaction. J Electroanal Chem 532: 165–170.

Head-Gordon M, Tully JC. 1993. Electric-field effects on chemisorption and vibrational-relaxation of CO on CU(100). Chem Phys 175: 37–51.

Henderson MA. 2002. The interaction of water with solid surfaces: fundamental aspects revisited. Surf Sci Rep 46: 1–308.

Henkelman G, Jonsson H. 2000. Improved tangent estimate in the nudged elastic band method for finding minimum energy paths and saddle points. J Chem Phys 113: 9978–9985.

Henkelman G, Uberuaga BP, Jonsson H. 2000. A climbing image nudged elastic band method for finding saddle points and minimum energy paths. J Chem Phys 113: 9901–9904.

Honkala K, Hellman A, Remediakis IN, Logadottir A, Carlsson A, Dahl S, Christensen CH, Nørskov JK. 2005. Ammonia synthesis from first-principles calculations. Science 307: 555–558.

Housmans THM, Wonders AH, Koper MTM. 2006. Structure sensitivity of methanol electro-oxidation pathways on platinum: An on-line electrochemical mass spectrometry study. J Phys Chem B 110: 10021–10031.

Hyman MP, Medlin JW. 2005. Theoretical study of the adsorption and dissociation of oxygen on Pt(111) in the presence of homogeneous electric fields. J Phys Chem B 109: 6304–6310.

Izvekov S, Mazzolo A, VanOpdorp K, Voth GA. 2001. Ab initio molecular dynamics simulation of the Cu(110)–water interface. J Chem Phys 114: 3248–3257.

Janik MJ, Neurock M. 2006. A first principles analysis of the electro-oxidation of CO over Pt(111). Electrochim Acta 52: 5517–5528.

Janik MJ, Taylor CD, Neurock M. 2007. First principles analysis of the electrocatalytic oxidation of methanol and carbon monoxide. Top Catal 46: 306–319.

Kresse G, Furthmuller J. 1996a. Efficiency of ab-initio total energy calculations for metals and semiconductors using a plane-wave basis set. Comput Mater Sci 6: 15–50.

Kresse G, Furthmuller J. 1996b. Efficient iterative schemes for ab initio total-energy calculations using a plane wave basis set. Phys Rev B 54: 11169–11186.

Kresse G, Hafner J. 1993. Ab initio molecular dynamics for liquid metals. Phys Rev B 47: 558–561.

Liu P, Logadottir A, Nørskov JK. 2003. Modeling the electro-oxidation of CO and $H_2.CO$ on Pt, Ru, PtRu, and Pt_3Sn. Electrochim Acta 48: 3731–3742.

Loffreda D, Simon D, Sautet P. 1999. Dependence of stretching frequency on surface coverage and adsorbate-adsorbate interactions: A density-functional theory approach of CO on Pd (111). Surf Sci 425: 68–80.

Lozovoi AY, Alavi A, Kohanoff J, Lynden-Bell RM. 2001. Ab initio simulation of charged slabs at constant chemical potential. J Chem Phys 115: 1661–1669.

Michaelides A, Alavi A, King DA. 2003. Different surface chemistries of water on Ru{0001}: From monomer adsorption to partially dissociated bilayers. J Am Chem Soc 125: 2746–2755.

Michaelides A, Ranea VA, de Andres PL, King DA. 2003b. General model for water monomer adsorption on close-packed transition and noble metal surfaces. Phys Rev Lett 90: 216102.

Mills G, Jonsson H, Schenter GK. 1995. Reversible work transition state theory: application to dissociative adsorption of hydrogen. Surf Sci 324: 305–337.

Nørskov JK, Rossmeisl J, Logadottir A, Lindqvist L, Kitchin JR, Bligaard T, Jonsson H. 2004. Origin of the overpotential for oxygen reduction at a fuel-cell cathode. J Phys Chem B 108: 17886–17892.

Okamoto Y, Sugino O, Mochizuki Y, Ikeshoji T, Morikawa Y. 2003. Comparative study of dehydrogenation of methanol at Pt(111)/water and Pt(111)/vacuum interfaces. Chem Phys Lett 377: 236–242.

Otani M, Sugino O. 2006. First-principles calculations of charged surfaces and interfaces: A plane-wave nonrepeated slab approach. Phys Rev B 73: 115407.

Panchenko A, Koper MTM, Shubina TE, Mitchell SJ, Roduner E. 2004. Ab initio calculations of intermediates of oxygen reduction on low-index platinum surfaces. J Electrochem Soc 151: A2016–A2027.

Pliego JR Jr, Riveros JM. 2000. On the calculation of the absolute solvation free energy of ionic species: Application of the extrapolation method to the hydroxide ion in aqueous solution. J Phys Chem B 104: 5155–5160.

Raghavan K, Foster K, Motakabbir K, Berkowitz ML. 1991. Structure and dynamics of water at the Pt(111) interface: molecular dynamics study. J Chem Phys 94: 2110–2117.

Rossmeisl J, Nørskov JK, Taylor CD, Janik MJ, Neurock M. 2006. Calculated phase diagrams for the electrochemical oxidation and reduction of water over Pt(111). J Phys Chem B 110: 21833–21839.

Sanchez CG, Lozovoi AY, Alavi A, 2004. Field-evaporation from first-principles. Mol Phys 102 (9–10): 1045–1055.

Shubina TE, Hartnig C, Koper MTM. 2004. Density functional theory study of the oxidation of CO by OH on Au(110) and Pt(111) surfaces. Phys Chem Chem Phys 6: 4215–4221.

Sidik RA, Anderson AB. 2006. Co_9S_8 as a catalyst for electroreduction of O_2: Quantum chemistry predictions. J Phys Chem B 110: 936–941.

Smith BB, Halley JW. 1994. Simulation study of the ferrous ferric electron transfer at a metal–aqueous electrolyte interface. J Chem Phys 101: 10915–10924.

Taylor CD, Neurock M. 2005. Theoretical insights into the structure and reactivity of the aqueous/metal interface. Curr Opin Solid State Mater Sci 9: 49–65.

Taylor C, Kelly RG, Neurock M. 2006a. First-principles calculations of the electrochemical reactions of water at an immersed Ni(111)/H_2O interface. J Electrochem Soc 153 (12): E207–E214.

Taylor CD, Wasileski SA, Filhol JS, Neurock M. 2006b. First principles reaction modeling of the electrochemical interface: Consideration and calculation of a tunable surface potential from atomic and electronic structure. Phys Rev B, 73.

Taylor CD, Janik MJ, Neurock M, Kelly RG. 2006c. Ab initio simulations of the electrochemical activation of water. Mol Sim 33: 429–436.

Taylor CD, Kelly RG, Neurock M. 2007a. A first-principles analysis of the chemisorption of hydroxide on copper under electrochemical conditions: A probe of the electronic interactions that control chemisorption at the electrochemical interface. J Electroanal Chem 607: 167–174.

Taylor CD, Kelly RG, Neurock M. 2007b. First-principles prediction of equilibrium potentials for water activation by a series of metals. J Electrochem Soc 154: F217–F221.

Taylor CD, Kelly RG, Neurock M. 2007c. Theoretical analysis of the nature of hydrogen at the electrochemical interface between water and a Ni(111) single-crystal electrode. J Electrochem Soc 154: F55–F64.

Taylor C, Kelly RG, Neurock M. 2009a. First principles analysis of the oxidatin of the Cu(111) surface under electrochemical conditions. Phys Rev B, submitted.

Taylor C, Kelly RG, Neurock M. 2009b. First principles modeling of the structure and reactivity of water at the metal/water interface. Submitted.

Toney MF, Howard JN, Richer J, Borges GL, Gordon JG, Melroy OR, Wiesler DG, Yee D, Sorensen LB. 1994. Voltage-dependent ordering of water-molecules at an electrode-electrolyte interface. Nature 368: 444–446.

van Santen RA, Neurock M. 2006. Molecular Heterogeneous Catalysis: A Conceptual and Computational Approach. Weinheim: Wiley-VCH.

Vassilev P, van Santen RA, Koper MTM. 2005. Ab initio studies of a water layer at transition metal surfaces. J Chem Phys 122.

Vassilev P, van Santen RA, Koper MTM. 2005. Ab initio studies of a water layer at transition metal surfaces. J Chem Phys 122: 54701–54712.

Vayner E, Sidik RA, Anderson AB, Popov B. 2007. Experimental and theoretical study of cobalt selenide as a catalyst for O_2 electroreduction. J Phys Chem C 111: 10508–10513.

Wagner FT. 1993. Structure of electrified interfaces. In: Lipkowski J, Ross P, eds. Structure of Electrified Interfaces. New York: VDH Publishers.

Zhang CJ, Hu P. 2000. Why must oxygen atoms be activated from hollow sites to bridge sites in catalytic CO oxidation? J Am Chem Soc 122: 2134–2135.

Zhang J, Vukmirovic MB, Sasaki K, Nilekar AU, Mavrikakis M, Adzic RR. 2005a. Mixed-metal Pt monolayer electrocatalysts for enhanced oxygen reduction kinetics. J Am Chem Soc 127: 12480–12481.

Zhang J, Vukmirovic MB, Xu Y, Mavrikakis M, Adzic RR. 2005b. Controlling the catalytic activity of platinum-monolayer electrocatalysts for oxygen reduction with different substrates. Angew Chem Int Ed 44: 2132–2135.

CHAPTER 5

Ab Initio Atomistic Thermodynamics for Fuel Cell Catalysis

TIMO JACOB

Fritz-Haber-Institut der Max-Planck-Gesellschaft, Faradayweg 4–6, Berlin, D-14915 Germany

5.1 INTRODUCTION

In recent years, there has been increasing interest in electrochemistry. Here, fuel cells have played an important role, triggered, for example, by the desire for ecologically sound and economically efficient use of energy resources. However, electrochemical reactions occur in a multicomponent environment (including, e.g., oxygen, water, and impurities) and under conditions of finite temperature, pressure, and electrode potential, leading to extremely complex systems. In particular, the presence of an electrode potential, which results in the formation of an electric double layer, influences the structure of the electrode/electrolyte interface. All of these factors make experimental studies and ab initio calculations most challenging.

Regarding the electrode/electrolyte interface, it is important to distinguish between two types of electrochemical systems: thermodynamically closed (and in equilibrium) and open systems. While the former can be understood by knowing the equilibrium atomic structure of the interface and the electrochemical potentials of all components, open systems require more information, since the electrochemical potentials within the interface are not necessarily constant. Variations could be caused by electrocatalytic reactions locally changing the concentration of the various species. In this chapter, we will focus on the former situation, i.e., interfaces in equilibrium with a bulk electrode and a multicomponent bulk electrolyte, which are both influenced by temperature and pressures/activities, and constrained by a finite voltage between electrode and electrolyte.

Experimentally, different structure- and surface-sensitive techniques such as in situ scanning tunnelling microscopy (STM), in situ X-ray diffraction (XRD), transition electron microscopy (TEM), and in situ infrared (IR) spectroscopy have been

---

Fuel Cell Catalysis. Edited by Marc T. M. Koper
Copyright © 2009 John Wiley & Sons, Inc.

developed to study electrochemical interfaces (see Abruna [1991], Kolb [1996, 2002], and Magnussen [2002], and references therein). Regarding theoretical studies, different attempts have been made to model and understand the structure and properties of systems under electrochemical conditions. An overview can be found in the reviews of Schmickler [1996a, 1999], Koper et al. [2003], Koper [2004] and the papers of Nazmutdinov and Shapnik [1996], Halley et al. [2000], Vassilev et al. [2001], Haftel and Rosen [2003], Feng et al. [2005], Kitchin et al. [2004], Gunnarsson et al. [2004], Taylor et al. [2006], and Jacob [2007a, b] and references therein. In these different works, mainly experimental input, semi-empirical approaches, or rather simplified models, have been used. The electrode potential has been either neglected or introduced by charging the electrode surface or applying an external electric field. While most of the theoretical studies have disregarded the electrode potential, some have attempted to consider its influence on catalytic reactions. For instance, Nørskov's group [Kitchin et al., 2004; Rossmeisl et al., 2006] studied the hydrogen evolution reaction (HER) and oxygen reduction reaction (ORR) on different electrodes, whose Fermi energies were shifted by the value of the electrode potential. Focusing more on the atomistic structure of the interface, Neurock's group [Taylor et al., 2006] performed ab initio molecular dynamics simulations on charged electrodes surrounded by water. For compensation, a counter-charge was located at a certain distance from the electrode surface, in an attempt to mimic the potential profile within the interfacial region.

While most theoretical studies have focused on electrochemical reactions by calculating the binding energies of particular adsorbates on the electrode, reaction barriers, and reaction mechanisms, the influence of the morphology has often been underestimated. Figure 5.1 schematically shows that there is a sensitive interplay between the morphology of the system, determined by its structure and composition, and the energetic of (electro)catalytic reactions. Furthermore, both are strongly influenced by the environment, which, when taking part in the reaction, could even be reactive, and the external parameters T, p/a, and electrostatic potential $\Delta\phi$. Therefore, prior to investigating detailed reactions, one has to understand the influence of these parameters and the environment on the morphology of the electrochemical system.

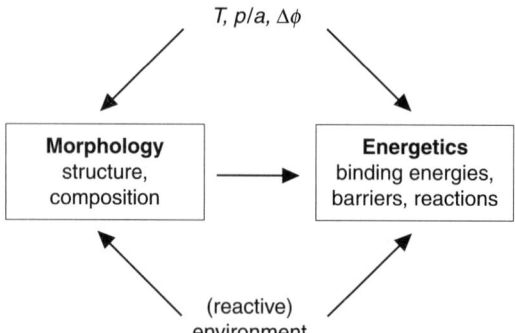

Figure 5.1 Interplay between the morphology of a system, its environment and energetics, and external parameters.

In this chapter, we will give a general description of electrochemical interfaces representing thermodynamically closed systems constrained by the presence of a finite voltage between electrode and electrolyte, which will then be taken as the basis for extending the ab initio atomistic thermodynamics approach [Kaxiras et al., 1987; Scheffler and Dabrowski, 1988; Qian et al., 1988; Reuter and Scheffler, 2002] to electrochemical systems. This will enable us to qualitatively and quantitatively investigate and predict the structures and stabilities of full electrochemical systems or single electrode/electrolyte interfaces as a function of temperature, activities/pressures, and external electrode potential.

The abilities of this approach will then be illustrated with two examples: (i) the potential-induced lifting of the Au(100) surface reconstruction; and (ii) the electrochemical oxidation of Pt(111).

5.2 THEORY

5.2.1 Electrochemical Potentials

Before we will discuss the electrochemical system, it is important to define the properties and characteristics of each component, especially the electrolyte. In the following, we assume macroscopic amounts of an electrolyte containing various ionic and nonionic components, which might be solvated. In the case that this *bulk electrolyte* is in thermodynamic equilibrium, each of the species present is characterized by its electrochemical potential, which is defined as the free energy change with respect to the particle number of species i:

$$\tilde{\mu}_i(T, c_i, \phi_S) = \left(\frac{\partial G}{\partial N_i}\right)_{T, c_i, N_j} + q_i \phi_S = \mu_i(T, c_i) + q_i \phi_S \tag{5.1}$$

Here G is the Gibbs free energy of the system without external electrostatic potential, and $q_i \phi_S$ refers to the energy contribution coming from the interaction of an applied constant electrostatic potential ϕ_S (which will be specified later) with the charge q_i of the species. The first term on the right-hand side of (5.1) is the usual chemical potential $\mu_i(T, c_i)$, which, for an ideal solution, is given by

$$\mu_i(T, c_i) = \bar{\mu}_i(T, c^0) + k_B T \ln\left(\frac{c_i}{c^0}\right) \tag{5.2}$$

where $\bar{\mu}_i$ is the chemical potential of species i at standard conditions, which for the solvent and the solute (electrolyte/ions) are defined as follows:

- solvent: standard temperature and pure solvent, which means the absence of electrolyte ions,
- solute (Henry's law standard state): standard temperature and concentration c^0 in the hypothetical state of an ideal solution (infinite dilution), which is reached at $\lim_{c_i \to 0} f_i = 1$.

All deviations from ideal behavior, which are directly (electrostatically) or indirectly (through the solvent) caused by ion–ion interactions are usually treated by scaling the concentration with a so-called activity coefficient f_i, which leads to the activity $a_i = f_i c_i$ [Atkins, 1990]. Inserting the activity into (5.2) gives

$$\mu_i(T, a_i) = \bar{\mu}_i(T, c^0) + k_B T \ln\left(\frac{c_i}{c^0}\right) + k_B T \ln(f_i) \tag{5.3}$$

In most electrochemical experiments, ion concentrations <1 M are used, and the last term in (5.3) contributes to the chemical potential by only a few percent. However, at higher concentrations (>5 M), this might increase even up to 10% [Bockris et al., 2000].

While different theoretical methods have been developed to calculate the deviations from ideal behavior, i.e., the activity coefficient, for each ionic species separately (anions: f_a; cations: f_c) [Bockris et al., 2000; Schmickler, 1996a, b; Abbas et al., 2002], only the mean activity coefficient is experimentally accessible, which is usually approximated by $f_\pm = \sqrt{f_a f_c}$. This is due to the fact that if only one of the ionic species were to be added to the electrolyte, the change in Gibbs free energy would result in an additional energy contribution from the interaction of the ionic species with an overall charged (electrified) solution. Since these two energy contributions cannot be separated unambiguously, and to avoid a charged electrolyte, one has to add an entire salt molecule to the electrolyte at a time.

The second term on the right-hand side of (5.1) is the energy required to transfer the charge q_i associated with each particle of the ith species from a reference potential ϕ_{ref} to the electrostatic potential of the bulk electrolyte (solution), ϕ_S. Since the reference potential can be chosen freely, throughout this chapter we use ϕ_{ref} as energy zero for the electrostatic potential. Further discussion on the importance of this aspect will be provided later.

5.2.2 The Non-Electrochemical Interface

Focusing on the example of a solid/gas interface, in the following, we will describe how to evaluate the stability of non-electrochemical interfaces, which are not influenced by a potential applied externally or caused by an inhomogeneous ion distribution within the system. In the case that both the solid and the gaseous phase are present in macroscopic quantities, we have already seen in the previous section that each of these reservoirs is characterized by its chemical potential $\mu_i(T, p_i)$, which for the non-electrochemical interface is a function of temperature and partial pressure.

If both reservoirs are brought into contact, different interfacial structures are possible (Fig. 5.2):

- phase separation;
- formation of an ordered adsorbate layer;
- formation of a stable or metastable surface compound (e.g., surface oxide or surface hydride);
- total phase mixing (e.g., bulk oxide or bulk hydride).

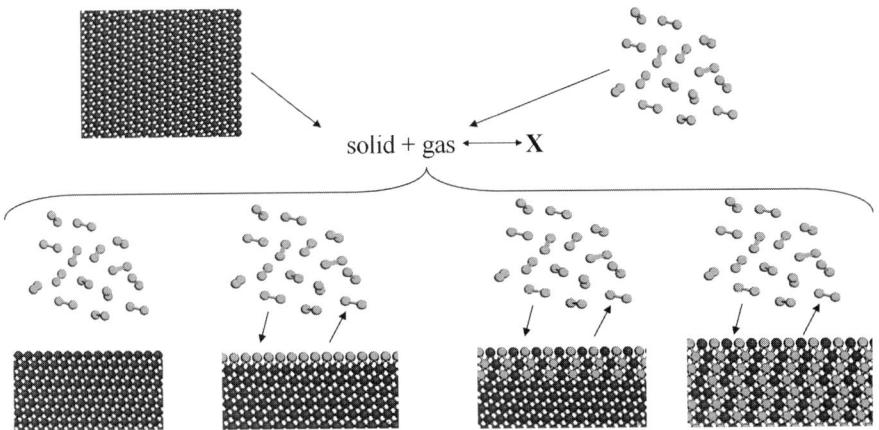

Figure 5.2 Schematic models of different solid/gas interface structures.

In order to evaluate which of these scenarios leads to the most stable interfacial structure, we have to analyze the relation between the chemical potentials of both reservoirs and the overall energy. Therefore, we begin with the Gibbs free energy of the interface,

$$G(T, \{p_i\}, \{N_i\}) = \sum_i N_i \mu_i(T, p_i) + A\gamma(T, \{p_i\}) \quad (5.4)$$

which is the sum of the energy required to extract N_i atoms or molecules of the ith species from their reservoir with $\mu_i(T, p_i)$ and the energy to build the interface with contact area A.

The Gibbs free energy G and the chemical potentials include contributions from the internal energy, vibrational free energy, and configurational entropy. Since most relevant structures will have a low surface free energy, we obtain from (5.4) that

$$\gamma(T, \{p_i\}) = \frac{1}{A}\left[G(T, \{p_i\}, \{N_i\}) - \sum_i N_i \mu_i(T, p_i)\right] \quad (5.5)$$

On the basis of this equation, we will now discuss different systems relevant to the examples presented later.

5.2.2.1 Pure Solid in Contact with a Single- or Multi-Component Environment Considering the example of a Pt surface in contact and in thermodynamic equilibrium with an oxygen atmosphere, the surface free energy (5.5) becomes

$$\gamma(T, p_{O_2}, p_{Pt}) = \frac{1}{A}[G(T, p_{O_2}, p_{Pt}, N_O, N_{Pt}) - N_{Pt}\mu_{Pt}(T, p_{Pt}) - N_{O_2}\mu_{O_2}(T, p_{O_2})] \quad (5.6)$$

While O_2 adsorbs molecularly only under specific circumstances or high coverages [Parker et al., 1989], on most transition metals adsorption of oxygen is accompanied by dissociation of the molecule, leading to atomic oxygen on the surface. In this case, the last term of (5.6) may be replaced by $N_O \mu_O(T, p_{O_2})$. Since the surface oxygens are in contact with the O_2 atmosphere, the chemical potential μ_O is then determined by the condition of thermodynamic equilibrium with the surrounding O_2 gas phase reservoir, meaning that $\mu_O = \frac{1}{2}\mu_{O_2}$ (taking the O—O dissociation energy as an arbitrary constant), and (5.6) becomes

$$\gamma(T, p_{O_2}, p_{Pt}) = \frac{1}{A}\left[G(T, p_{O_2}, p_{Pt}, N_O, N_{Pt}) - N_{Pt} g_{Pt}^{bulk}(T, p_{Pt}) - \frac{N_O}{2}\mu_{O_2}(T, p_{O_2})\right] \quad (5.7)$$

where, to be consistent with the existing literature, we have replaced the chemical potential of Pt by the Gibbs free energy of bulk Pt, g_{Pt}^{bulk}. In the case that the environment contains more than a single component, (5.7) contains additional terms similar to the oxygen term on the right-hand side.

5.2.2.2 Solid Compound in Contact with a Single- or Multi-Component Environment
Staying with the previous example of Pt in contact with an oxygen environment, at very low temperatures or high oxygen partial pressures, a bulk oxide (Pt_xO_y) will form. Since the Pt bulk oxide surface should be in thermodynamic equilibrium with the surrounding, the following condition must be fulfilled:

$$x g_{Pt}^{bulk} + y\mu_O = g_{Pt_xO_y}^{bulk} \quad (5.8)$$

Insertion into (5.7) finally gives

$$\gamma(T, p_{O_2}, p_{Pt_xO_y}) = \frac{1}{A}\left[G(T, p_{O_2}, p_{Pt_xO_y}, N_O, N_{Pt}) - \frac{N_{Pt}}{x} g_{Pt_xO_y}^{bulk}(T, p_{Pt_xO_y})\right.$$
$$\left. + \frac{1}{2}\left(y\frac{N_{Pt}}{x} - N_O\right)\mu_{O_2}(T, p_{O_2})\right] \quad (5.9)$$

5.2.3 Electric Double Layer

Compared with the non-electrochemical interface discussed in the previous section, where a solid was in contact with a surrounding gaseous atmosphere, the electrode/electrolyte interface is a multicomponent system and, besides temperature and partial pressures/concentrations, it is also influenced by the electrode potential. This results in greater complexity, which requires additional considerations prior to deriving an expression for the interfacial stability.

In the literature, one often finds different expressions or definitions for the various potentials relevant to the electrode/electrode interfaces. To provide a clear definition here and to better understand the role of the electrode potential, in the following we will first consider a system in which the electrolyte is absent, and afterwards address the electrochemical system.

Figure 5.3 shows a system of two electrodes of the *same* material, which is comparable to a capacitor in classical electrostatics, and the shape of the electrostatic potential before and after applying an external voltage bias, i.e., a charged capacitor. In the absence of an external voltage, the capacitor is uncharged. The electrostatic potential ϕ within each metallic electrode is determined by the charge of the nuclei, which gives rise to singularities at the positions of the atoms, and by the electrons. At the surface, the behavior is additionally influenced by the broken periodicity, leading to a so-called *surface dipole* [Horn and Scheffler, 2000]. Moving from the metal into the vacuum, the electrostatic potential finally approaches the vacuum level ϕ_{vac} (which is the same for each electrode of an uncharged capacitor). In the following, we define this value as reference potential ϕ_{ref}, which, without loss of generality, we have already chosen as the energy zero for the electrostatic potential.

The energy required to move an electron from deep inside the electrode into the vacuum is called the work function Φ. It allows us to define the (electro)chemical potential of the electrons, which is the energy change when removing one electron

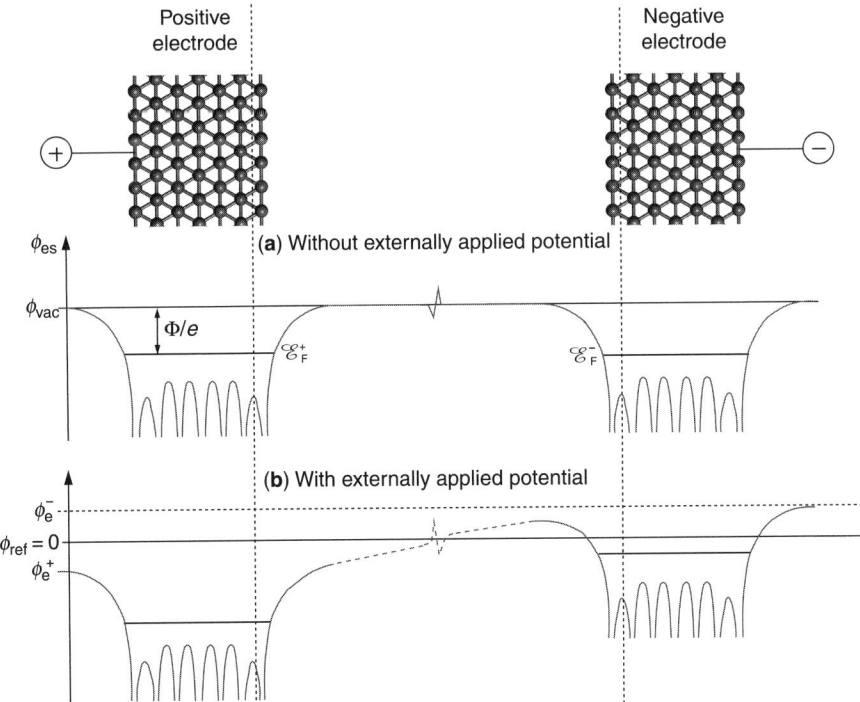

Figure 5.3 Electrostatic potential ϕ in a capacitor without (a) and with (b) an externally applied potential difference. Both electrodes are assumed to be of the same material.

from the electrode, as the difference between the work function Φ and the vacuum level ϕ_{vac}:

$$\tilde{\mu}_e = e\phi_{vac} - \Phi \tag{5.10}$$

Charging the capacitor is equivalent to applying an external potential difference between the two electrodes. In the system thus constrained (here by the applied potential), the vacuum levels of the electrodes are shifted to ϕ_e^- for the negative and ϕ_e^+ for the positive electrode, such that the difference between them corresponds exactly to the externally applied potential difference. This finally leads to an accumulation of charge (*excess charge*) at or near the electrode surfaces, which gives rise to a constant electric field between the electrodes as long as they are separated by vacuum or a solid dielectric. With respect to the excess charges, it should be noted that although the electric field tends to move all charge towards the surfaces of the electrodes, the Coulomb interactions between the excess charges and with the nuclei of the metal atoms result in a more diffuse charge layer. This might include several surface layers rather than producing surface-aligned excess charges. Thus, the only reasonable quantity to consider here is the surface charge density σ_e.

Since the externally applied potential difference influences the nuclei and electrons of the electrodes in the same way, not only the vacuum levels but the entire electrostatic potentials of both electrodes, and therefore the Fermi levels, are shifted, too. Consequently, the electrochemical potentials of the electrons become

$$\tilde{\mu}_e^\pm = e\phi_e^\pm - \Phi \tag{5.11}$$

In the case that the electrodes are of different materials, one has to distinguish between the corresponding work functions in this equation.

The reverse process of discharging the capacitor occurs when both electrodes are electrically connected. Through this connection, the system is allowed to reach its equilibrium state, and charge flows from the negative to the positive electrode until both Fermi levels are aligned again.

On the basis of the charged capacitor, we will now discuss the changes induced by filling the space between both electrodes with a liquid electrolyte. Because of its experimental relevance, we will consider a single electrode/electrolyte interface only, where the electrostatic potential of the electrode will simply be designated as ϕ_e (Fig. 5.4).

As described above, the electrolyte usually contains anions and cations, which are partially or fully solvated, water molecules and various species being involved in electrocatalytic reactions. The excess charge on the electrode surface is compensated by an accumulation of corresponding electrolyte counter-ions, leading to overall charge neutrality.

Since the capacitor without electrolyte shows a linear behavior of the electrostatic potential between the electrodes, $\phi(x)$, at the hypothetical moment when the electrolyte is added to the system ($t = 0$ s), the electrochemical potentials of the ions,

$$\tilde{\mu}_a = \mu_a + q_a\phi(x) \quad \text{and} \quad \tilde{\mu}_c = \mu_c + q_c\phi(x) \tag{5.12}$$

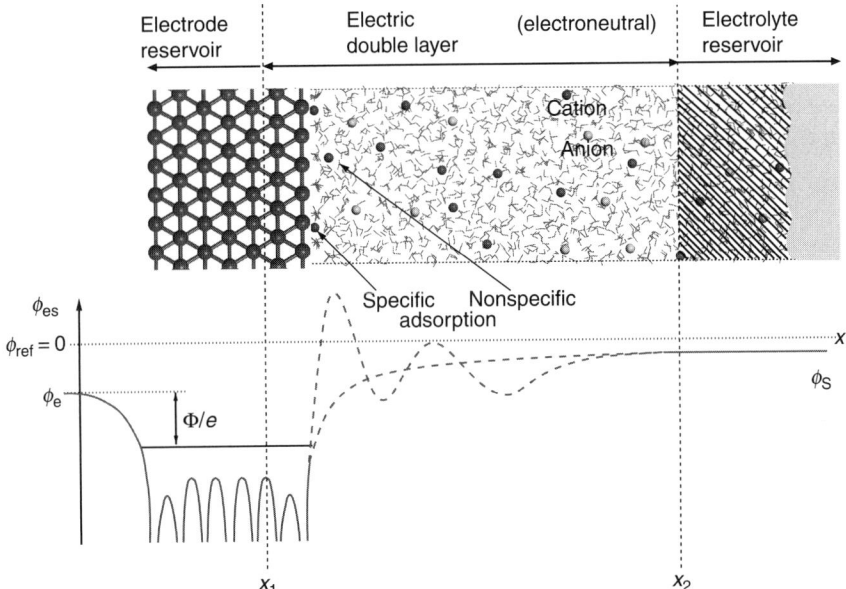

Figure 5.4 Atomistic model of the electrochemical half-cell, showing the electrode/electrolyte interface ($x_1 < x < x_2$), which is connected to the bulk electrode and electrolyte (reservoirs). The lower panel indicates the electrostatic potential within the electrode and the bulk electrolyte (solid lines), and possible shapes for the potential drop between them (dashed lines).

also show linear behavior as function of distance to the electrode x. At $t > 0$, the system tries to reach thermodynamic equilibrium by redistributing the ions within the electrolyte such that counter-ions in the electrolyte are accumulated at or near the electrode. Some ions might even lose parts of their solvation shell and adsorb at the electrode surface (*specific adsorption*) or just weakly interact as solvated ions (*nonspecific adsorption*). When thermodynamic equilibrium is reached, the electrochemical potentials of anions and cations are constant throughout the electrolyte. Furthermore, the electrolyte ions will exactly compensate the electrode excess charges, leading to a potential drop within the interfacial region.

Figure 5.4 shows schematically the electrochemical half-cell and the electrostatic potential after the system has reached equilibrium. It combines a single electrode/electrolyte interface in contact with two reservoirs: bulk electrode ($x < x_1$) and bulk electrolyte ($x > x_2$). Since, by definition, the excess charges on the electrode are fully shielded within the interface, far from the electrode, the electrostatic potential assumes the constant value ϕ_S with respect to the reference potential ϕ_{ref}. Regarding the dimensions, the width of such an interface might range from a few to several hundreds of Ångströms, which is rather small compared with the size of the bulk electrolyte region in realistic systems. Therefore, the latter region can be considered as an *electrolyte reservoir*.

Within the interfaces, the explicit profile of the electrostatic potential $\phi(x)$ as a function of electrode distance is unknown (possible shapes are indicated in Fig. 5.4 as dashed lines) and the only well-defined values are the asymptotic limits, which are the potential values of the electrode ϕ_e and the bulk electrolyte ϕ_S, respectively, and their difference (*electrode potential*)

$$\Delta\phi = \phi_e - \phi_S \tag{5.13}$$

A number of classical descriptions have been formulated to describe the shapes of the potential drop across the electrochemical interface [Schmickler, 1996b; Bockris et al., 2000], including the Helmholtz model, the Gouy–Chapman model, and the Gouy–Chapman–Stern model. The earliest model, formulated by Helmholtz in 1879, treats the interface mathematically as a simple capacitor formed by the electrode and a single layer of nonspecifically adsorbed (solvated) ions, leading to the so-called *inner layer*. Later, Gouy and Chapman (1910–1913) made significant improvements by introducing the diffuse character of the electrolyte to the electric double layer. In their model, the concentration of electrolyte ions, which are treated as point charges, reach their bulk concentrations only at larger distances from the electrode, leading to an exponential shape of the potential drop and a so-called *diffuse double layer*. A classical and widely applied description is the Gouy–Chapman–Stern model, which combines the Helmholtz inner layer with the Gouy–Chapman diffuse layer. In this model, the finite size of the ions is taken into account, so that they cannot approach the surface closer than their radius. The nonspecifically adsorbed ions of the Gouy–Chapman diffuse double layer are not at the surface, but at some distance (larger than their radius) away from the surface.

However, these classical models neglect various aspects of the interface, such as image charges, surface polarization, and interactions between the excess charges and the water dipoles. Therefore, the widths of the electrode/electrolyte interfaces are usually underestimated. In addition, the ion distribution within the interfaces is not fixed, which for short times might lead to much stronger electric fields near the electrodes.

5.2.4 Extended Ab Initio Atomistic Thermodynamics

Now having specified the bulk electrode, the bulk electrolyte, and the interface between them, our aim in this section is to quantify the atomistic structure of the interface and derive an expression that allows us to evaluate its stability. Based on (5.5), we will extend the ab initio atomistic thermodynamics approach to electrochemical systems.

For an electrode/electrolyte interface in equilibrium with the bulk electrode and the electrolyte reservoir, and constrained by the potential difference $\Delta\phi$, the most relevant structures are those with low interfacial free energies

$$\gamma(T, \{a_i\}, \{\phi_i\}) = \frac{1}{A}\left[G(T, \{a_i\}, \{N_i\}) - \sum_i N_i \tilde{\mu}_i(T, a_i, \phi_i)\right] \tag{5.14}$$

where A is the contact area and N_i is the number of atoms (or molecules) of the ith species, with $\tilde{\mu}_i(T, a_i, \phi_i)$ being the electrochemical potential of the corresponding reservoir at temperature T, activity a_i, and electrostatic potential ϕ_i (see above). Because of the presence of an electrostatic potential, here we have to use electrochemical potentials instead of chemical potentials. The sum over i involves metal atoms (me), excess electrons (e) at the electrodes, anions (a), cations (c), and neutral compounds (n). Then, (5.14) can be written as

$$\gamma(T, \{a_i\}, \phi_e, \phi_S) = \frac{1}{A}[G(T, \{a_i\}, \{N_i\}) - N_n\tilde{\mu}_n(T, a_n, \phi_S) - N_e\tilde{\mu}_e(\phi_e)$$
$$- N_{me}\tilde{\mu}_{me}(T, a_{me}, \phi_e) - N_a\tilde{\mu}_a(T, a_a, \phi_S) - N_c\tilde{\mu}_c(T, a_c, \phi_S)]$$
(5.15)

In order to evaluate different approximations in the context of modeling electrochemical systems [Haftel and Rosen, 2003; Kitchin et al., 2004; Gunnarsson et al., 2004; Feng et al., 2005; Rossmeisl et al., 2006; Taylor et al., 2006; Jacob, 2007a, b], in the following, we shall discuss each term of (5.15) separately:

- The term $G(T, \{a_i\}, \{N_i\})$ is the Gibbs free energy of the full electrochemical system ($x_1 < x < x_2$ in Fig. 5.4). It includes the electrode surface, which is influenced by possible reconstructions, adsorption, and charging, and the part of the electrolyte that deviates from the uniform ion distribution of the bulk electrolyte. The importance of these requirements becomes evident if we consider the theoretical modeling. If the interface model is chosen too small, then the excess charges on the electrode are not fully considered and/or, within the interface only part of the total potential drop is included, resulting in an electrostatic potential value at $x = x_2$ that differs from the required bulk electrolyte value ϕ_S. However, if we constrain such a model to reproduce the electrostatic potential ϕ_e at the electrode and ϕ_S at the far end of the electrolyte region, then the equilibrated system might differ geometrically and electronically from the realistic interface. Therefore, the modeled electrolyte region must be sufficiently large that, without any constraints, a certain "buffer zone" (see Fig. 5.4) at $x > x_2$ of the electrolyte region self-consistently reaches ϕ_S. Since, besides temperature and activities, the Gibbs energy also depends on the composition of the system (including electrolyte ions and the excess charges on the electrode surfaces), G implicitly depends on the electrode potential $\Delta\phi$.
- The term $N_n\tilde{\mu}_n(T, a_n, \phi_S)$ accounts for all neutral species, such as water or molecules involved in an electrochemical reaction. Since these are related to the electrolyte reservoir, the electrostatic potential that appears in $\tilde{\mu}_n$ is that of the bulk electrolyte. Although these species still might have dipole or higher multipole moments, the constancy of the potential within the reservoir allows $\tilde{\mu}_n(T, a_n, \phi_S)$ to be replaced by the corresponding chemical potential $\mu_n(T, a_n)$, which is potential-independent.
- The term $N_{me}\tilde{\mu}_{me}(T, a_{me}, \phi_e)$ gives the energy necessary to remove the metal atoms required to build the electrode from its corresponding bulk electrode

reservoir, which is at the electrostatic potential ϕ_e. However, since the electrode potential influences electrons and nuclei of the bulk electrode in the same way, there is no net effect on the neutral metal atoms. Consequently, $\tilde{\mu}_{me}(T, a_{me}, \phi_e)$ can be replaced by the potential-independent Gibbs free energies of the bulk electrode reservoir, $g_{me}^{bulk}(T, a_{me})$.

- The term $N_e\tilde{\mu}_e(\phi_e)$ accounts for the excess charges transferred from the reference electrode at the electrostatic potential ϕ_{ref} (which we have already defined as the energy zero) to the electrode. Thus, with (5.11) this term can be rewritten as $q_e(\phi_e - \Phi/e)$, where we have introduced the total excess charge at the electrode $q_e = N_e e$.

- The terms $N_a\tilde{\mu}_a(T, a_a, \phi_S)$ and $N_c\tilde{\mu}_c(T, a_c, \phi_S)$: for the bulk electrolyte reservoir, the condition

$$\tilde{\mu}_a(T, a_a, \phi_S) + \frac{y}{x}\tilde{\mu}_c(T, a_c, \phi_S) = \mu_{ac}(T, a_{ac}) \quad (5.16)$$

holds, where μ_{ac} describes entire salt molecules of the type $(c^{x+})_{y/x}(a^{y-})$ [e.g., $(H^+)_2SO_4^{2-}$], which are overall neutral and therefore potential-independent. Within the electrode/electrolyte interface, however, the amounts of anions and cations differ, leading to

$$N_a\tilde{\mu}_a(T, a_a, \phi_S) + N_c\tilde{\mu}_c(T, a_c, \phi_S)$$
$$= N_a\mu_{ac}(T, a_{ac}) + \left(N_c - \frac{y}{x}N_a\right)\tilde{\mu}_c(T, a_c, \phi_S) \quad (5.17)$$

Here, the last term accounts for the excess ions in the interfacial region, which compensate the excess charge q_e on the electrode surface and keep the overall interface electroneutral. What in electrochemical terms is often described as a polarizable active electrode and an unpolarizable reference electrode ensures that any change of the number of ions in the electrochemical half-cell under consideration, caused by an electrochemical reaction, is just compensated by a corresponding counter-reaction at the reference electrode.

Including these considerations, the interfacial free energy (5.15) becomes

$$\gamma(T, a_n, a_{me}, a_{ac}, a_c, \phi_e, \phi_S) = \frac{1}{A}\Big[G(T, \{a_i\}, \{N_i\}) - N_n\mu_n(T, a_n)$$
$$- N_{me}g_{me}^{bulk}(T, a_{me}) - q_e(\phi_e - \Phi/e)$$
$$- N_a\mu_{ac}(T, a_{ac}) - \left(N_c - \frac{y}{x}N_a\right)\tilde{\mu}_c(T, a_c, \phi_S)\Big]$$
$$(5.18)$$

As mentioned above, the last term in this equation represents the overall charge compensation within the interface. Therefore, with (5.13), we finally obtain the following expression for the interface stability:

$$\gamma(T, a_n, a_{me}, a_{ac}, a_c, \Delta\phi) = \frac{1}{A}\Big[G(T, \{a_i\}, \{N_i\}) - N_n\mu_n(T, a_n)$$
$$- N_{me}g_{me}^{bulk}(T, a_{me}) - q_e(\Delta\phi - \Phi/e)$$
$$- N_a\mu_{ac}(T, a_{ac}) - \Big(N_c - \frac{y}{x}N_a\Big)\mu_c(T, a_c)\Big] \quad (5.19)$$

The interfacial free energy $\gamma(T, a_n, a_{me}, a_{ac}, a_c, \Delta\phi)$ depends only on temperature, the activities, and the difference between the electrostatic potentials of electrode and electrolyte, $\Delta\phi$, which, apart from a_c, are all well-defined and experimentally accessible quantities. Therefore, the accurate calculation of γ depends on the accuracy of evaluating a_c or the corresponding activity coefficient f_-.

Finally, it should be remarked that, as long as the interfacial region is extended sufficiently to include all structural and electronic deviations from the reservoirs, (5.18) and (5.19) are valid for any type of connection between a metallic electrode and an electrolyte. They also include the cases of nonspecific and specific adsorption on the electrode.

5.3 APPLICATIONS

On the basis of (5.19), in this section we will evaluate the importance of different contributions to the overall interfacial free energy by applying the extended ab initio atomistic thermodynamics approach to two examples.

5.3.1 Potential-Dependent Surface Reconstruction of Au(100)

Surface reconstruction and relaxation can be understood as a deviation from the bulk-truncated structure on the atomic level, by which the surface minimizes its free energy [Kolb, 1996]. In particular, surface reconstruction usually involves a change in the periodicity of the surface and in some cases a change in symmetry as well, whereas surface relaxation is a (small) rearrangement of surface layers.

Under ultrahigh vacuum (UHV) conditions, some low index surfaces of the late $5d$ metals Au, Pt, and Ir show reconstruction of the first surface layer. In order to maximize the number of surface bonds, the ground state of the (100) face corresponds to a quasihexagonal (hex) close-packed structure, which results in an overall lower surface free energy. The reconstruction of Au(100) has been particularly well studied, both experimentally and theoretically. As nicely summarized in [Kolb, 1996] and [Feng et al., 2005], from low energy electron diffraction (LEED) and helium diffraction experiments, first a (1×5) and later a (20×15) reconstruction have been proposed [Fedak and Gjostein, 1966, 1967; Rieder et al., 1983]. However, further LEED studies [Wendelken and Zehner, 1978; Van Hove et al., 1981] suggested a $c(26 \times 68)$ reconstruction, while even more complex structures were found by scanning tunneling microscopy (STM) [Binnig et al., 1984]. Several theoretical attempts to explain the structure of the reconstructed surface can be found in the literature, ranging from tight-binding approaches [Tomanek and Bennemann, 1985], to simple glue models [Ercolessi et al., 1986], the application of embedded atom potentials [Dodson, 1987], and even first-principles calculations [Takeuchi et al., 1991].

Besides surface reconstructions induced by heat treatment, *potential-induced reconstruction* has recently become a topic of interest in electrochemistry. It has been observed that at potentials negative with respect to the potential of zero surface charge, ϕ_{pzc} [Kolb, 1996, 2002; Dakkouri, 1997], the reconstructions found under UHV conditions are also stable in contact with an electrolyte. Although all low index faces of Au and Pt undergo potential-induced reconstruction, it has been particularly well characterized for Au(100) (Fig. 5.5).

Analysis of STM images obtained for an Au(100) electrode in a 0.1M $HClO_4$ electrolyte solution at different electrode potentials indicate that below approximately -0.25 V (with respect to a saturated calomel electrode, SCE), the reconstruction of $(1 \times 1) \rightarrow$ hex starts by forming various "hex"-reconstructed domains with different surface orientations. At more negative electrode potentials (-0.3 to -0.4 V) large regions of (1×1) terraces are transformed into corrugated domains within a few minutes, and holding the potential at -0.35 V for about 10 minutes or longer results in a surface predominantly containing hexagonal reconstructed domains. Since the first surface layer is close-packed, the measured capacity curve was found to be comparable to that obtained for Au(111).

When the electrode potential is altered to positive potentials ($>+0.55$ V), a progressive removal ("*lifting*") of the reconstruction and a return of the dominant (1×1) structure is noticed by showing a pronounced current peak in the cyclic

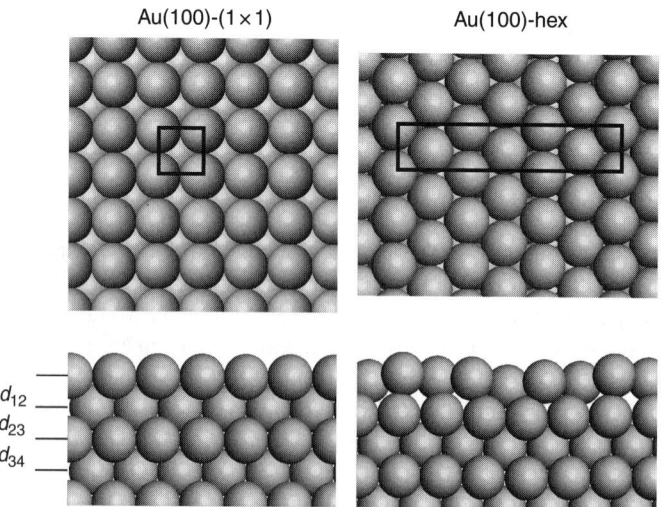

Figure 5.5 Top and side views of the unreconstructed Au(100)-(1×1) and the (5×1) hexagonal reconstructed Au(100)-hex surfaces. Unit cells are indicated by rectangles.

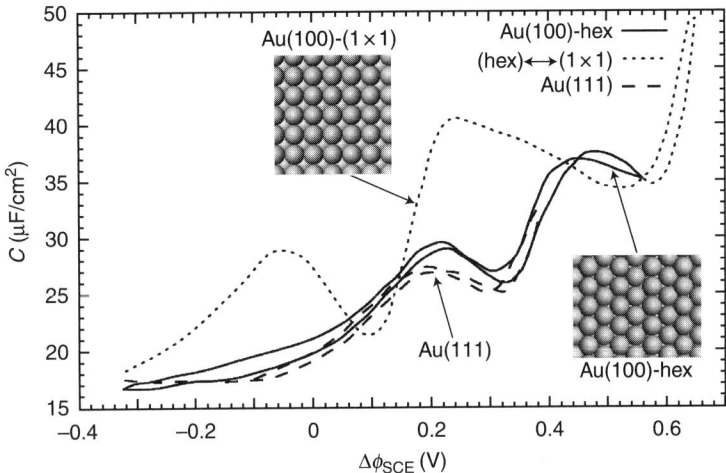

Figure 5.6 Capacity versus potential measurements on the lifting of surface reconstruction of Au(100) in 0.1 M H_2SO_4 [Kolb, 1996]. Whereas below 0.55 V the solid curve of Au(100)-hex more or less coincides with separate measurements on Au(111) (dashed curve), increasing the potential above +0.55 V lifts the reconstruction and gives Au(100)-(1 × 1) (dotted curve).

voltammogram [Gao et al., 1991] and an alteration in the capacity curve (Fig. 5.6). The potential-induced reconstruction therefore appears to be largely reversible under these conditions. Analysis of electrochemical data pertaining to the reconstruction of Au(100) suggests that the driving force for lifting the "hex" reconstruction [hex → (1 × 1)] is due to positive surface excess charge induced by the positive electrode potential. Such surface excess charges might lower the activation barrier for changes in the surface morphology.

Further STM and SXS studies [Wu et al., 1998] concerning this phenomenon indicated that the presence of specifically and nonspecifically adsorbing anions as well as organic molecules (e.g., pyridine, bipyridine, and uracil) may also lift the reconstructed surface by exhibiting a structural transition, and it has been extensively studied and reviewed in [Kolb, 1996].

Hence, the question arose *whether the lifting of reconstruction is due to surface charging (as a result of the electrode potential), adsorption of electrolyte ions (or molecules), or both*. Although various experimental and theoretical investigations have been carried out, the driving force for the lifting of reconstruction is not yet clear.

Assuming a simple capacitor model for the electric double layer, Santos and Schmickler [2004] used experimentally measured capacity curves for different electrolyte concentrations to evaluate the surface free energy γ for reconstructed and unreconstructed surfaces of Au(100) and Au(111). At around 0.6 V, they found an intersection between $\gamma_{Au(100)\text{-hex}}$ and $\gamma_{Au(100)\text{-}(1\times1)}$, and concluded that positive excess charge on the surface is responsible for the lifting of reconstruction in the case of weakly adsorbing electrolytes. They also evaluated the influence of the electrolyte concentration on the capacitance.

Haftel and Rosen used the embedded atom method (EAM) to study the structure and dynamics of metallic systems and investigated the surface reconstructions of various faces of Au and Pt theoretically with a number of different interaction potentials [Haftel and Rosen, 2001]. Later, they employed the surface embedded atom method (SEAM) to calculate the total energy of Au(100)-hex, Au(100)-(1 × 1), and Au(111)-(1 × 1) [Haftel and Rosen, 2003]. In order to account for the electrode potential, they added additional charge to the atoms at the electrode surface. From their studies, they concluded that surface charge plays an important role in the lifting of reconstruction. In addition, they estimated the reconstruction energy as function of change in surface strain, and found this factor to be important only for higher potentials.

While in previous ab initio studies the reconstructed surface was mostly simulated as Au(111), Feng et al. [2005] have recently performed periodic density functional theory (DFT) calculations on a realistic system in which they used a (5 × 1) unit cell and added an additional atom to the first surface layer. In their calculations, the electrode potential was included by charging the slab and placing a reference electrode (with the counter charge) in the middle of the vacuum region. From the surface free energy curves, which were evaluated on the basis of experimentally measured capacities, they concluded that there is no necessity for specific ion adsorption [Bohnen and Kolb, 1998] and that the positive surface charge alone would be sufficient to lift the reconstruction.

Although (5.19) is exact, the practical evaluation of the free energy G of the entire electrode/electrolyte interface with the accuracy required to tackle the effect of surface reconstruction is far beyond present capabilities. Therefore, instead of studying the influence of specific electrolyte ion adsorption directly, we focus on the electronic effects arising from the positive surface charge, and investigate whether these are already sufficient to lift the surface reconstruction. We further assume that the numbers of anions and cations within the interface stay constant over the entire range of electrode potentials and that the structure of the electrolyte does not change. These are certainly rather strong assumptions, but allow for quantum mechanical calculations of the electronic structure of Au(100)-hex and Au(100)-(1 × 1). As a consequence, the last two terms in (5.19) (hereafter denoted by K) become constant and the influence on the electrolyte on the electrode surface cancels out when we focus on relative stabilities only. Therefore, with

$$\gamma' = \gamma + N_a \mu_{ac}(T, a_{ac}) + \left(N_c - \frac{y}{x} N_a\right) \mu_c(T, a_c) \equiv \gamma + K \qquad (5.20)$$

we can rewrite (5.19) as

$$\gamma'(T, a_{Au}, \phi_e) = \frac{1}{A}\left[G(T, a_{Au}, N_{Au}, q_e) - N_{Au} g_{Au}^{bulk}(T, a_{Au}) - q_e(\Delta\phi - \Phi/e)\right] \qquad (5.21)$$

This expression has the advantages that the electrode potential only appears in the last term and that the Gibbs free energy G depends only on the temperature and quantities related to the electrode, allowing one to neglect the electrolyte part of the interface.

Again, the latter aspect is a direct consequence of the fact that we have assumed the electrolyte part of the interface to be fixed. Although the final reduction of the system size to the electrode surface alone allows the problem to be treated quantum mechanically, the Gibbs free energy G still depends on the amount of excess charge on the electrode surface, q_e, which changes as the electrode potential is varied. However, this brings us back to the problem of describing the shape of the potential drop within the electrode/electrolyte interface, for which, as we have learned already, no conclusive approach exists at present. In practical calculations, one way out of this dilemma is to revert to experimental measurements.

Therefore, instead of considering the electrode potential, the surface stability can be evaluated as a function of any quantity, which allows one to change the excess surface charge. When performing DFT calculations, this could be realized by adding an external electric field perpendicular to the surface plane, E_z. Since G and g_{Au}^{bulk} in (5.21) describe the Au surface and bulk, respectively, both of which have solid character, the temperature and activity (pressure) dependence will be rather small, which is why the DFT energies, which mimic $T = 0$, can be used instead. Therefore, G can be replaced by the total energy of the surface, leading to

$$\gamma'(E_z) = \frac{1}{A}\left[E_{tot}(E_z) - N_{Au}g_{Au}^{bulk}\right] - \sigma_e(E_z)\mu_e(E_z) \tag{5.22}$$

where we have introduced the excess surface charge density on the electrode surface, σ_e, which also depends on the applied electric field.

Figure 5.7 shows the modified surface free energies γ' calculated for unreconstructed and reconstructed Au(100) using DFT. The details can be found in [Jacob, 2007a]. The plot shows that the hexagonal reconstructed surface is more stable over

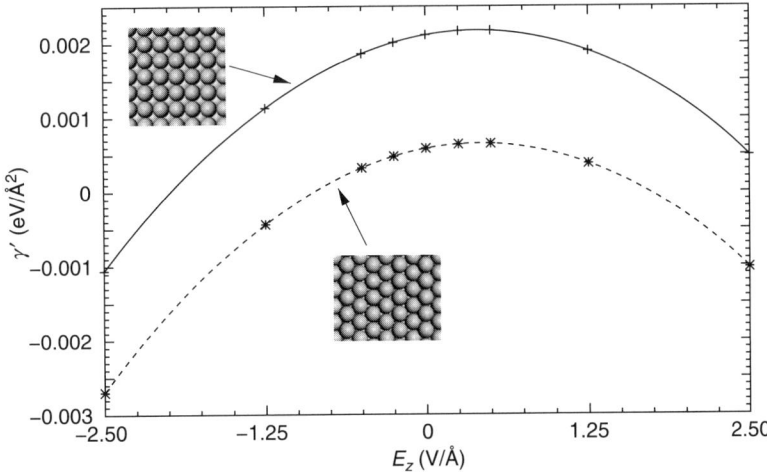

Figure 5.7 Calculated γ' versus electric field perpendicular to the surface, E_z, for unreconstructed and reconstructed Au(100).

the entire range of applied electric fields, and that the difference between the maxima (around $E_z \approx 0$) is only about 0.002 eV/Å^2 (0.04 J/m^2). Interestingly, the maxima are not exactly at $E_z = 0$, but are slightly shifted to positive E_z values. This reflects the fact that, without an electrode potential, there is still a surface dipole at the electrode surface, which requires the potential of zero charge, ϕ_{pzc}, to be compensated. Although the applied electric field induces relatively small variations in the surface free energy, there are major effects on the surface structure. Table 1.1 summarizes the surface charge density and layer distances for three different electrode potentials. At $E_z = \pm 3.85$ V/Å, the excess surface charge density σ_e is around ± 0.02 e/Å^2, which means an additional charge per surface atom of around ± 0.15 e. Without an electric field, the layer distances increase from 2.052 Å between the first two surface layers to 2.076 Å between the second and third layers and 2.085 Å between the third and fourth layers. The behavior is different in the presence of an external electric field. At the upper and lower electric field limits given in Table 5.1, the additional positive or negative surface charge causes repulsion between the surface layers. Consequently, d_{12} increases to 2.106 Å for $E_z = 3.85$ V/Å and even 2.155 Å for $E_z = -3.85$ V/Å. The behavior of the other layer spacings is comparable. Because of its relevance as the ground state, the hexagonal reconstructed surface is certainly more interesting. Without an external electric field, the additional surface atom per unit cell already leads to an overall buckled surface, with d_{12} in the range of 2.051–2.851 Å. On applying an electric field, the buckling is even more pronounced, and the upper limit of d_{12} increases to over 3 Å. This might already indicate a destabilization of the hexagonal surface layer and a lifting of the surface reconstruction.

However, as Fig. 5.7 shows, there is no crossing between the curves over the entire range of applied external electric fields, which would be necessary for a change in the preferred surface structure. This is because, instead of $\gamma'(\Delta\phi)$, which depends on the electrode potential and therefore allows comparison with experiments, we have used $\gamma'(E_z)$ from (5.22). As already mentioned, to solve this problem, we can use experimental information. While from our calculations we can extract the excess surface charge density as a function of the applied electric field $\sigma_e(E_z)$, integration of the experimental capacity measurements (Fig. 5.6) gives the same quantity, but as a function of the electrode potential $\sigma_e(\Delta\phi)$. By equating the calculated and experimental surface charge densities, we can extract a relation between the electric field applied in the calculations and the experimentally applied electrode potential, giving

TABLE 5.1 Calculated Excess Charge Densities and Layer Separations for the Unreconstructed and Reconstructed Au(100) Surfaces; In Each Case, the Values for Three Different Electric Fields are Given: -3.85, 0.00, and 3.85 V/Å

System	Au(100)-(1 × 1)			Au(100)-hex		
E_z (V/Å)	−3.85	0.00	3.85	−3.85	0.00	3.85
σ_e (e/Å^2)	−0.022	0.00	0.021	−0.019	0.00	0.018
d_{12} (Å)	2.155	2.052	2.106	2.073–3.026	2.051–2.851	2.061–3.067
d_{23} (Å)	2.132	2.076	2.149	2.126	2.057	2.148
d_{34} (Å)	2.135	2.085	2.149	2.133	2.071	2.161

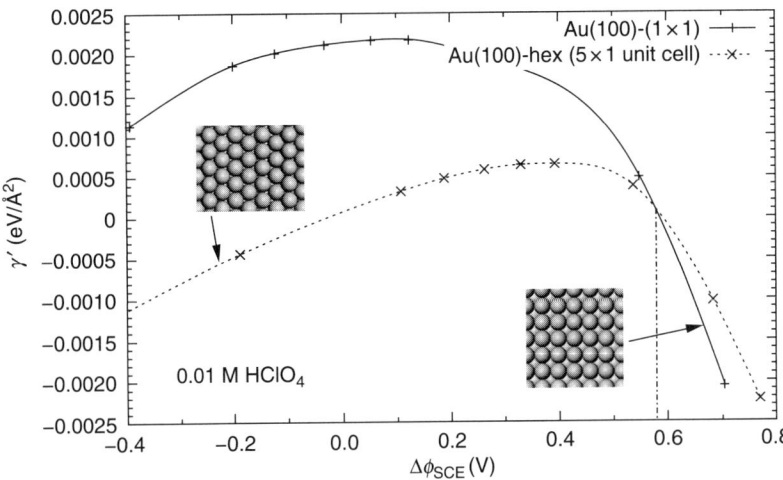

Figure 5.8 Calculated γ' versus electrode potential $\Delta\phi_{SCE}$ (referenced to an SCE) curves for Au(100) in 0.01 M HClO$_4$. The crossing between the curves indicates the electrode potential at which the surface reconstruction is lifted [Au(100)-hex $\rightarrow$ Au(100)-(1 × 1)].

$E_z(\Delta\phi)$. This relation can be used to plot $\gamma'(E_z)$ from Fig. 5.7 as a function of the electrode potential, $\gamma'[E_z(\Delta\phi)]$, for different electrolytes and concentrations, depending on which experimental capacity measurements have been used for the integration. Since these measurements were performed with an SCE, we have added a corresponding subscript to the electrode potential.

Figure 5.8 shows the resulting curves for 0.01 M HClO$_4$, which now exhibits a crossing at around 0.6 V, in good agreement with experiment. Since the reconstructed Au(100)-hex surface is stable below $\Delta\phi_{SCE} \approx 0.6$ V, the crossing of the curves indicates the lifting of surface reconstruction. Since only electronic effects and no specific ion adsorption have been considered in these calculations, from the agreement of the calculated transition potential around 0.6 V with the experimentally measured value of 0.55 V, one might conclude that the lifting of the Au(100)-hex surface reconstruction is exclusively caused by the surface charging. However, experimentally measured cyclic voltammetry curves were required as external input, which might reduce the predictive ability of the calculations. As the occurrence of any specific ion adsorption should influence the experimental cyclic voltammetry curves, from these kinds of studies, specific adsorption cannot be excluded. Instead, the results provide evidence that surface charging alone plays an important role in destabilizing the hexagonal reconstructed Au(100) surface at positive potentials.

The previous example of lifting the Au(100)-hex surface reconstruction showed that, even without specific adsorption of electrolyte ions, the electronic effects induced by the electrode potential might cause significant changes to the surface morphology of the electrode. Although the potential-induced changes to the surface free energy are only about 0.005 eV/Å^2, this is comparable to differences in the surface stabilities of clean metal surfaces. This might be of particular relevance for electrocatalytic reactions, since the surface morphology strongly influences reaction energetics and even entire reaction mechanisms.

In this example, we have focused on the surface excess charge term in (5.18) and (5.19); the next example will show that the potential is able to modify not only the electrode structure, but also its composition.

5.3.2 Electro-Oxidation of Pt Electrodes

Experimentally, cyclic voltammetry (CV) is one of the commonly used techniques to study the processes occurring at electrode surfaces as function of the electrode potential. With respect to the growth of oxides on metal electrodes by electro-oxidation, it allows one to distinguish between pronounced potential regions, which can be assigned to initial interface charging, followed by surface oxidation, and finally oxide formation at higher positive electrode potentials [Clavilier et al., 1996; Vielstich et al., 2003; Jerkiewicz et al., 2004; Climent et al., 2006]. Although different experimental techniques, such as CV, electrochemical quartz crystal nanobalance (EQCN), and Auger electron spectroscopy (AES) have been employed in order to obtain a better understanding of these structural changes, even for the "standard" system of a Pt electrode in an aqueous solution (e.g., H_2SO_4), there is still an ongoing debate regarding the geometry and composition on the atomic scale.

Figure 5.9 schematically shows the process of electro-oxidation of a polycrystalline Pt electrode with increasing electrode potential. Focusing on the part of the cyclic voltammogram that is marked by a dashed box, there is (almost) no current density at low positive electrode potentials, which indicates the absence of any charge transfer reaction at the electrode. Instead, the increasing electrode potential lowers the electronic Fermi level of the electrode, leading to an accumulation of positive excess charge at the electrode surface, accompanied by a structural reorganization within the electrolyte. As a consequence, the water molecules close to the electrode surface orient their dipoles according to the excess surface charge. This scenario is comparable to what we have already discussed in the previous example, where we have seen that surface excess charge densities around $0.02\,e/\text{Å}^2$ might cause significant structural changes at transition metal electrodes.

Further increase of the potential causes stronger interaction between the water dipoles and the electrode, finally leading to the onset of surface oxidation. Since this process involves charge transfer, in this potential range a finite current density can be observed (see Fig. 5.9). Whether this is caused by adsorption of OH^- or O^{2-} is still under debate. While the presence of OH^- is commonly accepted [Angerstein-Kozlowska et al., 1973; Dickinson et al., 1975], from recent EQCN and CV measurements Jerkiewicz et al. [2004] calculated a molecular weight of the adsorbing species of 15.8 g/mol and concluded the presence of atomic oxygen only. As a consequence of the surface oxidation, the electrolyte now sees a different (*shielded*) electrode surface, causing changes in the structure of the electrolyte. In this potential region, it is mainly the term $N_n\mu_n$ in (5.18) and (5.19) (with n = H_2O) that contributes to the stability, because water is the source for oxygen adsorbed on the surface.

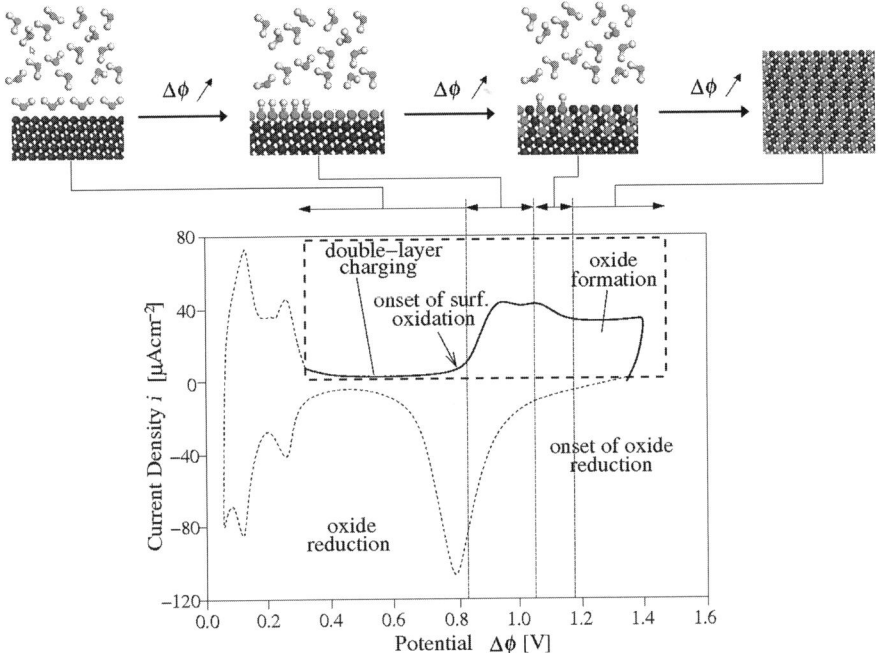

Figure 5.9 Schematic cyclic voltammogram showing the electro-oxidation of the electrode (dashed box). The curve was generated from measurements by Jerkiewicz et al. [2004] of Pt in 0.5 M H_2SO_4 with a reversible hydrogen reference electrode (RHE). For each separable potential range, an atomistic model of the electrode structure is shown above.

After forming an adsorbate layer of oxygen or an O-containing species on the electrode surface, the persistent current density that can be observed at higher electrode potentials is usually attributed to the formation of a surface oxide, which, after a certain time, continues growing to finally form the bulk oxide. While different electrochemical techniques [Gilroy and Conway, 1968; James, 1969; Kim et al., 1971; Allen et al., 1974] show evidence of the formation of an oxide, even for the *standard* system of Pt in contact with an aqueous solution, the exact structure and thickness of this oxide is still unclear [Dickinson et al., 1975; You et al., 1994; Clavilier et al., 1996; Jerkiewicz et al., 2004]. Here, the common view is that oxide growth first begins with the formation of a thin layer of PtO composition, onto which an oxide of PtO_2 composition continues growing.

Before we can apply the extended ab initio atomistic thermodynamics approach to the oxygen-covered surface or the surface/bulk oxide, we have to investigate the structure of the bulk electrode.

5.3.3 Morphology of the Bulk Electrode (Stability Condition)

While at low electrode potentials the bulk electrode will be purely platinum, at high positive electrode potentials the cyclic voltammogram shows that it becomes an oxide (PtO_2). In the following, we shall derive the condition under which the bulk oxide is the thermodynamically stable phase of the electrode.

For the non-electrochemical system of a Pt surface in contact with an O_2 atmosphere, the Pt_xO_y bulk electrode is only stable if

$$g^{bulk}_{Pt_xO_y} < xg^{bulk}_{Pt} + \frac{y}{2}\mu_{O_2} \qquad (5.23)$$

which for the electrochemical system becomes

$$g^{bulk}_{Pt_xO_y} < xg^{bulk}_{Pt} + y\left(\mu_{H_2O} + 2e\Delta\phi - \Delta G^{H_2O}_{O_2}\right) \qquad (5.24)$$

where $\Delta G^{H_2O}_{O_2}$ is the Gibbs energy required to form $\frac{1}{2}O_2$ out of liquid water (experimentally, 2.46 eV [Atkins, 1990]). Here, we have made use of the fact that water is the source of all the oxygen that, together with two electrons, is used to form the bulk oxide. Referencing the chemical potential of water by $\Delta\mu_{H_2O} = \mu_{H_2O} - \frac{1}{2}E^{tot}_{O_2}$, (5.24) can be rewritten as

$$\Delta\mu_{H_2O} > \frac{1}{y}\left[g^{bulk}_{Pt_xO_y} - xg^{bulk}_{Pt} - \frac{y}{2}E^{tot}_{O_2}\right] - 2e\Delta\phi + \Delta G^{H_2O}_{O_2} \qquad (5.25)$$

The expression in the square brackets is exactly the formation energy of the Pt bulk oxide out of bulk platinum and molecular O_2, which can easily be evaluated by DFT calculations [Jacob, 2007b]. Together with the experimental value for $\Delta G^{H_2O}_{O_2}$, we can finally formulate the following stability ranges at which the three known Pt bulk oxides are thermodynamically stable:

- α-PtO_2: $+1.74\,eV - 2e\Delta\phi < \Delta\mu_{H_2O} < 0\,eV$
- β-PtO_2: $+1.74\,eV - 2e\Delta\phi < \Delta\mu_{H_2O} < 0\,eV$ \qquad (5.26)
- PtO: $+1.97\,eV - 2e\Delta\phi < \Delta\mu_{H_2O} < 0\,eV$

Decomposition of the bulk oxide will occur if $\Delta\mu_{H_2O}$ is too negative to fulfill these conditions.

5.3.4 Pt Surface Oxidation

In this section, we will investigate the surface structure of the electrode in the potential range before a surface or bulk oxide starts forming, and will restrict ourselves to the adsorption of atomic oxygen only (not OH^-) [Jacob and Scheffler, 2007]. Furthermore, in our simulations, we assume a single-crystal Pt(111) electrode, which will be compared with the experimental CV curve (Fig. 5.9) for polycrystalline Pt. This simplification is motivated by the fact that our interest here is to describe the general behavior of the system only.

When calculating the interfacial free energies of particular surface structures, we are limited by the condition that the Pt bulk oxide is not the thermodynamically stable phase of the bulk electrode. Since α-PtO$_2$ and β-PtO$_2$ are the most stable Pt bulk oxides, this is fulfilled as long as $\Delta\mu_{H_2O} < 1.74\,\text{eV} - 2e\Delta\phi$. Keeping this condition in mind, similarly to the previous example, we again have to adapt (5.19) to the actual system in order to evaluate the stability of different surface structures. As already discussed for the Au(100) system, the exact evaluation of (5.19) requires a self-consistent calculation of the Gibbs free energy G of the entire electrode/electrolyte interface, which is far beyond present computer resources. However, under the following assumptions, the modeling can be drastically reduced:

- The structure and energetics of any adsorbate (here oxygen) is not influenced by the electrode potential, i.e., the adsorbed species have no surface dipole.
- A fixed electrolyte, respectively double-layer structure is assumed, which remains unchanged under potential variation.

The second assumption, which we have already made in the previous example, again allows us to consider the direct electrolyte contributions to the interfacial free energy as constant [see (5.20)]. Although the model constrained by these two strong assumptions certainly does not correspond to a realistic system, it represents the actual level at which electrochemical interfaces are usually calculated. However, as a consequence, the last three terms of (5.19) now give constant contributions, and the Gibbs free energy G becomes independent of the excess charges, respectively electrode potential, finally leading to

$$\gamma'' = \gamma' + q_e(\Delta\phi - \Phi/e) \approx \gamma + K' \qquad (5.27)$$

where we have used γ' from (5.20) and introduced the notation K' for all constant contributions. Applied to our system of Pt in contact with the aqueous electrolyte, this becomes

$$\gamma''(T, a_{H_2O}, a_{Pt}, \Delta\phi) = \frac{1}{A}\{G(T, a_{H_2O}, a_{Pt}, N_O, N_{Pt}) - N_{Pt} g_{Pt}^{bulk}(T, a_{Pt})$$
$$- N_O[\Delta\mu_{H_2O}(T, a_{H_2O}) + 2e\Delta\phi]\} \qquad (5.28)$$

The last term comes from the assumption that every oxygen atom adsorbing on the surface originates from a water molecule of the bulk electrolyte reservoir:

$$O_{adsorbed} + 2H^+ + 2e^- \longleftrightarrow H_2O(l) \qquad (5.29)$$

Since, by this reaction, two electrons are transferred from the reference electrode (which, for comparison with the experimental CV curves, we assume to be a reversible hydrogen electrode, giving $\phi_{ref} = 0$) to the electrode, the term $N_O(2e\Delta\phi)$ appears in (5.28).

In order to compare the stability of different oxygen overlayers, DFT calculations were performed on the energetic and structures of oxygen at different coverages. Since the electrode is present in the solid phase, it is reasonable to assume the T and a

dependence of G to be rather small, and the contributions from configurational and vibrational entropy to G will be mostly compensated by the g_{Pt}^{bulk} term on the right-hand side of (5.28). Therefore, the DFT energies, which mimic $T = 0$, can be used instead.

This finally leads to the $(T, a, \Delta\phi)$ phase diagram shown in Fig. 5.10 (Plate 5.1). The plot in Fig. 5.10a shows the modified interfacial free energy γ'' of different adsorbate overlayers as a function of the water chemical potential $\Delta\mu_{H_2O}$ and the electrode potential. Since the most stable structures have the lowest interfacial free energy, Fig. 5.10b shows the view to the bottom of the phase diagram. Each shaded area corresponds to a different stable adsorbate structure. Since, according to (5.2) and (5.3), $\Delta\mu_{H_2O}$ can be related to temperature and activity, for a water activity $a_{H_2O} = 1$ (ideal solution), we have added the corresponding temperature scale to Fig. 5.10b and marked $T = 298K$ by a horizontal dashed line. This line crosses the $\Delta\phi = 0$ line (vertical dashed line) in the area, which corresponds to the clean Pt(111) surface with no oxygen on the surface. Keeping these a_{H_2O} and T conditions and increasing the electrode potential shows that in the range from 0.95 to 1.20 V, a (2×2) oxygen overlayer structure with 0.25 monolayer (ML) coverage is dominant. In the case that the electrode potential exceeds 1.20 V, our calculations predict the formation of Pt bulk oxide. As discussed before and shown in Fig. 5.9, the experiments by Jerkiewicz et al. [2004] found the surface oxidation to occur from 0.85 to 1.1 V, followed by oxide formation above 1.1 V.

The agreement between our phase diagram and the experimental CV curve is surprisingly good, considering the simplicity of our interface model and the fact that polycrystalline Pt has been used experimentally. Furthermore, for our studies we have used Pt(111), whereas the experiments were performed with polycrystalline Pt. One of the reasons for the agreement might be the absence of specific ion adsorption on the electrode. Moreover, there might be a kinetic barrier to oxide formation,

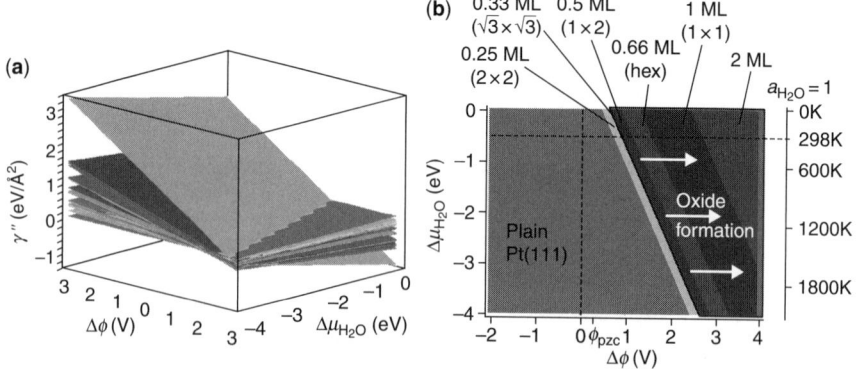

Figure 5.10 $(T, a, \Delta\phi)$ phase diagram for the electrochemical oxidation of Pt(111) in an aqueous electrolyte. (a) Modified interfacial free energy γ'' as a function of $\Delta\mu_{H_2O}$ and the electrode potential $\Delta\phi$. (b) View to the bottom of the phase diagram. In addition, the temperature scale corresponding to $a_{H_2O} = 1$ ($p = 1$atm) is given on the right side of the phase diagram.

necessitating higher oxygen coverages prior oxide formation, and higher electrode potentials to initiate this process.

While we have focused here on the oxygen adsorption at the electrode surface, in the following, we will discuss the part of the phase diagram shaded gray and labeled "Oxide formation" in Fig 5.10b.

5.3.5 Pt Oxide Growth

In the Pt bulk oxide range of the phase diagram, which is relevant provided that (5.26) is fulfilled, the bulk electrode is Pt oxide and no longer pure Pt. Therefore the corresponding term in (5.28) that accounts for the bulk electrode reservoir now has to involve $g_{Pt_xO_y}^{bulk}$. Since the bulk electrode should be in thermodynamic equilibrium with the surroundings,

$$x g_{Pt}^{bulk} + y\mu_O = g_{Pt_xO_y}^{bulk} \qquad (5.30)$$

holds [see (5.8)], which allows us to introduce the Gibbs energy of the bulk oxide. According to (5.29), each oxygen comes from a water molecule of the electrolyte, which means that $\mu_O = \mu_{H_2O} + 2e\Delta\phi$. Insertion of this condition into (5.30) can be used to obtain a modified interfacial free energy similar to (5.28):

$$\gamma''(T, a_{H_2O}, a_{Pt_xO_y}, \Delta\phi) = \frac{1}{A}\left\{G(T, a_{H_2O}, a_{Pt_xO_y}, N_O, N_{Pt}) - \frac{N_{Pt}}{x} g_{Pt_xO_y}^{bulk}(T, a_{Pt_xO_y})\right.$$
$$\left. + \left(y\frac{N_{Pt}}{x} - N_O\right)\left[\Delta\mu_{H_2O}(T, a_{H_2O}) + 2e\Delta\phi\right]\right\} \qquad (5.31)$$

In order to evaluate the structure of the electrode surface under high positive electrode potentials, we used the calculated DFT energies for all different slabs of low index Pt bulk oxide surfaces [Jacob, 2007b], together with (5.31), to obtain the corresponding $(T, a, \Delta\phi)$ phase diagrams for the two very stable Pt bulk oxides, α-PtO$_2$ and β-PtO$_2$. Similar to Fig. 5.10, that was for an oxygen overlayer on a pure Pt electrode, Fig. 5.11 (Plate 5.2) shows the stable bulk oxide surfaces for different $\Delta\mu_{H_2O}$ and $\Delta\phi$ conditions. To each plot we again have added the temperature scale that corresponds to an activity $a_{H_2O} = 1$, meaning an ideal solution.

For the α-PtO$_2$ system, we find that above an electrode potential of 1.2 V, the (001) surface with bulk composition is most stable and shows only minor relaxation effects (denoted as (001)-O in Fig. 5.11a). This surface structure corresponds to experimental UHV measurements of surface oxides on Pt(110), supported by DFT calculations [Li et al., 2004]. In the case of a very thin surface layer, the layer composition might even be PtO. Increasing the electrode potential above 2.0 V would cause stronger interactions with the surrounding water dipoles and lead to a α-PtO$_2$(011) surface with an enrichment of oxygen (as O^{2-}) on the surface.

For the β-PtO$_2$ system, we find the same general that increasing the electrode potential leads to surface structures with more oxygen, although the transition

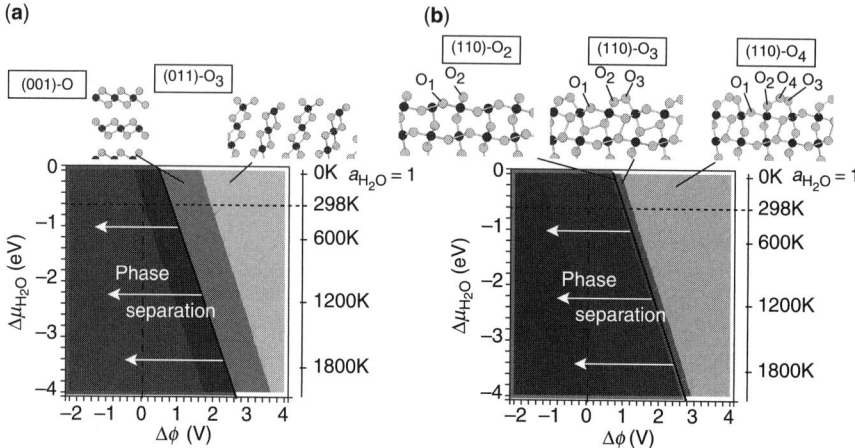

Figure 5.11 $(T, a, \Delta\phi)$ phase diagrams for all low index surfaces of α-PtO$_2$ (a) and β-PtO$_2$ (b). Each plot shows the thermodynamically most stable surface structures as functions of $\Delta\mu_{H_2O}$ and $\Delta\phi$. In addition, the temperature scale corresponding to $a_{H_2O} = 1$ ($p = 1$ atm) is shown on the right side of each phase diagram. The surface models are named with their orientations and surface terminations.

potentials are rather different. At room temperature and an electrode potential around 1.2 V, which corresponds exactly to the transition of the phase-separated system and the bulk oxide, the (110) surface with bulk composition is again most stable, showing only minor deviations from the corresponding bulk-truncated structure. A slight increase in the electrode potential stabilizes the same surface orientation, but with an additional oxygen atom per unit cell, until approximately 1.4 V. Induced by these excess atoms, which are not compensated by corresponding platinum atoms, structural changes on the surface can be observed such that adjacent oxygen atoms start forming bonds. The resulting O—O bond length is comparable to the distance in the OOH radical and H$_2$O$_2$ molecule [Jacob, 2006a, b], indicating a single covalent bond. Above 1.4 V, a second excess oxygen per surface unit cell is stabilized. As a result, two of now three excess oxygens form an O=O double bond.

This example describes how the extended ab initio atomistic thermodynamics approach can reveal further insights into the growth of the oxide. It has been shown that at electrode potentials that are not too high, oxide growth proceeds by first forming a thin α-PtO$_2$(001) surface oxide with a PtO composition. Further growth then results in an α-PtO$_2$(001) or β-PtO$_2$(110) bulk oxide. This finding agrees with experimental observations along the oxidation mode [Gilroy and Conway, 1968; Allen et al., 1974; Tremiliosi-Filho et al., 1992], but in addition specifies the corresponding surface structures. Along the reduction mode, Tremiliosi-Filho et al. [1992] found the opposite growth mechanism. Although this seems to be in contrast with our results, one has to consider that the surface structures discussed here correspond to thermodynamically stable phases. Thus, any energetic barriers that the system has to overcome in order to reach thermodynamic equilibrium are not included. Those studies would require extensive kinetic simulations, for which the results described here might be used as starting point.

5.4 SUMMARY

We have described the extended ab initio atomistic thermodynamics method, which allows structures and stabilities of electrode/electrolyte interfaces to be calculated quantitatively as functions of temperature, activities/pressures, and external electrode potential. Starting from the gas phase system, we have successively moved to the electrochemical system by first describing the properties and characteristics of the well-separated bulk electrolyte and bulk electrode independently, finally discussing the changes introduced by bringing both into contact. We have shown that when the electrochemical system is considered, i.e., the electrode/electrolyte interface, the interfacial free energy, which describes the stability of a particular interface structure, depends almost exclusively on well-defined quantities, even being accessible experimentally. Only the activity of one of the ionic electrolyte species depends on accurate theoretical modeling.

We have also discussed two applications of the extended ab initio atomistic thermodynamics approach. The first example is the potential-induced lifting of Au(100) surface reconstruction, where we have focused on the electronic effects arising from the potential-dependent surface excess charge. We have found that these are already sufficient to cause lifting of the Au(100) surface reconstruction, but contributions from specific electrolyte ion adsorption might also play a role. With the second example, the electro-oxidation of a platinum electrode, we have discussed a system where specific adsorption on the surface changes the surface structure and composition as the electrode potential is varied.

Although the extended ab initio atomistic thermodynamics approach provides an exact expression for the interfacial stability, the formalism requires self-consistent modeling of the entire electrochemical system, or electrode/electrolyte interface, exceeding presently available computational capabilities. Therefore, certain assumptions had to be made that reduce the effort to the calculation of the electrode surface only. Even with this simplified approach, which has been applied to the two examples discussed in this chapter, the qualitative behavior can be reproduced.

REFERENCES

Abbas Z, Gunnarsson M, Ahlberg E, Nordholm S. 2002. Corrected Debye-Hückel (CDH) analysis of surface complexation. J Phys Chem B 106: 1403–1420.

Abruna HD. 1991. Electrochemical Interfaces: Modern Techniques for In-Situ Interface Characterization. Weinheim: Wiley-VCH.

Allen GC, Tucker PM, Capon A, Parsons R. 1974. X-ray photoelectron-spectroscopy of adsorbed oxygen and carbonaceous species on platinum-electrodes. J Electroanal Chem 50: 335–343.

Angerstein-Kozlowska H, Conway BE, Sharp WBA. 1973. Real condition of electrochemically oxidized platinum surfaces: 1. Resolution of component processes. J Electroanal Chem 43: 9–36.

Atkins PW. 1990. Physikalische Chemie. Weinheim: Wiley-VCH.

Binnig GK, Rohrer H, Gerber C, Stoll E. 1984. Real-space observation of the reconstruction of Au(100). Surf Sci 104: 321–335.

Bockris JO'M, Reddy AKN, Gamboa-Aldeco ME. 2000. Modern Electrochemistry. 2nd ed. New York: Plenum Press.

Bohnen KP, Kolb DM. 1998. Charge versus adsorbate-induced lifting of the Au(100)-(hex) reconstruction in an electrochemical environment. Surf Sci 407: 629–632.

Clavilier J, Orts JM, Gómez R, Feliu JM, Aldaz A. 1996. Comparison of electrosorption at activated polycrystalline and Pt(531) kinked platinum electrodes. Surface voltammetry and charge displacement on potentiostatic CO adsorption. J Electroanal Chem 404: 281–289.

Climent V, Gómez R, Orts JM, Feliu JM. 2006. Thermodynamic analysis of the temperature dependence fo OH adsorption on Pt(111) and Pt(100) electrodes in acidic media in the absence of specific ion adsorption. J Phys Chem B 110: 11344–11351.

Dakkouri AS. 1997. Reconstruction phenomena at gold/electrolyte interfaces: An in-situ STM study of Au(100). Solid State Ionics 94: 99–114.

Dickinson T, Povey AF, Sherwood PMA. 1975. X-ray photoelectron spectroscopic studies of oxide-films on platinum and gold electrodes. J Chem Soc Faraday Trans I 71: 298–311.

Dodson BW. 1987. Atomistic study of structural instabilities in coherently strained Si-like layers. Phys Rev B 35: 5558–5562.

Ercolessi F, Tosatti E, Parrinello M. 1986. Au(100) surface reconstruction. Phys Rev Lett 57: 719–722.

Fedak DG, Gjostein NA. 1966. Structure and stability of (100) surfaces of gold. Phys Rev Lett 16: 171–172.

Fedak DG, Gjostein NA. 1967. On anomalous surface structures of gold. Surf Sci 8: 77–97.

Feng YJ, Bohnen KP, Chan CT. 2005. First-principles studies of Au(100)-hex reconstruction in an electrochemical environment. Phys Rev B 72: 125401.

Gao X, Hamelin A, Weaver MJ. 1991. Potential-dependent reconstruction at ordered Au(100)–aqueous interfaces as probed by atomic-resolution scanning tunneling microscopy. Phys Rev Lett 67: 618–621.

Gilroy D, Conway BE. 1968. Surface oxidation and reduction of platinum electrodes—coverage kinetic and hysteresis studies. Can J Chem 46: 875–876.

Gunnarsson M, Abbas Z, Ahlberg E, Nordholm S. 2004. Corrected Debye–Hückel analysis of surface complexation: III. Spherical particle charging including ion condensation. J Coll Interf Sci 274: 563–578.

Haftel MI, Rosen M. 2001. Surface embedded atom model of the electrolyte–metal interface. Phys Rev B 64: 195405.

Haftel MI, Rosen M. 2003. Surface-embedded-atom model of the potential-induced lifting of the reconstruction of Au(100). Surf Sci 523:118–124.

Halley JW, Schelling P, Duan Y. 2000. Simulation methods for chemically specific modeling of electrochemical interfaces. Electrochim Acta 46: 239–245.

Horn K, Scheffler M. 2000. Electronic Structure: 2 (Handbook of Surface Science). Amsterdam: North-Holland.

Jacob T. 2006a. The mechanism of forming H_2O from H_2 and O_2 over a Pt catalyst via direct oxygen reduction. Fuel Cells 6: 159–181.

Jacob T. 2006b. Water formation on Pt and Pt-based alloys: a theoretical description of a catalytic reaction. ChemPhysChem 7: 992–1005.

Jacob T. 2007a. Potential-induced lifting of the Au(100)-surface reconstruction studied with DFT. Electrochim Acta 52: 2229–2235.

Jacob T. 2007b. Theoretical investigations on the potential-induced formation of Pt-oxide surfaces. J Electroanal Chem 607: 158–166.

Jacob T, Scheffler M. 2007. Extended ab-initio atomistic thermodynamics for electrochemical systems. Submitted for publication.

James SD. 1969. Multilayer oxide films on anodized platinum. J Electrochem Soc 116: 1681–1682.

Jerkiewicz G, Vatankhah G, Lessard J, Soriaga MP, Park YS. 2004. Surface-oxide growth at platinum electrodes in aqueous H_2SO_4: Reexamination of its mechanism through combined cyclic-voltammetry, electrochemical quartz-crystal nanobalance, and Auger electron spectroscopy measurements. Electrochim Acta 49: 1451–1459.

Kaxiras E, Bar-Yam Y, Joannopoulos JD, Pandey KC. 1987. Ab-initio theory of polar semiconductor surfaces: 1. Methodology and the (2×2) reconstructions of GaAs(111). Phys Rev B 35: 9625–9635.

Kim KS, Winograd N, Davis RE. 1971. Electron spectroscopy of platinum-oxygen surfaces and application to electrochemical studies. J Am Chem Soc 93: 6296–6297.

Kitchin JR, Nørskov JK, Barteau MA, Chen JG. 2004. Modification of the surface electronic and chemical properties of Pt(111) by subsurface 3d transition metals. J Chem Phys 120: 10240–10246.

Kolb DM. 1996. Reconstruction phenomena at metal–electrolyte interfaces. Prog Surf Sci 51: 109–173.

Kolb DM. 2002. An atomistic view of electrochemistry. Surf Sci 500: 722–740.

Koper MTM. 2004. Ab initio quantum-chemical calculations in electrochemistry. In: Vayenas CG, Conway BE, White RE, Gamboa-Adelco ME, eds. Modern Aspects of Electrochemistry, No. 36, Berlin: Springer. pp. 51–130.

Koper MTM, van Santen RA, Neurock M. 2003. Theory and modeling of catalytic and electrocatalytic reactions. In: Savinova ER, Vayenas CG, Wieckowski A, eds. Catalysis and Electrocatalysis at Nanoparticle Surfaces. New York: Marcel Dekker. pp. 1–34.

Li WX, Österlund L, Vestergaard EK, Vang RT, Matthiesen J, Pedersen TM, Lægsgaard E, Hammer B, Besenbacher F. 2004. Oxidation of Pt(110). Phys Rev Lett 93: 146104.

Magnussen OM. 2002. Ordered anion adlayers on metal electrode surfaces. Chem Rev 102: 679–725.

Nazmutdinov RR, Shapnik MS. 1996. Contemporary quantum chemical modelling of electrified interfaces. Electrochim Acta 41: 2253–2265.

Parker DH, Bartram ME, Koel BE. 1989. Study of high coverages of atomic oxygen on the Pt(111) surface. Surf Sci 217: 489–510.

Qian GX, Martin RM, Chadi DJ. 1988. 1st-principles study of the atomic reconstructions and energies of Ga-stabilized and As-stabilized GaAs(100) surfaces. Phys Rev B 38: 7649–7663.

Reuter K, Scheffler M. 2002. Composition, structure, and stability of $RuO_2(110)$ as function of oxygen pressure. Phys Rev B 65: 035406.

Rieder KH, Engel T, Swendsen RH, Manninen M. 1983. A helium diffraction study of the reconstructed Au(100) surface. Surf Sci 127: 223–242.

Rossmeisl J, Nørskov JK, Taylor CD, Janik MJ, Neurock M. 2006. Calculated phase diagram for the electrochemical oxidation and reduction of water over Pt(111). J Phys Chem B 110: 21833–21839.

Santos E, Schmickler W. 2004. Changes in the surface energy during the reconstruction of Au(100) and Au(111) electrodes. Chem Phys Lett 400: 26–29.

Scheffler M, Dabrowski J. 1988. Parameter-free calculations of total energies, interatomic forces and vibrational entropies of defects in semiconductors. Phil Mag A 58: 107–121.

Schmickler W. 1996a. Electronic effects in the electric double layer. Chem Rev 96: 3177–3200.

Schmickler W. 1996b. Interfacial electrochemistry. New York: Oxford University Press.

Schmickler W. 1999. Recent progress in theoretical electrochemistry. Annu Rep Prog Chem Sect C 95: 117–162.

Takeuchi N, Chan CT, Ho KM. 1991. Reconstruction of the (100) surfaces of Au and Ag. Phys Rev B 43: 14363–14370.

Taylor CD, Wasileski SA, Filhol JS, Neurock M. 2006. First principles reaction modeling of the electrochemical interface: Consideration and calculation of a tunable surface potential from atomic and electronic structure. Phys Rev B 73: 165402.

Tomanek D, Bennemann KH. 1985. Electronic model for energies, relaxations and reconstruction trends at metal-surfaces. Surf Sci 163: 503–515.

Tremiliosi-Filho G, Jerkiewicz G, Conway BE. 1992. Characterization and significance of the sequence of stages of oxide film formation at platinum generated by strong anodic polarization. Langmuir 8: 658–667.

Van Hove MA, Koestner RJ, Stair PC, Biberian JP Kesmodel LL, Bartos I, Somorjai GA. 1981. The surface reconstructions of the (100) crystal faces of iridium, platinum and gold, 1. Experimental-observations and possible structural models. Surf Sci 103: 189–217.

Vassilev P, Hartnig C, Koper MTM, Frechard F, van Santen RA. 2001. Ab initio molecular dynamics simulation of liquid water and water–vapor interface. J Chem Phys 115: 9815–9820.

Vielstich W, Gasteiger HA, Lamm A. 2003. Handbook of Fuel Cells—Fundamentals, Technology and Applications. New York: Wiley.

Wendelken JF, Zehner DM. 1978. Au(100)—characterization of re-ordered and normal surfaces via LEED, AES, and ELS. Surf Sci 71: 178–184.

Wu S, Lipkowski J, Magnussen OM, Ocko BM, Wandlowski T. 1998. The driving force for $(p \times \sqrt{3}) \leftrightarrow (1 \times 1)$ phase transition of Au(111) in the presence of organic adsorption: A combined chronocoulometric and surface X-ray scattering study. J Electroanal Chem 446: 67–77.

You H, Zurawski DJ, Nagy Z, Yonco RM. 1994. In-situ X-ray reflectivity study of incipient oxidation of Pt(111) surface in electrolyte-solutions. J Chem Phys 100: 4699–4702.

CHAPTER 6

Mechanisms of the Oxidation of Carbon Monoxide and Small Organic Molecules at Metal Electrodes

MARC T.M. KOPER and STANLEY C.S. LAI

Leiden Institute of Chemistry, Leiden University, PO Box 9502, 2300 RA Leiden, The Netherlands

ENRIQUE HERRERO

Instituto de Electroquimica, Universidad de Alicante, Apartado 99, E-03080, Alicante, Spain

6.1 INTRODUCTION

The electrochemical oxidation of carbon monoxide (CO) and small organic molecules containing one or two carbon atoms is one of the central catalytic issues in the successful development of low temperature fuel cells. Methanol and ethanol are the most likely candidates for direct alcohol fuel cells, but the significant overpotential for their oxidation severely limits the fuel cell's efficiency. Understanding the nature of this overpotential is key to the development of better fuel cell catalysts, and also involves understanding the reactivity of intermediates in methanol and ethanol oxidation. CO is the most prominent intermediate, the oxidation of which is also of great significance for hydrogen–oxygen fuel cells in which the hydrogen stream is contaminated by small traces of CO.

This chapter summarizes our current understanding of the mechanisms of the oxidation of CO, methanol, and ethanol, as well as the important organic intermediates formaldehyde and formic acid, as gleaned from surface science-type studies employing well-defined monometallic electrode surfaces (primarily single crystals) combined with in situ spectroscopy and kinetic modeling. Computational methods such as density functional theory (DFT) are also beginning to contribute importantly to the unraveling of kinetic and thermodynamic factors, reaction pathways, and structure sensitivity issues in electrocatalysis research, and relevant results will be cited where appropriate. In the concluding section, we will also touch upon the relation

---

Fuel Cell Catalysis. Edited by Marc T. M. Koper
Copyright © 2009 John Wiley & Sons, Inc.

and relevance of these surface science studies employing single crystals with respect to the same reactions studied on more realistic catalytic surfaces consisting of small nanometer-sized particles on a carbon support. This review will not be exhaustive, and will mainly focus on results obtained in our own research groups. Nevertheless, we will point out the relation to similar work done in other groups as much as possible. The interested reader is also advised to consult literature reviews from other groups dealing with similar topics [Parsons and VanderNoot, 1988; Beden et al., 1992; Jarvi and Stuve, 1998; Sun, 1998; Markovic and Ross, 2002; Lai et al., 2008]. The electrode kinetics of the reactions considered in this chapter on bimetallic and nanoparticulate surfaces is discussed in detail in other chapters in this volume (e.g., Chapters 8, 10, and 15).

Perhaps the most important paradigm in research on the mechanism of the electrocatalytic oxidation of small organic molecules is the dual pathway mechanism introduced in Capon and Parsons [1973a, b], and reviewed in Parsons and VanderNoot [1988]. In terms of methanol oxidation, the dual pathway may be summarized in a simplified way by Fig. 6.1. The idea is that the complete oxidation of methanol to carbon dioxide may follow two different pathways:

- A so-called indirect pathway involving a strongly adsorbed intermediate. This intermediate is now commonly accepted to be CO. The sluggishness of this pathway has much to do with difficulty of oxidizing of this strongly adsorbed CO, explaining the importance of understanding the catalysis of CO oxidation. The oxidation of CO is the topic of Section 6.2.

- A so-called direct pathway involving a more weakly adsorbed perhaps even partially dissolved intermediate. Likely candidates for such intermediates are formaldehyde and formic acid. The oxidation mechanism of formic acid is discussed in Section 6.3. The idea is that the formation of a strongly adsorbed intermediate is circumvented in the direct pathway, though in practice this has appeared difficult to achieve (the dashed line in Fig. 6.1). Section 6.4 will discuss this in more detail in relation to the overall reaction mechanism for methanol oxidation.

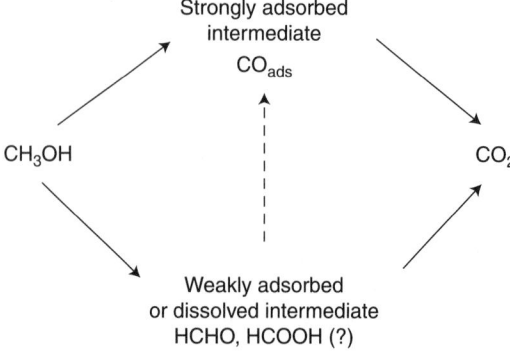

Figure 6.1 Schematic representation of the "dual pathway" mechanism for the electrocatalytic oxidation of methanol to carbon dioxide.

Similar considerations apply to the oxidation of ethanol, although one must take into account the breaking of the C—C bond, and the formation and involvement of different intermediates and side products. Ethanol oxidation will be the topic of Section 6.5. Section 6.6 will briefly summarize our main conclusions and discuss the relation of our "surface science approach" to real catalysts.

6.2 CARBON MONOXIDE OXIDATION

6.2.1 Carbon Monoxide Oxidation on Platinum

Poisoning of platinum fuel cell catalysts by CO is undoubtedly one of the most severe problems in fuel cell anode catalysis. As shown in Fig. 6.1, CO is a strongly bonded intermediate in methanol (and ethanol) oxidation. It is also a side product in the reformation of hydrocarbons to hydrogen and carbon dioxide, and as such blocks platinum sites for hydrogen oxidation. Not surprisingly, CO electro-oxidation is one of the most intensively studied electrocatalytic reactions, and there is a continued search for CO-tolerant anode materials that are able to either bind CO weakly but still oxidize hydrogen, or that oxidize CO at significantly reduced overpotential.

There are essentially two ways to study electrochemical CO oxidation. The first one, called stripping, involves pre-adsorbing a (saturated) monolayer of CO and then oxidizing it from a solution that does not contain CO, in either a voltammetric scan (stripping voltammetry) or a potential-step experiment (stripping chronoamperometry). Chronoamperometry has the advantage of studying the system at a fixed potential, and is generally the preferred method for investigating kinetics. The second way to study electrochemical CO oxidation is continuous or bulk CO oxidation, that is, with a fixed (often saturated) concentration of CO in solution. We will discuss here mainly CO stripping chronoamperometry on platinum single-crystal electrodes, as these experiments are relatively straightforward to interpret (Section 6.2.1.1). Stripping voltammetry of CO on platinum will be discussed in Section 6.2.1.2. Section 6.2.1.3 will deal with continuous CO oxidation.

All experiments to be described below are interpreted on the basis of the Langmuir–Hinshelwood (LH) mechanism for CO electro-oxidation suggested by Gilman more than 40 years ago [Gilman, 1964]. According to Gilman's model, water needs to be activated on a free site on the surface, leading to surface-bonded OH:

$$H_2O + {}^* \longleftrightarrow OH_{ads} + H^+ + e^- \qquad (6.1)$$

where the * denotes a free site on the platinum surface. Note that we write this reaction as reversible. The surface-bonded OH is the oxygen donor reacting with surface-bonded CO to form CO_2:

$$CO_{ads} + OH_{ads} \longrightarrow CO_2 + H^+ + e^- + 2^* \qquad (6.2)$$

162 MECHANISMS OF THE OXIDATION

This is an irreversible LH reaction (i.e., a second-order reaction between two surface adsorbates), and generates free sites for the adsorption of OH (Reaction 6.1) or, in the case of continuous CO oxidation, for the adsorption of CO:

$$CO + {}^* \longrightarrow CO_{ads} \tag{6.3}$$

6.2.1.1 CO Stripping Chronoamperometry

Before discussing experimental results, let us examine what the LH mechanism predicts for the chronoamperometric response of an experiment where we start at a potential at which the CO adlayer is stable and we step to a "final potential" E where the CO adlayer will be oxidized. We will also assume that the so-called mean field approximation applies, i.e., CO and OH are well mixed on the surface and the reaction rate can be expressed in terms of their average coverages θ_{CO} and θ_{OH}. The differential equation for the rate of change of θ_{CO} with time is

$$\Gamma_m \frac{d\theta_{CO}}{dt} = -k'(E)\theta_{CO}\theta_{OH} = -k(E)\theta_{CO}(1 - \theta_{CO}) \tag{6.4}$$

where Γ_m is the number of surface sites per unit area, and where in the last equality we have assumed that the formation of OH_{ads} through Reaction (6.1) is always in equilibrium, and its coverage is relatively small, such that $\theta_{OH} \propto (1 - \theta_{CO})$.

Equation (6.4) may be solved analytically assuming that at $t = 0$, $\theta_{CO} = \theta_{in}$, with θ_{in} the initial CO coverage (which should be <1), giving as a final expression for the current density j [Bergelin et al., 1999]

$$j(t) = \frac{Q(k/\Gamma_m)\exp[-k(t-t_{max})/\Gamma_m]}{\{1 + \exp[-k(t-t_{max})/\Gamma_m]\}^2} \tag{6.5}$$

where Q is the charge needed to oxidize the CO adlayer. In an experiment, this charge has two terms, $Q = Q_{CO} + Q_{anions}$; Q_{CO} is associated with the stripping of CO and equals $2e\Gamma_m\theta_{in}$, and Q_{anions} is associated with anion readsorption on the free sites created after the CO oxidation. It should be taken into account that anions are always adsorbed on a clean platinum electrode at the potentials where CO oxidation takes place. Thus, anions are immediately readsorbed when CO is stripped from the surface, and this readsorption process is always accompanied with a positive charge transfer. Equation (6.5) predicts that, for sufficiently high initial CO coverage, the current transient should exhibit a maximum at $t_{max} = (\Gamma_m/k)\ln[\theta_{in}/(1 - \theta_{in})]$. The reason for this maximum is the autocatalysis implied by Reactions (6.1) and (6.2), in which every free site on the surface creates two more. At high initial CO coverage, very few sites are available for OH adsorption, and the reaction rate is low. As the reaction proceeds, slowly new free sites are generated through the reaction of OH with CO, increasing the OH adsorption rate and consequently the overall CO oxidation rate and the current. The autocatalysis stops when $\theta_{CO} = 0.5$, for which the transient goes through a maximum. This autocatalytic mechanism for CO oxidation may lead to instabilities under conditions of continuous CO oxidation, as will be discussed in Section 6.2.1.3.

Chronoamperometric transients for CO stripping on polycrystalline platinum were measured by McCallum and Pletcher [1977]. Love and Lipkowski [1988] were the first to present chronoamperometric data for CO stripping on single-crystalline platinum. However, they interpreted their data on the basis of a different model than the one discussed above. Love and Lipkowski considered that the oxidation of the CO adlayer starts at holes or defects in the CO adlayer, where OH adsorbs. These holes act as nucleation centers for the oxidation reaction, and the holes grow as the CO at the perimeter of these holes is oxidized away by OH_{ads}. This "nucleation and growth" (N&G) mechanism is fundamentally different from the "mean field" model presented above, because it does not presume any kind of mixing of CO and OH [Koper et al., 1998]. Basically, it assumes complete surface immobility of the chemisorbed CO.

Mathematical expressions for the N&G model can be derived from the classical theory for the nucleation and growth of two-dimensional films [Schmickler, 1996]. Two regimes are distinguished:

- instantaneous nucleation, for which the nucleation is fast and holes are occupied instantaneously
- progressive nucleation, for which the nucleation rate is slow

The respective expressions are

$$j(t) \propto M k_G^2 t \exp(-\pi M k_G^2 t) \qquad (6.6)$$

$$j(t) \propto M k_N k_G^2 t^2 \exp(-\pi M k_N k_G^2 t^3 / 3) \qquad (6.7)$$

where M is the number of holes, k_N is the nucleation rate constant, and k_G is the growth rate constant. From their experiments, Love and Lipkowski argued that, at low potential, CO oxidation followed a progressive N&G mechanism, whereas, at high potential, CO oxidation followed an instantaneous N&G mechanism.

In order to assess the role of the platinum surface structure and of CO surface mobility on the oxidation kinetics of adsorbed CO, we carried out chronoamperometry experiments on a series of stepped platinum electrodes of $[n(111) \times (110)]$ orientation [Lebedeva et al., 2002c]. If the (110) steps act as active sites for CO oxidation because they adsorb OH at a lower potential than the (111) terrace sites, one would expect that for sufficiently wide terraces and sufficiently slow CO diffusion, the chronoamperometric transient would display a Cottrell-like tailing for longer times owing to slow diffusion of CO from the terrace to the active step site. The mathematical treatment supporting this conclusion was given in Koper et al. [2002].

Figure 6.2a shows chronoamperometric transients for CO oxidation recorded on three different stepped electrodes for the same final potential. Clearly, the electrode with the higher step density is more active, as it oxidizes the CO adlayer in a shorter period of time. Figure 6.2b shows a fit of a transient obtained on a Pt(15, 15, 14) electrode (terrace 30 atoms wide) by both the mean field model [(6.5), solid line] and the N&G model [(6.6), dashed line]. The mean field model gives a slightly better fit. More importantly, the mean field model gives a good fit of all transients on all electrodes,

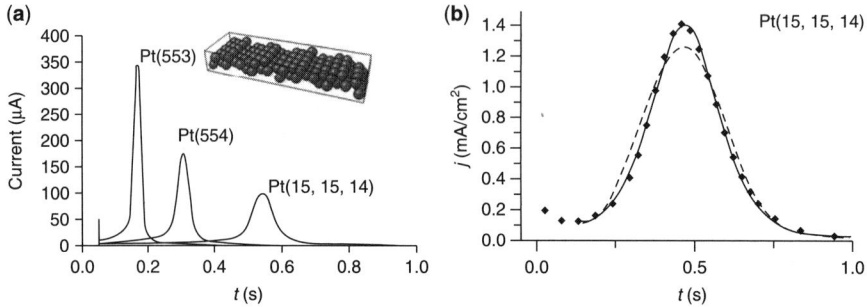

Figure 6.2 (a) Stripping chronoamperometry current transients for the oxidation of a saturated CO monolayer on Pt(553), Pt(554), and Pt(15, 15, 14) in 0.5 M H_2SO_4. The potential is 0.855 V with respect to RHE. (b) Fit of a current transient obtained at Pt(15, 15, 14) for $E = 0.88$ V. The diamonds are experimental data points, the solid line is the best fit by the mean field model, and the dashed line is the best fit by the N&G model.

and no tailing in the transient is observed on the electrodes with wide terraces, such as Pt(15, 15, 14) or Pt(111). Both observations strongly suggest that the mobility of CO on the (111) terraces must be considered high.

From the fit of the experimental transient, the rate constant $k_{hkl}(E)$ can be obtained for the various electrode surfaces at a series of electrode potentials. Figures 6.3 and 6.4 plot the rate constant as a function of the step density at various potentials, and the logarithm of the rate constant as a function of potential for the different surfaces (Tafel plot). The inset in Fig. 6.3 plots the "intrinsic" rate constant, i.e., the measured rate constant divided by the step density, and, together with the main plot in Fig. 6.3, these data demonstrate that the measured rate is proportional to the step density. Therefore, the reaction takes place exclusively at the steps, and, in the potential range studied, the Pt(111) terraces must be considered inactive for CO oxidation. The Tafel slopes obtained from Fig. 6.4 are all very close to 70–80 mV/dec, in good agreement with earlier measurements of the Tafel slope for CO oxidation [Love and Lipkowski, 1988; Palaikis et al., 1988; Bergelin et al., 1999]. We interpret this Tafel slope as an "EC" mechanism, in which an electron transfer reaction is in equilibrium, followed by an essentially potential-independent chemical step:

$$H_2O + * \longleftrightarrow OH_{ads} + H^+ + e^- \quad \text{equilibrium} \quad (6.8)$$

$$CO_{ads} + OH_{ads} \longrightarrow COOH_{ads} \quad \text{rate-determining step} \quad (6.9)$$

$$COOH_{ads} \longrightarrow CO_2 + H^+ + e^- + 2^* \quad (6.10)$$

Shubina and co-workers calculated the activation energy for the reaction between CO and OH on a Pt(111) surface in the absence of water, and obtained a value of about 0.6 eV [Shubina et al., 2004]. Janik and Neurock [2007] calculated the barrier for this reaction on Pt(111) in the presence of water and as a function of the surface charge of the Pt(111) electrode. They found a value of 0.50 eV in the absence of a surface

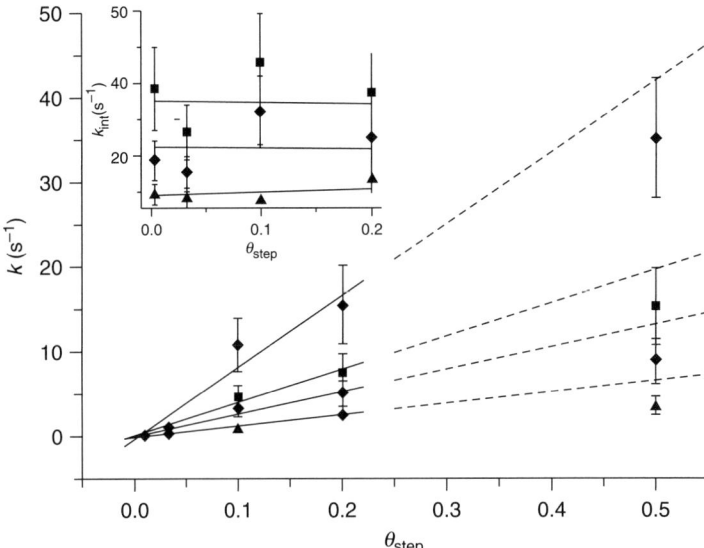

Figure 6.3 Dependence of the apparent rate constants, determined by fitting the experimental transients with (6.5), on the step fraction. The final potentials are 0.73 V (triangles), 0.755 V (diamonds), 0.78 V (squares), and 0.805 V (circles). The value of the step fraction for Pt(111) was estimated using a procedure described in [Lebedeva et al., 2002c]. The inset shows the independence of the apparent "intrinsic" rate constant per step.

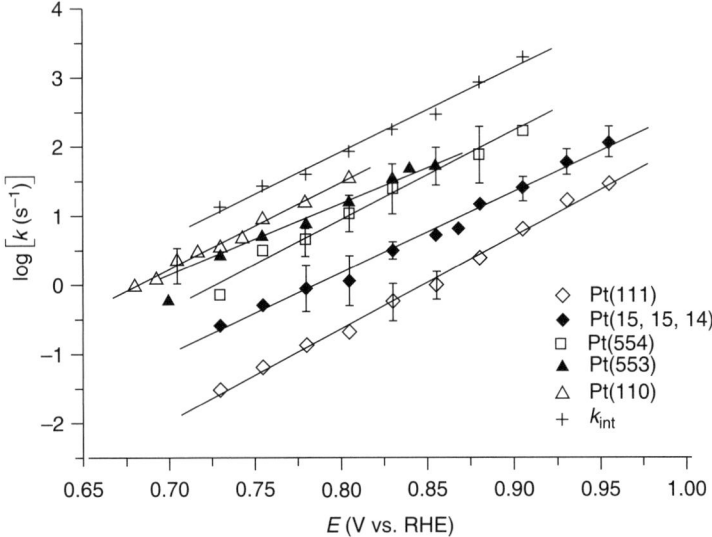

Figure 6.4 Dependence of the apparent rate constants and the apparent intrinsic rate constant on the potential ("Tafel plots"), determined by fitting the experimental transients with (6.5).

charge, decreasing to 0.30 eV at a surface charge corresponding to a potential about 1.5 V more positive than the potential at which the surface is uncharged. They also found that the reaction of CO with O in the presence of water has a much higher barrier than the reaction of CO with OH.

The above experiments show convincingly that the active sites for CO oxidation are the step sites and that the mobility of adsorbed CO on Pt(111) terraces should be considered high. The idea of a highly mobile CO chemisorbate is always inconsistent with a N&G model, because in the N&G model the assumption of immobile reaction partners is essential. This does not prove that the mean field model is quantitatively correct, although clearly the fit is very good, but merely that on the electrodes that we have studied, there is no physical basis for the key assumption in the N&G model, namely the supposed surface immobility of adsorbed CO. Interestingly, in later work on stepped rhodium single-crystal electrodes [Housmans et al., 2004; Housmans and Koper, 2005a, b], experiments do suggest a low CO mobility. The chronoamperometric transients for rhodium are very different from those obtained for platinum, and a consistent kinetic modeling is quite a bit more involved, as will be discussed in Section 6.2.2.

Close inspection of the experimental transients shows that the initial part of the transient cannot be modeled by a mean field or N&G model (see, e.g., Fig. 6.2b). After the double-layer charging has died out, the current shows an almost constant value for a certain period of time, before it begins to rise as predicted by the mean field and N&G models. We have ascribed this constant current (the "plateau") at the beginning of the transient to an incipient oxidation of the CO adlayer during which the CO adlayer relaxes such that no new free sites are created for OH adsorption [Bergelin et al., 1999; Lebedeva et al., 2002a, c]. This would explain the constant current and the implied zeroth-order reaction kinetics in θ_{CO}. Once the CO coverage is below a certain critical coverage, which we expect to be around 2–3% below the saturation coverage, the second-order LH mechanism sets in (6.2). The fact that the potential dependence of this "plateau" process, as discerned from a Tafel slope of 70–80 mV/dec for the various stepped electrodes [Lebedeva et al., 2002c], is similar to that of the potential dependence of the process associated with the chronoamperometric peak suggests that the mechanism is similar but that the reaction order is different. The current in the plateau was also found to depend on step density. A large increase in plateau current is observed when switching from Pt(111) to Pt(15, 15, 14), while further increases in the step density lead to much less significant current enhancements. Various authors have suggested that this CO oxidation initiation, which is also seen in voltammetry as a prewave (see Section 6.2.1.2), may be ascribed to a kind of Eley–Rideal mechanism in which the adsorbed CO is attacked directly by a water molecule from the double layer, and not by adsorbed OH. There are at least two reasons why we believe this interpretation is less likely. First, for such a reaction, a different potential dependence would be expected, and one would expect it to be less sensitive to the step density. Second, DFT calculations have shown that the activation energy for such a reaction would be unrealistically high [Dunietz et al., 2002].

6.2.1.2 CO Stripping Voltammetry

As discussed in the previous section, the chronoamperometric transients can be modeled using the LH mechanism. Using the

Reactions (6.8)–(6.10) corresponding to the proposed EC mechanism, and not taking into account the initiation step (which involves a small fraction of the charge), it is possible to obtain an analytical expression for the voltammetric CO stripping peak [Herrero et al., 2004]:

$$j = Q \frac{k_{app} \exp\left(\frac{FE}{RT}\right) \exp\left[-\frac{k_{app}RT}{Fv} \exp\left(\frac{FE}{RT}\right) + b\right]}{\left\{\exp\left[-\frac{k_{app}RT}{Fv} \exp\left(\frac{FE}{RT}\right) + b\right] + 1\right\}^2} \quad (6.11)$$

where Q is the charge associated to the oxidation of the adsorbed CO, k_{app} is the apparent rate constant for the process measured at 0 V (RHE), v is the scan rate, and b is a parameter that groups all the initial values for the process (essentially, the initial potential for the voltammetric scan, E_{ini}, and the initial coverage value, θ_{in}). The voltammetric peaks for CO oxidation on Pt(111) and Pt(100) were fitted using (6.11). A very good fitting was obtained for both electrodes, validating the approach used. It should be mentioned that the peak width depends strongly on the mechanism; the good fit obtained indicates that the proposed reaction mechanism is a good approximation to the process (Fig. 6.5). This approach was used to calculate the value of k_{app} for different electrodes under different conditions.

The variation of k_{app} with the temperature was used to determine the apparent activation energy of the process at 0 V for both electrodes. The values of the activation

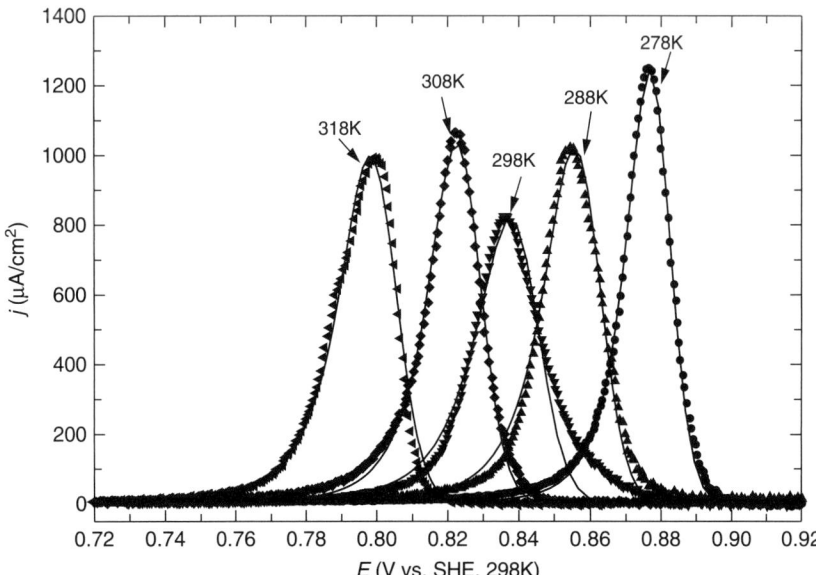

Figure 6.5 CO_{ads} voltammetric stripping peak (symbols) and the fits obtained from (6.11) (lines) for the Pt(111) electrode in 0.5 M H_2SO_4 at different temperatures. The scan rate was 50 mV/s.

energy obtained are 106 and 125 kJ/mol for the Pt(111) and Pt(100) electrodes, respectively, in 0.1 M $HClO_4$ [Herrero et al., 2000, 2004]. In order to compare these values with those obtained from DFT calculations [Anderson and Albu, 2000], it is necessary to take into account that the DFT values have been calculated for the elementary Reaction (6.9), whereas the apparent activation energy measured with voltammetry is a combination of the activation energies of the reversible Reaction (6.8) in both directions and the Reaction (6.9). When the values of the activation energies of the reversible Reaction (6.8) are taken into account, a good agreement is obtained between the values (approximately 100 kJ/mol) [Herrero et al., 2004].

The role of the anions of the solution in the oxidation process was also analyzed. Although the anions are not explicitly considered in the EC mechanism, they play an important role, since they compete with OH for the adsorption sites. Therefore, the presence of strongly adsorbed anions hinders CO oxidation. Thus, the stripping peak is displaced towards more positive potentials as the adsorption strength of the anion increases. The stripping peak for CO oxidation on the Pt(111) electrode in 0.5 M H_2SO_4 solution occurs at potentials about 120–150 mV more positive than in 0.1 M $HClO_4$. In the same way, the activation energy is 25 kJ/mol higher. The differences for the Pt(100) electrode are smaller, since (bi)sulfate adsorption on this electrode is weaker than on the Pt(111) electrode [Herrero et al., 2004].

The potential of the stripping peak, and hence the activity of the electrode for CO oxidation, also depends on the platinum surface structure in general and on the step density in particular. Based on the chronoamperometry experiments described in Section 6.2.1.1, one would expect the stripping peak to shift to lower potential with increasing step density. That this is indeed the case is shown in Fig. 6.6. Again, this

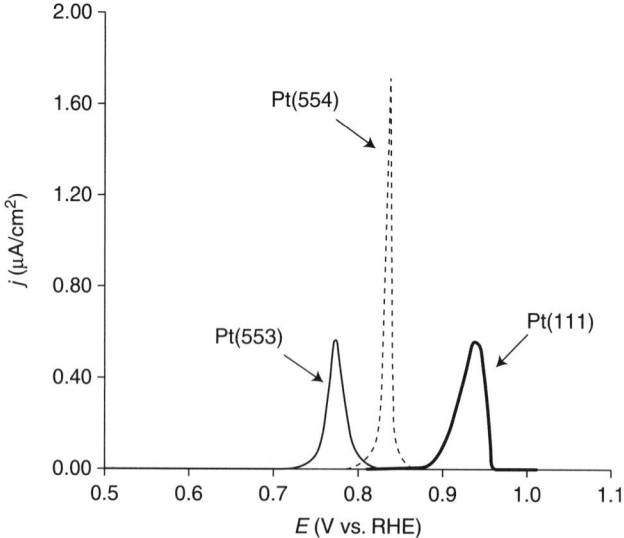

Figure 6.6 Stripping voltammetry of saturated CO adlayers on Pt(553) (thin full line), Pt(554) (dashed line) and Pt(111) (thick full line); sweep rate 50 mV/s.

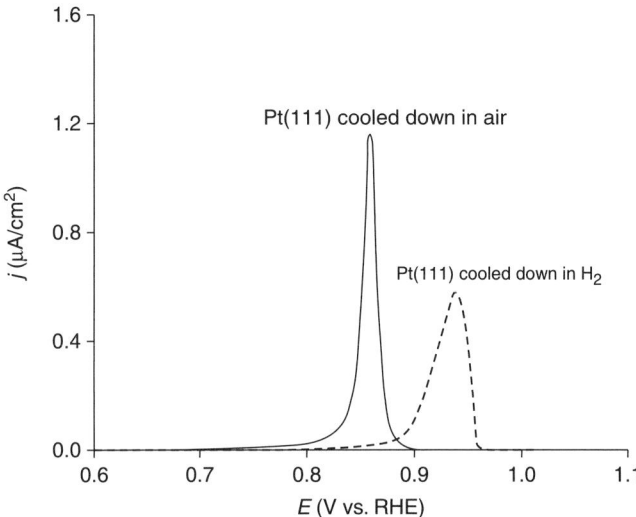

Figure 6.7 Stripping voltammetry of saturated CO adlayers on Pt(111) cooled in a hydrogen–argon atmosphere (dashed line), leading to a well-defined smooth surface, and on Pt(111) cooled in air (full line), leading to a less well-defined defect-rich surface; sweep rate 50 mV/s.

effect is easily explained by the steps being the active sites for CO oxidation, and a rapid mobility of the adsorbed CO on the terraces. In situ Fourier transform infrared (FTIR) spectroscopy showed that the terrace-bonded CO disappears first from the infrared spectrum, and the more strongly adsorbed step-bonded CO is oxidized last [Lebedeva et al., 2002b]. This also suggests that CO is being supplied from the terrace to the step during the oxidation. The high sensitivity to steps and defects was also demonstrated by an experiment in which the Pt(111) electrode was cooled down during the preparation procedure in either air or a hydrogen–argon atmosphere. The former leads to a less perfect Pt(111) surface, as evidenced by very small but significant changes in blank voltammetry. However, the CO stripping voltammetry was strongly different between the two surfaces, as shown in Fig. 6.7, with the air-cooled less perfect Pt(111) being the more active surface.

Under certain circumstances, the CO stripping profile on single-crystal platinum exhibits more than one peak. Lebedeva and co-workers showed that multiple stripping peaks may be observed on Pt(111) and stepped Pt surfaces if the initial CO coverage is below saturation (which is 0.68 monolayer (ML) on Pt(111) if no CO is present in solution) [Lebedeva et al., 2000a, b]. The origin of these multiple peaks has remained somewhat elusive, as they cannot be explained by the simple mean field models discussed above. A special case that has been studied in some detail is the so-called "pre-peak," "pre-wave," or "pre-ignition" peak that is often observed during CO stripping from a Pt(111) electrode (as well as other Pt surfaces) [Markovic and Ross, 2002]. López-Cudero and co-workers showed that a pre-wave is typically observed before the main oxidation peak if the CO is dosed onto the Pt(111) electrode while maintaining the potential during dosing within the hydrogen adsorption region (say 0.1 V vs. RHE)

[López-Cudero et al., 2005]. This was related to the observation that the highest CO coverage of 0.68 ML is obtained only if the dosing potential is below 0.25 V, i.e., in the hydrogen adsorption region. The pre-peak probably corresponds to the oxidation of CO at certain defect sites, and seems to be related to the initial plateau current observed in chronoamperometric transients (see the last paragraph of Section 6.2.1.1). Stamenkovic and co-workers have observed that in the pre-ignition region there is a blue shift (lowering) of the C—O stretching frequency corresponding to the linearly bonded CO, which they ascribe to bisulfate adsorbing onto the Pt sites that were previously occupied by CO, leading to a mild compression of the CO islands and an associated enhanced dipole–dipole coupling [Stamenkovic et al., 2005].

6.2.1.3 Continuous CO Oxidation on Platinum

The main difference between CO stripping and continuous CO oxidation is the CO (re-)adsorption Reaction (6.3). In contrast to CO stripping, this leads a steady-state CO oxidation current because of the continuous supply of CO. In modeling the continuous CO oxidation, we also need to consider the mass transport of CO from the bulk of the solution to the electrode surface. The temporal change in the CO coverage is now given by

$$\frac{d\theta_{CO}}{dt} = k_{ads}c_{CO,s}(1 - \theta_{CO}) - k(E)\theta_{CO}(1 - \theta_{CO}) \tag{6.12}$$

where $c_{CO,s}$ is the CO concentration in solution and k_{ads} is the CO adsorption rate constant. This equation must be coupled to an equation for the temporal evolution of the surface concentration of CO in which the diffusion/mass transport is taken into account. The equations for modeling this situation may be found in the original reference [Koper et al., 2001]. The important point here is that, in the case that CO adsorption is stronger than OH adsorption over a wide potential range, such that it leads to self-poisoning as is observed experimentally, the model predicts a very peculiar current–potential polarization curve.

The thick line in Fig. 6.8 shows the steady-state polarization curve predicted by this model. In contrast to the usual sigmoidal shape of a steady-state polarization curve for a first-order heterogeneous electrode reaction coupled to diffusion, the curve is S-shaped owing to the nonlinear autocatalytic nature of the CO oxidation reaction, as discussed in Section 6.2.1.1. The thin line in Fig. 6.8 is the predicted cyclic voltammetric curve obtained at a low scan rate. Starting from a low potential, the system follows the low current branch until it reaches $E = E_2$. At this potential, there is a very sharp current spike, which basically corresponds to a fast oxidation of the poisoning CO adlayer that was essentially stable for $E < E_2$, after which the current settles down on the high current branch, which corresponds to a diffusion-limited CO oxidation. After reversing the scan, the system stays on the high current branch until $E = E_1$. At this potential, the current quickly drops to the low current branch. The observed hysteresis between $E = E_1$ and $E = E_2$ cannot be removed by lowering the scan rate. It is intrinsic to the system, owing to the S-shaped nature of the polarization curve.

In general, the onset of CO oxidation occurs at higher potentials than those recorded for the stripping peak in the absence of CO in solution, i.e., E_2 values are

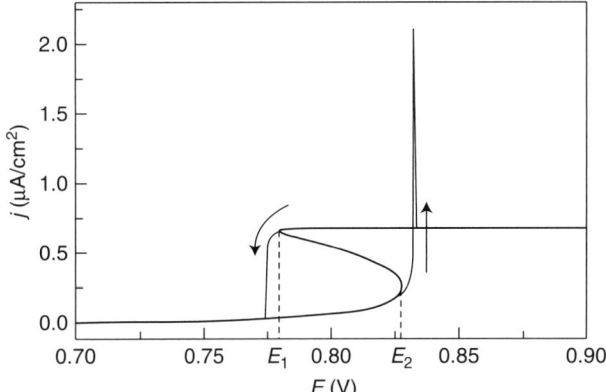

Figure 6.8 S-shaped polarization curve observed in the CO oxidation model (for the exact model parameters, see Koper et al. [2001]). The thin line shows the cyclic voltammetry observed at a low scan rate of 2 mV/s.

always higher than the values for the CO stripping peak. This effect is due to the quenching or self-poisoning effect of the CO incoming molecules in the autocatalytic process of CO oxidation; i.e., when the oxidation rate is low, the new free sites that are created in the initial stages of the oxidation are immediately covered by CO molecules from solution, thereby preventing a fast oxidation of the CO molecules. In order to achieve a sustained oxidation of the CO molecules, a higher CO oxidation rate constant is required, which necessarily implies higher potentials. An additional consequence of this is the dependence of the potential E_2 on the mass transport rate, as observed in Fig. 6.9 for different rotation rates of the electrode.

Figure 6.9 shows the experimental voltammetry on a Pt(110) electrode in $HClO_4$ in a rotating disk configuration. The experimental curves in Fig. 6.9 display all the qualitative characteristics of the model, and the influence of the scan rate and the disk rotation rate are well reproduced by the model. In the negative scan and at fast sweep rates, negative currents have been measured for $E < E_1$ (see, e.g., Fig. 6.9 at 200 mV/s). This negative current is associated with anion displacement from the surface. At positive potentials, where the stable oxidation current is obtained, the CO coverage is very low, and anions are adsorbed on the free electrode surface. When the potential is scanned in the negative direction, the CO coverage increases abruptly at $E = E_1$, where the CO oxidation rate becomes lower than the supply of the CO molecules transported to the surface. The adsorption of the CO on the surface displaces the anion from the surface, giving rise to the observed negative current.

In order to prove the S-shaped character of the polarization curve, the system was studied galvanostatically. The model predicts that the "sandwiched" branch of the polarization curve should be stable, and therefore measurable under galvanostatic conditions. Figure 6.10 shows the results of the experiment: depending on the scan rate, an S-shaped curve can be observed in the back scan, i.e., from high to low current. At low

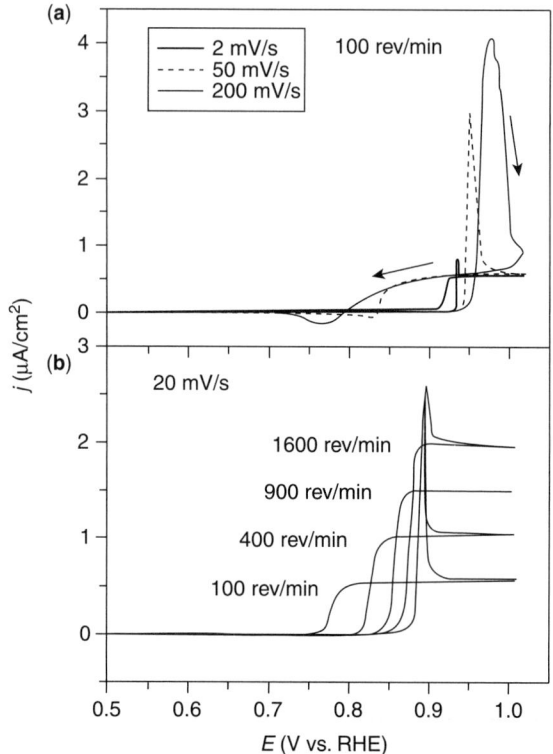

Figure 6.9 Cyclic voltammetry of a Pt(110) rotating disk electrode in a CO-saturated 0.1 M $HClO_4$ solution. (a) Influence of the voltage scan rate. (b) Influence of the disk rotation rate.

scan rates, however, the curve is not clearly S-shaped, and irregular potential oscillations are observed. These potential oscillations are most likely due to a spatial instability, which causes the electrode to exhibit patches of low potential and patches of high potential simultaneously. The instability would not have the time to fully manifest when the current is scanned rapidly, explaining why the S-shaped curve is observed under these conditions. The possibility of such a spatiotemporal instability in an electrochemical system with an S-shaped polarization curve was proved mathematically by Mazouz and Krischer [2000]. The same group has also studied this system experimentally (albeit using polycrystalline Pt), and indeed observed spatiotemporal phenomena that suggest that the system should not be considered as homogeneous [Bonnefont et al., 2003].

Similar studies have been carried out with Pt(111) and stepped surfaces with (111) terraces [Angelucci et al., 2007a, b]. The voltammetric profiles of these surfaces agree qualitatively with those depicted in Fig. 6.9. For the stepped surfaces, the potentials E_1 and E_2 depend linearly on the step density for terraces wider than 5 atoms. This linear dependence is a consequence of the dependence of the oxidation rate on the step density, as was observed in the chronoamperometric CO stripping experiments. In H_2SO_4

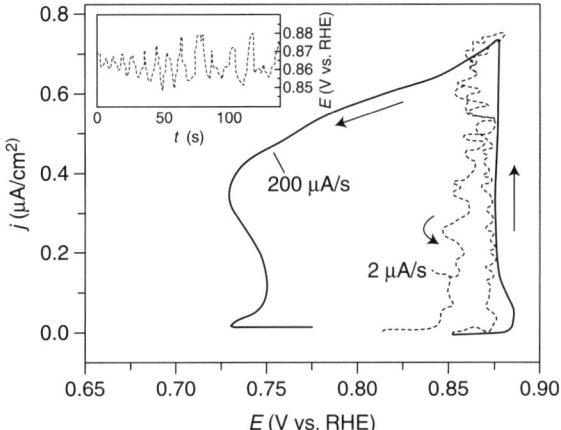

Figure 6.10 Galvanostatic scans of a Pt(110) rotating disk electrode in a CO-saturated 0.1 M HClO$_4$ solution at two different current scan rates (disk rotation rate 400 rev/min). The insert shows the potential fluctuations observed at an applied current density of 0.74 mA/cm^2 (disk rotation rate 900 rev/min).

solutions, Pt(111) deviates from the expected values extrapolated from the observed behavior of the stepped surfaces owing to the formation of an ordered bisulfate layer on the Pt(111) electrode.

6.2.2 Carbon Monoxide Oxidation on Rhodium

Given the results obtained on platinum electrodes discussed in some detail in the previous section, it is clearly of fundamental interest to study the mechanism of CO oxidation on other transition metal electrodes, and to compare the results with platinum. Rhodium has been the electrode material that has been studied in greatest detail after platinum, and results obtained with rhodium have provided some very significant insights into some of the general issues about the CO oxidation mechanism.

From earlier work on the CO electro-oxidation on single-crystal Rh, it is known that the reaction is much slower than on Pt, and that the stripping of a CO adlayer may take several cyclic voltammetric scans to completely oxidize CO$_{ads}$, from about 30 on Rh(111) to 1–2 on Rh(110) [Gomez et al., 1997; Housmans and Koper, 2004]. Housmans and Koper studied this reaction in H$_2$SO$_4$ by chronoamperometry [Housmans and Koper, 2005a]. Figure 6.11 shows a series of transients obtained on different stepped Rh electrodes, and it is clear that the results are very different from those obtained with Pt. Instead of one clear oxidation peak as on Pt, CO oxidation on single-crystal Rh exhibits two peaks: a pre-peak at short times (1–4 s in Fig. 6.11b) and a main peak at longer times (5–50 s in Fig. 6.11a). The charge associated with the pre-peak increases with step density, but the time of the pre-peak is not very sensitive to the step density. On the other hand, the main peak is very structure-sensitive: it rapidly shifts to shorter times with increasing step density, although it is virtually

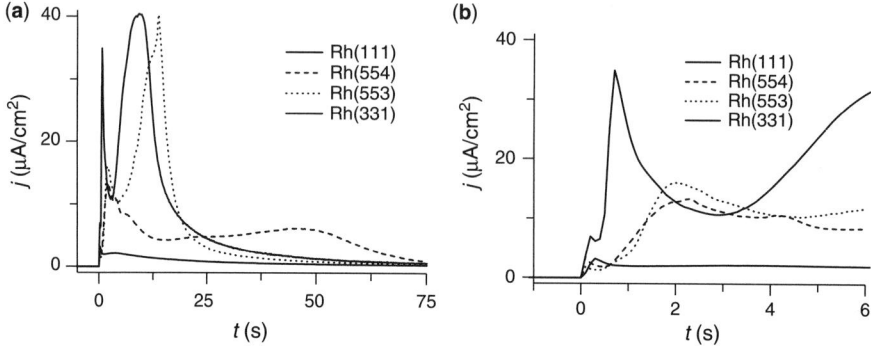

Figure 6.11 Stripping chronoamperometric current transients on stepped Rh single-crystal surfaces. (a) Full transients obtained on the four different surfaces at $E = 0.65$ V (with respect to an RHE), in 0.5 M H_2SO_4. (b) Initial parts of the transients in (a).

absent on Rh(111). Our interpretation of these transients is that the pre-peak corresponds to the oxidation of CO at or near the step sites, and that the main peak corresponds to CO oxidation on terraces. The fact that these processes can be observed separately suggests a low mobility of CO on the Rh(111) terraces. Kinetic Monte Carlo simulations of a model incorporating these different reactions indeed show that in the limit of low CO mobility, two separate peaks can be observed in the chronoamperometry, which behave qualitatively similar to the experimental transients [Housmans et al., 2007].

In a later communication [Housmans and Koper, 2005b], it was shown that the low CO mobility on Rh(111) is not an intrinsic feature of this surface, but is strongly influenced by the co-adsorbing anion. Whereas the experiments above were carried out in H_2SO_4, the same experiments in $HClO_4$ showed a very different behavior. CO oxidation on Rh in $HClO_4$ looks qualitatively similar to CO oxidation on Pt in H_2SO_4 or $HClO_4$. The chronoamperometry shows only one peak in $HClO_4$, in contrast to the two features observed in H_2SO_4. The difference is most clearly illustrated by comparing the CO stripping voltammetry on a series of Rh stepped single crystals

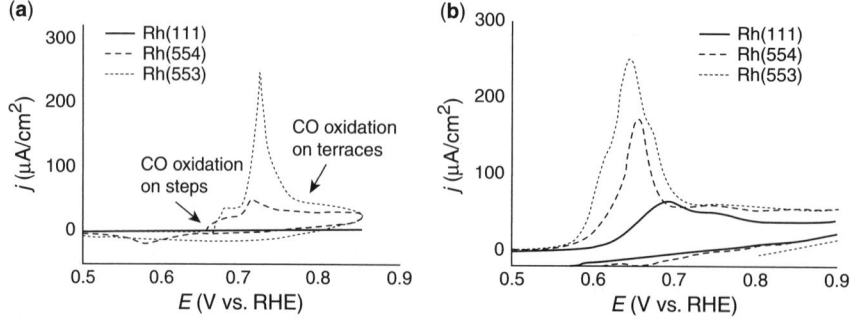

Figure 6.12 Stripping voltammetry of a saturated monolayer of adsorbed CO on Rh(111), Rh(554), and Rh(553) in (a) 0.5 M H_2SO_4 and (b) 0.1 M $HClO_4$; scan rate 20 mV/s.

in H_2SO_4 with that in $HClO_4$ (Fig. 6.12). In Fig. 6.12a, the stripping voltammetry in H_2SO_4 solution again shows two features: a pre-peak corresponding to CO oxidation at steps, and a main peak corresponding to CO oxidation at terraces. The potentials of these peaks do not shift with step density, which suggests that these different parts of the surface do not communicate and that therefore CO mobility is low. On the other hand, in $HClO_4$ (Fig. 6.12b), only a single main CO stripping peak is observed, the peak potential of which shifts to lower values with increasing step density, very similar to the situation on Pt (see Fig. 6.6). This latter observation can *only* be explained if we allow for a finite CO mobility on the terraces. The conclusion must be that the strong adsorption of the sulfate anion on Rh hampers CO surface mobility. Sulfate is less strongly adsorbed on Pt than on Rh, and therefore on Pt the influence of the anion on CO mobility is not noticeable. As a generic conclusion, the effective surface mobility of any species in an electrocatalytic reaction should always be considered in terms of the co-adsorption of other species, especially "spectator" species such as anions.

6.2.3 Carbon Monoxide Oxidation on Gold

In the catalysis community, there is considerable interest in the catalytic properties of oxide-supported nanocrystalline gold, which has been found to be remarkably active for the oxidation of CO [Haruta, 1997]. In electrochemistry, the ability of gold to oxidize CO, in the absence of an oxide support, has been known for many years [Roberts and Sawyer, 1964].

Weaver and co-workers have carried out extensive studies of CO electro-oxidation on Au single crystals [Chang et al., 1991; Edens et al., 1996]. Continuous oxidation of CO on Au starts at potentials where the formation of surface oxides or surface-bonded hydroxyl (OH) is not apparent from voltammetry. Weaver suggested the following mechanism:

$$CO + ^* \longrightarrow CO_{ads} \tag{6.13}$$

$$CO_{ads} + H_2O \longrightarrow COOH_{ads} + H^+ + e^- \tag{6.14}$$

$$COOH_{ads} \longrightarrow CO_2 + H^+ + e^- + ^* \tag{6.15}$$

CO adsorption is considered reversible, as the CO binding to Au is relatively weak (in the absence of CO in solution, CO does not bind to Au). The observed Tafel slope (120 mV/dec) is in agreement with Reaction (6.14) as the rate-determining step, but the pH dependence (60 mV/pH unit) is not, as the rate of the irreversible Reaction (6.14) should formally not depend on pH. Weaver and co-workers also reported CO electro-oxidation on gold to be structure-sensitive, with Au(110) being the most and Au(111) the least active [Edens et al., 1996]. This structure sensitivity does not reflect the CO adsorption capacity, however, since Au(210) was found to have a higher CO surface concentration than Au(110), but a lower oxidation capacity [Chang et al., 1991; Edens et al., 1996].

Markovic and co-workers [Blizanac et al., 2004a, b] have suggested that a small voltammetric feature near 0.5 V (vs. RHE) observed on clean Au(100) in $HClO_4$

may be related to the formation of "activated water" or surface-bonded hydroxyl that can oxidize CO. Note that this would correspond to a potential some 0.8 V negative of Au surface oxidation. Such small features are generally observed on Au single-crystal electrodes [Hamelin, 1996; Hamelin and Martins, 1996], although they have never been unequivocally been identified with OH_{ads} formation. On Au(100), surface reconstruction is known to occur in this potential region [Kolb, 1996], but whether it is associated with concomitant OH adsorption remains an open question. Using surface-enhanced Raman spectroscopy (SERS), Li and Gewirth [2003] observed a spectral feature at 790 cm^{-1} at potentials above 0.7 V (vs. RHE), which they ascribed to a bending vibration of surface-bonded OH, lending credibility to the claim by Markovic and co-workers that some level of OH formation may occur well below the Au surface oxidation potential (1.2–1.3 V).

DFT calculations showed that a reaction between surface-bonded CO and OH on Au(110) has a low activation barrier (approximately 0.2 eV) whereas the same reaction on Pt(111) has a much higher barrier [Shubina et al., 2004]. Both on Au and on Pt, the resulting COOH is relatively strongly bonded. This is evidence in favor of Weaver's model for CO oxidation on Au, in which adsorbed CO reacts directly with nearby water to form adsorbed COOH.

6.2.4 Carbon Monoxide Oxidation in Alkaline Solution

Alkaline solutions are generally known to lead to better catalytic activities than acidic solutions for many relevant electrode reactions. However, owing to the paucity in the development of suitable electrolyte materials, such as alkaline membranes, there has been much less fundamental work in the area of fuel cell catalysis in alkaline media. Nevertheless, there are a few hopeful developments in new alkaline polymer membranes [Varcoe and Slade, 2005] that are currently stirring up interest in studying fuel cell catalytic reactions in alkaline solution.

Spendelow, Wieckowski, and co-workers have published results on the mechanism of CO oxidation on a Pt(111) electrode in 0.1 M KOH [Spendelow et al., 2004, 2006]. In 0.1 M NaOH, the CO stripping potential lies about 100 mV lower than in 0.1 M H_2SO_4, on the reversible hydrogen potential scale. On slightly disordered Pt(111), a new CO oxidation peak appears between 0.4 and 0.6 V, whereas the one corresponding to well-ordered Pt(111) remains at about 0.8 V but loses intensity. The low potential oxidation peak is ascribed to small Pt islands, the edges of which are supposed to adsorb OH at low potential. Spendelow, Wieckowski, and co-workers attribute the higher CO electro-oxidation activity in alkaline media to the higher affinity of the Pt(111) surface, and especially the defects, for the adsorption of OH. This should be more than a trivial pH effect, as that has been taken into account by referring all the potentials to the RHE scale. They also conclude that CO oxidation in alkaline media follows the standard Gilman–Langmuir–Hinshelwood mechanism, in both potential regions.

More recently, García and Koper studied CO adlayer stripping on a series on stepped Pt electrodes in alkaline media [García and Koper, 2008]. They found that, in contrast to acidic media, CO mobility on the (111) terraces is very low, a conclusion based on the fact that the amount of CO oxidizing on the terrace depends on the scan rate following a diffusion law. The exact reason for this low surface mobility is not fully clear, but may be related to the strong adsorption of the final CO oxidation product in alkaline media, i.e., carbonate.

Also, under continuous CO oxidation conditions, alkaline media exhibit a much higher activity than acidic media. Markovic and co-workers observed a shift of about 150 mV of the main oxidation wave, and a pre-wave corresponding to CO oxidation at potentials as low as 0.2–0.3 V [Markovic et al., 2002]. Remarkably, the hysteresis that is so prominently observed in the diffusion-controlled CO oxidation wave in acidic media (see Fig. 6.9), is no longer present in alkaline media. Markovic and co-workers also attribute the high activity of alkaline media to a "pH-dependent" adsorption of OH_{ads} at defect/step sites.

6.3 FORMIC ACID OXIDATION

6.3.1 Formic Acid Oxidation on Platinum

Within the general mechanism for the oxidation of C_1 molecules, proposed by Bagotzky, formic acid is one of the simplest cases, since it requires only the transfer of two electrons for the complete oxidation to CO_2 [Bagotzky et al., 1977]. In fact, it has the same oxidation valency as CO; both require two electrons for complete oxidation to CO_2. When compared with CO, the reaction mechanism of formic acid is more complex although the catalysis of the oxidation reaction is much easier. In fact, formic acid can be readily oxidized at potentials as low as 0.2 V (vs. RHE). Its reaction mechanism takes place according to the well-established dual path mechanism [Capon and Parsons, 1973a, b]:

$$\text{HCOOH} \begin{array}{c} \nearrow \text{active intermediate} \xrightarrow{E_1} CO_2 \\ \\ \searrow \text{poisoning intermediate} \xrightarrow{E_2} CO_2 \end{array} \quad E_2 > E_1 \quad (6.16)$$

Earlier results on the oxidation of formic acid on Pt electrodes have been extensively reviewed [Parsons and VanderNoot 1988; Jarvi and Stuve, 1998; Sun, 1998; Vielstich, 2003; Feliu and Herrero, 2003]. Here, we will summarize previous results, but will focus on the most recent results.

The voltammetric profile for formic acid on Pt shows a clear hysteresis (Fig. 6.13). In the general case, currents in the positive-going scan are lower than in the negative-going one. Additionally, the hydrogen adsorption states, which appear at low potentials, are clearly blocked, indicating that there is a species adsorbing on the electrode surface, i.e., the poisoning intermediate. The hysteresis of the voltammetric profile is a consequence of the accumulation of the poisoning intermediate at low potentials, as

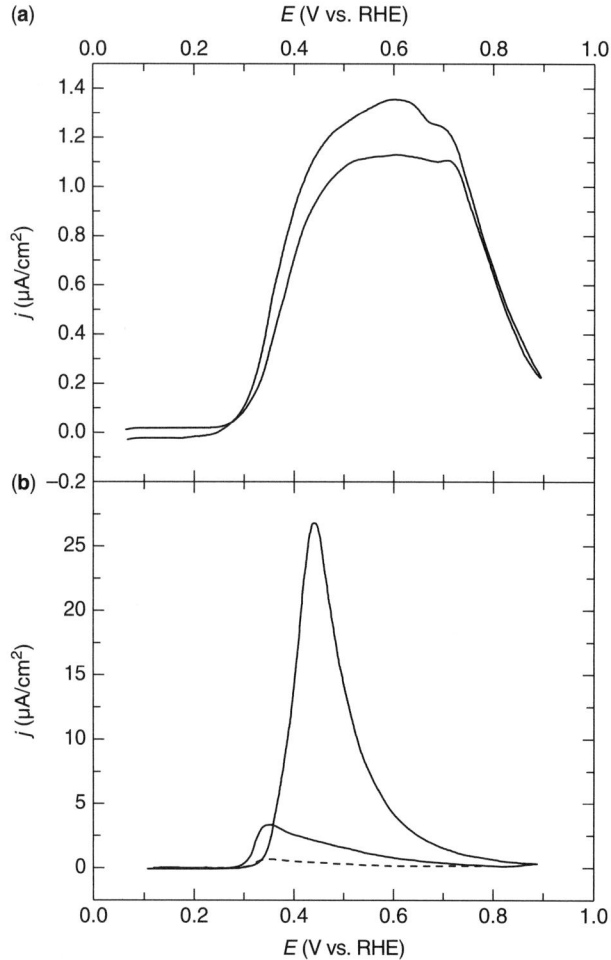

Figure 6.13 Voltammetric profile for (a) Pt(111) and (b) Pt(100) electrodes in 0.1 M HCOOH + 0.5 M H_2SO_4 on electrode. The full lines show the first cycle and the dashed line the second cycle. The scan rate was 50 mV/s.

revealed by the blocking of the hydrogen adsorption states, and its oxidation at high potentials. Thus, in the return scan, the platinum surface is completely clean, and therefore the catalytic activity of the surface is much higher than that recorded for the positive scan.

This general scheme for the oxidation reaction is very sensitive to the surface structure. The first studies with single-crystal electrodes revealed that the voltammetric profiles for the three basal planes, Pt(111), Pt(100), and Pt(110), were completely different (Fig. 6.13) [Clavilier et al., 1981; Lamy et al., 1983; Adzic et al., 1982]. The lowest currents are obtained for the Pt(111) electrode, which in turn has a very low poisoning rate, as suggested by the small hysteresis. In fact, the reaction on this electrode is completely dominated by the kinetics of the oxidation process. Experiments carried out with a Pt(111) hanging meniscus rotating disk electrode showed that the voltammogram is independent of the rotation rate (formic acid concentration 0.2 M), which indicated that the reaction is controlled by the electron transfer process [Maciá et al., 2003].

On the other hand, the Pt(100) electrode showed almost no currents in the positive-going scan, a clear indication that the surface is completely blocked by the poisoning intermediate, which is accumulated on the surface at low potentials. Once the poison is oxidized, above 0.7 V (vs. RHE), currents in the negative-going scan are almost one order of magnitude higher than those recorded for Pt(111) [Clavilier et al., 1981]. This indicates that both paths of the reaction mechanism are much faster for the Pt(100) electrode.

For the Pt(110) electrode, there are some contradictory results regarding its catalytic performance compared with Pt(100); some studies indicate that the activity is higher for Pt(110), whereas others suggest the opposite [Chang et al., 1990; Clavilier et al., 1981; Lamy et al., 1983]. The differences are probably associated with different surface states of the Pt(110) electrode. The actual surface structure of the Pt(110) electrode is strongly dependent on the electrode pretreatment. Since formic acid oxidation is a surface-sensitive reaction, different electrocatalytic behavior can be obtained for the same electrode after different treatments.

The studies with Pt single crystals showed some correlation between the measured current, which, obviously, is a measure of the total CO_2 produced, and the poisoning rate; i.e., the surfaces with higher catalytic activity showed higher poisoning rates. It may happen that the poisoning intermediate is the only reactive intermediate, and the reactivity through the active intermediate path would be negligible. However, the significant currents measured for the electrode at low potentials would indicate that the reactivity through the active intermediate is significant. In order to gain insight into the mechanism, we will discuss the two paths separately.

The first step towards disentangling the mechanism is identification of the poisoning intermediate. Several candidates were proposed—CO, COH, and CHO—but the only one identified by electrochemically modulated infrared spectroscopy (EMIRS) and FTIR spectroscopy was CO [Kunimatsu, 1986; Chang et al., 1990; Sun et al., 1988; Beden et al., 1983]. In fact, CO can be considered as the typical poisoning intermediate, since its oxidation to CO_2 on platinum takes place at significant rates only above 0.6 V (see Section 6.2). Furthermore, CO could be detected by IR spectroscopy at potentials as low as 0.2 V, and the total elimination occurs only above 0.7 V. The confirmation that CO was the poisoning intermediate and not an active intermediate was given by differential electrochemical mass spectroscopy (DEMS) [Willsau and Heitbaum, 1986; Wolter et al., 1985]. In these experiments (Fig. 6.14), $H^{13}COOH$ was initially put in contact with a platinum electrode until the maximum amount of CO was formed. Then, the solution was replaced with $H^{12}COOH$ and the potential swept positively. The formed CO_2 was analyzed with a mass spectrometer. In this case, $^{12}CO_2$ was detected prior to the formation of $^{13}CO_2$, which clearly indicates that there is a path going through an active intermediate different from CO, yielding CO_2. In fact, recent FTIR studies indicate that the contribution of the CO path to

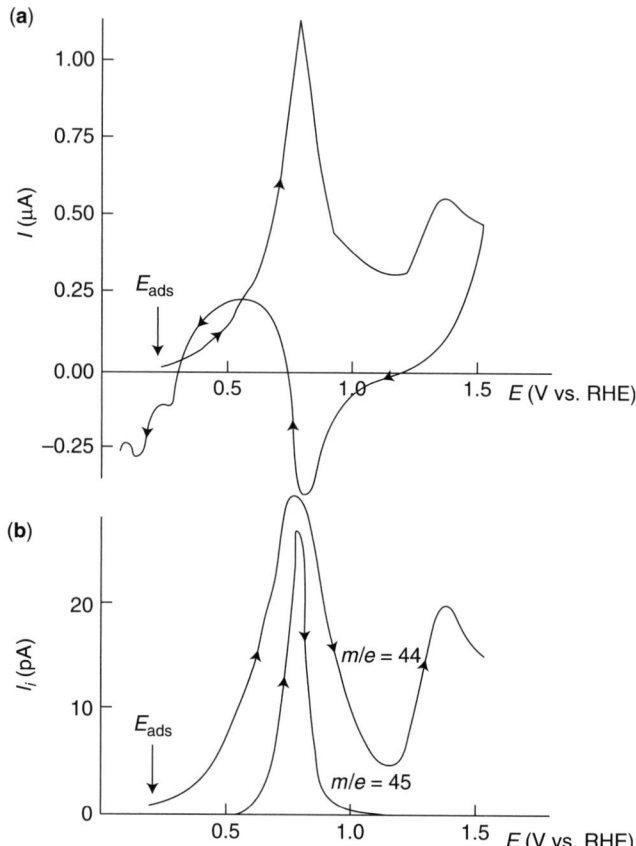

Figure 6.14 Oxidation of 0.01 M H^{12}COOH in 0.5 M H$_2$SO$_4$ on a polycrystalline electrode pre-covered with ^{13}CO. (a) Voltammetric currents. (b) Mass signals for ^{12}CO$_2$ ($m/e = 44$) and ^{13}CO$_2$ ($m/e = 45$). (Reproduced from Willsau and Heitbaum [1986].)

the total current is less that 5% in a wide range of temperatures [Chen et al., 2006b]. Therefore, the reaction through the poisoning intermediate can be written as

$$\text{HCOOH} + {^*} \longrightarrow \text{CO}_{\text{ads}} + \text{H}_2\text{O} \longrightarrow \text{CO}_2 + 2\text{H}^+ + 2e^- + {^*}$$

As can be seen, the first part of the reaction is chemical decomposition of the formic acid to yield CO and water, and the second part is the classical reaction for adsorbed CO, which has been reviewed extensively in the previous section. We shall focus on the first part of the reaction.

As shown in the reaction Scheme (6.16), the reaction that yields the poisoning intermediate is a chemical reaction, and no electrons are implied in the process. Additionally, the reaction takes place at open circuit, i.e., without potential control. This property has been used to measure the total amount of poisoning intermediate that can be accumulated on the surface. The amount and fraction of the surface blocked by CO are dependent on the surface structure [Clavilier and Sun, 1986]. The fraction of the surface blocked by the CO is always less than 1, indicating that there are some steric effects preventing further formation of the poison at high CO coverages. Experiments carried out with highly blocked surfaces have suggested that for the poison formation reaction to occur, there must be more than one adsorption site [Llorca et al., 1994a; Leiva et al., 1997].

When the poisoning reaction is analyzed under potential control, the formation rate is dependent on the electrode potential. The first experiments that clearly showed that the poison formation reaction was potential-dependent were performed by Clavilier using pulsed voltammetry [Clavilier, 1987] (Fig. 6.15). In this technique, a short pulse at high potential is superimposed on a normal voltammetric potential

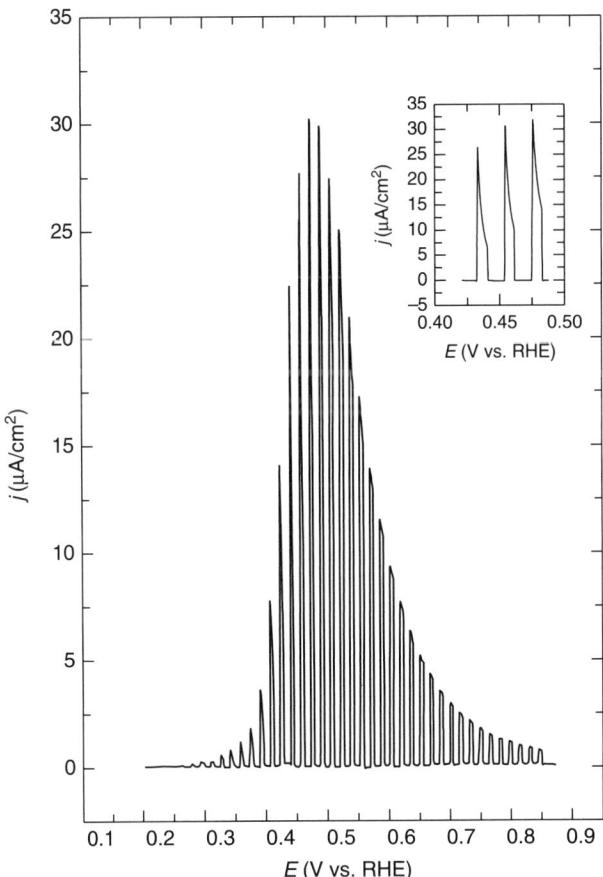

Figure 6.15 Pulse voltammetry of a Pt(100) electrode in 0.5 M H_2SO_4 + 0.25 M HCOOH. The inset is an expansion of the region between 0.4 and 0.5 V to show the decay in the current due to the poisoning of the surface. During the pulse at high potentials (0.9 V) to remove the poison, the current was not recorded (baseline between pulses). (Adapted from Clavilier [1987].)

perturbation. The final potential for the pulse is selected in such a way that the CO is completely oxidized, without perturbing the surface order. These experiments showed that poisoning rate was higher between 0.2 and 0.4 V, and almost negligible for potentials above 0.5 V [Clavilier, 1987; Fernández-Vega et al., 1991]. The poisoning rate measured by chronoamperometry indicated that the maximum rate is obtained at 0.3 V [Sun et al., 1994]. It has been suggested that the position of this maximum is influenced by hydrogen and anion adsorption; strong anion or hydrogen adsorption inhibits the poison formation reaction [Lu et al., 1999]. For that reason, the poisoning rate diminishes at very low or high potentials, where hydrogen and anions, respectively, are strongly adsorbed on the surface. In fact, the position of the maximum is close to the potential of zero total charge of the Pt electrode [Gómez et al., 2000], a potential for which the coverages of both species are low.

As already mentioned, the poison formation reaction is potential-dependent, and the poisoning rate for the basal planes is Pt(110) > Pt(100) > Pt(111) [Sun et al., 1994; Iwasita et al., 1996]. The case of Pt(111) is special, since the poisoning has been associated with the presence of defects on the surface. Selective covering of the defects on the Pt(111) electrode by some adatoms prevents the formation of CO on the electrode surface [Maciá et al., 1999, 2001; Smith et al., 2000].

The simplest way to study the reaction through the active intermediate is to measure the currents in absence of the poisoning intermediate. This effect can be achieved in two different ways: either by eliminating the CO prior to the studies or by suppressing the formation of CO with some surface modification. The first attempts were carried out by Clavilier using pulse voltammetry [Clavilier, 1987]. The current measured just after the pulse is the intrinsic activity of the surface at this potential, i.e., the maximum activity (current density) that can be achieved in absence of poison. These studies revealed that the Pt(100) electrode is very active for formic acid oxidation, its maximum intrinsic activity being around $30\,\text{mA}/\text{cm}^2$ at 0.46 V (vs. RHE) (Fig. 6.15), which means a turnover rate of more than 70 molecules per (100) site per second under these conditions. This can be compared with the nearly zero current obtained in the positive-going scan of the voltammetric experiments for which the surface is completely covered by the poison.

In order to obtain quantitative reaction rates, chronoamperometric experiments are normally used, because their analysis is always simpler. This strategy was used by Sun's group [Sun and Yang, 1999]. Assuming that formic acid oxidation can be described as a process controlled by both charge transfer and mass transport, this group was able to obtain the rate constant of the reaction from the analysis of the current–time transient obtained in potentiostatic experiments, as it would proceed in the absence of poison on the surface. In this way, activation energies and Tafel slopes were obtained. For Pt(100), a Tafel slope of about 120 mV has been obtained. Using the other strategy, namely, preventing CO formation by blocking all the defect sites on a Pt(111) electrode, the same Tafel slope was measured [Maciá et al., 2001]. Such a Tafel slope indicates that the rate-determining step for the formic acid oxidation reaction through the active intermediate is the first electron transfer. Since two electrons are exchanged in the oxidation of formic acid, the step that leads to the formation of the reactive intermediate is the slowest step. Once the active intermediate has

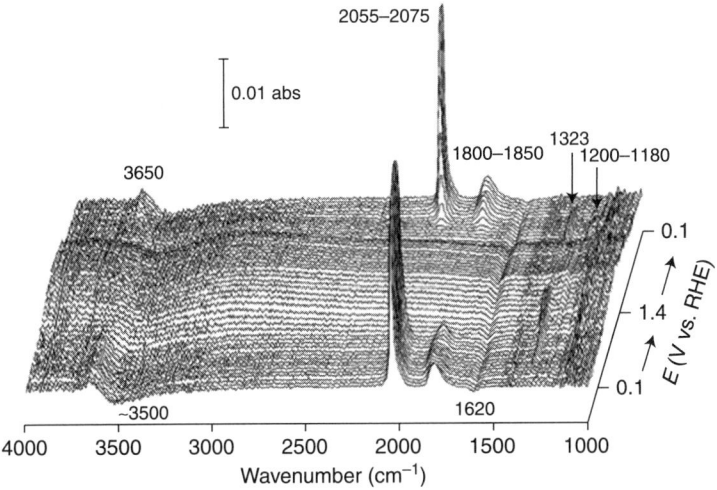

Figure 6.16 Attenuated total reflection surface enhanced infrared reflection absorption spectroscopy (ATR–SEIRAS) spectra for the oxidation of 0.1 M HCOOH in 0.5 M H_2SO_4 on a polycrystalline electrode. The bands at 2055–2075 and 1800–1850 cm^{-1} are assigned to linear- and bridge-bonded CO, whereas the band at 1323 cm^{-1} corresponds to adsorbed formate. (Reproduced from Samjeske et al. [2006].)

formed, the following reactions will proceed much more rapidly. Therefore, the active intermediate will be short-lived and its detection will be difficult. That is the reason why the active intermediate has been so elusive.

The first proposed active intermediate was Pt–COOH, as suggested by EMIRS experiments on Pt(111) and Pt(100) electrodes [Sun et al., 1988]. On the other hand, adsorbed formate has been detected by Osawa and co-workers during formic acid oxidation on a polycrystalline electrode [Samjeske et al., 2005, 2006; Samjeske and Osawa, 2005; Mukouyama et al., 2006] (Fig. 6.16). Therefore, they proposed that this species was the elusive reactive intermediate. However, the issue is not yet settled, because combined FTIR and electrochemical methods under controlled flow rate indicate that there is another active intermediate not detected that accounts at least for the 85% of the total CO_2 formed through the active intermediate [Chen et al., 2006a]. These results are corroborated by isotope effects in the kinetics of the oxidation reaction, which suggest that the rate-limiting step of the process is the excision of the H—C bond [Chen et al., 2007]. This would indicate that the active intermediate cannot be the adsorbed formate.

6.3.2 Formic Acid Oxidation on Other Electrodes

Palladium electrodes are also very active for formic acid oxidation, with higher current densities than platinum electrodes [Capon and Parsons, 1973c]. Oxidation occurs almost exclusively through the active intermediate path, without poison formation. The reaction is also very sensitive to the surface structure, and the activity of the basal planes decreases as follows: Pd(100) > Pd(111) and Pd(110) [Hoshi et al., 2006]. The activity of Pd nanoparticles has also been studied, and it has been shown that there is a small dependence of their catalytic activity on size [Zhou et al., 2006]. However, the most active electrode material for formic acid oxidation is 1–2 layers of Pd deposited on a Pt(111) or PtRu(111) electrode [Llorca et al., 1994b; Baldauf and Kolb, 1996; Kibler et al., 2005]. In this case, high currents are obtained at very low potentials (as low as 0.2 V) without poisoning.

Although gold normally shows very little activity for the oxidation of small organic molecules, it is able to oxidize formic acid to CO_2 in acidic media. Interestingly, as Hamelin, Weaver, and co-workers demonstrated, there is no indirect pathway on Au and no formation of a CO poisoning intermediate [Hamelin et al., 1992]. They observed a relatively weak but significant crystal face dependence of the formic acid oxidation current on Au: the Au(111) surface was the most active of the low index planes, and Au(110) the least active. Combining SERS and DFT, Beltramo and co-workers identified the low frequency mode between 280 and 320 cm^{-1} observed during formic acid oxidation on polycrystalline Au with adsorbed formate, $HCOO^-$ [Beltramo et al., 2005].

6.4 METHANOL OXIDATION

6.4.1 Methanol Oxidation on Platinum

In the oxidation of methanol to CO_2, six electrons are involved. This high number of electrons implies that the mechanism is inevitably very complex, with several intermediate species participating in the mechanism. In spite of its complexity, it has been proposed that the oxidation mechanism follows the same general scheme as the oxidation of formic acid, i.e., a dual path mechanism with active and poisoning intermediates (see the reaction Scheme 6.16) [Parsons and VanderNoot, 1988]. For that reason, we will compare the behavior with that of formic acid to highlight the similarities and differences.

The qualitative voltammetric behavior of methanol oxidation on Pt is very similar to that of formic acid. The voltammetry for the oxidation of methanol on Pt single crystals shows a clear hysteresis between the positive- and negative-going scans due to the accumulation of the poisoning intermediate at low potentials and its oxidation above 0.7 V (vs. RHE) [Lamy et al., 1982]. Additionally, the reaction is also very sensitive to the surface structure. The order in the activity of the different low index planes of Pt follows the same order than that observed for formic acid. Thus, the Pt(111) electrode has the lowest catalytic activity and the smallest hysteresis, indicating that both paths of the reaction are slow, whereas the Pt(100) electrode displays a much higher catalytic activity and a fast poisoning reaction. As before, the activity of the Pt(110) electrode depends on the pretreatment of the surface (Fig. 6.17).

The first IR studies detected the formation and adsorption of CO, and therefore CO was proposed as the poisoning intermediate [Beden et al., 1981; Nichols and Bewick, 1988; Corrigan and Weaver, 1988]. The formation of CO is structure-dependent and takes place at open circuit, and the maximum amount accumulated on the electrode

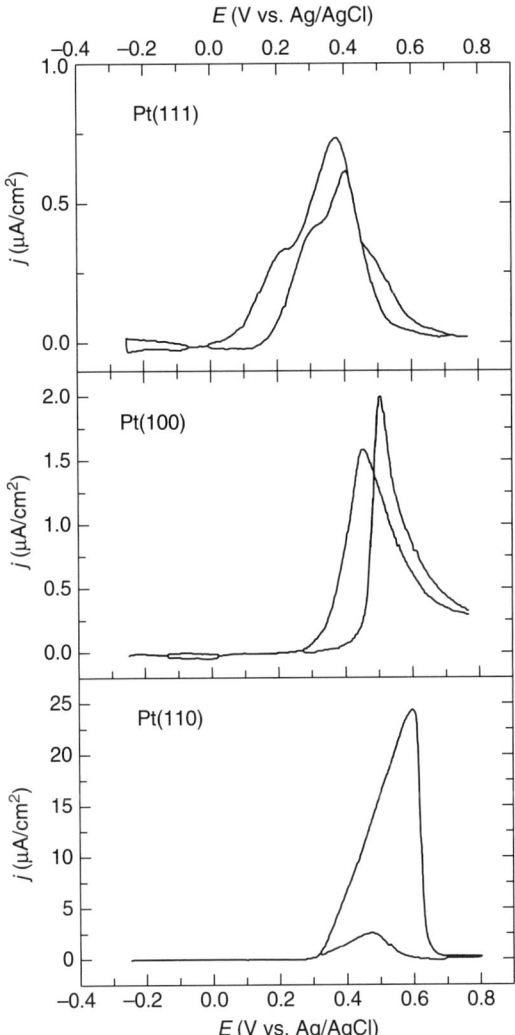

Figure 6.17 Voltammetric profile for Pt(111), Pt(100), and Pt(110) electrodes in 0.2 M CH_3OH + 0.5 M H_2SO_4 on electrode; scan rate 50 mV/s. (Data taken from Herrero et al. [1994].)

surface is very similar to that obtained with formic acid [Sun and Clavilier, 1987]. Owing to the similarities with formic acid, the following mechanism was proposed for the oxidation of methanol:

$$\text{methanol} \longrightarrow \text{formaldehyde} \longrightarrow \text{formic acid} \begin{array}{c} \nearrow \text{active intermediate} \searrow \\ \\ \searrow \qquad \qquad \nearrow \\ \text{CO} \end{array} CO_2 \qquad (6.17)$$

The first part of the mechanism is a sequential reaction yielding formic acid, and from that point the typical dual path mechanism for formic acid occurs. In fact, it has been proposed that the mechanisms of formic acid and methanol oxidation consist of the same dominating elemental steps [Okamoto et al., 2005]. However, experiments have revealed that the mechanism is much more complicated than that.

The initial DEMS studies showed the first differences [Willsau and Heitbaum, 1986]. The surface was initially covered with ^{13}CO coming from $^{13}CH_3OH$ dissociation. Next, the solution was exchanged with one containing $^{12}CH_3OH$ and the oxidation of methanol was followed with voltammetry. The CO_2 thus formed, analyzed with a mass spectrometer, revealed that $^{12}CO_2$ and $^{13}CO_2$ were formed simultaneously, i.e., the oxidation of methanol only took place when the CO present on the electrode surface oxidized. As already mentioned, the same experiment with formic acid detected the formation of $^{12}CO_2$ prior to $^{13}CO_2$. This fact indicates that steric requirements for methanol to oxidize are higher than those needed for formic acid oxidation, since formic acid oxidation can take place on a surface pre-covered with the poisoning intermediate. In fact, it has been proposed that the formation of CO from methanol requires at least three contiguous Pt sites [Cuesta, 2006].

Another important difference in the poison formation reaction is observed when studying this reaction on Pt(111) electrodes covered with different adatoms. On Pt(111) electrodes covered with bismuth, the formation of CO ceased at relatively high coverages only when isolated Pt sites were found on the surface [Herrero et al., 1993]. For formic acid, the formation takes place only at defects; thus, small bismuth coverages are able to stop poison formation [Herrero et al., 1993; Maciá et al., 1999]. Thus, an ideal Pt(111) electrode would form CO from methanol but not from formic acid. This important difference indicates that the mechanism proposed in (6.17) is not valid. It should be noted that the most difficult step in the oxidation mechanism of methanol is probably the addition of the oxygen atom required to yield CO_2. In the case of formic acid, this step is not necessary, since the molecule has already two oxygen atoms. For that reason, the adatoms that enhance formic acid oxidation, such as bismuth or palladium, do not show any catalytic effect for methanol oxidation.

The FTIR studies revealed that the formation of CO_2 is only detected when the CO starts to be oxidized (Fig. 6.18). Therefore, it was proposed that the mechanism has only one path, with CO as the CO_2-forming intermediate [Chang et al., 1992; Vielstich and Xia, 1995]. This has two important and practical consequences. First, methanol oxidation will be catalyzed by the same adatoms that catalyze CO oxidation, mainly ruthenium. Second, since the steric requirements for CO formation from methanol are quite high, the catalytic activity of small (<4 nm) nanoparticles diminishes [Park et al., 2002].

Aside from CO, other intermediate species have been detected. The formation of formic acid was detected by DEMS [Jusys and Behm, 2001; Wang and Baltruschat, 2007], whereas formaldehyde was found by fluorescence and DEMS [Korzeniewski and Childers, 1998; Jusys and Behm, 2001; Wang and Baltruschat, 2007]. The presence of formic acid clearly indicates that the mechanism should always have a parallel path, although its contribution to the total CO_2 could be minor. In fact, only

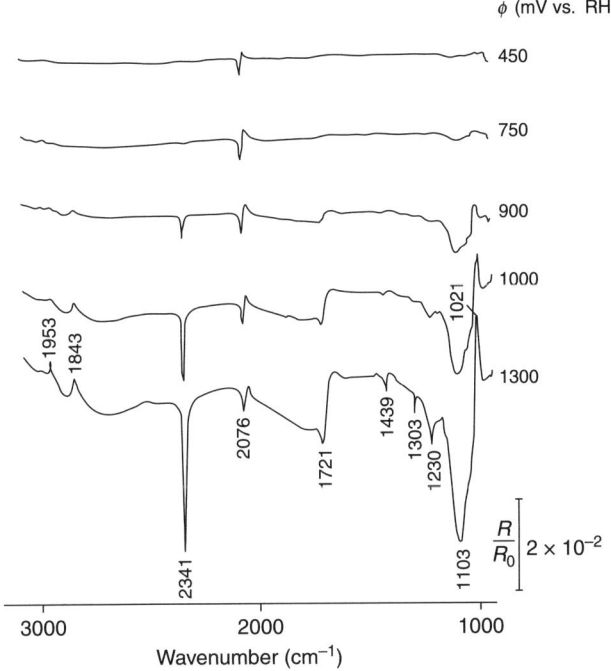

Figure 6.18 Subtractively normalized interfacial Fourier transform infrared spectroscopy (SNIFTIRS) spectra of a polished polycrystalline Pt electrode, immersed in 0.1 M $HClO_4$, + 5 M CH_3OH electrolyte. All spectra were normalized to the base spectrum collected at 0 mV vs. RHE. (Reproduced from Iwasita and Vielstich [1988].)

for Pt(111) electrodes is this contribution through the non-CO path important [Herrero et al., 1995]. The formation of formic acid on a polycrystalline Pt electrode has also been detected indirectly by attenuated total reflection (ATR) spectroscopy, since its adsorption product, formate, has been detected during the oxidation of methanol, which supports the dual pathway mechanism for any Pt electrode [Chen et al., 2003]. The formation of intermediate oxidation products (formaldehyde and formic acid) reveals that the oxidation of methanol is not complete, and this is one of the reasons for the low currents detected, since the average number of electrons exchanged per methanol molecule is always smaller than 6.

The amounts of formaldehyde and formic acid have been measured for several Pt single crystals [Batista et al., 2003, 2004; Housmans et al., 2006; Wang and Baltruschat, 2007]. The fraction of methanol molecules oxidized to CO_2 is about 20% at 0.60 V and 32% at 0.80 V, with formic acid and formaldehyde being the remaining oxidation products [Wang and Baltruschat, 2007].

Increasing the temperature seems to facilitate the oxidation of CO, since lower amounts are accumulated on the surface [Kardash and Korzeniewski, 2000]. The presence of strongly adsorbed anions in the electrolytic solution affects the distribution of products. It has been shown that adsorbed (bi)sulfate diminishes the production of CO, whereas formic acid and formaldehyde yields are not affected [Batista et al., 2003, 2004]. A detailed study of the product distribution on a series of stepped single-crystal Pt electrodes using online electrochemical mass spectrometry revealed that steps influence not only the activity of methanol oxidation, but also the reaction paths chosen [Housmans et al., 2006] (Fig. 6.19). In $HClO_4$ (i.e., in the absence of strongly adsorbing anions), steps promote the formation of methyl formate, which is produced by the reaction of methanol with the oxidation intermediate formic acid. Pt(110) produces about 4–5 times more methyl formate than Pt(111). This suggests that the direct path via formic acid is catalyzed by steps and defects. In H_2SO_4, the same trend with step density is observed, but Pt(111) is nevertheless the most active surface in producing methyl formate. Pt(111) is also the only surface that produces more methyl formate in H_2SO_4 than in $HClO_4$. This suggests that anion adsorption on the (111) terrace also favors the direct pathway.

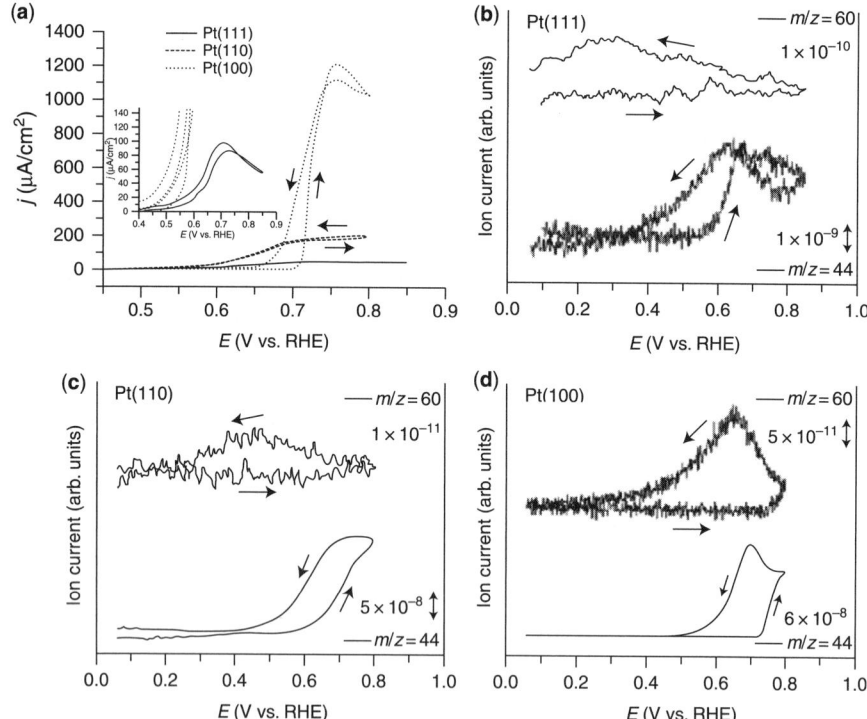

Figure 6.19 (a) Cyclic voltammograms of Pt(111) (solid line), Pt(110) (dashed line), and Pt(100) (dotted line) in 0.5 M CH_3OH and 0.5 M H_2SO_4 at a scan rate of 2 mV/s. The inset shows a zoom of the Pt(111) cyclic voltammogram. The top mass shown in the associated mass spectroscopic cyclic voltammograms for (b) Pt(111), (c) Pt(110), and (d) Pt(100) displays the methyl formate signal, which is associated with the formic acid produced in the reaction ($m/z = 60$), and the bottom one displays the CO_2 signal ($m/z = 44$). (Reproduced from Housmans et al. [2006].)

The kinetics of the reaction has been studied generally with chronoamperometry, since the analysis of the data is simpler. The first studies with Pt single crystals analyzed the kinetics in the absence of CO [Franaszczuk et al., 1992; Herrero et al., 1994]. An important isotopic effect was detected when CD_3OH was used instead of CH_3OH; currents in the former case in the absence of poison were about 3–4 times lower than those recorded for CH_3OH. Although the effect is smaller than that expected for a typical C—H bond breaking during the rate-determining step (RDS) (a 6-fold effect), it clearly indicates that the breaking of a C—H bond is taking place in the RDS. The smaller isotopic effect has been explained by the partial contribution of other steps to the total kinetics [Jusys and Behm, 2001]. Tafel slopes were also measured; approximately 120 mV was obtained for Pt(111) and Pt(110) electrodes, whereas 60 mV was measured for Pt(100) electrodes (Fig. 6.20) [Herrero et al., 1994]. The value 120 mV suggests that the RDS is the first electron transfer, whereas 60 mV is associated with a chemical step after the first electron transfer. The combination of the isotopic effect and the Tafel slopes indicates that the first step in methanol oxidation is the formation of adsorbed CH_2OH. This intermediate is different from that found in ultrahigh vacuum (UHV) environments, where adsorbed CH_3O is detected [Franaszczuk et al., 1992]. The difference in the behavior between the UHV and electrochemical environments has been explained by the presence of water, which affects the interactions of methanol with the surface.

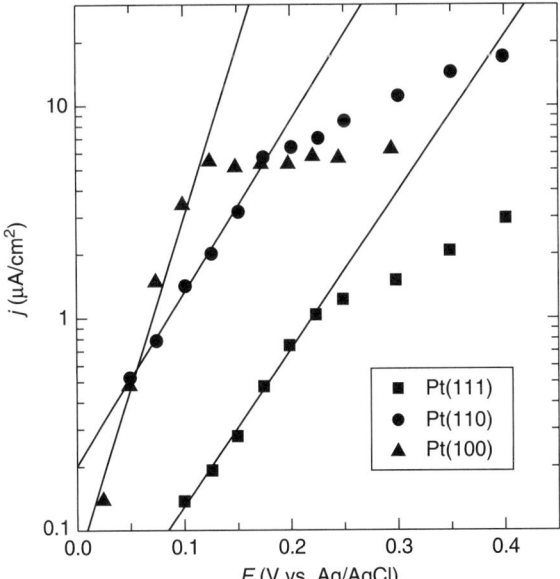

Figure 6.20 Extrapolated current density at $t = 0$ obtained from chronoamperometric experiments for Pt(111), Pt(100), and Pt(110) electrodes in 0.2 M HCOOH + 0.5 M H_2SO_4 on electrode. The straight lines show the regions where the Tafel behavior is observed. (Data taken from Herrero et al. [1994].)

A more complete analysis of the kinetics on the Pt(111) electrode has been carried out by Stuve's group [Sriramulu et al., 1999]. In this model, the first step is a dehydrogenation reaction. From this point, two paths emerge: one proceeds forming CO, and the second one involves an active intermediate. The model fits the experimental data well, and kinetic constants are obtained. Activation energies have been determined for polycrystalline electrodes. The apparent activation energies measured in acid media range between 24 and 74 kJ/mol, depending on the applied potential [Cohen et al., 2007].

Theoretical studies on Pt(111) surfaces in the presence of water have validated the above model [Hartnig and Spohr, 2005]. DFT calculations indicated that the oxidation of methanol starts by the formation of a hydrogen bond between the OH group of the methanol and a water molecule, followed by the cleavage of a C—H bond close to the Pt surface. In the next step, a rapid cleavage of the O—H bond leads to the formation of formaldehyde as a stable intermediate. Other studies indicate that the dehydrogenation reaction to yield CO takes place at low potentials, whereas the other paths are active at higher potentials [Cao et al., 2005]. They also suggest that dehydrogenation takes place by the formation of an adsorbed CH_2OH, whereas the paths through an active intermediate have an adsorbed CH_3O species. This latter path would be more favorable on steps and defect sites.

Housmans and Koper [2003] have carried out a detailed analysis of the chronoamperometric transients on a series of stepped single-crystal electrodes using reaction models very similar to those used in Franaszczuk et al. [1992] and Jarvi and Stuve [1998]. The expression that they used to model the current transient was

$$j(t) = 4eN_{Pt}k_{dec}[1 - \theta_{CO}(t)]^2 + 2eN_{Pt}k_{ox}\theta_{CO}(t)[1 - \theta_{CO}(t)] + j_d[1 - \theta_{CO}(t)] \quad (6.18)$$

with

$$\theta_{CO}(t) = \frac{1 - \exp(k_{ox}t)}{1 - (1 + k_{ox}/k_{dec})\exp(k_{ox}t)} \quad (6.19)$$

where the first term is rate of methanol decomposition to surface-bonded CO, the second term is the rate of CO oxidation, and the third term is a very general expression for the direct oxidation path through formaldehyde and formic acid. The expression for the time dependence of the CO coverage is obtained by solving the differential equation for the CO coverage as it follows from the expression for its formation and subsequent oxidation.

Without the direct pathway contribution, this equation may either yield an increasing or decreasing current transient, depending on the value of k_{ox}/k_{dec}. If this ratio is larger than 4, i.e., if methanol decomposition is slow compared with CO oxidation, then the current is predicted to increase with time. Experimentally, this situation has been observed for a low methanol concentration and an almost perfect Pt(111) electrode [Housmans and Koper, 2003], which both lead to a low methanol decomposition rate. Typically, however, current transients decrease with time, suggesting that the rate of poisoning increases with increasing step density. Fitting of transients with (6.18) indeed suggested that both k_{dec} and k_{ox} depend on step density, with Pt(111) having a much lower decomposition rate constant than Pt(554) and Pt(553), although the model is probably not accurate enough to establish a clear linear relationship between rate constant and step density. It was also found that decomposition is typically faster at low potentials (<0.6 V vs. RHE) and oxidation is faster at high potentials (>0.6 V vs. RHE).

From the results described above, a more detailed reaction scheme has been suggested, which is summarized in Fig. 6.21. As in the mechanism of Cao and co-workers [Cao et al., 2005], the decision between the direct and indirect pathway is made at the initial dehydrogenation step in the approach of the methanol to the surface. The indirect path (Reactions 1–7 in Fig. 6.21) is initiated by the dehydrogenation of methanol to hydroxymethyl, which is further dehydrogenated to a C/O/H species, the nature of which is still unclear, and eventually to CO_{ads}, which acts as a surface poison at low potentials. Online electrochemical mass spectrometry (OLEMS) results [Housmans et al., 2006] indicate that the indirect pathway is favored on (111) terraces in the absence of a strongly adsorbing anions and on the Pt(100) plane. The direct pathway, reactions 8–15, is initiated by O—H scission to form methoxy, a reaction known to occur readily under UHV conditions [Kizhakevariam and Stuve, 1993]. Under electrochemical conditions, the OLEMS data suggest that methoxy formation is preferred on Pt(111) in H_2SO_4 media, on Pt surfaces with (110) steps in the absence of (bi)sulfate and on Pt(110) regardless of the electrolyte. The DFT calculations by Cao and co-workers indicate that dehydrogenation to methoxy is also preferred on (100) steps [Cao et al., 2005]. The formed methoxy is further dehydrogenated to

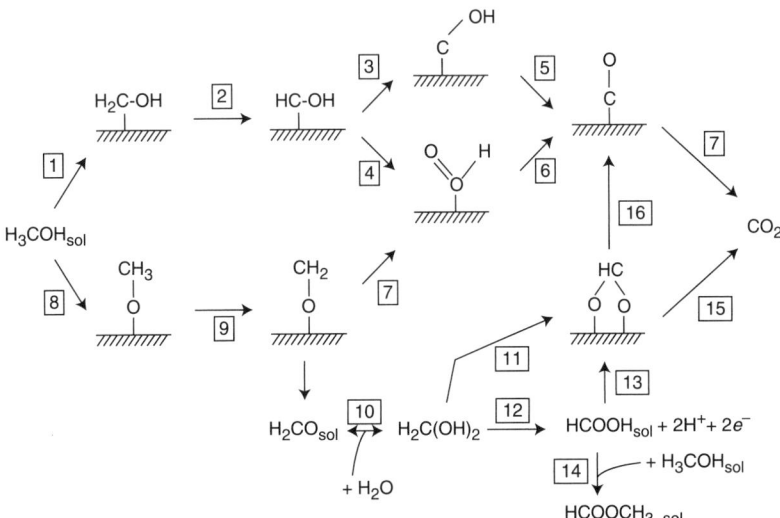

Figure 6.21 Suggested reaction scheme for the electrochemical oxidation of methanol on metal electrodes. (After Housmans et al. [2006].)

adsorbed H$_2$CO (Reaction 9), which may desorb as formaldehyde. Formaldehyde is then nearly completely hydrated to methylene glycol, which will be oxidized to formic acid in solution, either directly (Reaction 12) or through adsorbed formate (Reactions 11 and 13). Finally, this scheme also explains why formate was observed on Pt surfaces during the electrochemical oxidation of methanol (Reactions 8–12), formic acid (Reaction 13), and formaldehyde (Reactions 10–13), as well as why CO$_{ads}$ was found during the oxidation of formaldehyde and of formic acid (Reactions 10, 11, and 16) by Osawa and co-workers [Chen et al., 2003].

6.4.2 Methanol Oxidation on Other Transition Metals

Methanol oxidation on any other transition metal than platinum is rather sluggish, and none of these metals displays significant catalytic oxidation activity. Generally, current densities are very small in the region of interest. The only case where significant activity is found is for the oxidation of methanol on Au nanoparticles or rough surfaces in alkali solutions [Borkowska et al., 2004b; Hernández et al., 2006]. For these electrodes, significant currents can be recorded at potentials as low as 0.2 V (vs. RHE) [Hernández et al., 2006]. This situation contrasts with that observed for single-crystal electrodes, where the onset of the oxidation is always above 0.8 V [Borkowska et al., 2004a]. The activity of these electrodes has to be associated to the presence of a high number of low coordinated atoms on the surface, which are probably the responsible for this unusual catalytic activity. The final product of methanol oxidation on Au in alkaline media has been suggested to be formate [Hernandez et al., 2006], formed through a formaldehyde intermediate, similar to the pathway suggested in Fig. 6.21. Interestingly, formate cannot be further oxidized to CO$_2$ in alkaline solution.

6.5 ETHANOL OXIDATION

6.5.1 Ethanol Oxidation on Platinum

Ethanol, being a renewable fuel, is often mentioned as one of the potential candidates for low temperature fuel cell applications. Besides practical advantages in the employment of ethanol, such as the ease of transport of large quantities of ethanol and its nontoxicity, the interest is justified by the high energy content of ethanol (6.09 kWh/kg), corresponding to 12 electrons per molecule for total oxidation. Furthermore, ethanol is the smallest alcohol containing a C—C bond, and can therefore serve as a model for the electro-oxidation of compounds containing C—C bonds. Although the C—C bonds are, theoretically, the weakest bond in small alcohols [Tsang, 1976], they are difficult to access (electro)catalytically, posing a major challenge for the complete oxidation of ethanol.

It is well established that the main products of ethanol electro-oxidation on Pt in acidic media are acetaldehyde and acetic acid, partial oxidation products that do not require C—C bond breaking, with their relative yields depending on the experimental conditions [Iwasita and Pastor, 1994]. Apart from the loss of efficiency associated with the partial oxidation, acetic acid is also unwanted, as it constitutes a catalyst poison.

The complete oxidation product CO_2 is only found to a minor extent [Chang et al., 1990; Xia et al., 1997]. It is generally assumed that these products are formed by the following general reaction scheme:

$$C_2H_5OH \rightleftharpoons C_2H_5OH_{ads} \tag{6.20}$$

$$C_2H_5OH_{ads} \longrightarrow CO_{ads} + CH_{x,ads} + (6-x)H^+ + (6-x)e^- \tag{6.21a}$$

$$C_2H_5OH_{ads} \longrightarrow CH_3CHO_{ads} + 2H^+ + 2e^- \tag{6.21b}$$

$$H_2O \longrightarrow OH_{ads} + H^+ + e^- \tag{6.22}$$

$$CH_3CHO_{ads} \rightleftharpoons CH_3CHO \tag{6.23a}$$

$$CH_3CHO_{ads} \longrightarrow CO_{ads} + CH_{x,ads} + (4-x)H^+ + (4-x)e^- \tag{6.23b}$$

$$CH_3CHO_{ads} + OH_{ads} \longrightarrow CH_3COOH + H^+ + e^- \tag{6.23c}$$

$$CO_{ads} + OH_{ads} \longrightarrow CO_2 + H^+ + e^- \tag{6.24}$$

$$CH_x + 2\,OH_{ads} \longrightarrow CO_2 + (2+x)H^+ + (2+x)e^- \tag{6.25}$$

It should be noted that this scheme does not represent the elementary steps of the ethanol oxidation mechanism. The reaction is initiated by the adsorption of an ethanol molecule on a vacant surface site (6.20). After adsorption, ethanol can decompose (6.21a) in strongly adsorbed CO and CH_x, or be oxidized to weakly adsorbed acetaldehyde (6.21b), which can desorb and diffuse into the bulk of the solution (6.23a). Alternatively, acetaldehyde can decompose into strongly adsorbed fragments (6.23b). In the presence of adsorbed OH (6.22), the weakly adsorbed acetaldehyde and the strongly absorbed CO fragments can be oxidized to acetic acid (6.23c) and CO_2 (6.24), respectively. The CH_x fragment resulting from the dissociative adsorption of ethanol or acetaldehyde can also be oxidized to CO_2 (6.25), most likely through a CO-like intermediate [Shin et al., 1996].

The first molecular-level study of products and intermediates of ethanol electrooxidation was provided by Willsau and Heitbaum with the use of DEMS [Willsau and Heitbaum, 1985]. They identified CO_2 and acetaldehyde as the primary reaction products, although it should be noted that acetic acid is not volatile enough to be detected by DEMS. By employing deuterium and ^{18}O labeling, this study showed that acetaldehyde is formed by the cleavage of hydrogen atoms from the hydroxyl group and the α-carbon of ethanol. Furthermore, it was found that CO_2 is formed through a strongly bound intermediate, which is oxidized to give two CO_2 molecules, one originating from the methyl group and the other from the alcohol group and still containing the alcoholic O. Subsequent DEMS measurements [Iwasita and Pastor, 1994; Schmiemann et al., 1995] studied the C—C bond breaking in more detail. By isotopic labeling of the carbons in ethanol, methane and ethane originating from the methyl group of ethanol were detected during slow cathodic potential sweeps into the hydrogen adsorption region after ethanol adsorption at potentials before the onset of CO_2 evolution, indicating that C—C bond breaking and CO_2 production are decoupled processes.

Most studies on ethanol oxidation on Pt single crystal electrodes have focused on the structure sensitivity of the formation of the different products, mainly on the basal planes of Pt. By studying the early stages of adsorption and oxidation of ethanol on Pt with in situ FTIR spectroscopy, Xia and co-workers found that the onset potential for ethanol dissociation, as followed by the CO_{ads} infrared absorption intensities, coincides with the hydrogen desorption potential [Xia et al., 1997]. In addition, the presence of adsorbed CO strongly inhibited further oxidation on all surfaces. By relating the spectroscopic measurements with the voltammetric data, it was concluded that the number of surface sites covered with CO_{ads}, and thus the capacity to cleave the C—C bond, is higher on Pt(100) than on Pt(111) or Pt(110). It is also worth noting that the amount of CO_2 produced keeps increasing after the surface becomes free of CO_{ads}, indicating the presence of other strongly adsorbed species, likely CH_x, which is not observable by infrared spectrometry.

More recently, the dissociation of ethanol was studied by SERS [Lai et al., 2008]. By employing isotopically labelled ethanol, it was found that C—C bond breaking already occurs at low potentials, resulting in chemisorbed CH and CO. Upon oxidation the CH fragments are converted to CO at a potential below that of CO oxidation, suggesting that, at least on platinum, the potential limiting step in the oxidation of the adsorbed C species is the oxidation of CO.

Infrared spectroscopy has also been employed to follow the formation of acetaldehyde and acetic acid on Pt during ethanol electro-oxidation. On the basal planes, acetaldehyde could be observed starting at about 0.4 V (vs. RHE), well before the onset of CO oxidation, while the onset of acetic acid formation closely follows CO_2 formation [Chang et al., 1990; Xia et al., 1997]. This is readily explained by the fact that both CO oxidation and acetic acid formation require a common adsorbed co-reactant, OH_{ads}, whereas the formation of acetaldehyde from ethanol merely involves a relatively simple proton–electron transfer.

So far, few studies have focused on the effect of (the density of) defect sites, as modeled by using Pt single-crystal electrodes with varying step density. A notable exception is a study by Tarnowski and Korzeniewski, who followed the quantities of acetate formed at different potentials in potential step experiments using ion chromatography on Pt(111), Pt(755) $\equiv$ Pt[6(111) $\times$ (100)] and Pt(533) $\equiv$ Pt[4(111) $\times$ (100)] [Tarnowski and Korzeniewski, 1997]. It was shown that, although the maximum currents increased with step density, the relative contribution of acetic acid formation decreased. Since step sites are assumed to facilitate C—C bond breaking in ethanol oxidation [Shin et al., 1996], this decreased activity was partly attributed to increased surface poisoning, blocking sites for water adsorption and thereby inhibiting acetic acid formation. Since the maximum currents do increase with step density, it is likely that other processes, such as acetaldehyde formation, become more pronounced on (partially blocked) stepped surfaces [Leung et al., 1989].

Since several steps in the ethanol oxidation mechanism require the presence of an OH_{ads} species, the use of an alkaline medium has also attracted some attention, owing to the ubiquitous hydroxide ions leading to significantly higher oxidation currents. However, for fuel cell purposes, one has to consider the progressive carbonation of the electrolyte solution due to CO_2, which is produced as a product of the oxidation reaction, leading to a decrease in the pH of the solution. A study on the electrooxidation of C_1–C_4 alcohols by Tripković and co-workers on Pt(111), Pt(755) ≡ Pt[6(111) × (100)] and Pt(332) ≡ Pt[5(111) × (110)] in an alkaline electrolyte has shown that the potentials at which ethanol can be oxidized are mainly affected by the ability of the surface to adsorb OH^- anions [Tripkovic et al., 2001]. On all three surfaces, the onset potential of ethanol oxidation coincides with the completion of hydrogen desorption, and accelerates in the potential region in which the hydroxide anions can adsorb reversibly. The peak potential in all cases was found to be coupled to the onset of the irreversible hydroxide adsorption or oxide layer formation, which occurs at a lower potential on the stepped surfaces than on Pt(111).

6.5.2 Ethanol Oxidation on Gold

Gold is generally considered a poor electro-catalyst for oxidation of small alcohols, particularly in acid media. In alkaline media, however, the reactivity increases, which is related to that fact that no poisoning CO-like species can be formed or adsorbed on the surface [Nishimura et al., 1989; Tremiliosi-Filho et al., 1998]. Similar to Pt electrodes, the oxidation of ethanol starts at potentials corresponding to the onset of surface oxidation, emphasizing the key role of surface oxides and hydroxides in the oxidation process. The only product observed upon the electrooxidation of ethanol on Au in an alkaline electrolyte is acetate, the deprotonated form of acetic acid. The lack of carbon dioxide as a reaction product again suggests that adsorbed CO-like species are an essential intermediate in CO_2 formation.

In acidic media, the reactivity of ethanol on Au electrodes is much lower than in alkaline media. The main product of the oxidation of ethanol on Au in an acidic electrolyte was found to be acetaldehyde, with small amounts of acetic acid [Tremiliosi-Filho et al., 1998]. The different reactivities and the product distributions in different media were explained by considering the interactions between the active sites on Au, ethanol, and active oxygen species absorbed on or near the electrode surface. In acidic media, surface hydroxide concentrations are low, leading to relatively slow dehydrogenation of ethanol to form acetaldehyde as the main oxidation pathway. In contrast, in alkaline media, ethanol, adsorbed as an ethoxy species, reacts with a surface hydroxide, forming adsorbed acetate, leading to acetate (acetic acid) as the main reaction product.

6.5.3 Ethanol Oxidation on Rhodium

Another metal that has attracted interest for use as electrode material is rhodium, inspired by its high activity in the catalytic oxidation of CO in automotive catalysis. It is found that Rh is a far less active catalyst for the ethanol electro-oxidation reaction than Pt [de Souza et al., 2002; Leung et al., 1989]. Similar to ethanol oxidation on Pt, the main reactions products were CO_2, acetaldehyde, and acetic acid. Rh, however, presents a significant better CO_2 yield relative to the C_2 compounds than Pt, indicating a relatively higher C—C bond breaking activity. This increased ability to activate the ethanol C—C bond is also reflected in cathodic scans, during which the sole reduction product was methane.

The lower total activity for Rh electrodes may be partly due to increased CO poisoning and slower CO electro-oxidation kinetics compared with Pt electrodes, as demonstrated by the number of voltammetric cycles required to oxidize a saturated CO adlayer from Rh electrodes (see Section 6.2.2) [Housmans et al., 2004]. In addition, it is argued that the barrier to dehydrogenation is higher on Rh than on Pt, leading to a lower overall reaction rate [de Souza et al., 2002]. These effects may also explain the lower product selectivity towards acetaldehyde and acetic acid, which require the dehydrogenation of weakly adsorbed species.

6.5.4 Acetaldehyde Oxidation on Platinum

Since acetaldehyde has repeatedly been observed as one of the main products of ethanol electro-oxidation, it has become clear that investigations on the mechanism of the oxidation of acetaldehyde may help to understand the processes involved in ethanol oxidation. In contrast to the large number of papers on ethanol oxidation, relatively little has been published on acetaldehyde oxidation. By following the electro-oxidation of acetaldehyde with in situ FTIR spectroscopy at varying concentrations, Farias and co-workers concluded that acetaldehyde is oxidized in two parallel pathways [Farias et al., 2007]. The pathway producing CO_2 requires acetaldehyde to be adsorbed dissociatively on the (polycrystalline) platinum electrode. From the dependence of the amount of CO_2 formed on the acetaldehyde concentrations, it was suggested that CO_2 is formed through a Langmuir–Hinshelwood mechanism. At low acetaldehyde concentrations (<0.05 M), CO_2 production is the dominant pathway. At higher concentrations, the main reaction product is acetic acid, generated through an Eley–Rideal-like mechanism involving (weakly adsorbed) acetaldehyde reacting directly with OH_{ads} to form acetic acid. Similar conclusions were drawn for Pt(111) and Pt(110) electrodes [Rodriguez et al., 2000].

Recently, a systematic study of the influence of electrode structure on acetaldehyde electro-oxidation was performed in Leiden [Lai and Koper, 2009]. By employing a series of $Pt[n(111) \times (111)] \equiv Pt[(n-1)(111) \times (110)]$ single-crystal electrodes (Pt(111), Pt(15, 15, 14), Pt(554), and Pt(553), with $n = 200–500, 30, 10$, and 5, respectively) in acidic media, the density of step sites could be controlled in a precise manner. The results of this investigation are shown in Fig. 6.22. Surprisingly, in contrast to ethanol electro-oxidation, oxidation activity was found to decrease with increasing step density, although this decrease is relatively small. This decrease in activity is reflected both in decreasing maximum current densities and increasing peak potentials with decreasing terrace width. The dependences of both peak parameters on the step density (defined as $\theta_{step} = 1/(n - \frac{2}{3})$ [Clavilier et al., 1990]) were found to be linear, suggesting a cumulative effect of steps on the oxidation behavior of acetaldehyde. One possible explanation for this behavior could be the faster decomposition of acetaldehyde on step sites, compared with ethanol, owing to a relatively easy-to-access C—C bond. Combined with the preferential poisoning of step sites, this restricts oxidation to the terrace sites, explaining the observed behavior.

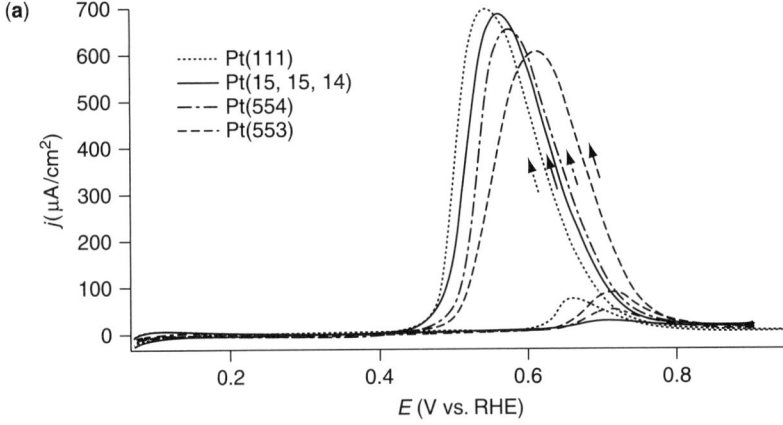

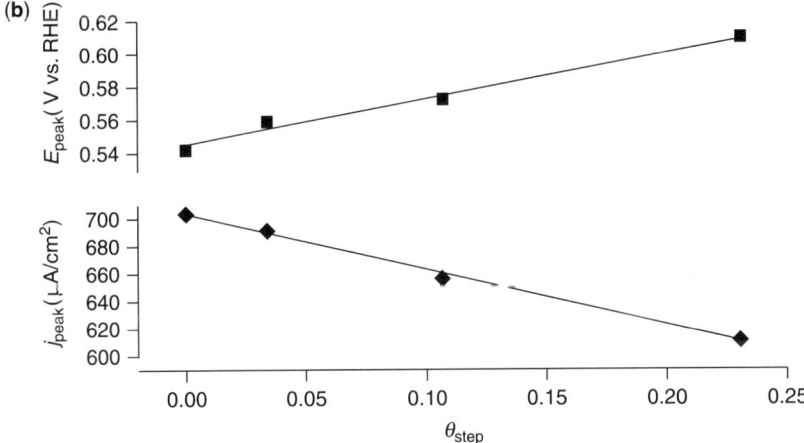

Figure 6.22 (a) Cyclic voltammograms of the oxidation of 0.1 M acetaldehyde on Pt single-crystal electrodes in 0.5 M H_2SO_4 at a scan rate of 10 mV/s. (b) Dependence of the peak potential (upper) and maximum current density (lower) of acetaldehyde oxidation on the step density. The solid lines are the least squares fit of the data.

6.6 CONCLUSIONS

In this chapter, we have summarized (recent) progress in the mechanistic understanding of the oxidation of carbon monoxide, formic acid, methanol, and ethanol on transition metal (primarily Pt) electrodes. We have emphasized the "surface science approach" employing well-defined electrode surfaces, i.e., single crystals, in combination with surface-sensitive techniques (FTIR and online DEMS), kinetic modeling and first-principles DFT calculations.

CO oxidation is a highly structure-sensitive reaction that needs steps and defect sites. Mobility of CO on the electrode surface does not seem to play a role on

Pt single crystals, but was found to be essential on Rh in the presence of strongly adsorbing anions. Methanol oxidation is also highly structure-sensitive, and is typically enhanced by steps and defects, which is partly due to the prominent role of the poisoning intermediate CO. However, owing to the more complicated reaction scheme, the effects are not that clear-cut or easy to model. Both defect density and strongly adsorbing anions have a significant influence not only on the overall methanol oxidation activity but also on the selectivity of the reaction. Soluble intermediates such as formic acid and formaldehyde are favored by the presence of steps and defects. Formic acid oxidation is typically poisoned faster in the presence of steps and defects, as they promote the formation of chemisorbed CO. Although adsorbed formate has been detected as an intermediate in formic acid oxidation (and also methanol oxidation), it is not yet fully clear whether adsorbed formate is the precursor to CO_2 formation. The oxidation of ethanol presents an additional complication in the sense that a C—C bond must also be cleaved in order to accomplish full oxidation to CO_2. Intermediates in ethanol oxidation are adsorbed CO, CH_x, acetaldehyde, and acetic acid. Acetic acid cannot be oxidized further, and CO and CH_x act as surface poisons. Steps enhance the oxidation of ethanol, and are assumed to facilitate C—C bond breaking, although clear direct proof for this is still missing. Interestingly, steps have the opposite effect on the oxidation of acetaldehyde.

Real fuel cell catalysts hardly ever consist of pure metals, and are typically based on (bi)metallic nanoparticles supported on high surface area carbon. It is not always straightforward to translate results obtained on (stepped) single-crystal surfaces to the kinetics observed on nanoparticles, in particular size effects. For example, if smaller particles were to exhibit a higher density of active defects or low coordination sites, then the trend observed with CO oxidation on stepped single crystals would predict that CO oxidation should be enhanced on smaller particles. Experimentally, the opposite is observed: smaller particles are less active for CO oxidation—and also for methanol oxidation, for that matter. This apparent paradox probably has to do with identifying the actual active sites on nanoparticles. A detailed discussion of this important issue is given in Chapter 15. Also, as is well known, certain reactions in the overall reaction scheme may be catalyzed by the addition of a second metal to the Pt catalyst, CO oxidation by PtRu and PtSn alloys being prominent examples. We have not discussed bimetallic catalysis in this chapter, but the matter is dealt with in detail in Chapters 7, 8, and 11. Nevertheless, we strongly believe that the approaches described in this chapter are a necessary and indispensable step in obtaining a molecular-level mechanistic view of the relevant reactions, which may ultimately help in finding or designing a superior real (bi)metallic nanoparticulate catalyst.

REFERENCES

Adzic RR, Tripkovic AV, O'Grady W. 1982. Structural effects in electrocatalysis. Nature 296: 137.

Anderson AB, Albu TV. 2000. Catalytic effect of platinum on oxygen reduction—An ab initio model including electrode potential dependence. J Electrochem Soc 147: 4229–4238.

Angelucci CA, Nart FC, Herrero E, Feliu JM. 2007a. Anion re-adsorption and displacement at platinum single crystal electrodes in CO-containing solutions. Electrochem Commun 9: 1113–1119.

Angelucci CA, Herrero E, Feliu JM. 2007b. Bulk CO oxidation on platinum electrodes vicinal to the Pt(111) surface. J Solid State Electrochem 11: 1532–1539.

Bagotzky VS, Vassiliev YB, Khazova OA. 1977. Generalized scheme of chemisorption, electrooxidation and electroreduction of simple organic compounds on platinum group metals. J Electroanal Chem 81: 229–238.

Baldauf M, Kolb DM. 1996. Formic acid oxidation on ultrathin Pd films on Au(hkl) and Pt(hkl) electrodes. J Phys Chem 100: 11375–11381.

Batista EA, Malpass GRP, Motheo AJ, Iwasita T. 2003. New insight into the pathways of methanol oxidation. Electrochem Commun 5: 843–846.

Batista EA, Malpass GRP, Motheo AJ, Iwasita T. 2004. New mechanistic aspects of methanol oxidation. J Electroanal Chem 571: 273–282.

Beden B, Lamy C, Bewick A, Kunimatsu K. 1981. Electrosorption of methanol on a platinum electrode. IR spectroscopic evidence for adsorbed carbonyl species. J Electroanal Chem 121: 343–347.

Beden B, Bewick A, Lamy C. 1983. A study by electrochemically modulated infrared reflectance spectroscopy of the electrosorption of formic acid at a platinum electrode. J Electroanal Chem 148: 147–160.

Beden B, Lamy C, Léger JM. 1992. Electrocatalytic oxidation of oxygenated aliphatic organic compounds at noble metal electrodes. In: Bockris JO'M, Conway BE, White RE, eds. Modern Aspects of Electrochemistry. Volume 22. New York: Plenum Press. p 97–264.

Beltramo G, Shubina TE, Koper MTM. 2005. Oxidation of formic acid and carbon monoxide on gold electrodes studied by surface-enhanced Raman spectroscopy and DFT. ChemPhysChem 6: 2597–2606.

Bergelin M, Herrero E, Feliu JM, Wasberg M, 1999. Oxidation of CO adlayers on Pt(111) at low potentials: an impinging jet study in H_2SO_4 electrolyte with mathematical modeling of the current transients. J Electroanal Chem 467: 74–84.

Blizanac BB, Lucas CA, Gallagher ME, Arenz M, Ross PN, Markovic NM. 2004a. Anion adsorption, CO oxidation, and oxygen reduction reaction on a Au(100) surface: The pH effect. J Phys Chem B 108: 625–634.

Blizanac BB, Arenz M, Ross PN, Markovic NM. 2004b. Surface electrochemistry of CO on reconstructed gold single crystal surfaces studied by infrared reflection absorption spectroscopy and rotating disk electrode. J Am Chem Soc 126: 10130–10141.

Bonnefont A, Varela H, Krischer K. 2003. Stationary small and large amplitude patterns during bulk CO electrooxidation on platinum. ChemPhysChem 4: 1260–1263.

Borkowska Z, Tymosiak-Zielinska A, Shul G. 2004a. Electrooxidation of methanol on polycrystalline and single crystal gold electrodes. Electrochim Acta 49: 1209–1220.

Borkowska Z, Tymosiak-Zielinska A, Nowakowski R. 2004b. High catalytic activity of chemically activated gold electrodes towards electro-oxidation of methanol. Electrochim Acta 49: 2613–2621.

Cao D, Lu GQ, Wieckowski A, Wasileski SA, Neurock M. 2005. Mechanisms of methanol decomposition on platinum: A combined experimental and ab initio approach. J Phys Chem B 109: 11622–11633.

Capon A, Parsons R. 1973a. Oxidation of formic acid at noble metal electrodes. I. Review of previous work. J Electroanal Chem 44: 1–7.

Capon A, Parsons R. 1973b. Oxidation of formic acid at noble metal electrodes. III. Intermediates and mechanism on platinum electrodes. J Electroanal Chem 45: 215–231.

Capon A, Parsons R. 1973c. The oxidation of formic acid on noble metal electrodes. I. A comparison of the behaviour of pure electrodes. J Electroanal Chem 44: 239–254.

Chang SC, Leung LWH, Weaver MJ. 1990. Metal crystallinity effects in electrocatalysis as probed by real-time FTIR spectroscopy: electrooxidation of formic acid, methanol, and ethanol on ordered low-index platinum surfaces. J Phys Chem 94: 6013–6021.

Chang SC, Hamelin A, Weaver MJ. 1991. Dependence of the electrooxidation rates of carbon monoxide at gold on the surface crystallographic orientation: A combined kinetic-surface infrared spectroscopy study. J Phys Chem 95: 5560–5567.

Chang SC, Ho Y, Weaver MJ. 1992. Applications of real-time infrared spectroscopy to electrocatalysis at bimetallic surfaces. I. Electrooxidation of formic acid and methanol on bismuth-modified platinum (111) and platinum (100). Surf Sci 265: 81–94.

Chen YX, Miki A, Ye S, Sakai H, Osawa M. 2003. Formate, an active intermediate for direct oxidation of methanol on Pt electrode. J Am Chem Soc 125: 3680–3681.

Chen YX, Heinen M, Jusys Z, Behm RJ. 2006a. Bridge-bonded formate: Active intermediate or spectator species in formic acid oxidation on a Pt film electrode? Langmuir 22: 10399–10408.

Chen YX, Ye S, Heinen M, Jusys Z, Osawa M, Behm RJ. 2006b. Application of in-situ attenuated total reflection—Fourier transform infrared spectroscopy for the understanding of complex reaction mechanism and kinetics: Formic acid oxidation on a Pt film electrode at elevated temperatures. J Phys Chem B 110: 9534–9544.

Chen YX, Heinen M, Jusys Z, Behm RJ. 2007. Kinetic isotope effects in complex reaction networks: Formic acid electro-oxidation. ChemPhysChem 8: 380–385.

Clavilier J. 1987. Pulsed linear sweep voltammetry with pulses of constant level in a potential scale, a polarization demanding condition in the study of platinum single crystal electrodes. J Electroanal Chem 236: 87–94.

Clavilier J, Sun SG. 1986. Electrochemical study of the chemisorbed species formed from formic acid dissociation at platinum single crystal electrodes. J Electroanal Chem 199: 471–480.

Clavilier J, Parsons R, Durand R, Lamy C, Leger JM. 1981. Formic acid oxidation on single crystal platinum electrodes. Comparison with polycrystalline platinum. J Electroanal Chem 124: 321–326.

Clavilier J, El Achi K, Rodes A. 1990. In situ probing of step and terrace sites on Pt(s)-[n(111)x(111)] electrodes. Chem Phys 141: 1–14.

Cohen JL, Volpe DJ, Abruña HD. 2007. Electrochemical determination of activation energies for methanol oxidation on polycrystalline platinum in acidic and alkaline electrolytes. Phys Chem Chem Phys 9: 49–77.

Corrigan DS, Weaver MJ. 1988. Mechanisms of formic acid, methanol, and carbon monoxide electrooxidation at platinum as examined by single potential alteration infrared spectroscopy. J Electroanal Chem 241: 143–162.

Cuesta A. 2006. At least three contiguous atoms are necessary for CO formation during methanol electrooxidation on platinum. J Am Chem Soc 128: 13332–13333.

de Souza JPI, Queiroz SL, Bergamaski K, Gonzalez ER, Nart FC. 2002. Electro-oxidation of ethanol on Pt, Rh, and PtRh electrodes. A study using DEMS and in-situ FTIR techniques. J Phys Chem B 106: 9825–9830.

Dunietz BD, Markovic NM, Ross Jr PN, Head-Gordon M. 2004. Initiation of electro-oxidation of CO on Pt based electrodes at full coverage conditions simulated by ab initio electronic structure calculations. J Phys Chem B 108: 9888–9892.

Edens GJ, Hamelin A, Weaver MJ. 1996. Mechanism of carbon monoxide electrooxidation on monocrystalline gold surfaces: Identification of a hydroxycarbonyl intermediate. J Phys Chem 100: 2322–2329.

Farias MJS, Camara GA, Tanaka AA, Iwasita T. 2007. Acetaldehyde electrooxidation: The influence of concentration on the yields of parallel pathways. J Electroanal Chem 600: 236.

Feliu JM, Herrero E. 2003. Formic acid oxidation. In: Vielstich W, Lamm A, Gasteiger H, eds. Handbook of Fuel Cells. Volume 2. New York; Wiley-VCH. p 625–638.

Fernández-Vega A, Feliu JM, Aldaz A, Clavilier J. 1991. Heterogeneous electrocatalysis on well-defined platinum surfaces modified by controlled amounts of irreversibly adsorbed adatoms: Part IV. Formic acid oxidation on the Pt(111)–As system. J Electroanal Chem 305: 229–240.

Franaszczuk K, Herrero E, Zelenay P, Wieckowski A, Wang J, Masel RI. 1992. A comparison of electrochemical and gas-phase decomposition of methanol on platinum surfaces. J Phys Chem 96: 8509–8516.

García G, Koper MTM. 2008. Stripping voltammetry of carbon monoxide oxidation on stepped platinum single-crystal electrodes in alkaline solution. Phys Chem Chem Phys 10: 3802–3811.

Gilman S. 1964. The mechanism of electrochemical oxidation of carbon monoxide and methanol on platinum. II. The "reactant-pair" mechanism for electrochemical oxidation of carbon monoxide and methanol. J Phys Chem 68: 70–80.

Gomez R, Orts JM, Feliu JM, Clavilier J, Klein LH. 1997. The role of surface crystalline heterogeneities in the electrooxidation of carbon monoxide adsorbed on Rh(111) electrodes in sulphuric acid solutions. J Electroanal Chem 432: 1–5.

Gómez R, Climent V, Feliu JM, Weaver MJ. 2000. Dependence of the potential of zero charge of stepped platinum (111) electrodes on the oriented step-edge density: Electrochemical implications and comparison with work function behavior. J Phys Chem B 104: 597–605.

Hamelin A. 1996. Cyclic voltammetry at gold single-crystal surfaces. Part 1. Behaviour at low-index faces. J Electroanal Chem 407: 1–11.

Hamelin A, Martins AM. 1996. Cyclic voltammetry at gold single-crystal surfaces. Part 2. Behaviour of high-index faces. J Electroanal Chem 407: 13–21.

Hamelin A, Ho Y, Chang SC, Gao X, Weaver MJ. 1992. Surface crystallographic dependence of voltammetric oxidation of polyhydric alcohols and related systems at monocrystalline gold–acidic aqueous interfaces. Langmuir 8: 975–981.

Hartnig C, Spohr E. 2005. The role of water in the initial steps of methanol oxidation on Pt(111). Chem Phys 319: 185–191.

Haruta M. 1997. Size- and support-dependency in the catalysis of gold. Catalysis Today 36: 153–166.

Hernández J, Solla-Gullón J, Herrero E, Aldaz A, Feliu JM. 2006. Methanol oxidation on gold nanoparticles in alkaline media: Unusual electrocatalytic activity. Electrochim Acta 52: 1662–1669.

Herrero E, Fernandez-Vega A, Feliu JM, Aldaz A. 1993. Poison formation reaction from formic acid and methanol on platinum (111) electrodes modified by irreversibly adsorbed bismuth and arsenic. J Electroanal Chem 350: 73–88.

Herrero E, Franaszczuk K, Wieckowski A. 1994. Electrochemistry of methanol at low index crystal planes of platinum: An integrated voltammetric and chronoamperometric study. J Phys Chem 98: 5074–5083.

Herrero E, Chrzanowski W, Wieckowski A. 1995. Dual path mechanism in methanol electro-oxidation on a platinum electrode. J Phys Chem 99: 10423–10424.

Herrero E, Feliu JM, Blais S, Jerkiewicz G. 2000. Temperature dependence of CO chemisorption and its oxidative desorption on the Pt(111) electrode. Langmuir 16: 4779–4783.

Herrero E, Álvarez B, Feliu JM, Blais S, Radovic-Hrapovic Z, Jerkiewicz G. 2004. Temperature dependence of the CO_{ads} oxidation process on Pt(111), Pt(100), and Pt(110) electrodes. J Electroanal Chem 567: 139–149.

Hoshi N, Kida K, Nakamura M, Nakada M, Osada K. 2006. Structural effects of electrochemical oxidation of formic acid on single crystal electrodes of palladium. J Phys Chem B 110: 12480–12484.

Housmans THM, Koper MTM. 2003. Methanol oxidation on stepped $Pt[n(111) \times (110)]$ electrodes: A chronoamperometric study. J Phys Chem B 107: 8557–8567.

Housmans THM, Koper MTM. 2005a. CO oxidation on stepped $Rh[n(111) \times (111)]$ single-crystal electrodes: A chrono-amperometric study. J Electroanal Chem 575: 39–51.

Housmans THM, Koper MTM. 2005b. CO oxidation on stepped $Rh[n(111) \times (111)]$ single-crystal electrodes: Anion effects on CO surface mobility. Electrochem Commun 7: 581–588.

Housmans THM, Feliu JM, Koper MTM. 2004. CO oxidation on stepped $Rh[n(111) \times (111)]$ single-crystal electrodes: A voltammetric study. J Electroanal Chem 572: 79–91.

Housmans THM, Wonders AH, Koper MTM. 2006. Structure sensitivity of methanol electro-oxidation pathways on platinum: An on-line electrochemical mass spectrometry study. J Phys Chem B 110: 10021–10031.

Housmans THM, Hermse CGM, Koper MTM. 2007. CO oxidation on stepped single-crystal electrodes: A dynamic Monte Carlo study. J Electroanal Chem 607: 67–82.

Iwasita T, Pastor E. 1994. A DEMS and FTIR spectroscopic investigation of adsorbed ethanol on polycrystalline platinum. Electrochim Acta 39: 531–537.

Iwasita T, Vielstich W. 1988. New in-situ IR results on adsorption and oxidation of methanol on platinum in acidic solution. J Electroanal Chem 250: 451–456.

Iwasita T, Xia X, Herrero E, Liess HD. 1996. Early stages during the oxidation of HCOOH on single-crystal Pt electrodes as characterized by infrared spectroscopy. Langmuir 12: 4260–4265.

Janik MJ, Neurock M. 2007. A first-principles analysis of the electro-oxidation of CO over Pt(111). Electrochim Acta 52: 5517–5528.

Jarvi TD, Stuve EM. 1998. Fundamental aspects of vacuum and electrocatalytic reactions of methanol and formic acid on platinum surface. In: Lipkowski J, Ross PN, eds. Electrocatalalysis. New York: Wiley-VCH. p 75–154.

Jusys Z, Behm RJ. 2001. Methanol oxidation on a carbon-supported Pt fuel cell catalyst—A kinetic and mechanistic study by differential electrochemical mass spectrometry. J Phys Chem B 105: 10874–10883.

Kardash D, Korzeniewski C. 2000. Temperature effects on methanol dissociative chemisorption and water activation at polycrystalline platinum electrodes. Langmuir 16: 8419–8425.

Kibler LA, El-Aziz AM, Hoyer R, Kolb DM. 2005. Tuning reaction rates by lateral strain in a palladium monolayer. Angew Chem Int. Ed 44: 2080–2084.

Kizhakevariam N, Stuve EM. 1993. Promotion and poisoning of the reaction of methanol on clean and modified platinum (100). Surf Sci 286: 246–260.

Kolb DM. 1996. Reconstruction phenomena at metal–electrolyte interfaces. Prog Surf Sci 51: 109–173.

Koper MTM, Jansen APJ, van Santen RA, Lukkien JJ, Hilbers PAJ. 1998. Monte Carlo simulations of a simple model for the electrocatalytic CO oxidation on platinum. J Chem Phys 109: 6051–6062.

Koper MTM, Schmidt TJ, Markovic NM, Ross PN Jr. 2001. Potential oscillations and S-shaped polarization curve in the continuous electro-oxidation of CO on platinum single-crystal electrodes. J Phys Chem B 105: 8381–8386.

Koper MTM, Lebedeva NP, Hermse CGM, 2002. Dynamics of CO at the solid/liquid interface studied by modeling and simulation of CO oxidation on Pt and PtRu electrodes. Faraday Discuss 121: 301–311.

Korzeniewski C, Childers CL. 1998. Formaldehyde yields from methanol electrochemical oxidation on platinum. J Phys Chem B 102: 489–492.

Kunimatsu K. 1986. Infrared spectroscopic study of methanol and formic acid adsorbates on a platinum electrode: Part I. Comparison of the infrared absorption intensities of linear CO(a) derived from CO, CH_3OH and HCOOH. J Electroanal Chem 213: 149–157.

Lai SCS, Koper MTM. 2009. Electro-oxidation of ethanol and acetaldehyde on platinum single-crystal electrodes. Faraday Discuss 140: 399–416.

Lai SCS, Kleyn SEF, Rosca V, Koper MTM. 2008. Mechanism of the dissociation and electro-oxidation of ethanol and acetaldehyde on platinum as studied by SERS. J Phys Chem C 112: 19080–19087.

Lai SCS, Lebedeva NP, Housmans THM, Koper MTM. 2007. Mechanisms of carbon monoxide and methanol oxidation at single-crystal electrodes. Top Catalysis 46: 320–333.

Lamy C, Leger JM, Clavilier J. 1982. Structural effects in the electrooxidation of methanol in alkaline medium. Comparison of platinum single crystal and polycrystalline electrodes. J Electroanal Chem 135: 321–328.

Lamy C, Leger JM, Clavilier J, Parsons R. 1983. Structural effects in electrocatalysis: A comparative study of the oxidation of CO, HCOOH and CH_3OH on single crystal Pt electrodes. J Electroanal Chem 150: 71–77.

Lebedeva NP, Koper MTM, Herrero E, Feliu JM, van Santen RA. 2000a. CO oxidation on stepped $Pt[n(111) \times (111)]$ electrodes. J Electroanal Chem 487: 37–44.

Lebedeva NP, Koper MTM, Feliu JM, van Santen RA. 2000b. The effect of the cooling atmosphere in the preparation of flame-annealed Pt(111) electrodes on CO adlayer oxidation. Electrochem Commun 2: 487–490.

Lebedeva NP, Koper MTM, Feliu JM, van Santen RA. 2002a. Mechanism and kinetics of the electrochemical CO adlayer oxidation on Pt(111). J Electroanal Chem 524/525: 242–251.

Lebedeva NP, Rodes A, Feliu JM, Koper MTM, van Santen RA. 2002b. Role of crystalline defects in electrocatalysis: CO adsorption and oxidation on stepped platinum electrodes as studied by *in situ* infrared spectroscopy. J Phys Chem B 106: 9863–9872.

Lebedeva NP, Koper MTM, Feliu JM, van Santen RA. 2002c. Role of crystalline defects in electrocatalysis: Mechanism and kinetics of CO adlayer oxidation on stepped platinum electrodes. J Phys Chem B 106: 12938–12947.

Leiva E, Iwasita T, Herrero E, Feliu JM. 1997. Effect of adatoms in the electrocatalysis of HCOOH oxidation. A theoretical model. Langmuir 13: 6287–6293.

Leung LWH, Chang SC, Weaver MJ. 1989. Real-time FTIR spectroscopy as an electrochemical mechanistic probe—Electrooxidation of ethanol and related species on well-defined Pt(111) surfaces. J Electroanal Chem 266: 317–336.

Li X, Gewirth AA. 2003. Peroxide electroreduction on Bi-modified Au surfaces: Vibrational spectroscopy and density functional calculations. J Am Chem Soc 125: 7086–7099.

Llorca MJ, Herrero E, Feliu JM, Aldaz A. 1994a. Formic acid oxidation on Pt(111) electrodes modified by irreversibly adsorbed selenium. J Electroanal Chem 373: 217–225.

Llorca MJ, Feliu JM, Aldaz A, Clavilier J. 1994b. Formic acid oxidation on Pd + Pt(100) and Pd + Pt(111) electrodes. J Electroanal Chem 376: 151–160.

López-Cudero A, Cuesta A, Gutiérrez C. 2005. Potential dependence of the saturation CO coverage of Pt electrodes: The origin of the pre-peak in CO-stripping voltammograms. Part 1: Pt(111). J Electroanal Chem 579: 1–12.

Love B, Lipkowski J. 1988. Effect of surface crystallography on electrocatalytic oxidation of carbon monoxide on Pt electrodes. ACS Symp Ser 378: 484.

Lu GQ, Crown A, Wieckowski A. 1999. Formic acid decomposition on polycrystalline platinum and palladized platinum electrodes. J Phys Chem B 103: 9700–9711.

McCallum C, Pletcher D, 1978. An investigation of the mechanism of the oxidation of carbon monoxide adsorbed onto a smooth Pt electrode in aqueous acid. J Electroanal Chem 70: 277.

Maciá MD, Herrero E, Feliu JM, Aldaz A. 1999. Formic acid self-poisoning on bismuth-modified Pt(755) and Pt(775) electrodes. Electrochem Commun 1: 87–89.

Maciá MD, Herrero E, Feliu JM, Aldaz A. 2001. Formic acid self-poisoning on bismuth-modified stepped electrodes. J Electroanal Chem 500: 498–509.

Maciá MD, Herrero E, Feliu JM. 2003. Formic acid oxidation on Bi–Pt(111) electrode in perchloric acid media. A kinetic study. J Electroanal Chem 554/555: 25–34.

Markovic NM, Ross PN Jr. 2002. Surface science studies of model fuel cell electrocatalysts. Surf Sci Rep 45: 117–229.

Markovic NM, Lucas CA, Rodes A, Stamenkovic V, Ross PN. 2002. Surface electrochemistry of CO on Pt(111): Anion effects. Surf Sci 499: L149–L158.

Mazouz N, Krischer K. 2000. A theoretical study on Turing patterns in electrochemical systems. J Phys Chem B 104: 6081–6090.

Mukouyama Y, Kikuchi M, Samjeske G, Osawa M, Okamoto H. 2006. Potential oscillations in galvanostatic electrooxidation of formic acid on platinum: A mathematical modeling and simulation. J Phys Chem B 110: 11912–11917.

Nichols RJ, Bewick A. 1988. SNIFTIRS with a flow cell: the identification of the reaction intermediates in methanol oxidation at platinum anodes. Electrochim Acta 33: 1691–1694.

Nishimura K, Kunimatsu K, Enyo M. 1989. Electrocatalysis on Pd + Au alloy electrodes: Part III. IR spectroscopic studies on the surface species derived from CO and CH_3OH in NaOH solution. J Electroanal Chem 260: 167.

Okamoto H, Kon W, Mukouyama Y. 2005. Five current peaks in voltammograms for oxidations of formic acid, formaldehyde, and methanol on platinum. J Phys Chem B 109: 15659–15666.

Palaikis L, Zurawski D, Hourani M, Wieckowski A. 1988. Surface electrochemistry of carbon monoxide adsorbed from electrolytic solutions at single crystal surfaces of Pt(111) and Pt(100). Surf Sci 199: 183–198.

Park S, Xie Y, Weaver MJ. 2002. Electrocatalytic pathways on carbon-supported platinum nanoparticles: Comparison of particle-size-dependent rates of methanol, formic acid, and formaldehyde electrooxidation. Langmuir 18: 5792–5798.

Parsons R, VanderNoot T. 1988. The oxidation of small organic molecules. A survey of recent fuel cell related research. J Electroanal Chem 257: 9–45.

Roberts JL, Sawyer DT. 1964. Voltammetric determination of carbon monoxide at gold electrodes. J Electroanal Chem 7: 315–319.

Rodriguez JL, Pastor E, Xia XH, Iwasita T. 2000. Reaction intermediates of acetaldehyde oxidation on Pt(111) and Pt(100). An in situ FTIR study. Langmuir 16: 5479–5486.

Samjeske G, Osawa M. 2005. Current oscillations during formic acid oxidation on a Pt electrode: Insight into the mechanism by time-resolved IR spectroscopy. Angew Chem 44: 5694–5698.

Samjeske G, Miki A, Ye S, Yamakata A, Mukouyama Y, Okamoto H, Osawa M. 2005. Potential oscillations in galvanostatic electrooxidation of formic acid on platinum: A time-resolved surface-enhanced infrared study. J Phys Chem B 109: 23509–23516.

Samjeske G, Miki A, Ye S, Osawa M. 2006. Mechanistic study of electrocatalytic oxidation of formic acid at platinum in acidic solution by time-resolved surface-enhanced infrared absorption spectroscopy. J Phys Chem B 110: 16559–16566.

Schmickler W. 1996. Interfacial Electrochemistry. New York: Oxford University Press.

Schmiemann U, Muller U, Baltruschat H. 1995. The influence of the surface-structure on the adsorption of ethene, ethanol and cyclohexene as studied by DEMS. Electrochim Acta 40: 99–107.

Shin J, Tornquist WJ, Korzeniewski C, Hoaglund CS. 1996. Elementary steps in the oxidation and dissociative chemisorption of ethanol on smooth and stepped surface planes of platinum electrodes. Surf Sci 364: 122.

Shubina TE, Hartnig C, Koper MTM. 2004. Density functional theory study of the oxidation of CO by OH on Au(110) and Pt(111) surfaces. Phys Chem Chem Phys 6: 4215–4221.

Smith PE, Ben-Dor KF, Abruña HD. 2000. Poison formation upon the dissociative adsorption of formic acid on bismuth-modified stepped platinum electrodes. Langmuir 16: 787–794.

Spendelow J, Lu GQ, Kenis PJA, Wieckowski A. 2004. Electrooxidation of adsorbed CO on Pt(111) and Pt(111)/Ru in alkaline media and comparison with results from acidic media. J Electroanal Chem 568: 215–224.

Spendelow J, Goodpaster JD, Kenis PJA, Wieckowski A. 2006. Mechanism of CO oxidation on Pt(111) in alkaline media. J Phys Chem B 110: 9545–9555.

Sriramulu S, Jarvi TD, Stuve EM. 1999. Reaction mechanism and dynamics of methanol electrooxidation on platinum (111). J Electroanal Chem 467: 132–142.

Stamenkovic V, Chou KC, Somorjai GA, Ross PN, Markovic NM. 2005. Vibrational properties of CO at the Pt(111)-solution interface: The anomalous Stark–Tuning slope. J Phys Chem B 109: 678–680.

Sun SG. 1998. Studying electrocatalytic oxidation of small organic molecules with in-situ infrared spectroscopy. In: Lipkowski J, Ross PN, eds. Electrocatalalysis. New York: Wiley-VCH. p 243–291.

Sun SG, Clavilier J. 1987. Electrochemical study on the poisoning intermediate formed from methanol dissociation at low index and stepped platinum surfaces. J Electroanal Chem 236: 95–112.

Sun SG, Yang YY. 1999. Studies of kinetics of HCOOH oxidation on Pt(100), Pt(110), Pt(111), Pt(510) and Pt(911) single crystal electrodes. J Electroanal Chem 467: 121–131.

Sun SG, Clavilier J, Bewick A. 1988. The mechanism of electrocatalytic oxidation of formic acid on Pt(100) and Pt(111) in sulphuric acid solution: An EMIRS study. J Electroanal Chem 240: 147–159.

Sun SG, Lin Y, Li NH, Mu JQ. 1994. Kinetics of dissociative adsorption of formic acid on Pt(100), Pt(610), Pt(210) and Pt(110) single-crystal electrodes in perchloric acid solutions. J Electroanal Chem 370: 273–280.

Tarnowski DJ, Korzeniewski C. 1997. Effects of surface step density on the electrochemical oxidation of ethanol to acetic acid. J Phys Chem B 101: 253.

Tremiliosi-Filho G, Gonzalez ER, Motheo AJ, Belgsir EM, Leger JM, Lamy C. 1998. Electro-oxidation of ethanol on gold: Analysis of the reaction products and mechanism. J Electroanal Chem 444: 31–39.

Tripkovic AV, Popovic KD, Lovic JD. 2001. The influence of the oxygen-containing species on the electrooxidation of the C-1–C-4 alcohols at some platinum single crystal surfaces in alkaline solution. Electrochim Acta 46: 3163–3173.

Tsang W. 1976. Thermal stability of alcohols. Int J Chem Kin 8: 173–192.

Varcoe JR, Slade RCT. 2005. Prospects for alkaline anion-exchange membranes in low temperature fuel cells. Fuel Cells 5: 187–200.

Vielstich W. 2003. CO, formic acid, and methanol oxidation in acid electrolytes—mechanisms and electrocatalysis. In: Bard AJ, Stratmann M, Calvo EJ, eds. Encyclopedia of Electrochemistry. Volume 2. New York: Wiley. p 466–511.

Vielstich W, Xia XH. 1995. Comments on "Electrochemistry of methanol at low index crystal planes of platinum: An integrated voltammetric and chronoamperometric study." J Phys Chem 99: 10421–10422.

Wang H, Baltruschat H. 2007. DEMS study on methanol oxidation at poly- and monocrystalline platinum electrodes: The effect of anion, temperature, surface structure, Ru adatom, and potential. J Phys Chem C 111: 7038–7048.

Willsau J, Heitbaum J. 1985. Elementary steps of ethanol oxidation on Pt in sulfuric-acid as evidenced by isotope labeling. J Electroanal Chem 194: 27–35.

Willsau J, Heitbaum J. 1986. Analysis of adsorbed intermediates and determination of surface potential shifts by DEMS. Electrochim Acta 31: 943–948.

Wolter O, Willsau J, Heitbaum J. 1985. Reaction pathways of the anodic oxidation of formic acid on platinum evidenced by oxygen-18 labeling—A DEMS study. J Electrochem Soc 132: 1635–1638.

Xia XH, Liess HD, Iwasita T. 1997. Early stages in the oxidation of ethanol at low index single crystal platinum electrodes. J Electroanal Chem 437: 233–240.

Zhou WP, Lewera A, Larsen R, Masel RI, Bagus PS, Wieckowski A. 2006. Size effects in electronic and catalytic properties of unsupported palladium nanoparticles in electrooxidation of formic acid. J Phys Chem B 110: 13393–13398.

CHAPTER 7

Clues for the Molecular-Level Understanding of Electrocatalysis on Single-Crystal Platinum Surfaces Modified by *p*-Block Adatoms

V. CLIMENT, N. GARCÍA-ARÁEZ, and J.M. FELIU

Instituto de Electroquímica, Universidad de Alicante, Ap. 99, E-03080 Alicante, Spain

7.1 INTRODUCTION

The development of new electrocatalytic materials by deposition of foreign adatoms on a host metal is currently a very active area of research in surface electrochemistry. It has been demonstrated that the simple modification of the surface composition of noble metal electrodes constitutes a convenient approach to enhance their electrocatalytic properties in a controlled way, as an alternative to the formation of bulk surface alloys. The deposition of submonolayer amounts of foreign metal adatoms can enhance the electrocatalytic activity of a given electrode or improve its selectivity. The localized modification of the surface composition offers several advantages, especially for the investigation from a fundamental point of view of the mechanism of electrocatalytic enhancement. It should be kept in mind that the electrocatalytic activity produced by submonolayer modification will not be stable in long-term practical applications, unless periodic regeneration of the adlayers is performed. Still, this approach will be useful to understand the reactivity of alloys and intermetallic compounds in an easy and controllable way. Normally, a given electrode can be modified with several surface ad-species, allowing easy systematic study of a broad range of compositions. This approach is especially advantageous when the initial electrode substrate can be recovered by removing the surface modifier, and, thus, a new electrode can be easily prepared with a different surface composition, a requisite that cannot be accomplished with bulk alloys. Also, from a fundamental point of view, this methodology allows one to perform studies with proper control of the surface structure, a

---

Fuel Cell Catalysis. Edited by Marc T. M. Koper
Copyright © 2009 John Wiley & Sons, Inc.

property that has been revealed in the last few decades as a fundamental parameter with paramount influence on the reactivity of a metal. In some cases, not only can the surface structure of the substrate material be finely controlled, but so can the location of the deposited species, achieving a regular distribution of adatoms that allows the study of structural parameters in great detail.

There are two main approaches to modifying the surface composition of an electrode with foreign metal adatoms. First, underpotential deposition occurs when there is equilibrium between the interphase and the solution containing the metallic precursor [Herrero et al., 2001; Szabo, 1991]. Underpotential deposition of a metal differs from bulk deposition in that the former corresponds to the formation of only one layer (sometimes two layers) of the metal on a foreign substrate. This process can take place at potentials more anodic than the Nernst potential for bulk deposition, owing to the existence of different bonding interactions between the metal and the substrate, stronger than those between the metal atoms in the bulk material. The existence of equilibrium between the precursor in solution and the adatom on the surface implies the existence of a relationship between electrode potential and surface coverage. This can be either an advantage or a disadvantage, depending on the desired experiment. On the one hand, this relationship allows precise modification of surface composition simply by selecting the electrode potential. In addition, the existence of equilibrium allows the use of thermodynamic relationships to characterize the composition of the adlayer. The clear disadvantage is that the surface composition depends on the potential, and therefore it is not easy to separate the effect of potential from the effect of coverage when electrocatalytic activity is investigated in the presence of the precursor in solution. Moreover, coverage by adatoms will depend on solution composition, and, although it can be well defined in a solution of simple composition, characterization of coverage is usually very difficult in the presence of other reductant or oxidant compounds in solution. This is especially complex when these compounds adsorb on the electrode, which is the most familiar case with compounds that undergo electrocatalytic processes.

The second procedure for surface modification of an electrode is irreversible adsorption of adatoms. Under certain conditions, some adatoms are adsorbed, and then remain on the surface of the electrode even in the absence of the precursor in solution [Clavilier et al., 1988, 1989a, 1990a, b; Evans and Attard, 1993; Feliu et al., 1988, 1991, 1993a, b; Gómez et al., 1992; Sung et al., 1997, 1998]. In this case, there is no equilibrium between the species on the surface and the solution—hence the terminology *irreversible*. The advantage of this method is that coverage and potential can be varied independently. This constitutes a distinctive advantage over underpotential deposition, and therefore most of studies of electrocatalysis with modified surfaces that have been performed recently are based on this method [Clavilier et al., 1989b, c; Climent et al., 1998; Fernández-Vega et al., 1989, 1991; Gomez et al., 1993, 1997; Herrero et al., 1993, 1995a–d, 1996; Leiva et al., 1997; Llorca et al., 1994; Maciá et al., 1999, 2001, 2002, 2003; Schmidt et al., 2000a, 2001; Smith and Abruña, 1999a, b; Smith et al., 1999, 2000].

Because of the very high reactivity of platinum in many reactions of technological interest, such as those in fuel cells, and the well-demonstrated requirement for control of the surface structure for fundamental studies (see, e.g., Chapter 6), surface modification of platinum single-crystal electrodes by irreversible adsorption of adatoms has been extensively studied in the literature, and it will be the central subject of this chapter. First, methods of preparation and electrochemical characterization will be reviewed in the next section. Then, the effect of surface modification on electrocatalysis will be discussed, and some reasonable explanations for the enhancement of electrocatalytic activity by adatom deposition will be proposed.

7.2 ELECTROCHEMICAL CHARACTERIZATION OF MODIFIED ELECTRODES

7.2.1 Procedure for Irreversible Adsorption of the Adatom

Many elements of the *p*-block of the periodic table spontaneously adsorb on the surface of a platinum electrode when this is immersed in a solution containing a soluble salt of the element, without an external supply of electricity [Clavilier et al., 1988, 1989a, b, 1990a, b; Evans and Attard, 1993; Feliu et al., 1988, 1991, 1993a, b; Gómez et al., 1992; Sung et al., 1997, 1998]. The electrode can then be rinsed and transferred to an electrochemical cell that does not contain the corresponding ion of the deposited element, which remains on the surface, irreversibly adsorbed.

The process for this irreversible adsorption has not been investigated in detail. The mechanism by which the metal is deposited has not been unambiguously elucidated, and several possibilities have been proposed. One possibility is the formation of local cells, with the ion of the adatom being reduced and either hydrogen [Szabo and Nagy, 1978] or platinum [Clavilier et al., 1988] being oxidized:

$$\begin{aligned} \text{cathode} \quad & \text{Pt} + \text{M}_{\text{sol}}^{z+} + (z-z')e^- \longrightarrow \text{Pt} - \text{M}_{\text{ads}}^{z'+} \\ \text{anode} \quad & \text{Pt-H} \longrightarrow \text{Pt} + \text{H}^+ + e^- \\ & \text{Pt} + \text{H}_2\text{O} \longrightarrow \text{Pt-OH} + \text{H}^+ + e^- \end{aligned} \quad (7.1)$$

where z', the charge retained by the adsorbed adatom, can be zero in some cases. The potential of the anodic reaction should be lower than the potential for the reduction of the adatom, if the local cell is to be spontaneous. For hydrogen oxidation, this condition is plausible, although spontaneous deposition of adatoms also occurs even when the surface is not expected to be pre-covered by hydrogen. In this regard, it is worth mentioning that adsorbed Pb and Ge species on Pt electrodes do not remain stable on the surface unless the rinsing step is avoided or it is done using water in equilibrium with H_2/Ar [Clavilier et al., 1990b; Gómez et al., 1992]. On the other hand, the possibility that Pt-OH is formed at low enough potentials is unexpected, especially in the case of Pt(111) surfaces. In this regard, it has been proposed that the surface oxidation could take place at defect sites [Clavilier et al., 1988].

In most cases, the adatom deposition is reductive, as sketched in (7.1), because the adatom precursor in the solution is in an oxidized state. However, oxidative adsorption has also been reported, for instance, in the case of sulfur [Sung et al., 1997, 1998]. In this case, oxygen reduction seems the more likely cathodic reaction:

$$Pt + S^{2-} + 2H^+ + \tfrac{1}{2}O_2 \longrightarrow Pt\text{-}S + H_2O \tag{7.2}$$

Alternatively, a disproportionation reaction has been proposed for the case of Sn [Rodes et al., 1988]:

$$Pt + 2Sn^{2+} \longrightarrow Pt\text{-}Sn + Sn^{4+} \tag{7.3}$$

Finally, another possibility is an initial physisorption (adsorption without charge transfer) of the adatom when the electrode is put in contact with the solution, the physisorbed species remaining on the surface after rinsing, and the reduction taking place after the first negative scan of the potential.

7.2.2 Voltammetric Characterization of the Modified Electrode

The presence of the adatom on the surface after irreversible deposition is evidenced by a significant suppression of the hydrogen and anion adsorption processes characteristic of clean Pt electrodes. In addition, a new redox process takes place, resulting in two new voltammetric peaks. The voltammetric profile remains stationary over a wide potential range (typically from 0 to 0.8–1.0 V vs. RHE), indicating that the adatom remains stable on the surface. However, desorption of the adsorbed adatom takes place if the upper potential limit is increased further, as evidenced from the decrease of the redox peaks associated with the adatom and the progressive recovery of the hydrogen and anion adsorption processes on the free Pt sites. Figures 7.1 and 7.2 show some representative cyclic voltammograms corresponding to adatom-modified Pt(111) and Pt(100) electrodes. It is evident from these figures that the peak potential and the shape of the voltammograms for the new redox process are characteristic of the nature of the adatom. Comparison of Figs. 7.1 and 7.2 reveals that the surface process is also very sensitive to the crystallographic orientation of the electrode surface. The new redox peak has been attributed to the oxidation/reduction of the deposited adatoms according to

$$Pt_m\text{-}M \longrightarrow Pt_m\text{-}M^{n+} + ne^- \tag{7.4}$$

From the pH dependence of the peak potential (close to 60 mV/decade), it was proposed that the oxidation involves adsorption of oxygenated species, leading to the formation of either the oxide or the hydroxide:

$$Pt_m\text{-}M + nH_2O \rightleftharpoons Pt_m\text{-}M(OH)_n + nH^+ + ne^- \tag{7.5a}$$
$$Pt_m\text{-}M + \tfrac{n}{2}H_2O \rightleftharpoons Pt_m\text{-}MO_{n/2} + nH^+ + ne^- \tag{7.5b}$$

where m is the number of Pt sites blocked by the adatom and n is the number of electrons transferred in the oxidation of one adatom. It has been observed that the charge density associated with this surface process and the blockage of the hydrogen and

7.2 ELECTROCHEMICAL CHARACTERIZATION OF MODIFIED ELECTRODES

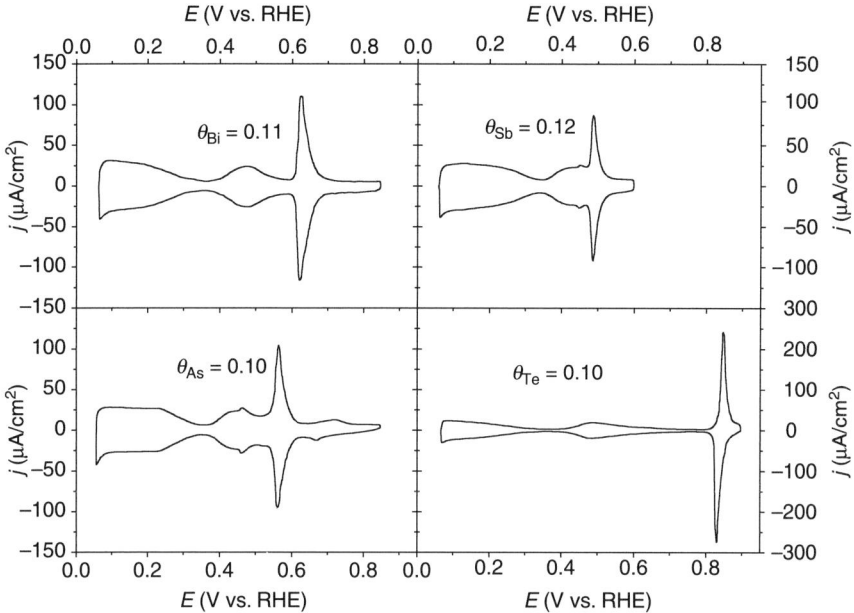

Figure 7.1 Cyclic voltamograms of Pt(111) electrodes modified by Bi, Sb, As, and Te deposition at intermediate coverages, as indicated, in 0.5 M H_2SO_4 solution. Sweep rate: 50 mV/s.

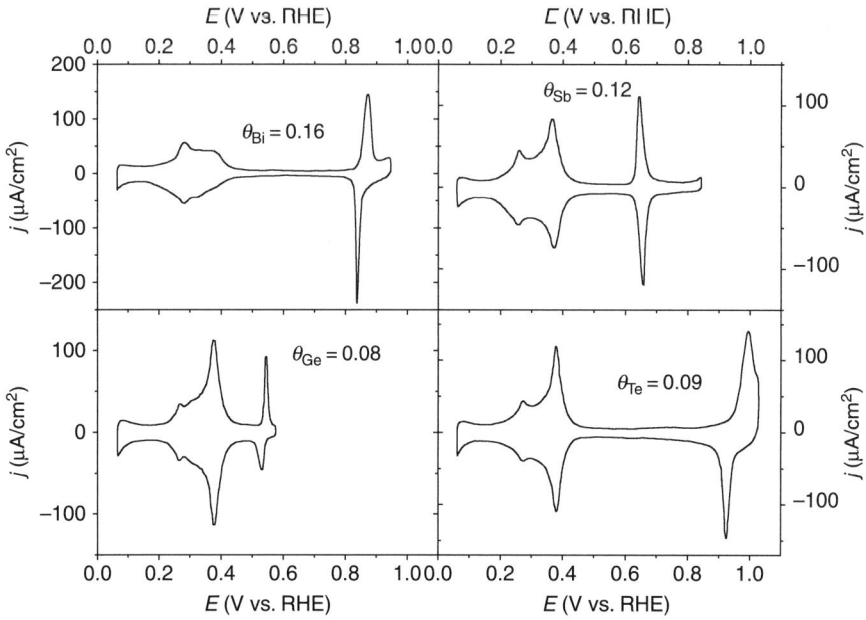

Figure 7.2 Cyclic voltamograms of Pt(100) electrodes modified by Bi, Sb, Ge, and Te deposition at intermediate coverages, as indicated, in 0.5 M H_2SO_4 solution. Sweep rate: 50 mV/s.

anion adsorption reactions on the bare Pt sites are directly proportional. From the relation between the charge magnitudes, it is possible to obtain the ratio between m and n (usually referred to as the stoichiometry of the adatom reaction), as discussed in the following paragraphs.

The adatom coverage is given by

$$\theta = \frac{q_{Ad}}{n q_{Pt(hkl)}} \tag{7.6}$$

where q_{Ad} is the charge density involved in the adatom oxidation process and $q_{Pt(hkl)}$ is the charge density corresponding to the transfer of one electron per Pt atom on the surface [241 $\mu C/cm^2$ for Pt(111) and 209 $\mu C/cm^2$ for Pt(100)].

The charge density corresponding to hydrogen adsorption on the free Pt sites is given by

$$q_H^\theta = q_H^{\theta=0}(1 - m\theta) \tag{7.7}$$

where $q_H^{\theta=0}$ is the maximum hydrogen charge density attained in the absence of the adatom. From (7.6) and (7.7), it is possible to eliminate θ, obtaining the following relation between hydrogen and adatom charge densities:

$$q_H^\theta = q_H^{\theta=0} - \frac{m}{n} \frac{q_H^{\theta=0}}{q_{Pt(hkl)}} q_{Ad} \tag{7.8}$$

Then, the maximum adatom charge density corresponding to full blockage of hydrogen adsorption is

$$q_{Ad} = \frac{n}{m} q_{Pt(hkl)} \tag{7.9}$$

Since in many cases it is not possible to unambiguously separate hydrogen from anion adsorption processes, it is convenient to write another equation similar to (7.7) for the charge density corresponding to anion adsorption on the free Pt sites:

$$q_{An}^\theta = q_{An}^{\theta=0}(1 - m\theta) \tag{7.10}$$

At the time when most of the studies on the characterization of adatom-modified Pt surfaces were done, it was customary to integrate the total Pt voltammetric charge, corresponding to the sum of the hydrogen and anion adsorption processes on the free Pt sites [Clavilier et al., 1989a, 1990a; Evans and Attard, 1993; Feliu et al., 1991, 1993a, b; Gómez et al., 1992]:

$$q_{Pt}^\theta = q_H^\theta + q_{An}^\theta = (q_H^{\theta=0} + q_{An}^{\theta=0})\left(1 - m\frac{q_{Ad}}{nq_{Pt(hkl)}}\right) = q_{Pt}^{\theta=0} - \frac{m}{n}\frac{q_{Pt}^{\theta=0}}{q_{Pt(hkl)}} q_{Ad} \tag{7.11}$$

In this equation, it has been assumed that the adatom exerts the same blockage on hydrogen and anion adsorption (i.e., the stoichiometric number m is the same for both processes). Although, a priori, this assumption seems arbitrary, it has been proven correct in many situations. Besides, in the particular case of Pt(111) in H_2SO_4 solutions, it has been shown that $q_{Pt}^{\theta=0} = (q_H^{\theta=0} + q_{An}^{\theta=0}) = q_{Pt(hkl)}$ [Feliu et al., 1994], and hence (7.11) simplifies to

$$q_{Pt}^{\theta} = q_{Pt}^{\theta=0} - \frac{m}{n} q_{Ad} \Rightarrow q_{Pt}^{\theta=0} = q_{Pt}^{\theta} + \frac{m}{n} q_{Ad} \quad (7.12)$$

On the other hand, in certain cases (e.g., Ad/Pt(111) in $HClO_4$ solutions), the adatom oxidation and anion adsorption on the free sites (OH adsorption in $HClO_4$) overlap. Then, another refinement to this analysis would be to use (7.7) and (7.10) to calculate q_{An}^{θ} according to

$$q_{An}^{\theta} = q_{An}^{\theta=0} \frac{q_H^{\theta}}{q_H^{\theta=0}} \quad (7.13)$$

and then subtract this charge density from the total charge density measured under the adatom peak [Climent et al., 2006].

According to (7.8) and (7.12), the stoichiometry (m/n) can be extracted from the slope of the plots of adatom charge density versus hydrogen (or hydrogen plus anion) charge density. Some representative plots are shown in Fig. 7.3. The conclusions extracted from this kind of analysis are summarized in Tables 7.1 and 7.2 for Pt(111) and Pt(100) modified surfaces, respectively. The extension of this analysis

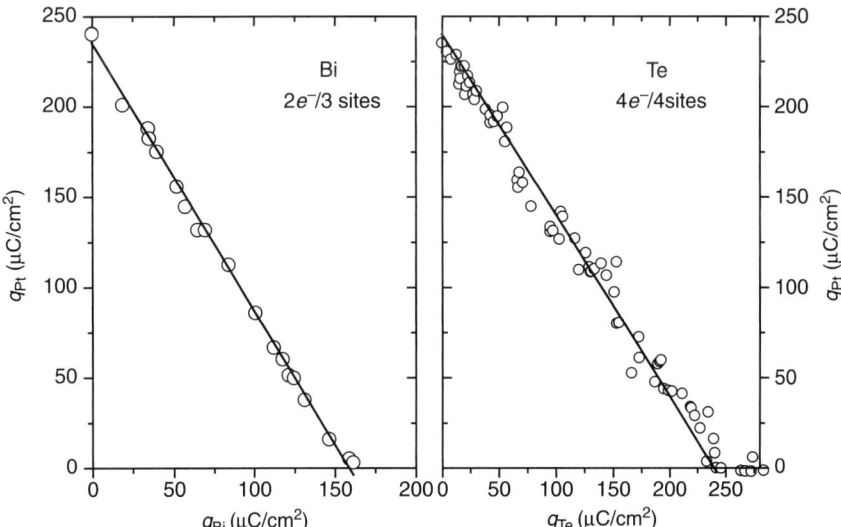

Figure 7.3 Plot of the platinum (hydrogen plus anion) charge density versus the charge density associated with the adatom redox process (Bi or Te, as indicated) on a Pt(111) electrode in 0.5 M H_2SO_4 solution. Straight lines represent the expected behavior for the stoichiometry indicated in the figure.

TABLE 7.1 Summary of Redox Properties of Adatom-Modified Pt(111) Electrodes in 0.5 M H_2SO_4 Solution (Except Where Otherwise Stated). All Potentials are Given vs. RHE and Measured at 50 mV/s

		S [Batina et al., 1989] $E_p > 0.8$ V in 10 mM NaF at pH 9–10[a] $S(0) \rightleftharpoons S(VI) + 6e^-$ $\theta_{max} = 0.33$[a]
Ge [Gómez et al., 1992] $E_{an} = 0.73$ V[b] $Ge(0) \rightleftharpoons Ge(IV) + 4e^-$	As [Feliu et al., 1988] $E_p = 0.56$ V[c] $q_{Pt}^0 = q_{Pt} + q_{As}$ $As(0) \rightleftharpoons As(III) + 3e^-$ $\theta = q_{As}/(3 \times 241)$ $\theta_{max} = 0.33$	Se [Feliu et al., 1993a] $E_{an} = 1.08$ and 0.98 V[d] $E_{cat} = 0.79$ V $q_{Pt}^0 = q_{Pt} + \frac{3}{4} q_{Se}$ $Se(0) \rightleftharpoons Se(IV) + 4e^-$ $\theta = q_{Se}/(4 \times 241)$ $\theta_{max} = 0.33$
Sn [Xiao et al., 2002] Broad peaks in 0.05 M H_2SO_4[e] $Sn(0) \rightleftharpoons Sn(IV) + 4e^-$ $\theta = q_{Sn}/(4 \times 241)$ $\theta_{max} = 0.32$	Sb [Climent et al., 1998; Feliu et al., 1988] $E_p = 0.485$ V[f] $q_{Pt}^0 = q_{Pt} + \frac{3}{2} q_{Sb}$ $Sb(0) \rightleftharpoons Sb(II) + 2e^-$ $\theta = q_{Sb}/(2 \times 241)$ $\theta_{max} = 0.33$	Te [Feliu et al., 1993b] $E_p = 0.83$ V[g] $q_{Pt}^0 = q_{Pt} + q_{Te}$ $Te(0) \rightleftharpoons Te(IV) + 4e^-$ $\theta = q_{Te}/(4 \times 241)$ $\theta_{max} = 0.25$
Pb [Feliu et al., 1991] $E_p =$ several peaks between 0.5 and 0.6 V[h] $q_{Pt}^0 = q_{Pt} + \frac{3}{2} q_{Pb}$ $Pb(0) \rightleftharpoons Pb(II) + 2e^-$ $\theta = q_{Pb}/(2 \times 241)$ $\theta_{max} = 0.33$	Bi [Feliu et al., 1991] $E_p = 0.62$ V[i] $q_{Pt}^0 = q_{Pt} + \frac{3}{2} q_{Bi}$ $Bi(0) \rightleftharpoons Bi(II) + 2e^-$ $\theta = q_{Bi}/(2 \times 241)$ $\theta_{max} = 0.33$	

[a]The S oxidation process involves desorption of S adatoms, and the formation of H_2SO_4 species has been tentatively proposed. The maximum coverage was determined from Auger electron spectroscopy (AES) and low energy electron diffraction (LEED) measurements. When the irreversible adsorption of sulfur is performed in an oxygen-free Na_2S solution, the formation of more compact S adlayers on Pt(111) has been identified from AES and LEED measurements [Sung et al., 1997, 1998].
[b]The Ge adatoms do not remain adsorbed after oxidation, and the characteristic voltammetric profile of the blank is recovered after about 2–3 cycles up to 1.2 V. The Ge reduction process does not take place in a well-defined peak.
[c]At near-saturation coverages, the As redox process splits, with $E_{p2} = 0.54$ V. Partial desorption of As can be achieved by increasing the positive potential limit to values higher than 0.85 V.
[d]The second Se oxidation peak at 0.98 V appears at $\theta_{Se} > 0.17$. At near-saturation coverages, this peak becomes broader and shifts to positive potentials, leading to an apparent increase in current density of the peak at 1.08 V. In order to desorb the Se adatoms, the upper potential limit has to be increased to 1.3 V, where a new, small oxidation peak is observed, tentatively assigned to the formation of Se(VI) species.
[e]At saturation coverage, Sn oxidation starts at 0.54 V and shows a broad wave, centered at 0.86 V, up to 1.3 V. The reduction process is centered at 0.56 V. At lower Sn coverages, a new peak appears at 0.61 V, which becomes sharper and is displaced to 0.56 V as the coverage is further decreased. This second process has been tentatively assigned to $Sn \rightarrow Sn(I)$.
[f]At high coverages ($\theta_{Sb} \geq 0.2$), the formation of a Pt–Sb alloy has been proposed, concomitant with a broadening of the Sb process at 0.49 V and the appearance of a new peak around 0.43 V.
[g]Only at saturation coverage is there a certain degree of irreversibility, which is reflected in $\Delta E_p = E_{an} - E_{cat} = 45$ mV. A new oxidation process is observed at 1.01 V that leads to progressive dissolution of the adsorbed adlayer. This process has been tentatively ascribed to the formation of soluble Te(VI) species. The corresponding reduction process is observed around 0.9 V.
[h]A continuous decrease in the amount of adsorbed Pb is observed in successive voltammetric cycles up to 0.6 V. For this reason, the stoichiometry of the Pb process has been inferred from measurements in 0.1 M NaOH solutions.
[i]The dissolution of the adsorbed Bi only takes place at potentials around 1 V.

TABLE 7.2 Summary of Redox Properties of Adatom-Modified Pt(100) Electrodes in 0.5 M H_2SO_4 Solution (Except Where Otherwise Stated). All Potentials are Given vs. RHE and Measured at 50 mV/s

		S [Batina et al., 1989] $E_p > 0.8$ V in 10 mM NaF at pH 9–10[a] $S(0) \rightleftharpoons S(VI) + 6e^-$ $\theta_{max} = 0.5$[a]
Ge [Gómez et al., 1992] $E_p = 0.54$ V[b] $q_{Pt}^0 = q_{Pt} + 2q_{Ge}$ $Ge(0) \rightleftharpoons Ge(IV) + 2e^-$ $\theta = q_{Ge}/(2 \times 209)$ $\theta_{max} = 0.25$	As [Feliu et al., 1988] Multiple peaks[c]	Se [Feliu et al., 1993a] $E_{an} = 1.15$ V $E_{cat} \approx 0.82$ V[d]
Sn	Sb [Feliu et al., 1988] $E_p = 0.65$ V[e] $q_{Pt}^0 = q_{Pt} + q_{Sb}$ $Sb(0) \rightleftharpoons Sb(II) + 2e^-$ $\theta = q_{Sb}/(2 \times 209)$ $\theta_{max} = 0.50$	Te [Feliu et al., 1993b] $E_{an} = 1.0$ V $E_{cat} = 0.93$ V[f] $q_{Pt}^0 = q_{Pt} + \frac{1}{2}q_{Te}$ $Te(0) \rightleftharpoons Te(IV) + 4e^-$ $\theta = q_{Te}/(4 \times 209)$ $\theta_{max} = 0.50$
Pb [Clavilier et al., 1990b] $E_{an} = 0.830$ V, $E_{cat} = 0.815$ V in 0.1 M NaOH[g]	Bi [Clavilier et al., 1989a] $E_{an} = 0.88$ V, $E_{cat} = 0.84$ V[h] $q_{Pt}^0 = q_{Pt} + q_{Bi}$ $Bi(0) \rightleftharpoons Bi(II) + 2e^-$ $\theta = q_{Bi}/(2 \times 209)$ $\theta_{max} = 0.50$	

[a] The S oxidation process involves desorption of S adatoms, and the formation of H_2SO_4 species has been tentatively proposed. The maximum coverage was determined from AES and LEED measurements.
[b] At near-saturation coverages, the Ge process becomes less reversible and the reduction process is split to lower potential values. There is a second Ge process at $E_{an2} = 0.72$ V, $E_{cat2} = 0.62$ V.
[c] Two broad peaks at 0.55 and 0.75 V. Partial desorption of As can be achieved by increasing the positive potential limit to values higher than 0.9 V.
[d] The Se oxidation process also implies partial dissolution of the adatoms and oxidation of the bare Pt sites, thus hindering the acquisition of any quantitative information of the adatom process on well-ordered Pt(100) electrodes.
[e] Electrochemical desorption of Sb occurs above 0.9 V.
[f] At near-saturation coverages, the oxidation peak potential is displaced to approximately 1.03 V. There is a second oxidation process at 1.1 V that involves desorption of Te adatoms, which has been ascribed to the displacement of the Te(IV) oxide by surface Pt oxide.
[g] At saturation coverage, the Pb oxidation process is split into three peaks, and the charge under the Pb redox process amounts to approximately 300 $\mu C/cm^2$, indicating the formation of a close-packed layer.
[h] A second oxidation peak at 0.96 V can be observed, ascribable to the oxidative desorption of Bi(II) adsorbed on defects, to form Bi(III) species in solution.

to the Pt(110) electrode is complicated by the fact that the adatom oxidation processes on this electrode take place at relatively high potentials, and severe interference from the surface oxidation of the substrate is expected. Moreover, oxidation/reduction cycles induce disordering of the surface, precluding a study with proper control of the surface structure [Clavilier et al., 1988; Hayden et al., 1996].

In some cases, the maximum coverage of the adatom and the stoichiometry of the reaction have been corroborated by independent techniques, normally involving the transfer of the modified surface to ex situ conditions.

Ex situ scanning tunneling microscopy (STM) measurements demonstrated that the maximum As coverage on Pt(111) is 0.33, corresponding to a $(\sqrt{3} \times \sqrt{3})R30°$ structure [Orts et al., 1997]. Moreover, in situ infrared measurements showed that the participation of adsorbed anions during the surface redox process could be ignored [Orts et al., 1997]. This last result validates the above explained analysis of charge densities, since the participation of anion adsorption/desorption, coupled to adatom oxidation, could lead to some uncertainty in the estimation of surface coverages from coulometric measurements. Another system that has been characterized by STM, this time in situ, is Te on Pt(111) [Rhee and Kim, 2001]. In this case, two structures were found for the reduced state at saturation coverage, namely (2×2) and (11×8), corresponding to coverages 0.25 and 0.284, respectively. These values of the Te maximum coverage agree well with the conclusions extracted from the coulometric data, $\theta_{Te,max} = 0.25$ [Feliu et al., 1993b]. In the oxidized state, a square $(\sqrt{2} \times \sqrt{2})$ lattice was detected, and interpreted as the intermixing of two $(2 \times \sqrt{3})$ rectangular structures, for Te and O. The same structures could also be identified by ex situ STM experiments for Pt(111) and Pt(111) stepped surfaces [Rodriguez et al., 2006]. For the Pt(100) basal plane and vicinal surfaces, a $c(2 \times 2)$ square structure was imaged with ex situ STM, which also confirms the maximum coverage of 0.5 obtained by coulometry [Rodriguez et al., 2006].

The valence state of different adatoms and its dependence on the electrode potential has been monitored in several studies for As, Bi, and Te by means of X-ray photoelectron spectroscopy (XPS) measurements [Hamm et al., 1998; Schmidt et al., 2000b; Zhou et al., 2002, 2004]. A clear limitation of this approach is the uncertainty that arises during the emersion of the double layer and subsequent transfer to the ultrahigh vacuum (UHV) environment. It was recognized by the authors of these works that some decomposition of the oxidized species can take place during the emersion, and therefore, some caution should be applied in the quantitative analysis of the results. However, the technique has proven valuable in demonstrating that a change of the oxidation state of the adatoms takes place during the redox process observed in the voltammetric profile. For As on Pt(111) [Zhou et al., 2004], it was confirmed that the valence state of the ad-species changes during the voltammetric redox process from As(0) to As(III). Similarly, for Te on Pt(111) [Zhou et al., 2002], a change from Te(0) to Te(IV) was observed, in accordance with the electrochemical measurements. On the other hand, contrasting results were obtained for Bi on Pt(111) [Hamm et al., 1998; Schmidt et al., 2000b], since for this system no change in the valence state of the adatom was observed and only elemental Bi was measured, regardless of the emersion potential. Clearly, this result has to be taken with great caution, since a change of the redox state of the adatom can take place during the collapse of the double layer in UHV. The question could be solved with careful measurements of the work function changes after emersion, since it is expected that the work function of the emersed electrode should be proportional to the emersion potential, while variations from this trend would indicate that the electrode no longer behaves as ideally polarizable [Stuve et al., 1995].

7.3 ENERGETICS OF THE SURFACE REDOX PROCESS

There are conceptual similarities between the linear sweep voltammetry and the temperature programmed desorption (TPD) experiments. The latter are commonly performed in gas phase studies of adsorption to determine binding energies of adsorbed species. In this regard, the variation of the electrode potential in the voltammetric experiment can be understood as the imposition of an increasing driving force towards desorption, analogous to the effect of the temperature increase in TPD experiments. Hence, the potential at which desorption takes place contains energetic information about the adsorption/desorption process. In the case of the redox processes exhibited by irreversibly adsorbed adatoms, the reaction driven by the potential is the adsorption/desorption of oxygenated species, and the peak potential must then be related to the energy of adsorption of oxygen or hydroxyl species on the adatom-modified surface. In some particular cases, however, the adatom oxidation also involves the desorption of the oxidized adatom species. In this case, the peak potential value will also be influenced by the interaction energy of the adatom with the surface.

Applying the common equations for the thermodynamics of reversible cells, it is possible to extract energetic parameters for the adatom redox reaction. This approach requires the measurement of voltammograms at different temperatures. If we consider that the adatom oxidation reaction involves the formation of the hydroxide, we can write the following equation for the overall cell reaction:

$$\text{Pt-M} + n\text{H}_2\text{O} \longrightarrow \text{Pt-M(OH)}_n + \tfrac{1}{2}n\text{H}_2 \qquad (7.14)$$

Then, the equilibrium potential (peak potential) is related to ΔG^0 for this reaction by:

$$\Delta G^0 = nFE \qquad (7.15)$$

(Since the reaction in the working electrode is an oxidation when the overall reaction is (7.14), the cell potential in (7.15) is defined as $E = E_{\text{work}} - E_{\text{ref}} = E_{\text{anode}} - E_{\text{cathode}}$ and the sign in this equation is opposite to that obtained with the more common convention that defines the cell potential as $E = E_{\text{cathode}} - E_{\text{anode}}$.) From the temperature variation of the cell potential, the following equation can be written for the entropy of the overall reaction:

$$\Delta S^0 = -nF\left(\frac{\partial E}{\partial T}\right)_\theta \qquad (7.16)$$

Finally, the enthalpy of the reaction can be extracted from

$$\Delta H^0 = nF \left(\frac{\partial (E/T)}{\partial (1/T)} \right)_\theta \quad (7.17)$$

It is interesting to compare these thermodynamic magnitudes with the bulk formation of the hydroxide, given by:

$$M + \tfrac{n}{2} H_2 + \tfrac{n}{2} O_2 \longrightarrow M(OH)_n \quad (7.18)$$

Clearly, (7.14) and (7.18) are related by the reaction of water formation:

$$H_2 + \tfrac{1}{2} O_2 \longrightarrow H_2O \quad (7.19)$$

Consequently, it is convenient to convert the thermodynamic results obtained in the electrochemical media, corresponding to the reaction (7.14), to those corresponding to the analogous reaction (7.18). This can easily be done by adding the tabulated values of the energetic parameters of water formation: ($\Delta G^0_{H_2O} = -237.1$ kJ/mol, $\Delta H^0_{H_2O} = 285.8$ kJ/mol, and $\Delta S^0_{H_2O} = -163.3$ J/mol K) [Lide and Frederikse, 1998].

It is worth mentioning that, in some cases, the adatom oxidation reaction may involve the formation of oxide species. In particular, the oxidation of electronegative adatoms, such as Te and Se, involves the interchange of four electrons, giving rise to Te(IV) and Se(IV) species. The formation of tetravalent hydroxide species seems very unlikely, both for steric reasons and because of the strong polarizing character of the cation, and hence the following equation for the overall cell reaction can be proposed:

$$Pt\text{-}M + \tfrac{n}{2} H_2O \longrightarrow Pt\text{-}MO_{n/2} + \tfrac{n}{2} H_2 \quad (7.20)$$

In this case, the values of the thermodynamic magnitudes can be also determined from (7.15)–(7.17), but these results should be compared with the bulk formation of the oxide:

$$M + \tfrac{n}{4} O_2 \longrightarrow MO_{n/2} \quad (7.21)$$

Unfortunately, the above thermodynamic approach has only been followed for two systems, namely As-Pt(111) and Bi-Pt(111) [Blais et al., 2001, 2002]. Table 7.3 summarizes the main results. Thermodynamic data about the bulk formation of As(OH)$_3$ and Bi(OH)$_2$ are not available for comparison. The only data available is the standard enthalpy for the bulk formation of Bi(OH)$_3$ ($\Delta H^0_f = -711.3$ kJ/mol). Considering that the standard enthalpies of formation of various oxides, sulfides, and halides of As, Sb, and Bi are very close to each other, the similarity between this value and the standard enthalpy for the surface formation of As(OH)$_3$, $\Delta H^0_f = -680 \pm 20$ kJ/mol, was taken as an indication that, indeed, the As redox process on Pt(111) involves the formation of hydroxide species [Blais et al., 2001]. For comparison with ΔH^0_f for the surface formation of Bi(OH)$_2$, the strategy followed

TABLE 7.3 Energetic Data for the Bi-Pt(111) [Blais et al., 2002] and As-Pt(111) [Blais et al., 2001] Oxidation Processes in 0.5 M H_2SO_4 Solution

	E_p (V vs. SHE)	dE_{an}/dT (mV/K)	dE_{cat}/dT (mV/K)	ΔS^0 (J/mol K)	ΔG^0 (kJ/mol)	ΔH^0 (kJ/mol)[a]	ΔH_f^0 (kJ/mol)[a]
Bi-Pt(111)	0.60	−0.288	−0.237	56, 46	116 (SHE)	133, 129	−439, −443
As-Pt(111)[b]	0.55	0.355	0.543	−103, −155	160 (SHE)	191, 184	−666, −674
		0.536	0.726	−157, −210		174, 166	−684, −691

[a] ΔH^0 and ΔH_f^0 are the enthalpies of formation of the surface hydroxide according to (7.14) and (7.18), respectively.
[b] For As, the different values of dE/dT are due to the splitting of the peak into two peaks with slightly different temperature dependences.

in [Blais et al., 2002] was to correct the value of $Bi(OH)_3$ with a factor of 2/3. In this way, reasonable coincidence is observed between the enthalpy of formation of the surface hydroxide in the electrochemical environment ($\Delta H_f^0 = -440$ kJ/mol) and the formation of the bulk hydroxide ($\Delta H_f^0 = -474$ kJ/mol), supporting the interpretation of the voltammetric data.

Although, for other adatoms, the study of the temperature dependence of the peak potential has not yet been performed, it is still worth analyzing the relationship between the peak potential of the adatom redox process and the energy of the bond between the element and the oxygen or hydroxyl species. Figure 7.4 plots peak potential values of adatom redox processes on Pt(111) and Pt(100) electrodes against the enthalpy of formation of the corresponding bulk oxide [Dean, 1999]. In this graph, the distinction between the formations of oxide or hydroxide species has not been considered, in order to compare results for different adatoms. Following the procedure in [Blais et al., 2002], when the oxidation state of the adatom on the Pt surface is not stable in the bulk solid state, the enthalpy values have been corrected by the ratio between the oxidation state in the electrochemical environment and in the bulk oxide. This is the case for Bi_2O_3, Sb_2O_3, Pb_3O_4, and GeO_2.

It is observed that higher potential values for the adatom redox process are correlated with a lower energy of the M—O bond, i.e., lower (less negative) enthalpy of formation of the adatom oxygenated species. In this regard, the discrepant behavior of Ge-Pt(100) may be related to the dilute nature of this adlayer, with a maximum coverage of only 0.25.

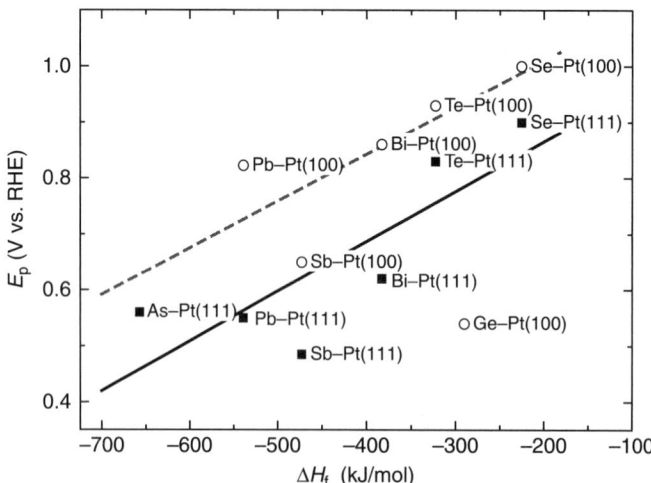

Figure 7.4 Peak potential values of adatom redox processes on Pt(111) and Pt(100) electrodes in 0.5 M H_2SO_4 solution, as labeled, plotted against the enthalpy of formation of the corresponding bulk oxide. Lines are included to indicate the tendency (the full line corresponds to the filled squares, and the dashed line to the open circles).

TABLE 7.4 Summary of Work Function Values [Haas, 1972; Michaelson, 1977; Trasatti, 1972] and Electronegativities [Lide and Frederikse, 1998] of the Adatoms Studied in the Present Work. The Corresponding Values for Pt are $\chi \approx 2.2$ [Lide and Frederikse, 1998] and $\Phi \approx 5.40$ eV [Trasatti, 1972]

		S $\chi = 2.58$ $\Phi \approx 6.2$ eV[a]
Ge [Michaelson, 1977] $\chi = 2.01$ $\Phi \approx 5.0$ eV	As [Haas, 1972] $\chi = 2.18$ $\Phi \approx 4.66$–4.72 eV	Se [Michaelson, 1977] $\chi = 2.55$ $\Phi \approx 5.9$ eV
Sn [Trasatti, 1972] $\chi = 1.96$ $\Phi \approx 4.35$ eV	Sb [Trasatti, 1972] $\chi = 2.05$ $\Phi \approx 4.56$ eV	Te [Trasatti, 1972] $\chi = 2.1$ $\Phi \approx 4.70$ eV
Pb [Trasatti, 1972] $\chi = 1.8$ $\Phi \approx 4.18$ eV	Bi [Trasatti, 1972] $\chi = 1.9$ $\Phi \approx 4.36$ eV	

[a]Work function values calculated from electronegativities [Gordy and Thomas, 1956].

On the other hand, it is also observed that adatom redox processes take place, in most cases, at higher potential values on a Pt(100) substrate than on a Pt(111) substrate. This behavior is unexpected from the point of view that oxidation of the Pt(100) substrate takes place at lower potentials than on Pt(111), owing to the more open surface structure and lower work function of the former.

Finally, it is worth noting that the peak potential value of the adatom redox process and the enthalpy of formation of the oxide are also correlated with the electronegativity and work function of the adatom, more electronegative adatoms having higher peak potential. Electronegativity and work function values of the different adatoms are summarized in Table 7.4 for comparison purposes.

7.4 STEPPED SURFACES

It has been often stressed that low coordinated atoms (defects, steps, and kink sites) play an important role in surface chemistry. The existence of dangling bonds makes steps and kinks especially reactive, favoring the adsorption of intermediate species on these sites. Moreover, studies of single-crystal surfaces with a complex geometry have been demonstrated very valuable to link the gap between fundamental studies of the basal planes [Pt(111), Pt(100), and Pt(110)] and applied studies of nanoparticle catalysts and polycrystalline materials. In this context, it is relevant to mention results obtained with adatom-modified Pt stepped surfaces, prior to discussing the effect of adatom modification on electrocatalysis.

Interestingly, it has been shown that some adatoms can be selectively deposited on step sites, taking advantage of the enhanced reactivity of these sites. Figure 7.5 shows the voltammogram of a Pt(775) surface in 0.5 M H_2SO_4. The hard sphere model for

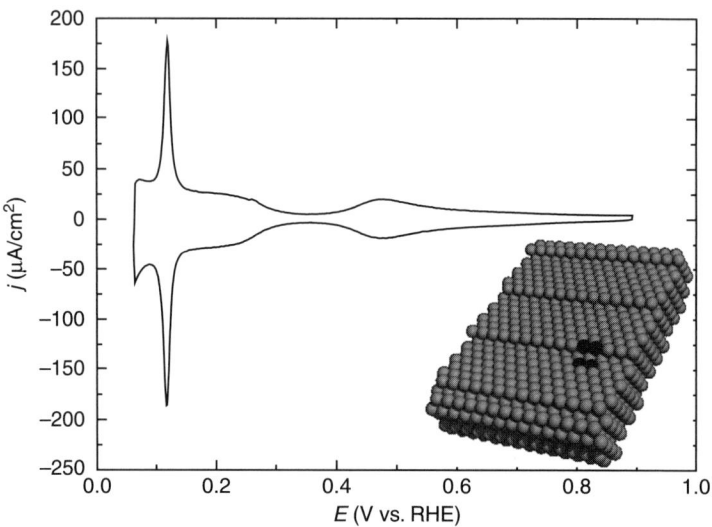

Figure 7.5 Cyclic voltammogram of a Pt(775) electrode in 0.5 M H_2SO_4 solution and a hard sphere model of this surface. Sweep rate: 50 mV/s. In the hard sphere model, four atoms forming the (110) step site have been identified in black.

this surface is shown as an inset. This surface can be considered as composed of (111) terraces separated by (110) monoatomic steps. The voltammogram shows three clearly differentiated regions: (i) a sharp peak at 0.12 V, corresponding to hydrogen adsorption on steps; (ii) a flat region between 0.06 and 0.35 V (overlapping with the peak at 0.12 V), corresponding to hydrogen adsorption on terrace sites; and (iii) a broad wave between 0.4 and 0.6 V, corresponding to anion adsorption on terrace sites. Figure 7.6

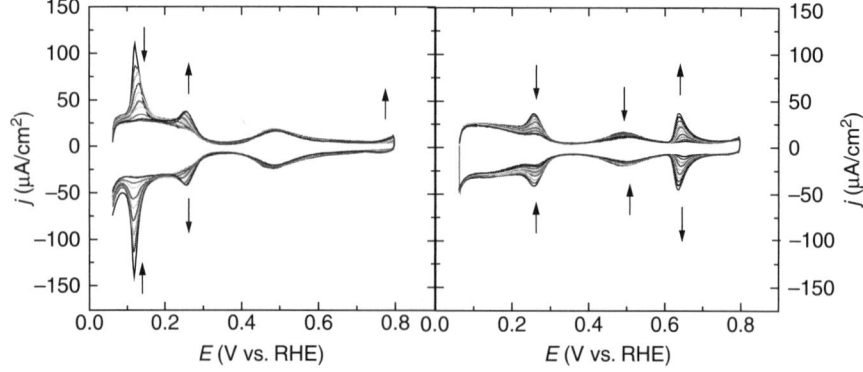

Figure 7.6 Bismuth deposition on a Pt(775) electrode from 5 μM Bi^{3+} + 0.5 M H_2SO_4 solution. The results have been divided into two graphs for the sake of clarity. Sweep rate: 50 mV/s.

shows the evolution of the cyclic voltammogram recorded with this electrode in a solution containing a small concentration of Bi(III) cations (approximately 5 μM). The adatoms progressively deposit on the surface, a process that can be monitored by the simultaneous measurement of cyclic voltammograms [Climent et al., 2001; Herrero et al., 2000]. In the first stage of the deposition, the voltammetric peak at 0.12 V decreases, while the wave corresponding to anion adsorption on the terraces remains essentially unaltered, signifying that the adatom is selectively being deposited on the steps. During this stage, a new peak at 0.25 V grows, linked to the decrease of the peak of the step, a fact that has been assigned to induced anion adsorption on Pt sites adjacent to the adatom-covered step site, in view of its marked dependence on the nature of the anion in solution (e.g., in $HClO_4$ solution, this peak is not observed). In a second stage, which starts when the step peak is completely blocked, adatom deposition on the terrace sites takes place, as evidenced by the progressive blockage of the hydrogen and anion adsorption processes on the terraces, the appearance of the redox peak characteristic of Bi oxidation on Pt(111) surfaces at 0.63 V, and the progressive disappearance of the adatom-induced anion peak at 0.25 V. The redox process corresponding to adatoms on steps is not observed, because it lies above the potential range selected in this experiment.

It is noteworthy that the same qualitative behavior is observed with Te deposition. Similar behavior also occurs with As and Sb, although, in these cases, deposition on the terrace starts slightly before the step is completely blocked, indicating that the selectivity for step sites is smaller with these adatoms. Another important difference observed in the decoration of stepped surfaces with As is that the adatom on the step features a redox process at 0.57 V. This case is illustrated in Fig. 7.7, where successive voltammetric cycles recorded with a Pt(775) electrode in 1 μM As_2O_3 + 0.1 M H_2SO_4 solution are shown. The selective deposition of As on step sites is evidenced by the progressive attenuation of the peak at 0.12 V, while the region for anion adsorption on the terraces, between 0.40 and 0.50 V, remains unaffected. However, a new peak grows at 0.57 V, corresponding to the surface oxidation of the adatom. The possible oxidation of the adatom in this relatively low potential range will mark a difference for the electrocatalysis of CO oxidation, as will be discussed in Section 7.6.

Completely different behavior is observed with S and Se, as shown in Fig. 7.8. With these adatoms, deposition on the terrace starts from the very beginning and no selectivity towards the step is observed. Additionally, deposition of the adatom changes the hydrogen adsorption energy on the (110) step sites, as reflected by the progressive shift of the peak at 0.12 V towards higher potential values.

The different behavior exhibited by the studied adatoms has been rationalized in terms of their different work function or electronegativity values, relative to the Pt substrate (see Table 7.4). Those elements with lower electronegativity tend to create a surface dipole with excess of positive charge on the adatom. In contrast, more-electronegative adatoms will retain negative charge. The second basis for the observed behavior is the particular charge distribution on step sites predicted by the Smoluchowski effect or the inability of the soft electron cloud to follow the sharp step edge defined by the atomic nuclei [Smoluchowski, 1941; Zangwill, 1988]. Accordingly, negative charge will accumulate on the lower part of the step, while the upper part of the step will

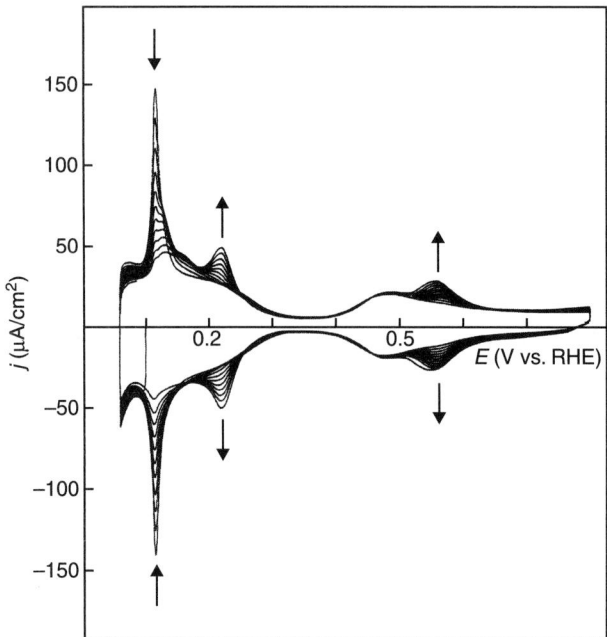

Figure 7.7 Arsenic deposition on a Pt(775) electrode from 1 µM As_2O_3 + 0.1 M H_2SO_4 solution. Sweep rate: 50 mV/s.

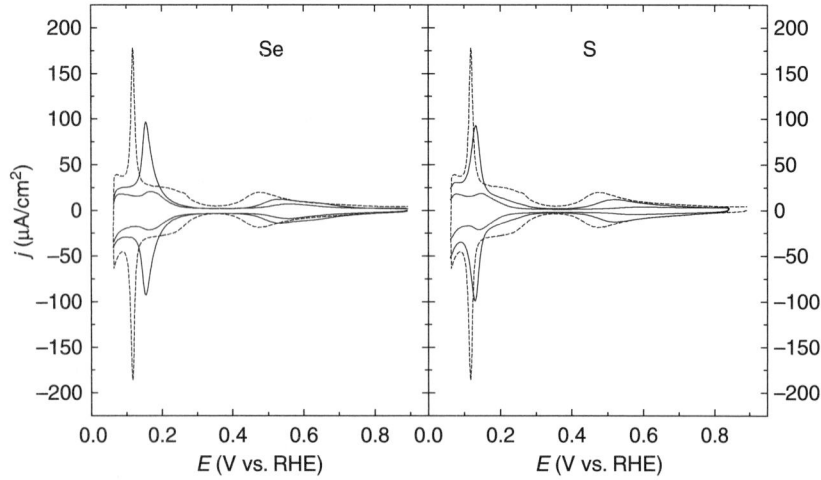

Figure 7.8 Different stages in Se and S deposition on a Pt(775) electrode in 0.5 M H_2SO_4 solution containing 5 µM SeO_2 and Na_2S, respectively. The dashed line is the first scan, the full line is the third scan, and the dotted line is the 32nd scan. Sweep rate: 50 mV/s.

become positively charged. Now, it is easily understood that electropositive adatoms will tend to interact more strongly with the lower part of the step, while electronegative adatoms will deposit on the terrace, more likely near the upper part of the step. Although it must be recognized that this explanation should be viewed as a simplified picture, since other aspects need consideration, such as water adsorption on steps or specific (chemical) interactions between the adatom and the d-orbitals of surface Pt atoms, it does serve to rationalize the observed experimental behavior.

7.5 DOUBLE-LAYER EFFECTS OF ADATOM MODIFICATION

The careful analysis of the double-layer effects involved in adatom modification provides valuable information for the understanding of the consequent electrocatalytic effects, discussed in the next section. One key parameter for understanding electrochemical reactivity is the potential of zero charge (PZC). The notion of electrode charge and PZC are unambiguously defined only for an ideal polarized interface. When adsorption phenomena with charge transfer processes are present, as in the case of Pt electrodes, it is necessary to distinguish the total charge, which includes the charge involved in the adsorption process, from the true excess electronic charge on the metal, i.e., the free charge [Frumkin and Petrii, 1975]. It is the latter that is mainly responsible for the electric field at the interphase and therefore it is a very important parameter for understanding the electronic effects caused by the adatom. Unfortunately, none of the methods suitable for the determination of the PZC can be applied to adatom-modified platinum surfaces.

The CO charge displacement method has been applied to the determination of the potential of zero total charge (PZTC) of Pt surfaces, and the effect of the surface structure and the composition of the solution have been investigated in this way [Climent et al., 1997, 1999, 2000; Gómez et al., 2000]. This method involves measurement of the total charge flow during potentiostatic adsorption of CO. It is assumed that the charge remaining on the CO-covered surface is negligible, and therefore that the charge flowing during the adsorption process (displaced charge) equals that initially present at the interphase at the potential of the experiment:

$$q_{dis} = q_f - q_i \approx -q_i \quad (7.22)$$

where q_{dis} is the displaced charge, q_i is the initial charge on the CO-free interphase, and q_f is the final charge at the CO-covered electrode. The experiment can be performed at different potentials to locate the particular value where the displaced charge equals zero. Alternatively, one particular value of q_{dis} at a given potential is enough to provide the integration constant to integrate the voltammetric current density to construct the total charge versus potential curve. From this curve, the value of the PZTC can be read as the intersection at $q = 0$. Moreover, from these results and under some assumptions, it has been possible in some cases to estimate the potential of zero free charge (PZFC) [Climent et al., 1997; Gómez et al., 2000]. However, this method cannot be applied to adatom-modified surfaces in general, because CO does not adsorb on the adatom, and hence the charge associated to the adatom will not be probed by the CO adsorption, and also because CO adsorption causes, in most cases, desorption of the adatom.

The other approach that has been proven valuable for the determination of the PZTC of Pt electrodes is the kinetic study of N_2O reduction [Attard and Ahmadi, 1995; Attard et al., 2004; Climent et al., 2002a]. This molecule adsorbs very weakly on the Pt surface. As a result, the maximum rate of reduction coincides with the maximum availability of surface sites for adsorption, and this, in turn, coincides with the PZTC. For heterogeneous surfaces, such as Pt stepped surfaces, various peaks in the N_2O reduction current have been attributed to the existence of various local PZTCs. Unfortunately, N_2O reduction on adatom-modified surfaces is inhibited, and therefore this method is not suitable for the PZTC determination of these surfaces. Only the particular case of decorated stepped surfaces vicinal to Pt(111) has been studied by these methods [Attard et al., 2004; Climent et al., 2001]. Under these conditions, CO adsorption does not induce the desorption of the adatom and the catalysis of the N_2O reduction is still sizeable. Figure 7.9a, b illustrates the results of a CO charge displacement study for the case of Pt(775) in 0.1 M $HClO_4$. These plots present the total charge curves for the bare and the step Bi-decorated surfaces, as obtained from integration of the voltammograms, in combination with the total CO displaced charge at 0.1 and 0.2 V. For the bare stepped surface, the presence of steps causes a decrease in the PZTC that has been explained in terms of the formation of dipoles that lower the surface

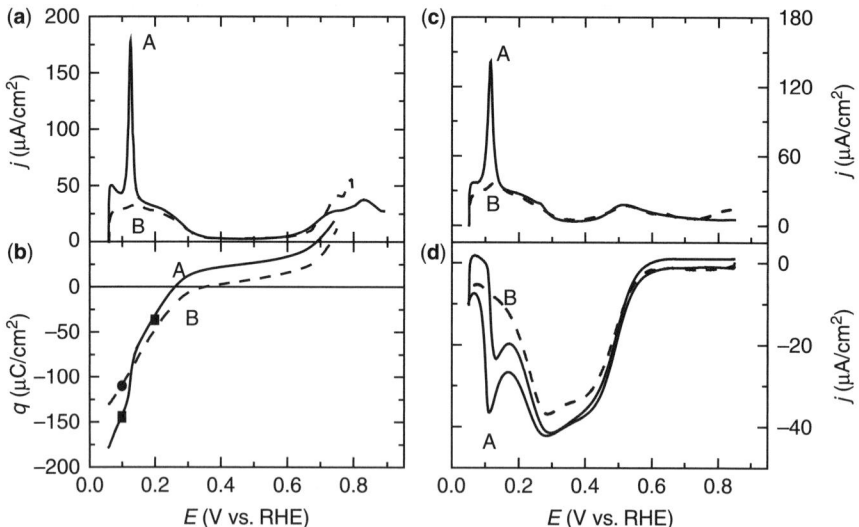

Figure 7.9 (a, b) Anodic voltammetric scan (a) and total charge density (b) of a Pt(775) electrode in 0.1 M $HClO_4$ solution. The filled squares and circle represent the opposite of the CO displaced charge. (c, b) Voltammograms corresponding to a Pt(332) stepped surface in 0.1 M H_2SO_4 solution without (c) and with (d) N_2O. Curve A corresponds to the bare stepped surface, while curve B corresponds to the step decorated surface by Bi deposition. Sweep rate: (a–c) 50 mV/s; (d) 10 mV/s.

potential owing to the Smoluchowski effect mentioned above [Climent et al., 1999; Gómez et al., 2000]. The effect of adatom decoration is to quench the local charge associated with step sites, resulting in a positive shift of the overall PZTC towards that of the terrace. The N_2O reduction on the bare stepped surface confirms the existence of two local PZTCs of terrace and step sites, the local PZTC of the step being lower than that of the terrace, and the PZTC of the overall surface being a weighted average of the two local values [Attard et al., 2004; Climent et al., 2002a]. Figure 7.9d shows this result for Pt(332) in 0.1 M H_2SO_4. Curve A in this figure corresponds to the unmodified surface, showing the two local minima at 0.12 and 0.30 V, due to step and terrace contributions, respectively. Curve B corresponds to the Bi-decorated surface, showing only the local minima due to terrace sites at the same potential value.

A novel approach to investigating important interfacial properties of Pt single-crystal electrodes is the laser-induced temperature jump method [Climent et al., 2002b, 2004, 2006]. In this technique, a nanosecond laser pulse is used to suddenly increase the temperature of the interphase. The coulostatic response to this temperature change allows estimation of the sign and magnitude of the thermal coefficient of the potential drop at the interphase. Remarkably, since the main contribution to this parameter comes from the effect of the temperature on the dipolar contribution from the solvent, these experiments provide unique information about the interaction of water with the metallic surface. At low enough potentials, the thermal coefficient is negative, indicating that interfacial water molecules exhibit a net orientation with their positive end (hydrogen) towards the metal. As the potential is increased, the oxygen-towards-the-metal orientation is favored and the thermal coefficient changes to positive values. The particular potential where the thermal coefficient becomes zero can be identified with the potential at which the net orientation of water changes. Moreover, it can be demonstrated from thermodynamic considerations that this particular potential coincides with the potential of maximum entropy of double-layer formation (PME).

As a result of the electrostatic interaction of the water dipoles with the electric field at the interphase, the PME is related to the PZC. Then, if the surface is negatively charged, we can expect the water molecules to be polarized with the positive end closer to the metal, while the opposite will be true when the surface is positively charged. The picture is complicated by the existence of a chemical interaction between the water and the metal surface that tends to orientate the water molecules even in the absence of any electric field. This natural orientation is with the oxygen closer to the metal surface in the case of gold and mercury electrodes, as reflected in the fact that the PME is located slightly negative to the PZC [Benderskii and Velichko, 1982; Climent et al., 2002c]. The decrease in the work function upon water dosage supports the same picture for Pt [Villegas and Weaver, 1996]. Consequently, the study of the shift of the PME of adatom-modified Pt surfaces provides insight on the direction and magnitude of the surface dipole induced by adatom deposition, and also on the specific interactions of the water molecules with the metal surface.

Figure 7.10 plots values of the PME as a function of adatom coverage for different adatoms [Garcia-Araez et al., 2008]. The medium–high coverage region will be discussed first. In this region, two different behaviors are observed, represented by S and Se, on the one hand, which displace the PME towards higher values, and Bi

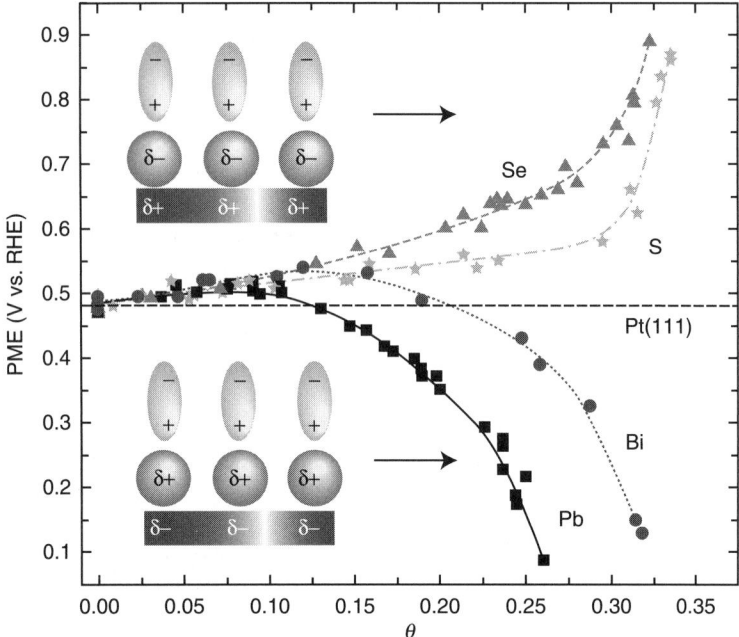

Figure 7.10 Potential of maximum entropy (PME) of a Pt(111) electrode modified by Bi, Pb, Se, and S deposition in 1 mM $HClO_4$ + 0.1 M $KClO_4$ solution, as a function of adatom coverage. The dashed, zero-slope line corresponds to the averaged reference PME value of unmodified Pt(111). The cartoons show the schematic interpretation for the effect of the adatoms at high coverage on the potential transients. (Reprinted with permission from Garcia-Araez et al. [2008].)

and Pb, on the other hand, which clearly decrease the PME. These trends can be easily explained by considering the differences in work function values and electronegativities (see Table 7.4) between the adatom and the Pt substrate: more-electropositive adatoms (Bi and Pb) shift the PME towards lower potential values, while more-electronegative adatoms (Se and S) displace it towards higher potential values. This is consistent with the discussion given above to explain the ability of adatoms to decorate step sites, and reinforces the idea that electropositive adatoms will retain partial positive charge while electronegative adatoms will remain partially negatively charged. The consequence of these charge distributions is the formation of surface dipoles that will affect the electrostatic interaction of the water molecules with the surface (see the cartoons in Fig. 7.10): water will remain with the oxygen towards the metal in a broader potential range when the surface is modified with electropositive adatoms, while the opposite orientation will be favored in the case of electronegative adatoms.

The formation of surface dipoles by adatom deposition can also be inferred from work function measurements in UHV. In this regard, Bi and Pb deposition causes a marked decrease in the work function of a Pt(111) surface [Mazinangokoudi and

Argile, 1992; Paffett et al., 1986]. This decrease is consistent with the formation of dipoles with negative charge on the metal and positive charge on the adatom, in agreement with the model proposed to explain both the decoration of stepped surfaces and the results from laser-induced potential transients. Unfortunately, detailed work function data in this regard are very scarce. Another approach that can be used to indirectly infer changes in the electronic structure induced by adatom modification is based on the analysis of vibrational frequencies of coadsorbed species sensitive to the local electronic state of the surface. In this way, the CO stretching frequency is sensitive to the ability of the substrate to back-donate electronic density towards the antibonding $2\pi^*$ orbital. This backdonation results in a weakening of the C$\equiv$O bond and decreases the vibrational frequency. However, this analysis should be performed with caution, since the frequency of the CO stretching band may depend also on other factors such as electrode potential and local coverage (through dipole–dipole coupling). Besides, spectroscopic results will be relevant for the present purpose only if both adlayers are properly intermixed.

It has been shown that Bi [Chang and Weaver, 1991; Herrero et al., 1995a, d; Lin et al., 1994], Pb [Hoshi et al., 2000], Sb [Kizhakevariam and Weaver, 1994] and S [Gracia et al., 2005; Lin et al., 1994] form mixed adlayers with CO. Adsorption of electropositive adatoms causes a displacement of the CO stretching frequency to lower values, indicating a reinforcement of the backdonation mechanism. On the other hand, S causes a displacement of the CO vibrational frequency to higher values, suggesting that the withdrawal of electronic density caused by the electronegative adatom decreases the degree of backdonation. These explanations are in agreement with quantum chemical calculations performed with model cluster systems that demonstrate that Bi donates electrons to Pt, causing an increase in the extent of $d\pi_{Pt} \rightarrow 2\pi^*_{CO}$ backdonation, whereas S gains electrons from Pt, decreasing the extent of backdonation from Pt to CO [Lin et al., 1994].

Coming back to the results obtained with the laser-induced temperature jump method, a small, but measurable, increase in the PME is observed at low coverages. This increase in the PME caused by adatom deposition is similar for all studied adatoms, regardless of their chemical nature. This observation has been explained by the disruptive effect of adatom deposition on the structure of interfacial water. Optimization of hydrogen bonding is a key factor determining the structure of the water network. It is reasonable to propose that hydrogen bonding will stabilize a configuration that possesses a net orientation of water molecules with the oxygen towards the metal. The disruption of the water network caused by the adatoms would facilitate the turnover of the water dipoles to the hydrogen-toward-the-metal orientation, which would take place at less negative potentials, i.e., shifting the PME towards positive potentials.

7.6 ADATOMS AND ELECTROCATALYSIS

In general, the effect of surface modifiers on the electrocatalytic activity of a bimetallic material can be classified intro three main categories [Bligaard and

Norskov, 2007; Jarvi and Stuve, 1998; Motoo and Watanabe, 1980; Parsons and VanderNoot, 1988]:

- *Change in the Electronic Properties of the Substrate.* The new electronic properties alter the adsorption energy and intramolecular bond energy in adsorbed reactants and intermediates. Besides the ligand effect that occurs when an atom changes the electronic state of neighboring atoms, foreign metal adatoms can also induce or suffer strain effects, when either the adatoms or the host metal atoms are forced to adopt positions different from the equilibrium position in the bulk materials.
- *Ensemble or Third-Body Effects.* These effects concern the selective blockage of a particular adsorption site by adatom deposition. This can be advantageous when the reaction mechanism contains parallel paths that can be affected differently by blocking particular sites. In some cases, the undesired reaction needs more than one free adjacent site (ensemble), and can be inhibited by blocking particular sites without decreasing the reactivity of the surface for the catalyzed reaction.
- *Bifunctional Catalysts.* The adatoms provide suitable adsorption sites for a second reactant necessary for the reaction to proceed, while the main reactant still adsorbs on the free sites of the substrate.

A group of reactions extensively studied in electrocatalysis, because of their great interest in the context of fuel cell technology, involve oxidation of small oxygenated organic compounds. Most of these reactions proceed through a dual-path mechanism. This means that at least two parallel pathways exists: one leading to the formation of a poisoning intermediate and another producing a stable and soluble oxidized compound, ideally CO_2. The main poison identified spectroscopically is CO. Adatom modification can affect each of these pathways independently, the effect normally sought being inhibition of the poisoning reaction and enhancement of the direct oxidation reaction.

The simplest reactions in this group are oxidations of molecules containing only one carbon (C_1 molecules), and, among these, the two most widely studied reactions are methanol and formic acid oxidation. The interest in the latter arises not only from its possible application as a noncontaminant fuel but also because this simple reaction has been considered as a prototype among the oxidation of C_1 molecules. Surface modifications with sp^n elements have greater effects on the electrocatalysis of formic acid [Clavilier et al., 1989b, c; Fernández-Vega et al., 1989, 1991] than on methanol oxidation [Chang et al., 1992]. Electrocatalysis for the oxidation of CO has also been studied, as this molecule constitutes the main poison in the oxidation of small organic molecules.

7.6.1 Electrocatalysis of CO Oxidation

CO forms intermixed adlayers with most of the p-block adatoms. CO oxidation from mixed adlayers with Bi [Chang and Weaver, 1991; Herrero et al., 1995a, d], As [Herrero et al., 1995d], Sb [Kizhakevariam and Weaver, 1994], Se [Herrero et al., 1996], and Te [Herrero et al., 1996] on Pt(111) has been studied. The formation of mixed adlayers is inferred from voltammetric and coulometric analysis of CO oxidation and corroborated by Fourier transform infrared (FTIR) measurements. Two possible effects have to be considered. First, there is the electronic effect that would alter the stability of the CO adlayer. This is related to the modification of the electronic backdonation from the metal to the CO molecule and also to the existence of lateral interactions between CO and the adatom. On the other hand, the existence of a bifunctional mechanism with the adatoms acting as donors of oxygenated species also needs to be considered. The changes that the adatom exerts on the CO adlayer on Pt(111) can be summarized as follows:

- In general, the adatoms exert a significant effect on the CO adlayer structure, displacing it from multifolded adsorption sites to on-top sites. The fact that, for the same adatom coverage, the degree of blocking of hydrogen adsorption is more important than that of CO adsorption suggests that the adatoms adsorb on hollows, blocking hydrogen multifolded adsorption but allowing on-top adsorption of CO [Chang and Weaver, 1991; Herrero et al., 1995a, d; Herrero et al., 1996; Kizhakevariam and Weaver, 1994].
- In the presence of Bi or Te, the C≡O bond is weakened, as concluded from the displacement of the CO stretching band to lower wavenumbers. There is also a change in the dependence of the band frequency on electrode potential, with the slope dv/dE increasing for the adatom-modified surfaces. These changes indicate that the adatom alters the electronic properties of the surface, increasing the amount of electronic backdonation and stabilizing the adsorbed CO molecule. No catalytic enhancement is expected from this effect.
- In the presence of As, a displacement of the vibrational band towards higher wavenumbers indicates a weakening of the backdonation mechanism, decreasing the stability of CO on the electrode surface.
- Almost no displacement of the vibrational band is observed in the presence of Sb or Se, indicating that electronic effects are very weak in these cases.
- For Bi, As, and Sb, the CO oxidation undergoes bifunctional catalysis through an adatom-mediated oxygen transfer. This conclusion can be obtained from the following observations based on experiments similar to that shown in Fig. 7.11. When CO oxidation on a Bi-modified and on a clean surface are compared, a slight decrease in the onset potential for CO oxidation is observed. The oxidation of the adatom is also severely affected, and takes place coupled with the CO oxidation. The reduction of the adatom in the first negative scan after the CO stripping attests that the adatoms were oxidized together with CO oxidation. For Bi, however (Fig. 7.11), the charge under the first negative scan is lower than in subsequent cycles. This has been taken as an indication that adatoms donate oxygenated species to adjacent CO molecules to promote their oxidation. In this way, after CO oxidation, the Bi adlayer ends up partially reduced. Chronoamperometric measurements of CO oxidation at different constant potentials demonstrate that the catalytic enhancement only takes place in the potential region where the adatom is oxidized.

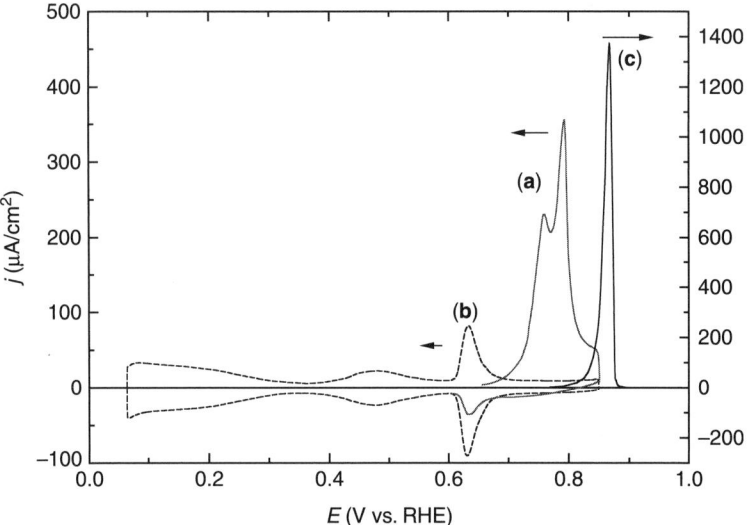

Figure 7.11 CO stripping experiment for a Pt(111) electrode with $\theta_{Bi} = 0.14$ (a) and recovery of the electrode surface after CO removal (b). Curve (c) shows the CO stripping for the unmodified surface. Sweep rate: 50 mV/s.

- At low CO coverages, the adatom oxidation peak can be distinguished from the CO oxidation peak. Lateral interactions between CO and the adatom stabilize the elemental Bi state, increasing the potential of the adatom redox peak. For As, a displacement of the redox peak to lower potentials is observed, indicating an stabilization of the As(III) state on the CO-As mixed adlayer.
- For Se and Te, oxidation of the adatom takes place at potentials higher than that of CO oxidation. The adatom is always in its reduced state, and no bifunctional catalysis through the transfer of oxygen from the adatom to the CO molecule can take place.

Fewer studies have been carried out for electrodes with other crystallographic orientations, since normally the oxidation of the adatom takes place at potentials higher than that of CO oxidation, and hence no bifunctional enhancement is expected. For Pt(100), formation and oxidation of mixed CO adlayers with Bi, Te, and Sb adatoms have been studied [Feliu et al., 1996]. In all three cases, CO forms mixed adlayers, although some segregation of the two species was observed at low CO coverages. While for Bi and Te no electrocatalytic effect is observed, for Sb the CO stripping peak potential is reduced by approximately 60 mV. This observation agrees with the hypothesis that a bifunctional mechanism operates in the electrocatalysis of CO oxidation on Pt electrodes modified by sp^n elements, since only adsorbed Sb has an oxidation potential lower than CO. In this case, a displacement of the Sb redox peak towards lower potentials due to CO coadsorption is observed, indicating a destabilization of the elemental Sb state. CO oxidation has also been studied on

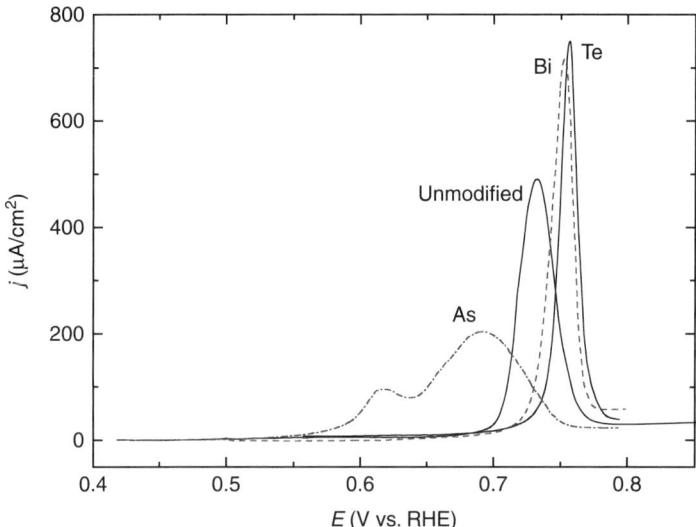

Figure 7.12 CO stripping experiment on a Pt(775) electrode decorated with different adatoms, as labelled, in 0.1 M HClO$_4$. The blank voltammogram is also included for comparison. Sweep rate: 50 mV/s.

Pt(S)[n(111) × (110)] stepped surfaces decorated with adatoms [Climent et al., 2001]. Figure 7.12 shows the voltammetric stripping of CO on a Pt(775) stepped surface that has been modified by selectively depositing different adatoms on step sites. As expected, no enhancement of CO oxidation is observed for Bi and Te, since oxidation of these adatoms adsorbed on step sites takes place at potentials much higher than that of CO oxidation. In these cases, the CO stripping peak is displaced towards positive values, owing to the blocking effect of the step sites, which are the most reactive for this reaction [Lebedeva et al., 2002]. However, as mentioned above, As on steps undergoes a surface oxidation process at 0.57 V, and therefore is capable of providing the oxygenated species needed for CO oxidation, acting as a bifunctional catalyst. This is reflected in a significant displacement of the CO oxidation peak towards lower potential values.

7.6.2 Electrocatalysis of the Poison Formation Reaction

The effect of adatom deposition on the poison formation reaction from formic acid dissociative adsorption has been extensively studied [Climent et al., 1998; Herrero et al., 1993, 1994, 1995b, c; Llorca et al., 1994]. The poisoning reaction takes place spontaneously under open circuit conditions when the electrode is immersed in a solution containing formic acid. Then, the poisoned surface can be transferred to an electrochemical cell containing a solution without formic acid, and the amount of poison formed can be easily characterized by cyclic voltammetry, through the stripping charge. In this way, the poisoning reaction can be easily separated from the direct

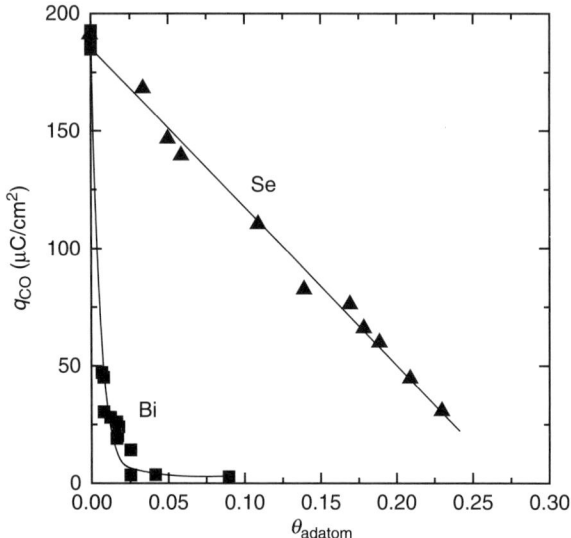

Figure 7.13 Variation of the CO stripping charge formed from formic acid dissociative adsorption as a function of adatom coverage for a Pt(111) electrode modified with Bi and Se, as indicated, in 0.5 M H_2SO_4 solution.

oxidation and the effect of the adatom on this particular path of the overall mechanism can be studied. Figure 7.13 shows some selected experimental results, plotting the amount of poison formed as a function of adatom coverage on Pt(111) surfaces. Two different kinds of behavior are observed. For Se-Pt(111) surfaces, the amount of poison decreases linearly with the adatom coverage. The same behavior is observed with S-modified Pt(111) electrodes. This behavior can be explained by arguing a third-body effect, the amount of poison decreasing merely because there are fewer Pt sites available for the poisoning reaction. The interesting point here is that the amount of poison falls to zero at a coverage lower than saturation, indicating that a certain number of adjacent Pt atoms are necessary for the poisoning reaction. At a coverage close to 0.25, the probability of finding two adjacent unblocked Pt atoms drops to zero, and therefore no poison is formed at this coverage, although other reactions might proceed if the ensemble requirements are less demanding. Similar behavior is also observed with Sb-, Bi-, and Te-modified Pt(100) electrodes, although in these cases the linearity is lost at high coverages.

The second behavior observed in the results plotted in Fig. 7.13 is that of Bi-Pt(111) electrodes. In this case, a very small amount of adatom on the surface can produce a significant effect on the amount of poison formed. In fact, no poison can be detected when the Bi coverage is as low as 0.04. It has been calculated that if this effect were due to an extended electronic modification, it would mean that one adatom inhibits the poison formation in the surrounding Pt atoms, extending its influence as far as to 7 neighbors. This is quite unexpected, and contrary to theoretical calculations. An alternative explanation for this behavior is given below. Similar behavior has been observed with Sb, Te, As, Sn, and Pb on Pt(111).

If we recall now what has been discussed before about the different abilities of adatoms to decorate step sites and the relation to the induced surface dipole, a correlation between these effects and electrocatalytic behavior becomes clear: the adatoms that decorate steps on Pt(111) are the same ones that give the extended electrocatalytic effect. More insight into this can be obtained from study of the poisoning reaction on adatom-modified stepped surfaces [Macia et al., 1999, 2001]. With sufficiently wide terraces, the amount of poison remains essentially constant until Bi completely covers the step. Once the step sites are completely covered and Bi starts to deposit on the terraces, a strong inhibition of the poisoning reaction is observed. This result was interpreted as evidence that poison formation only takes place on step sites, and when those are completely blocked by adatoms, the poisoning reaction no longer takes place on the modified surface. This argument was also used to explain the "long-range effect" observed for Bi on Pt(111) surfaces [Macia et al., 2001]. It was considered that, even on the most perfect surface, there are always a certain number of defects. These defects would be, under this hypothesis, the only active sites for poison formation. The amount of poison is not limited to the step sites, because of the significant mobility of adsorbed CO, which can migrate from the step where it is formed to the adjacent terrace. Then, when the Pt(111) surface is modified with a small amount of an electropositive adatom, defect sites will be preferentially covered, rendering a surface inactive for the poison formation reaction. On the other hand, electronegative adatoms such as Se and S will randomly cover both defect and terrace, giving a third-body effect.

A different mechanism seems to operate in the case of poison formation from methanol [Herrero et al., 1993]. In this case, modification of the Pt(111) surface by Bi deposition only causes a linear decrease in the amount of poison formed, indicating the existence of a mere third-body effect. Complete inhibition of the poisoning reaction is achieved for $\theta_{Bi} > 0.23$, i.e., before the surface is completely covered. This suggests the existence of ensemble requirements for this reaction, which need enough free contiguous Pt sites to take place.

7.6.3 Electrocatalysis of the Direct Reaction

In the previous section, it has been shown how it is possible to isolate and measure the amount of poison formed from formic acid dissociation by transferring the poisoned electrode to a solution free of formic acid. Separation of the direct path from the poisoning reaction is more difficult, since this reaction can only be studied in solutions containing formic acid, and hence both reaction pathways will take place simultaneously in this case. However, the capacity of Bi to virtually stop the poisoning reaction on Pt(111) for any coverage higher than 0.04 allows one to study the effect of surface modification on the direct formic acid oxidation without the interference of the poisoning reaction. Figure 7.14a plots the maximum current density measured in the anodic scan during formic acid oxidation as a function of adatom coverage on Pt(111). The curve features a roughly linear increase at low coverages and a

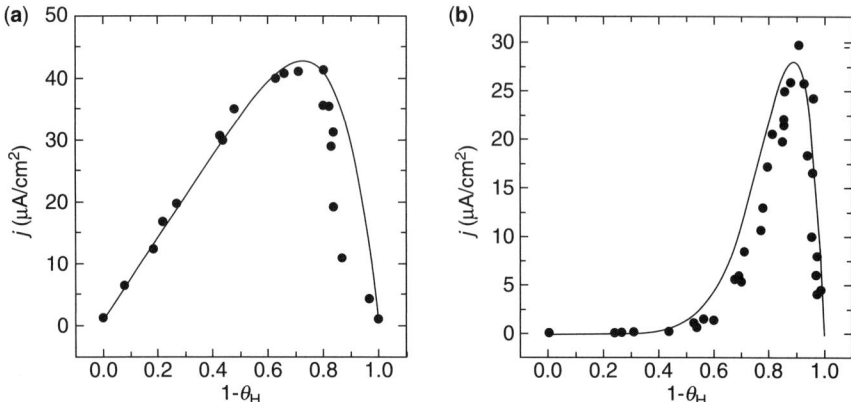

Figure 7.14 Current density for formic acid oxidation as a function of the fraction of Pt surface atoms blocked by adatoms on two different electrodes: (a) Bi/Pt(111); (b) Sb/Pt(100). (Reprinted with permission from Leiva et al. [1997].)

very steep decrease at coverages near saturation, with a maximum current density for a blockage of around 70%. This shape has been explained with a very simple model [Leiva et al., 1997] that considers that the adatoms play an electronic effect in such a way that the current density for a given coverage is proportional to the number of Bi–Pt pairs (i.e., the number of Bi atoms that have at least one adjacent unblocked Pt atom). The initial linear increase supports this model, since, for low adatom coverage, each deposited adatom contains at least one neighboring free Pt atom. In this model, the activity of each adatom is the same, regardless of the number of unoccupied sites next to it. For high adatom coverages, the availability of free sites starts to decrease and, with it, the activity towards the direct oxidation reaction. The position of maximum activity can be calculated from probabilistic considerations, being located at 72% occupied Pt sites. The agreement between the model and the experimental results is excellent.

The situation is more complex when the poisoning reaction takes place at the same time as the direct oxidation and the poison acts as a blocking agent for the direct pathway. In this case, the result depends on the degree of poison formation. One extreme case is when all available sites are blocked by the poison. This is the case with Sb-Pt(100), depicted in Fig. 7.14b. The shape of the curve in this figure has been explained by considering that adatoms exert an ensemble effect and that poison formation can only take place if at least two adjacent Pt atoms are available [Leiva et al., 1997]. Then, free Pt sites completely surrounded by adatoms will not be active for the poison formation but will be active for the direct oxidation. For low coverages, the activity remains very low, since the probability of having a Pt site completely surrounded by adatoms will be very low. The current only starts to increase for blockage degrees higher than 0.5, featuring a maximum at about 0.9, in excellent agreement with the theoretical model.

The good coincidence between the model and the experimental data depicted in Fig. 7.14 supports the idea that this model captures the essential aspects of the effect of different adatoms on the electrocatalysis of formic acid oxidation.

ACKNOWLEDGMENTS

Financial support from the MEC (Spain) through Project CTQ 2006-04071/BQU is gratefully acknowledged.

REFERENCES

Attard GA, Ahmadi A. 1995. Anion–surface interactions. Part 3: N_2O reduction as a chemical probe of the local potential of zero total charge. J Electroanal Chem 389: 175–190.

Attard GA, Hazzazi O, Wells PB, Climent V, Herrero E, Feliu JM. 2004. On the global and local values of the potential of zero total charge at well-defined platinum surfaces: Stepped and adatom modified surfaces. J Electroanal Chem 568: 329–342.

Batina N, Mccargar JW, Salaita GN, Lu F, Lagurendavidson L, Lin CH et al. 1989. Structure and composition of Pt(111) and Pt(100) surfaces as a function of electrode potential in aqueous sulfide solutions. Langmuir 5: 123–128.

Benderskii VA, Velichko GI. 1982. Temperature jump in electric double-layer study. Part I. Method of measurements. J Electroanal Chem 140: 1–22.

Blais S, Jerkiewicz G, Herrero E, Feliu JM. 2001. Temperature-dependence of the electro-oxidation of the irreversibly chemisorbed As on Pt(111). Langmuir 17: 3030–3038.

Blais S, Jerkiewicz G, Herrero E, Feliu JM. 2002. New insight into the electro-oxidation of the irreversibly chemisorbed bismuth on Pt(111) through temperature-dependent research. J Electroanal Chem 519: 111–122.

Bligaard T, Norskov JK. 2007. Ligand effects in heterogeneous catalysis and electrochemistry. Electrochim Acta 52: 5512–5516.

Chang S-C, Weaver MJ. 1991. Influence of coadsorbed bismuth and copper on carbon monoxide adlayer structures at ordered low-index platinum–aqueous interfaces. Surf Sci 241: 11–24.

Chang S-C, Ho Y, Weaver MJ. 1992. Applications of real-time infrared spectroscopy to electrocatalysis at bimetallic surfaces. I. Electrooxidation of formic acid and methanol on bismuth-modified Pt(111) and Pt(100). Surf Sci 265: 81–94.

Clavilier J, Feliu JM, Aldaz A. 1988. An irreversible structure sensitive adsorption step in bismuth underpotential deposition at platinum electrodes. J Electroanal Chem 243: 419–433.

Clavilier J, Feliu J, Fernández-Vega A, Aldaz A. 1989a. Electrochemical behaviour of irreversibly adsorbed bismuth on Pt(100) with different degrees of crystalline surface order. J Electroanal Chem 269: 175–189.

Clavilier J, Fernández-Vega A, Feliu JM, Aldaz A. 1989b. Heterogeneous electrocatalysis on well-defined Pt surfaces modified by controlled amounts of irreversible adsorbed adatoms. Part I. Formic acid oxidation on Pt(111)-Bi system. J Electroanal Chem 258: 89–100.

Clavilier J, Fernández-Vega A, Feliu JM, Aldaz A. 1989c. Heterogeneous electrocatalysis on well-defined Pt surfaces modified by controlled amounts of irreversible adsorbed adatoms. Part III. Formic acid oxidation on Pt(100)-Bi system. J Electroanal Chem 261: 113–125.

Clavilier J, Feliu JM, Fernandez-Vega A, Aldaz A. 1990a. Electrochemical behaviour of the Pt(111)-As system in acidic medium: Adsorbed hydrogen and hydrogen reaction. J Electroanal Chem 294: 193–208.

Clavilier J, Orts JM, Feliu JM, Aldaz A. 1990b. Study of the conditions for irreversible adsorption of lead at Pt(*hkl*) electrodes. J Electroanal Chem 293: 197–208.

Climent V, Gómez R, Orts JM, Aldaz A, Feliu JM. 1997. The potential of zero total charge of single-crystal electrodes of platinum group metals. In: Korzeniewski C, Conway BE, eds. The Electrochemical Society Proceedings (Electrochemical Double Layer). Pennington, NJ: The Electrochemical Society. pp. 222–237.

Climent V, Herrero E, Feliu JM. 1998. Electrocatalysis of formic acid and CO oxidation on antimony-modified Pt(111) electrodes. Electrochim Acta 44: 1403–1414.

Climent V, Gómez R, Feliu JM. 1999. Effect of increasing amount of steps on the potential of zero total charge of Pt(111) electrodes. Electrochim Acta 45: 629–637.

Climent V, Gómez R, Orts JM, Aldaz A, Feliu JM. 2000. Potential of zero total charge of platinum single crystal electrodes. In: Jerkiewicz G, Feliu JM, Popov BN, eds. The Electrochemistry Society Proceedings (Hydrogen at Surface and Interfaces). Pennington, NJ: The Electrochemical Society. pp. 12–30.

Climent V, Herrero E, Feliu JM. 2001. Positive shift of the potential of zero total charge of stepped Pt(111) electrodes decorated by irreversibly adsorbed bismuth. Electrochem Comm 3: 590–594.

Climent V, Attard GA, Feliu JM. 2002a. Potential of zero charge of platinum stepped surfaces: a combined approach of CO charge displacement and N_2O reduction. J Electroanal Chem 532: 67–74.

Climent V, Coles BA, Compton RG. 2002b. Coulostatic potential transients induced by laser heating of a Pt(111) single-crystal electrode in aqueous acid solutions. Rate of hydrogen adsorption and potential of maximum entropy. J Phys Chem B 106: 5988–5996.

Climent V, Coles BA, Compton RG. 2002c. Laser-induced potential transients on a Au(111) single-crystal electrode. Determination of the potential of maximum entropy of double layer formation. J Phys Chem B 106: 5258–5265.

Climent V, Coles BA, Compton RG, Feliu JM. 2004. Coulostatic potential transients induced by laser heating of platinum stepped electrodes: Influence of steps on the entropy of double layer formation. J Electroanal Chem 561: 157–165.

Climent V, Garcia-Araez N, Compton RG, Feliu JM. 2006. Effect of deposited bismuth on the potential of maximum entropy of Pt(111) single-crystal electrodes. J Phys Chem B 110: 21092–21100.

Dean JA. 1999. Lange's Handbook of Chemistry, 15th ed. New York: McGraw-Hill.

Evans RW, Attard GA. 1993. The redox behavior of compressed bismuth overlayers irreversibly adsorbed on Pt(111). J Electroanal Chem 345: 337–350.

Feliu JM, Fernández-Vega A, Aldaz A, Clavilier J. 1988. New observation of a structure sensitive electrochemical behaviour of irreversibly adsorbed As and Sb from acidic solution on Pt(111) and Pt(100) orientations. J Electroanal Chem 256: 149–163.

Feliu JM, Fernandez-Vega A, Orts JM, Aldaz A. 1991. The behavior of lead and bismuth adatoms on well-defined platinum surfaces. J Chim Phys Phys Chim Biol 88: 1493–1518.

Feliu JM, Gómez R, Llorca MJ, Aldaz A. 1993a. Electrochemical behavior of irreversible adsorbed selenium dosed from solution on Pt(*hkl*) single crystal electrodes in sulphuric and perchloric acid media. Surf Sci 289: 152–162.

Feliu JM, Gómez R, Llorca MJ, Aldaz A. 1993b. Electrochemical behavior of irreversible adsorbed tellurium dosed from solution on Pt(*hkl*) single crystal electrodes in sulphuric and perchloric acid media. Surf Sci 297: 209–222.

Feliu JM, Orts JM, Gómez R, Aldaz A, Clavilier J. 1994. New information on the unusual adsorption states of Pt(111) in sulphuric acid solutions from potentiostatic adsorbate replacement by CO. J Electroanal Chem 372: 265–268.

Feliu JM, Herrero E, Orts JM, Rodes A. 1996. CO adsorption and oxidation on Pt(100) single crystals modified by irreversibly adsorbed adatoms. Proc Electrochem Soc 96/98: 68–82.

Fernández-Vega A, Feliu JM, Aldaz A, Clavilier J. 1989. Heterogeneous electrocatalysis on well-defined Pt surfaces modified by controlled amounts of irreversible adsorbed adatoms. Part II. Formic acid oxidation on Pt(100)-Sb system. J Electroanal Chem 258: 101.

Fernández-Vega A, Feliu JM, Aldaz A, Clavilier J. 1991. Heterogeneous electrocatalysis on well-defined Pt surfaces modified by controlled amounts of irreversible adsorbed adatoms. Part IV. Formic acid oxidation on Pt(111)-As system. J Electroanal Chem 305: 229.

Frumkin AN, Petrii OA. 1975. Potentials of zero total charge and zero free charge of platinum group metals. Electrochim Acta 20: 347–359.

Garcia-Araez N, Climent V, Feliu JM. 2008. Evidence of water reorientation on model electrocatalytic surfaces from nanosecond-laser-pulsed experiments. J Am Chem Soc 130: 3824–3833.

Gómez R, Llorca MJ, Feliu JM, Aldaz A. 1992. The behavior of germanium adatoms irreversibly adsorbed on platinum single crystals. J Electroanal Chem 340: 349–355.

Gomez R, Fernandez-Vega A, Feliu JM, Aldaz A. 1993. Hydrogen evolution on Pt single-crystal surfaces—Effects of Irreversibly adsorbed bismuth and antimony on hydrogen adsorption and evolution on Pt(100). J Phys Chem 97: 4769–4776.

Gómez R, Feliu JM, Aldaz A. 1997. Effects of irreversibly adsorbed bismuth on hydrogen adsorption and evolution on Pt(111). Electrochim Acta 42: 1675–1683.

Gómez R, Climent V, Feliu JM, Weaver MJ. 2000. Dependence of the potential of zero charge of stepped platinum (111) electrodes on the oriented step-edge density: Electrochemical implications and comparison with work function behavior. J Phys Chem B 104: 597–605.

Gordy W, Thomas WJO. 1956. Electronegativities of the elements. J Chem Phys 24: 439–444.

Gracia FJ, Guerrero S, Wolf EE, Miller JT, Kropf A. 2005. Kinetics, operando FTIR, and controlled atmosphere EXAFS study of the effect of sulfur on Pt-supported catalysts during CO oxidation. J Catal 233: 372–387.

Haas GA. 1972. Work Function and Secondary Emission. American Institute of Physics Handbook. New York: McGraw-Hill. pp. 172–181.

Hamm UW, Kramer D, Zhai RS, Kolb DM. 1998. On the valence state of bismuth adsorbed on a Pt(111) electrode—An electrochemistry, LEED and XPS study. Electrochim Acta 43: 2969–2978.

Hayden BE, Murray AJ, Parsons R, Pegg DJ. 1996. UHV and electrochemical transfer studies on Pt(110)-(1 × 2): The influence of bismuth on hydrogen and oxygen adsorption, and the electro-oxidation of carbon monoxide. J Electroanal Chem 409: 51–63.

Herrero E, Fernández-Vega A, Feliu JM, Aldaz A. 1993. Poison formation reaction from formic acid and methanol on Pt(111) electrodes modified by irreversibly adsorbed Bi and As. J Electroanal Chem 350: 73–88.

Herrero E, Feliu JM, Aldaz A. 1994. Poison formation reaction from formic-acid on Pt(100) electrodes modified by irreversibly adsorbed bismuth and antimony. J Electroanal Chem 368: 101–108.

Herrero E, Feliu JM, Aldaz A. 1995a. CO adsorption and oxidation on Pt(111) electrodes modified by irreversibly adsorbed bismuth in sulfuric acid medium. J Catal 152: 264–274.

Herrero E, Llorca MJ, Feliu JM, Aldaz A. 1995b. Oxidation of formic-acid on Pt(100) electrodes modified by irreversibly adsorbed tellurium. J Electroanal Chem 383: 145–154.

Herrero E, Llorca MJ, Feliu JM, Aldaz A. 1995c. Oxidation of formic-acid on Pt(111) electrodes modified by irreversibly adsorbed tellurium. J Electroanal Chem 394: 161–167.

Herrero E, Rodes A, Perez JM, Feliu JM, Aldaz A. 1995d. CO adsorption and oxidation on Pt(111) electrodes modified by irreversibly adsorbed arsenic in sulfuric-acid medium—Comparison with bismuth-modified electrodes. J Electroanal Chem 393: 87–96.

Herrero E, Rodes A, Pérez JM, Feliu JM, Aldaz A. 1996. CO adsorption and oxidation on Pt(111) electrodes modified by irreversibly adsorbed selenium and tellurium. J Electroanal Chem 412: 165–174.

Herrero E, Climent V, Feliu JM. 2000. On the different adsorption behavior of bismuth, sulfur, selenium and tellurium on a Pt(775) stepped surface. Electrochem Commun 2: 636–640.

Herrero E, Buller LJ, Abruna HD. 2001. Underpotential deposition at single crystal surfaces of Au, Pt, Ag and other materials. Chem Rev 101: 1897–1930.

Hoshi N, Bae IT, Scherson DA. 2000. In situ infrared reflection absorption spectroscopic studies of coadsorption of CO with underpotential-deposited lead on Pt(111) in an aqueous acidic solution. J Phys Chem B 104: 6049–6052.

Jarvi TD, Stuve EM. 1998. Fundamental aspects of vacuum and electrocatalytic reactions of methanol and formic acid on platinum surfaces. In: Lipkowski J, Ross PN, eds. Electrocatalysis. New York: Wiley-VCH. pp. 75–153.

Kizhakevariam N, Weaver MJ. 1994. Structure and reactivity of bimetallic electrochemical interfaces: Infrared spectroscopy studies of carbon monoxide adsorption and formic acid electrooxidation on antimony-modified Pt(100) and Pt(111). Surf Sci 310: 183–197.

Lebedeva NP, Koper MTM, Feliu JM, van Santen RA. 2002. Role of crystalline defects in electrocatalysis: Mechanism and kinetics of CO adlayer oxidation on stepped platinum electrodes. J Phys Chem B 106: 12938–12947.

Leiva E, Iwasita T, Herrero E, Feliu JM. 1997. Effect of Adatoms in the Electrocatalysis of HCOOH oxidation—A theoretical-model. Langmuir 13: 6287–6293.

Lide DR, Frederikse HPR, eds. 1998. CRC Handbook of Chemistry and Physics, 79th ed. Boca Raton: CRC Press.

Lin W-F, Sun S-G, Tian Z-W. 1994. Investigations of coadsorption of carbon monoxide with S or Bi adatoms at a platinum electrode by in-situ FTIR spectrocopy and quantum chemistry analysis. J Electroanal Chem 364: 1–7.

Llorca MJ, Herrero E, Feliu JM, Aldaz A. 1994. Formic acid oxidation on Pt(111) electrodes modified by irreversible adsorbed selenium. J Electroanal Chem 373: 217–225.

Macia MD, Herrero E, Feliu JM, Aldaz A. 1999. Formic acid self-poisoning on bismuth-modified Pt(755) and Pt(775) electrodes. Electrochem Commun 1: 87–89.

Macia MD, Herrero E, Feliu JM, Aldaz A. 2001. Formic acid self-poisoning on bismuth-modified stepped electrodes. J Electroanal Chem 500: 498–509.

Macia MD, Herrero E, Feliu JM. 2002. Formic acid self-poisoning on adatom-modified stepped electrodes. Electrochim Acta 47: 3653–3661.

Macia MD, Herrero E, Feliu JM. 2003. Formic acid oxidation on Bi-Pt(111) electrode in perchloric acid media. A kinetic study. J Electroanal Chem 554: 25–34.

Mazinangokoudi M, Argile C. 1992. Adsorption and coadsorption of lead and 1,3-butadiene on platinum (111)—An AES, LEED and DF study. Surf Sci 262: 307–317.

Michaelson HB. 1977. Work function of elements and its periodicity. J Appl Phys 48: 4729–4733.

Motoo S, Watanabe M. 1980. Electrocatalysis by ad-atoms: Part VII. Enhancement of CO oxidation on platinum by As ad-atoms. J Electroanal Chem 111: 261–268.

Orts JM, Rodes A, Feliu JM. 1997. Irreversibly adsorbed As at full blockage on Pt(111) electrodes—Surface stoichiometry. J Electroanal Chem 434: 121–127.

Paffett MT, Campbell CT, Taylor TN. 1986. Adsorption and growth modes of Bi on Pt(111). J Chem Phys 85: 6176–6185.

Parsons R, VanderNoot T. 1988. The oxidation of small organic molecules. A survey of recent fuel cell related research. J Electroanal Chem 257: 9–45.

Rhee CK, Kim D-K. 2001. Electrochemical scanning tunneling microscope study of irreversibly adsorbed Te on a Pt(111) single crystal electrode surface. J Electroanal Chem 506: 149–154.

Rodes A, Feliu JM, Aldaz A, Clavilier J. 1988. Irreversible tin adsorption on polyoriented gold electrodes. J Electroanal Chem 256: 455–462.

Rodriguez P, Herrero E, Aldaz A, Feliu JM. 2006. Tellurium adatoms as an in-situ surface probe of (111) two-dimensional domains at platinum surfaces. Langmuir 22: 10329–10337.

Schmidt TJ, Behm RJ, Grgur BN, Markovic NM, Ross PN. 2000a. Formic acid oxidation on pure and Bi-modified Pt(111): Temperature effects. Langmuir 16: 8159–8166.

Schmidt TJ, Grgur BN, Behm RJ, Markovic NM, Ross PN. 2000b. Bi adsorption on Pt(111) in perchloric acid solution: A rotating ring-disk electrode and XPS study. Phys Chem Chem Phys 2: 4379–4386.

Schmidt TJ, Stamenkovic VR, Lucas CA, Markovic NM, Ross PN. 2001. Surface processes and electrocatalysis on the Pt(hkl)/Bi–solution interface. Phys Chem Chem Phys 3: 3879–3890.

Smith SPE, Abruña HD. 1999a. Structural effects on the oxidation of HCOOH by bismuth modified Pt(111) electrodes with (110) monatomic steps. J Electroanal Chem 467: 43–49.

Smith SPE, Abruña HD. 1999b. The coadsorption of UPD copper and irreversibly adsorbed bismuth on Pt(111) and Pt(100) electrodes. J Phys Chem B 103: 6764–6769.

Smith SPE, Bendor KF, Abruña HD. 1999. Structural effects on the oxidation of HCOOH by bismuth-modified Pt(111) electrodes with (100) monatomic steps. Langmuir 15: 7325–7332.

Smith SPE, Ben-Dor KF, Abruña HD. 2000. Poison formation upon the dissociative adsorption of formic acid on bismuth-modified stepped platinum electrodes. Langmuir 16: 787–794.

Smoluchowski R. 1941. Anisotropy of the electronic work function of metals. Phys Rev 60: 661–674.

Stuve EM, Krasnopoler A, Sauer DE. 1995. Relating the in-situ, ex-situ, and non-situ environments in surface electrochemistry. Surf Sci 335: 177–185.

Sung YE, Chrzanowski W, Zolfaghari A, Jerkiewicz G, Wieckowski A. 1997. Structure of chemisorbed sulfur on a Pt(111) electrode. J Am Chem Soc 119: 194–200.

Sung YE, Chrzanowski T, Wieckowski A, Zolfaghari A, Blais S, Jerkiewicz G. 1998. Coverage evolution of sulfur on Pt(111) electrodes: From compressed overlayers to well-defined islands. Electrochim Acta 44: 1019–1030.

Szabo S. 1991. Underpotential depostion of metals on foreing metal substrates. Int Rev Phys Chem 10: 207–248.

Szabo S, Nagy F. 1978. Investigations of bismuth adsorption via the ionization of hydrogen adsorbed on platinized platinum in hydrochloric acid solutions. J Electroanal Chem 87: 261–265.

Trasatti S. 1972. Electronegativity, work function, and heat of adsorption of hydrogen on metals. J Chem Soc Faraday Trans I 68: 229–236.

Villegas I, Weaver MJ. 1996. Infrared spectroscopy of model electrochemical interfaces in ultrahigh vacuum: Evidence for coupled cation-anion hydration in the Pt(111)/K^+,Cl^- system. J Phys Chem 100: 19502–19511.

Xiao XY, Tillmann S, Baltruschat H. 2002. Scanning tunneling microscopy of Sn coadsorbed with Cu and CO on Pt(111) electrodes. Phys Chem Chem Phys 4: 4044–4050.

Zangwill A. 1988. Physics at Surfaces. Cambridge: Cambridge University Press.

Zhou WP, Kibler LA, Kolb DM. 2002. Evidence for a change in valence state for tellurium adsorbed on a Pt(111) electrode. Electrochim Acta 47: 4501–4510.

Zhou WP, Kibler LA, Kolb DM. 2004. XPS study of irreversibly adsorbed arsenic on a Pt(111) electrode. Electrochim Acta 49: 5007–5012.

CHAPTER 8

Electrochemistry at Well-Characterized Bimetallic Surfaces

VOJISLAV R. STAMENKOVIC and NENAD M. MARKOVIC

Argonne National Laboratory, Materials Science Division, University of Chicago, Argonne, IL 60439, USA

8.1 INTRODUCTION

Interest in electrochemistry on well-defined (usually single-crystalline) surfaces has risen steadily over the last four decades, and coincides with the development of surface structural techniques for the microscopic-level characterization of metal single-crystal surfaces in ultrahigh vacuum (UHV). Not surprisingly, an early emphasis was placed on developing the UHV-based experimental strategies applicable to electrochemical interfaces. This ex situ approach, pioneered during the 1970s by Hubbard (1998), Yeager [Hagans et al., 1979], and in Ross (1982), provided the first successful means of obtaining atomic-level structural information for electrochemical adsorbates on metal single-crystal surfaces. Although such an ex situ strategy opened some new directions in surface electrochemistry, establishing the relationship between the structure of the interface in an electrolyte and that observed in UHV was always problematic, and had to be carefully examined on a case-by-case basis. More sophisticated microscopic-level descriptions of electrochemical systems became possible in the early 1990s through the emergence of new in situ surface-sensitive probes, most notably synchrotron-based surface X-ray scattering (SXS) [Ocko et al., 1990; Lucas et al., 1996; Tidswell et al., 1993] and scanning tunneling microscopy (STM) [Bard et al., 1991; Kolb, 1996; Itaya, 1998]. While these two techniques are markedly different in character and exhibit contrasting strengths, both are providing information on potential-dependent surface structures at a level of sophistication that is on a par with (or even beyond) that obtained for surfaces in traditional (vacuum-based) surface science [Somorjai, 1985]. Of the various systems examined, a decided emphasis has been placed on elucidating the surface electrochemistry of bimetallic surfaces, which play a crucial role in a number of technologically important areas, including

---

Fuel Cell Catalysis. Edited by Marc T. M. Koper
Copyright © 2009 John Wiley & Sons, Inc.

electrocatalysis, corrosion, and electrochemical synthesis. Although this field is still in its early years, a great deal has already been learned, and trends are beginning to emerge that provide some predictive ability with respect to the surface structure/composition assumed by bimetallic surfaces and their corresponding chemisorption properties and activities towards simple molecules. The aim of this chapter is to review some of the recent advances in bimetallic surface electrochemistry and to give selected examples that will demonstrate that it is indeed possible to prepare and characterize ordered bimetallic systems in an electrochemical environment, assess the adsorption and electrochemical reactivity of these systems, and use knowledge from well-defined surfaces in nanoscale catalyst design. A brief summary of experimental techniques that have been commonly used for the characterization of bimetallic systems is provided. However, this overview covers techniques that have been employed at the authors' laboratories, and the reader should be aware that this is not a complete summary.

8.2 SURFACE CHARACTERIZATION TECHNIQUES: EX SITU VERSUS IN SITU

8.2.1 Surface Structure

In heterogeneous catalysis, chemical reactions are taking place at solid surfaces, and therefore it is of paramount importance to obtain detailed insight into fundamental surface properties such as structure, composition, and surface electronic profile. UHV surface-sensitive tools have been extensively used for determination of these critical parameters [Somorjai, 1992]. For instance, low energy electron diffraction (LEED) is a widely adopted tool to study the surface crystallography of various materials [Van Hove et al., 1986]. In LEED experiments, an incident low energy electron beam that is in the range from about 20 to 500 eV is diffracted from the surface of the sample. The emerging diffracted beams are usually detected on a phosphorescent screen as a characteristic pattern, which represents the reciprocal space. From this pattern, conclusions can be drawn about the symmetry of the surface structure and the size of characteristic periodicities on the surface. LEED is usually applied to control the surface quality of a sample after preparation in UHV by establishing the exact surface structure, eventual presence of reconstruction, impurities, etc. Moreover, LEED has been extensively used to study chemisorption and catalytic properties in UHV, i.e., correlations between surface structure and adsorption/desorption processes [Somorjai, 1993].

As an example, Fig. 8.1 shows a result from Bardi and co-workers obtained on a bimetallic $Au_3Pd(100)$ single-crystal alloy [Kuntze et al., 1999]. The LEED pattern indicates a sharp (1×1) unit cell that corresponds to the bulk-truncated structure of the substitutionally disordered Au_3Pd alloy. Additionally, the authors determined the composition of the first outermost layer to be pure Au. These findings revealed that the (100) oriented surface of Au over Au_3Pd alloy is not reconstructed, which is unique, since pure Au, Pt, and Ir (100) crystals are all known to be reconstructed in similar ways [Van Hove et al., 1981; Ritz et al., 1997]. In this case, the presence

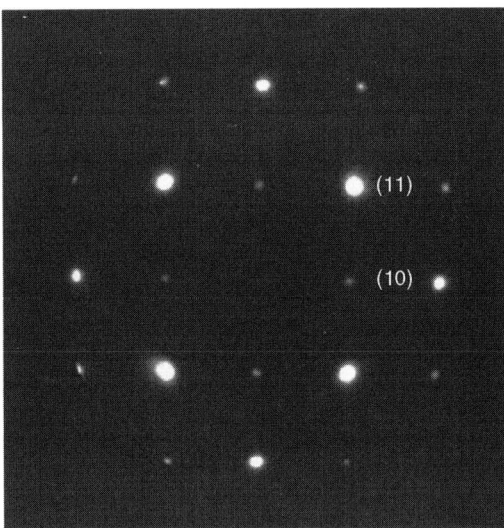

Figure 8.1 LEED pattern of $Au_3Pd(100)$, showing a sharp (1×1) unit cell. (Reprinted with permission from Kuntze et al. [1999]. Copyright 1999. The American Physical Society.)

of the other metal Pd in the alloy stabilizes the unreconstructed surface. Obviously, LEED provides an important insight into the arrangements of surface atoms of bimetallic systems; therefore, surface structure, its stability, and surface adsorption properties, as well as adsorption sites, can be revealed. Besides studies in UHV, LEED has also been successfully employed to analyze adsorbates as an ex situ probe for surfaces previously exposed to an electrochemical environment [Wagner and Ross, 1983].

Furthermore, LEED is an important tool to study thin film growth, and can reveal the mechanism of thin film formation as well as the final structure of mono/multi topmost layers versus substrate [Attard et al., 1995].

The sharpness of the spots and the amount of diffuse background are good indicators of the long-range order of the sample surface. By measuring the spot intensity as a function of the incoming beam voltage ($I-V$ curve) and comparing this with simulations, models of the surface structure can be tested [Heinz et al., 1992]. Also from an $I-V$ analysis, the surface segregation profiles have been estimated for a number of bimetallic alloys, and found to be in good agreement with theoretical predictions [Gauthier et al., 1985; Vasiliev, 1997].

Overall, LEED in UHV provides the exclusive ability to study structure–function relationships in heterogeneous catalysis, and for that reason it has become a routine surface analysis tool.

Nevertheless, in applications relevant for electrocatalysis and reactions that occur at solid–liquid interfaces, it has been essential to develop a methodology that can provide detailed insight into the surface and near-surface structure during the course of reaction. For that purpose, the in situ SXS diffraction technique, depicted in

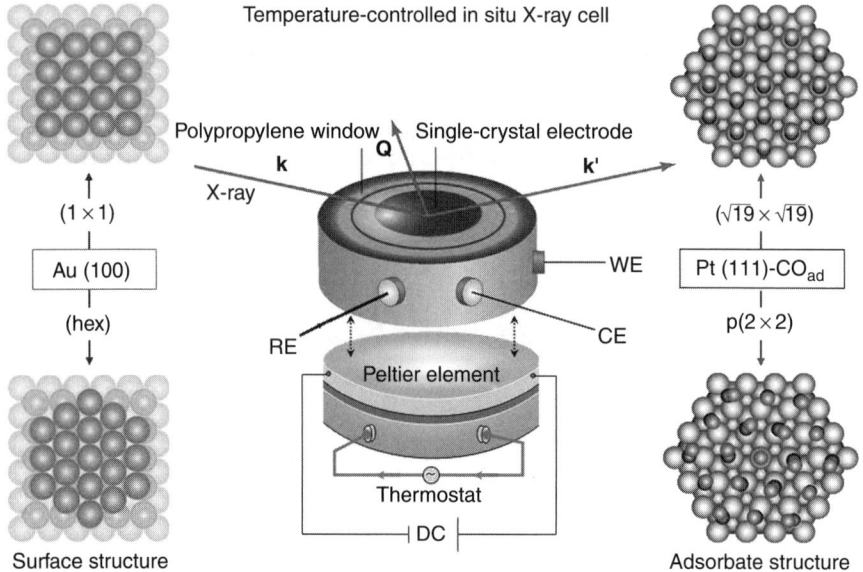

Figure 8.2 In situ SXS electrochemical cell: WE, working electrode; CE, counter-electrode; RE, reference electrode. On the left is shown the transition from (1 × 1) to (hex) for a Au(100) surface and on the right the characteristic adsorbate structures of CO on Pt(111) commonly observed by SXS.

Fig. 8.2, takes advantage of the ability of X-rays to penetrate a thin solution layer. Together with a simple theory of backscattering, this allows very detailed spatial information to be obtained [Lucas and Markovic, 2006].

For instance, one of the most relevant systems to study is reconstruction of Au(100), and its (hex) ↔ (1 × 1) transition. This transition can be followed by SXS [Ocko et al., 1990; Tidswell et al., 1993], as the reconstruction gives rise to X-ray scattering that is separated in reciprocal space from the scattering due to the bulk of the crystal. Figure 8.2 shows ball models of the square planar (1 × 1) surface that is stable at positive potentials, and the hexagonal (hex) reconstructed phase, which is commonly detected at more negative potentials. On the right-hand side in Fig. 8.2, schematic structures of the Pt(111)-CO system are given. The potential-induced p(2 × 2) ↔ $\sqrt{19}$ CO phase transition of an ordered CO adlayer has been successfully followed by SXS in acid and alkaline electrolytes [Tolmachev et al., 2004; Wang et al., 2005; Strmcnik et al., 2007]. Both examples in Fig. 8.2 are associated with a new SXS cell design, which allowed the exploration of temperature effects on surface reconstructions of monometallic surfaces both in nonadsorbing electrolytes and in the presence of different adsorbates, and for different electrolyte pH values. Great progress has been achieved with SXS in studies of bimetallic systems such as $Pt_3Sn(111)$ [Stamenkovic et al., 2003] and $Pt_3Ni(hkl)$ [Stamenkovic et al., 2007a] surfaces, which will be discussed in more detail in the following sections.

8.2.2 Surface Structure and Surface Composition

Besides SXS, another matching technique that has become available to surface scientists in the last two decades is scanning tunneling microscopy (STM). This technique has proved to be a valuable tool in different environments such as UHV, ambient pressure, and liquids. A large number of studies have been published on results obtained with STM, providing deep insight into atomic structure [Binning et al., 1982], corrosion [Zhang and Stimming, 1990; Oppenheim et al., 1991], surface defects, and thin film growth [Hosler et al., 1986; Sieradzki et al., 1999], for example. In electrocatalysis, STM became established in the early 1990s [Bard et al., 1991], and is now widely accepted as a routine tool for the characterization of solid–liquid electrified interfaces [Bard et al., 1993]. In addition, STM has been successfully used for bimetallic surfaces in order to obtain atomic resolution with the ability to distinguish between different elements [Varga and Schmmid, 1999]. Two examples are mentioned here: the work on $Pt_{25}Rh_{75}(100)$ by Varga's group [Hebenstreit et al., 1999] and the study of Pd-Au surfaces by Behm's group [Maroun et al., 2001]. Varga and co-workers used STM images with chemical contrast that allowed direct determination of the surface composition and short-range order of various $Pt_{25}Rh_{75}$ single-crystal surfaces in vacuum (Fig. 8.3). For all crystallographic orientations, they found a strong Pt enrichment at the surface. In the other example, Behm and co-workers described the evaluation of catalytic properties of defined atomic ensembles in atomically flat PdAu(111). Catalytic behavior was correlated with the concentration and distribution of Pd ad-atoms, which were determined by in situ STM with chemical contrast.

The latter report demonstrated the unique ability of this technique to resolve surface structure as well as surface composition at the electrified solid–liquid interfaces. In particular, STM has become an important tool for ex situ and in situ characterization of surfaces at the atomic level, in spite its significant limitations regarding surface composition characterization for bimetallic systems, such as the lack of contrast for different elements and the scanned surface area being too small to be representative for the entire surface. To avoid these limitations, STM has been mostly used as a complementary tool in surface characterization.

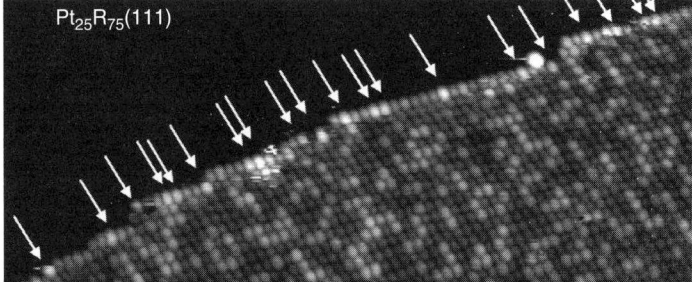

Figure 8.3 STM image with chemical contrast of the $Pt_{25}Rh_{75}(100)$ surface. Arrows point at brighter Rh atoms at the step edge. (Reprinted with permission from Hebenstreit et al. [1999]. Copyright 1999. Elsevier.)

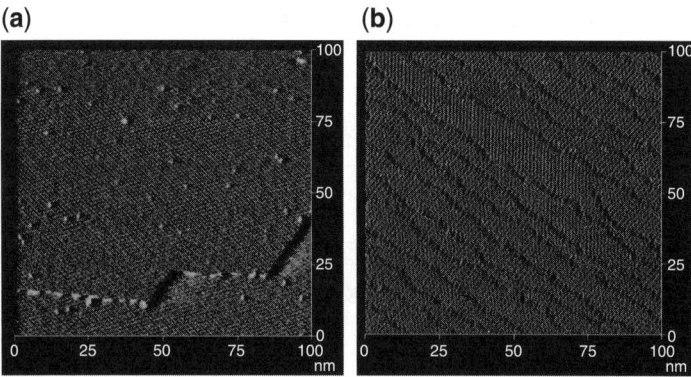

Figure 8.4 STM images (100 nm × 100 nm size) of Pt(111) covered by CO for (a) as-prepared electrode and (b) CO-annealed electrode (20 consecutive cycles) to 0.95 V. The images ($U_{tip} = 0.15$ V; $I_{tip} = 1$ nA) illustrate the presence of islands and steps on the surface.

For instance, in our laboratory, we have recently successfully coupled ex situ STM experiments with electrochemical treatment of Pt single crystals, and we have been able to assign certain changes in surface morphology to electrochemical pretreatment [Strmcnik et al., 2008].

The image of the as-prepared Pt(111) electrode (Fig. 8.4a) reveals that the surface is composed of a flat terrace that is covered by a small number of ad-islands with monatomic height. These features are even more evident on the as-prepared Pt(100) electrode [Strmcnik et al., 2008]. The origin of such islands is likely due to lifting of reconstructions in the thermally annealed surfaces. However, after CO annealing to 0.95 V [Arenz et al., 2005], i.e., cycling the potential in the CO saturated electrolyte, the ad-islands completely disappear (Fig. 8.4b) and are replaced by a series of smooth terrace–step structures. Since this transformation only occurs above the ignition potential for CO oxidation, conclusion is that the CO oxidation reaction, as opposed to simple CO adsorption, leads to step formation on the surface. Both surface features can have a dramatic influence on catalytic activity, since the topmost atoms have different coordination numbers in each case [Strmcnik et al., 2008; Arenz et al., 2005; Lebedeva et al., 2002]. This illustrative example emphasizes the capability of STM in the characterization and control of the active sites.

8.2.3 Surface Composition

Another very important technique for fundamental consideration of multicomponent systems is low energy ion scattering (LEIS) [Taglauer and Heiland, 1980; Brongersma et al., 2007]. This is a unique tool in surface analysis, since it provides the ability to define the atomic composition of the topmost surface layer under UHV conditions. The signal does not interfere with the subsurface atomic layers, and therefore the results of LEIS analysis represent exclusively the response from the outer surface. In LEIS, a surface is used as a target that scatters a noble gas ion beam (He^+, Ne^+,

8.2 SURFACE CHARACTERIZATION TECHNIQUES: EX SITU VERSUS IN SITU

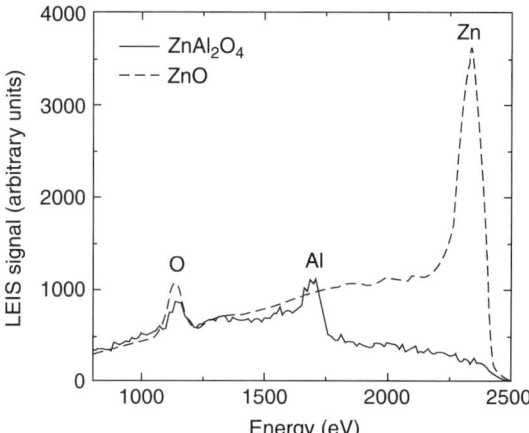

Figure 8.5 LEIS spectra for $ZnAl_2O_4$ (solid line) and ZnO (dashed line) obtained with a 3 keV He^+ beam. (Reprinted with permission from Brongersma and Jacobs [1994]. Copyright 1999. The American Physical Society.)

or Ar^+) with energies between 0.5 and 3 keV. The energy of the scattered ions is in direct correlation with the mass of the target atoms, and therefore can be interpreted as a "mass spectrum" of the surface.

Figure 8.5 shows the LEIS spectra of $ZnAl_2O_4$ and ZnO as a characteristic example of a multicomponent system analyzed by this technique [Brongersma and Jacobs, 1994]. Since only the surface peaks of Al and O were detected for $ZnAl_2O_4$, the Zn atoms must be located in the subsurface layers. The onset of the tail agrees between the spectra, indicating that Zn is present in the second and deeper layers. This example illustrates the strength of the LEIS technique, in that characteristic peaks from different elements can be used to selectively analyze the atomic composition of the topmost surface. In addition, the shape of the tails could provide information on the in-depth distribution of the elements.

This has been used, for instance, to follow the formation of palladium silicide in a silicon wafer for thicknesses up to 6 nm [Vanleerdam et al., 1990]. More recently, investigation of the tails in LEIS has been used as a tool for high resolution nondestructive in-depth composition analysis of ultrathin layers [Brongersma et al., 2003] and shallow interfaces [Janssen et al., 2004].

Although it is, in principle, feasible to quantify LEIS results, the determination of absolute ion fractions is far from trivial. In LEIS experiments, the surface composition is therefore generally obtained in a relative measurement by comparison of signals of the element in the sample of interest with those of reference samples, assuming that the matrix does not have a significant role and that the surface roughness is similar.

8.2.4 Chemical Nature of Metal–Electrolyte Interfaces

In addition to the structure and composition of the electrode, evaluation of an electrochemical interface in terms of the nature of adsorbates has long been recognized as a key requirement for understating reactivity of the electrochemical interface. In situ infrared reflection absorption spectroscopy (IRAS) in a thin liquid film cell mode has been successfully applied for characterization of electrified solid–liquid interfaces [Iwasita and Nart, 1997].

This method provides the unique ability to follow changes at a molecular level during the course of reaction. For instance, for each potential change, characteristic infrared spectra can be obtained, and therefore the surface electrochemistry of reactants, products, and spectator molecular species can be followed. Figure 8.6 shows spectra of adsorbed CO on Pt(111) and Au(*hkl*) surfaces obtained in an infrared cell that has been designed in our laboratory. Besides the possibility of working under temperature-controlled conditions, this configuration offers the opportunity to substantially decrease the recording time for each potential step. This has led to the development of a so-called rapid scan mode of operation, which delivers a continuous mode for recording of spectra for potential sweep rates between 1 and 10 mV/s.

For fundamental studies of nanoparticles in IRAS measurements, a very important issue is how to attach metal nanoparticles onto a conducting substrate without changing their physical and chemical properties. Recently, we have developed a new method for anchoring metallic nanoparticles on reflective substrates of gold and/or glassy carbon, which we have termed a temperature-induced deposition (TID) method [Stamenkovic et al., 2004]. A key advantage of this method is that the catalysts

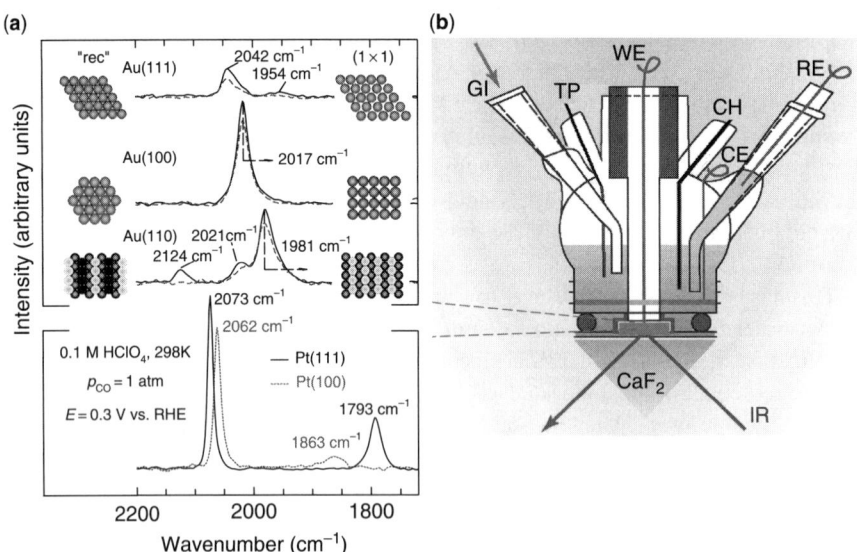

Figure 8.6 (a) Infrared spectra of adsorbed CO on Pt(111) and Au(*hkl*) surfaces obtained in a spectroelectrochemical cell (b) for in situ IRAS measurements at elevated temperatures (the initial design of the cell was described by Iwasita [1997]): GI, gas inlet (bubbler); TP, thermocouple; WE, working electrode; CH, cartridge heate; CE, counter-electrode; RE, reference electrode; CaF_2, CaF_2 prism. (Reprinted with permission from Stamenkovic et al. [2004]. Copyright 2004. The American Chemical Society.)

are uniquely spread over the substrate by instantaneous evaporation of the aqueous matrix, forming a thin and uniform film. With this approach, for instance, the commonly detected negative v_{CO} absorbance is completely eliminated, so that the v_{CO} spectra can routinely be measured from nanoparticles and compared with the corresponding spectra from single-crystal surfaces. The IRAS technique described here has demonstrated its ability to provide invaluable details about the true morphology of multimetallic well-defined extended surfaces as well as nanoparticle systems, as will be discussed in the following sections.

8.3 RULES GOVERNING ELECTROCATALYSIS AT BIMETALLIC SYSTEMS

A major focus in electrocatalysis is on the development of a fundamental understanding of the catalytic activity of bi- and/or multimetallic systems in order to gain unprecedented control of reactivity and stability during the transformation of the chemical energy of hydrogen, hydrocarbons, and oxygen into electrical energy. The primary interest is in controlling the key parameters that could lead to a catalytic system with advanced properties. It is well known that an enhancement of catalytic properties could be achieved by the addition of a second metal [Appleby, 1970; Kinoshita, 1992], and the mechanism of improvement may occur through a change in the local bonding geometry (structure effects), a bifunctional mechanism, the distribution of active sites (ensemble effects), or directly by modifying the reactivity of Pt surface atoms (electronic effects) [Ross, 1998]. In a real system, all of these factors will, in general, operate simultaneously, and separating their effects and assessing their relative importance for catalytic activity and reaction mechanism is a very challenging problem [Kinoshita, 1992; Markovic et al., 2003]. This lack of fundamental knowledge about the mechanisms responsible for the catalytic activity of bi/multimetallic systems, in turn, makes a further improvement of the materials in a rational, *science-based* manner very difficult. In what follows, we will emphasize several illustrative mechanisms for improved electrocatalysis on bimetallic systems.

8.3.1 Bimetallic Surfaces

Well-characterized Pt-bimetallic surfaces can be prepared by conventional metallurgy, as bulk alloys. Pt_xRu_y [Gasteiger et al., 1993], Pt_3Mo [Grgur,], and $Pt_3Sn(hkl)$ [Gasteiger et al., 1995] alloys have been thoroughly characterized in our laboratory, and we have been able to determine by UHV and in situ electrochemical techniques that both metals are present and stable at the outermost surface layer under relevant conditions. The $Pt_3Sn(111)$ system, for instance, is the most active catalyst for CO oxidation [Stamenkovic et al., 2003; Gasteiger et al., 1996]. The structural properties of the $Pt_3Sn(111)$ surface in an electrolyte have been thoroughly examined by in situ SXS, including both the stability of the UHV-prepared $p(2 \times 2)$ structure in H_2SO_4 solution and the potential-dependent relaxation of Pt and Sn atoms in the near-surface region (Fig. 8.7).

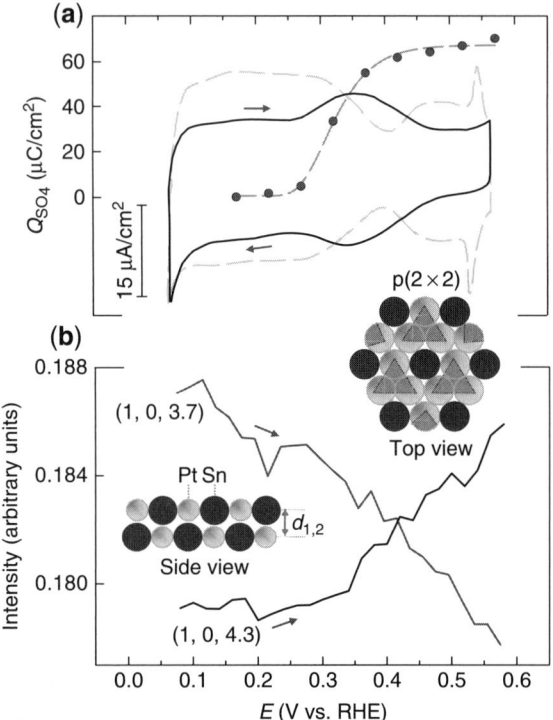

Figure 8.7 (a) Cyclic voltammograms of Pt(111) (dashed gray line) and Pt$_3$Sn(111) (solid black line) in 0.5 M H$_2$SO$_4$; scan rate 50 mV/s. Potential-dependent integrated charges for the adsorption of (bi)sulfate anions on the Pt$_3$Sn(111) surface are represented by circles. (b) Measured X-ray intensities at (1, 0, 3.7) and (1, 0, 4.3) as functions of electrode potential. The ball models show top and side views of the p(2 × 2) structure: gray circles, Pt atoms; black circles, Sn atoms; triangles, (bi)sulfate anions adsorbed on Pt sites. The side view indicates the surface normal spacing that is derived from the crystal truncation rod (CTR) measurements. (Reprinted with permission from Stamenkovic et al. [2003]. Copyright 1999. The American Chemical Society.)

The results indicate two important characteristics of the surface structure at 0.05 and 0.55 V: (i) at 0.05 V, expansion of the Pt surface atoms induced by the adsorption of hydrogen is very similar to that observed on Pt(111) [Janssen et al., 2004; Iwasita and Nart, 1997] ($\Delta d^1_{Pt} = +2\%$); (ii) at 0.55 V, the p(2 × 2) structure remains stable but, while the Pt surface atoms are unrelaxed, the Sn atoms in the topmost layer expand up to ($\Delta d^1_{Sn} = 8.5\%$) of the lattice spacing. At potentials higher than 0.55 V, this expansion is even more pronounced, indicating that before Sn dissolution, the Sn surface atoms are expanded by approximately 12% of the bulk lattice spacing. Cyclic voltammetry confirms the presence of non-Pt atoms on the surface by a significantly attenuated hydrogen adsorption/desorption region, as well as characteristic adsorption of anion (sulfate) on Pt sites, which is represented by a reversible voltammetric feature at about 0.35 V (Fig. 8.7a). These in situ results confirmed that the surface structure and composition previously determined under UHV conditions were stable in an electrochemical environment.

In addition, the properties of adsorbed CO were examined by in situ IRAS, and the origin of the high catalytic activity was fully explained in combination with ex situ UHV measurements (Fig. 8.8). In contrast to the near-invariant bands of atop CO_{ad} on Pt(111) [Markovic et al., 2002], changes in the band shape (splitting of the band) and frequency are clearly visible on the $Pt_3Sn(111)$ surface in the same potential range. The splitting and the early onset of CO electro-oxidation have been explained by the presence of oxygenated species on every second Sn atom, which was first confirmed by ex situ LEED, i.e., a sharp p(4 × 4) pattern was obtained by LEED on a $Pt_3Sn(111)$ surface previously exposed to oxygen at 4×10^{-8} Torr at 300 °C (Fig. 8.8c, d), having a Sn/Pt Auger electron spectroscopy (AES) ratio of 4.8 compared with 4.5 for the p(2 × 2) pattern [Stamenkovic et al., 2003]. These results also indicate that chemisorption of oxygen occurs without changing the underlying structure or composition of the $Pt_3Sn(111)$ alloy surface.

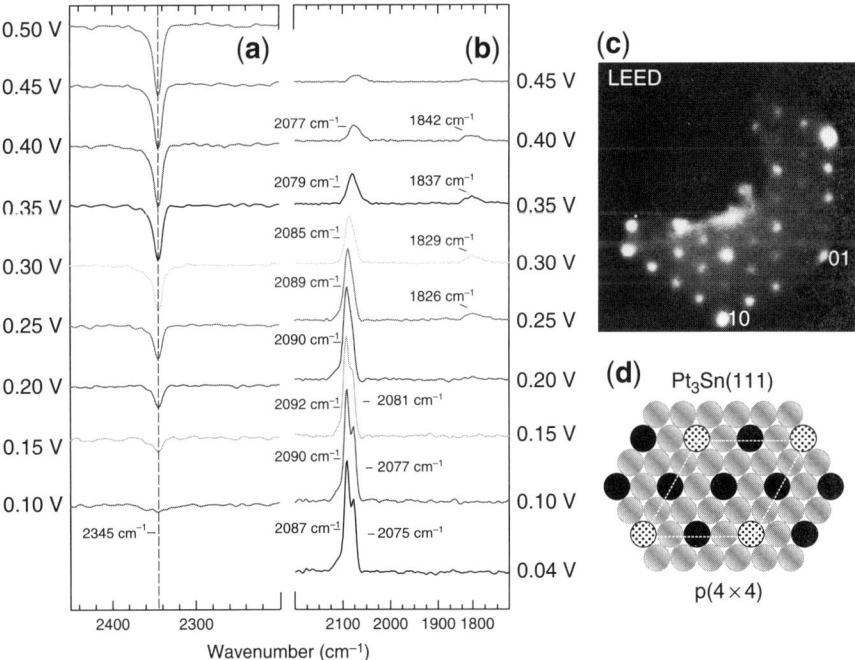

Figure 8.8 Series of infrared spectra during (a) CO_2 production and (b) progressive oxidation of CO_{ad} on $Pt_3Sn(111)$ in 0.5 M H_2SO_4 saturated with CO; each spectrum was accumulated from 50 interferometers at the potential indicated. (c, d) LEED pattern and schematic representation of the p(4 × 4) structure observed on $Pt_3Sn(111)$ after exposing the surface to O_2 and electrolyte. The gray circles are Pt surface atoms, the black circles are Sn atoms covered with OH, and the dotted circles are Sn atoms that are chemically different from Sn atoms modified with OH. (Reprinted with permission from Stamenkovic et al. [2003]. Copyright 1999. The American Chemical Society.)

The same approach was used for the Pt$_3$Sn(111) surface exposed to an electrochemical environment, where the oxygenated species are assumed to be OH$^-$. The existence of two types of Sn sites (oxide-covered and oxide-free) causes compression of CO$_{ad}$ molecules, resulting in higher local CO coverage, blue-shifted CO frequencies due to enhanced dipole–dipole coupling, and onset of CO oxidation at 0.1 V. Therefore, the oxidative removal of CO$_{ad}$ on Pt$_3$Sn(111) at low potentials (based on CO$_2$ production in Fig. 8.9 at about 0.1 V) proceeds on the surface sites where local (microscopic) CO$_{ad}$ coverage remains high and thus CO$_{ad}$ is weakly adsorbed on Pt. In contrast to the polarization curves obtained from rotating disk electrode measurements [Gasteiger et al., 1996], spectroscopic data unambiguously show the true onset potential for CO oxidation. This, in turn, emphasizes that classical electrochemical methods are not capable of measuring the reaction rates with very low turnover frequency (TOF), such as the oxidative removal of CO$_{ad}$ from the Pt$_3$Sn(111) surface at low overpotential. Nevertheless, linking the microscopic and macroscopic levels of characterization, the so-called weakly adsorbed (CO$_{ad}^w$) [Markovic et al., 1999] state on this surface may correspond microscopically to the formation of disordered but compressed CO$_{ad}$ patches with a characteristic high frequency (2090 cm^{-1}) for atop CO$_L$. The remaining fraction of CO$_L$ with frequency of 2077 cm^{-1} may correspond to more strongly adsorbed CO$_{ad}$, which requires higher overpotentials to be oxidized. For $E >$ 0.25 V, a pair of bands is transformed into a single peak shifted towards 2077 cm^{-1} (Fig. 8.8b). This transition is accompanied by the appearance of a bridge-bonded CO$_B$ stretching band (centered at about 1820 cm^{-1}). As noted by many authors [Lebedeva et al., 2000; Markovic and Ross, 2002], the oxidation of CO$_{ad}$ occurs

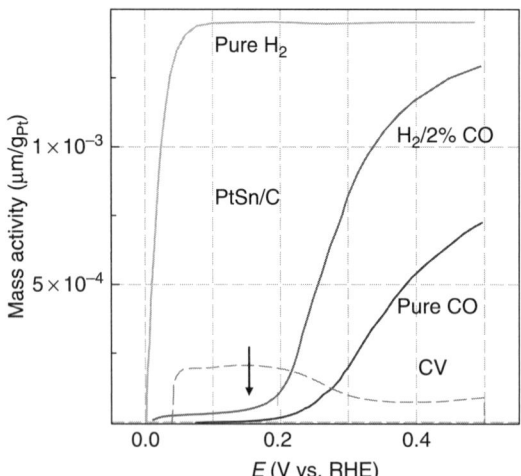

Figure 8.9 Polarization curves for a PtSn/C catalyst recorded by a rotating disk electrode in 0.5 M H$_2$SO$_4$ saturated with either pure hydrogen, a H$_2$/2% CO mixture, and pure CO (the arrow points to the onset of CO oxidation) at 60 °C with 1 mV/s and 2500 rev/min; the dashed curve is the cyclic voltammogram (in arbitrary units) in an argon-purged solution at 60 °C with 50 mV/s. (Reprinted with permission from Arenz et al. [2005]. Copyright 2005. Elsevier.)

through a Langmuir–Hinshelwood (L–H)-type reaction and follows a reaction scheme in which oxygenated species can be adsorbed on Sn or Pt at low potentials:

$$\text{Pt-CO}_{ad} + \text{Sn-OH}_{ad} \longrightarrow \text{CO}_2 + \text{H}^+ + e^- \tag{8.1}$$

It should be mentioned here that Sn sites are not considered to be the solitary source for OH_{ad}, which could be adsorbed on Pt sites owing to the influence of adjunct Sn atoms [Stamenkovic et al., 2005]. The promotional effect of Sn was later confirmed on a PtSn/C nanocatalyst [Arenz et al., 2005], which exhibits similar behavior that was assigned primarily to the formation of reactive OH species at much lower potential than on pure Pt catalysts. Based on these findings, the *bifunctional effect* was unambiguously confirmed for Pt-Sn surfaces, where Sn sites serve as a source of oxygenated species that boost CO oxidation at low potentials and allow these surfaces to be employed as CO-tolerant catalysts.

8.3.2 Bimetallic Systems with Nanosegregated Profile

In our early work with bimetallic systems, we noticed that, depending on the preparation procedure in UHV, different surface compositions could be produced over the same bulk material owing to the phenomenon of surface segregation [Stamenkovic et al., 2002]. It was essential, then, to establish a methodology for transferring a well-defined bimetallic system into an electrochemical environment for further electrochemical characterization.

The first results reporting a relation between electrocatalytic activity and surface composition determined by LEIS were reported by Gasteiger et al. (1993), for a series of PtRu alloys. After annealing in UHV, these systems showed evidence of Pt enrichment on the surface, and a clear link between the surface composition and activity for electrooxidation of methanol was established. More recently, our initial work on polycrystalline Pt_3M alloys (M = Ni, Co) [Stamenkovic et al., 2002, 2003] shown in Fig. 8.10, revealed a complete segregation of Pt after annealing in UHV at 1000K, i.e., the outermost surface layer contains 100 at% of Pt. A pure topmost atomic layer of Pt has been designated as the *Pt-skin* surface. The same results have been found for all Pt_3M (M = Co, Ni, Fe, V, Ti) alloys studied by our group, i.e., formation of a Pt-skin composition upon annealing, and a surface composition after mild sputtering that corresponds to the bulk ratio of alloying components (75 at% of Pt and 25 at% of Co) [Stamenkovic et al., 2007b]. However, in contrast to Pt_3M systems, no enrichment in Pt has been found for the annealed PtCo alloy surface, suggesting that Pt segregation thermodynamics depends strongly on the bulk ratio of alloying components [Stamenkovic et al., 2006a]. In addition, we have challenged the chemical stability of well-defined surfaces that were exposed to electrochemical environment utilizing AES and LEIS spectroscopy. Following (electro)chemical pretreatment, each alloy surface was rinsed with ultrapure triply distilled water, dried in an argon stream, and then transferred back to the UHV environment.

Comparison between data obtained on surfaces treated in water and acid unambiguously confirms a dramatic change in surface composition; in fact, complete dissolution

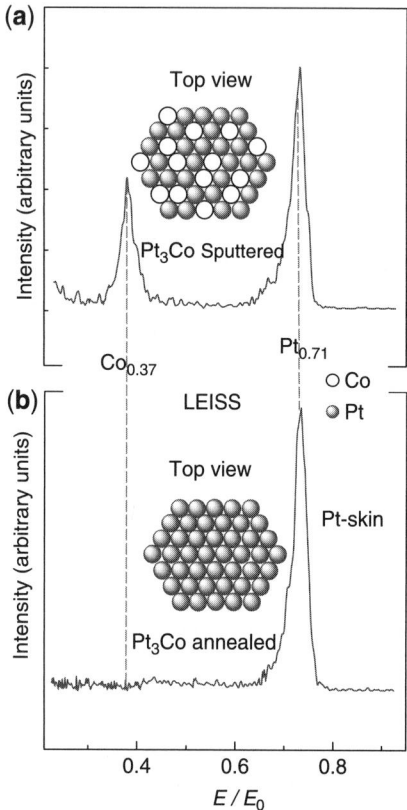

Figure 8.10 LEIS spectra of Pt$_3$Co: (a) lightly sputtered surface with 0.5 keV Ar$^+$ ions; (b) annealed surface at 1000K. (Reprinted with permission from Stamenkovic et al. [2002]. Copyright 2002. The American Chemical Society.)

of surface Co atoms was revealed by the LEIS spectra shown in Fig. 8.11d, while Co atoms are stable and covered with oxides (Fig. 8.11a, b), i.e., upon exposure of a PtCo surface to an acidic solution, Co surface atoms are instantaneously dissolved from the near-surface layers. The remaining surface, which consists only of Pt atoms we term a *Pt-skeleton*. This type of surface had been described previously in the literature by Watanabe and co-workers [Toda et al., 1999], but it was designated as a skin of Pt with thickness of 3–5 atomic layers determined by X-ray photoelectron spectroscopy (XPS) measurements. The conclusions drawn from this study are, however, equally valid for the Pt$_3$Co, Pt$_3$Fe, Pt$_3$Ni, Pt$_3$Ti, and Pt$_3$V sputtered surfaces. In each case, whenever transition metal atoms are exposed to the acidic environment, non-Pt atoms are dissolved, forming a Pt-skeleton surface. There may be some differences between the alloys in the depth profile of transition metal atoms in the subsurface layers, but that determination was not performed on polycrystalline systems. The same set of ex situ analyses was performed on the Pt-skin surfaces formed over all three Pt$_3$M alloys, showing that the Pt-skin surface is a stable formation in an

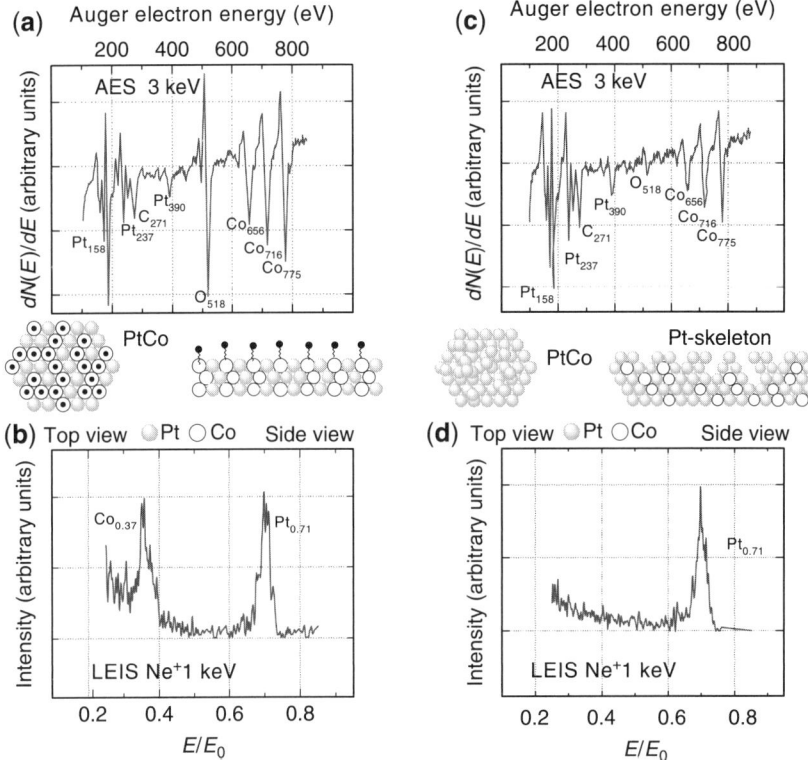

Figure 8.11 (a, b) PtCo surface after exposure to H_2O: (a) Auger spectra reveal the formation of stable oxides (black spheres); (b) LEIS spectra confirm that the Co surface concentration remains the same upon transfer to UHV. (c, d) PtCo surface after exposure to 0.1 M $HClO_4$: (c) Auger spectra show a decrease in intensity of the Co peaks; (d) LEIS spectra reveal that the PtCo surface contains only Pt atoms in the topmost atomic layer after exposure to the electrolyte, surface Co atoms are being dissolved forming a Pt-skeleton surface. (Reprinted with permission from Stamenkovic et al. [2006a]. Copyright 2006. The American Chemical Society.)

(electro)chemical environment. Overall, we concluded that in both the Pt-skin and Pt-skeleton surfaces, the pure Pt outermost layer protects the subsurface transition metals atoms from (further) dissolution.

This allows a direct influence of the alloying component on the electronic properties of these unique Pt near-surface formations from subsurface layers, which is the crucial difference in these materials. In addition, the electronic and geometric structures of skin and skeleton were found to be different; for example, the skin surface is "smoother" and the d-band center position with respect to the metallic Fermi level is downshifted for skin surfaces (Fig. 8.12) [Stamenkovic et al., 2006a] owing to the higher content of non-Pt atoms in the second layer. On both types of surface, the relationship between the specific activity for the oxygen reduction reaction (ORR) and the d-band center position exhibits a volcano-shape, with the maximum

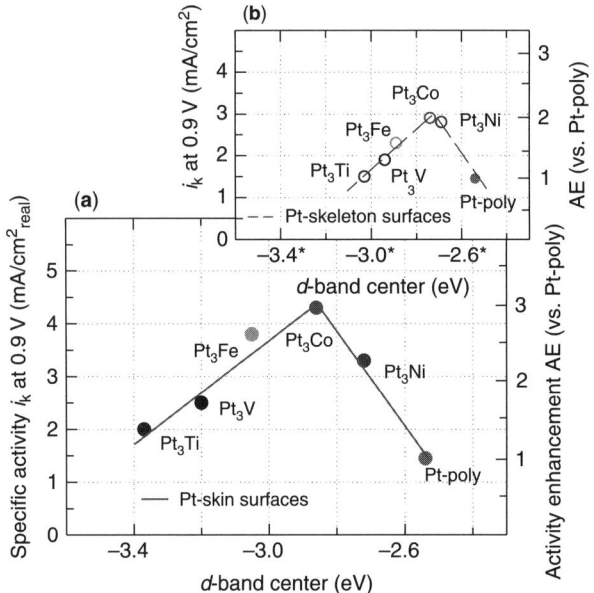

Figure 8.12 Relationships between the catalytic properties and electronic structure of Pt_3M alloys: correlation between the specific activity for the oxygen reduction reaction measured experimentally by a rotating disk electrode on Pt_3M surfaces in 0.1 M $HClO_4$ at 333 K and 1600 rev/min versus the d-band center position for (a) Pt-skin and (b) Pt-skeleton surfaces. (Reprinted with permission from Stamenkovic et al. [2007b]. Copyright 2007. Nature Publishing Group.)

catalytic activity being obtained for Pt_3Co (Fig. 8.12) [Stamenkovic et al., 2007b]. The overall consequence of this trend is that an active catalyst for the ORR should counterbalance two opposing effects, namely, adsorption energy of O_2 and reaction intermediates (O_2^-, O_2^{2-}, H_2O_2, etc.), against the adsorption strength for spectator oxygenated species. For metal surfaces that bind oxygen too strongly, as in the case of Pt, the d-band center is too close to the Fermi level, and the rate of the ORR is limited by the availability of "spectator-free" (e.g., OH_{ad}) Pt sites. On the other hand, when the d-band center is too far from the Fermi level, as in the case of Pt_3V and Pt_3Ti, the surface binds intermediates and O_2 too weakly to significantly promote the ORR.

The success and great future potential of this approach was recently demonstrated in our collaboration with Norskov's group, where our experimental results (summarized in Fig. 8.12) and Norskov's componential screening of the same binary alloys [Stamenkovic et al., 2006b] converge to the same optimal composition. This may serve as a textbook example that the design of stable and catalytically active materials for ORR electrocatalysis requires fundamental breakthroughs that come only from basic research on well-characterized surfaces. However, even though it was tempting to conclude that the rationale for the variation in activity depends exclusively on the position of the metal d-states relative to the Fermi level, there is no proof that this is always the case, because the values of the d-band center position for sputtered surfaces were obtained in UHV, whereas, in the electrolyte, there were substantial changes in surface composition and morphology. Therefore, these values (Fig. 8.12b) are just *virtual*, and the results imply that the catalytic activity depends also on the detailed morphology of the alloy surface. Unfortunately, it was impossible to determine this morphology in even an approximate manner on polycrystalline samples, and therefore, in the most recent work [Stamenkovic et al., 2007a], we used well-defined single-crystal surfaces to elucidate structural versus electronic effects. As for polycrystalline systems, in UHV, all $Pt_3Ni(hkl)$ surfaces show a strong enrichment of Pt and the formation of a Pt-skin layer with different geometry of surface atoms, but also with significant variations in the density of states (DOS) near the Fermi level. Although the phenomenon of surface segregation at bimetallic systems has been well studied under UHV conditions, especially with the progress of the LEIS technique [Brongersma et al., 2007] there has been no study examining segregation in situ during reaction conditions.

Great progress has been made, however, in our later study using in situ SXS [Stamenkovic et al., 2007a], where, by simultaneously fitting the intensity ratio between two different sets of crystal truncation rod (CTR) data that constrain the fit to the full CTR data [Robinson, 1986; Warren, 1990], it was possible to reveal the elemental concentration profile at the surface (Fig. 8.13c). Based on the in situ SXS results depicted in Fig. 8.13a, the topmost surface layer is confirmed to be 100 at%

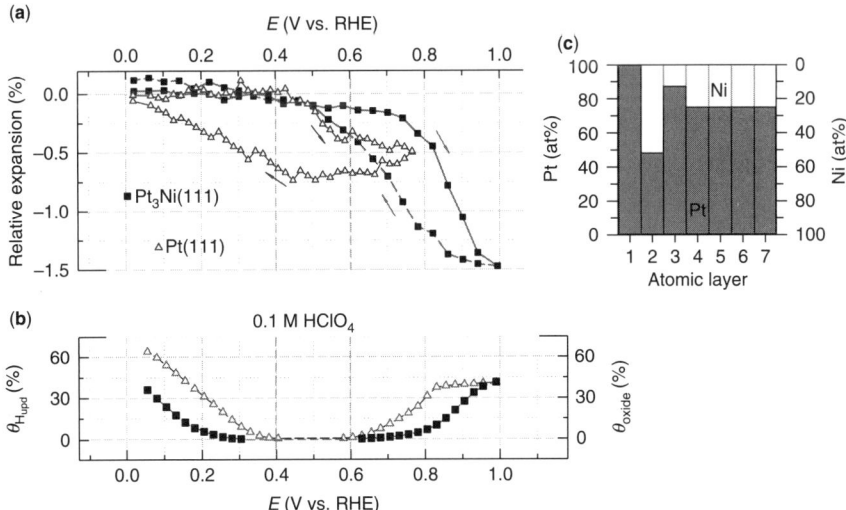

Figure 8.13 In situ electrochemical SXS characterization of $Pt_3Ni(111)$ and $Pt(111)$ surfaces: (a) XRV measurements for $Pt_3Ni(111)$ at the (0, 0, 2.7) (filled squares) and $Pt(111)$ at (1, 0, 3.6) (open triangles); (b) surface coverage by underpotentially deposited hydrogen (H_{upd}) and hydroxyl species (OH_{ad}) calculated from the cyclic voltammograms; (c) segregation profile ascertained from the SXS measurements. (Reprinted with permission from Stamenkovic et al. [2007a]. Copyright 2007. American Association for the Advancement in Science.)

Pt, the second atomic layer is found to be depleted in Pt (48 at% of Pt), the third layer is again enriched in Pt (87 at%), and, beyond that, the bulk value of 75 at%. It is clear from these results that the Pt-skin(111) surface is stable in UHV, during transfer and in the electrochemical environment. It is also apparent from X-ray voltammetry (XRV) (Fig. 8.13a) that the relaxation of both surfaces is induced by OH_{ad} and hence the potential shift is consistent with the cyclic voltammogram. The Pt-skin surface is stable over a wider range than the Pt(111) surface, perhaps owing to the increased contraction and the lower coverage by OH_{ad}, which is the precursor to irreversible oxide formation as well as dissolution. On this basis, it was possible to draw the conclusion that the Pt-skin surface is more stable owing to a less pronounced interaction with surface oxides. The stability of such surfaces in an electrochemical environment was relatively unknown, and so these findings were crucially important for potential applications of these surfaces as electrocatalysts.

The observed oscillatory concentration profile is the first instance of segregation profiles being monitored at different potentials in electrochemical environment and it was in excellent agreement with the results previously obtained on similar single-crystalline systems in UHV by Gautier and co-workers [Gauthier,]. Since this type of segregation is characteristic of the first four atomic layers, we term these systems surfaces with a *nanosegregated profile*. Further electrochemical characterization revealed the unique behavior of these surfaces. In Fig. 8.14, a representative cyclic voltammogram of $Pt_3Ni(111)$-skin is compared with a Pt(111) surface. The two electrodes have the same surface composition and structure of the topmost layer, but different compositions of the subsurface layer and hence different electronic structures. This leads to dramatic differences in their adsorption properties, which are nicely visible

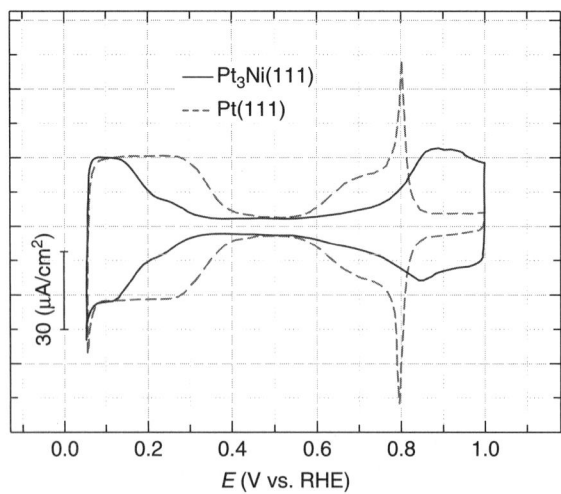

Figure 8.14 Cyclic voltammetry of $Pt_3Ni(111)$ and Pt(111) surfaces in 0.1 M $HClO_4$, at 30 mV/s. (Reprinted with permission from Stamenkovic et al. [2007a]. Copyright 2007. American Association for the Advancement in Science.)

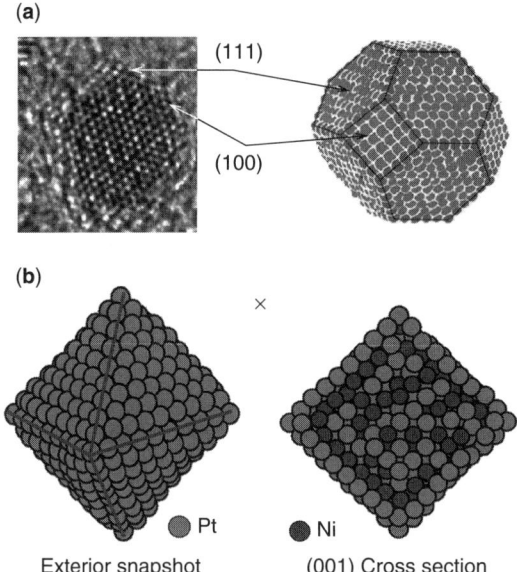

Figure 8.15 (a) TEM micrograph of a Pt nanoparticle supported on carbon and the corresponding cubo-octahedral model. (b) Snapshot of equilibrium Pt_3Ni octahedral nanoparticles containing 586 atoms simulated at 600K, exterior and cross-sectional views with (111) facets.

as a change in the surface coverage of underpotentially deposited hydrogen as well as oxide species (Figs. 8.13b and 8.14).

The attenuated oxide formation boosts the activity for the ORR, and for $Pt_3Ni(111)$-skin we found the highest activity that has ever been observed on cathode catalysts [Stamenkovic et al., 2007a], with a specific activity *10-fold higher than Pt(111) and 90-fold higher than state-of-the-art Pt/C catalysts*. Therefore, a critical goal in ORR electrocatalysis is to prepare PtM nanoparticles with electronic and morphological properties similar to those of the nanosegregated $Pt_3Ni(111)$ structure.

Given that the contribution of the (111) plane is maximized in an octahedral particle (Fig. 8.15) consisting of 8 (111) facets, 12 (111)–(111) step-edges, and 6 vertices, synthesis of uniform alloys of $Pt_{75}M_{25}$ octahedra with the Pt-skin structure is the ultimate goal [Wang et al., 2005]. These procedures are difficult, but, based on indications obtained from Monte Carlo simulations, such syntheses with metals may eventually be possible. An additional challenge involves preserving the Pt-skin-like surface composition in the particles.

8.3.3 Bimetallic Systems by Thin Film Deposition

In addition to the establishment and understanding of activity trends on nanosegregated surfaces, it has been anticipated that finding relationships between chemical and electronic properties of thin metal films of Pt group metals deposited over $3d$ and $5d$ elements has the potential to open up new opportunities in the quest to create novel, stable, and more efficient multimetallic catalysts. A number of procedures can be employed for thin film (0.1–1 monatomic layers: 0.1–1 ML) depositions; however, the major division is between deposition in vacuum and in electrochemical environments [Sieradzki et al., 1999; Tomellini and Attard, 1991].

In our recent work, we have used Pd deposited on Pt(hkl) in order to characterize thin metal films [Arenz et al., 2002] and to test their catalytic activity (Fig. 8.16). We have employed both methodologies: thermal evaporation in UHV and electrochemical deposition. For the Pd/Pt(111) system, in situ SXS measurements have been used to show that after the formation of 1 ML of pseudomorphic Pd film, three-dimensional pseudomorphic crystalline islands of pure Pd begin to aggregate.

For the purpose of demonstrating the effects of surface coverage by Pd, θ_{Pd}, on the rate of electro-oxidation of formic acid and the ORR, Fig. 8.17 reveals that the i versus θ_{Pd} relationship again has a volcano-like form, with the maximum catalytic activity being exhibited for 1 ML of Pd. The examples that we have given indicate that volcano relationships are the rule rather than the exception, emphasizing the importance of a systematic evaluation of the catalyst factors that control catalytic activity. A thorough

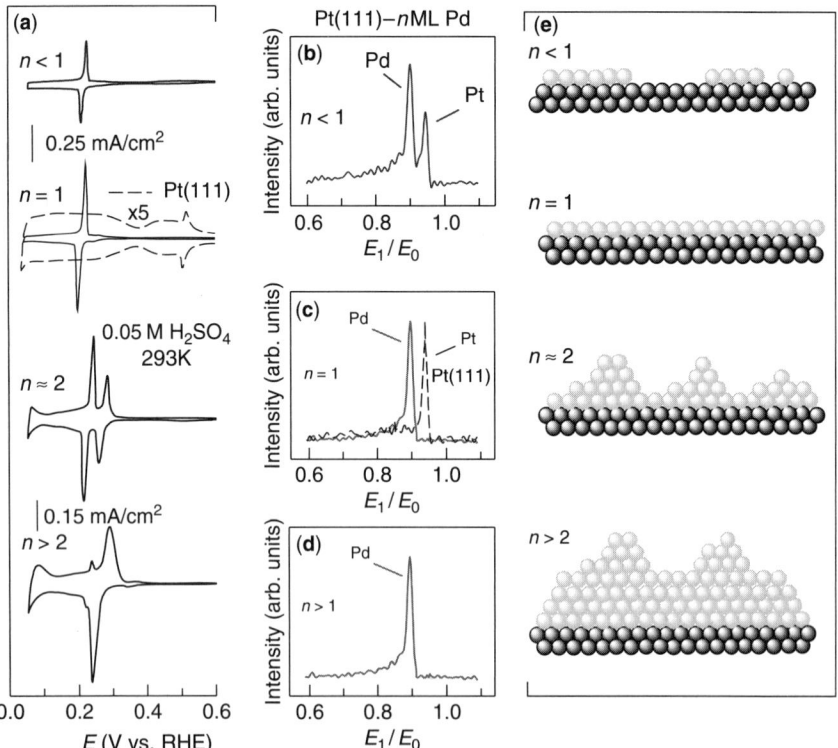

Figure 8.16 (a) Cyclic voltammetry for Pt(111)-nPd deposited in UHV and in 0.05 M H_2SO_4 with 50 mV/s. (b–e) corresponding LIES spectra after nML Pd deposition ($0 < n < 1$); He^+ energy = 1 keV. (e) Ball models assessed from the in situ SXS analysis.

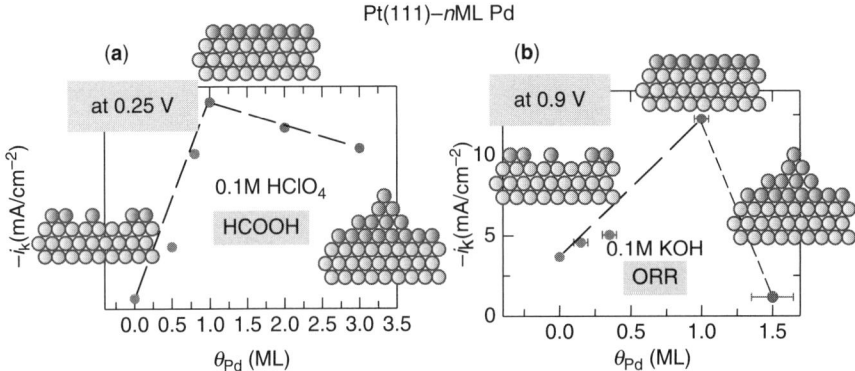

Figure 8.17 Activities of Pt(111)-nML Pd electrodes from rotating disk electrode measurements, with corresponding ball models: (a) electro-oxidation of formic acid in 0.1 M $HClO_4$; (b) oxygen reduction reaction in 0.1 M KOH.

exploration of all of these factors has allowed us to understand how to tune the catalytic activity of bi/multimetallic systems by systematic changes in the electronic/structural properties of the catalysts and ultimately to minimize the amount of precious metals needed by maximizing the catalytic properties of Pt-based multimetallic catalysts.

8.4 CONCLUSION

In spite of the important insights obtained from well-defined bimetallic surfaces, many fundamental aspects of the structure–reactivity relationship are not understood even in these idealized systems. Hence, much more needs to be learned to permit the development of a truly rational, *science-based* approach to tailoring and making multimetallic surfaces with targeted reactivity properties. With few exceptions (such as the work from our team), the field of electrocatalysis on multimetallic surfaces is still in its early years, and there is little knowledge available concerning the structure and composition of electrochemical interfaces under in situ conditions or of trends in the reactivity of simple molecules with multimetallic surfaces. In the particular case of the crucially important ORR—the cathodic half-cell reaction in fuel cells—bimetallic PtNi alloys with nanosegregated surfaces, for example, were demonstrated to be 90 times more active than the existing state-of-the-art Pt/C catalysts. The use of well-defined systems has enabled us to determine that this remarkable activity is due, in part, to the presence of a *Pt-skin* on the alloy surfaces, and to the effect on the surface electronic structure of the near-surface composition.

Overall, this chapter aimed to emphasize and demonstrate the great potential of utilizing a multidisciplinary approach to bimetallic systems that combines computational methods with a number of highly sophisticated in situ and ex situ surface-sensitive techniques at electrified solid–liquid interfaces. Advances in the understanding of fundamental properties that govern catalytic processes at well-defined multimetallic systems could form the basis for producing tailor-made (nano)catalysts with desired and tunable reactive properties for reactions relevant to energy conversion.

ACKNOWLEDGMENTS

The authors would like to acknowledge collaborators who were an integral part of the work described in this chapter: Philip Ross, Chris Lucas, Hubert Gasteiger, Thomas Schmidt, Matthias Arenz, Berislav Blizanac, Guofeng Wang, Karl Mayrhofer, Jeff Greeley, and Simon Mun. This work was supported by a contract (DE-AC02-06CH11357) between the University of Chicago and Argonne, LLC, and the US Department of Energy.

REFERENCES

Appleby AJ. 1970. Electrocatalysis and fuel cells. Catal Rev 4: 221–244.

Arenz M, Mayrhofer KJJ, Stamenkovic V, Blizanac BB, Tomoyuki T, Ross PN, Markovic NM. 2005. The effect of the particle size on the kinetics of CO electrooxidation on high surface area Pt catalysts. J Am Chem Soc 127: 6819–6829.

Arenz M, Stamenkovic V, Blizanac BB, Mayrhofer KJJ, Markovic NM, Ross PN. 2005. Carbon-supported Pt-Sn electrocatalysts for the anodic oxidation of H_2, CO, and H_2/CO mixtures. Part II: The structure–activity relationship. J Catal 232: 402–410.

Arenz M, Stamenkovic V, Wandelt K, Ross PN, Markovic NM. 2002. CO adsorption and kinetics on well-characterized Pd films on Pt(111) in alkaline solutions. Surf Sci 506: 287–296.

Attard GA, Price R, Alakl A. 1995. Electrochemical and ultra-high-vacuum characterization of rhodium on Pt(111)—A temperature-dependent growth mode. Surf Sci 335: 52–62.

Bard AJ, Abruna HD, Chidsey CE, Faulkner LR, Feldberg SW, Itaya K, Majda M, Melroy O, Murray RW, Porter MD, Soriaga MP, White HS. 1993. The electrode–electrolyte interface—A status-report. J Phys Chem 97: 7147–7173.

Bard AJ, Fan FRF, Pierce DT, Unwin PR, Wipf DO, Zhou FM. 1991. Chemical imaging of surfaces with the scanning electrochemical microscope. Science 254: 68–74.

Binning G, Rohrer H, Gerber C, Weibel E. 1982. Surface Studies by scanning tunneling microscopy. Phys Rev Lett 49: 57–61.

Brongersma HH, Draxler M, de Ridder M, Bauer P. 2007. Surface composition analysis by low-energy ion scattering. Surf Sci Rep 62: 63–109.

Brongersma HH, Jacobs JP. 1994. Application of low-energy ion-scattering to studies of growth. Appl Surf Sci 75: 133–138.

Brongersma HH, de Ridder M, Gildenpfennig A, Viitanen MM. 2003. Insight in the outside: Materials science at the atomic level using LEIS. J Eur Ceramic Soc 23: 2761–2767.

Gasteiger HA, Markovic NM, Ross PN. 1995. Electrooxidation of CO and H_2/CO mixtures on a well-characterized Pt_3Sn electrode surface. J Phys Chem 99: 8945–8949.

Gasteiger HA, Markovic NM, Ross PN Jr. 1996. Structural effects in electrocatalysis: Electrooxidation of carbon monoxide on Pt_3Sn single-crystal surfaces. Catal Lett 36: 1–8.

Gasteiger HA, Ross PN, Cairns EJ. 1993. LEIS and AES on sputtered and annealed polycrystalline Pt-Ru bulk alloys. Surf Sci 293: 67–80.

Gauthier Y, Joly Y, Baudoing R, Rundgren J. 1985. Surface-sandwich segregation on nondilute bimetallic alloys—$Pt_{50}Ni_{50}$ and $Pt_{78}Ni_{22}$ Probed by low-energy electron-diffraction. Phys Rev B 31: 6216–6218.

Hagans P, Homa A, Yeager E. 1979. In situ LEED–Auger–thin layer electrochemical studies of the underpotential deposition of Pb on Au single-crystals. J Electrochem Soc 126 (3): C147.

Hebenstreit ELD, Hebenstreit W, Schmid M, Varga P. 1999. $Pt_{25}Rh_{75}$(111), (110), and (100) studied by scanning tunnelling microscopy with chemical contrast. Surf Sci 441: 441–453.

Heinz K, Starke U, Vanhove MA, Somorjai GA. 1992. The angular-dependence of diffuse LEED intensities and its structural information-content. Surf Sci 261: 57–63.

Hosler W, Behm RJ, Ritter E. 1986. Defects on the Pt(100) surface and their influence on surface-reactions—A scanning tunneling microscopy study. IBM J Res Dev 30: 403–410.

Hubbard AT. 1998. Electrochemistry at well-characterized surfaces. Chem Rev 88: 633–656.

Itaya K. 1998. In-situ scanning tunneling microscopy in electrolyte solutions. Prog Surf Sci 58: 121–247.

Iwasita T, Nart FC. 1997. In situ infrared spectroscopy at electrochemical interfaces. Prog Surf Sci 55: 271–340.

Janssen FJJ, van IJzendoorn LJ, van der Gon AWD, de Voigt MJA, Brongersma HH. 2004. Interface formation between metal and poly-dialkoxy-*p*-phenylene vinylene. Phys Rev B 70: 165425.

Kinoshita K. 1992. Electrochemical Oxygen Technology. New York: Wiley.

Kolb DM 1996. Reconstruction phenomena at metal–electrolyte interfaces. Prog Surf Sci. 51: 109–173.

Kuntze J, Speller S, Heiland W, Atrei A, Rovida G, Bardi U. 1999. Surface structure and composition of the alloy Au_3Pd(100) determined by LEED and ion scattering spectroscopy. Phys Rev B 60: 1535–1538.

Lebedeva NP, Koper M, Herrero E, Feliu JM, van Santen RA. 2000. CO oxidation on stepped Pt(111) × (111) electrodes. J Electroanal Chem 487: 37–44.

Lebedeva NP, Rodes A, Feliu JM, Koper MTM, van Santen RA. 2002. Role of crystalline defects in electrocatalysis: CO adsorption and oxidation on stepped platinum electrodes as studied by in situ infrared spectroscopy. J Phys Chem B 106: 9863–9872.

Lucas C, Markovic NM, Ross PN. 1996. Surface structure at the Pt(110)/electrolyte interface. Phys Rev Lett 77: 4922–4925.

Lucas CA, Markovic NM. 2006. In-situ X-ray diffraction studies of the electrode/solution interface. In: Alkire RC, Kolb DM, Lipkowski J, Ross PN, editors. Advances in Electrochemical Science and Engineering. Volume 9. New York: Wiley-VCH, pp. 1–45.

Markovic NM, Grgur BN, Lucas CA, Ross PN. 1999. Electrooxidation of CO and H_2/CO mixtures on Pt(111) in acid solutions. J Phys Chem B 103: 487–495.

Markovic NM, Lucas CA, Rodes A, Stamenkovic V, Ross PN. 2002. Surface electrochemistry of CO on Pt(111): Anion effects. Surf Sci Lett 499: L149–L158.

Markovic NM, Radmilovic V, Ross PN. 2003. Physical and electrochemical characterization of bimetallic nanoparticle electrocatalysts. In: Wieckowski A, Savinova E, Vayenas C, eds. Catalysis and Electrocatalysis at Nanoparticle Surfaces. New York: Marcel Dekker, pp. 311–342.

Markovic NM, Ross PN. 2002. Surface science studies of model fuel cell electrocatalysts. Surf Sci Rep 45: 117–230.

Maroun F, Ozanam F, Magnussen OM, Behm RJ. 2001. The role of atomic ensembles in the reactivity of bimetallic electrocatalysts. Science 293: 1811–1814.

Ocko BM, Wang J, Davenport A, Isaacs H. 1990. In situ X-ray reflectivity and diffraction studies of the Au(001) reconstruction in an electrochemical cell. Phys Rev Lett 65: 1466–1469.

Oppenheim IC, Trevor DJ, Chidsey CED, Trevor PL, Sieradzki K. 1991. In situ scanning tunneling microscopy of corrosion of silver–gold alloys. Science 254: 687–689.

Ritz G, Schmid M, Varga P, Borg A, Ronning M. 1997. Pt(100) quasihexagonal reconstruction: A comparison between scanning tunneling microscopy data and effective medium theory simulation calculations. Phys Rev B 56: 10518–10525.

Robinson IK. 1986. Crystal truncation rods and surface roughness. Phys Rev B 33: 3830–3836.

Ross PN Jr. 1982. Studies of adsorption at well-ordered electrode surfaces using low-energy electron diffraction. In: Vanselow R, Howe R, editors. Chemistry and Physics of Solid Surfaces IV. Berlin: Springer-Verlag, pp. 173–201.

Ross PN. 1998. The science of electrocatalysis on bimetallic surfaces. In: Lipkowski J, Ross PN Jr, editors. Electrocatalysis. New York: Wiley-VCH. pp. 43–74.

Sieradzki K, Brankovic SR, Dimitrov N. 1999. Electrochemical defect-mediated thin-film growth. Science 284: 138–141.

Somorjai GA. 1985. Surface science and catalysis. Science 227: 902–908.

Somorjai GA. 1992. The frontiers of surface-structure analysis. Surf Interface Anal 19: 493–507.

Somorjai GA. 1993. Introduction to Surface Chemistry and Catalysis. New York: Wiley.

Stamenkovic V, Arenz M, Blizanac BB, Mayrhofer KJJ, Ross PN, Markovic NM. 2005. In situ CO oxidation on well characterized $Pt_3Sn(hkl)$ surfaces: A selective review. Surf Sci 576: 145–157.

Stamenkovic V, Arenz M, Lucas C, Gallagher M, Ross PN, Markovic NM. 2003. Surface chemistry on bimetallic alloy surfaces: adsorption of anions and oxidation of CO on $Pt_3Sn(111)$. J Am Chem Soc 125: 2736–2745.

Stamenkovic V, Arenz M, Ross PN, Markovic NM. 2004. Temperature-induced deposition method for anchoring metallic nanoparticles onto reflective substrates for in situ electrochemical infrared spectroscopy. J Phys Chem B 108: 17915–17920.

Stamenkovic VR, Fowler D, Mun BS, Wang G, Ross PN Jr, Lucas C, Markovic NM. 2007a. Improved oxygen reductionactivity on Pt + Ni(111) via increased surfrce site avalability. Science 315: 493–497.

Stamenkovic VR, Mun BS, Arenz M, Mayrhofer KJJ, Lucas C, Wang G, Ross PN, Markovic NM. 2007b. Trends in electrocatalysis on extended and nanoscale Pt-bimetallic surfaces. Nature Mater 6: 241–247.

Stamenkovic V, Mun BS, Blizanac BB, Mayrhofer KJJ, Ross PN Jr, Markovic NM. 2006a. The effect of surface composition on electronic structure, stability and electrocehmical properties of Pt-transition metal alloys; Pt-skin vs. Pt-skeleton surfaces. J Am Chem Soc 137: 1.

Stamenkovic V, Mun BS, Mayrhofer KJJ, Ross PN, Markovic NM, Rossmeisl J, Greeley J, Norskov JK. 2006b. Changing the activity of electrocatalysts for oxygen reduction by tuning the surface electronic structure. Angew Chem. 4: 1.

Stamenkovic V, Schmidt TJ, Markovic NM, Ross PN Jr. 2002. Surface composition effects in electrocatalysis: Kinetics of oxygen reaction on well defined Pt_3Ni and Pt_3Co alloy surfaces. J Phys Chem B 106: 11970–11979.

Stamenkovic V, Schmidt TJ, Ross PN, Markovic NM. 2003. Surface segregation effects in electrocatalysis: Kinetics of oxygen reduction reaction on polycrystalline Pt_3Ni alloy surfaces. J Electroanal Chem 554: 191–199.

Strmcnik D, Rebec P, Gaberscek M, Tripkovic D, Stamenkovic V, Lucas C, Markovic N. 2007. Relationship between the surface coverage of spectator spacies and the rate of the electrochemical reactions. J Phys Chem C 111: 18672–18678.

Strmcnik D, Tripkovic D, van der Vliet D, Chang K-C, Karapetrov G, Komanicky V, You H, Greeley J, Stamenkovic V, Markovic NM. 2008. J Am Chem Soc 130: 15332–15340.

Taglauer E, Heiland W. 1980. Ion scattering spectroscopy. Appl Surf Anal 669(ASTM STP): 111–124.

Tidswell IM, Markovic NM, Lucas C, Ross PN. 1993. In-situ X-ray scattering study of the Au(100) reconstructuin in alkaline and acid solution. Phys Rev B 47: 16542.

Toda T, Igarashi H, Watanabe M. 1999. Enhancement of the electrocatalytic O_2 reduction on Pt-Fe alloys. J Electroanal Chem 460: 258–262.

Tolmachev YV, Menzel A, Tkachuk A, Chu SY, You H. 2004. Electrochem Solid-State Lett 7: E23.

Tomellini M, Attard GA. 1991. On the question of determining thin-film growth using Auger-electron spectroscopy. Surf Sci 245: L179–L184.

Van Hove MA, Weinberg WH, Chan CM. 1986. Low Energy Electron Diffraction. Berlin: Springer-Verlag.

Van Hove MA, Koestner RJ, Stair PC, Biberian JP, Kesmodel LL, Bartos I, Somorjai GA. 1981. The surface reconstructions of the (100) crystal faces of iridium, platinum and gold. 1. Experimental-observations and possible structural models. Surf Sci 103: 189–217.

Vanleerdam GC, Lenssen KMH, Brongersma HH. 1990. Charge-exchange processes in low-energy He^+ ion-scattering from Si and Pd_2Si Surfaces. Nucl Instrum Meth B 45: 390–393.

Varga P, Schmid M. 1999. Chemical discrimination on atomic level by STM. Appl Surf Sci 141: 287–293.

Vasiliev MA. 1997. Surface effects of ordering in binary alloys. J Phys D Appl Phys 30: 3037–3070.

Wagner FT, Ross PN. 1983. LEED analysis of electrode surfaces—Structural effects of potentiodynamic cycling on Pt single-crystals. J Electroanal Chem 150: 141–164.

Wang G, Van Hove MA, Ross PN, Baskes MI. 2005. Quantitative prediction of surface segregation in bimetallic Pt-M alloy nanoparticles (M = Ni, Re, Mo). Prog Surf Sci 79: 28–45.

Wang J, Robinson IK, Ocko BM, Adzic RR. 2005. Adsorbate-geometry specific subsurface relaxation in the CO/Pt(111) system. J Phys Chem B 109: 24–26.

Warren BE. 1990. X-Ray Diffraction. New York: Dover.

Zhang XOG, Stimming U. 1990. Scanning tunneling microscopy of copper corrosion in aqueous perchloric-acid. Corrosion Sci 30: 951–954.

CHAPTER 9

Recent Developments in the Electrocatalysis of the O_2 Reduction Reaction

YE XU

Center for Nanophase Materials Sciences, Oak Ridge National Laboratory, Oak Ridge, TN 37831, USA

MINHUA SHAO

Chemistry Department, Brookhaven National Laboratory, Upton, NY 11973, USA

MANOS MAVRIKAKIS

Department of Chemical and Biological Engineering, University of Wisconsin–Madison, Madison, WI 53705, USA

RADOSLAV R. ADZIC

Chemistry Department, Brookhaven National Laboratory, Upton, NY 11973, USA

9.1 NEW INSIGHTS INTO THE OXYGEN REDUCTION REACTION FROM EXPERIMENT AND THEORY

The oxygen reduction reaction (ORR) is the primary electrochemical reaction occurring at the cathode of fuel cells, and is central to this promising technology for efficient and clean energy generation. In low temperature proton exchange membrane fuel cells (PEMFCs), the ORR is traditionally catalyzed by platinum because it has the highest activity of the elemental metals. A multi-electron reaction, the ORR appears to occur in two pathways (represented in acidic electrolytes):

(i) A "direct" four-electron pathway, wherein four electrons are transferred in concert:

$$O_2 + 4H^+ + 4e^- \longrightarrow 2H_2O$$

Fuel Cell Catalysis. Edited by Marc T. M. Koper
Copyright © 2009 John Wiley & Sons, Inc.

(ii) A "series" pathway, wherein electrons are transferred consecutively:

$$O_2 + 2H^+ + 2e^- \longrightarrow H_2O_2 + 2H^+ + 2e^- \longrightarrow 2H_2O$$

Unfortunately, oxygen reduction kinetics on all-Pt catalysts contributes significantly to the overpotential loss [Adzic, 1998], thus requiring high loadings of Pt, a very expensive metal, to compensate. This, together with catalyst and membrane stability issues, is among the factors hindering the commercialization of PEMFCs, and has motivated intense research for a more active, inexpensive, and stable replacement for Pt. Promising directions include metal alloys, notably Pt- and Pd-based [Anderson et al., 2005; Mukerjee and Srinivasan, 1993; Shao et al., 2006a; Stamenkovic et al., 2006, 2007a, b; Zhang et al., 2005a, b] inorganic compounds such as chalcogenides [Alonso-Vante et al., 2002], and organic compounds such as transition metal macrocycles [Bashyam and Zelenay, 2006]. This chapter presents a partial review of the recent work that has led to the development of a class of highly promising monolayer-type Pt- and Pd-based alloy electrocatalysts with competitive or higher oxygen reduction activity and lower Pt loading than commercial Pt electrodes, with emphasis on the experimental results and, in particular, the corresponding mechanistic underpinnings provided by electronic structure theory.

9.1.1 The *d*-Band Theory of Surface Reactivity and Rational Approach to Catalyst Design

The last decade has seen significant development of the *theory* of surface reactivity. That is, the theoretical inquiry into the fundamental, atomic-level factors that determine catalytic activity and selectivity. One prominent example is the *d*-band theory for metal surfaces developed by Hammer and Nørskov [Hammer and Nørskov, 1995a, b]. The *d*-band theory states that the electronic states in the entire valence band of a metal surface are responsible for the reactivity of the surface. For the transition and noble metals, the contribution from the metal *sp*-states is dominant but relatively constant, so the coupling between an adsorbate and the metal *d*-states is mainly responsible for the variation in the interaction energy on different metals. The key properties of the *d*-band are its energy-weighted center (ε_d), its filling, and the coupling matrix element [Hammer and Nørskov, 1995a, b].

An important corollary of the *d*-band theory is that, for a given metal and adsorbate species, the position of the *d*-band center determines the binding energy of the adsorbate. Subsequent studies have demonstrated a strong correlation between these two variables for a great number of systems involving small atoms and molecules interacting with similar metal surfaces. In particular, the theory has proved very useful when applied to explaining metal alloy reactivity and the strain, ligand, and ensemble effects [Hammer and Nørskov, 1995a, b; Jacobsen et al., 2001; Nørskov et al., 2002]. The identification of ε_d as a key factor in the reactivity of metal surfaces is significant. Previous explanations of surface reactivity were adapted from prevalent concepts in chemistry, which focused on isolated aspects of the electronic structure, such as the density of states at the Fermi level and *d*-band vacancy, and which dominated the electrocatalysis community until recently but fell short in explaining, for example, the enhanced ORR activity of bulk Pt alloy catalysts. It is worth pointing out that the d-band theory inherently assumes that surface reactivity is a local property, which differs from another earlier perspective that bulk properties determine catalytic properties.

This new atomistic and mechanistic perspective rooted in theoretical solid state physics has brought fresh fundamental insights to heterogeneous catalysis: The activity for a number of reactions often peaks on a certain metal when correlated with some measure of reactivity across a series of metals, thus displaying a "volcano" trend. Using extensive density functional theory (DFT) calculations, Nørskov and co-workers have convincingly demonstrated that this behavior is the result of key competing processes in a reaction network connected via a common intermediate [Nørskov et al., 2002], just as described by the Sabatier principle, which observed that catalytic activity is often maximized when the reactants interact with the catalysts with intermediate strength.

Heterogeneous catalysis and electrocatalysis research has traditionally been based on empirical approaches. These new theoretical insights, in combination with sophisticated experimental techniques that give increasingly greater imaging resolution and control over matter, and aided by ever-growing computational power, open the door to the rational design and screening of catalytic materials for desired properties (e.g., activity, selectivity, and stability). Examples have already emerged in which hitherto untested metal alloys have been identified to possess activity superior to traditional monometallic catalysts for such reactions as ammonia synthesis, methanation, water-gas shift, and electrocatalytic hydrogen evolution [Greeley et al., 2006; Jacobsen et al., 2001; Knudsen et al., 2007; Sehested et al., 2007]. As demonstrated below, this possibility has not been lost on researchers investigating ways to improve the oxygen reduction electrocatalysts for PEMFCs.

9.2 CHEMICAL SPECIES AFFECTING THE ORR: NEW ATOMIC-LEVEL INFORMATION

9.2.1 Adsorption of Oxygen-Containing Species

9.2.1.1 Oxygen Adsorption Formation of adsorbed oxygen-containing species in the ORR can occur from two sources: molecular oxygen (O_2) and water molecules that are oxidized above a certain potential to form hydroxyl (OH) and eventually atomic O. As one of the main reactants of the ORR, O_2 must of course interact with the electrode in order for the net reaction to proceed. The thermochemical interaction of O_2 with Pt surfaces under ultrahigh vacuum (UHV) conditions has been extensively studied [Avery, 1983; Gland, 1980; Gland et al., 1980; Outka et al., 1987; Sexton, 1981]. DFT calculations have identified several adsorbed di-σ O_2 states on Pt(111) (Fig. 9.1) [Eichler et al., 2000; Gambardella et al., 2001; Shao et al., 2006b; Sljivancanin and Hammer, 2002; Xu et al., 2004]. Based on the O–O vibrational frequency and electronic properties, the top–fcc/hcp–top (t-f/h-b) states are identified as peroxide states (O_2^{2-}), whereas the identity of the top–bridge–top (t-b-t)

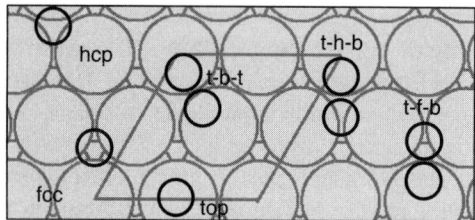

Figure 9.1 High symmetry adsorption sites for O and O_2 on Pt(111). Large gray circles represent Pt atoms, and small open circles represent O atoms. "t," "b," "h," and "f" stand for top, bridge, hcp, and fcc, respectively. The surface unit cell is delineated. (Reproduced with permission from Xu et al. [2004].)

state is ambiguous and is suggested by some authors to be a peroxide state and by others to be a superoxide state (O_2^-). The t-b-t state is the most stable, with a binding energy of -0.6 eV at $\frac{1}{4}$ monolayer (ML).

To gain information on O_2 adsorption on Pt(111) during the course of ORR, Adzic and Wang used the adsorption of other metal atoms (Ag) as a probe [Adzic and Wang, 1998]. The inhibition of the ORR on Pt(111) by sub- and monolayer coverages of Ag was studied using electrochemical and in situ surface X-ray scattering techniques. The analysis of the extent of the inhibition of the ORR as a function of the Ag coverage showed that the data are best interpreted with O_2 occupying a bridge site. More recently, Shao and co-workers used surface-enhanced infrared reflection and

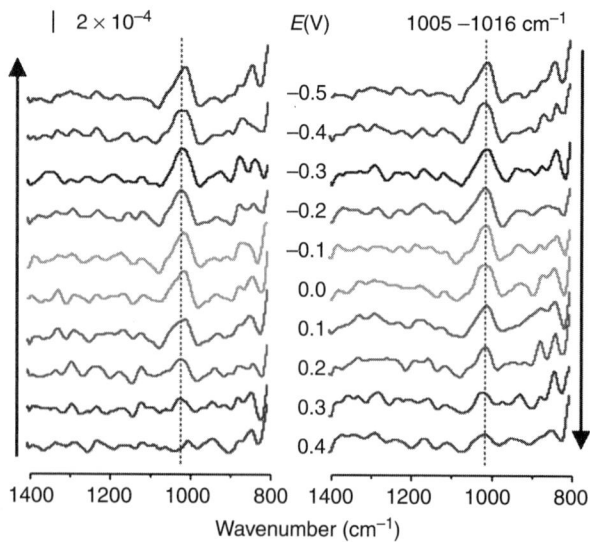

Figure 9.2 SEIRAS spectra for a Pt thin film electrode in O_2-saturated 0.1 M $NaClO_4$ + NaOH (pH 11). The reference spectrum at 0.4 V was taken before the potential sweep started. The scan rate was 10 mV/s. (Reproduced with permission from Shao et al. [2006a].)

9.2 CHEMICAL SPECIES AFFECTING THE ORR: NEW ATOMIC-LEVEL INFORMATION

TABLE 9.1 Calculated Parameters for Free O_2^- Molecule and Adsorbed O_2^- on Pt(111) at Different Coverages. Reprinted with Permission from Shao et al. [2006b]

	ML	BE (eV)	d_{O-O} (Å)	Site	ν_{O-O} (cm^{-1})
O_2^-			1.27		1123
	1/9	−0.60	1.39	b-f-t	796
	1/9	−0.76	1.37	t-b-t	875
O_2^-/Pt(111)	1/4	−0.68	1.36	t-b-t	896
	1/2	−0.39	1.34	t-b-t	1051
	1	>0	1.27	t-b-t	1292

absorption spectroscopy (SEIRAS) in an attenuated total reflection mode (ATR) to study the ORR on a Pt electrode in alkaline solution [Shao et al., 2006a]. Spectral and voltammetry data (Fig. 9.2), together with the vibrational frequencies calculated using DFT (Table 9.1), provide evidence for a superoxide intermediate (O_2^-) present at medium to high coverage. Further evidence supporting this assignment includes similar spectra in acetonitrile solutions and a lack of the ν_{O-O} signature in the 1200–1000 cm^{-1} range in the absence of O_2. Gewirth and co-workers, using surface-enhanced Raman spectroscopy (SERS), assigned a band at 1162 cm^{-1} to adsorbed HO_2 (peroxyl), the superoxide intermediate in acid solutions, on a BiOH-modified Au(111) surface [Li and Gewirth, 2005].

Experimentally, the activation energy for the thermal dissociation of adsorbed O_2 has been estimated to be about 0.30 eV, which means that O_2 dissociates on Pt(111) at low temperature. However, this barrier increases significantly with increasing oxygen coverage. Using the climbing-image nudged elastic band (cNEB) method [Henkelman and Jónsson, 2000; Henkelman et al., 2000] the activation energy for O_2 dissociation is calculated to be 0.77 eV at $\frac{1}{4}$ ML of O_2 [Xu et al., 2004]. A cartoon of the reaction sequence is shown in Fig. 9.3. Thus, a high coverage of co-adsorbed O or OH would substantially hinder O_2 dissociation. The dissociated product, atomic O, prefers to adsorb in the fcc threefold hollow site, with a binding energy of −3.88 eV on Pt(111), followed by the hcp hollow, which is 0.4 eV less stable. The binding

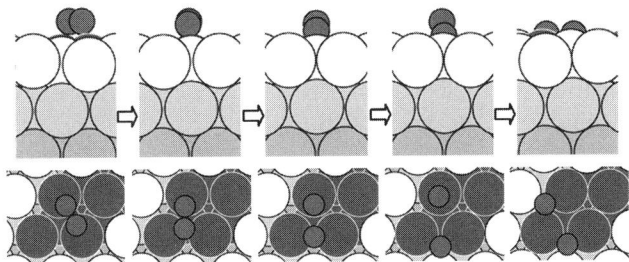

Figure 9.3 Cross-section and top views of selected states along the O_2 dissociation path on Pt(111), from the initial state (t-b-t) to the dissociated product state (fcc × 2). The third image is the transition state. (Reproduced with permission from Xu et al. [2004].)

energy of an O atom on Pt(111) was estimated to be -3.6 eV experimentally [Campbell et al., 1981; Gland et al., 1980].

9.2.1.2 Inhibition Effect of Adsorbed OH and Pt Oxidation

The question about the equivalence of adsorbed OH species (OH_{ads}) formed through dissociative adsorption of molecular oxygen followed by protonation and via water oxidation has been recurring, although a consensus has emerged that high OH_{ads} coverage can reduce the ORR rate. In the potential region of 0.75–1.10 V, OH_{ads} is thought to form from the oxidation of H_2O:

$$Pt + H_2O \longrightarrow Pt-OH + H^+ + e^- \quad (9.1)$$

OH_{ads} formation has a clear voltammetric signature on a number of surfaces, including the (111)-oriented surfaces of platinum group metals, Pt(111) in alkaline and acid electrolytes of non-adsorbing anions [Markovic and Ross, 2002], and Au(111), Au(100), and Ag(111) in neutral and alkaline electrolytes [Savinova et al., 2002]. On these surfaces, the reaction has a reversible character. Anderson and co-workers calculated the reversible potential of Reaction (9.1) on Pt to be 0.62 V with respect to a reversible hydrogen electrode (RHE) [Anderson, 2002]. The Pt(111)–OH bond energy has been estimated to be about -1.4 eV in an alkaline electrolyte [Markovic and Ross, 2002].

Wang and co-workers have analyzed the site-blocking and electronic effects of OH_{ads} (in $HClO_4$) and bisulfate (in H_2SO_4) with the adsorption isotherms incorporated in the model for a Pt(111) surface [Wang et al., 2004]. The best fits yielded the intrinsic Tafel slope in the range from -118 to -130 mV/dec, supporting the interpretation of single electron transfer in the rate-determining step, with the corresponding transfer coefficients equal to 0.50 and 0.45, respectively. In addition to site blocking, a negative electronic effect on ORR kinetics was found for both OH and bisulfate, with the effect of the latter being much stronger. The deviation of the apparent Tafel slope in $HClO_4$ from its intrinsic value can be fully accounted for by the site-blocking and electronic effects of OH_{ads}, which vary with coverage over the mixed kinetic–diffusion-controlled region. Since the anion adsorption effects are proportional to the coverage, they affect the slope of the polarization curve only when the coverage changes in the mixed kinetic–diffusion-controlled region of ORR. That appears to be the reason why the apparent Tafel slope deviates from its intrinsic value in $HClO_4$ but not in H_2SO_4.

The quantitative evaluation of the effects of adsorbed species on the ORR kinetics was obtained from the following equation:

$$j_k(E) = -j_0^*[1 - \gamma\theta(E)]^m \exp\left[2.3\frac{E - E^\circ - \varepsilon\theta(E)}{b^*}\right] \quad (9.2)$$

where j_0^* and b^* are the *intrinsic* exchange current and Tafel slope for an adsorbate-free Pt(111) surface. The $(1 - \gamma\theta)^m$ term accounts for the geometric site-blocking effect, whereas the electronic effect of anion adsorption on ORR kinetics is described by a

9.2 CHEMICAL SPECIES AFFECTING THE ORR: NEW ATOMIC-LEVEL INFORMATION

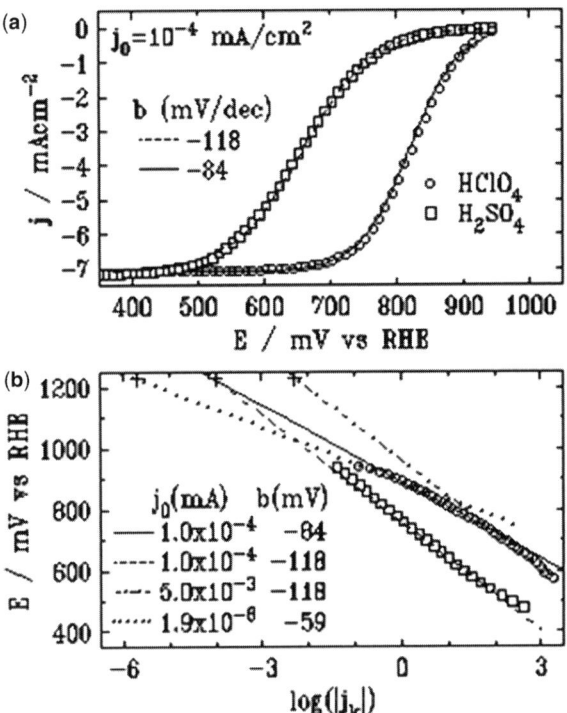

Figure 9.4 (a) Rotating disk electrode polarization curves for ORR on Pt(111) at 2500 rev/min with a sweep rate of 50 mV/s in oxygen-saturated 0.1 M $HClO_4$ (circles) and 0.05 M H_2SO_4 (squares) solution, obtained by averaging the absolute currents in both positive and negative sweep directions. The solid and dashed lines are fitted curves using j_k in (9.2). (b) The Tafel plots for the data in (a), with the straight lines corresponding to different Tafel slopes and exchange currents. (Reproduced with permission from Wang et al. [2004].)

coverage-dependent potential shift, $-\varepsilon\theta$, through the exponential term, where γ and ε are the site-blocking and energy coefficients for either OH or the bisulfate anion (represented by OH and A, respectively) and m is the number of Pt sites involved in the rate-determining step. The potential-dependent coverages $\theta_{OH}(E)$ and $\theta_A(E)$ were obtained by integrating the currents measured by linear sweep voltammetry in nitrogen-saturated solutions. Very good fitting of the experimental curves was obtained (Fig. 9.4).

The relevance of Pt–OH formation to the change in the Tafel slope has been demonstrated by varying the content of water in the electrolyte [Murthi et al., 2004]. The experiments were performed in H_2O/trifluoromethanesulfonic acid (TFMSA) mixtures with several water/acid molar ratios. Whereas at high water contents the usual change in the Tafel slope from -112 to -59 mV/dec observed in aqueous solutions of H_2SO_4 and $HClO_4$ took place, at low water contents no change in the Tafel slope was observed. This corroborates the involvement of water in the formation

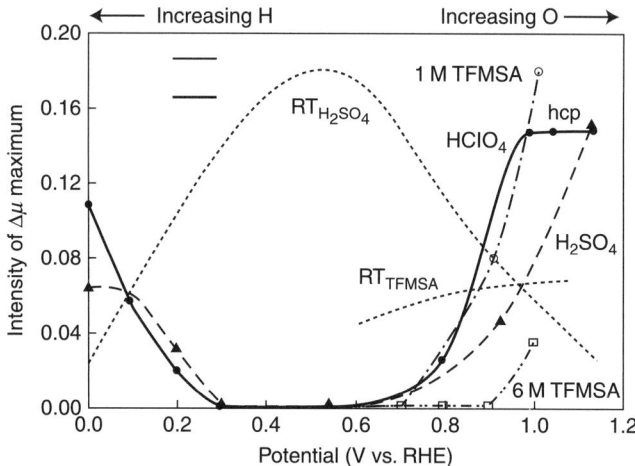

Figure 9.5 Anion adsorption on Pt in several electrolytes: bisulfates and TFMSA as determined by radiotracers (RT), and OH and H as reflected in $\Delta\mu$, the amplitude of the XANES peak. The effect of 6 M TFMSA suppressing OH formation can be seen. (Reproduced with permission from Teliska et al. [2007].)

of oxides on the Pt surface. Figure 9.5 compares anion adsorption on Pt in several electrolytes as determined by radiotracers showing the adsorption of bisulfates [Zelenay and Wieckowski, 1991] and TFMSA, and OH and H as reflected in $\Delta\mu$, the amplitude of the X-ray absorption near-edge structure spectroscopy (XANES) peak [Teliska et al., 2007]. The effect of 6 M TFMSA suppressing OH_{ads} formation is seen. Enhanced ORR kinetics in this solution indicates that the suppression of OH is not due to impurities that are often observed in concentrated TMFSA.

Earlier work on Pt oxidation was summarized in [Conway, 1995]. Recent EXAFS studies, by Teliska and co-workers have revealed new information on Pt oxidation [Teliska et al., 2007]. For Pt/C nanoparticles in a fuel cell electrode at 80 °C, the peak corresponding to the Pt–O bond occurred concomitantly with the decrease in the intensity of the peak corresponding to the Pt–Pt bond. This is indicative of the restructuring of a Pt surface. The adsorption of O and OH on a Pt/C electrode comprising 1.5–3 nm Pt nanoparticles in 0.1 M $HClO_4$ or H_2SO_4 electrolytes were studied in situ with extended X-ray absorption fine structure (EXAFS) and XANES techniques. Differences in the L3 XANES spectra at different potentials were used to separate the effects of O or OH adsorption in the XANES spectra. At low coverages, OH adsorbs primarily in 1-fold coordinated atop sites. As the coverage increases, O binds in the bridge/fcc sites, and at still higher coverages and further oxidation above 1.05 V (vs. RHE), O adsorbs into a higher coordinated n-fold or subsurface site, which is apparently the result of Pt–O site exchange and oxide formation [Sasaki et al., 2008].

Studies have shown that the Pt oxides are not hydrated [Birss et al., 1993; Harrington, 1997; Jerkiewicz et al., 2004]. Electrochemical quartz crystal microbalance [Birss et al., 1993] and nanobalance [Jerkiewicz et al., 2004] experiments

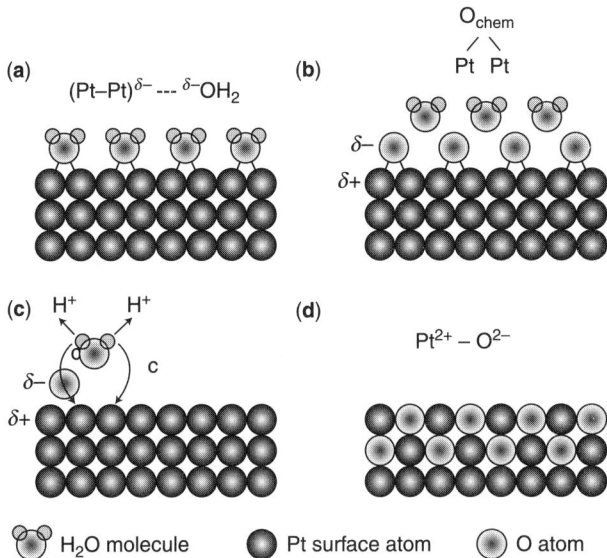

Figure 9.6 Visual representation of the platinum oxide growth mechanism. (a) Interaction of H_2O molecules with the Pt electrode occurring in the $0.27\ V \leq E \leq 0.85\ V$ range. (b) Discharge of $\frac{1}{2}$ ML of H_2O molecules and formation of $\frac{1}{2}$ ML of chemisorbed oxygen (O_{chem}). (c) Discharge of the second $\frac{1}{2}$ ML of H_2O molecules; the process is accompanied by the development of repulsive interactions between $(Pt-Pt)^{\delta+}-O^{\delta-}_{chem}$ surface species that stimulate an interfacial place exchange of O_{chem} and Pt surface atoms. (d) Quasi-3D surface PtO lattice, comprising Pt^{2+} and O^{2-} moieties, that forms through the place-exchange process. (Reproduced with permission from Jerkiewicz et al. [2004].)

indicated that a $\frac{1}{2}$ ML of chemisorbed O_{ads}, instead of OH_{ads}, was formed at 0.85–1.15-V through the discharge of H_2O. Further discharge of H_2O results in the formation of the second $\frac{1}{2}$ ML of O_{ads}, with the first $\frac{1}{2}$ ML of O_{ads} undergoing place-exchange with interfacial Pt atoms to form a quasi-3D near-surface Pt–O lattice [Jerkiewicz et al., 2004], as shown in Fig. 9.6. You and co-workers found that only 0.3 ML of Pt atoms exchanged places with oxygen-containing species [Nagy and You, 2002; You et al., 2000]. The place-exchange mechanism exposes Pt to the electrolyte and allows further oxidation and dissolution [Wang XP et al., 2006]. Furthermore, Nagy and You showed that PtO is mobile and can diffuse to energetically favorable sites on the Pt surface, further exposing underlying Pt atoms to the electrolyte [Nagy and You, 2002]. At higher potentials, the PtO film can be oxidized to PtO_2, which is also mobile [You et al., 2000]. This was studied using surface X-ray scattering (SXS) by You and co-workers, who showed that the reversible oxidation involved a place-exchange mechanism up to 1.25 V, while irreversible oxidation (restructuring) occurred above that potential [You et al., 1994].

OH adsorption on Ru is a key factor that makes this metal the major component of various bimetallic catalysts for anode reactions. Ru–OH causes a significant inhibition of the ORR [Inoue et al., 2002]. In situ SXS data for the oxidation of Ru(0001) in acid

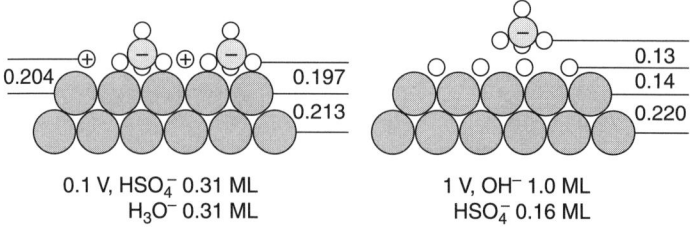

Figure 9.7 Proposed structural models for Ru(0001) oxidation, where the O, S, and Ru atoms are represented by the open, lightly shaded, and heavily shaded circles, respectively. The interlayer distances are in nanometers and coverage in monolayers (ML). (Reproduced with permission from Wang et al. [2001].)

solutions below the onset of bulk oxide formation show that the spacing between the top two Ru layers is 0.213 nm at 0.1 V, and 0.220 nm at 1.0 V in 1 M H_2SO_4 solution, similar to the values in the gas phase for bare Ru and for 1 ML of oxygen on Ru (0.210 and 0.222 nm, respectively) [Wang et al., 2001]. At low potentials, the specular reflectivity data support a model involving the co-adsorption of bisulfate and hydronium ions on Ru(0001). The coverage of bisulfate is close to $\frac{1}{3}$ ML at potentials below 0.57 V. Figure 9.7 shows the proposed structural models. In contrast to the behavior of Pt(111) and Au(111) surfaces, no place exchange is involved in Ru(0001) surface oxidation. The formation of a monolayer of Ru oxide induces partial desorption of bisulfate, in agreement with the Fourier transform infrared (FTIR) data. These properties of Ru–OH affect the activity of mixed Pt–M monolayers for the ORR, as discussed later in this chapter.

9.2.2 Adsorption of Anions

The effects of anions on the kinetics of the ORR are of fundamental interest to developing an understanding of the relationship between surface structure and composition and surface activity. Like OH_{ads}, strongly (also termed "specifically") adsorbed anions can inhibit the ORR and significantly reduce its reaction rate. The effects are surface structure-dependent, with the inhibition ranging from low to very high for the same metal. The most pronounced inhibition is caused by halides (chlorides, bromides, and iodides). Bisulfates/sulfates and phosphates have weaker adsorption and cause less inhibition; nitrates adsorb more weakly still, while, for all practical purposes, perchlorates, fluorides, and perfluoroacid anions do not adsorb [Adzic and Wang, 2000]. Traces of Cl^- impurities often affect the results with nonadsorbing anions.

There are several factors through which anions can influence the pathway and O_2 reduction kinetics. The main factors are competition with O_2 for surface sites; changes in the activity coefficients of the reactants, intermediates, and transition states; and the acidity and dielectric properties of the electrolyte side of the interface [Adzic, 1998]. For example, perfluoro acids have higher O_2 solubility and lower adsorbability than electrolytes with specifically adsorbed anions. They also decrease the effective dielectric constant adjacent to the electrode surface, which favors the adsorption of the relatively nonpolar O_2 molecule.

9.2.2.1 Inhibition Effect of Anions For an adequate understanding of the effect of anions on ORR kinetics, it is necessary to obtain information on the adsorbed anion structure during reaction. Considerable progress in structural determination of adlayers on electrode surfaces in the last decade and a half has inspired the application of these techniques to the anion effects on ORR kinetics [Wang et al., 2000]. Obtaining atomic-scale information on the structure of catalytic surfaces during reaction has been one of the major goals of studies in both heterogeneous catalysis and electrocatalysis. SXS techniques can provide detailed structural information on long-range ordered adlayers during the course of electrocatalytic reactions. Obtaining information on the anion inhibition mechanism of the ORR on an atomic level has been demonstrated by Adzic and Wang, who used SXS to monitor the structure of adsorbed Br^- during the course of the ORR [Adzic and Wang, 2000]. Structures of Br^- adlayers are determined on Pt(111) and Au(100) electrodes in the absence and in the presence of O_2 reduction in 0.1 M $HClO_4$ solutions. On Au(100), the $c(\sqrt{2} \times 2\sqrt{2})$-R45° and $c(\sqrt{2} \times p)$R45° phases are found above 0.35 and 0.68 V, respectively. On Pt(111), the commensurate (3×3)-Br^- adlayer is found between 0.3 and 0.8 V. Sharp peaks are due to phase transitions. As shown in Figs. 9.8 and 9.9, the $c(\sqrt{2} \times 2\sqrt{2})$-R45° adlayer of Br^- on Au(100) and the (3×3)-Br^- adlayer on Pt(111) inhibit the ORR. The reaction takes place only at potentials more negative than those of the ordered Br^- phases. The stability of Br^- adlayers is monitored in situ by measuring X-ray intensity at a particular position in reciprocal space corresponding to the diffraction peak for the adlayer as a function of potential during O_2 reduction. Adsorbed disordered Br^- anions change the mechanism of ORR on Pt(111) from a four-electron into a two-electron reaction.

For Ag(100) in 0.1 M $NaClO_4$ solutions, the adlayer $c(2 \times 2)$-Br^- precludes the adsorption of O_2 in the bridge state, and the reduction current originates solely from O_2 adsorbing in the end-on configuration through the four-fold symmetry holes in the $c(2 \times 2)$-Br^- adlayer. The experiment demonstrates the site sensitivity of the blocking effect. Markovic and co-workers reported the same structure of Br^- on Pt(111) in the absence of O_2 and the inhibition of the ORR measured using a rotating disk electrode. They also studied halide adsorption on Pt(100) and Pt(110) [Markovic and Ross, 2002].

The activity of Pt for the ORR in H_2SO_4 solutions is highly dependent on the structure of electrode surface. Bisulfate anions show the most pronounced effect on the ORR on Pt(111). The strong adsorption of bisulfate is a consequence of its structure and its adsorption through the three unprotonated oxygen atoms. The inhibition was ascribed to the same symmetry of the anion and the adsorption sites. The exact nature of the adsorbed species has not yet been completely determined. Yeager's group identified it as bisulfate using FTIR spectroscopy [Faguy et al., 1990], but more recent data [Faguy et al., 1996] are more consistent with an adsorbed

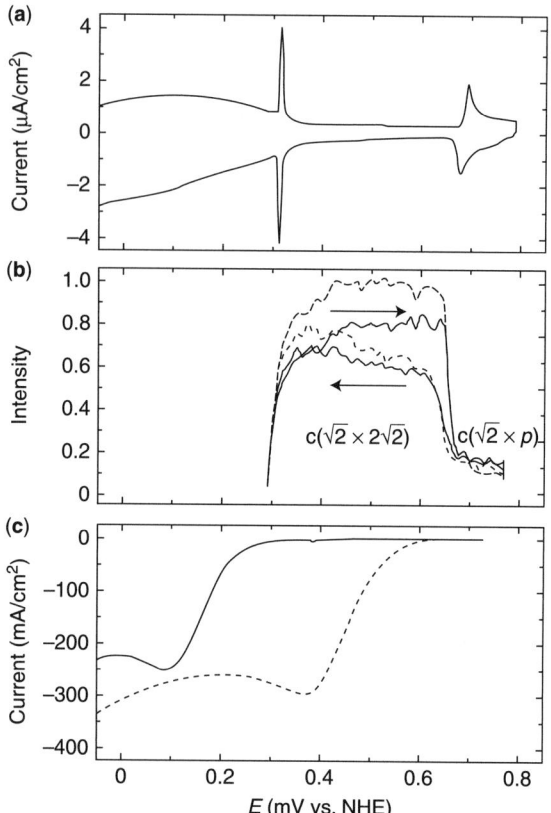

Figure 9.8 (a) Voltammetry curve of a Au(100) electrode in 0.1 M $HClO_4$ with 20 mM Br^-, at 10 mV/s. (b) X-ray intensity at (1:2, 1:2, 0.1) position as a function of potential for Br:Au(100) in the absence (dashed line) and presence (full line) of O_2, at 1 mV/s. (c) O_2 reduction on Au(100) (dashed line) and Au /Br (full line), at 20 mV/s. (Reproduced with permission from Adzic and Wang [2000].)

H_3O^+–SO_4^{2-} ion pair, possibly with the three unprotonated sulfate oxygen atoms interacting with Pt sites. A weak band at 940 cm^{-1} is assigned to the S–O(H_3O^+) stretch of the ion pair, and a band at 1040 cm^{-1}, which disappears with increasingly more positive sample potentials, is assigned to the symmetric stretching mode for solution phase bisulfate ions. Chronocoulommetry data also indicate the adsorption of bisulfates [Savich et al., 1995]. Although there are no structural studies of bisulfate adsorption during the ORR, the structure of the ordered bisulfate adlayer on submersed Pt(111) has been found to be ($\sqrt{3} \times \sqrt{3}$)-R30° [Thomas et al., 1996]. This adlayer can effectively block the ORR on the Pt(111) surface. Using Monte Carlo methods, Koper and Lukkien modeled the butterfly peak associated with bisulfate adsorption on Pt(111) with the latter structure [Koper and Lukkien, 2000].

9.2 CHEMICAL SPECIES AFFECTING THE ORR: NEW ATOMIC-LEVEL INFORMATION

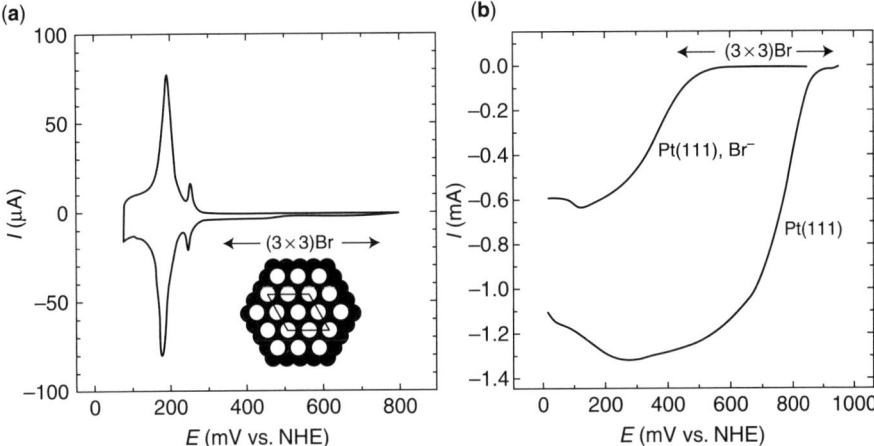

Figure 9.9 (a) Voltammetry curve for Pt(111) in 0.1 M HClO$_4$ with 20 mM Br$^-$, at 20 mV/s. The insert shows a real space model of a (3 × 3) Br adlayer; the shaded circles are Pt atoms and the open circles are the Br adatoms. (b) O$_2$ reduction on a rotating Pt(111) disk electrode in 0.1 M HClO$_4$, without and with 20 mM Br$^-$, at 625 rev/min and 20 mV/s. (Reproduced with permission from Adzic and Wang [2000].)

The specific adsorption of bisulfate anions is observed in H$_2$SO$_4$ in both EXAFS and XANES data and, in agreement with voltammetry, is seen to impede oxygen adsorption. Significant specific anion adsorption was found in 6 M TFMSA, but not in 1 M TFMSA [Teliska et al., 2007]. As mentioned above, this specific anion adsorption suppresses OH adsorption (particularly the formation of subsurface O), causes the Pt nanoparticle to become more round, and weakens the Pt–Pt bonding at the surface. The specific anion adsorption becomes site-specific only after lateral interactions from other chemisorbed species such as OH force the anions to adsorb into specific sites.

Based on the available atomic-level information and other relevant literature data, factors determining the effect of anion adsorption on the ORR include the following:

- Anion adsorbability: this increases in the sequence F$^-$ ~ ClO$_4^-$ ~ TFMSA < HSO$_4^-$ ~ PO$_4^{3-}$ < Cl$^-$ < Br$^-$ < I$^-$.
- The symmetry of the anions and the surface structure: tetrahedral bisulfates bond very strongly to threefold symmetry sites of (111), but weakly to (100) and (110) surfaces. Halides adsorb more strongly on (100) surfaces.
- The structure and coverage of anion adlayers: close-packed adlayers usually cause a complete inhibition.
- The anion electronic effect: in a recent analysis of the inhibition effect of bisulfate and comparison with the effect of OH adsorption (Pt–OH formation), Wang and co-workers found that the electronic effect must also be operative [Wang et al., 2004].

9.3 IMPROVING ORR KINETICS AND PLATINUM UTILIZATION

9.3.1 Platinum Alloy Electrocatalysts

It was known as early as the 1970s that some Pt bulk alloys possessed higher activity for catalyzing the ORR. Typically, one of the $3d$ transition metals (including Cr, Mn, Fe, Co, and Ni) was alloyed with Pt at an atomic concentration of less than 50%, as higher concentration diminished the beneficial effects. Enhancements of ORR kinetics of from 2–3 times [Mukerjee et al., 1995; Paulus et al., 2002; Stamenkovic et al., 2002] up to 10–25 times [Toda et al., 1999a, b] were reported. The original explanation of the alloying effect came in the form of the hypothesized dual-site mechanism for breaking the O–O bond: Jalan and Taylor suggested that the distance between neighboring Pt atoms was too large for the dissociative adsorption of O_2 species [Jalan and Taylor, 1983]. Alloying with base metals was beneficial, they argued, because it caused the Pt lattice constant to contract, a phenomenon well documented with X-ray diffraction (XRD), while maintaining the generally favorable properties of Pt. The strong correlation between Pt–Pt distance and ORR activity was repeatedly confirmed [Mukerjee et al., 1995].

Subsequently, it was discovered by using surface-sensitive techniques such as X-ray photoelectron spectroscopy (XPS) that Pt alloy catalysts lost all base metal content from their surfaces in electrochemical experiments, leaving behind a pure Pt shell a few atomic layers thick [Igarashi et al., 2001; Paffett et al., 1988; Toda et al., 1999a, b; Wan et al., 2002; Watanabe et al., 1994]. Because of the strong tendency of Pt to segregate to the surface when alloyed with most metals [Ruban et al., 1999], annealing the alloy catalysts at high temperature also resulted in a very thin outer shell of pure Pt, most likely only one atomic layer thick (Pt "skin") [Atli et al., 1994; Stamenkovic et al., 2002].

The discovery of the Pt "skin" led some researchers to reject the geometric argument [Toda et al., 1998] and instead explain the enhanced ORR activity of the alloys in terms of the bulk electronic structure. Interpretation of in situ XANES and EXAFS results suggested 10–20% greater Pt d-band vacancy for the alloys compared with pure Pt catalysts [Mukerjee et al., 1995]. It was thus suggested that increased Pt d-band vacancy strengthened the metal–O_2 interaction and thus facilitated O–O bond scission [Toda et al., 1998, 1999a, b]. Others emphasized effects on surface-blocking species such as OH and anions [Arenz et al., 2003; Mukerjee et al., 1995; Stamenkovic et al., 2007a, b] and pointed out the importance of high surface area [Paffett et al., 1988; Stamenkovic et al., 2007a, b] and specific crystalline facets [Markovic and Ross, 2002]. It became apparent that the way in which the ORR was affected by alloying Pt with the base metals could not be unequivocally explained with current experimental techniques.

To shed light on the origin of the enhanced ORR activity, Xu and co-workers performed extensive DFT calculations to investigate the reactivity of the Pt skin [Xu et al., 2004], in particular how oxygen interacts in vacuum with the ordered Pt_3Co alloy and with a monolayer of Pt formed on the alloy as a model for Pt skin. Figure 9.10 identifies the various adsorption sites for O and O_2. Experiments have shown that up to four layers of Pt could sustain a 2.5% compressive strain without creating any surface

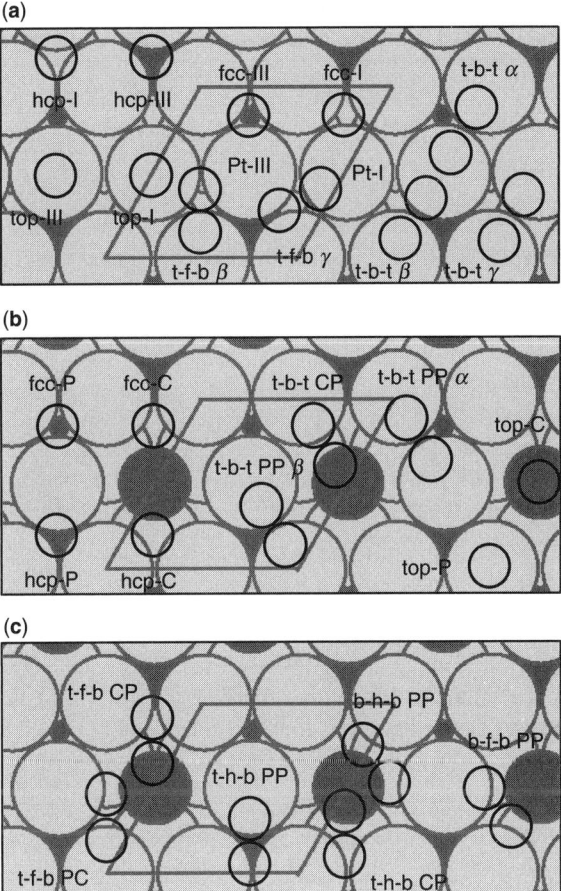

Figure 9.10 Like Pt, the ordered Pt$_3$Co phase has the fcc structure and is represented by the thermodynamically most stable (111) facet. These diagrams show high symmetry adsorption sites identified for O and O$_2$ on (a) monolayer Pt skin on Pt$_3$Co(111) and (b, c) Pt$_3$Co(111). Large circles represent Pt atoms, medium gray circles represent Co atoms, and small open circles represent O atoms. The surface unit cells are delineated. (Reproduced with permission from Xu et al. [2004].)

defects when deposited on Ru(0001) [Schlapka et al., 2003], which suggests that the observed Pt skin could be compressed so as to conform to the slightly contracted lattices of the underlying alloys (2% for Pt$_3$Co and 1.5% for Pt$_3$Fe vs. bulk Pt). In an attempt to decouple the possible ligand effect (electronic interaction with the substrate) from the strain effect (conforming to the substrate geometry) [Kitchin et al., 2004], a 2% compressed Pt(111) surface was separately investigated.

The stoichiometric Pt$_3$Co(111) surface has one-quarter of the Pt atoms replaced with Co atoms. Bulk Co is very reactive toward oxygen, on which O$_2$ dissociation is probably non-activated. The Co atoms on the alloy surface are diluted by Pt, but

TABLE 9.2 Perdew-Wang (PW91) Binding Energies of O and O_2 on the Pt and Pt_3Co Alloy Surfaces. Reprinted with Permission from Xu et al. [2004]

	Pt(111)	Pt(111) −2%	Pt skin on $Pt_3Co(111)$			$Pt_3Co(111)$		
			I	III		C	P	
Atomic oxygen (eV/O)								
top	−2.49	−2.36	−2.31	−2.28		−3.50	n.s.	
hcp	−3.49	−3.28	−3.13	−3.12		−4.02	−3.46	
fcc	−3.88	−3.63	−3.50	−3.20		−4.29	−3.52	
Molecular oxygen (eV/O_2)			α	β	γ	CP	PP	PC
t-b-t	−0.62	−0.50	−0.24	−0.34	−0.34	−0.92	−0.53	←α
							−0.65	←β
t-f-b	−0.61	−0.46	n.s.	−0.25	−0.28	−0.92	n.s.	−0.77
t-h-b	−0.45	−0.32	n.s.	n.s.	n.s.	−0.67	−0.48	n.s.
b-f-b	n.s.	n.s.	n.s.	n.s.	n.s.	n.s.	−0.70	n.s.
b-h-b	n.s.	n.s.	n.s.	n.s.	n.s.	n.s.	−0.63	n.s.

The binding energies of O and O_2 are referenced to $O_{(g)}$ and $O_{2(g)}$, respectively. Surface coverage is 1/4 ML for each species. For comparison, the binding energy of O on Co(0001) is −5.44 eV. The bond energy of a gas-phase O_2 molecule is 5.64 eV. See Fig. 3.1 for site labeling.
n.s. indicates that the corresponding state is not stable.

remain the reaction centers: those sites that involve a Co atom always bind O and O_2 more strongly than the corresponding Pt-only sites (Table 9.2). O_2 also dissociates more easily at Co centers, where the transition states of O_2 dissociation are stabilized and the activation energy is lowered considerably compared with Pt-only sites of $Pt_3Co(111)$ or Pt(111) (Fig. 9.11). However, given that the base metal

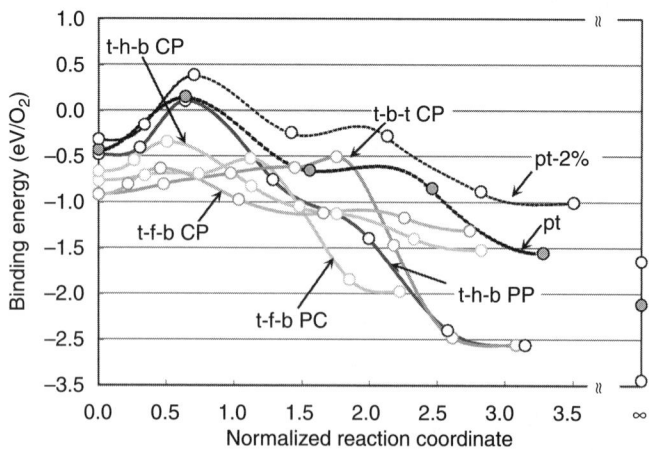

Figure 9.11 Several minimum energy paths for O_2 dissociation on $Pt_3Co(111)$ (labeled by respective initial states), generating $\frac{1}{2}$ ML of atomic O, compared with equilibrium and 2% compressed Pt(111). The points on each path are the "images" or states used to discretize the path with the climbing-image nudged elastic band method. The zero of the energy axis corresponds to an O_2 molecule and the respective clean surfaces at infinite separation. The points located on the right vertical axis represent atomic O at $\frac{1}{4}$ ML. (Reproduced with permission from Xu et al. [2004].)

can leach out under reaction conditions and that it binds oxygen so strongly, the stoichiometric Pt$_3$Co surface cannot be directly responsible for enhanced ORR activity.

In contrast to the alloy surface, the Pt skin on Pt$_3$Co(111) binds atomic and molecular oxygen *less strongly* than pure Pt(111) by about 0.3 eV. Moreover, the activation barrier for O$_2$ dissociation is *increased* by 0.16 eV. A comparison of the Pt skin with the 2% compressed pure Pt surface shows that, in terms of the binding energies of O and O$_2$ and the activation energy for O$_2$ dissociation, the 2% compressed Pt falls squarely in between the equilibrium Pt and the Pt skin on Pt$_3$Co. Thus, the strain effect and the ligand effect appear to be additive, and each contributes to rendering the Pt skin less reactive than a normal Pt surface.

The key parameters of the electronic structure of these surfaces are summarized in Table 9.3. The calculated *d*-band vacancy of Pt shows no appreciable increase. Instead, there is a slight charge transfer from Co to Pt, which may be attributable to the difference in electronegativity of Pt and Co, in apparent contradiction with the substantial increase in Pt *d*-band vacancy previously reported [Mukerjee et al., 1995]. What does change systematically across these surfaces is the *d*-band center (ε_d) of Pt, which, as Fig. 9.12 demonstrates, systematically affects the reactivity of the surfaces. This correlation is consistent with the previous successes [Greeley et al., 2002; Mavrikakis et al., 1998] of the *d*-band model in describing the reactivity of various bimetallic surfaces and the effect of strain. Compressive strain lowers ε_d, which, in turn, leads to weaker adsorbate–surface interaction, whereas expansive strain has the opposite effect.

In summary, the Pt skin structure clearly modifies the reactivity of what is otherwise a pure Pt surface. These findings contradict the original geometric argument that the Pt–Pt distance needs to contract to be better at activating the O–O bond: a compressed Pt surface with a shorter Pt–Pt distance is in fact less active for dissociating O$_2$. The ligand effect in this case further destabilizes adsorption and makes O$_2$ dissociation more activated. These findings provide seemingly counterintuitive but important clues for the ORR mechanism: more facile O$_2$ dissociation does not necessarily

TABLE 9.3 The Electronic Structure of the Pt and Pt$_3$Co Alloy Surfaces. Reprinted with Permission from Xu et al. [2004]

		$\varepsilon_d - \varepsilon_F$ (eV)	σ	f (%)
Pt(111)		−2.52	5.93	93.3
Pt(111) −2%		−2.63	6.20	93.2
Pt skin on Pt$_3$Co(111)	Pt-I	−2.58	6.14	93.4
	Pt-III	−2.79	6.13	93.3
Pt$_3$Co(111)	Pt	−2.69	5.88	93.5
	Co	−1.45	5.56	79.0
Co(0001)		−1.48	5.57	81.3

ε_d is the center of the d-band (with respect to the Fermi level ε_F) or the first moment of the d-band; σ, the spread of the d-band or its second moment; and f, the filling of the d-band. Co(0001) is included for comparison.

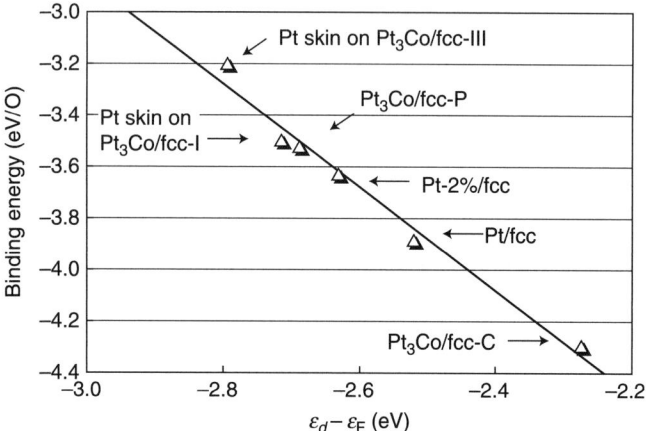

Figure 9.12 Binding energies of O versus d-band center (relative to the Fermi level, $\varepsilon_d - \varepsilon_F$) of the Pt and Pt$_3$Co alloy surfaces. Labels identify the adsorption sites. The line is the best linear fit. (Reproduced with permission from Xu et al. [2004].)

entail higher ORR activity; subsequent events (i.e., the hydrogenation/protonation of oxygen) likely play an important role as well.

9.3.2 Platinum Monolayer Electrocatalysts

The successful use of Pt sub-monolayer on Ru as the electrocatalyst for hydrogen oxidation, with high stability and high CO tolerance, led to extension of the monolayer catalyst concept to the ORR [Brankovic et al., 2001a, b; Sasaki et al., 2003]. The Pt monolayers are formed on other pure metals or alloys. The electronic structure of Pt skins could be altered significantly upon bonding with the substrate metal [Hammer and Nørskov, 2000; Rodriguez, 1996]. In addition to the possibility of tailoring the catalytic properties of the resultant Pt surface [Kibler et al., 2005], this approach also allows for a substantial reduction in Pt loading. The catalytic properties of bimetallic surfaces consisting of metal monolayers on single-crystal metal surfaces have been extensively studied in UHV [Rodriguez and Goodman, 1992], but to a lesser extent in electrochemical systems [Adzic, 2002; Baldauf and Kolb, 1996; Naohara et al., 2000; Schmidt et al., 2002]. Using the technique of displacing the metal adlayer deposited at underpotentials [Brankovic et al., 2001a], Zhang and co-workers prepared Pt monolayers on Ru(0001), Ir(111), Rh(111), Au(111), and Pd(111) single crystals and studied their ORR activity using rotating disk measurements and DFT calculations [Zhang et al., 2005b].

The experiments showed that the properties of the Pt monolayer were modified differently by the different supporting metal (Fig. 9.13). Pt$_{ML}$/Pd(111) and Pt$_{ML}$/Ru(0001) are the most and least active of these surfaces, respectively. When the ORR kinetic currents obtained from Koutecky–Levich plots are plotted against the

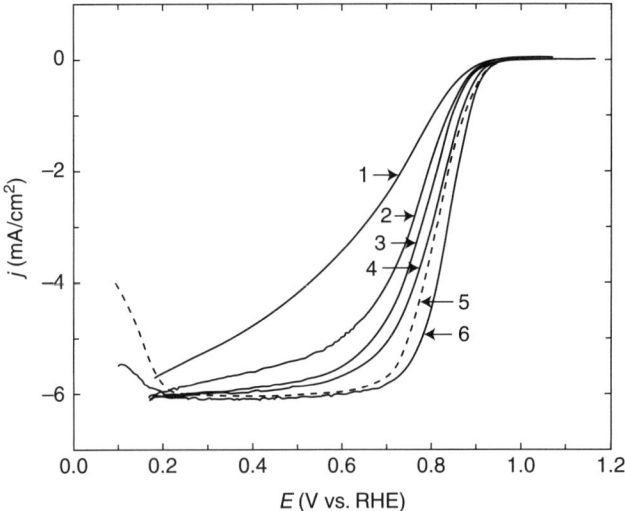

Figure 9.13 Polarization curves for ORR on Pt monolayers supported on Ru(0001), Ir(111), Rh(111), Au(111), and Pd(111) in a 0.1 M HClO$_4$ solution on a disk electrode. The curve for Pt(111) was obtained from [Markovic et al., 1999] and is included for comparison. The rotation rate was 1600 rev/min and the sweep rate was 20 mV/s (50 mV/s for Pt(111)). Key: 1, Pt$_{ML}$/Ru(0001); 2, Pt$_{ML}$/Ir(111); 3, Pt$_{ML}$/Rh(111); 4, Pt$_{ML}$/Au(111); 5, Pt(111); 6, Pt$_{ML}$/Pd(111). (Reproduced with permission from Zhang et al. [2005a].)

DFT-calculated ε_d, a familiar "volcano" plot emerges (Fig. 9.14). Like the Pt$_3$Co alloy systems, a linear correlation also exists between the calculated O binding energy and ε_d. The binding energy is higher on Pt$_{ML}$/Au(111) and lower on Pt$_{ML}$/Ir(111), Pt$_{ML}$/Ru(0001) and Pt$_{ML}$/Rh(111). Strain partially explains the observed modifications to the Pt monolayer properties: compared with equilibrium Pt(111), the Pt monolayer is compressed on Ir(111), Ru(0001), and Rh(111), whereas it is stretched by almost 4% on Au(111). There is also a ligand effect, for example: Au, being a less reactive metal, affects the Pt layer to a lesser extent than the more reactive Ru. Because of the similar lattice constants and chemical properties of Pt and Pd, Pt(111) and Pt$_{ML}$/Pd(111) bind oxygen with similar strength, and yet Pd support causes an approximately 0.1 eV destabilization of O and OH, compared with Pt(111).

Although extensive work has been done to resolve the detailed mechanism of the ORR, important details remain to be clarified, including the rate-determining step (RDS) [Anderson et al., 2005; Li and Gewirth, 2005; Sidik and Anderson, 2002; Wang and Balbuena, 2004, 2005a, b]. Some investigators have proposed the dissociative adsorption of O$_2$ as the RDS, while others have identified the first electron transfer step as the RDS and estimated its activation energy at the reversible electrode potential [Anderson et al., 2005; Grgur et al., 1997; Sidik and Anderson, 2002]. Despite the lack of consensus, the earlier study on Pt$_3$Co alloys suggested that both O–O bond scission and O–H bond formation types of elementary steps are important in the ORR, and that the optimum ORR catalyst must facilitate both types of transformations without

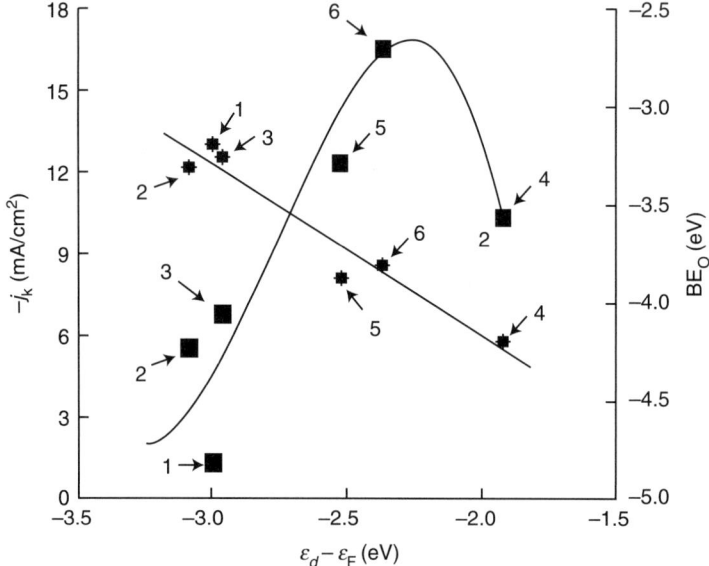

Figure 9.14 Kinetic current density (squares) at 0.8 V for O_2 reduction on the Pt monolayer deposited on various metal single-crystal surfaces in a 0.1 M $HClO_4$ solution, and calculated binding energies (circles) of atomic oxygen (BE_O), as a function of calculated d-band center (relative to the Fermi level, $\varepsilon_d - \varepsilon_F$) of the respective surfaces. The data for Pt(111) were obtained from [Markovic et al., 1999] and are included for comparison. Key: 1, Pt_{ML}/Ru(0001); 2, Pt_{ML}/Ir(111); 3, Pt_{ML}/Rh(111); 4, Pt_{ML}/Au(111); 5, Pt(111); 6, Pt_{ML}/Pd(111). (Reproduced with permission from Zhang et al. [2005a].)

unduly hindering either. To identify the fundamental reasons behind the volcano-type behavior shown in Fig. 9.14, DFT calculations were performed to study the following elementary steps:

$$O_2 \longrightarrow O + O \tag{9.3}$$
$$O + H \longrightarrow OH \tag{9.4}$$

These two steps were chosen because they are the most activated versions of O–O bond scission and O–H bond formation, respectively. There is evidence that the reactivity of other hydrogenated forms of oxygen follows a similar trend to that of atomic oxygen [Shao et al., 2007a].

The calculated activation energy barrier E_a for the elementary steps (9.3) and (9.4) are plotted against O binding energy in Fig. 9.15. The binding energy of O is emphasized because atomic oxygen is the common intermediate for (9.3) and (9.4), and interacts more strongly than any other ORR intermediate with the surface, thus making O adsorption more sensitive to surface properties. Previous studies have shown that surfaces that bind an adsorbate strongly tend to enhance the kinetics of bond scission steps in which the adsorbate is a product. On the other hand, surfaces that bind an

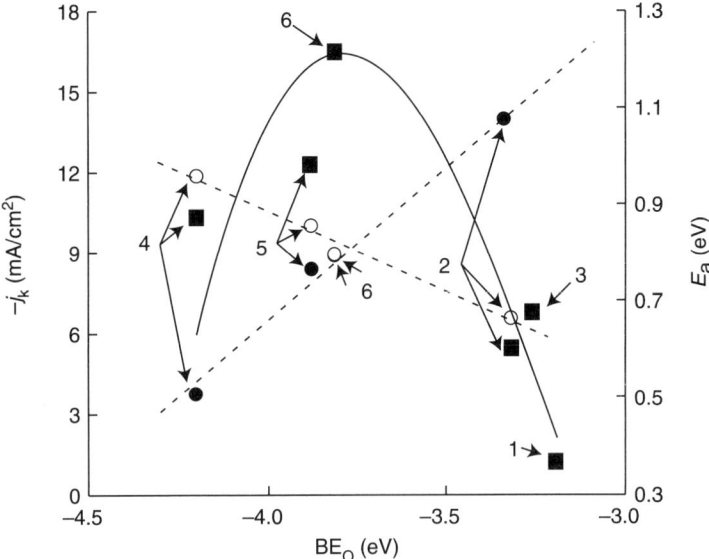

Figure 9.15 Kinetic current density (squares) at 0.8 V for O_2 reduction on supported Pt monolayers in a 0.1 M $HClO_4$ solution, and the calculated activation energy barriers for O_2 dissociation (filled circles) and OH formation (open circles) on $Pt_{ML}/Au(111)$, Pt(111), $Pt_{ML}/Pd(111)$, and $Pt_{ML}/Ir(111)$, as a function of the calculated binding energy of atomic oxygen (BE_O). The current density data for Pt(111) were obtained from [Markovic et al., 1999] and are included for comparison. Key: 1, $Pt_{ML}/Ru(0001)$; 2, $Pt_{ML}/Ir(111)$; 3, $Pt_{ML}/Rh(111)$; 4, $Pt_{ML}/Au(111)$; 5, Pt(111); 6, $Pt_{ML}/Pd(111)$. Surface coverage is $\frac{1}{4}$ ML O_2 in O_2 dissociation and $\frac{1}{4}$ ML each for O and H in OH formation. (Reproduced with permission from Zhang et al. [2005a].)

adsorbate weakly tend to facilitate the kinetics of bond formation steps in which the adsorbate is a reactant [Mavrikakis et al., 1998; Wintterlin et al., 2003; Xu et al., 2004]. Not surprisingly, E_a for O_2 dissociation is smallest on $Pt_{ML}/Au(111)$ and largest on $Pt_{ML}/Ir(111)$. The exact opposite trend is found for E_a of oxygen hydrogenation, which is most facile on $Pt_{ML}/Ir(111)$ and most difficult on $Pt_{ML}/Au(111)$. In accord with the Sabatier principle, the volcano-type dependence of ORR activity on ε_d as seen in Fig. 9.14 is a consequence of two opposing demands on the catalyst connected via a common intermediate. Pt(111) lies close to the intersection of the two E_a trend lines, suggesting that the optimum compromise between the two opposite trends is well represented by this intersection. The only Pt monolayer in the vicinity of the intersection of the two trend lines is $Pt_{ML}/Pd(111)$, which immediately raises the possibility that this bimetallic system is a promising ORR catalyst. Indeed, according to experiments, $Pt_{ML}/Pd(111)$ performs better than all the other Pt monolayers, and even shows an approximately 30% increase in current density compared with Pt(111). Although the exact crossing point may depend on factors including surface coverage, electric field and potential, and solvent, the binding energy of O appears to be a robust *reactivity descriptor* for the ORR on these surfaces. More recently, detailed theoretical calculations of free energy changes in all possible ORR elementary steps, including potential bias, entropy, and zero-point energy corrections according to the methodology developed by Nørskov and co-workers [Nørskov et al., 2004], suggest that OH removal from Pt and from supported Pt$_{ML}$ surfaces and O–O bond scission are the limiting factors to the ORR performance of these catalysts [Nilekar and Maurikakis, 2008]. Clearly, regardless the level of detail used in the analysis, and despite the additional complexity that the electrochemical environment presents, the intrinsic reactivity of the metal surface remains a crucial factor in determining its catalytic properties.

An additional factor favoring Pt$_{ML}$/Pd(111) is the reduced binding energy of OH on the Pt$_{ML}$ surface compared with Pt(111) and Pd(111), since a high OH coverage may inhibit the ORR. Voltammetry and in situ XANES [Zhang et al., 2004] have confirmed that Pt–OH adsorption occurs at more positive potentials on carbon-supported Pt$_{ML}$/Pd nanoparticles (Pt$_{ML}$/Pd/C) than on Pt/C, which in turn occurs at higher potential than Pd–OH on Pd/C, in line with the calculations. Increase in the potential of OH formation on Pt skin on Pt$_3$Co(111) versus Pt(111) has also been reported [Roques et al., 2005]. It stands to reason that catalyst performance may be further improved if ways to destabilize OH can be devised.

One possibility would be to replace some of the Pt atoms on the surface with atoms of other metals that oxidize more easily than Pt. At low potentials, these metal atoms should attract oxygen-containing species (e.g., O and OH) to themselves and, through the electronic modification of the surface plus the enhanced lateral repulsion among OH groups, destabilize OH on adjacent Pt sites and decrease the lifetime and coverage of OH on those sites. As a test, Zhang and co-workers deposited a mixed Pt-Ru and Pt-Ir monolayer of varying compositions on Pd(111) and tested the ORR reactivity of these surfaces by performing rotating disk experiments [Zhang et al., 2005b]. The kinetic current density was observed to increase substantially with the Ru mole fraction up to a maximum at a Pt : Ru ratio of 4 : 1, after which it decreased as the Pt content of the surface diminished. Similar results were obtained for (Pt–Ir)$_{ML}$/Pd(111) (Fig. 9.16). In fact, the kinetic current density for the Pt : M = 4 : 1 composition is significantly enhanced compared with that on the Pt$_{ML}$/Pd(111) surface, which already has a higher ORR activity than Pt(111).

To put these observations into a broader perspective, additional DFT calculations on (Pt$_3$M)$_{ML}$/Pd(111) and rotating disk experiments on (Pt$_{0.8}$M$_{0.2}$)$_{ML}$/Pd(111) with M = Au, Pd, Rh, Re, or Os were carried out [Zhang et al., 2005b]. The calculated interaction energies between OH groups and the geometries of the most stable OH states at $\frac{1}{2}$ ML coverage on all the mixed Pt-M monolayer surfaces are shown in Fig. 9.17a. Negative (positive) interaction energy implies attractive (repulsive) interaction between two neighboring OH groups. For metals with oxidation potential comparable to or higher than that of Pt (M = Au, Pt, and Pd), the OH groups have an attractive interaction and prefer to adsorb alternating on bridge and top Pt sites. For surfaces where M is Rh, Ru, and Ir, all of which have lower oxidation potential than Pt, the OH groups experience an increased destabilization, with half of the OH groups being adsorbed atop M atoms and half atop Pt atoms. The repulsive interaction is the largest among all the Pt-M monolayers for M = Re and Os.

9.3 IMPROVING ORR KINETICS AND PLATINUM UTILIZATION

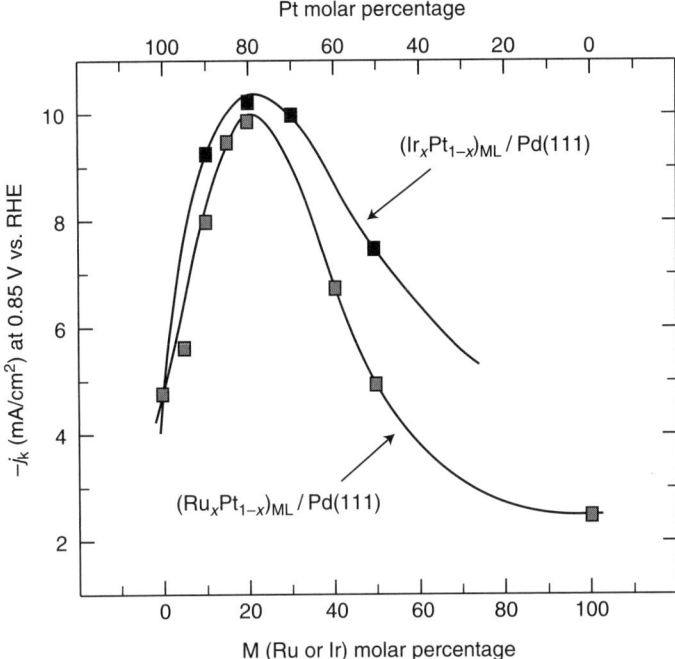

Figure 9.16 ORR activity of two mixed-metal monolayer electrocatalysts supported on Pd(111), expressed as the kinetic current density at 0.85 V as a function of the M:Pt ratio in the Pd-supported Pt-M monolayer. (Reproduced with permission from Zhang et al. [2005b].)

In fact, for these two metals (M), adjacent OH groups react spontaneously to yield an H_2O molecule and an O atom adsorbed on M.

When the kinetic current densities measured on the $(Pt-M)_{ML}/Pd(111)$ surface are plotted against the effective repulsion energies (relative to the repulsion energy on $Pt_{ML}/Pd(111)$), an excellent linear correlation is found between the two variables (Fig. 9.17b), which shows that the increased ORR activity of these Pt-M/Pd(111) surfaces can be attributed to an increase in repulsive interaction between surface-bound OH species, or, alternatively, an increased destabilization of OH species on the Pt sites of the surface. This supports the original hypothesis that easily oxidized metal atoms attract OH to them at lower potentials, thereby destabilizing OH on adjacent Pt sites and reducing OH lifetime and coverage on those sites. Therefore, DFT calculations have been critical for the identification of this key reactivity descriptor (OH–OH and OH–O interaction), which is responsible for reduced OH coverage on these ternary alloy surfaces.

Additional experiments were carried out to study the behavior of Pd nanoparticles coated with Pt or with Pt plus M, which more closely reflects the morphology of actual catalyst particles. Figure 9.18a displays the polarization curves for the ORR on commercial carbon-supported Pt nanoparticles (Pt/C), Pd nanoparticles (Pd/C), a monolayer of Pt on Pd/C ($Pt_{ML}/Pd/C$), and mixed $(Pt_{0.8}Ir_{0.2})_{ML}/Pd/C$ and

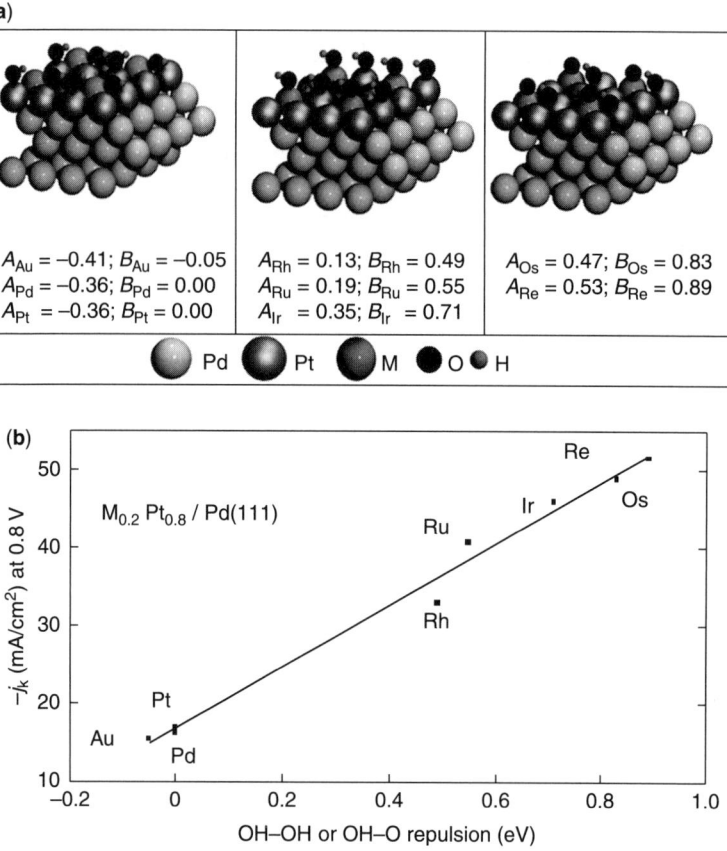

Figure 9.17 (a) Calculated most stable configurations for OH or OH + O ($\frac{1}{2}$ ML total coverage) on seven different $(Pt_3M)_{ML}/Pd(111)$ surfaces and on a $Pt_{ML}/Pd(111)$ surface, which is used as reference for the data in (b). In the left and middle panes, A is the attractive (when negative) or repulsive (when positive) interaction between two OH groups in the unit cell, referenced to the binding energy of OH at $\frac{1}{4}$ ML coverage; B is the same quantity, but referenced to the corresponding value on $Pt_{ML}/Pd(111)$. All energies are in eV. In the right panel, A is the repulsive interaction between adsorbed O and OH; B is the the same quantity rescaled to the repulsion between two OH groups on the $Pt_{ML}/Pd(111)$ surface. (b) Measured kinetic current densities at 0.80 V as a function of the calculated repulsion energy between two OH groups or between OH and O in a (2 × 2) unit cell. (Reproduced with permission from Zhang et al. [2005b].)

$(Pt_{0.8}Re_{0.2})_{ML}/Pd/C$. Although comparing the activity of the Pt/C electrocatalyst with an average particle size of 3.1 nm with that of $Pt_{ML}/Pd/C$ electrocatalyst of 9 nm [Zhang et al., 2004] does not provide an adequate activity assessment because of their different surface areas, this does not adversely affect the main conclusions derived from experiments on single-crystal surfaces.

The ORR activity of $Pt_{ML}/Pd/C$ is much higher compared to that of Pd/C, but, more importantly, it is indeed also higher than that of Pt/C. The amounts of Pt in

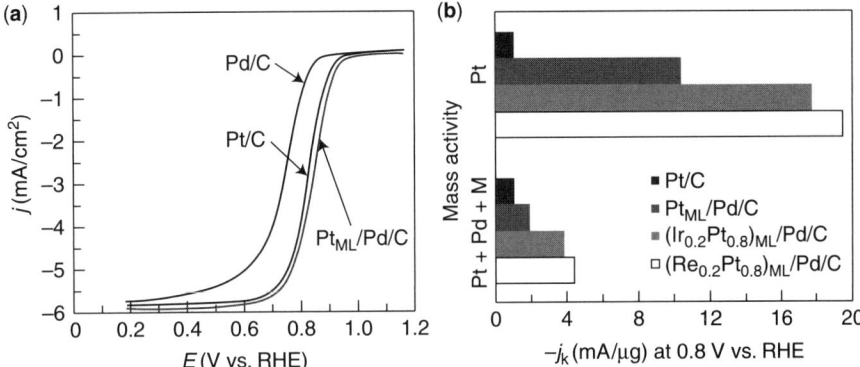

Figure 9.18 (a) Polarization curves for the ORR on Pd/C (10 nmol), Pt/C (10 nmol), and $Pt_{ML}/Pd/C$ nanoparticles (10 nmol Pd) in 0.1 M $HClO_4$ solution. The electrode geometric area was 0.164 cm². The rotation rate was 1600 rev/min and the sweep rate was 10 mV/s. (b) The Pt and total noble metal mass activities for the ORR of Pt/C (10 nmol), $Pt_{ML}/Pd/C$ (10 nmol Pd), and $(Ir_{0.2}Pt_{0.8})_{ML}/Pd/C$ and $(Re_{0.2}Pt_{0.8})_{ML}/Pd/C$ (20 nmol Pd) nanoparticles expressed as a current at 0.8 V. (Reproduced with permission from Vukmirovic et al. [2007].)

$Pt_{ML}/Pd/C$ and Pt/C are 1.7 and 12 $\mu g_{Pt}/cm^2$, respectively. The amount of Pt corresponding to $Pt_{ML}/Pd/C$ can be obtained by calculating the charge associated with the deposition of a Cu monolayer at underpotentials (after correcting for the double-layer charging) on Pd/C, assuming that there is a one-to-one ratio between the Cu and Pd atoms. Therefore, even though Pt/C has a higher surface area, and a Pt loading that is seven times larger, its ORR activity is lower than that of $Pt_{ML}/Pd/C$.

When the ORR activity is compared on a Pt mass basis, $Pt_{ML}/Pd/C$ is eight times more active than commercial Pt electrodes, whereas $(Pt_{0.8}Ir_{0.2})_{ML}/Pd/C$ and $(Pt_{0.8}Re_{0.2})_{ML}/Pd/C$ are nearly 20 times more active than Pt/C (Fig. 9.18b). This finding underscores the promise of monolayer-based electrocatalysts in significantly reducing the amount of Pt in fuel cell electrodes. The positive comparisons with Pt/C nanoparticles remain valid even when the total precious metal content ($m_{Pt} + m_M + m_{Pd}$) is considered instead of Pt mass alone: the total noble metal mass activity of Pt-Ir and Pt-Re monolayer electrocatalysts is 4- and 4.5-fold higher than Pt/C. This is still very satisfactory, particularly since the main constituent of the ternary alloys is Pd, which is considerably less expensive than Pt.

XANES and in situ voltammetry [Zhang et al., 2005b] experiments show that, in the case of $(Pt_{0.8}Ir_{0.2})_{ML}/Pd/C$, Pt-OH formation is suppressed until potentials as high as 1.17 V, which also supports the original hypothesis that an appropriate choice of metal M can keep the Pt sites from being poisoned by OH by destabilizing interactions, and thereby facilitating the ORR. For comparison, the onset of Pt-OH formation on the Pt/C catalysts was found to be less than 0.6 V. Thus, it is likely that the ternary nanoparticles are more stable under potential cycling regimes than Pt nanoparticles, resulting in more sinter-resistant catalysts.

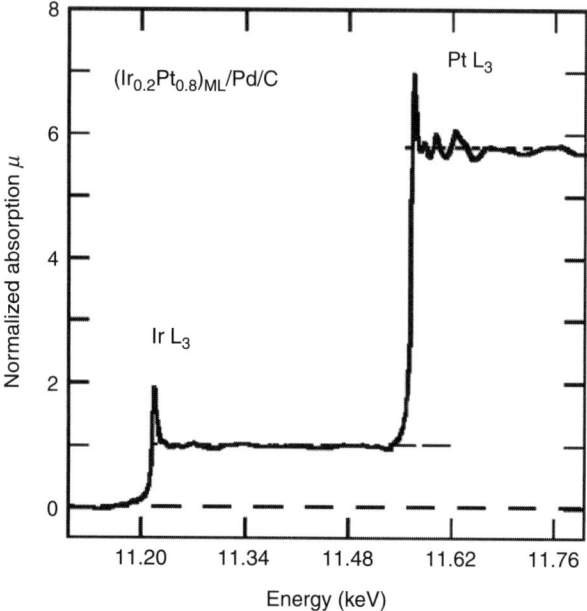

Figure 9.19 The Pt and Ir L_3 edge in situ XANES spectra for $Pt_{0.8}Ir_{0.2}/Pd/C$ (20 nmol Pd) in 1 M $HClO_4$ at 0.47 V. The Pd loading in these nanoparticles has been doubled to enhance the Ir signal in the specific experiment. (Reproduced with permission from Vukmirovic et al. [2007].)

To verify that intended amounts of Pt and the second metal are indeed deposited on the substrate, XANES spectra are taken in situ for $(Pt_{0.8}Ir_{0.2})_{ML}/Pd/C$ (20 nmol Pd) in 1 M $HClO_4$ at 0.47 V. The L_3 edge for Pt and Ir (Fig. 9.19) indicates that the Pt : Ir ratio is 4.66 : 1, or 82% Pt and 18% Ir, in very good agreement with the 80 : 20 ion concentration used in the Cu-displacement preparation method.

9.3.3 Mechanistic Insights for the ORR

To further substantiate these insights for the ORR mechanism, Wang and co-workers analyzed the four-electron ORR in acidic media using an intrinsic kinetic equation with the free energies of activation and adsorption as the kinetic parameters [Wang JX et al., 2007]. The kinetic model consisted of four essential elementary reactions:

1. dissociative adsorption (DA) and
2. reductive adsorption (RA),
   which yield two reaction intermediates, O and OH;
3. reductive transition (RT) O → OH; and
4. reductive desorption (RD) of OH as H_2O.

Analytic expressions were obtained for the O and OH adsorption isotherms by solving the steady state rate equations.

9.3 IMPROVING ORR KINETICS AND PLATINUM UTILIZATION 297

The resulting free energy diagram at zero overpotential and the activity-and-barrier plots (Fig. 9.20) (Plate 9.1) indicate that reductive adsorption ($\Delta G_{RA}^o = 0.46$ eV) is apparently not the RDS for the ORR on Pt, because dissociative adsorption ($\Delta G_{DA}^o = 0.26$ eV) offers a more favorable pathway at high potentials. This is at variance with the conclusions in [Sidik and Anderson, 2002]. Above the reversible potential for O → OH, the O coverage is high, causing severe inhibition of ORR activity. As the potential decreases, the O coverage falls, making the sites available for OH adsorption and the kinetic current increases with the decreasing reductive desorption activation barrier. Nonetheless, the activation barriers for both O → OH ($\Delta G_{RT}^o = 0.50$ eV) and OH desorption ($\Delta G_{RD}^o = 0.45$ eV) are high, causing

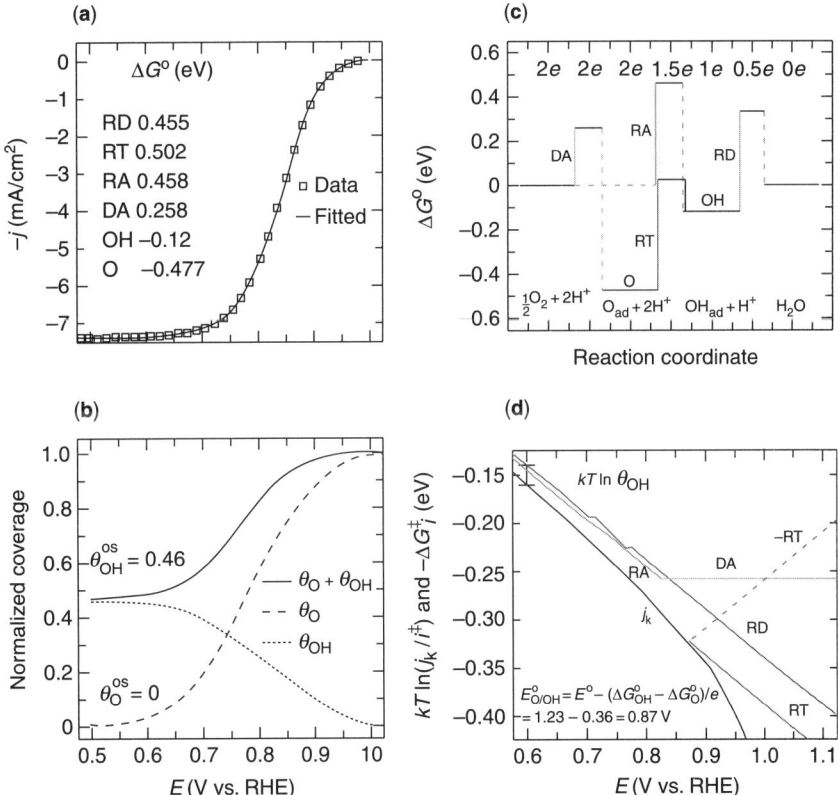

Figure 9.20 (a) ORR polarization curve for Pt(111) in 0.1 M HClO$_4$ solution. The fitted parameters were used to calculate the adsorption isotherms for O and OH in (b) and to construct the free energy diagrams in (c), where the lengths of the vertical lines represent the activation free energies of the forward (solid) and backward (dashed) reactions. (d) Activity-and-barriers plot that shows the kinetic current (black line) and the activation free energies (colored lines) on an equivalent energy scale. Above the reversible potential for O → OH transition (0.87 V), the apparent Tafel slope is half of that at low potentials. (Reproduced with permission from Wang JX et al. [2007].)

a large potential loss for the ORR. The activity-and-barrier plots illustrate why the RDS may vary with reaction conditions, or may not even exist. They also hint at the important role that the elementary steps (9.3) and (9.4) play directly in the ORR.

9.3.4 Palladium Alloy Electrocatalysts

As already mentioned, considerable ongoing research is focused on replacing Pt altogether without sacrificing ORR activity. Recent reports [Fernandez et al., 2005a, b; Lee et al., 2006; Mustain et al., 2007; Raghuveer et al., 2005; Savadogo et al., 2004; Shao et al., 2006a, c, 2007b; Tarasevich et al., 2007; Wang W et al., 2007; Zhang L et al., 2007] have indicated that binary Pd-M alloys (M = Cr, Co, Fe, Ni, Ti) and ternary ones were comparable to, or slightly better than commercial Pt catalysts. These Pd-M catalysts have attracted considerable attention because they are comparatively inexpensive and have very high methanol tolerance, making them promising cathode catalytic materials for hydrogen PEMFCs and especially direct methanol fuel cells (DMFCs).

Theoretical calculations and experimental data demonstrated that, upon annealing at elevated temperatures, the Pd-M alloys, like the Pt-M alloys, undergo Pd segregation, in which Pd atoms migrate to the surface to form a pure Pd skin on the bulk alloys [Bozzolo et al., 2003; Rousset et al., 1996; Ruban et al., 1999]. The Pd lattice also contracts upon alloying with a $3d$ metal, which generates compressive strains in the Pd skin. A good correlation was found between specific activity and the Pd–Pd interatomic distance expressed as a function of Fe content (Fig. 9.21). The highest ORR activity was obtained for Pd_3Fe, which has a Pd–Pd bond distance of 0.273 nm, smaller than that in Pd by 2.3%. Similar results were obtained for Pd–Co alloys [Suo et al., 2007; Wang W et al., 2007].

DFT calculations show that, like Pt_3Co alloy surfaces, the electronic structure of the Pd skin is modified by strain and the underlying alloy, which in turn modifies the

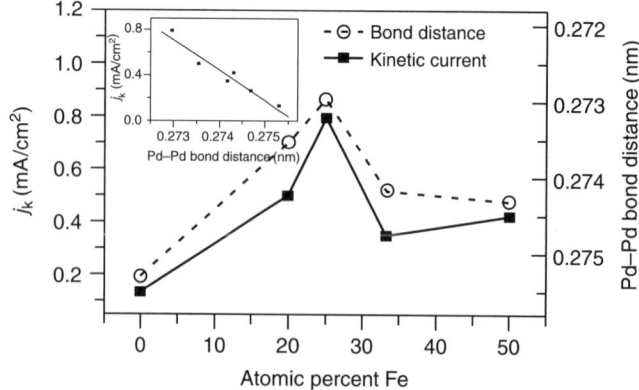

Figure 9.21 Specific activity and Pd–Pd bond distance calculated from XRD data against the concentration of Fe in Pd-Fe/C electrocatalysts treated at 500 °C. (Reproduced with permission from Shao et al. [2006c].)

reactivity of the surface [Shao et al., 2007a, b; Suo et al., 2007]. Compressive strain alone accounts for a 0.1 eV destabilization of the Pd–O bond. $Pd_3Fe(111)$ and PdFe(111) substrates contribute further destabilization of O by about 0.25 and 0.35 eV, respectively. The native $Pd_3Fe(111)$ and PdFe(111) surfaces bind oxygen about 1 eV *more strongly* than equilibrium Pd, and are therefore not expected to be good ORR catalysts. Of these surfaces, Pd skin on $Pd_3Fe(111)$ is most similar to Pt(111) in terms of both the d-band center position and the oxygen binding energy, which corroborates the findings in Fig. 9.21.

Similarly to the Pt monolayer catalysts, a series of Pd monolayers deposited on different metal single crystals were tested for ORR activity. The results are shown in Fig. 9.22. The ORR activity increases in the order $Pd_{ML}/Ru(0001) < Pd_{ML}/Ir(111) < Pd_{ML}/Rh(111) < Pd_{ML}/Au(111) < Pd(111) < Pd_{ML}/Pt(111) < Pt(111)$. The activity of the $Pd_{ML}/Pt(111)$ surface is higher than that of Pd(111), but somewhat lower than that of Pt(111).

Some thermodynamic guidelines have been proposed for designing non-Pt alloy ORR electrocatalysts. Bard and co-workers suggested that for Pd-M alloys, the reactive metal M constitutes the site for breaking the O–O bonds, forming O_{ads} that migrates to the hollow sites dominated by Pd atoms, where it is readily reduced to water [Fernandez et al., 2005a, b, 2006]. Based on this mechanism, the alloy surface should consist of a relatively reactive metal such as Co, and the atomic ratio of this metal should be 10–20% so that there are sufficient sites for reactions of O–O bond breaking on M and O_{ads} reduction at hollow sites formed by Pd atoms. DFT calculations indicated that one of the O atoms diffused to the Pd hollow site while

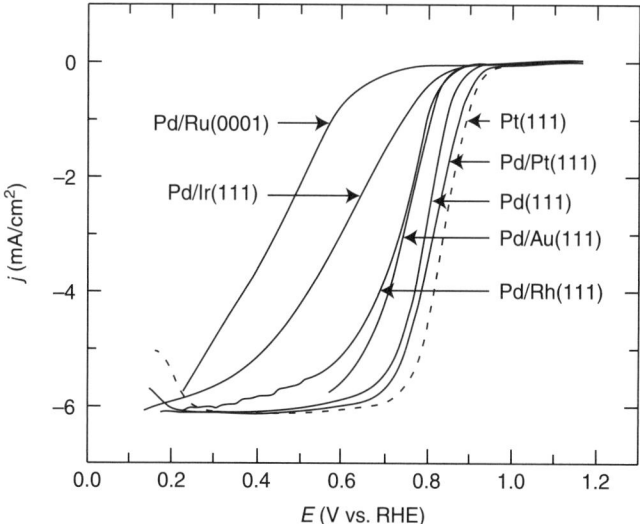

Figure 9.22 Comparison of polarization curves for the ORR on Pd monolayers on different substrates, and on Pd(111) and Pt(111), in 0.1 M $HClO_4$ solution; sweep rate 10 mV/s; room temperature. (Reproduced with permission from Shao et al. [2006a].)

the other still adsorbed on the hollow site near Co after the dissociative adsorption of the O_2 molecule [Fernandez et al., 2006]. The second O_2 could dissociate on Co with an O atom pre-bound on the hollow site near it. Balbuena and co-workers proposed a similar thermodynamic guideline for designing Pd alloy catalysts [Wang and Balbuena, 2005a, b]. For Pd with fully occupied valence d-orbitals, alloying with transition metals such as Co with unoccupied valence d-orbitals reduces significantly the Gibbs free energy both for the first charge-transfer step and for the steps involving the reduction of intermediates. However, it is not clear whether O_2 can still easily dissociate after the reactive metal centers are fully occupied by O; besides, the reactive metals on the alloy surface are unstable and leach out rapidly during electrochemical measurements. Thus, these arguments for ORR electrocatalysis on bimetallic surfaces cannot explain the relatively good stability of Pd-M alloys in acidic media [Tarasevich et al., 2007]. The Pd-enriched skin, on the other hand, can account for both the good activity and the stability [Lamas and Balbuena, 2006; Shao et al., 2007a; Suo et al., 2007].

Pd ternary alloys, including Pd-Co-Au [Fernandez et al., 2005a, b] and Pd-Co-Mo [Raghuveer et al., 2005] have been developed to further improve the stability of the catalyst. The addition of 10% Au to the Pd-Mo mixture improved catalyst stability. Another promising way to improve the activity and durability of Pd-M alloys is to deposit a Pt monolayer on them. Recently, a Pt monolayer deposited on Pd_3Fe/C was found to possess higher activity than that of Pt/C [Shao et al., 2007b].

Methanol tolerance is a very important property of Pd-M alloys. In particular, methanol tolerance was demonstrated for Pd_3Fe/C and for Pd-Co based alloys [Mustain et al., 2007; Raghuveer et al., 2005; Shao et al., 2006c; Zhang L et al., 2007]. The high ORR activity in the presence of a high concentration of methanol indicates that the Pd-Co and Pd-Fe electrocatalysts are not active for methanol oxidation.

9.4 CATALYST STABILITY

One of the critical issues with regard to low temperature fuel cells is the gradual loss of performance due to the degradation of the cathode catalyst layer under the harsh operating conditions, which mainly consist of two aspects: electrochemical surface area (ECA) loss of the carbon-supported Pt nanoparticles and corrosion of the carbon support itself. Extensive studies of cathode catalyst layer degradation in phosphoric acid fuel cells (PAFCs) have shown that ECA loss is mainly caused by three mechanisms:

- Pt dissolution and redeposition (Ostwald ripening)
- Pt particle agglomeration due to crystallite migration on carbon supports
- Pt particle agglomeration due to carbon corrosion

Similar ECA loss phenomena have been observed in PEMFCs. Understanding ECA loss and carbon corrosion mechanisms may help with designing more durable catalysts for the ORR. In this section, we will discuss the ECA loss mechanisms. The carbon corrosion mechanism is outside the scope of this review.

9.4.1 Platinum Dissolution

The thermodynamic behavior of Pt bulk is described by potential–pH diagrams (Pourbaix diagrams) [Pourbaix, 1974]. The main pathways for Pt dissolution involve either direct dissolution of the metal,

$$Pt \longrightarrow Pt^{2+} + 2e^- \qquad E_0 = 1.19 + 0.029 \ \log [Pt^+] \qquad (9.5)$$

or formation of an oxide film and subsequent chemical reaction,

$$Pt + H_2O \longrightarrow PtO + 2H^+ + 2e^- \qquad E_0 = 0.98 - 0.59 \ \text{pH} \qquad (9.6)$$

$$PtO + 2H^+ \longrightarrow Pt^{2+} + H_2O \qquad \log [Pt^{2+}] = -7.06 - 2 \ \text{pH} \qquad (9.7)$$

The dissolution through Pt oxides is expected to be slow owing to the self-passivation effect. The solubility of Pt at different potentials, pH, and temperatures has been extensively studied. In acidic solutions, Pt solubility increases with decreasing pH, suggesting an acidic dissolution mechanism [Mitsushima et al., 2007a, b]. Pt solubility depends strongly on temperature, following an Arrhenius relationship [Dam and de Bruijn, 2007; Mitsushima et al., 2007a, b]. The effect of potential on Pt solubility has been extensively studied in acidic solutions [Bindra et al., 1979; Dam and de Bruijn, 2007; Ferreira et al., 2005; Mitsushima et al., 2007a, b; Wang XP et al., 2006] and summarized in [Borup et al., 2007]. The equilibrium concentration of dissolved Pt increases with potential up to 1.1 V [Bindra et al., 1979; Ferreira et al., 2005; Wang XP et al., 2006] and decreases thereafter, which is ascribed to the formation of a PtO film [Wang XP et al., 2006]. At higher potential, Wang and co-workers found that Pt dissolution was alleviated by the formation of a PtO_2 film. From the slope of the logarithmic plot of Pt solubility against potential in hot concentrated H_3PO_4 solution, Bindra and co-workers concluded that Pt dissolution underwent a two-electron reaction pathway (9.5) [Bindra et al., 1979], which is consistent with the Pourbaix diagram [Pourbaix, 1974]. However, in H_2SO_4 (80 °C) [Ferreira et al., 2005] and $HClO_4$ (23 °C) [Wang X et al., 2006] solutions and lower temperatures, the slopes are much smaller, indicating that Pt dissolution involves the formation of dissolved species other than Pt^{2+}.

Potential cycling has been found to accelerate Pt dissolution compared with potentiostatic conditions. The dissolution mechanisms and dissolved species involved in this process are unclear [Johnson et al., 1970; Kinoshita et al., 1973; Ota et al., 1988; Rand and Woods, 1972]. Darling and Meyers have developed a mathematical model based on (9.5)–(9.7) to study Pt dissolution and movement in a PEMFC during potential cycling from 0.87 to 1.2 V [Darling and Meyers, 2003, 2005]. Severe Pt dissolution occurs when the potential switches to the upper limit potential (1.2 V), and then stops once a monolayer of PtO has formed. The charge difference between the anodic and cathodic cycles was found to be consistent with the amount of Pt dissolved. Thus, an anodic dissolution mechanism was suggested either via (9.5) or via (9.6) and (9.7). A cathodic dissolution mechanism is also possible according to Johnson and co-workers, who detected Pt^{2+} in a rotating ring–disk electrode study during the negative-going potential scan in 0.1 M $HClO_4$ solution [Johnson et al., 1970]. The Pt^{2+} species was formed as a result of the reduction of PtO_2 [Johnson et al., 1970; Mitsushima et al., 2007a, b]:

$$PtO_2 + 4H^+ + 2e^- \longrightarrow Pt^{2+} + 2H_2O \quad E_0 = 0.84 + 0.12\,pH + 0.029\log[Pt^{2+}] \quad (9.8)$$

The dissolution rate during potential cycling was reported to be around 3.0–5.5 ng/cm^2 per cycle, with the upper potential limit between 1.2 and 1.5 V and various potential scanning rates [Johnson et al., 1970; Kinoshita et al., 1973; Rand and Woods, 1972; Wang XP et al., 2006]. The predominant forms of dissolved Pt were Pt^{2+} species [Johnson et al., 1970; Wang XP et al., 2006] and Pt^{4+} species [Mitsushima et al., 2007a, b; Rand and Woods, 1972] in $HClO_4$ and H_2SO_4 solution, respectively, which may be due to the ability of H_2SO_4 to form a more stable complex with Pt^{4+} [Rand and Woods, 1972].

Dissolved Pt species (Pt^{z+}) at the cathode can either redeposit on existing Pt particles, resulting in particle growth, or diffuse into the membrane (PEMFC) or matrix (PAFC). The concentrations of these species increased upon the aging of the membrane electrode assembly (MEA), suggesting that mobile Pt species move into the membrane [Guilminot et al., 2007a, b]. A Pt band of large particles forms in the membrane near the interface of membrane/cathode during cycling with H_2/N_2 [Ferreira et al., 2005; Ferreira and Shao-Horn, 2007; Yasuda et al., 2006a, b], and somewhere away from the cathode with H_2/O_2(air) [Bi et al., 2007; Patterson, 2002; Yasuda et al., 2006a, b; Zhang J et al., 2007a]. Studies combining experimental data and mathematical modeling suggested that the location of the Pt band in the membrane under open circuit voltage and cycling conditions depends on the partial pressures of H_2 and O_2 and on the permeability of those species and Pt^{z+} through the membrane [Bi et al., 2007; Zhang J et al., 2007a] (Fig. 9.23). These results suggest that Pt^{z+} in the membrane is chemically reduced by H_2. Pt can also move into the anode in the absence of H_2 [Yasuda et al., 2006a, b].

The driving force for the crossover of dissolved Pt species into the membrane/ matrix can be electro-osmotic drag and/or concentration gradient diffusion [Guilminot et al., 2007a, b]. The identity of the counteranions of Pt^{z+}, however, is not yet clear [Johnson et al., 1970; Kinoshita et al., 1973; Ota et al., 1988; Rand and Woods, 1972]. Membrane degradation products, such as fluoride [Healy et al., 2005; Xie et al., 2005a, b] and sulfate [Teranishi et al., 2006; Xie et al., 2005a, b] anions, have been detected during the operation of PEMFCs, and may be the complexing ligands for Pt^{z+}. In fact, Guilminot and co-workers presented strong evidence that the concentration of fluoride around the Pt nanoparticles in an aged membrane was higher than that in a new one [Guilminot et al., 2007a, b]. Other halide ions, such as chloride and bromide, left on carbon and Pt surfaces during catalyst synthesis represent another possibility [Guilminot et al., 2007a, b].

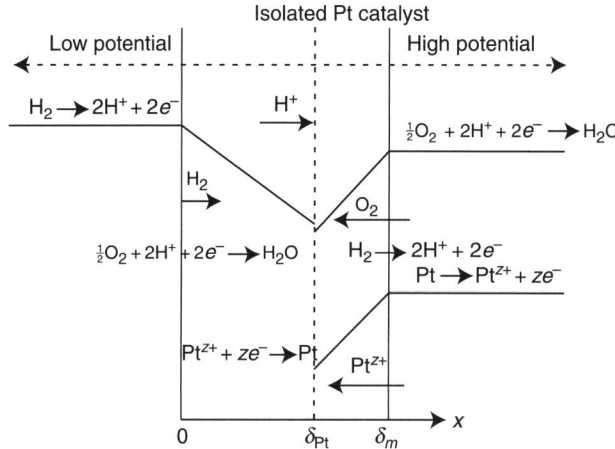

Figure 9.23 Schematic representation of the various electrochemical and chemical reactions occurring in a membrane electrode assembly and the concentration gradients of O_2, H_2, and Pt ions. The location where the local O_2 molar flux equals one-half of the local H_2 molar flux is marked by δ_{Pt}. (Reproduced with permission from Zhang J et al. [2007a].)

Pt dissolution mechanisms also depend on the surface crystallographic orientation. Edges and corners on Pt nanoparticles, which have low coordinated sites, are suspected to have higher dissolution tendency compared with the terrace facets, even though no experimental or theoretical data are available to support this claim. Komanicky and co-workers studied Pt dissolution on different low index facets as a function of potential in 0.6 M $HClO_4$ solution (Fig. 9.24) [Komanicky et al., 2006]. Even at potentials as low as 0.65 V, dissolution was observed on edges and pits on Pt(111) surfaces, while terraces remained stable. Surprisingly, the dissolution of Pt(111) was inhibited at 0.95 V owing to the formation of oxide films at edges, but started again at 1.15 V, where terraces were irreversibly roughened, with a large number of pits on the surface. For Pt(100) and (110) facets, the dissolution was alleviated with increasing potential, owing to the formation of a passivation layer on these surfaces. A (111)–(100) nanofaceted surface, which was used to simulate the edges of Pt nanoparticles, dissolved quickly in the potential range of 0.65–1.15 V.

9.4.2 Platinum Particle Growth Mechanisms

Potential-dependent particle growth rates in PAFCs [Honji et al., 1988] and the existence of Pt particles in the matrix [Aragane et al., 1988] and membrane [Patterson, 2002] after fuel cell operation suggest a Pt dissolution/redeposition mechanism, i.e., "3D Ostwald ripening" [Honji et al., 1988; Tseung and Dhara, 1975]. The mechanism of redeposition of mobile Pt^{z+} on the existing Pt nanoparticles on carbon supports in the cathode has not, however, been studied extensively. Based on their finding that no particle growth was observed on nonconductive Al_2O_3, Virkar and

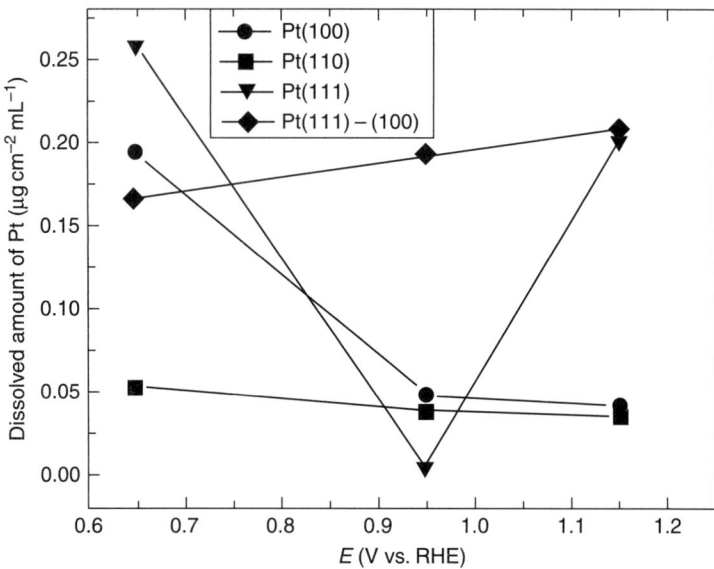

Figure 9.24 Amounts of dissolved platinum, normalized to 1 cm^2 and 1 mL of solution, plotted against the dissolution potentials for all the investigated surfaces. (Reproduced with permission from Komanicky et al. [2006].)

Zhou [2007] have recently suggested that electronically conductive supports, such as carbon black, are critical for Pt re-precipitation. Therefore, particle growth by Ostwald ripening involves coupled transport, for example, Pt ions and electron transport via water/ionomer and conductive supports, respectively [Virkar and Zhou, 2007]. Yasuda and co-workers performed an interesting study using double-layer cathodes. Pt particles were not observed in the membrane with a Pt black layer sandwiched between the membrane and a normal Pt/C catalyst layer, while no substantial difference was found between cathode layers with and without sandwiching an extra carbon black layer. The authors concluded that dissolved Pt species were easier to deposit on unsupported Pt black, which has a much higher surface energy than carbon [Yasuda et al., 2006a, b]. If these results are correct, then a new door to minimizing the ECA loss may be opened through capturing the dissolved Pt species by materials with higher surface energy than Pt [Tseung and Dhara, 1975].

However, minimization of the surface energy can drive crystallite migration/coalescence as well as Ostwald ripening, independent of an electric potential. There were indeed reports that ECA loss during the operation of PAFCs was insensitive to the potential [Blurton et al., 1978; Gruver et al., 1980]. The crystallite migration and coalescence mechanism would result in the particle size distribution tailing toward large particles, which in fact has been observed in studies of both PAFCs [Aragane et al., 1988; Bett et al., 1976; Blurton et al., 1978] and PEMFCs [Wilson et al., 1993]. Bett and co-workers found that particle coalescence did not become more severe with higher Pt loading on carbon supports, as would be expected from this mechanism [Bett et al., 1976]. They instead proposed a "2D Ostwald ripening" mechanism involving surface transport of Pt atoms (not soluble Pt species) from one crystallite to another. In such a mechanism, the particle size distribution would broaden and shift toward larger particles owing to the consumption of smaller particles. However, a bimodal particle size distribution was observed during potential cycling in some cases where some small Pt particles remained [Garzon et al., 2006; More et al., 2006; Xie et al., 2005a, b], suggesting a combination of crystallite coalescence and Ostwald ripening processes [Borup et al., 2007].

9.4.3 Stabilization Effects of Au Clusters on Pt Electrocatalysts

Transportation is expected to be an especially important application of fuel cells, since their uniquely high energy conversion efficiency may result in a substantial decrease in

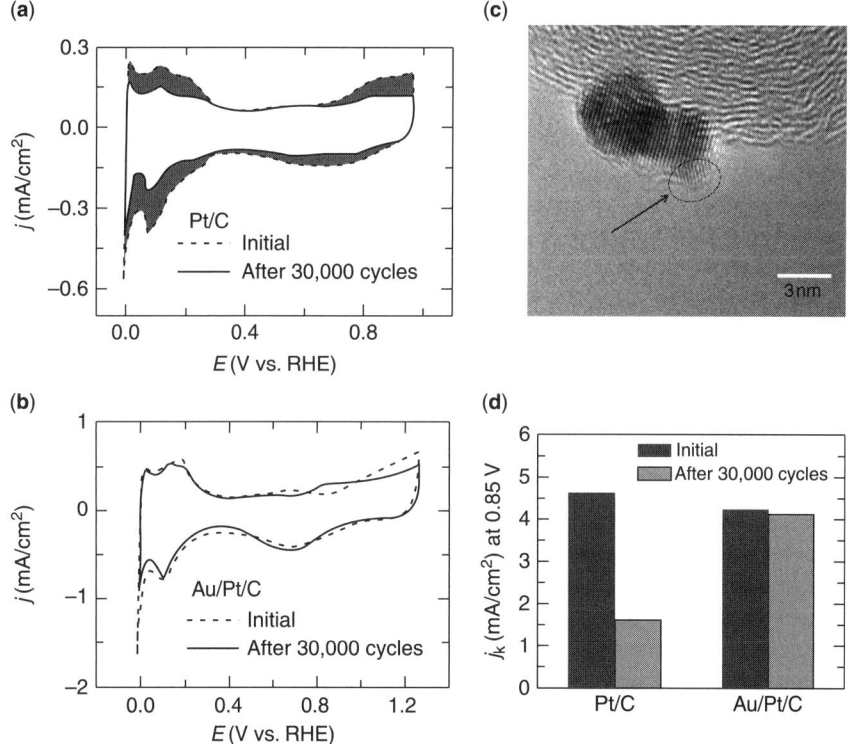

Figure 9.25 (a, b) Voltammetry curves for (a) Pt/C and (b) Au/Pt/C catalysts before and after 30,000 cycles; the sweep rate was 50 and 20 mV/s, respectively. The potential cycles were from 0.6 to 1.1 V in an O_2-saturated 0.1 M $HClO_4$ solution at room temperature. For all electrodes, the Pt loading was 1.95 μg (10 nmol) Pt on a 0.164 cm^2 glassy carbon rotating disk electrode. The shaded area in (a) indicates the lost Pt area. (c) Au-modified Pt. Different interference fringes (circled) indicate clusters of Au on Pt. (d) Catalytic activities of Pt/C and Au/Pt/C before and after 30,000 cycles (normalized to geometric area). (Reproduced with permission from Bi et al. [2007] and Zhang et al. [2007b].)

TABLE 9.4 A Comparison of Surface Area and the Catalytic Activity Data for Pt/C and Au/Pt/C Before and After 30,000 Potential Cycles from 0.6 to 1.1 V under the Oxidizing Conditions of the O_2 Reduction Reaction. Reprinted with Permission from Zhang et al. [2007b]

Catalyst / Kinetic Data	Pt Dispersion (m^2/g_{Pt})	Half-Wave Potential at 1600 rpm (V)	Kinetic Current Density at 0.85 V (mA/cm^2)	Specific Kinetic Current Density at 0.85 V (A/m^2_{Pt})
Pt/C initial	65.5	0.841	4.56	5.80
Pt/C after 30,000 cycles	35.5	0.802	1.60	3.72
Au/Pt/C initial	63.1	0.838	4.23	5.64
Au/Pt/C after 30,000 cycles	60.6	0.833	4.10	5.69

the adverse environmental effects of using fossil fuels. A major obstacle to this application has been observed in recent studies that recorded a substantial loss of Pt surface area over time—and it is Pt that is the electrocatalyst for the cathodic reduction of oxygen in PEMFCs during the stop-and-go driving of an electric car. Such regimes generate large electrode potential excursions (typically 0.6–1.0 V), resulting in Pt dissolution, which is thus a major challenge for fuel cell vehicles and the science of electrocatalysis.

Zhang and co-workers demonstrated that Pt oxygen reduction fuel cell electrocatalysts can be stabilized against dissolution under potential cycling regimes by modifying the Pt nanoparticles with small Au clusters [Zhang J et al., 2007b]. The stabilizing effect of Au clusters was assessed in an accelerated stability test by continuously applying linear potential sweeps from 0.6 to 1.1 V. After 30,000 cycles, Pt/C suffered a loss of 39 mV half-wave potential and 45% in active surface area under the same conditions (Fig. 9.25). The same experiment with the Au/Pt/C electrode at 60 °C showed negligible change in the Pt surface area and electrocatalytic activity. In situ XANES and voltammetry data suggest that the Au clusters confer stability by raising the oxidation potential of Pt [Zhang et al., 2007b]. Possibly, Au atoms block the kink and step sites of Pt where dissolution starts. The same effect was observed with the aforementioned Pt monolayer-on-Pd nanoparticle electrocatalysts (Table 9.4). This finding has the potential to solve one of the major problems for fuel cell application in transportation. Apart from fuel cells, the ability of Au clusters to alter the properties of other metals may have other scientific and technological applications, although further research is needed to explore such effects.

ACKNOWLEDGMENTS

Work at BNL was supported by the US Department of Energy, Divisions of Chemical and Material Sciences, under Contract DE-AC02-98CH10886. Work at ORNL was conducted at the Center for Nanophase Materials Sciences and was sponsored by the Division of

Scientific User Facilities, DOE-BES. Work at UW was supported by DOE-BES, Chemical Sciences Division, and by the NSF. Supercomputing resources at NERSC, PNNL, and ORNL were utilized.

REFERENCES

Adzic RR. 1998. Recent advances in the kinetics of oxygen reduction. In: Lipkowski J, Ross PN, eds. Electrocatalysis. New York: Wiley-VCH.

Adzic RR. 2002. In: Bard A, Stratmann M, eds. Encyclopedia of Electrochemistry, Volume 1. New York: Wiley-VCH.

Adzic RR, Wang JX. 1998. Configuration and site of O_2 adsorption on the Pt(111) electrode surface. J Phys Chem B 102: 8988–8993.

Adzic RR, Wang JX. 2000. Structure and inhibition effects of anion adlayers during the course of O_2 reduction. Electrochim Acta 45: 4203–4210.

Alonso-Vante N, Malakhov IV, Nikitenko SG, Savinova ER, Kochubey DI. 2002. The structure analysis of the active centers of Ru-containing electrocatalysts for the oxygen reduction. An in situ EXAFS study. Electrochim Acta 47: 3807–3814.

Anderson AB. 2002. O_2 reduction and CO oxidation at the Pt–electrolyte interface. The role of H_2O and OH adsorption bond strengths. Electrochim Acta 47: 3759–3763.

Anderson AB, Roques J, Mukerjee S, Murthi VS, Markovic NM, Stamenkovic V. 2005. Activation energies for oxygen reduction on platinum alloys: Theory and experiment. J Phys Chem B 109: 1198–1203.

Aragane J, Murahashi T, Odaka T. 1988. Change of Pt distribution in the active components of phosphoric acid fuel cell. J Electrochem Soc 135: 844–850.

Arenz M, Schmidt TJ, Wandelt K, Ross PN, Markovic NM. 2003. The oxygen reduction reaction on thin palladium films supported on a Pt(111) electrode. J Phys Chem B 107: 9813–9819.

Atli A, Abon M, Beccat P, Bertolini JC, Tardy B. 1994. Carbon monoxide adsorption on a $Pt_{80}Fe_{20}$(111) single-crystal alloy. Surf Sci 302: 121–125.

Avery NR. 1983. An EELS and TDS study of molecular-oxygen desorption and decomposition on Pt(111). Chem Phys Lett 96: 371–373.

Baldauf M, Kolb DM. 1996. Formic acid oxidation on ultrathin Pd films on Au(*hkl*) and Pt(*hkl*) electrodes. J Phys Chem 100: 11375–11381.

Bashyam R, Zelenay P. 2006. A class of non-precious metal composite catalysts for fuel cells. Nature 443: 63–66.

Bett JAS, Kinoshita K, Stonehart P. 1976. Crystallite growth of platinum dispersed on graphitized carbon black: II. Effect of liquid environment. J Catal 41: 124–133.

Bi W, Gray GE, Fuller TF. 2007. PEM fuel cell Pt/C dissolution and deposition in nafion electrolyte. Electrochem Solid State Lett 10: B101–B104.

Bindra P, Clouser SJ, Yeager E. 1979. Platinum dissolution in concentrated phosphoric-acid. J Electrochem Soc 126: 1631–1632.

Birss VI, Chang M, Segal J. 1993. Platinum oxide film formation reduction—An in-situ mass measurement study. J Electroanal Chem 355: 181–191.

Blurton KF, Kunz HR, Rutt DR. 1978. Surface area loss of platinum supported on graphite. Electrochim Acta 23: 183–190.

Borup R, Meyers J, Pivovar B, Kim YS, Mukundan R, Garland N, Myers D, Wilson M, Garzon F, Wood D, Zelenay P, More K, Stroh K, Zawodzinski T, Boncella J, McGrath JE, Inaba M, Miyatake K, Hori M, Ota K, Ogumi Z, Miyata S, Nishikata A, Siroma Z, Uchimoto Y, Yasuda K, Kimijima Ki, Iwashita N. 2007. Scientific aspects of polymer electrolyte fuel cell durability and degradation. Chem Rev 107: 3904–3951.

Bozzolo G, Noebe RD, Khalil J, Morse J. 2003. Atomistic analysis of surface segregation in Ni-Pd alloys. Appl Surf Sci 219: 149–157.

Brankovic SR, Wang JX, Adzic RR. 2001a. Metal monolayer deposition by replacement of metal adlayers on electrode surfaces. Surf Sci 474: L173–L179.

Brankovic SR, Wang JX, Adzic RR. 2001b. Pt submonolayers on Ru nanoparticles—A novel low Pt loading, high CO tolerance fuel cell electrocatalyst. Electrochem Solid State Lett 4: A217–A220.

Campbell CT, Ertl G, Kuipers H, Segner J. 1981. A molecular beam study of the adsorption and desorption of oxygen from a Pt(111) surface. Surf Sci 107: 220–236.

Conway BE. 1995. Electrochemical oxide film formation at noble metals as a surface-chemical process. Prog Surf Sci 49: 331–452.

Dam VAT, de Bruijn FA. 2007. The stability of PEMFC electrodes—Platinum dissolution vs. potential and temperature investigated by quartz crystal microbalance. J Electrochem Soc 154: B494–B499.

Darling RM, Meyers JP. 2003. Kinetic model of platinum dissolution in PEMFCs. J Electrochem Soc 150: A1523–A1527.

Darling RM, Meyers JP. 2005. Mathematical model of platinum movement in PEM fuel cells. J Electrochem Soc 152: A242–A247.

Eichler A, Mittendorfer F, Hafner J. 2000. Precursor-mediated adsorption of oxygen on the (111) surfaces of platinum-group metals. Phys Rev B 62: 4744–4755.

Faguy PW, Markovic N, Adzic RR, Fierro C, Yeager E. 1990. A study of bisulfate adsorption on Pt(111) single crystal electrodes using in-situ Fourier transform infrared spectroscopy. J Electroanal Chem 289: 245–262.

Faguy PW, Marinkovic NS, Adzic RR. 1996. Infrared spectroscopic analysis of anions adsorbed from bisulfate-containing solutions on Pt(111) electrodes. J Electroanal Chem 407: 209–218.

Fernandez JL, Raghuveer V, Manthiram A, Bard AJ. 2005a. Pd-Ti and Pd-Co-Au electrocatalysts as a replacement for platinum for oxygen reduction in proton exchange membrane fuel cells. J Am Chem Soc 127: 13100–13101.

Fernandez JL, Walsh DA, Bard AJ. 2005b. Thermodynamic guidelines for the design of bimetallic catalysts for oxygen electroreduction and rapid screening by scanning electrochemical microscopy. M-Co (M: Pd, Ag, Au). J Am Chem Soc 127: 357–365.

Fernandez JL, White JM, Sun YM, Tang WJ, Henkelman G, Bard AJ. 2006. Characterization and theory of electrocatalysts based on scanning electrochemical microscopy screening methods. Langmuir 22: 10426–10431.

Ferreira PJ, Shao-Horn Y. 2007. Formation mechanism of Pt single-crystal nanoparticles in proton exchange membrane fuel cells. Electrochem Solid State Lett 10: B60–B63.

Ferreira PJ, La O GJ, Shao-Horn Y, Morgan D, Makharia R, Kocha S, Gasteiger HA. 2005. Instability of Pt/C electrocatalysts in proton exchange membrane fuel cells—A mechanistic investigation. J Electrochem Soc 152: A2256–A2271.

Gambardella P, Sljivancanin Z, Hammer B, Blanc M, Kuhnke K, Kern K. 2001. Oxygen dissociation at Pt steps. Phys Rev Lett 87: 056103.

Garzon FH, Davey JR, Borup R. 2006. Fuel cell catalyst particle size growth characterized by X-ray scattering methods. ECS Transactions 1(8): 153.

Gland JL. 1980. Molecular and atomic adsorption of oxygen on the Pt(111) and Pt(s)-12(111) × (111) surfaces. Surf Sci 93: 487–514.

Gland JL, Sexton BA, Fisher GB. 1980. Oxygen interactions with the Pt(111) surface. Surf Sci 95: 587–602.

Greeley J, Nørskov JK, Mavrikakis M. 2002. Electronic structure and catalysis on metal surfaces. Annu Rev Phys Chem 53: 319–348.

Greeley J, Jaramillo TF, Bonde J, Chorkendorff IB, Nørskov JK. 2006. Computational high-throughput screening of electrocatalytic materials for hydrogen evolution. Nature Mater 5: 909–913.

Grgur BN, Markovic NM, Ross PN. 1997. Underpotential deposition of lead on Pt(111) in perchloric acid solution: $RRD_{Pt(111)}E$ measurements. Langmuir 13: 6370–6374.

Gruver GA, Pascoe RF, Kunz HR. 1980. Surface Area loss of platinum supported on carbon in phosphoric acid electrolyte. J Electrochem Soc 127: 1219–1224.

Guilminot E, Corcella A, Charlot F, Maillard F, Chatenet M. 2007a. Detection of Pt^{z+} ions and Pt nanoparticles inside the membrane of a used PEMFC. J Electrochem Soc 154: B96–B105.

Guilminot E, Corcella A, Chatenet M, Maillard F, Charlot F, Berthome G, Iojoiu C, Sanchez JY, Rossinot E, Claude E. 2007b. Membrane and active layer degradation upon PEMFC steady-state operation. J Electrochem Soc 154: B1106–B1114.

Hammer B, Nørskov JK. 1995a. Electronic factors determining the reactivity of metal surfaces. Surf Sci 343: 211–220.

Hammer B, Nørskov JK. 1995b. Why gold is the noblest of all the metals. Nature 376: 238–240.

Hammer B, Nørskov JK. 2000. Theoretical surface science and catalysis: Calculations and concepts. Adv Catal 45: 71–129.

Harrington DA. 1997. Simulation of anodic Pt oxide growth. J Electroanal Chem 420: 101–109.

Healy J, Hayden C, Xie T, Olson K, Waldo R, Brundage A, Gasteiger H, Abbott J. 2005. Aspects of the chemical degradation of PFSA ionomers used in PEM fuel cells. Fuel Cells 5: 302–308.

Henkelman G, Jónsson H. 2000. Improved tangent estimate in the nudged elastic band method for finding minimum energy paths and saddle points. J Chem Phys 113: 9978–9985.

Henkelman G, Uberuaga BP, Jónsson H. 2000. A climbing image nudged elastic band method for finding saddle points and minimum energy paths. J Chem Phys 113: 9901–9904.

Honji A, Mori T, Tamura K, Hishinuma Y. 1988. Agglomeration of platinum particles supported on carbon in phosphoric acid. J Electrochem Soc 135: 355–359.

Igarashi H, Fujino T, Zhu Y, Uchida H, Watanabe M. 2001. CO tolerance of Pt alloy electrocatalysts for polymer electrolyte fuel cells and the detoxification mechanism. Phys Chem Chem Phys 3: 306–314.

Inoue H, Brankovic SR, Wang JX, Adzic RR. 2002. Oxygen reduction on bare and Pt monolayer-modified Ru(0001), Ru(101-0) and Ru nanostructured surfaces. Electrochim Acta 47: 3777–3785.

Jacobsen CJH, Dahl S, Clausen BS, Bahn S, Logadottir A, Nørskov JK. 2001. Catalyst design by interpolation in the periodic table: Bimetallic ammonia synthesis catalysts. J Am Chem Soc 123: 8404–8405.

Jalan V, Taylor EJ. 1983. Importance of interactomic spacing in catalytic reduction of oxygen in phosphoric acid. J Electrochem Soc 130: 2299–2301.

Jerkiewicz G, Vatankhah G, Lessard J, Soriaga MP, Park YS. 2004. Surface-oxide growth at platinum electrodes in aqueous H_2SO_4: Reexamination of its mechanism through combined cyclic-voltammetry, electrochemical quartz-crystal nanobalance, and Auger electron spectroscopy measurements. Electrochim Acta 49: 1451–1459.

Johnson DC, Napp DT, Bruckenstein S. 1970. A ring-disk electrode study of the current/potential behaviour of platinum in 1.0 M sulphuric and 0.1 M perchloric acids. Electrochim Acta 15: 1493–1509.

Kibler LA, El-Aziz AM, Hoyer R, Kolb DM. 2005. Tuning reaction rates by lateral strain in a palladium monolayer. Angew Chem Int Ed 44: 2080–2084.

Kinoshita K, Lundquist JT, Stonehart P. 1973. Potential cycling effects on platinum electrocatalyst surfaces. J Electroanal Chem 48: 157–166.

Kitchin JR, Nørskov JK, Barteau MA, Chen JG. 2004. Modification of the surface electronic and chemical properties of Pt(111) by subsurface $3d$ transition metals. J Chem Phys 120: 10240–10246.

Knudsen J, Nilekar AU, Vang RT, Schnadt J, Kunkes EL, Mavrikakis M, Dumesic JA. 2007. A Cu/Pt near-surface alloy for water-gas shift catalysis. J Am Chem Soc 129: 6485–6490.

Komanicky V, Chang KC, Menzel A, Markovic NM, You H, Wang X, Myers D. 2006. Stability and dissolution of platinum surfaces in perchloric acid. J Electrochem Soc 153: B446–B451.

Koper MTM, Lukkien JJ. 2000. Modeling the butterfly: the voltammetry of $(\sqrt{3} \times \sqrt{3})R30°$ and p(2 × 2) overlayers on (111) electrodes. J Electroanal Chem 485: 161–165.

Lamas EJ, Balbuena PB. 2006. Oxygen reduction on $Pd_{0.75}Co_{0.25}(111)$ and $Pt_{0.75}Co_{0.25}(111)$ surfaces: An ab initio comparative study. J Chem Theory Comput 2: 1388–1394.

Lee K, Savadogo O, Ishihara A, Mitsushima S, Kamiya N, Ota K-I. 2006. Methanol-tolerant oxygen reduction electrocatalysts based on Pd-3D transition metal alloys for direct methanol fuel cells. J Electrochem Soc 153: A20–A24.

Li X, Gewirth AA. 2005. Oxygen electroreduction through a superoxide intermediate on Bi-modified Au surfaces. J Am Chem Soc 127: 5252–5260.

Markovic NM, Ross PN. 2002. Surface science studies of model fuel cell electrocatalysts. Surf Sci Rep 45: 121–229.

Markovic NM, Gasteiger HA, Grgur BN, Ross PN. 1999. Oxygen reduction reaction on Pt(111): Effects of bromide. J Electroanal Chem 467: 157–163.

Mavrikakis M, Hammer B, Nørskov JK. 1998. Effect of strain on the reactivity of metal surfaces. Phys Rev Lett 81: 2819.

Mitsushima S, Kawahara S, Ota K-I, Kamiya N. 2007a. Consumption rate of Pt under potential cycling. J Electrochem Soc 154: B153–B158.

Mitsushima S, Koizumi Y, Ota K, Kamiya N. 2007b. Solubility of platinum in acidic media (I)—In sulfuric acid. Electrochemistry 75: 155–158.

More KL, Bentley L, Reeves KS. 2006. DOE Hydrogen Program Annual Progress Report. Available at: http://www.hydrogen.energy.gov/annual_progress06.html.

Mukerjee S, Srinivasan S. 1993. Enhanced electrocatalysis of oxygen reduction on platinum alloys in proton exchange membrane fuel cells. J Electroanal Chem 357: 201–224.

Mukerjee S, Srinivasan S, Soriaga M, McBreen J. 1995. Role of structural and electronic properties of Pt and Pt alloys on electrocatalysis of oxygen reduction. J Electrochem Soc 142: 1409–1422.

Murthi VS, Urian RC, Mukerjee S. 2004. Oxygen reduction kinetics in low and medium temperature acid environment: Correlation of water activation and surface properties in supported Pt and Pt alloy electrocatalysts. J Phys Chem B 108: 11011–11023.

Mustain WE, Kepler K, Prakash J. 2007. CoPd$_x$ oxygen reduction electrocatalysts for polymer electrolyte membrane and direct methanol fuel cells. Electrochim Acta 52: 2102–2108.

Nagy Z, You H. 2002. Applications of surface X-ray scattering to electrochemistry problems. Electrochim Acta 47: 3037–3055.

Naohara H, Ye S, Uosaki K. 2000. Electrocatalytic reactivity for oxygen reduction at epitaxially grown Pd thin layers of various thickness on Au(111) and Au(100). Electrochim Acta 45: 3305–3309.

Nilekar AU, Mavrikakis M. 2008. Improved oxygen reduction reactivity of platinum monolayers on transition metal surfaces. Surf Sci 602: L89–L94.

Nørskov JK, Bligaard T, Logadottir A, Bahn S, Hansen LB, Bollinger M, Bengaard H, Hammer B, Sljivancanin Z, Mavrikakis M, Xu Y, Dahl S, Jacobsen CJH. 2002. Universality in heterogeneous catalysis. J Catal 209: 275–278.

Nørskov JK, Rossmeisl J, Logadottir A, Lindqvist L, Kitchin JR, Bligaard T, Jónsson H. 2004. Origin of the overpotential for oxygen reduction at a fuel-cell cathode. J Phys Chem B 108: 17886–17892.

Ota K-I, Nishigori S, Kamiya N. 1988. Dissolution of platinum anodes in sulfuric acid solution. J Electroanal Chem 257: 205–215.

Outka DA, Stohr J, Jark W, Stevens P, Solomon J, Madix R. 1987. Orientation and bond length of molecular oxygen on Ag(110) and Pt(111): A near-edge X-ray-absorption fine-structure study. Phys Rev B 35: 4119–4122.

Paffett MT, Beery JG, Gottesfeld S. 1988. Oxygen Reduction at $Pt_{0.65}Cr_{0.35}$, $Pt0_{.2}Cr_{0.8}$ and roughened platinum. J Electrochem Soc 135: 1431–1436.

Patterson TW. 2002. In: Igwe GJ, Mah D, eds. Fuel Cell Technology Topical Conference Proceedings. AIChE Spring National Meeting 2002; New Orleans, LA. New York: AIChE. p. 313.

Paulus UA, Wokaun A, Scherer GG, Schmidt TJ, Stamenkovic V, Radmilovic V, Markovic NM, Ross PN. 2002. Oxygen reduction on carbon-supported Pt-Ni and Pt-Co alloy catalysts. J Phys Chem B 106: 4181–4191.

Pourbaix M. 1974. Atlas of Electrochemical Equilibria. 2nd ed. Houston: NACE.

Raghuveer V, Manthiram A, Bard AJ. 2005. Pd-Co-Mo electrocatalyst for the oxygen reduction reaction in proton exchange membrane fuel cells. J Phys Chem B 109: 22909–22912.

Rand DAJ, Woods R. 1972. A study of the dissolution of platinum, palladium, rhodium and gold electrodes in 1 M sulphuric acid by cyclic voltammetry. J Electroanal Chem 35: 209–218.

Rodriguez JA. 1996. Physical and chemical properties of bimetallic surfaces. Surf Sci Rep 24: 225–287.

Rodriguez JA, Goodman DW. 1992. The nature of the metal–metal bond in bimetallic surfaces. Science 257: 897–903.

Roques J, Anderson AB, Murthi VS, Mukerjee S. 2005. Potential shift for OH(ads) formation on the Pt skin on Pt_3Co-(111) electrodes in acid. J Electrochem Soc 152: E193–E199.

Rousset JL, Bertolini JC, Miegge P. 1996. Theory of segregation using the equivalent-medium approximation and bond-strength modifications at surfaces: Application to fcc Pd-X alloys. Phys Rev B 53: 4947.

Ruban AV, Skriver HL, Nørskov JK. 1999. Surface segregation energies in transition-metal alloys. Phys Rev B 59: 15990–16000.

Sasaki K, Mo Y, Wang JX, Balasubramanian M, Uribe F, McBreen J, Adzic RR. 2003. Pt submonolayers on metal nanoparticles—Novel electrocatalysts for H_2 oxidation and O_2 reduction. Electrochim Acta 48: 3841–3849.

Sasaki K, Zhang L, Adzic RR. 2008. Niobium oxide-supported platinum ultra-low amount electrocatalysts for oxygen reduction. Phys Chem Chem Phys 10: 159–167.

Savadogo O, Lee K, Oishi K, Mitsushimas S, Kamiya N, Ota K-I. 2004. New palladium alloys catalyst for the oxygen reduction reaction in an acid medium. Electrochem Commun 6: 105–109.

Savich W, Sun SG, Lipkowski J, Wieckowski A. 1995. Determination of the sum of Gibbs excesses of sulfate and bisulfate adsorbed at the Pt(111) electrode surface using chronocoulometry and thermodynamics of the perfectly polarized electrode. J Electroanal Chem 388: 233–237.

Savinova ER, Scheybal A, Danckwerts M, Wild U, Pettinger B, Doblhofer K, Schlogl R, Ertl G. 2002. Structure and dynamics of the interface between a Ag single crystal electrode and an aqueous electrolyte. Faraday Discuss 121: 181–198.

Schlapka A, Lischka M, Groß A, Käsberger U, Jakob P. 2003. Surface strain versus substrate interaction in heteroepitaxial metal layers: Pt on Ru(0001). Phys Rev Lett 91: 016101.

Schmidt TJ, Stamenkovic V, Arenz M, Markovic NM, Ross PN. 2002. Oxygen electrocatalysis in alkaline electrolyte: Pt(*hkl*), Au(*hkl*) and the effect of Pd-modification. Electrochim Acta 47: 3765–3776.

Sehested J, Larsen KE, Kustov AL, Frey AM, Johannessen T, Bligaard T, Andersson MP, Nørskov JK, Christensen CH. 2007. Discovery of technical methanation catalysts based on computational screening. Top Catal 45: 9–13.

Sexton BA. 1981. Identification of adsorbed species at metal-surfaces by electron-energy loss spectroscopy (EELS). Appl Phys A 26: 1–18.

Shao MH, Huang T, Liu P, Zhang J, Sasaki K, Vukmirovic MB, Adzic RR. 2006a. Palladium monolayer and palladium alloy electrocatalysts for oxygen reduction. Langmuir 22: 10409–10415.

Shao MH, Liu P, Adzic RR. 2006b. Superoxide is the intermediate in the oxygen reduction reaction on platinum electrode. J Am Chem Soc 128: 7408–7409.

Shao MH, Sasaki K, Adzic RR. 2006c. Pd-Fe nanoparticles as electrocatalysts for oxygen reduction. J Am Chem Soc 128: 3526–3527.

Shao M, Liu P, Zhang J, Adzic RR. 2007a. Origin of enhanced activity in palladium alloy electrocatalysts for oxygen reduction reaction. J Phys Chem B 111: 6772–6775.

Shao M, Sasaki K, Liu P, Adzic RR. 2007b. Pd_3Fe and Pt monolayer-modified Pd_3Fe. Z Phys Chem 221: 1175–1190.

Sidik RA, Anderson AB. 2002. Density functional theory study of O_2 electroreduction when bonded to a Pt dual site. J Electroanal Chem 528: 69–76.

Sljivancanin Z, Hammer B. 2002. Oxygen dissociation at close-packed Pt terraces, Pt steps, and Ag-covered Pt steps studied with density functional theory. Surf Sci 515: 235–244.

Stamenkovic V, Schmidt TJ, Ross PN, Markovic NM. 2002. Surface composition effects in electrocatalysis: Kinetics of oxygen reduction on well-defined Pt_3Ni and Pt_3Co alloy surfaces. J Phys Chem B 106: 11970–11979.

Stamenkovic V, Mun BS, Mayrhofer KJJ, Ross PN, Markovic NM, Rossmeisl J, Greeley J, Nørskov JK. 2006. Changing the activity of electrocatalysts for oxygen reduction by tuning the surface electronic structure. Angew Chem Int Ed 45: 2897–2901.

Stamenkovic VR, Fowler B, Mun BS, Wang G, Ross PN, Lucas CA, Markovic NM. 2007a. Improved oxygen reduction activity on $Pt_3Ni(111)$ via increased surface site availability. Science 315: 493–497.

Stamenkovic VR, Mun BS, Arenz M, Mayrhofer KJJ, Lucas CA, Wang G, Ross PN, Markovic NM. 2007b. Trends in electrocatalysis on extended and nanoscale Pt-bimetallic alloy surfaces. Nature Mater 6: 241–247.

Suo YG, Zhuang L, Lu JT. 2007. First-principles considerations in the design of Pd-alloy catalysts for oxygen reduction. Angew Chem Int Ed 46: 2862–2864.

Tarasevich MR, Zhutaeva GV, Bogdanovskaya VA, Radina MV, Ehrenburg MR, Chalykh AE. 2007. Oxygen kinetics and mechanism at electrocatalysts on the base of palladium–iron system. Electrochim Acta 52: 5108–5118.

Teliska M, Murthi VS, Mukerjee S, Ramaker DE. 2007. Site specific vs specific adsorption of anions on Pt and Pt-based alloys. J Phys Chem C 111: 9267–9274.

Teranishi K, Kawata K, Tsushima S, Hirai S. 2006. Degradation mechanism of PEMFC under open circuit operation. Electrochem Solid State Lett 9: A475–A477.

Thomas S, Sung Y-E, Kim KS, Wieckowski A. 1996. Specific adsorption of a bisulfate anion on a Pt(111) electrode. Ultrahigh vacuum spectroscopic and cyclic voltammetric study. J Phys Chem 100: 11726–11735.

Toda T, Igarashi H, Watanabe M. 1998. Role of electronic property of Pt and Pt alloys on electrocatalytic reduction of oxygen. J Electrochem Soc 145: 4185–4188.

Toda T, Igarashi H, Uchida H, Watanabe M. 1999a. Enhancement of the electroreduction of oxygen on Pt alloys with Fe, Ni, and Co. J Electrochem Soc 146: 3750–3756.

Toda T, Igarashi H, Watanabe M. 1999b. Enhancement of the electrocatalytic O_2 reduction on Pt-Fe alloys. J Electroanal Chem 460: 258–262.

Tseung ACC, Dhara SC. 1975. Loss of surface area by platinum and supported platinum black electrocatalyst. Electrochim Acta 20: 681–683.

Virkar AV, Zhou YK. 2007. Mechanism of catalyst degradation in proton exchange membrane fuel cells. J Electrochem Soc 154: B540–B547.

Vukmirovic MB, Zhang J, Sasaki K, Nilekar AU, Uribe F, Mavrikakis M, Adzic RR. 2007. Platinum monolayer electrocatalysts for oxygen reduction. Electrochim Acta 52: 2257–2263.

Wan L-J, Moriyama T, Ito M, Uchida H, Watanabe M. 2002. In situ STM imaging of surface dissolution and rearrangement of a Pt-Fe alloy electrocatalyst in electrolyte solution. Chem Commun 1: 58–59.

Wang JX, Robinson IK, DeVilbiss JE, Adzic RR. 2000. Structural trends among ionic mental-halide adlayers on electrode surfaces. J Phys Chem B 104: 7951–7959.

Wang JX, Marinkovic NS, Zajonz H, Ocko BM, Adzic RR. 2001. In situ X-ray reflectivity and voltammetry study of Ru(0001) surface oxidation in electrolyte solutions. J Phys Chem B 105: 2809–2814.

Wang JX, Markovic NM, Adzic RR. 2004. Kinetic analysis of oxygen reduction on Pt(111) in acid solutions: Intrinsic kinetic parameters and anion adsorption effects. J Phys Chem B 108: 4127–4133.

Wang JX, Zhang JL, Adzic RR. 2007. Double-trap kinetic equation for the oxygen reduction reaction on Pt(111) in acidic media. J Phys Chem A 111: 12702–12710.

Wang W, Zheng D, Du C, Zou Z, Zhang X, Xia B, Yang H, Akins DL. 2007. Carbon-supported Pd-Co bimetallic nanoparticles as electrocatalysts for the oxygen reduction reaction. J Power Sources 167: 243–249.

Wang X, Li WZ, Chen ZW, Waje M, Yan YS. 2006. Durability investigation of carbon nanotube as catalyst support for proton exchange membrane fuel cell. J Power Sources 158: 154–159.

Wang XP, Kumar R, Myers DJ. 2006. Effect of voltage on platinum dissolution relevance to polymer electrolyte fuel cells. Electrochem Solid State Lett 9: A225–A227.

Wang YX, Balbuena PB. 2004 Roles of proton and electric field in the electroreduction of O_2 on Pt(111) surfaces: Results of an ab-initio molecular dynamics study. J Phys Chem B 108: 4376–4384.

Wang YX, Balbuena PB. 2005a. Design of oxygen reduction bimetallic catalysts: Ab-initio-derived thermodynamic guidelines. J Phys Chem B 109: 18902–18906.

Wang YX, Balbuena PB. 2005b. Potential energy surface profile of the oxygen reduction reaction on a Pt cluster: Adsorption and decomposition of OOH and H_2O_2. J Chem Theory Comput 1: 935–943.

Watanabe M, Tsurumi K, Mizukami T, Nakamura T, Stonehart P. 1994. Activity and stability of ordered and disordered Co-Pt alloys for phosphoric acid fuel cells. J Electrochem Soc 141: 2659–2668.

Wilson MS, Garzon FH, Sickafus KE, Gottesfeld S. 1993. Surface area loss of supported platinum in polymer electrolyte fuel cells. J Electrochem Soc 140: 2872–2877.

Wintterlin J, Zambelli T, Trost J, Greeley J, Mavrikakis M. 2003. Atomic-scale evidence for an enhanced catalytic reactivity of stretched surfaces. Angew Chem Int Ed 42: 2850–2853.

Xie J, Wood DL, More KL, Atanassov P, Borup RL. 2005a. Microstructural changes of membrane electrode assemblies during PEFC durability testing at high humidity conditions. J Electrochem Soc 152: A1011–A1020.

Xie J, Wood DL, Wayne DM, Zawodzinski TA, Atanassov P, Borup RL. 2005b. Durability of PEFCs at high humidity conditions. J Electrochem Soc 152: A104–A113.

Xu Y, Ruban AV, Mavrikakis M. 2004. Adsorption and dissociation of O_2 on Pt-Co and Pt-Fe alloys. J Am Chem Soc 126: 4717–4725.

Yasuda K, Taniguchi A, Akita T, Ioroi T, Siroma Z. 2006a. Characteristics of a platinum black catalyst layer with regard to platinum dissolution phenomena in a membrane electrode assembly. J Electrochem Soc 153: A1599–A1603.

Yasuda K, Taniguchi A, Akita T, Ioroi T, Siroma Z. 2006b. Platinum dissolution and deposition in the polymer electrolyte membrane of a PEM fuel cell as studied by potential cycling. Phys Chem Chem Phys 8: 746–752.

You H, Zurawski DJ, Nagy Z, Yonco RM. 1994. In-situ X-ray reflectivity study of incipient oxidation of Pt(111) surface in electrolyte-solutions. J Chem Phys 100: 4699–4702.

You H, Chu YS, Lister TE, Nagy Z, Ankudiniv AL, Rehr JJ. 2000. Resonance X-ray scattering from Pt(1 1 1) surfaces under water. Physica B: Cond Matt 283: 212–216.

Zelenay P, Wieckowski A. 1991. Anion radiotracers. In: Abruna HD, ed. Electrochemical Interfaces. New York: VCH. p. 479.

Zhang J, Mo Y, Vukmirovic MB, Klie R, Sasaki K, Adzic RR. 2004. Platinum monolayer electrocatalysts for O_2 reduction: Pt monolayer on Pd(111) and on carbon-supported Pd nanoparticles. J Phys Chem B 108: 10955–10964.

Zhang J, Vukmirovic MB, Xu Y, Mavrikakis M, Adzic RR. 2005a. Controlling the catalytic activity of platinum-monolayer electrocatalysts for oxygen reduction with different substrates. Angew Chem Int Ed 44: 2132–2135.

Zhang J, Vukmirovic MB, Sasaki K, Nilekar AU, Mavrikakis M, Adzic RR. 2005b. Mixed-metal Pt monolayer electrocatalysts for enhanced oxygen reduction kinetics. J Am Chem Soc 127: 12480–12481.

Zhang J, Litteer BA, Gu W, Liu H, Gasteiger HA. 2007a. Effect of hydrogen and oxygen partial pressure on Pt precipitation within the membrane of PEMFCs. J Electrochem Soc 154: B1006–B1011.

Zhang J, Sasaki K, Sutter E, Adzic RR. 2007b. Stabilization of platinum oxygen reduction electrocatalysts using gold clusters. Science 315: 220–222.

Zhang L, Lee K, Zhang JJ. 2007. The effect of heat treatment on nanoparticle size and ORR activity for carbon-supported Pd-Co alloy electrocatalysts. Electrochim Acta 52: 3088–3094.

CHAPTER 10

Electrocatalysis at Platinum and Bimetallic Alloys

MASAHIRO WATANABE and HIROYUKI UCHIDA

Clean Energy Research Center, University of Yamanashi, Takeda 4, Kofu 400-8510, Japan

10.1 INTRODUCTION

Polymer electrolyte fuel cells (PEFCs) have attracted great interest as a primary power source for electric vehicles or residential co-generation systems. However, both the anode and cathode of PEFCs usually require platinum or its alloys as the catalyst, which have high activity at low operating temperatures (<100 °C). For large-scale commercialization, it is very important to reduce the amount of Pt used in fuel cells for reasons of cost and limited supply.

In order to obtain high mass activity of Pt, it is essential to disperse Pt or alloy nanoparticles on high surface area supports. Some questions then arise. What kind of alloys and composition should we choose? Is there any good parameter for screening the catalysts? What size of catalyst particles should we prepare to obtain the maximum performance? Unfortunately, there has been much controversy about such issues in the literature.

In order to establish a clear strategy, we have examined the properties of Pt-based catalysts for both the hydrogen oxidation reaction (HOR) and the oxygen reduction reaction (ORR) systematically and comprehensively using various techniques, the results of which complement each other. We have also developed a standard method to evaluate the real activity. In this chapter, we summarize our recent research results.

Fuel Cell Catalysis. Edited by Marc T. M. Koper
Copyright © 2009 John Wiley & Sons, Inc.

10.2 ANODE CATALYSTS FOR CO-TOLERANT HOR

When pure hydrogen is used as the fuel, the overpotential for the hydrogen oxidation reaction (HOR) at the Pt anode is negligibly small.

$$H_2 \longrightarrow 2H^+ + 2e^- \qquad (10.1)$$

However, the Pt anode is seriously poisoned by trace amounts of carbon monoxide in reformates (fuel gas reformed from hydrocarbon), because CO molecules strongly adsorb on the active sites and block the HOR [Lemons, 1990; Igarashi et al., 1993]. Therefore, extensive efforts have been made to develop CO-tolerant anode catalysts and cell operating strategies to suppress CO poisoning, such as anode air-bleeding or pulsed discharging.

To evaluate the catalytic activity or to investigate the reaction mechanism, planar electrodes with well-defined characteristics such as surface area, surface and bulk compositions, and crystalline structure have often been examined in acidic electrolyte solutions. An appreciable improvement in CO tolerance has been found at Pt with adatoms such as Ru, Sn, and As [Watanabe and Motoo, 1975a, 1976; Motoo and Watanabe, 1980; Motoo et al., 1980; Watanabe et al., 1985], Pt-based alloys Pt-M (M = Ru, Rh, Os, Sn, etc.) [Ross et al., 1975a, b; Gasteiger et al., 1994, 1995; Grgur et al., 1997; Ley et al., 1997; Mukerjee et al., 2004], and Pt with oxides (RuO_xH_y) [Gonzalez and Ticianelli, 2005; Sugimoto et al., 2006].

We have found new CO-tolerant catalysts by alloying Pt with a second, nonprecious, metal (Pt-Fe, Pt-Co, Pt-Ni, etc.) [Fujino, 1996; Watanabe et al., 1999; Igarashi et al., 2001]. In this section, we demonstrate the properties of these new alloy catalysts together with Pt-Ru alloy, based on voltammetric measurements, electrochemical quartz crystal microbalance (EQCM), electrochemical scanning tunneling microscopy (EC-STM), in situ Fourier transform infrared (FTIR) spectroscopy, and X-ray photoelectron spectroscopy (XPS).

10.2.1 Screening Tests of Sputtered Pt Alloys by Rotating Disk Electrode

We prepared thin film Pt alloy electrodes by Ar-sputtering Pt and the second metal targets simultaneously onto a disk substrate at room temperature (thickness approximately 200 nm). The resulting alloy composition was determined by gravimetry and X-ray fluorescent analysis (EDX). Grazing incidence ($\vartheta = 1°$) X-ray diffraction patterns of these alloys indicated the formation of a solid solution with a face-centered cubic (fcc) crystal structure.

A standard rotating disk electrode (RDE) setup with a gas-tight Pyrex cell was used for the experiment on CO adsorption and the HOR. A Pt wire was used as counterelectrode. A reversible hydrogen electrode, RHE(t), kept at the same temperature as that of the cell (t, in °C), was used as the reference. All the electrode potentials in this chapter will be referenced to RHE(t). The electrolyte solution of 0.1 M $HClO_4$

was prepared from reagent grade chemicals and Milli-Q water and purified in advance with conventional pre-electrolysis methods [Watanabe and Motoo, 1975b; Uchida et al., 1997].

Before the measurement of HOR activity, a pretreatment of the alloy electrode was carried out by potential sweeps ($10\,V\,s^{-1}$) of 10 cycles between 0.05 and 1.20 V in N_2-purged 0.1 M $HClO_4$. The cyclic voltammograms (CVs) at all the alloys resembled that of pure Pt. As described below, these alloy electrodes were electrochemically stabilized by the pretreatment. Hydrodynamic voltammograms for the HOR were then recorded in the potential range from 0 to 0.20 V with a sweep rate of $10\,mV\,s^{-1}$ in 0.1 M $HClO_4$ saturated with pure H_2 or 100 ppm CO/H_2 at room temperature. The kinetically controlled current I_k for the HOR at 0.02 V was determined from Levich–Koutecky plots [Bard and Faulkner, 1994].

Figure 10.1 shows I_k at various electrodes as a function of CO poisoning time at 26 °C. For the pure Pt electrode, the value of I_k decreases and reaches nearly zero after 30 minutes. In contrast, the Pt-Fe, Pt-Ni, Pt-Co, and Pt-Mo alloys retain high HOR activity for a prolonged period of time; the reduction in I_k is negligibly small. Such CO tolerance of these alloys was found to be almost independent of the composition; for example, alloying Pt with only 5 at% Fe resulted in excellent tolerance. However, Pt alloys with Ti, Cr, Cu, Ge, Nb, Pd, In, Sb, W, Au, Pb, or Bi showed complete CO poisoning after a short time, while the combination of Pt with Mn, Zn, Ag, or Sn exhibited only limited CO tolerance.

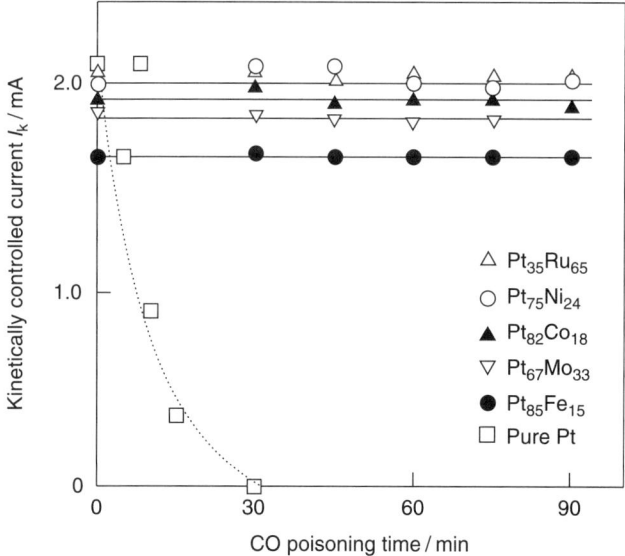

Figure 10.1 Time courses of kinetically controlled currents I_k for the HOR at 0.02 V and 26 °C on various electrodes in 0.1 M $HClO_4$ saturated with 100 ppm CO (H_2 balance). CO was adsorbed on each electrode at 0.02 V under the rotation rate of 1500 rev min^{-1}. (From Igarashi et al. [2001], reproduced by permission of the PCCP Owner Societies.)

We calculated the CO coverage θ_{CO} as the site occupation by the following equation, regardless of the type of adsorbed CO (CO_{ad}), e.g., linear (on-top), bridged, etc.:

$$\theta_{CO} = 1 - \frac{\Delta Q_H}{\Delta Q_H^o} \quad (10.2)$$

where ΔQ_H and ΔQ_H^o are the hydrogen-desorption charges (from 0.05 V to 0.40 V) in the CVs with and without CO_{ad}, respectively. It was found that the values of θ_{CO} on all of the CO tolerant alloys were suppressed to less than 0.6 even after 90 minutes, while the less CO-tolerant alloys and pure Pt were almost completely covered with CO.

The most essential question is why the CO-free sites are secured for H_2 adsorption and oxidation. Watanabe and Motoo proposed a so-called "bifunctional mechanism" originally found at Pt electrodes with various oxygen-adsorbing adatoms (e.g., Ru, Sn, and As), which facilitate the oxidation of adsorbed CO_{ad} at Pt sites [Watanabe and Motoo, 1975a; Watanabe et al., 1985]. This mechanism has been adopted for the explanation of CO-tolerant HOR on Pt-Ru, Pt-Sn, and Pt-Mo alloys [Gasteiger et al., 1994, 1995], and recently confirmed by in situ FTIR spectroscopy [Yajima et al., 2004]. To investigate the role of such surface sites, we examined the details of the alloy surface states by various methods.

10.2.2 Surface States of CO-Tolerant Pt Alloys

10.2.2.1 EQCM Analyses It was suggested by XPS analysis that the nonprecious metals were leached out from the alloy surface during CVs in acidic electrolyte solution. Therefore, we used an EQCM to measure the mass change at the electrode. The resolution and stability of our 10 MHz EQCM was ± 0.1 Hz, i.e., ± 0.44 ng cm^{-2} without any signal averaging.

Figure 10.2 shows the CVs and concomitant mass changes at a working $Pt_{49}Co_{51}$ alloy EQCM electrode during repetitive potential sweeps in 0.1 M $HClO_4$ solution. The CV approaches a steady state after a noticeable decrease in the current density. The final CV shows a feature similar to that of polycrystalline Pt. The electrode mass decreases steeply during the several initial sweeps, and reaches a steady value at the 35th sweep. These results indicate that the Co component was dissolved out from the alloy and the dissolution was suppressed after several potential sweeps, i.e., the electrode was electrochemically stabilized. We observed similar behavior for a Pt-Fe alloy EQCM [Uchida et al., 2002].

If only nonprecious atoms (e.g., Co or Fe) were selectively dissolved from the alloy surface, the roughness of the resulting Pt surface layer would be increased with the number of sweeps. However, as seen in Fig. 10.2a, the value of ΔQ_H^o (a measure of the electrochemical surface area) rather decreases gradually and reaches a steady value.

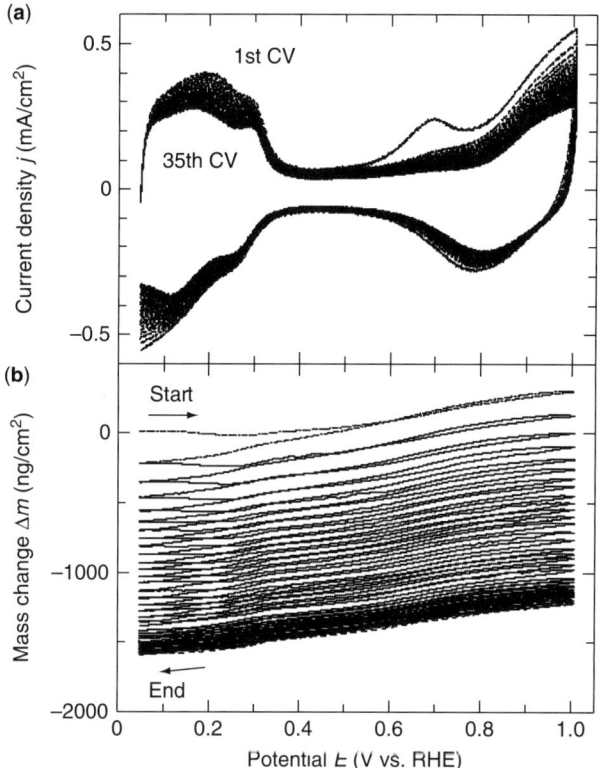

Figure 10.2 (a) Cyclic voltammograms and (b) the simultaneous changes in mass during repetitive potential scans on $Pt_{49}Co_{51}$ alloy EQCM at 25 °C in 0.1 M $HClO_4$ purged with He. Sweep rate 20 mV s^{-1}.

10.2.2.2 EC-STM Observation

This puzzle of the change in the surface state of Pt alloys during the potential sweeps was clearly resolved by EC-STM [Wan et al., 2002]. When 0.1 M $HClO_4$ solution was injected into the EC-STM cell, the surface morphology of the Pt-Co or Pt-Fe film was found to change, resulting in a rough surface with cauliflower-like features that indicated dissolution of the surface alloy layer. However, after several potential sweeps, we observed STM images of a terrace–step structure, distinct from the rough surface. The step lines were found to cross each other at 60° or 120°. The width of the terraces was from several nanometers to more than 10 nm. Figure 10.3a shows an example of an atomic resolution STM image on the terrace. A hexagonal structure can be seen in the image. The interatomic distance was measured to be approximately 0.28 ± 0.02 nm, which is close to the lattice spacing of Pt(111). Hence, it is clear that the dissolution of surface alloy layer was followed by a rearrangement of the Pt layer, resulting in well-ordered Pt(111)-(1×1) structural domains of the Pt skin layer, which can protect underlying bulk alloy.

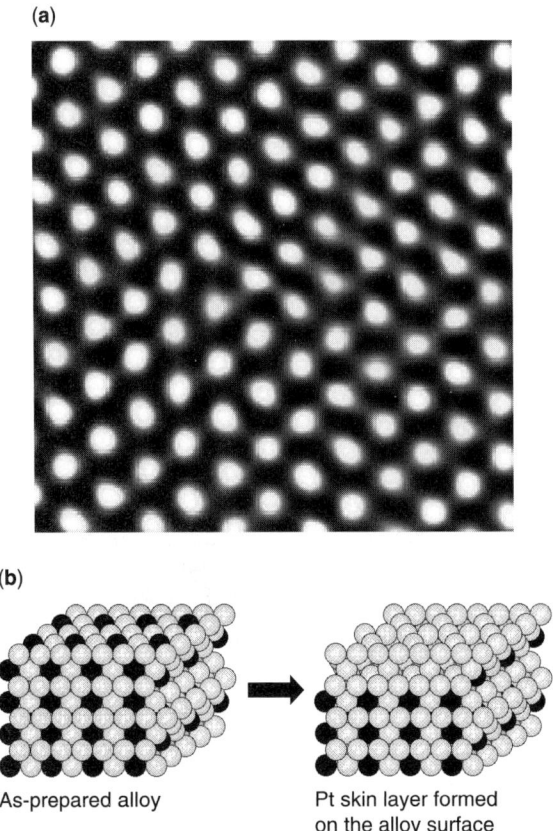

Figure 10.3 (a) An atomic-resolution STM image (2.5 nm × 2.5 nm) of a Pt skin layer formed on $Pt_{55}Co_{45}$ after electrochemical stabilization. The image was observed at 0.05 V in N_2-purged 0.1 M $HClO_4$. The tunneling current and bias voltage were 42.6 nA and −4 mV, respectively. (b) Scheme of formation of Pt skin layer on Pt alloyed with nonprecious metal.

10.2.2.3 Thickness of Pt Skin Layer Analyzed by ICP and EQCM

We determined the thickness of the Pt skin layer formed on the alloy, based on the mass changes from EQCM and the amount of each ion dissolved in the electrolyte solution from an inductively coupled plasma (ICP) analysis. The amount of Pt ions detected by ICP remained at a very low level compared with those of Fe or Co ions. This indicates that most of the Pt ions that might have dissolved out from the alloy were re-deposited on the surface, and the Pt atoms were rearranged during the potential sweeps (Fig. 10.3b). The thickness of the Pt skin layer calculated was about 1–3 nm, corresponding to 4–11 monolayers, respectively, if the layer is oriented preferentially to Pt(111). This is consistent with the order of the value estimated from XPS measurements [Watanabe et al., 1999; Igarashi et al., 2001].

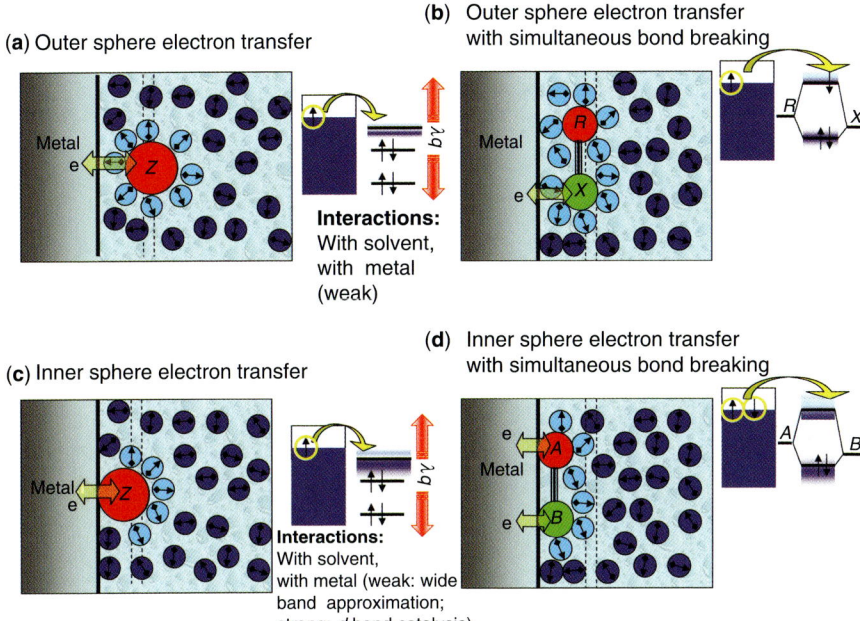

Figure 2.1 Classification of electrochemical electron transfer reaction on metal electrodes.

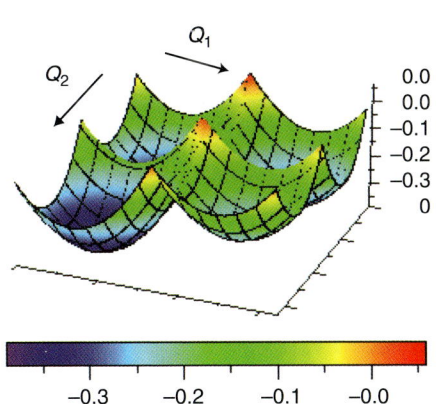

Figure 2.9 Model potential energy surface for combined electron and proton transfer. Q_1 is the solvent coordinate for electron transfer and Q_2 that for proton transfer.

Note: Several figures in this work that appear in black and white throughout this print version may be viewed in color free of charge on ftp://ftp.wiley.com/public/sci_tech_med/advanced_biomaterials

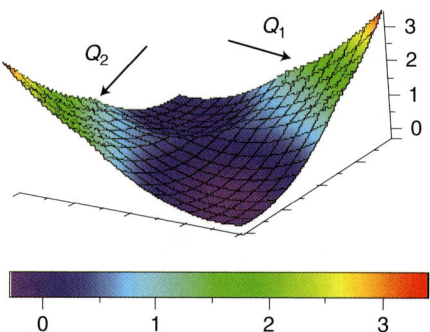

Figure 2.10 Potential energy surface for combined electron and proton transfer.

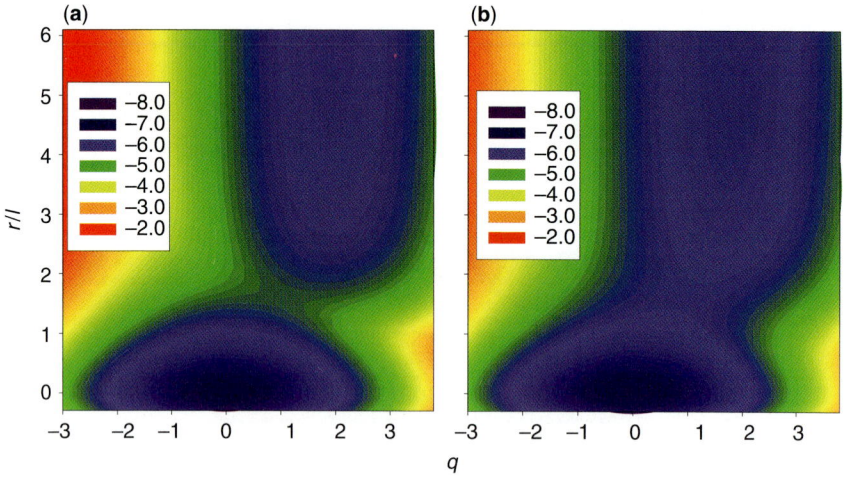

Figure 2.16 Potential energy surfaces for the reduction of a diatomic molecule: (a) without a d-band; (b) in the presence of a d-band. The parameters are $\Delta_{sp}^0 = 0.1$ eV, $\lambda = 0.5$ eV, $A = 4$ eV; in (b), $\Delta_d^0 = 1.0$ eV, $\varepsilon_c = -0.5$ eV, and $w_d = 1$ eV.

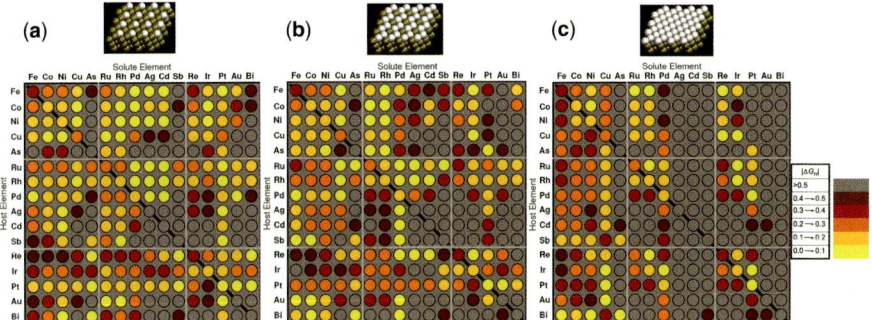

Figure 3.17 Computational high throughput screening for 736 pure metals and surface alloys. The rows indicate the identity of the pure metal substrates, and the columns indicate the identity of the solute embedded in the surface layer of the substrate. The solute coverage is (a) $\frac{1}{3}$ ML, (b) $\frac{2}{3}$ ML, and (c) 1 ML, and the adsorbed hydrogen coverage is also $\frac{1}{3}$ ML. The diagonals of the plots correspond to the hydrogen adsorption free energy on the pure metals. Adapted from [Greeley et al., 2006]; see this reference for more details.

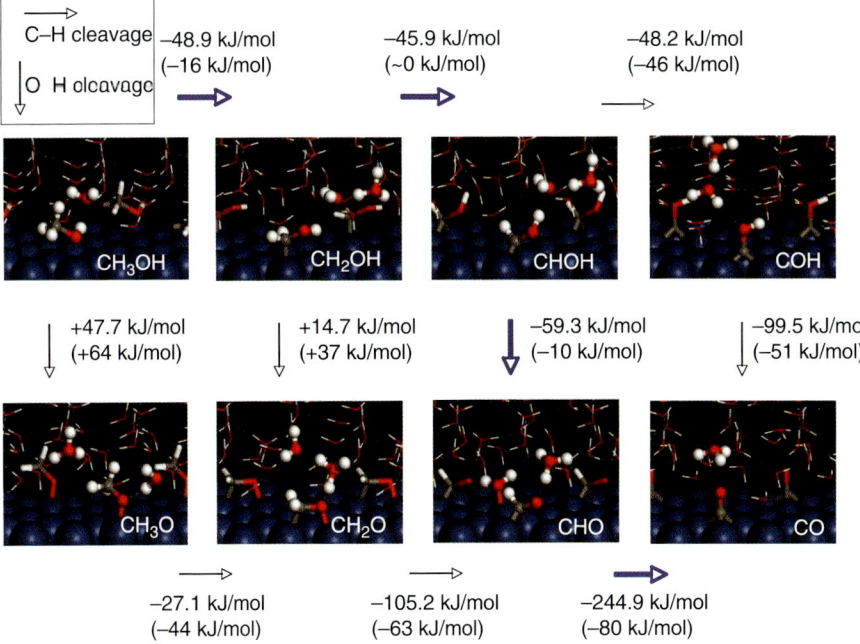

Figure 4.11 Optimized structures of CH_xO species, as indicated, over aqueous-solvated Pt(111) as determined by DFT in Cao et al. [2005]. Horizontal and vertical arrows indicate C—H and O—H cleavage steps, respectively. Reaction energies are included for the aqueous phase [Cao et al., 2005] and the vapor phase (in parentheses) [Desai et al., 2002]. The thermodynamically preferred aqueous phase pathway is indicated by bold arrows (in blue).

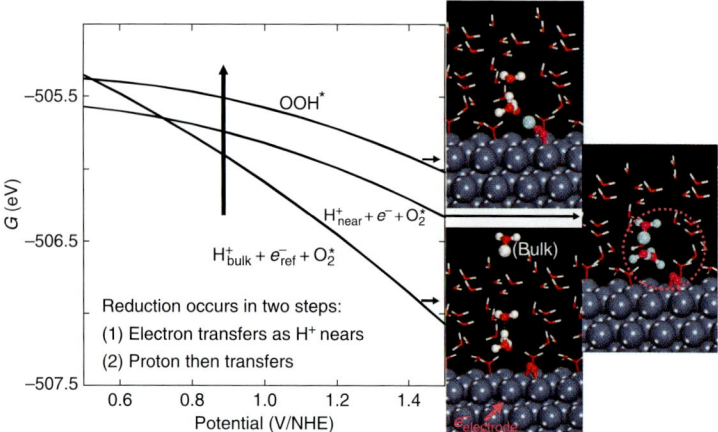

Figure 4.14 Free energy as a function of electrode potential for the species considered in the reduction of O_2^* to form OOH^*. In the atomistic diagrams, purple coloring is used to illustrate where the electron involved in reduction resides in each species.

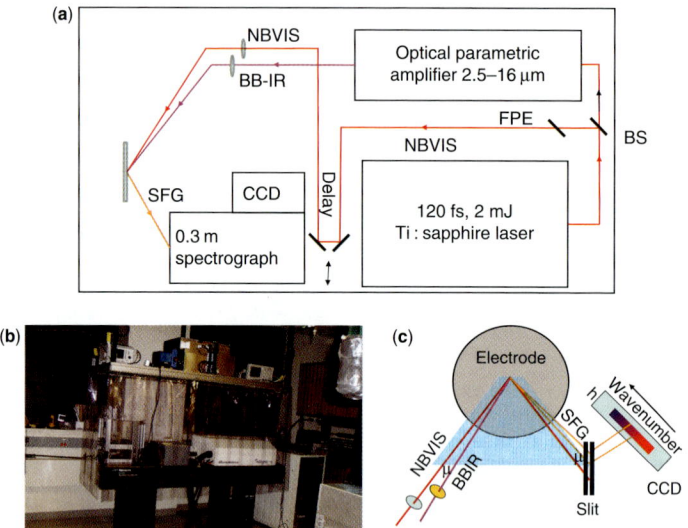

Figure 12.3 (a) Block diagram of the third-generation SFG spectrometer. A beam splitter (BS) directs $\frac{1}{4}$ of the output through a Fabry–Perot etalon (FPE) to produce the narrowband visible (NBVIS) pulses. The remaining output pumps an optical parametric amplifier to generate the broadband IR pulses (BBIR). (b) Photograph of the SFG apparatus on a 2 m × 0.8 m optical table. The laser, spectrograph, and sample chamber are shown in the foreground. (c) Schematic of the arrangement for SFG phase-matching. Different BBIR frequencies result in slightly different SFG output angles, but all SFG signals pass through the spectrograph slit and illuminate slightly different height (h) regions of the CCD. Output spectra are obtained by integrating along the CCD h-direction.

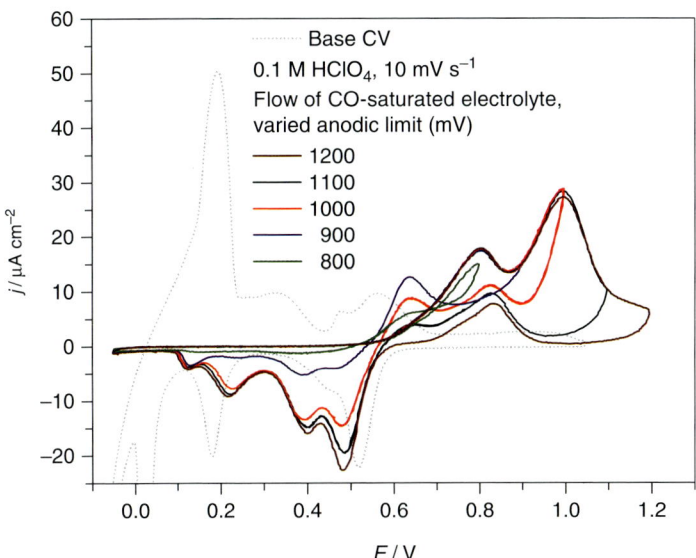

Figure 14.9 CO bulk electro-oxidation at bare Ru(0001) in flow cell; dotted line, CO free electrolyte; solid lines: flow of CO saturated electrolyte, with varied upper scan limits (see key on figure).

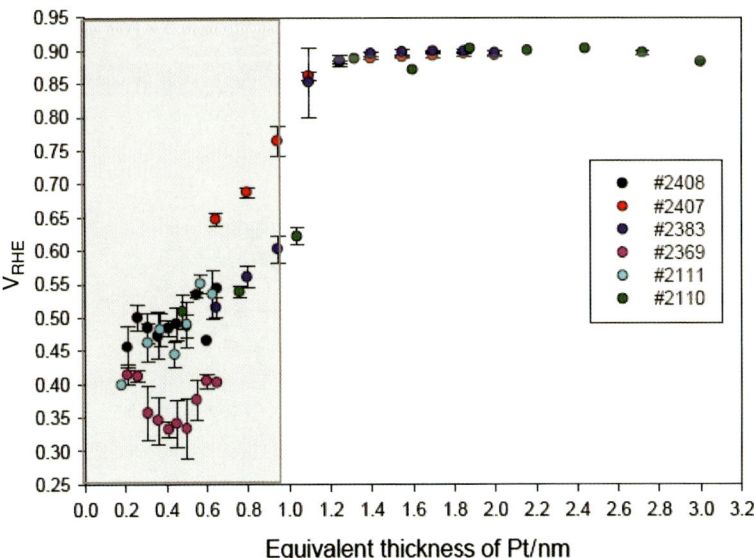

Figure 16.8 Pt/TiO$_x$-catalyzed oxygen reduction potential, where 0.01 mA cm^{-2} is reached during the negative scan in a cyclic voltammetry experiment (scan rate 20 mV s^{-1}) in oxygen-saturated 0.5 M HClO$_4$ at 25 °C.

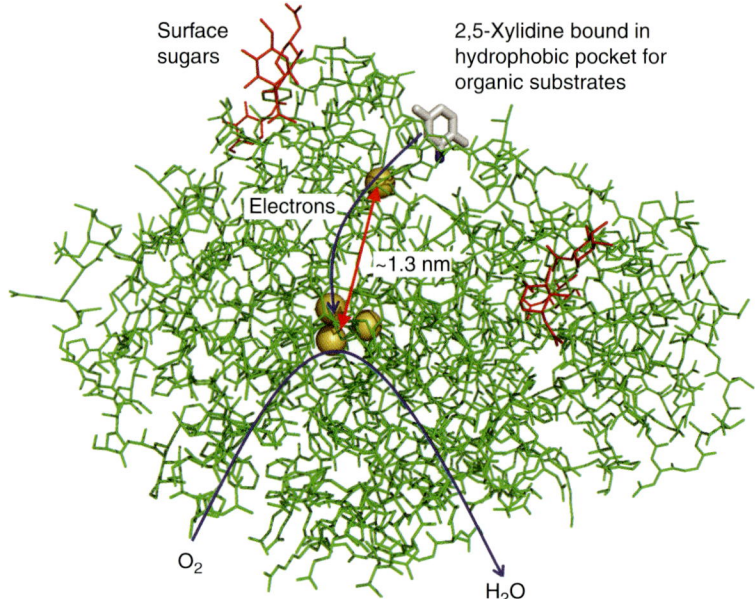

Figure 17.3 Anatomy of a redox enzyme: representation of the X-ray crystallographic structure of *Trametes versicolor* laccase III (PDB file 1KYA) [Bertrand et al., 2002]. The protein is represented in green lines and the Cu atoms are shown as gold spheres. Sugar moieties attached to the surface of the protein are shown in red. A molecule of 2,5-xylidine that co-crystallized with the protein (shown in stick form in elemental colors) is thought to occupy the broad-specificity hydrophobic binding pocket where organic substrates are oxidized by the enzyme. Electrons from substrate oxidation are passed to the mononuclear "blue" Cu center and then to the trinuclear Cu active site where O_2 is reduced to H_2O.

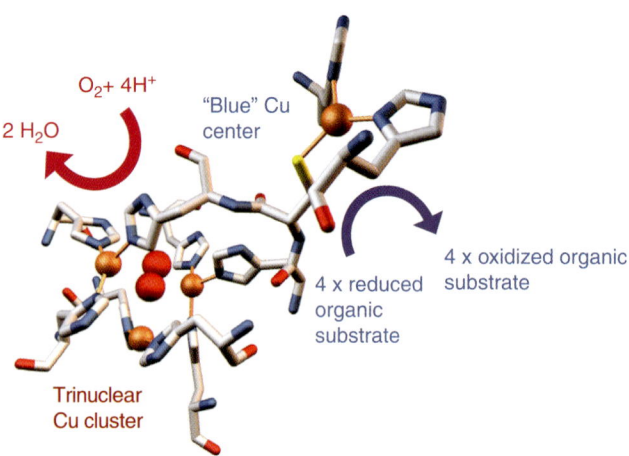

Figure 17.5 The protein environment around the Cu centers (gold spheres) of laccase from *Melanocarpus albomyces* (PDB file 1GW0) showing a substrate O_2 molecule bound in the trinuclear Cu site [Hakulinen et al., 2002]. The protein is depicted in stick representation with atoms in their conventional coloring. (Courtesy of Armand W. J. W. Tepper.)

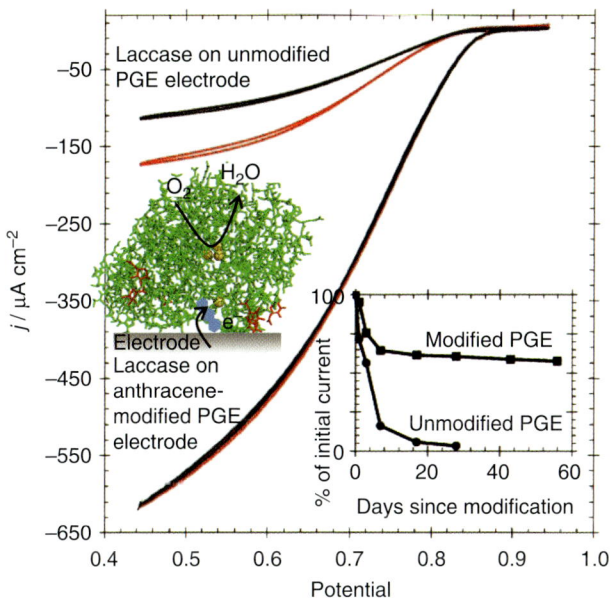

Figure 17.7 Electrocatalysis of O_2 reduction by *Pycnoporus cinnabarinus* laccase on a 2-aminoanthracene-modified pyrolytic graphite edge (PGE) electrode and an unmodified PGE electrode at 25 °C in sodium citrate buffer (200 mM, pH 4). Red curves were recorded immediately after spotting laccase solution onto the electrode, while black curves were recorded after exchanging the electrochemical cell solution for enzyme-free buffer solution. Insets show the long-term percentage change in limiting current (at 0.44 V vs. SHE) for electrocatalytic O_2 reduction by laccase on an unmodified PGE electrode (●) or a 2-aminoanthracene modified electrode (■) after storage at 4 °C, and a cartoon representation of the probable route for electron transfer through the anthracene (shown in blue) to the "blue" Cu center of laccase. Reproduced by permission of The Royal Society of Chemistry from Blanford et al., 2007.

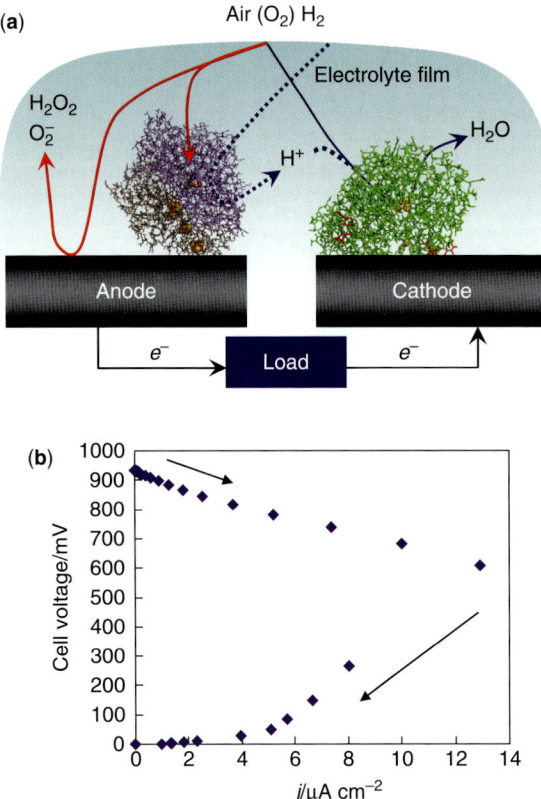

Figure 17.20 Use of enzyme catalysts in a membraneless fuel cell that operates on low levels of H_2 in air [Vincent et al., 2006]. (a) Schematic representation of the fuel cell. PGE electrodes modified with hydrogenase and laccase are inserted into a shallow tray of aqueous electrolyte (pH 5 citrate solution) in contact with an atmosphere of 3% H_2 in air. Blue lines indicate productive reactions: electrocatalytic H_2 oxidation at the anode and O_2 reduction at the cathode. Red lines indicate unproductive reactions: consumption of electrons at the anode by direct reduction of O_2 at bare graphite, generating radical species that may damage the enzymes, or reversible inhibition of hydrogenase by access of O_2 to the active site. (b) Typical cell voltage/current plot for this fuel cell obtained by applying a variable load, showing the rapid drop in current observed at high cell voltages, consistent with high potential inactivation of the hydrogenase at the anode.

10.2.3 Modified Electronic Structure of Pt Skin Layer and Pt-Ru Alloy

According to the EQCM and EC-STM results, we cannot invoke the bifunctional mechanism for CO tolerance of the present alloys, because of the absence of atoms of the second metal in the surface layer. It should also be noted that CO_{ad} cannot be oxidized at the HOR potential $E < 0.1$ V even on a Pt-Ru alloy, as shown below. In contrast, we found very interesting features in the CO adsorption on these alloys distinct from those of pure Pt. The first is the type of CO_{ad}. Using in situ FTIR spectroscopy at electrodes with $\theta_{CO} < 0.6$, linearly adsorbed CO was found to be dominant at Pt alloys, whereas a large fraction of CO adsorbed on pure Pt is in the bridged form, by which the HOR active sites were blocked more dramatically [Watanabe et al., 2000; Igarashi et al., 2001]. The same tendency was also confirmed at 30–70 °C based on CV analysis [Uchida et al., 2006]. Another important point is the deceleration of the CO adsorption rate on the alloys [Uchida et al., 2006]. These results strongly suggest a modified electronic structure of these alloys.

Here, we demonstrate clear and direct evidence for the modified electronic structures of surface Pt atoms in Pt-Co and Pt-Ru by using EC-XPS [Wakisaka et al., 2006]. The sample electrode was transferred between an XPS chamber and an electrochemical (EC) chamber without exposure to air (to minimize contamination of the surface). All photoelectron spectra, including the valence level region) were taken by using a monochromatic Al Kα ($h\nu = 1486.58$ eV). The uncertainty of binding energy measurement was less than ± 0.03 eV.

10.2.3.1 Electronic Structures of Pt in As-Prepared Alloys

Figure 10.4a shows Pt$4f_{7/2}$ core level (CL) spectra for pure Pt, Pt$_{58}$Co$_{42}$, and Pt$_{60}$Ru$_{40}$ as-prepared alloys (before electrochemical stabilization). The CLs for both Pt-Co and Pt-Ru alloys clearly shift to higher binding energy with respect to pure Pt. The magnitude of the CL shift for the Pt-Ru alloy was 0.39 eV, which is larger than the 0.19 eV for the Pt-Co alloy. A positive CL shift of an atom has been generally interpreted as electron loss from the atom [Briggs and Seah, 1990]. However, since the order of the work functions is Pt > Ru > Co, the electron transfer must occur from Ru or Co to Pt. The positive Pt$4f_{7/2}$ CL shift in the XPS is rather explained by re-hybridization of the d-band as well as the sp-band [Weinert and Watson, 1995] as follows. The change in the work function leads to a reference level (E_F) shift in the XPS measurement. The negative shift of E_F results in a positive shift of the Pt$4f_{7/2}$ CL as well as the d-band center. The observed values of the CL shifts for Pt$_{58}$Co$_{42}$ and Pt$_{60}$Ru$_{40}$ were in good agreement with those calculated from the electronic parameters of the pure elements given in the literature [Miedema et al., 1980].

10.2.3.2 Electronic Structures of Pt after Electrochemical Stabilization

Each test electrode was transferred to the EC chamber and subjected to electrochemical stabilization. The EC chamber was then evacuated rapidly by two sorption pumps and a cryopump to transfer the electrode to the XPS chamber again.

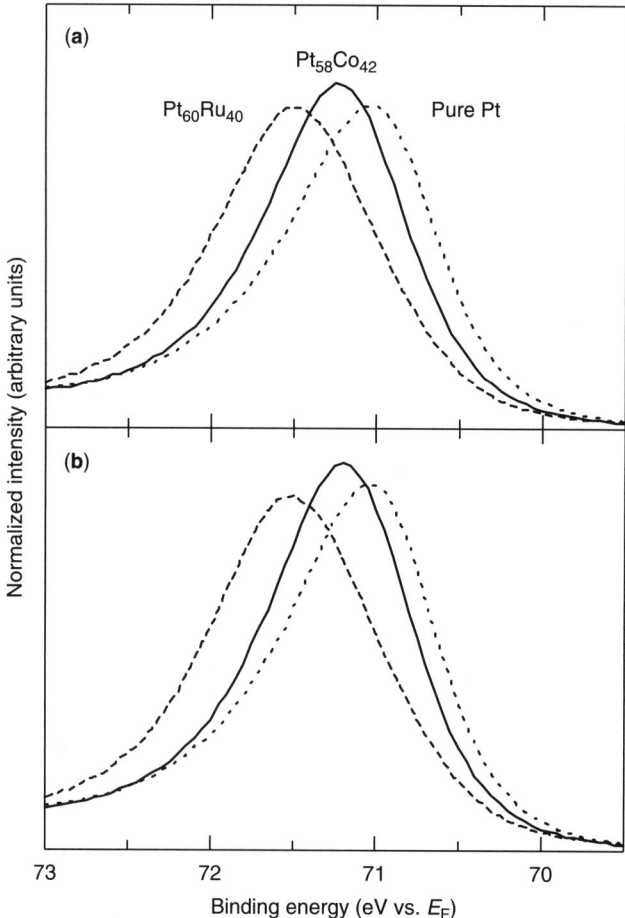

Figure 10.4 Area-normalized CL spectra of Pt$4f_{7/2}$ for the pure Pt (dotted line), Pt$_{58}$Co$_{42}$ (solid line), and Pt$_{60}$Ru$_{40}$ (dashed line) alloys with respect to E_F: (a) as-prepared; (b) after electrochemical stabilization. The samples were thin film pure Pt or Pt-based alloys (diameter 8 mm and thickness 80 nm) prepared on Au disks by DC sputtering. Electrochemical stabilization of Pt$_{58}$Co$_{42}$ was performed by repeated potential cycling between 0.075 and 1.00 V at a sweep rate of 0.10 V s^{-1} in 0.1 M HClO$_4$ under ultrapure N$_2$ (99.9999%) until CV showed a steady state. Pt$_{60}$Ru$_{40}$ was stabilized by several potential cycling between 0.075 and 0.80 V at 0.10 V s^{-1} in 0.05 M H$_2$SO$_4$ under ultrapure N$_2$. (From Wakisaka et al. [2006], reproduced by permission of the American Chemical Society.)

It was found that the intensity of Co$2p_{3/2}$ decreased significantly (by a factor of 2.5), supporting the concept of Co dissolution from the alloy and formation of the Pt skin layer on the electrode surface during electrochemical stabilization. As shown in Fig. 10.4b, a clear CL shift was still observed in the Pt$4f_{7/2}$ spectrum for the stabilized Pt-Co, in spite of the dissolution of Co, although the CL shift after stabilization was slightly smaller (0.15 eV) than in the as-prepared alloy (0.19 eV). Thus, we have confirmed the CL shift for Pt-Co, even after electrochemical stabilization, by using EC-XPS under the cleanest conditions.

For the $Pt_{60}Ru_{40}$ alloy, however, the intensity of the $Ru3d_{5/2}$ and the binding energy of the $Pt4f_{7/2}$ CL were unchanged after the potential sweeps. Therefore, it can be concluded that the surface composition of the $Pt_{60}Ru_{40}$ alloy was maintained without any Ru dissolution in 0.05 M H_2SO_4 solution during the potential sweeps up to 0.80 V.

10.2.3.3 Change in the Electronic Structure of Pt by CO Adsorption

Figure 10.5 shows CO stripping voltammograms at these electrodes in N_2-purged electrolyte solutions measured in the EC chamber. The complete suppression of the hydrogen desorption current at all the electrodes indicated a surface saturated with CO. In the positive-going potential scan at the pure Pt electrode, a pre-oxidation peak is seen between 0.50 and 0.70 V before a major CO oxidation peak at 0.73 V. At the stabilized Pt-Co alloy electrode, the pre-oxidation peak is seen between 0.35 and 0.60 V, but the major CO oxidation peak appeared at the same potential as that on the pure Pt. The onset potential for CO oxidation at the Pt-Ru alloy (0.35 V) is the same as at Pt-Co, but the current rises steeply to reach the major peak at a lower potential (0.52 V) than on Pt-Co or pure Pt. This is because CO oxidation at Pt sites is facilitated by oxygen species adsorbed at Ru sites, i.e., through the bifunctional mechanism. However, for the CO-tolerant HOR in the low potential region ($E <$ 0.1 V), the CO_{ad} molecules are never oxidized, even at Pt-Ru, indicating that the activity for the CO_{ad} oxidation cannot be directly related to the CO-tolerant HOR activity at such a low potential. Therefore, we focus on the interaction between CO_{ad} and these electrodes.

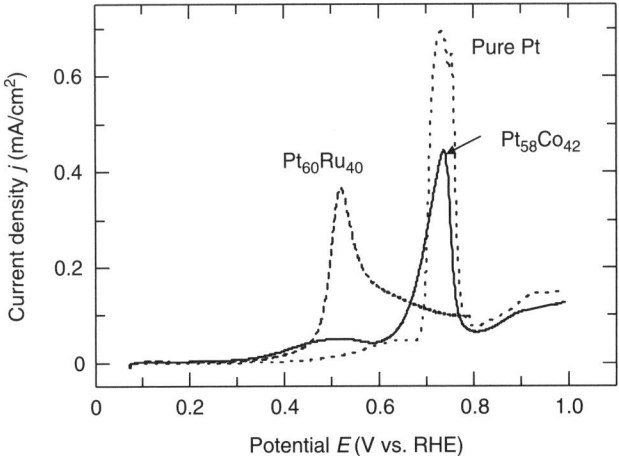

Figure 10.5 CO-stripping voltammograms (sweep rate 0.10 V s^{-1}) at pure Pt and stabilized $Pt_{58}Co_{42}$ in 0.1 M $HClO_4$, and $Pt_{60}Ru_{40}$ in 0.05 M H_2SO_4 at room temperature. The CO pre-adsorption was performed in CO-saturated solution at 0.075 V for 5 minutes. (From Wakisaka et al. [2006], reproduced by permission of the American Chemical Society.)

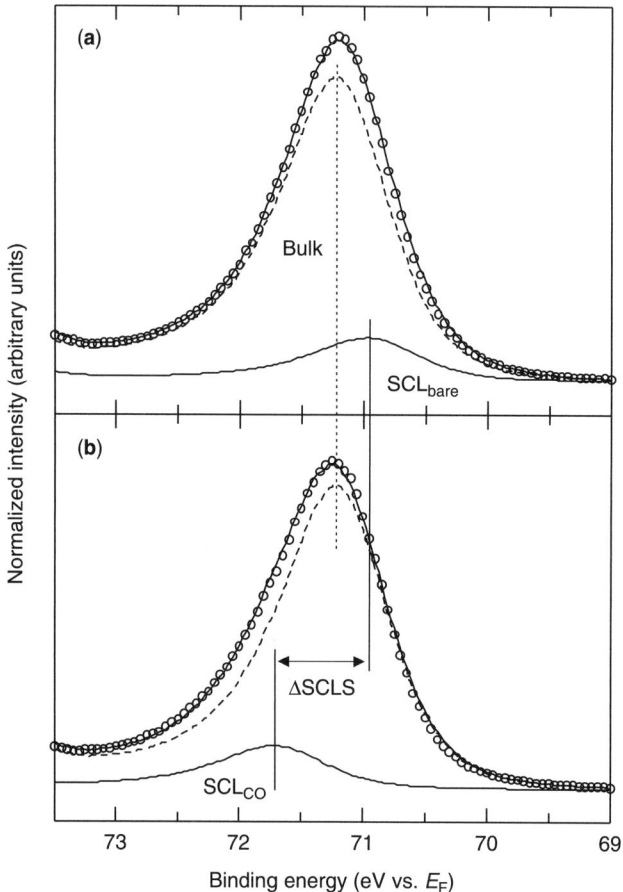

Figure 10.6 Decomposition of Pt$4f_{7/2}$ spectra (open circles) into surface and bulk core levels (solid and dashed lines), respectively, for stabilized Pt$_{58}$Co$_{42}$ alloy (a) before and (b) after CO adsorption. (From Wakisaka et al. [2006], reproduced by permission of the American Chemical Society.)

Figure 10.6 shows CL spectra of Pt$4f_{7/2}$ for stabilized Pt-Co with and without CO$_{ad}$. The CO adsorption induced both a shift in the Pt$4f_{7/2}$ CL to higher binding energy and an increase in the full width at half maximum (FWHM). Such changes can be explained by surface core level (SCL) shifts of Pt$4f_{7/2}$ by CO$_{ad}$, whereas the bulk CL is not affected by CO$_{ad}$. In order to extract the change in SCL shift (ΔSCLS) by CO$_{ad}$, we decomposed the Pt$4f_{7/2}$ spectra into two components: the bulk CL and SCL. It was found that the value of ΔSCLS decreased in the order pure Pt > stabilized Pt-Co > Pt-Ru.

The CO adsorption energy (the energy required to break the Pt–CO bond), E_{ads}(CO/Pt), was related to ΔSCLS by the following equation [Treglia et al., 1981;

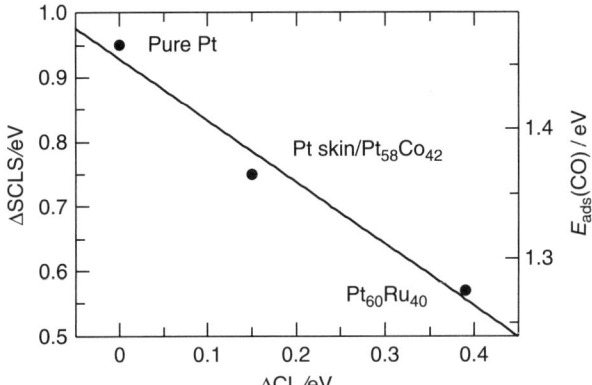

Figure 10.7 Linear relation between CL shifts (ΔCL) and ΔSCLS or CO adsorption energy E_{ads}(CO). (From Wakisaka et al. [2006], reproduced by permission of the American Chemical Society.)

Shek et al., 1982; Björneholm et al., 1994; Cabeza et al., 1999]:

$$E_{ads}(CO/Pt) = \Delta SCLS + E_{ads}(CO/Au) \tag{10.3}$$

Figure 10.7 shows a plot of ΔSCLS, as a measure of E_{ads}(CO), as a function of ΔCL. E_{ads}(CO) is found to decrease linearly with increasing ΔCL. The Pt–CO bond strength is evidently weaker on the alloy electrodes than on pure Pt. Thus, the HOR cannot be hindered by such a weakly bonded CO at the alloys. It should be noted that Ru is a more effective element in weakening the bond strength than Co. Because Pt-Ru exhibited a larger ΔCL than as-prepared Pt-Co (with Co remaining near the Pt site), as evidenced in Fig. 10.4a, the exact location of such second metals is not essential for the modification, as long as the skin layer is thin enough.

This is the first experimental demonstration of changes in the strength of CO adsorption at Pt-based alloy electrodes. Nørskov and co-workers theoretically predicted a similar linear relation between changes in E_{ads}(CO) and shifts in the d-band center [Hammer et al., 1996; Hammer and Nørskov, 2000; Ruban et al., 1997]. Because the Pt$4f_{7/2}$ CL shift due to alloying can be more easily measured by XPS than the d-band center can, this should be one of the most important parameters to aid in discovering CO-tolerant anode catalysts among Pt-based alloys or composites.

10.2.4 Temperature Dependence of CO-Tolerant HOR Activity at Pt, Pt-Co, and Pt-Ru

Besides using Pt-Ru or Pt-Co alloy anodes, CO poisoning can be mitigated by elevating the operating temperature. However, temperature dependencies of the HOR rates in the presence of CO with relevance to PEFC operation have been scarcely reported. One of the difficulties is correction of the change in H_2 concentration [H_2] in the electrolyte solution at an RDE cell open to the atmosphere at elevated temperatures. In this section, we report the kinetically controlled current density j_k for the HOR at Pt, Pt-Co, and Pt-Ru planar electrodes with and without CO poisoning as a function of temperature from 30 to 90 °C [Uchida et al., 2006]. j_k is considered as a clear target that one can achieve in the ideal case.

We employed a channel flow cell, which can be operated under a closed system with controlled [H_2] and can provide precise j_k from hydrodynamic voltammograms. The working electrode of Pt, Pt-Co, or Pt-Ru was prepared on a mirror-finished Au substrate (1 mm × 4 mm) by DC sputtering. The active area of the working electrode S_{Pt}^o was determined from the ΔQ_H^o in the CV in N_2-purged 0.1 M $HClO_4$ solution, where the superscript "o" denotes the CO-free surface. The CO coverage θ_{CO} was calculated from Equation (10.2) after CO was adsorbed at 0.05 V by supplying a solution saturated with 0.1% CO (H_2 balance) for various time intervals. Hydrodynamic voltammograms for the HOR at various working electrodes were recorded under a flow of H_2-saturated (CO-free) solution (mean flow rate $U_m = 10-50$ cm s^{-1}) from 0 to 0.10 V at the sweep rate of 0.5 mV s^{-1}.

The kinetically controlled current I_k at 0.02 V was determined from a well-defined equation [Levich, 1962; Gerischer et al., 1965], i.e., plotting the inverse of the current I^{-1} against $U_m^{-1/3}$ and extrapolating $U_m^{-1/3}$ to zero. j_k was obtained as I_k/S_{Pt}^o. The value of the apparent rate constant k_{app} for the HOR was calculated from the equation

$$j_k/2F = k_{app}[H_2] \tag{10.4}$$

where [H_2] was determined from the limiting current and the diffusion constant. Figure 10.8 shows Arrhenius plots of k_{app} at 0.02 V for CO-free electrodes. The

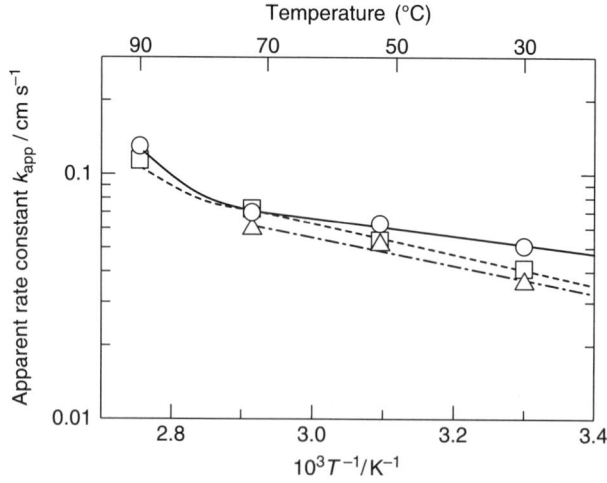

Figure 10.8 Arrhenius plots for the apparent rate constant k_{app} for the HOR (CO-free) at Pt (O), Pt$_{51}$Co$_{49}$ (Δ), and Pt$_{54}$Ru$_{46}$ ($\square$) working electrodes at 0.020 V vs. RHE(t). (From Uchida et al. [2006], reproduced by permission of the American Chemical Society.)

value of k_{app} at a pure Pt electrode is the highest in the whole temperature range. Pt-Co and Pt-Ru exhibited almost comparable k_{app} at 30–50 °C. The apparent activation energies at Pt, Pt-Co, and Pt-Ru at 30–70 °C were as low as 3.0, 4.2, and 5.3 kJ mol^{-1}, respectively. Hence, elevating the operation temperature does not lead to significant enhancement in HOR kinetics. For example, k_{app} for the HOR on Pt at 90 °C was two times higher than that at 30 °C, whereas k_{app} for the ORR on Pt increased by about one order of magnitude from 30 to 90 °C [Wakabayashi et al., 2005a], as described in the next section.

Next, the j_k values in the presence of CO_{ad} at these electrodes are compared. Figure 10.9 shows the current per remaining hydrogen adsorption site free from CO_{ad}, $j_k^{HA} = j_k/(1 - \theta_{CO})$, as a function of θ_{CO}. If the HOR active sites are tightly blocked by the immobile CO_{ad}, the value of j_k^{HA} should be a constant independent of θ_{CO}. j_k^{HA} at pure Pt increases slightly up to $\theta_{CO} = 0.6$, but it decreases steeply at high θ_{CO}. Such an increase in j_k^{HA} at low θ_{CO} suggests that Pt sites were blocked not rigidly by CO_{ad} at least for $\theta_{CO} < 0.6$, resulting in apparent active sites more than that expected from θ_{CO}. The steep decrease in j_k or j_k^{HA} at high $\theta_{CO} > \frac{2}{3}$ is certainly due to a decrease in nearest neighbor CO-free pair sites required for H_2 dissociative adsorption in the rate-determining Tafel step [Conway and Tilak, 2002; Uchida et al., 2006].

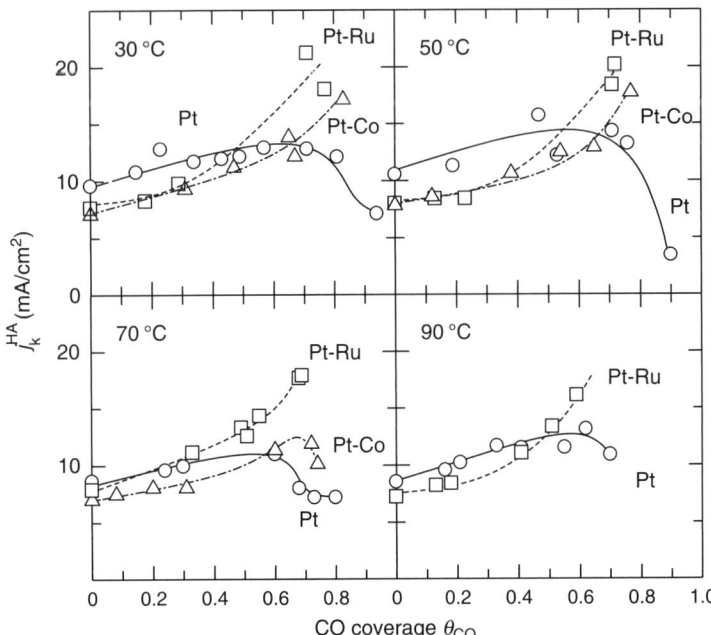

Figure 10.9 Dependence of $j_k^{HA} = j_k/(1 - \theta_{CO})$ at 0.020 V on θ_{CO} for Pt, $Pt_{51}Co_{49}$, and $Pt_{54}Ru_{46}$ electrodes in H_2-saturated 0.1 M $HClO_4$ solution. (From Uchida et al. [2006], reproduced by permission of the American Chemical Society.)

It is striking that j_k^{HA} at Pt-Ru increases with θ_{CO} at all temperatures, doubling in value at $\theta_{CO} = 0.7$ compared with $\theta_{CO} = 0$. The dependence of j_k^{HA} on θ_{CO} at Pt-Co is also similar to that at Pt-Ru up to 50 °C. This indicates that the HOR sites on Pt-Ru and Pt-Co are not so rigidly blocked by CO_{ad}, owing to its enhanced mobility, as indicated by EC-XPS experiments (Section 10.2.3).

However, the CO tolerance at Pt-Co degraded at 70 °C. As seen in Fig. 10.9, the HOR activity of Pt-Co at a given θ_{CO} is close to that of pure Pt, although the deceleration effect on the CO adsorption rate was still observed to some extent at 70 °C. Such a "deactivated" electrode cannot recover the original CO tolerance. This can certainly be ascribed to a severe dealloying of the nonprecious metal component (Co) in hot acid solution. We will discuss this in Section 10.3.2.

10.3 CATHODE CATALYSTS FOR THE ORR

It is very important to develop a high performance cathode catalyst, because a sluggish ORR causes a large overpotential at low temperatures. With respect to the total performance of activity and stability, the cathode catalyst material is limited to Pt or its alloys at present. In acidic media such as Nafion electrolyte or aqueous acid solutions, four-electron reduction is dominant at Pt-based electrodes:

$$O_2 + 4H^+ + 4e^- \longrightarrow 2H_2O \tag{10.5}$$

However, two-electron reduction occurs fractionally to produce H_2O_2, which may cause deterioration of the polymer electrolyte membrane or gaskets:

$$O_2 + 2H^+ + 2e^- \longrightarrow H_2O_2 \tag{10.6}$$

The ORR has been studied with a rotating ring–disk electrode (RRDE), which can provide the j_k and the H_2O_2 yield $P(H_2O_2)$ at around room temperature. However, for improving the ORR activity, PEFCs should be operated at high temperature (>80 °C). In this section, we demonstrate the temperature dependencies of ORR activity and $P(H_2O_2)$ at pure Pt (both bulk and supported catalyst) and bulk Pt alloys (Pt-Ni, Pt-Co, and Pt-Fe).

10.3.1 Particle Size Effect on the ORR at Pt/CB Catalysts

The mass activity MA (in A g^{-1}) of the Pt catalyst is, of course, the product of the specific activity j_s (in A m^{-2}) and the specific surface area S_{mass} (in m^2 g^{-1}): MA = $j_s S_{mass}$. Because S_{mass} is inversely proportional to the particle diameter d_{Pt}, the use of supported Pt nanoparticles is effective for increasing MA, if j_s is a constant independent of d_{Pt}. However, even at pure Pt, conflicting results on the values of j_s and $P(H_2O_2)$ have been reported, suggesting the presence of differences in electrochemical properties between bulk and supported nanoparticles. For example, Bregoli [1978] stated that j_s for the ORR decreased with increasing S_{mass} from 20 to 80 m² g⁻¹ (corresponding to a decrease in d_{Pt} from about 14 nm to 3.5 nm). Ross and co-workers observed a similar but more pronounced effect in the region of S_{mass} greater than 80 m² g⁻¹ with $d_{Pt} < 3$ nm [Ross, 1986; Sattler and Ross, 1986]. This so-called "particle size effect" might be attributed to an electronic effect [Mukerjee and McBreen, 1989; Takasu et al., 1996] or to a morphological effect (exposed facet dependent on d_{Pt}) [Kinoshita, 1990]. In contrast, Watanabe and co-workers found that j_s was independent of d_{Pt}, but did depend on the interparticle distance [Watanabe et al., 1988, 1989]. We consider that this inconsistency is partly due to inappropriate evaluation techniques for j_s or MA. We have therefore made a great effort to establish standard methods of evaluating the real activities [Higuchi et al., 2005; Wakabayashi et al., 2005a, b; Yano et al., 2006a].

In this section, we demonstrate the real ORR activities (apparent rate constant per real active surface area, k_{app}) and $P(H_2O_2)$ at bulk Pt and nanosized Pt catalysts dispersed on carbon black (Pt/CB) with $d_{Pt} = 1.6 \pm 0.4$, 2.6 ± 0.7, and 4.8 ± 1.0 nm in the practical temperature range 30–110 °C [Yano et al., 2006b]. The use of a channel flow double-electrode (CFDE) cell allowed us to evaluate k_{app} and $P(H_2O_2)$ precisely.

Details of the experimental set-up for the ORR measurements using the CFDE cell and a flow circuit of electrolyte solution are described in the literature [Wakabayashi et al., 2005a, b; Yano et al., 2006a]. The working electrode consisted of Pt/CB catalysts perfectly dispersed on a gold substrate electrode (1 mm × 4 mm) at a constant loading of the carbon support, 5.45 μg cm⁻², which corresponds to approximately a monolayer of CB particles. Nafion solution 0.2 wt% was put on top of the catalyst layer to yield an average film thickness of 0.1 μm, which is thinner than the critical value determined for the ORR [Higuchi et al., 2005]. Finally, Nafion–Pt/CB on gold was heated at 130 °C for 30 minutes in air. Hydrodynamic voltammograms at the working electrode under a flow of O_2-saturated 0.1 M HClO₄ solution ($U_m = 10$–50 cm s⁻¹) were recorded by scanning its potential from 0.3 to 1.0 V at 0.5 mV s⁻¹.

10.3.1.1 H₂O₂ Production Rate

Based on the collection efficiency N at the Pt collecting electrode located downstream from the working electrode, we calculated the ratio of the H₂O₂ production rate to the overall ORR rate, $P(H_2O_2)$:

$$P(H_2O_2) = \frac{2I_C}{NI_W + I_C} \times 100\% \quad (10.7)$$

where I_W and I_C are the currents at the working and collecting electrodes, respectively. Because we found that Nafion-coated carbon black (without Pt) and Au substrate produced H₂O₂ with $P(H_2O_2) = 98\%$ only at potentials lower than 0.65 V, the H₂O₂ production at $E > 0.7$ V was ascribed completely to the Nafion–Pt/CB interface. As shown in Fig. 10.10, the values of $P(H_2O_2)$ at all Nafion–Pt/CB are almost the same as those at Nafion–Pt(bulk) within experimental error, irrespective of d_{Pt} and Pt loading. $P(H_2O_2)$ increases slightly on lowering the potential, i.e., 0.6% and 1.0% at 0.80 and 0.70 V, respectively, and is almost independent of the temperature from 50 to 110 °C. Considering $P(H_2O_2) = 0$ at Pt(bulk) (without

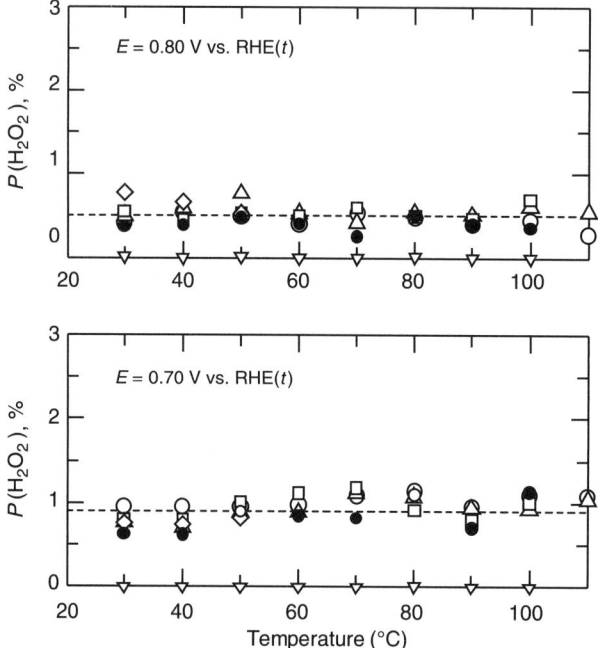

Figure 10.10 Temperature dependence of H_2O_2 yield $P(H_2O_2)$ at Nafion–Pt(4.8 nm)/CB (○), Nafion–Pt(2.6 nm)/CB (△), Nafion–Pt(1.6 nm)/CB (◇), Nafion–Pt(2.6 nm)/CB with the same Pt loading as 1.6 nm catalyst (□), Nafion–Pt(bulk) (●), and Pt(bulk) without Nafion coating (▽). $U_m = 50$ cm s^{-1}. (From Yano et al. [2006b], reproduced by permission of the PCCP Owner Societies.)

Nafion) [Wakabayashi et al., 2005a], the Nafion coating on Pt is the major reason for triggering H_2O_2 production in a practical potential region $E > 0.70$ V. It has been reported that ΔQ_H^o at Nafion–Pt(bulk) decreases by about 10% compared with a Pt(bulk) (without Nafion) electrode [Lawson et al., 1988; Chu et al., 1989; Zecevic et al., 1997; Chu, 1998]. This reduction is probably due to a specific adsorption of sulfonate groups in Nafion, as observed by in situ FTIR spectroscopy [Ayato et al., 2006]. Therefore, sulfonate groups in Nafion could be the species strongly adsorbed on Pt surface, modifying the surface property.

10.3.1.2 Arrhenius Plots of k_{app} The value of I_k at a given potential E is determined in the same manner as described in Section 10.2.4. Because the contribution of two-electron reduction to the production of H_2O_2 (compared with the overall ORR) was very low, we can evaluate an apparent rate constant k_{app} at a constant applied potential from the following equation [Wakabayashi et al., 2005a, b]:

$$\frac{I_k}{4FS_{Pt}^o} = -k_{app}[H^+][O_2] \tag{10.8}$$

where [H$^+$] is the bulk concentration of H$^+$ (0.1 M) and S_{Pt}^o is the active area of Pt, electrochemically determined with CV in N$_2$-purged 0.1 M HClO$_4$ at 30 °C. It was found that the values of S_{Pt}^o of d_{Pt} = 4.6 and 2.6 nm catalysts were unchanged even after measurement at 110 °C. Although S_{Pt}^o at the 1.6 nm catalyst was almost constant after measurement at 50 °C, it decreased significantly once heated to a temperature higher than 60 °C. It was also observed by transmission electron microscopy (TEM) that the Pt particles had grown to about 1.9 nm from the original size of 1.6 nm. Hence, we will consider the ORR properties of the 1.6 nm catalyst at 30–50 °C, because the effect of particle size on stability is not within the scope of this chapter.

Figure 10.11 shows Arrhenius plots for k_{app} at various electrodes. Linear relationships between log k_{app} and $1/T$ are observed at all the electrodes. Each solid line is the least squares fit of all the data at the constant applied potential. From the slope, we obtained an apparent activation energy E_a = 37 kJ mol^{-1} at −0.525 V vs. E^o, which is in very good agreement with that for bulk Pt electrodes [Wakabayashi et al., 2005a]. It is very striking that the values of k_{app} (per real surface area) at Nafion–Pt/CB with different d_{Pt} agreed beautifully with those at Nafion–Pt(bulk) in the whole temperature range and in the practical potential range. Hence, the ORR rate constant and its activation energy are independent of Pt particle size.

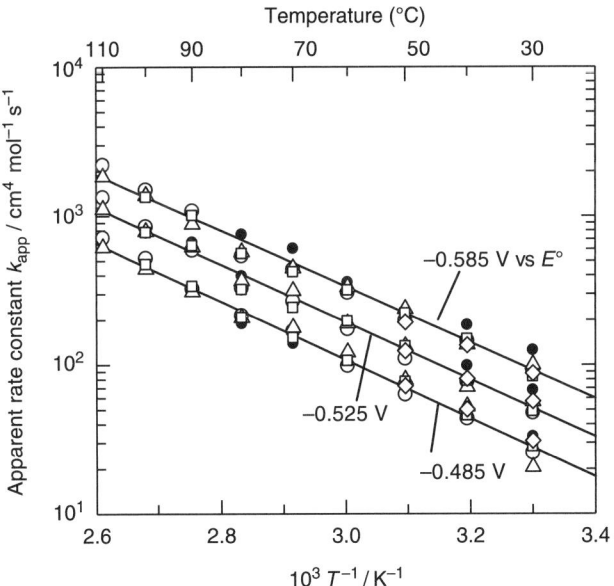

Figure 10.11 Arrhenius plots of the ORR rate constants k_{app} obtained at various electrodes. The symbols are the same as those in Fig. 10.10. Each solid line is the least squares fit of all the data at the constant applied potential. Since the standard potential E^o and $E[\text{RHE}(t)]$ shift to less positive values in a different manner, the corrected potential E is applied so as to keep a constant overpotential for the ORR at each temperature. The applied potentials of −0.485, −0.525, and −0.585 V vs. E^o correspond to 0.80, 0.76, and 0.70 V vs. RHE, respectively, at 30 °C. (From Yano et al. [2006b], reproduced by permission of the PCCP Owner Societies.)

In contrast, Paulus and co-workers reported that j_k (per real surface area) for the ORR at Nafion–Pt/CB were lower than those at bulk Pt at 60 °C in O_2-saturated 0.5 M H_2SO_4, namely, 190 μA cm^{-2} (20 wt% Pt/Vulcan XC72) compared with 400 μA cm^{-2} [Pt(bulk)] at 0.85 V vs. RHE [Paulus et al., 2001]. Such a noticeable reduction in activity should be ascribed to the stacked multilayer (about 10 monolayers) of Pt/CB catalyst coated on the glassy carbon disk substrate. Hence, it is essential for the evaluation of the true activity of electrocatalysts supported on CB to disperse them uniformly as in our experimental procedure. The present results have been supported by ^{195}Pt electrochemical nuclear magnetic resonance spectroscopy measurements (EC-NMR) on the same set of Pt/CB catalysts, which showed that the surface peak position and the spin–lattice relaxation time of surface Pt atoms exhibited practically no change with d_{Pt} [Yano et al., 2006b].

Because both the ORR activity per real active surface area and $P(H_2O_2)$ are the same, we can reduce the mass of Pt in the catalyst layer by using smaller particles, as long as their durability can be ascertained. It is also found that, at −0.525 V vs. $E°$, the k_{app} values for both bulk Pt and Nafion–Pt/CB at 60 and 110 °C are 5 and 35 times higher, respectively, than at 30 °C. This clearly shows the possibility of significant reduction of Pt loading in PEFCs by such factors by elevating the operating temperature, if the catalyst layer could properly be designed to supply sufficient O_2 gas to Pt active sites.

10.3.2 Enhanced ORR Activity at Pt Alloy Electrodes

Enhanced ORR activities of Pt alloyed with nonprecious metals such as Fe, Co, Ni, Mn, Cr, and V have been reported in acidic electrolyte solutions [Paffett et al., 1988; Mukerjee and Srinivasan, 1993; Mukerjee et al., 1995; Thamizhmani and Capuano, 1994; Toda et al., 1999; Min et al., 2000; Neergat et al., 2001; Stamenkovic et al., 2002, 2003; Antolini et al., 2002; Drillet et al., 2002; Yang et al., 2004]. In this section, we investigate the temperature dependencies of the real ORR activities k_{app} and $P(H_2O_2)$ at Pt-Fe, Pt-Co, and Pt-Ni sputtered film electrodes at 20–90 °C by the CFDE method [Wakabayashi et al., 2005b].

As described in the previous sections, a stable Pt skin of a few nanometers is formed on the Pt-Fe, Pt-Co, and Pt-Ni alloy surfaces after electrochemical stabilization. Figure 10.12 shows Arrhenius plots of k_{app} on the alloy electrodes at −0.525 V vs. $E°$ in comparison with that of a pure Pt electrode. In the low temperature region (20–50 °C for $Pt_{54}Fe_{46}$, 20–60 °C for $Pt_{68}Co_{32}$ and $Pt_{63}Ni_{37}$), linear relationships between log k_{app} and $1/T$ are observed at all the electrodes, corresponding to the following Arrhenius equation:

$$k_{app} = Z \exp\left(-\frac{\varepsilon_a}{RT}\right) \qquad (10.9)$$

The apparent activation energy ε_a on each alloy was found to be 41 kJ mol^{-1}, which is almost comparable to that on the pure Pt electrode. The values of k_{app} on the alloys are larger than that on the Pt electrode by factors of 4.0 ($Pt_{54}Fe_{46}$), 3.1

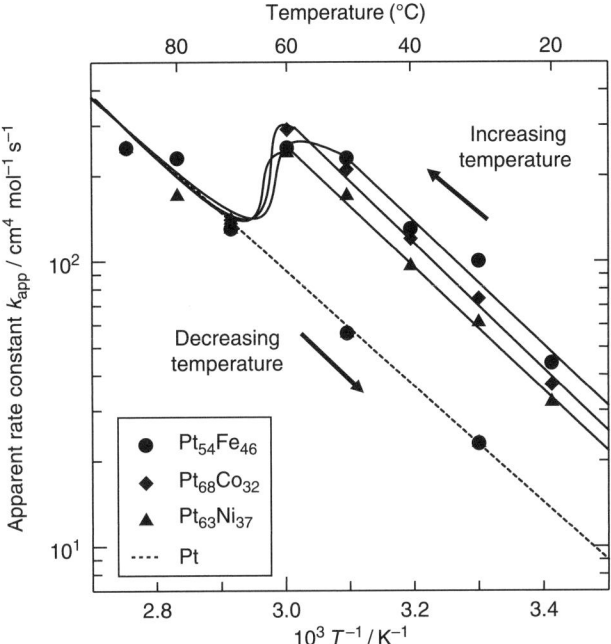

Figure 10.12 Arrhenius plots of k_{app} for the ORR at $Pt_{54}Fe_{46}$ (●), $Pt_{68}Co_{32}$ (♦), $Pt_{63}Ni_{37}$ (▲), and Pt (····) electrodes in 0.1 M $HClO_4$ at -0.525 V vs. $E°$. (From Wakabayashi et al. [2005b], reproduced by permission of the American Chemical Society.)

($Pt_{68}Co_{32}$), and 2.4 ($Pt_{63}Ni_{37}$), respectively. However, the k_{app} values on these alloys decrease with increasing temperature above 60 °C, and settle at almost the same value as for the pure Pt electrode. The degraded ORR activity after heating at high temperature in 0.1 M $HClO_4$ solution was never recovered, even in the low temperature region.

The values of $P(H_2O_2)$ were found to be almost independent of potential between $E = 0.8$ and 0.3 V, namely, 1.2% at $Pt_{54}Fe_{46}$ (20–50 °C) and 0.16% at $Pt_{68}Co_{32}$ and $Pt_{63}Ni_{37}$ (20–60 °C). However, further elevation of temperature decreases H_2O_2 production below the limit of detection (<0.05%). After these alloy electrodes experienced high temperatures (>60 °C) in the solution, H_2O_2 production was never detected over the whole temperature range. It should be noted that the critical temperature of about 60 °C for loss of the high ORR activity and H_2O_2 production activity corresponds well to that of a sharp decrease in the S_{Pt}^o. This deactivation in the ORR as well as the HOR is certainly attributable to a dealloying of the nonprecious metal component in hot acid solution, resulting in a thick Pt layer, the electronic state of which is no longer affected by the underlying alloy.

Finally, we briefly discuss the mechanism of enhanced specific ORR activities (k_{app}) at the Pt skin layer on the alloys. The Tafel slope in the high current density region at the Pt skin layer was found to be -120 mV/decade [Toda et al., 1999; Wakabayashi et al., 2005b], indicating that the enhancement of ORR activities at these alloys is dominantly brought about by promoting the same rate-determining step (*rds*) of the first one-electron-transfer step as that of pure Pt. Considering the first-order dependence of the ORR rate on [O_2] [Wakabayashi et al., 2005a] and pH [Sepa et al., 1981], the *rds* can be written as

$$x\text{Pt} + O_2 + H^+ + e^- \longrightarrow \text{Pt}_x - \dot{O}_2 - H \quad (10.10)$$

This reaction may consist of the following elementary steps:

$$\text{Pt}_x - yA_{ad} + O_2 \rightleftarrows \text{Pt}_x - O_2 + yA_{ad} \quad (A_{ad} = H_2O, \text{ adsorbed anions, etc.}) \quad (10.11)$$

$$\text{Pt}_x - O_2 + H^+(H_2O)_z \rightleftarrows [\text{Pt}_x - O_2 \cdots H^+ \cdots (H_2O)_z]^* \quad (10.12)$$

$$[\text{Pt}_x - O_2 \cdots H^+ \cdots (H_2O)_z]^* + e^- \longrightarrow \text{Pt}_x - \dot{O}_2 - H + z(H_2O) \quad (10.13)$$

where the asterisk denotes the activated complex. Because the activation energy ε_a on both alloys and pure Pt is the same, the high ORR activities at the alloys are ascribed to a large pre-exponential factor Z in the rate constant of (10.9). As stated in the previous sections, the electronic state at the Pt skin is strongly modified by underlying alloy. This may increase the coverage of adsorbed oxygen, for example in the reaction step (10.11), probably assisted by a decrease in the blocking of A_{ad} (H_2O, adsorbed anions, etc.). The production of H_2O_2 during the ORR at all the alloy electrodes at low temperatures also supports such a mechanism [Wakabayashi et al., 2005b]. While an inhibition of "OH poisoning" at the alloy electrodes has been proposed [Stamenkovic et al., 2002; Paulus et al., 2002; Murthi et al., 2004], it is necessary to identify the adsorbed species in the ORR on both pure Pt and Pt alloys to find a clue for the design of high performance cathode catalysts. Such studies are in progress in our laboratory [Wakisaka et al., 2008].

10.4 CONCLUSIONS

We have demonstrated enhanced ORR activities and excellent CO tolerances in the HOR at Pt alloyed with nonprecious metals such as Fe, Co, and Ni. It has been shown that dissolution of the surface alloy layer is followed by rearrangement of the Pt layer, resulting in a Pt skin layer, which can protect the underlying bulk alloy. The CO-tolerant HOR activity at Pt skins or Pt-Ru at $E < 0.10$ V vs. RHE is ascribed to weaker CO bond strength due to the modification of the Pt electronic structures, observed as a positive core level shift in Pt$4f_{7/2}$. It has been suggested that such a modified electronic structure at the Pt skin layer also induces enhanced ORR activity. To clarify such a mechanism in detail, a number of investigations of the ORR at these alloys and pure Pt are under progress in our group.

We have also clarified for Pt/CB that ORR activity is not affected by differences in the particle size. It is very important for highly dispersed alloy catalysts to examine the size dependences of both activities and stabilities in the ORR and the CO-tolerant HOR. For this purpose, we have recently developed a nanocapsule method to prepare monodispersed Pt alloy particles supported on CB with well-controlled composition [Yano et al., 2007, 2008]. This method, of course, can be applied to prepare various kind of alloys (binary, ternary, or more), which can contribute to the exploration of highly active and stable electrocatalysts for PEFCs.

ACKNOWLEDGMENT

This work was partly supported by the fund for "Leading Project; Next Generation Fuel Cells" of the Ministry of Education, Science, Culture, Sports, and Technology of Japan.

REFERENCES

Antolini E, Passos RR, Ticianelli EA. 2002. Electrocatalysis of oxygen reduction on a carbon supported platinum–vanadium alloy in polymer electrolyte fuel cells. Electrochim Acta 48: 263–270.

Ayato Y, Kunimatsu K, Osawa M, Okada T. 2006. Study of Pt electrode/Nafion ionomer interface in $HClO_4$ by in situ surface-enhanced FTIR spectroscopy. J Electrochem Soc 153: A203–A209.

Bard AJ, Faulkner LF. 1994. Electrochemical Methods. New York: Wiley. p. 288.

Bjorneholm O, Nilsson A, Tillborg H, Bennich P, Sandell A, Hernnas B, Puglia C, Mårtensson N. 1994. Overlayer structure from adsorbate and substrate core level binding energy shifts: CO, CCH_3 and O on Pt(111). Surf Sci 315: L983–L989.

Bregoli LJ. 1978. The influence of platinum crystallite size on the electrochemical reduction of oxygen in phosphoric acid. Electrochim Acta 23: 489–492.

Briggs D, Seah MP, eds. 1990. Practical Surface Analysis. Volume 1. Auger and X-ray Photoelectron Spectroscopy. New York: Wiley.

Cabeza GF, Castellani NJ, Légaré P. 1999. Theoretical study of CO adsorption on Ni(111), Pt(111) and Pt/Ni(111) surfaces. Surf Rev Lett 6: 369–381.

Chu D. 1998. O_2 reduction at the Pt/Nafion interface in 85% concentrated H_3PO_4. Electrochim Acta 43: 3711–3718.

Chu D, Tryk D, Gervasio D, Yeager EB. 1989. Examination of the ionomer/electrode interface using the ferric/ferrous redox couple. J Electroanal Chem 272: 277–284.

Conway BE, Tilak BV. 2002. Interfacial processes involving electrocatalytic evolution and oxidation of H_2, and the role of chemisorbed H. Electrochim Acta 47: 3571–3594.

Drillet J-F, Ee A, Friedemann J, Kötz R, Schnyder B, Schmidt VM. 2002. Oxygen reduction at Pt and $Pt_{70}Ni_{30}$ in H_2SO_4/CH_3OH solution. Electrochim Acta 47: 1983–1988.

Fujino T. 1996. Development of anode catalysts for PEFCs. MEng Thesis, Yamanashi University.

Gasteiger HA, Markovic N, Ross PN Jr, Cairns EJ. 1994. Carbon monoxide electrooxidation on well-characterized platinum–ruthenium alloys. J Phys Chem 98: 617–625.

Gasteiger HA, Markovic N, Ross PN Jr. 1995. Electrooxidation of CO and H_2/CO mixtures on a well-characterized Pt_3Sn electrode surface. J Phys Chem 99: 8945–8949.

Gerisher H, Mattes I, Braun R. 1965. Electrolyse in Strömungskanal. J Electroanal Chem 10: 553–567.

Gonzalez ER, Ticianelli EA. 2005. Studies of CO tolerance on modified gas diffusion electrodes containing ruthenium dispersed on carbon. J Electroanal Chem 575: 53–60.

Grgur BN, Zhuang G, Markovic N, Ross PN Jr. 1997. Electrooxidation of H_2/CO mixtures on a well-characterized $Pt_{75}Mo_{25}$ alloy surface. J Phys Chem B 101: 3910–3913.

Hammer B, Nørskov JK. 2000. Theoretical surface science and catalysis—calculations and concepts. Adv Catal 45: 71–129.

Hammer B, Morikawa Y, Nørskov JK. 1996. CO chemisorption at metal surfaces and overlayers. Phys Rev Lett 76: 2141–2144.

Higuchi E, Uchida H, Watanabe M. 2005. Effect of loading level in platinum-dispersed carbon black electrocatalysts on oxygen reduction activity evaluated by rotating disk electrode. J Electroanal Chem 583: 69–76.

Igarashi H, Fujino T, Watanabe M. 1993. Hydrogen electro-oxidation on platinum catalysts in the presence of trace carbon monoxide. J Electroanal Chem 391: 119–123.

Igarashi H, Fujino T, Zhu Y, Uchida H, Watanabe M. 2001. CO tolerance of Pt alloy electrocatalysts for polymer electrolyte fuel cells and the detoxification mechanism. Phys Chem Chem Phys 3: 306–314.

Kinoshita K. 1990. Particle size effects for oxygen reduction on highly dispersed platinum in acid electrolytes. J Electrochem Soc 137: 845–848.

Lawson DR, Whiteley LD, Martin CR. 1988. Oxygen reduction at Nafion film-coated platinum electrodes: Transport and kinetics. J Electrochem Soc 135: 2247–2253.

Lemons RA. 1990. Fuel cells for transportation. J Power Sources 29: 251–264.

Levich VG. 1962. Physicochemical Hydrodynamics. Englewood Cliffs, NJ: Prentice Hall. pp. 112–116.

Ley KL, Liu RX, Pu C, Fan QB, Leyarovska N, Segre C, Smotkin ES. 1997. Methanol oxidation on single-phase Pt-Ru-Os ternary alloys. J Electrochem Soc 144: 1543–1548.

Miedema AR, De Chatel PF, De Boer FR. 1980. Cohesion in alloys—Fundamentals of a semiempirical model. Physica B 100: 1–28.

Min M, Cho J, Cho K, Kim H. 2000. Particle size and alloying effects of Pt-based alloy catalysts for fuel cell applications. Electrochim Acta 45: 4211–4217.

Motoo S, Watanabe M. 1980. Electrocatalysis by ad-atoms. Part VII. Enhancement of CO oxidation on platinum by As ad-atoms. J Electroanal Chem 111: 261–268.

Motoo S, Shibata M, Watanabe M. 1980. Electrocatalysis by ad-atoms. Part VI. Enhancement of CO oxidation on Pt(subs) and Pt-Au(subs) electrodes by Sn ad-atoms. J Electroanal Chem 110: 103–109.

Mukerjee S, McBreen J. 1989. Effect of particle size on the electrocatalysis by carbon-supported Pt electrocatalysts: an in situ XAS investigation. J Electroanal Chem 448: 163–171.

Mukerjee S, Srinivasan S. 1993. Enhanced electrocatalysis of oxygen reduction on platinum alloys in proton exchange membrane fuel cells. J Electroanal Chem 357: 201–224.

Mukerjee S, Srinivasan S, Soriaga MP, McBreen J. 1995. Role of structural and electronic properties of Pt and Pt alloys on electrocatalysis of oxygen reduction. J Electrochem Soc 142: 1409–1422.

Mukerjee S, Urian RC, Lee SJ, Ticianelli EA, McBeen J. 2004. Electrocatalysis of CO tolerance by carbon-supported PtMo electrocatalysts in PEMFCs. J Electrochem Soc 151: A1094–A1103.

Murthi VS, Urian RC, Mukerjee S. 2004. Oxygen reduction kinetics in low and medium temperature acid environment: correlation of water activation and surface properties in supported Pt and Pt alloy electrocatalysts. J Phys Chem B 108: 11011–11023.

Neergat N, Shukla AK, Gandhi KS. 2001. Platinum-based alloys as oxygen-reduction catalysts for solid-polymer-electrolyte direct methanol fuel cells. J Appl Electrochem 31: 373–378.

Paffett MT, Berry JG, Gottesfeld S. 1988. Oxygen reduction at $Pt_{0.65}Cr_{0.35}$, $Pt_{0.2}Cr_{0.8}$ and roughened platinum. J Electrochem Soc 135: 1431–1436.

Paulus UA, Schmidt TJ, Gasteiger HA, Behm RJ. 2001. Oxygen reduction on a high-surface area Pt/Vulcan carbon catalyst: A thin-film rotating ring-disk electrode study. J Electroanal Chem 495: 134–145.

Paulus UA, Wokaun A, Scherer GG, Schmidt TJ, Stamenkovic V, Markovic NM, Ross PN. 2002. Oxygen reduction on carbon-supported Pt-Ni and Pt-Co alloy catalysts. J Phys Chem B 106: 4181.

Ross PN. 1986. In: Rao UV, ed. Precious Metals 1986. Allentown, PA: International Precious Metals Institute. p. 355.

Ross PN, Kinoshita K, Scarpellino AJ, Stonehart P. 1975a. Electrocatalysis on binary alloys: I. Oxidation of molecular hydrogen on supported Pt-Rh alloys. J Electroanal Chem 59: 177–189.

Ross PN, Kinoshita K, Scarpellino AJ, Stonehart P. 1975b. Electrocatalysis on binary alloys: II. Oxidation of molecular hydrogen on supported Pt + Ru alloys. J Electroanal Chem 63: 97–110.

Ruban A, Hammer B, Stoltze P, Skriver HL, Nørskov JK. 1997. Surface electronic structure and reactivity of transition and noble metals. J Mol Catal A 115: 421–429.

Sattler ML, Ross PN. 1986. The surface structure of Pt crystallites supported on carbon black. Ultramicroscopy 20: 21–28.

Sepa DB, Vojnovic MV, Damjanovic A. 1981. Reaction intermediates as a controlling factor in the kinetics and mechanism of oxygen reduction at platinum electrodes. Electrochim Acta 26: 781–793.

Shek ML, Stefan PM, Binns C, Lindau I, Spicer WE. 1982. Chemisorption-induced Pt $4f$ surface core level shifts. Surf Sci 115: L81–L85.

Stamenkovic V. Schmidt TJ, Markovic NM, Ross PN Jr. 2002. Surface composition effects in electrocatalysis: Kinetics of oxygen reduction on well-defined Pt_3Ni and Pt_3Co alloy surfaces. J Phys Chem B 106: 11970–11979.

Stamenkovic V, Schmidt TJ, Ross PN Jr, Markovic NM. 2003. Surface segregation effects in electrocatalysis: Kinetics of oxygen reduction reaction on polycrystalline Pt_3Ni alloy surfaces. J Electroanal Chem 554/555: 191–199.

Sugimoto W, Saida T, Takasu Y. 2006. Co-catalytic effect of nanostructured ruthenium oxide towards electro-oxidation of methanol and carbon monoxide. Electrochem Commun 8: 411–415.

Takasu Y, Ohashi N, Zhang X-G, Murakami Y, Yahikozawa K. 1996. Size effects of platinum particles on the electroreduction of oxygen. Electrochim Acta 41: 2595–2600.

Thamizhmani G, Capuano GA. 1994. Improved electrocatalytic oxygen reduction performance of platinum ternary alloy-oxide in solid-polymer-electrolyte fuel cells. J Electrochem Soc 141: 968–975.

Toda T, Igarashi H, Uchida H, Watanabe M. 1999. Enhancement of the electroreduction of oxygen on Pt alloys with Fe, Ni, and Co. J Electrochem Soc 146: 3750–3756.

Treglia G, Desjonqueres MC, Spanjaard D, Lassailly Y, Guillot C, Jugnet Y, Duc TM, Lecante J. 1981. Study of the W (Ta) core level shifts induced by the adsorption of oxygen on tungsten (tantalum) (110). J Phys C 14: 3463–3474.

Uchida H, Ikeda N, Watanabe M. 1997. Electrochemical quartz crystal microbalance study of copper ad-atoms on gold electrodes. Part II. Further discussion on the specific adsorption of anions from the solutions. J Electroanal Chem 424: 5–12.

Uchida H, Ozuka H, Watanabe M. 2002. Electrochemical quartz crystal microbalance analysis of CO-tolerance at Pt-Fe alloy electrodes. Electrochim Acta 47: 3629–3636.

Uchida H, Izumi K, Watanabe M. 2006. Temperature dependence of CO-tolerant hydrogen oxidation reaction activity at Pt, Pt-Co, and Pt-Ru electrodes. J Phys Chem B 110: 21924–21930.

Wakabayashi N, Takeichi M, Itagaki M, Uchida M, Watanabe M. 2005a. Temperature-dependence of oxygen reduction activity at a platinum electrode in an acidic electrolyte solution investigated with a channel flow double electrode. J Electroanal Chem 574: 339–346.

Wakabayashi N, Takeichi M, Uchida H, Watanabe M. 2005b. Temperature dependence of oxygen reduction activity at Pt-Fe, Pt-Co, and Pt-Ni alloy electrodes. J Phys Chem B 109: 5836–5841.

Wakisaka M, Mitsui S, Hirose H, Kawashima K, Uchida H, Watanabe M. 2006. Electronic structures of Pt-Co and Pt-Ru alloys for CO-tolerant anode catalysts in polymer electrolyte fuel cells studied by EC-XPS. J Phys Chem B 110: 23489–23496.

Wakisaka M, Suzuki H, Mitsui S, Uchida H, Watanabe M. 2008. Increased oxygen coverage at Pt-Fe alloy cathode for the enhanced oxygen reduction reaction studied by EC-XPS. J Phys Chem C 112: 2750–2755.

Wan L, Moriyama T, Ito M, Uchida H, Watanabe M. 2002. In situ STM imaging of surface dissolution and rearrangement of a Pt-Fe alloy electrocatalyst in electrolyte solution. Chem Commun 1: 58–59.

Watanabe M, Motoo S. 1975a. Electrocatalysis by ad-atoms. Part III. Enhancement of the oxidation of carbon monoxide on platinum by ruthenium ad-atoms. J Electroanal Chem 60: 275–283.

Watanabe M, Motoo S. 1975b. Electrocatalysis by ad-atoms. Part I. Enhancement of the oxidation of methanol on platinum and palladium by gold ad-atoms. J Electroanal Chem 60: 259–266.

Watanabe M, Motoo S. 1976. Electrocatalysis by Sn and Ge ad-atoms. J Electroanal Chem 69: 429–431.

Watanabe M, Shibata M, Motoo S. 1985. Electrocatalysis by ad-atoms. Part XII. Enhancement of carbon monoxide oxidation on platinum electrodes by oxygen adsorbing ad-atoms (Ge, Sn, Pb, As, Sb and Bi). J Electroanal Chem 187: 161–174.

Watanabe M, Saegusa S, Stonehart P. 1988. Electro-catalytic activity on supported platinum crystallites for oxygen reduction in sulphuric acid. Chem Lett 17: 1487–1490.

Watanabe M, Sei H, Stonehart P. 1989. The influence of platinum crystallite size on the electro-reduction of oxygen. J Electroanal Chem 261: 375–387.

Watanabe M, Igarashi H, Fujino T. 1999. Design of CO tolerant anode catalysts for polymer electrolyte fuel cell. Electrochemistry 67: 1194–1196.

Watanabe M, Zhu Y, Uchida H. 2000. Oxidation of CO on a Pt-Fe alloy electrode studied by surface enhanced infrared reflection–absorption spectroscopy. J Phys Chem B 104: 1762–1768.

Weinert M, Watson RE. 1995. Core-level shifts in bulk alloys and surface adlayers. Phys Rev B 51: 17168–17180.

Yajima T, Uchida H, Watanabe M. 2004. In-situ ATR-FTIR spectroscopic study of electro-oxidation of methanol and adsorbed CO at Pt-Ru alloy. J Phys Chem B 108: 2654–2659.

Yang H, Vogel W, Lamy C, Alonso-Vante N. 2004. Structure and electrocatalytic activity of carbon-supported Pt-Ni alloy nanoparticles toward the oxygen reduction reaction. J Phys Chem B 108: 11024–11034.

Yano H, Higuchi E, Uchida H, Watanabe M. 2006a. Temperature dependence of oxygen reduction activity at Nafion-coated bulk Pt and Pt/carbon black catalysts. J Phys Chem B 110: 16544–16549.

Yano H, Inukai J, Uchida H, Watanabe M, Babu PK, Kobayashi T, Chung JH, Oldfield E, Wieckowski A. 2006b. Particle-size effect of nanoscale platinum catalysts in oxygen reduction reaction: an electrochemical and ^{195}Pt EC-NMR study. Phys Chem Chem Phys 8: 4932–4939.

Yano H, Kataoka M, Yamashita H, Uchida H, Watanabe M. 2007. Oxygen reduction activity of carbon supported Pt-M (M = V, Ni, Cr, Co and Fe) alloys prepared by nanocapsule method. Langmuir 23: 6438–6445.

Yano H, Song JM, Uchida H, Watanabe M. 2008. Temperature dependence of oxygen reduction activity at carbon-supported Pt_XCo (X = 1, 2, and 3) alloy catalysts prepared by the nano-capsule method. J Phys Chem C 112: 8372–8380.

Zecevic SK, Wainright JS, Litt MH, Gojkovic SL, Savinell RF. 1997. Kinetics of O_2 reduction on a Pt electrode covered with a thin film of solid polymer electrolyte. J Electrochem Soc 144: 2973–2982.

CHAPTER 11

Electrocatalysis for the Direct Alcohol Fuel Cell

J.-M. LEGER, C. COUTANCEAU, and C. LAMY

Laboratoire de Catalyse en Chimie Organique, UMR 6503 CNRS, Equipe Electrocatalyse, Université de Poitiers, 40 Avenue du Recteur Pineau, F-86022 Poitiers, France

11.1 INTRODUCTION

Fuel cells are electrochemical devices transforming the heat of combustion of a fuel (hydrogen, natural gas, methanol, ethanol, hydrocarbons, etc.) directly into electricity. The fuel is electrochemically oxidized at the anode, whereas the oxidant (oxygen from the air) is reduced at the cathode. This process does not follow Carnot's theorem, so that higher energy efficiencies are expected: up to 40–50% in electrical energy and 80–85% in total energy (heat production in addition to electricity).

Typically, pure hydrogen or hydrogen-rich gases are used as fuel for proton exchange membrane fuel cells (PEMFC), allowing the highest electrical efficiency. However, the production, storage, and distribution of hydrogen are still strong limitations on the development of such techniques [Gosselink, 2002; Ströbel et al., 2002]. In this context, the use of hydrogen carriers such as alcohols (methanol, ethanol, etc.), in a direct alcohol fuel cell (DAFC), appears particularly convenient for two main reasons: they are liquids (allowing easy storage) and their theoretical mass energy density is rather high (6.1 and 8.0 kWh kg^{-1} for methanol and ethanol, respectively), close to that of gasoline (10.5 kWh kg^{-1}) [Lamy and Léger, 1994].

Methanol [Bett et al., 1998; Wasmus and Küver, 1999; Coutanceau et al., 2004; Batista et al., 2004] and ethanol [Xia et al., 1997; Vigier et al., 2004a; Lamy et al., 2004] are the most studied alcohols for DAFC application (in DMFCs and DEFCs, respectively). As a consequence of the acidic environment of the ionomeric conducting membrane (Nafion) and of the low working temperatures of DAFCs (60–120 °C), the use of platinum is impossible to avoid, owing to its catalytic properties to activate C–H bond cleavage during the first adsorption steps, although this leads to rather poor

---

Fuel Cell Catalysis. Edited by Marc T. M. Koper
Copyright © 2009 John Wiley & Sons, Inc.

electro-oxidation kinetics. This is due to the formation of strongly bonded species limiting the number of active sites. In order to improve the reaction kinetics, a deep understanding of the mechanisms of the electrocatalytic reactions is a key issue. As several different steps are generally necessary to carry out the complete electro-catalytic reaction, an optimized catalyst should be multifunctional.

At Pt catalysts, methanol is adsorbed, with the formation of poisoning species (adsorbed carbon monoxide) [Beden et al., 1987]. On the other hand, owing to the chemical structure of ethanol, its electro-oxidation is more difficult because of the necessary cleavage of the C–C bond to form the final product (carbon dioxide). However, even if such bond breaking is difficult, adsorbed CO is also observed by in situ infrared reflectance spectroscopy [Vigier et al., 2004a]. In both cases, the formation of such poisoning species leads to poor activity, and the challenge is to enhance the activity of Pt. Because of the different steps likely involved in methanol and ethanol adsorption and oxidation at Pt, the design of multi-metallic electrocatalysts is essential. The composition of the catalysts (nature and proportion of the metals involved) and the structure (size of particles, atomic arrangement, superficial structure, etc.) are crucial, and their tolerance to ageing should also be taken into account.

11.2 THERMODYNAMICS AND KINETICS OF REACTIONS IN A DAFC: ETHANOL OXIDATION

Considering as an example a DEFC, electro-oxidation of ethanol takes place at the anode (negative pole of the cell),

$$CH_3CH_2OH + 3H_2O \longrightarrow 2CO_2 + 12H^+ + 12e^- \qquad E_1^\circ = 0.085 \text{ V vs. SHE} \tag{11.1}$$

while at the cathode (positive pole) oxygen undergoes electro-reduction,

$$O_2 + 4H^+ + 4e^- \longrightarrow 2H_2O \qquad E_2^\circ = 1.229 \text{ V vs. SHE} \tag{11.2}$$

where E_1° and E_2° are the electrode potentials versus the standard hydrogen (reference) electrode (SHE). This corresponds to the overall combustion reaction of ethanol in oxygen:

$$CH_3CH_2OH + 3O_2 \longrightarrow 2CO_2 + 3H_2O \tag{11.3}$$

with the thermodynamic data, under standard conditions:

$$\Delta G^\circ = -1325 \text{ kJ mol}^{-1}, \quad \Delta H^\circ = -1366 \text{ kJ mol}^{-1} \tag{11.4}$$

This gives a standard electromotive force (EMF) at equilibrium

$$E_{eq}^\circ = \frac{-\Delta G^\circ}{nF} = \frac{1325 \times 10^3}{12 \times 96{,}485} = E_2^\circ - E_1^\circ = 1.144 \text{ V} \tag{11.5}$$

11.2 THERMODYNAMICS AND KINETICS OF REACTIONS IN A DAFC

and a theoretical energy efficiency

$$\varepsilon_{eq}^{rev} = \frac{-\Delta G^o}{-\Delta H^o} = \frac{1325 \times 10^3}{1366 \times 10^3} = 97\% \quad (11.6)$$

with $F = 96,485$ C mol^{-1} the Faraday constant and $n = 12$ the number of electrons exchanged per molecule for complete oxidation to CO_2.

However, under working conditions, with a current density j, the cell voltage $E(j)$ decreases greatly as the result of three limiting factors: the charge transfer overpotentials $\eta_{a,act}$ and $\eta_{c,act}$ at the two electrodes due to slow kinetics of the electrochemical processes (η_i is defined as the difference between the working electrode potential $E(j)_i$ and the equilibrium potential $E_{eq,i}$), the ohmic drop $R_e|j|$, with R_e the ohmic resistance of the electrolyte and interface, and the mass transfer limitations for reactants and products. The cell voltage can thus be expressed as

$$E(|j|) = E_2(|j|) - E_1(|j|) = E_2^o + \eta_c - (E_1^o + \eta_a) - R_e|j|$$
$$= E_{eq}^o - (|\eta_a| + |\eta_c| + R_e|j|) \quad (11.7)$$

where the overvoltages η_a (>0 for an anodic reaction, i.e., here the oxidation of the fuel) and η_c (<0 for a cathodic reaction, i.e., here the reduction of the oxidant) take into account both the slow rate of the electrochemical reactions (activation polarization) and the limiting rate of mass transfer (concentration polarization). Thus, the energy efficiency can be expressed as

$$\varepsilon_{cell}^{C_2H_5OH/O_2} = \frac{n_{exp}FE(j)}{-\Delta H^o} = \frac{n_{exp}}{n} \frac{E(j)}{E_{eq}^o} \frac{nFE_{eq}^o}{-\Delta H^o}$$
$$= \frac{n_{exp}}{n} \frac{E(j)}{E_{eq}^o} \frac{\Delta G^o}{\Delta H^o} = \varepsilon_F \varepsilon_E \varepsilon_{eq}^{rev} \quad (11.8)$$

Therefore, for a DEFC working at 0.5 V at 50 mA cm^{-2} and leading either to CO_2 (complete oxidation) or to acetic acid (partial oxidation), the overall efficiency would be

$$\varepsilon_{cell}^{C_2H_5OH/O_2} = \frac{12}{12} \times \frac{0.5}{1.144} \times 0.97 \approx 42.4\% \quad (11.9)$$

$$\varepsilon_{cell}^{C_2H_5OH/O_2} = \frac{4}{12} \times \frac{0.5}{1.144} \times 0.97 \approx 14\% \quad (11.10)$$

respectively. Since ε_{rev}^{eq} is given by the thermodynamics (one can increase it slightly by changing the operating pressure and temperature conditions), this expression shows that the only ways to increase significantly the overall energy efficiency are to increase ε_E, i.e., to decrease the overvoltages η, the ohmic drop $R_e|j|$, and to increase ε_F, i.e., to favor the complete oxidation of alcohol to CO_2 (avoiding the formation of aldehyde and carboxylic acid). The decrease in $|\eta|$ is directly related to the increase in the rate of the electrochemical reactions occurring at both electrodes. This is typical in electrocatalysis, where the actions of the electrode potential and the catalytic electrode material will synergistically increase the reaction rate. The current intensity j is proportional to the rate of reaction v, i.e., $j = nFv$. For a first-order electrochemical reaction (the rate of which is proportional to the reactant concentration c) the current density can be expressed as

$$j = nFv = nFk(T, E)c = nFk_0 c e^{-\Delta G_0^+/RT} e^{\alpha nFE/RT} = j_0 e^{\alpha nFE/RT} \quad (11.11)$$

This last equation contains the two essential activation terms met in electrocatalysis: an exponential function of the electrode potential E and an exponential function of the chemical activation energy ΔG_0^+ (defined as the activation energy at the standard equilibrium potential). By modifying the nature and structure of the electrode material (the catalyst), one may decrease ΔG_0^+, thus increasing j_0, as a result of the catalytic properties of the electrode. This leads to an increase in the reaction rate j.

11.3 EFFECT OF CATALYST STRUCTURE AND COMPOSITION: METHANOL ELECTRO-OXIDATION

Platinum–ruthenium systems are known to be the most effective bimetallic catalysts in terms of activity with regard to methanol oxidation and selectivity with regard to complete oxidation to CO_2 [Kabbabi et al., 1998; Schmidt et al., 1999; Hamnett, 1999; Dinh et al., 2000]. Radioactive labeling experiments provided some evidence for the effects of Ru on methanol adsorption and oxidation. The presence of Ru increases the rate of methanol adsorption, the maximum CO coverage of the surface being reached more rapidly at a Pt-Ru than at a Pt catalyst (Fig. 11.1) at a given adsorption potential. Moreover, the presence of Ru leads to a higher rate of CO oxidative desorption (Fig. 11.2), i.e., CO_2 evolution, than at a pure Pt catalyst. The reason for such an enhancement of activity with regard to methanol electro-oxidation is generally related to the higher CO tolerance of Pt-Ru. This is the bifunctional mechanism according to which adsorbed CO species are oxidized by OH species generated on Ru surface atoms at lower anode potentials [Watanabe and Motoo, 1975a, b; Iwasita et al., 2000]. Another explanation for the enhanced CO tolerance is the electronic effect [Waszczuk et al., 2001a, 2002] by which the presence of Ru brings about a change in the electronic state of Pt, leading to a change in CO binding. A mechanism combining both effects is also possible. However, Lu et al. showed on the basis of UHV, electrochemical NMR and electrochemical studies that the bifunctional mechanism for the enhancement of CO oxidation at Pt/Ru fuel cell catalysts was four times larger than the ligand effect [Lu et al., 2002].

A bifunctional catalyst should be able to activate two different reaction steps (methanol and water adsorption and surface reaction between adsorbed species), and so active sites with different properties are necessary. As an example, investigations of possibility of enhancing activity with regard to methanol electro-oxidation with Pt-Ru-based electrodes are of great interest with regard to improving the electrical efficiency of DMFCs. Several approaches have been considered: the effect of Pt-Ru

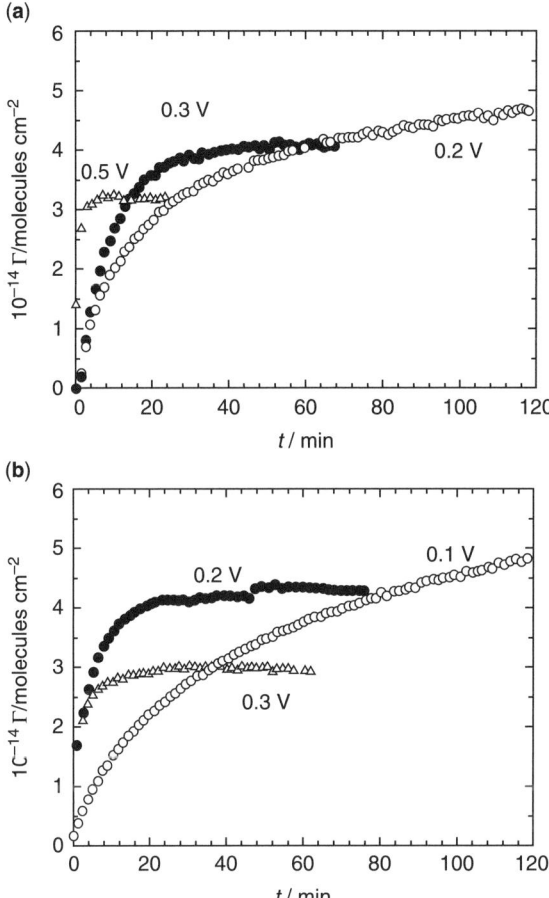

Figure 11.1 Kinetics of adsorption of CO at a Pt catalyst from a 0.01 M methanol solution at different potentials. (a) Pt black catalyst, with Pt loading 0.8 mg cm^{-2}. (b) Pt$_{0.5}$Ru$_{0.5}$ black catalyst, with catalyst loading 0.8 mg cm^{-2} (0.1 M H$_2$SO$_4$, $T = 298$K).

catalyst structure [Dubau et al., 2003b; Waszczuk et al., 2001b; Brankovic et al., 2002a, b]; optimization of the Pt/Ru atomic ratio [Kabbabi et al., 1998; Iwasita et al., 2000; Dinh et al., 2000; Watanabe et al., 1987; Gasteiger et al., 1994; Ianniello et al., 1994]; and the use of a third alloying metal [Lima et al., 2001]. A crucial aspect is the method used to prepare nanostructured catalysts, leading to different catalyst structures and compositions, in a controlled way by varying the experimental conditions only slightly. The colloidal route for catalyst preparation derived from the method developed by Bönnemann and co-workers [Bönnemann et al., 1991, 1996] is very convenient for obtaining Pt-Ru compounds with controlled size, composition, and structure [Dubau et al., 2003a, b; Dubau, 2002]. Catalytic powders can be obtained in the following ways: (i) Pt/C and Ru/C, from deposition of a Pt colloid or a Ru colloid solution on carbon; (ii) Pt-Ru/C, from deposition of alloyed Pt-Ru

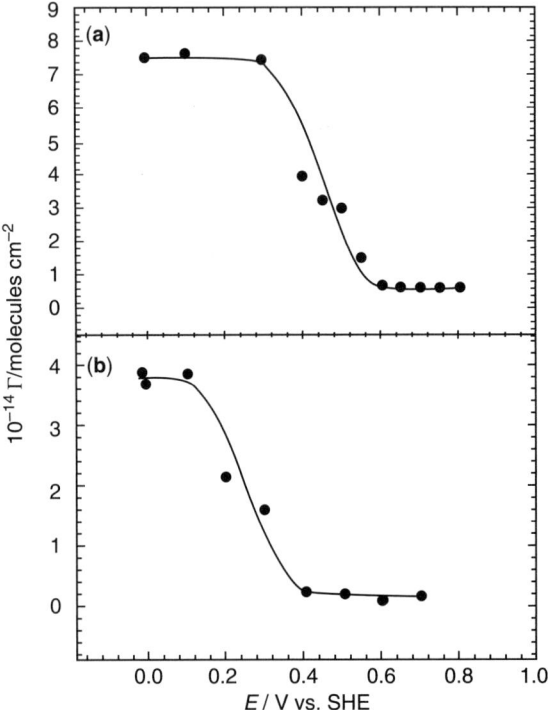

Figure 11.2 Adsorbed CO coverage at Pt black (a) and Pt-Ru black (b) catalysts as a function of electrode potential in the presence of methanol in the solution. Methanol was adsorbed from a 0.01 M CH_3OH solution in the supporting electrolyte. The Pt loading was 0.8 mg cm^{-2} (0.1 M H_2SO_4, $T = 298$K).

colloid particles (obtained by co-reduction of Pt and Ru salts) on carbon; and (iii) Pt + Ru/C, from the deposition of a mixture of Pt and Ru colloid solutions on carbon. The amounts of each reagent can be adjusted to obtain different Pt/Ru atomic ratios and various metal loadings on carbon. Characterizations by transmission electron microscopy (TEM), X-ray diffraction (XRD), and energy-dispersive X-ray spectroscopy (EDX) were carried out, and results are presented in Table 11.1. Analysis of the XRD patterns led to the conclusion that catalysts prepared by method (ii) were true Pt-Ru alloys and those prepared according to method (iii) were not alloyed compounds but appeared to be Pt particles decorated by small Ru particles, or at least Pt particles in very strong interaction with smaller Ru particles [Dubau et al., 2003a; Dubau, 2002].

An important conclusion was that the best catalyst is not the alloyed one as expected, nor the mixture of Pt/XC 72 and Ru/XC 72 powders, but one consisting of a dispersion of Pt colloid and Ru colloid on the same carbon support, i.e., the Pt + Ru/XC 72 catalyst. The latter leads to higher current densities for the electro-oxidation of methanol than the other catalysts with the same atomic ratio for potentials lower than 0.5 V versus a reversible hydrogen electrode (RHE) (Fig. 11.3). This result

11.3 EFFECT OF CATALYST STRUCTURE AND COMPOSITION

TABLE 11.1 TEM and XRD Characterization of Different Catalysts Prepared by the Colloidal Route[a]

Catalyst	Average Size/nm	Structure
Pt/C	2.2	fcc
$Pt_{0.8} + Ru_{0.2}/C$	2.1	Pt fcc + Ru hc
$Pt_{0.8}Ru_{0.2}/C$	1.9	fcc (alloy)
Pt/C + Ru/C (80 at% Pt and 20 at% Ru)	Pt 2.1; Ru 1.5	Pt fcc + Ru hc
Ru/C	1.5	hc

[a]Experimental details are given in Dubau et al. [2003a, b] and Dubau [2002].

is in agreement with the works of Waszczuk and Brankovic, who demonstrated an electrocatalytic enhancement of methanol oxidation at Pt particles decorated by Ru compared with the alloy compounds of the same composition [Waszczuk et al., 2001b; Brankovic et al., 2002a, b].

The problem of the optimum Pt/Ru atomic ratio is of great interest. The differences in activity as a function of Ru content are due to the balance between the initial step of adsorption–dehydrogenation of methanol at Pt sites and the following step of adsorbed CO species oxidation [Gasteiger et al., 1993]. Some authors claim that a Pt/Ru atomic ratio of 50/50 is the best for methanol electro-oxidation [Dinh et al., 2000; Watanabe et al., 1987]. Others propose that a ratio close to 80/20 is optimal [Kabbabi et al., 1998; Gasteiger et al., 1993; Ianniello et al., 1994; Dubau et al., 2003a]. The discrepancy in the literature likely comes from a lack of knowledge of the surface composition. The work of Wieckowski's group on Pt-Ru catalysts prepared by spontaneous deposition of Ru at Pt nanoparticles is very interesting in this context

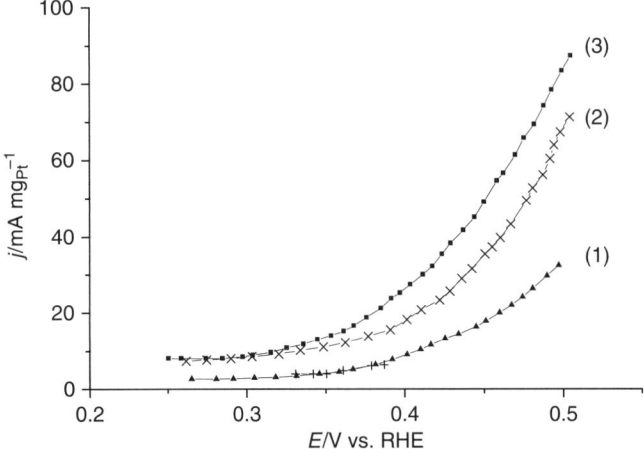

Figure 11.3 $j(E)$ polarization curves for methanol oxidation on different colloid catalysts prepared in different ways: (1) Pt/C; (2) $Pt_{0.8}Ru_{0.2}/C$; (3) $Pt_{0.8} + Ru_{0.2}/C$ (0.5 M H_2SO_4, 1.0 M MeOH; sweep rate 1 mV s^{-1}, $T = 298$K).

[Waszczuk et al., 2001b; Tong et al., 2002]. Because Ru is deposited as nanosized Ru islands of monoatomic height, the Ru coverage of Pt could be determined accurately. In that case, the best activity with regard to methanol oxidation was found for a Ru coverage close to 40–50% at 0.3 and 0.5 V vs. RHE. However, the structure of such catalysts and the conditions of study are far from those used in DMFCs. Moreover, the surface composition of a bimetallic catalyst likely depends on the method of preparation of the catalyst [Caillard et al., 2006] and on the potential [Blasini et al., 2006].

Using the colloidal $Pt_{(1-x)} + Ru_x/C$ catalysts described above, the optimal atomic ratio depends upon methanol concentration, cell temperature, and applied potential, as shown by the Tafel plots recorded with methanol concentrations of 1.0 and 0.1 M at $T = 298K$ (Fig. 11.4) and 318K (Fig. 11.5). Some authors have stated that for potentials between 0.35 and 0.6 V vs. RHE, the slow reaction rate between adsorbed CO and adsorbed OH species must be responsible for the rate of the overall process [Iwasita et al., 2000]. From these results, it can be underlined that, at a given constant potential lower than 0.45–0.5 V vs. RHE, an increase in temperature requires an increase in Ru content to enhance the rate of methanol oxidation, and that, at a given constant potential greater than 0.5 V vs. RHE, an increase in temperature requires a decrease in Ru content to enhance the rate of methanol oxidation.

Increasing the temperature above 40 °C gives to Ru the ability to adsorb and dehydrogenate methanol [Chu and Gilman, 1996]. Moreover, Ru allows the activation of water molecules at lower potentials than Pt. According to the bifunctional theory of electrocatalysis for the complete oxidation of methanol [Watanabe and Motoo, 1975b], the presence of a large amount of Ru in the $Pt_{0.5} + Ru_{0.5}/C$ catalyst can explain the higher activity of this catalyst at lower potentials. At higher potentials, above 0.5 V vs. RHE, Pt-rich catalysts become more active. In the limiting current range, the catalytic surface is blocked by adsorbed oxygen species [Watanabe and Motoo, 1975a, b], which makes the adsorption of organic species more difficult. Because Ru adsorbs oxygen species at more negative potentials, increasing the content of Ru causes the limiting current to decrease. On the other hand, some authors have shown that in the temperature range 25–60 °C, pure Pt displays greater activity than Ru for methanol oxidation for potentials higher than 0.5 V vs. RHE [Gasteiger et al., 1994]. The combination of both effects may explain the decrease in limiting current with an increase in the Ru atomic ratio.

Arrhenius plots recorded with different catalyst compositions (Fig. 11.6) allow determination of the apparent activation energy ΔH^* at 0.5 V vs. RHE. Although radioactive labeling measurements indicated that the presence of Ru leads to an increase in the rate of methanol adsorption/dehydrogenation at Pt for potentials lower than 0.5 V vs. RHE (Fig. 11.1), infrared studies have shown that the coverage by CO-adsorbed species remains small for Ru-rich catalysts [Bett et al., 1998], which indicates that the limiting step may still be the adsorption/dehydrogenation of methanol, in agreement with the proposition of Lei and co-workers from deuterium isotope analysis of methanol oxidation on a $Pt_{0.5}Ru_{0.5}$ catalyst [Lei et al., 2002]. Therefore, a value close to 60 kJ mol^{-1} of the apparent activation energy is reasonable. For Ru-poor catalysts, another rate-determining step than that proposed for Ru-rich catalysts seems to be involved because of the low value of the apparent energy of

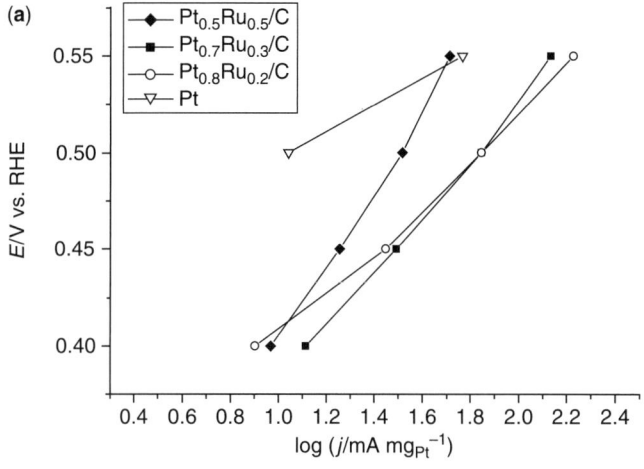

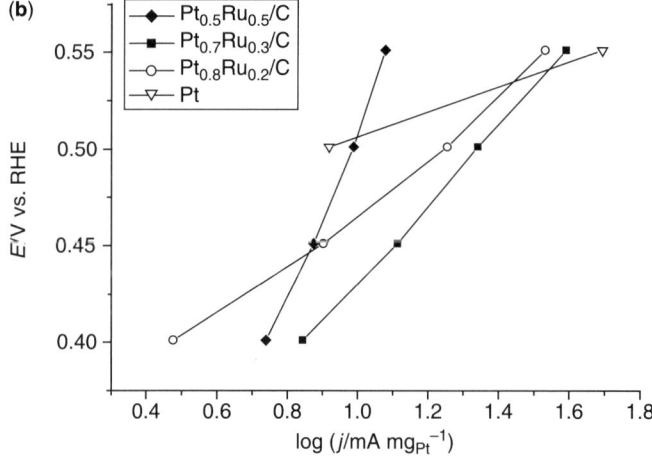

Figure 11.4 Tafel plots for methanol oxidation on Pt + Ru/C colloid catalysts with different atomic compositions; at $T = 298$ K: (a) 1.0 M MeOH; (b) 0.1 M MeOH (0.5 M H_2SO_4; sweep rate 1 mV s^{-1}).

activation (30 kJ mol^{-1}). Electrochemical NMR studies [Tong et al., 2002] have shown that CO diffusion at a pure Pt surface was slow compared with that near Pt/Ru islands. Then, the slow diffusion of adsorbed CO from Pt sites away from Ru to Pt sites close to Ru where the oxygenated species on Ru oxidize the CO species can be proposed as the rate-determining step [Gasteiger et al., 1994]. Finally, the intermediate value obtained with the Pt catalyst (45 kJ mol^{-1}) may be attributed to the oxidation of CO_{ads} to CO_2, which is the rate-determining step, according to the reaction

$$CO_{ads} + OH_{ads} \longrightarrow CO_2 + H^+ + e^- \tag{11.12}$$

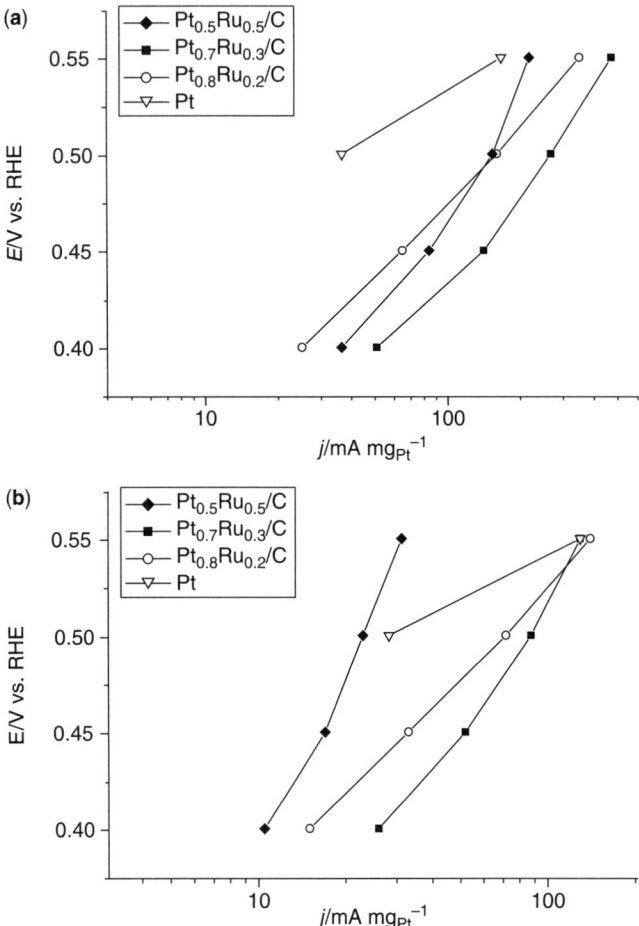

Figure 11.5 Tafel plots for methanol oxidation on Pt + Ru/C colloid catalysts with different atomic compositions; at $T = 318$ K: (a) 1.0 M MeOH; (b) 0.1 M MeOH (0.5 M H_2SO_4; sweep rate 1 mV s^{-1}).

For potentials higher than 0.5 V vs. RHE, the formation of adsorbed oxygen species at Ru as well as at Pt will block the catalytic surface, leading to a decrease in the methanol adsorption kinetics. Therefore, in a potential range higher than 0.5 V vs. RHE, the kinetics of methanol oxidation is optimized at a Ru-poor catalyst, because methanol adsorption is not blocked and because the presence of Ru provides the extra oxygen atom needed to complete the oxidation of adsorbed CO to CO_2.

Finally, trimetallic compounds have been developed to enhance the electroactivity of Pt-based catalysts, for either methanol or ethanol electro-oxidation. A long time ago, it was reported that adsorption of molybdates (Na_2MoO_4) at a Pt black electrode

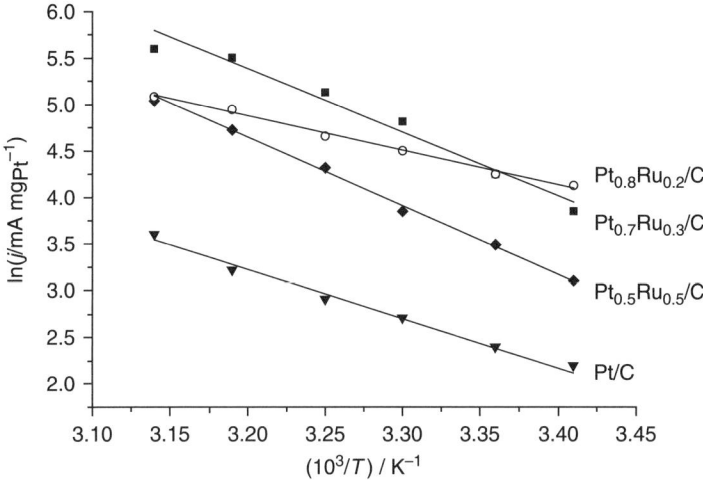

Figure 11.6 Arrhenius plots obtained at 0.5 V on Pt + Ru/C colloid catalysts with different atomic compositions (0.5 M H_2SO_4, 1.0 M MeOH; sweep rate 1 mV s^{-1}).

before adding the fuel (HCHO or CH_3OH) resulted in a decrease of 0.3 V in the oxidation onset potential with respect to pure Pt [Shropshire, 1965]. Moreover, when added to Pt-Ru catalysts, molybdenum led to higher activity with regard to methanol electro-oxidation at low potentials [Lima et al., 2001; Jusys et al., 2002]. Such anodes (prepared by electrodeposition of metals on a carbon gas diffusion layer [Coutanceau et al., 2004]) led to an improvement in the electrical performance of a single DMFC at 90 °C (with the electrode characteristics given in the caption of Fig. 11.7). The trimetallic $Pt_{0.7}Ru_{0.26}Mo_{0.04}$/C catalyst led to a higher cell performance than a $Pt_{0.8}Ru_{0.2}$/C anode prepared in the same way (Fig. 11.7a). For ethanol electro-oxidation, a trimetallic $Pt_{0.86}Sn_{0.1}Ru_{0.04}$/C catalyst made by the colloidal route [Rousseau et al., 2006] leads to a cell performance twice as high as that obtained with a $Pt_{0.9}Sn_{0.1}$/C catalyst under the same experimental conditions (Fig. 11.7b). In the case of molybdenum, the increase in the open circuit voltage (OCV) of the single cell of about 100 mV can be explained by a decrease in surface poisoning or by an effect on the composition of chemisorbed species [Podlovchenko et al., 1966; Smirnova et al., 1988]. Moreover, the higher capacity of this third metal to form oxy-hydroxyl species at low potentials was often proposed as an explanation for the enhancement of methanol oxidation. In the case of ethanol electro-oxidation at PtSnRu and PtSn, the OCV is similar for both catalysts, and it was shown that the presence of Ru does not lead to a noticeable change in the product distribution in the anode outlet [Rousseau et al., 2006]. Here, the role of Ru seems more difficult to explain. Its main role may be to limit strong adsorption of surface poisons by diluting adsorption sites [Adžič, 1984], or to provide adsorbed OH species as soon as tin starts to form higher oxide species that are not catalytically active [Rousseau, 2004].

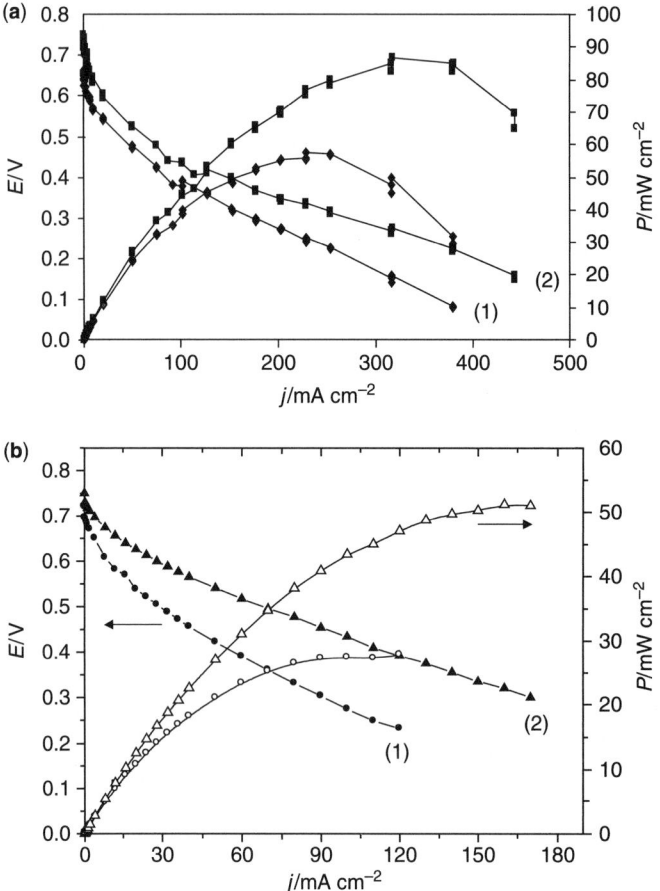

Figure 11.7 (a) Electric characteristics of a single 5 cm^2 DMFC recorded at 90 °C: (1) Pt$_{0.8}$Ru$_{0.2}$/C anode; (2) Pt$_{0.7}$Ru$_{0.26}$Mo$_{0.04}$/C anode. The catalyst loading was 2.0 mg cm^{-2}; O$_2$: 3 bar, 120 mL min^{-1}; MeOH 2.0 M: 2 bar, 2 mL min^{-1}. The membrane electrode assembly (MEA) was prepared by hot pressing at 130 °C, 35 kg cm^{-2} for 3 minutes, the home-made anode and the cathode (E-TEK, 2.0 mg cm^{-2} Pt loading, 40 wt% Pt on carbon) against a Nafion 117 membrane. (b) Electric cell characteristics of a single 25 cm^2 DEFC recorded at 80 °C: (1) Pt$_{0.9}$Sn$_{0.1}$/C anode; (2) Pt$_{0.86}$Sn$_{0.1}$Ru$_{0.04}$/C anode (3.0 mg cm^{-2} Pt loading, 60 wt% catalyst on carbon). O$_2$: 3 bar, 120 mL min^{-1}; EtOH 2.0 M: 1 bar; 2 mL min^{-1}. The MEA was prepared as in (a).

However, in the case of multimetallic catalysts, the problem of the stability of the surface layer is crucial. Preferential dissolution of one metal is possible, leading to a modification of the nature and therefore the properties of the electrocatalyst. Changes in the size and crystal structure of nanoparticles are also possible, and should be checked. All these problems of ageing are crucial for applications in fuel cells.

11.4 EFFECT OF FOREIGN METALS ALLOYED TO Pt: COMPARISON BETWEEN ETHANOL AND METHANOL

In situ infrared spectroscopy has shown that the dissociative adsorption of both methanol and ethanol leads to the formation of strongly adsorbed CO species at low potentials [Beden et al., 1987; Perez et al., 1989; Iwasita and Nart, 1991]. Indeed, in both cases, one can see the existence of an absorption band located close to 2050 cm^{-1} in the subtractively normalized interfacial Fourier transform infrared (SNIFTIR) spectra (Fig. 11.8). With Pt/C, this species is linearly bonded to the Pt sites and plays the role of a poison for the electrocatalytic reaction. The surface reaction between this adsorbed CO species and adsorbed OH (coming from the dissociation of water) occurs at potentials around 0.4–0.5 V for methanol and ethanol with Pt/C. The formation of the final product (CO_2 for both alcohols detected at 2345 cm^{-1} [Vigier et al., 2004a; Dubau et al., 2003b, Léger et al., 2005]) occurs mainly when the coverage in adsorbed CO begins to decrease (Fig. 11.9). With Pt-Ru/C, the maximum coverage in CO_{ads} (in the case of methanol) appears at lower potentials in comparison with Pt/C. With Pt-Sn/C, the formation of CO_2 (in the case of ethanol) is clearly visible after the disappearance of adsorbed CO, the coverage of which is higher in the low potential range. Bands corresponding to the formation of acetaldehyde and acetic acid are clearly observed during ethanol oxidation at Pt-Sn/C electrodes (Fig. 11.10). The bands located close to 1720, 1370, and 1280 cm^{-1} are respectively attributed to the formation of products containing a carbonyl group: a CO stretching mode from –COOH or –CHO (1370 and 1280 cm^{-1}), –C–O symmetric stretching in adsorbed CH_3COO^-, and coupled C O stretching and OH deformation from –COOH [Xia et al., 1997]. The bands located between 1280 and 1400 cm^{-1} probably correspond to the symmetric CH_3 bending mode of acetic acid [Iwasita et al., 1989]. The absorption band located close to 1720 cm^{-1} originates from acetic acid and/or acetaldehyde [Iwasita and Pastor, 1994a; Chang et al., 1990; Beden, 1984].

All these points show clearly that in the case of the oxidation of methanol, the adsorbed CO species, which is without any doubt a poisoning species at low potentials, is also a reactive intermediate when Pt/C or Pt-Ru/C is used as the electrocatalyst. The main advantage in using Ru to modify Pt is the possibility of activating water at lower potentials and consequently oxidizing adsorbed CO earlier. During ethanol oxidation, on Pt/C, linearly adsorbed CO is also a poisoning species at low potentials. Cleavage of the C–C bond is clearly possible in this potential range. The formation of CO_2 is clearly related to the removal of CO_{ads} from the Pt/C surface, as it coincides with the maximum of CO_{ads} coverage (similarly to the oxidation of methanol). Pt-Sn is known to be the best bimetallic catalyst for ethanol electro-oxidation [Delime et al., 1999; Zhou et al., 2004a, b, 2005]. When Pt-Sn/C is used, the formation of CO_2 seems to be, at least partially, disconnected from the coverage in adsorbed CO. The amount of CO_2 formed with Pt-Sn/C, for potentials greater than 0.4 V, should be related not only to the removal of adsorbed CO, but also to the further oxidation of acetaldehyde. Two main routes then exist in the overall mechanism of ethanol oxidation with Pt-Sn/C: the first through the formation of adsorbed CO, involving electrode poisoning (which occurs strongly with Pt/C); the second via the formation

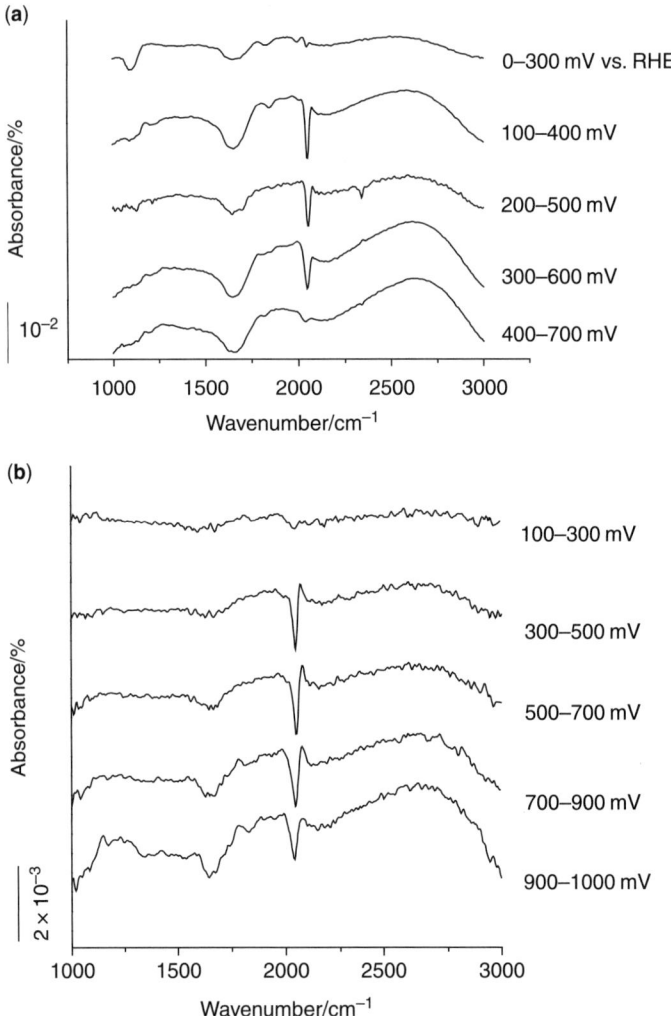

Figure 11.8 (a) SNIFTIR spectra of the species coming from methanol adsorption and oxidation at a Pt/C electrode; 0.1 M $HClO_4$ + 0.1 M CH_3OH; 25 °C. (b) SNIFTIR spectra of the species coming from ethanol adsorption and oxidation on a Pt/C electrode; 0.1 M $HClO_4$ + 0.1 M C_2H_5OH; 25 °C.

of an adsorbed acetyl species (CH_3CHO_{ads}), which can then lead to the formation of acetaldehyde and acetic acid as final products. The acetyl species can also be dissociated at the electrode surface, and, after reaction with adsorbed OH coming from water, may produce CO_2. This acetyl species has already been proposed in the case of bulk Pt [Perez et al., 1989].

The addition of Sn to Pt increases the activity of ethanol electro-oxidation and single DEFC performance [Rousseau et al., 2006; Zhou et al., 2004a; Vigier

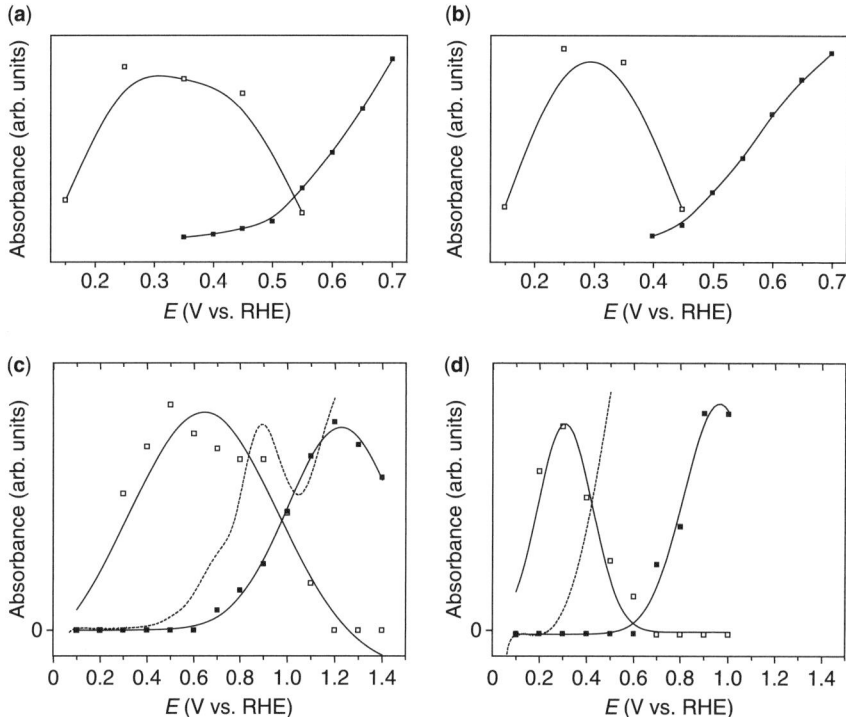

Figure 11.9 Intensities of the CO_L (□) and CO_2 (■) bands as functions of potential. (a, b) From the spectra of the species coming from methanol adsorption and oxidation (0.1 M $HClO_4$ + 0.1 M CH_3OH, 25 °C): (a) Pt/C electrode; (b) $Pt_{0.8}$ + $Ru_{0.2}$/C electrode. (c, d) From the spectra of the species coming from ethanol adsorption and oxidation (0.1 M $HClO_4$ + 0.1 M C_2H_5OH, 25 °C): (c) Pt/C electrode; (d) $Pt_{0.9}Sn_{0.1}$/C electrode. The dashed curves in (c) and (d) show $I(E)$.

et al., 2004b]. Actually, this can be explained by the fact that modification of Pt by Sn induces two important effects: first, it leads to a decrease in the yield of CO_2 (which is twice as high with a Pt/C catalyst as with a Pt-Sn/C catalyst); second, it greatly favors the formation of acetic acid compared with acetaldehyde, as was shown by analyzing the reaction products in the anode outlet of a DAFC (Table 11.2) [Rousseau et al., 2006]. The dilution of surface Pt sites by addition of Sn atoms is responsible for the first effect. The introduction of Sn atoms between Pt atoms decreases, by a statistical effect, the ability of the catalytic surface to cleave the C–C bond in the adsorption reaction of ethanol (which needs at least three adjacent Pt atoms). The bifunctionnal mechanism [Watanabe et Motoo, 1975a] is responsible for the second effect. In this mechanism, ethanol is adsorbed dissociatively at Pt sites, either via an O-adsorption or a C-adsorption process [Iwasita and Pastor, 1994a, b; Rightmire et al., 1964] to form acetaldehyde species, according to the following reactions:

$$Pt + CH_3-CH_2OH \longrightarrow Pt-(OCH_2-CH_3)_{ads} + e^- + H^+ \qquad (11.13)$$

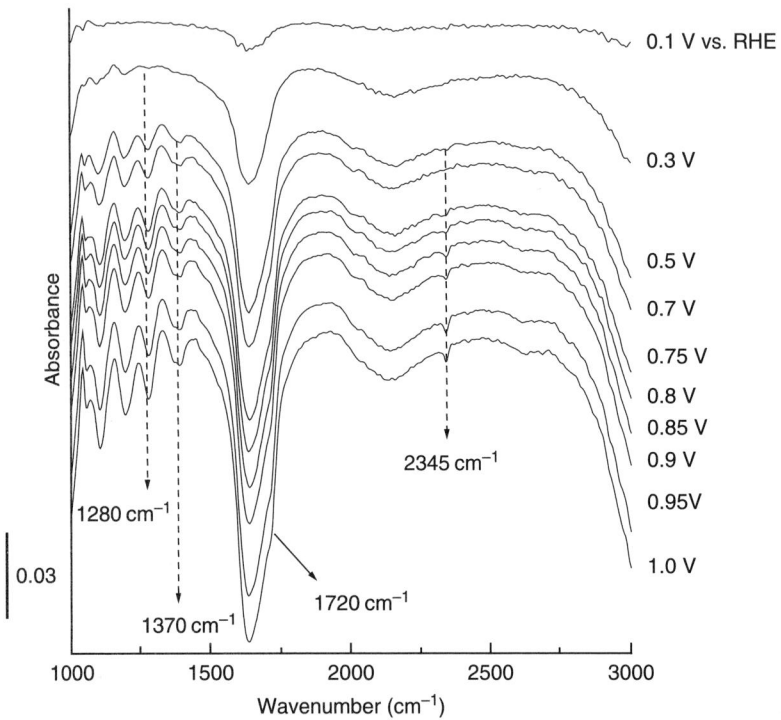

Figure 11.10 Single potential alteration infrared (SPAIR) spectra of the species coming from ethanol adsorption and oxidation on a Pt-Sn (90:10)/C electrode (0.1 M $HClO_4$ + 0.1 M C_2H_5OH, 25 °C). The reference spectrum was taken at 50 mV vs. RHE.

or

$$Pt + CH_3-CH_2OH \longrightarrow Pt-(CHOH-CH_3)_{ads} + e^- + H^+ \quad (11.14)$$

$$Pt-(OCH_2-CH_3)_{ads} \longrightarrow Pt + CHO-CH_3 + e^- + H^+ \quad (11.15)$$

or

$$Pt-(CHOH-CH_3)_{ads} \longrightarrow Pt + CHO-CH_3 + e^- + H^+ \quad (11.16)$$

TABLE 11.2 Chemical Yields in Acetic Acid, Acetaldehyde, and CO_2 for Pt/C and Pt-Sn (9:1)/C Catalysts during Ethanol Electro-oxidation under DEFC Operating Conditions at 80 °C for 4 Hours[a]

	Acetic Acid/Products/%	Acetaldehyde/Products/%	CO_2/Products/%
Pt/C 60 wt%	32.5	47.5	20
Pt-Sn (9:1)/ C 60 wt%	76.9	15.4	7.7

[a]Experimental conditions are described in Rousseau et al. [2006].

As soon as acetaldehyde is formed, it can adsorb on Pt sites, leading to a Pt–CO–CH$_3$ species:

$$Pt + CHO-CH_3 \longrightarrow Pt-(CO-CH_3)_{ads} + e^- + H^+ \quad (11.17)$$

Because Sn is known to activate water at lower potentials than Pt, OH species can be formed at low potentials on Sn sites according to the reaction

$$Sn + H_2O \longrightarrow Sn-(OH)_{ads} + e^- + H^+ \quad (11.18)$$

and the adsorbed acetaldehyde species can react with adsorbed OH species to produce acetic acid according to the reaction

$$Pt-(CO-CH_3)_{ads} + Sn-(OH)_{ads} \longrightarrow Pt + Sn + CH_3-COOH \quad (11.19)$$

At this stage, it should be pointed out that modification of a Pt-Sn catalyst by Ru atoms increases cell performance (and hence catalytic activity with regard to ethanol electro-oxidation), but has no effect on the OCV or on product distribution [Rousseau et al., 2006]. It seems, then, that the oxidation mechanism is the same on Pt-Sn and Pt-Sn-Ru, which supports the proposition that Ru allows OH species to be produced when the anode potential is increased and noncatalytically active tin oxides are formed.

11.5 ELECTROCATALYTIC ASPECTS OF OXYGEN REDUCTION

One of the main issues concerning the improvement of DAFC efficiency is to limit or avoid the effect of crossover of the alcohol, i.e., crossing of the electrolytic membrane by alcohol from the anode side to the cathode side due to intrinsic permeability of the membrane and electro-osmosis. This phenomenon causes a great depolarization of the cathode and hence a decrease in fuel cell electrical performance [Oedergaard et al., 2004; Hogarth and Ralph, 2002; Yang, 2004]. Several cathode catalysts have been studied in order to improve the oxygen reduction reaction (ORR) in the absence or in the presence of methanol, including Pt-alloy particles [Paffett et al., 1988; Beard and Ross, 1990; Toda et al., 1999a, b; Neergat et al., 2001]; transition metal chalcogenides [Alonso-Vante et al., 2000]; macrocycles, either heat-treated [Lalande et al., 1995, 1996] or not [Zagal et al., 1980; Coutanceau et al., 1994, 1995]; and N$_2$-Co or N$_2$-Fe based catalysts [Bashyam and Zelenay, 2006]. Other non-Pt catalysts have recently been investigated for acid membrane electrolyte fuel cell application, such as chalcogenides (CoSe or FeSe [Sidik and Anderson, 2006]; CoS or FeS [Campbell, 2006]), and CoN$_2$-type catalysts [Bashyam and Zelenay, 2006]. These catalysts displayed relatively good activity, although lower than that of Pt, and good tolerance to the presence of alcohol. However, for most of these catalysts, their structure, and therefore the reason for their activity and stability, is still unknown.

Numerous studies have shown that Pt-based binary alloy electrocatalysts such as Pt-Fe, Pt-Co, Pt-Ni, and Pt-Cr exhibit a higher catalytic activity for the ORR in an acid electrolyte than pure Pt [Paffett et al., 1988; Beard and Ross, 1990; Toda et al., 1999a, b; Neergat et al., 2001]. In that case, to achieve economic metal loadings, fuel cell electrodes must be prepared from Pt-based nanoparticles ($d < 5$ nm) supported on high surface area carbon black. For the ORR, the specific activity SA (in $\mu A\, cm_{Pt}^{-2}$) decreases with decreasing particle size, while it remains approximately the same as on Pt electrodes for particles larger than 5 nm [Kinoshita, 1992]. As a result, plots of the mass activity MA (in $A\, g_{Pt}^{-1}$) versus particle size for oxygen reduction in a sulfuric or phosphoric acid electrolyte exhibit a marked maximum for particle sizes close to 3.5 nm [Kinoshita, 1992].

As an example, Fig. 11.11 shows the oxygen reduction polarization curves obtained at $Pt_{1-x}Cr_x/C$ catalysts prepared using the carbonyl route [Yang et al., 2004]. The presence of Cr leads to a shift in the onset of the reduction wave towards higher potential, whereas the limiting current density is the same as that at Pt nanoparticles, indicating that the final product is likely the same (in agreement with a four-electron process of oxygen electro-reduction). The observed electrocatalytic enhancement was interpreted either by an electronic factor, namely, the change in the d-band vacancy in Pt upon alloying, and/or geometric effects (Pt coordination number and Pt–Pt distance). On the basis of X-ray absorption near-edge structure (XANES) and extended X-ray absorption fine structure (EXAFS) spectroscopic investigations of the ORR at Pt/C and Pt-X/C catalysts, Mukerjee and co-workers showed that Pt/C exhibited a significant increase in the number of Pt d-vacancies per atom compared with Pt alloys [Mukerjee et al., 1995]. They showed that this increase was due to the adsorption of OH species from the electrolyte, resulting in a decrease in electrocatalytic activity with regard to the ORR compared with Pt alloys. Both

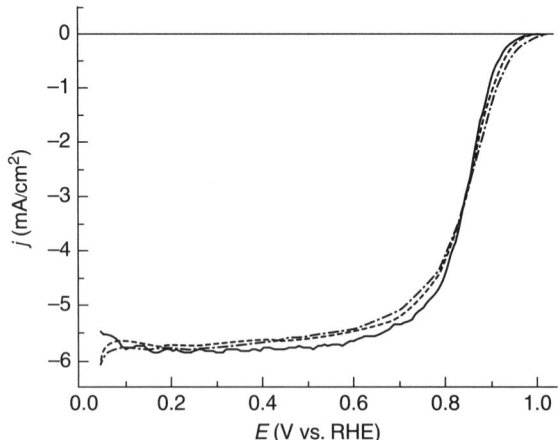

Figure 11.11 Linear cyclic voltammograms of carbon-supported nanosized Pt and Pt-Cr alloy catalysts with different atomic ratios (prepared using the carbonyl route [Yang et al., 2004]) recorded in 0.5 M $HClO_4$ saturated with pure oxygen at a scan rate of 5 mV s^{-1} and a rotation speed of 2000 rev min^{-1}. Current densities are normalized to the geometric surface area. ———, Pt/C; – – – –, $Pt_{0.7}Cr_{0.3}/C$; — - — -, $Pt_{0.5}Cr_{0.5}/C$.

effects should enhance the reaction rates of oxygen adsorption and of O–O bond breaking during the reduction reaction. For example, the lattice parameter a_0 in the case of cubic Pt-X (X = Fe, Co, Ni) decreases with increasing content of the alloying component X, leading to a variation in catalytic behavior. In the case of Pt-Ni alloys, the maximum electrochemical activity for ORR is obtained with 30 at% Ni [Toda et al., 1999a, b]. The presence of highly uncoordinated atoms is also very important, as shown by Lemire and co-workers for CO adsorption at Au nanoparticles [Lemire et al., 2004]. Finally, some authors have proposed that the role of the foreign metal is to protect the Pt surface from oxidation, by the presence of more easily oxidizable species [Shukla et al., 2001].

Such bimetallic alloys display higher tolerance to the presence of methanol, as shown in Fig. 11.12, where Pt-Cr/C is compared with Pt/C. However, an increase in alcohol concentration leads to a decrease in the tolerance of the catalyst [Koffi et al., 2005; Coutanceau et al., 2006]. Low power densities are currently obtained in DMFCs working at low temperature [Hogarth and Ralph, 2002] because it is difficult to activate the oxidation reaction of the alcohol and the reduction reaction of molecular oxygen at room temperature. To counterbalance the loss of performance of the cell due to low reaction rates, the membrane thickness can be reduced in order to increase its conductance [Shen et al., 2004]. As a result, methanol crossover is strongly increased. This could be detrimental to the fuel cell's electrical performance, as methanol acts as a poison for conventional Pt-based catalysts present in fuel cell cathodes, especially in the case of mini or micro fuel cell applications, where high methanol concentrations are required (5–10 M).

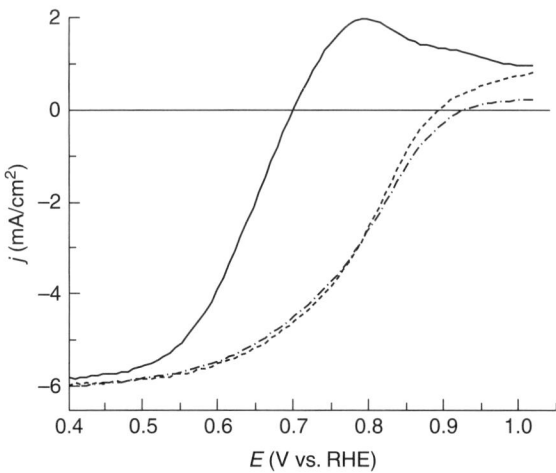

Figure 11.12 Linear cyclic voltammograms of carbon-supported nanosized Pt and Pt-Cr alloy catalysts with different atomic ratios (prepared using the carbonyl route [Yang et al., 2004]) recorded in 0.5 M $HClO_4$ + 0.5 M CH_3OH saturated with pure oxygen at a scan rate of 5 mV s^{-1} and a rotation speed of 2000 rev min^{-1}. ——, Pt/C; – – – –, $Pt_{0.7}Cr_{0.3}$/C; — - — -, $Pt_{0.5}Cr_{0.5}$/C.

For this reason, the development of cathode electrocatalysts insensitive to alcohols is of great interest. Among such catalysts, metal macrocycles, and most particularly iron phthalocyanines (FePcs) and porphyrins (FePPs), have been extensively studied [Lalande et al., 1995, 1996; Zagal et al., 1980; Coutanceau et al., 1994, 1995]. These catalysts were shown to be totally insensitive to the presence of methanol [Jiang and Chu, 2000; Convert et al., 2001; Baranton et al., 2005], and present a good alternative to Pt, which is a rare and expensive material. FePc/C catalysts displayed better catalytic activity with regard to the ORR than Pt in the presence of methanol (Fig. 11.13), and very good stability when used under working conditions close to that of a DMFC at room temperature [Baranton et al., 2005] (Fig. 11.14). It was proposed that using a solid proton exchange membrane as electrolyte, avoiding the presence of "free" protons as in liquid acid electrolytes, avoids demetalation of the macrocycle [Baranton et al., 2005]. These characteristics made FePc a good candidate for a cathode catalyst, at least for a fuel cell working at or close to room temperature, for example for mini and micro fuel cells developed for portable applications [Lu et al., 2004; Yamazaki, 2004]. However, FePc catalysts do not match Pt catalyst performances in methanol-free oxygen-saturated solutions [Baranton et al., 2005]. It is therefore very important to obtain a better understanding of the oxygen reduction mechanism at such macrocyclic catalysts in order to propose some ways to improve their activity and their selectivity. For example, a co-facial structure (Fig. 11.15) of the macrocycles with a metal–metal distance close to 0.4 nm seems to improve the electroactivity with regard to oxygen reduction as well as the selectivity with regard to water formation via a four-electron process [Collman et al., 1980; Baranton et al., 2006].

For example, the α- and β-phases of FePc, which correspond to co-facial structures of the macrocycles, were studied and compared. The structure of MPcs (M = metal)

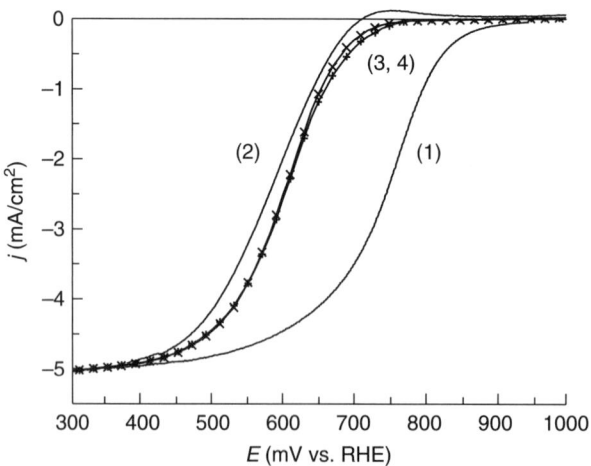

Figure 11.13 Linear cyclic voltammograms of different carbon-supported catalysts recorded in an O_2-saturated electrolyte (0.5 M H_2SO_4): (1) Pt/C catalyst; (2) Pt/C catalyst in the presence of 1.0 M methanol; (3) FePc/C catalyst; (4) FePc/C catalyst in the presence of 1.0 M methanol (temperature 20 °C, scan rate 5 mV s^{-1}, rotation speed 2500 rev min^{-1}).

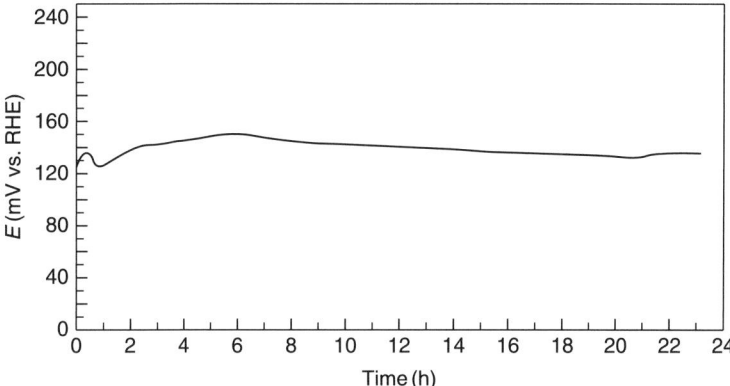

Figure 11.14 E versus time curve recorded during the ORR on a FePc/C gas diffusion electrode ($j = -25\,\text{mA cm}^{-2}$). The FePc electrode was fed with ambient air without any convection (temperature 20 °C) [Baranton et al., 2005].

was studied by Kirner's and Ballirano's groups [Kirner et al., 1976; Ballirano et al., 1998]. These authors found that in the α-phase of the MPc, the macrocycles were disposed in parallel stacks in the cell, whereas in the β-form, they were disposed in perpendicular stacks. Figure 11.16 gives a representation of the structures viewed perpendicularly with respect to the (100) crystallographic plane. The authors also pointed out that in the case of CoPc, the transition between the α- and β-forms led to increase the Co Co distance in the unit crystallographic cell: the Co Co distance close to 0.375 nm for α-CoPc increases to 0.479 nm for the β-phase. A similar value of 0.479 nm between Fe centers was found for β-FePc. Moreover, Ballirano and

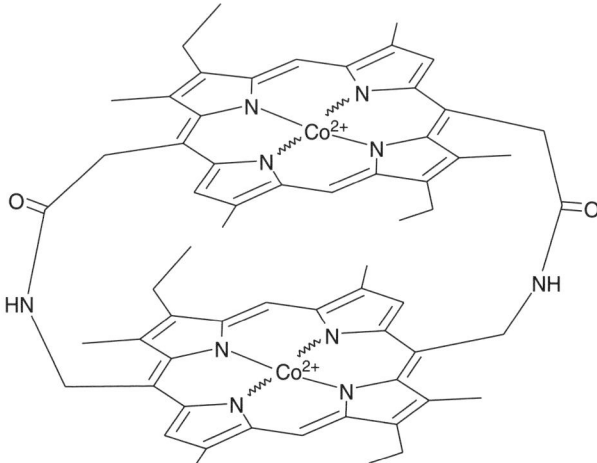

Figure 11.15 Schematic representation of co-facial cobalt porphyrins $Co_2FTF\,4\text{-}2,1\text{-NH}$ synthesized by Collman and co-workers [Collman et al., 1980].

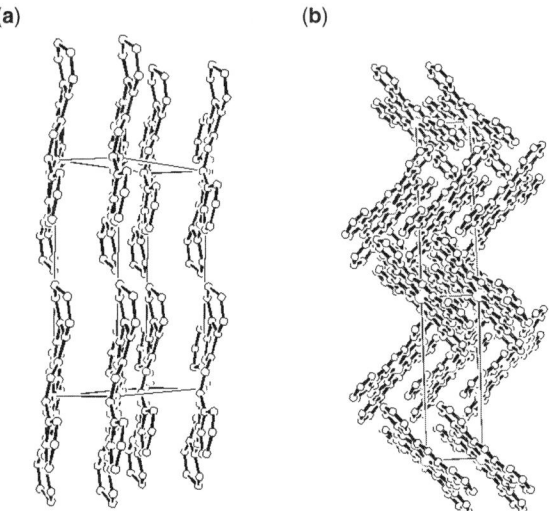

Figure 11.16 Schematic orientation of the planar FePc molecules in the crystallographic cell: (a) α-FePc calculated from the data obtained by Ballirano and co-workers [Ballirano et al., 1998]; (b) β-FePc calculated from the data obtained by Kirner and co-workers [Kirner et al., 1976].

co-workers pointed out the similarity of the structure of the α- and β-forms between CoPc and FePc [Ballirano et al., 1998].

α-FePc led to a shift of the reduction wave of 100 mV towards higher potentials than for β-FePc, and to higher absolute values of current densities in the diffusion plateau (Fig. 11.17). Tafel plots derived from these polarization curves give the following:

- For an α-FePc/C electrode, there is a first slope for potentials between 850 and 700 mV vs. RHE, with a value of -65 mV/decade ($2.3\,RT/F$), and a second one for potentials lower than 700 mV, with a value of -121 mV/decade ($2.3\,RT/0.5F$). These values are close to those observed at Pt electrodes [Damjanovic and Bockris, 1966; Gnanamuthu and Petrcelli, 1967; Damjanovic et al., 1967; Sepa et al., 1986a, b; Toda et al., 1999a, b].
- For β-FePc, only one Tafel slope could be observed, with a value close to -63 mV/decade.

These facts indicate that two mechanisms are involved, depending on the electrode potential in the case of α-FePc whereas only one mechanism seems to be involved in the case of β-FePc.

At Pt electrodes, adsorption of oxygen species is supposed to be controlled by the Temkin isotherm in the low overpotential region [Damjanovic and Bockris, 1966], whereas in the higher overpotential region, the absence of an oxide layer leads to

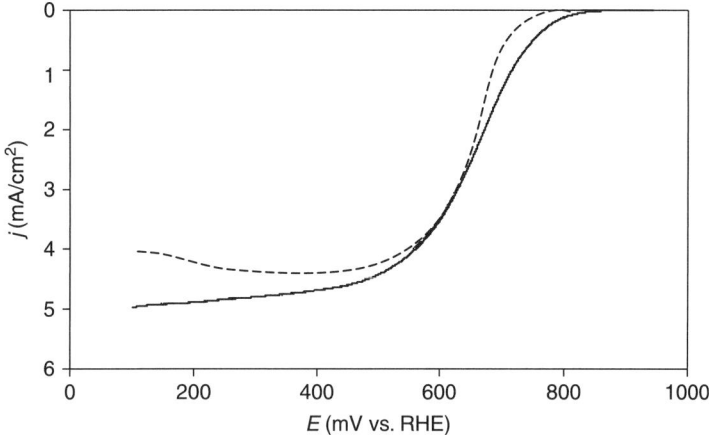

Figure 11.17 Linear cyclic voltammograms of α-FePc/C (——) and β-FePc (- - -) recorded in an O_2-saturated electrolyte (0.5 M H_2SO_4, temperature 20 °C, sweep rate 5 mV s^{-1}, rotation speed 2500 rev min^{-1}).

control of the adsorption of the oxygen species by the Langmuir isotherm [Sepa et al., 1986a]. The explanation for the existence of two Tafel slopes is somewhat different at α-FePc; however, different adsorption modes of molecular oxygen are certainly involved. One explanation could be the capability of α-FePc to form both PcFe–O_2–FePc μ-oxo species, allowing breaking of the O–O bond, and FePc–O_2 peroxo species [Baranton et al., 2006]. The difference in catalytic behavior between the two forms of FePc crystals can then be explained by the catalyst structure. It is

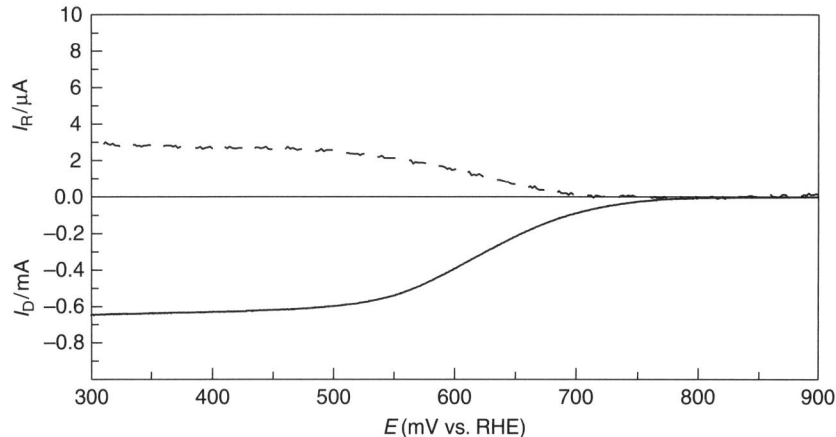

Figure 11.18 Linear cyclic voltammograms of a FePc/C disk electrode and corresponding oxidation current of a Pt ring electrode maintained at 1.2 V vs. RHE, recorded at 2500 rev min^{-1} in an O_2-saturated 0.5 M H_2SO_4 electrolyte (temperature 20 °C, sweep rate 5 mV s^{-1}).

likely that the structure of β-FePc in perpendicular stacks in the unit cell [Kirner et al., 1976] leads either to an increase in the metal–metal distance, which favors the adsorption of oxygen as FePc–O_2 peroxo species rather than as PcFe–O_2–FePc μ-oxo species, or to a decrease in the accessibility of molecular oxygen to the catalytic centers. α-FePc appears then as the better of the catalysts for fuel cell applications in terms of higher current densities achieved at higher cathode potentials. Moreover, rotating ring–disk electrode experiments (Fig. 11.18) indicated that α-FePc catalyzes a four-electron process of oxygen reduction to form water only for potentials higher than 0.8 V vs. RHE (which correspond to the cathode potentials of fuel cells under operating conditions) and with an efficiency not less than 96% for higher overpotentials [Baranton et al., 1995, 1996]. H_2O_2 formation at β-FePc was not evaluated; however, considering the limited current density achieved (which is lower than that obtained with α-FePc), it can reasonably be assumed that water and hydrogen peroxide are formed at this catalyst.

11.6 CONCLUSIONS

This chapter has presented the main features concerning electrocatalysts for DAFCs in acidic media. To achieve the best electrical efficiency, it is necessary to develop electrocatalysts able to oxidize the alcohol completely to CO_2 and to reduce oxygen completely to water with high reaction rates. For the ORR, Pt-based bimetallic catalysts and non-noble metal catalysts have shown higher activity than pure Pt, and, moreover, exhibit a higher tolerance or even total insensitivity to the presence of alcohol. On the other hand, in a strongly acidic medium as in a DEFC using a proton exchange membrane, Pt is necessary to realize the dissociative adsorption of the alcohol, but is easily poisoned by adsorbed species such as CO. To reduce the effect of poisoning, it is necessary to add to the Pt anode other metals, such as Sn, Ru, or Mo. For example, Pt-Ru catalysts exhibit higher activity for methanol electro-oxidation, whereas Pt-Sn catalysts exhibit higher activity for ethanol electro-oxidation, both in an electrochemical half-cell as well as in a DAFC. However, in the case of ethanol, selectivities with regard both to CO_2 and to acetaldehyde are lowered, whereas selectivity with regard to acetic acid is increased. Therefore, it seems that the possibility of obtaining a total faradaic efficiency ε_F (selectivity of the catalyst) and a higher potential efficiency ε_E (poisoning of Pt) has to be abandoned. Still, one goal could be to achieve a total selectivity of the electrocatalyst with regard to the production of acetic acid, which is a liquid and easy to manage, but it would provide only one-third of the faradaic efficiency ($\varepsilon_F = 0.33$), and thus leads to a decrease in the overall efficiency and power density by two-thirds.

Solid alkaline membrane fuel cells (SAMFCs) can be a good alternative to PEMFCs. The activation of the oxidation of alcohols and reduction of oxygen occurring in fuel cells is easier in alkaline media than in acid media [Wang et al., 2003; Yang, 2004]. Therefore, less Pt or even non-noble metals can be used owing to the improved electrode kinetics. For example, Ag/C catalytic powder can be used as an efficient cathode material [Demarconnay et al., 2004; Lamy et al., 2006]. It has also been shown that both doped and undoped manganese oxides [Chatenet et al., 2006] display good activity and selectivity with regard to a four-electron reduction process, and therefore could be used as cathode catalysts in SAMFCs. In alkaline media, MN_4 catalysts such as porphyrins and phthalocyanines display higher activity with regard to ORR and higher selectivity with regard to water production than in acidic media. Moreover, in such systems, crossover of fuel is limited owing to the transfer of hydroxyl ions from the cathode side to the anode side by electro-osmosis. For the oxidation reaction of alcohol in this system, Pt-based anodes are mainly used [Matsuoka et al., 2005; Varcoe and Slade, 2006; Yu and Scott, 2004; Demarconnay et al., 2007]. Bianchini's group has developed a non-noble Co-based catalyst [Barbaro et al., 2006], and has claimed that these catalysts are active for ethanol electro-oxidation in an alkaline medium, and could be used in a DEFC. Finally, in a SAMFC, the use of a solid polymer as electrolyte [Matsuoka et al., 2005; Varcoe and Slade, 2006; Yu and Scott, 2004; Demarconnay et al., 2007] instead of KOH gel, which is very corrosive, avoids the problem of safety. It also limits the carbonation of the electrolyte by gaseous impurities or reaction products.

ACKNOWLEDGEMENT

Part of this chapter is based on thesis works by Laetitia Dubau, Stève Baranton, and Séverine Rousseau. We also acknowledge Emmanuel Garnier for his helpful contributions to determining the crystallographic structure of catalysts.

REFERENCES

Adžič RR. 1984. Electrocatalytic properties of surfaces modified by foreign metal adatoms. In: Gerisher H, Tobias CW, eds. Advances in Electrochemistry and Electrochemical Engineering. Volume 13. New York: Wiley-Interscience. pp. 159–260.

Alonso-Vante N, Cattarin S, Musiani M. 2000. Electrocatalysis of O_2 reduction at polyaniline + molybdenum-doped ruthenium selenide composite electrodes. J Electroanal Chem 481: 200–207.

Ballirano P, Caminiti R, Ercolani C, Maras A, Orrù MA. 1998. X-ray powder diffraction structure reinvestigation of the α and β forms of cobalt phthalocyanine and kinetics of the $\alpha \rightarrow \beta$ phase transition. J Am Chem Soc 120: 12798–12807.

Baranton S, Coutanceau C, Roux C, Hahn F, Léger JM. 2005. Oxygen reduction reaction in acid medium at iron phthalocyanine dispersed on high surface area carbon substrate: tolerance to methanol, stability and kinetics. J Electroanal Chem 577: 223–234.

Baranton S, Coutanceau C, Garnier E, Léger JM. 2006. How does α-FePc catalysts dispersed onto high specific surface carbon support work towards oxygen reduction reaction (orr)? J Electroanal Chem 590: 100–110.

Barbaro P, Bert P, Bianchini C, Giambastiani G, Tampucci A, Vizza F. 2006. Catalysts for fuel cells electrodes based on cobalt and its alloys, their preparation and use and fuel cells containing them. European Patent PCT/EP/2005/053576; World Patent WO 2006/008319.

Bashyam R, Zelenay P. 2006. A class of non-precious metal composite catalysts for fuel cells. Nature 443: 63–66.

Batista EA, Malpass GRP, Motheo AJ, Iwasita T. 2004. New mechanistic aspects of methanol oxidation. J Electroanal Chem 571: 273–282.

Beard BC, Ross PN. 1990. The structure and activity of Pt-Co alloys as oxygen reduction electrocatalysts. J Electrochem Soc 137: 3368–3374.

Beden B. 1984. In situ infra-red reflection spectroscopy actual developments at the electrode–solution interface. II. Applications. Spectra 2000 12: 31–39.

Beden B, Juanto S, Léger JM, Lamy C. 1987. Infrared spectroscopy of the methanol adsorbates at a platinum electrode: Part III. Structural effects and behaviour of the polycrystalline surface. J Electroanal Chem 238: 323–331.

Bett JS, Kunz HR, Aldykiewicz AJ Jr, Fenton JM, Bailey WS, Mc Grath DV. 1998. Platinum-macrocycle co-catalysts for the electrochemical oxidation of methanol. Electrochim Acta 43: 3645–3655.

Blasini DR, Rochefort D, Fachini E, Alden LR, DiSalvo FJ, Cabrera CR, Abruña HD. 2006. Surface composition of ordered intermetallic compounds PtBi and PtPb. Surf Sci 600: 2670–2680.

Bönnemann H, Brijoux W, Brinkmann R, Dinjus E, Joussen T, Korall B. 1991. Formation of colloidal transition metals in organic phases and their application in catalysis. Angew Chem Int Ed 30: 1312–1314.

Bönnemann H, Braun G, Brijoux W, Brinkmann R, Schultze Tilling A, Seevogel K, Siepen K. 1996. Nanoscale colloidal metals and alloys stabilized by solvents and surfactants. Preparation and use as catalyst precursors. J Organomet Chem 520: 143–162.

Brankovic SR, Wang JX, Zhu Y, Sabatini R, McBreen J, Adzic RR. 2002a. Electrosorption and catalytic properties of bare and Pt modified single crystal and nanostructured Ru surfaces. J Electroanal Chem 524/525: 231–241.

Brankovic SR, Marinkovic NS, Wang JX, Adzic RR. 2002b. Carbon monoxide oxidation on bare and Pt-modified Ru($10\bar{1}0$) and Ru(0001) single crystal electrodes. J Electroanal Chem 532: 57–66.

Caillard A, Coutanceau C, Brault P, Mathias J, Léger JM. 2006. Structure of Pt/C and PtRu/C catalytic layers prepared by plasma sputtering and electric performance in direct methanol fuel cells (DMFC). J Power Sources 162: 66–73.

Campbell S. 2006. Ballard Power System. Development of transition metal/chalcogen based cathode catalysts for PEM fuel cells. DOE Hydrogen Program Review, May 16–19, Washington, DC. Available at: http://www.hydrogen.energy.gov/annual_review06_fuelcells.html (click on "catalysts" section).

Chang SC, Leung LW, Weaver MJ. 1990. Metal crystallinity effects in electrocatalysis as probed by real-time FTIR spectroscopy: Electrooxidation of formic acid, methanol, and ethanol on ordered low-index platinum surfaces. J Phys Chem 94: 6013–6021.

Chatenet M, Micoud F, Roche I, Chainet E, Vondràk J. 2006. Kinetics of sodium borohydride direct oxidation and oxygen reduction in sodium hydroxide electrolyte. Part II. O_2 reduction. Electrochim Acta 51: 5452–5458.

Chu D, Gilman S. 1996. Methanol electro-oxidation on unsupported Pt-Ru alloys at different temperatures. J Electrochem Soc 143: 1685–1690.

Collman JP, Denisevich P, Konai Y, Marrocco M, Koval C, Anson FC. 1980. Electrode catalysis of the four-electron reduction of oxygen to water by dicobalt face-to-face porphyrins. J Am Chem Soc 102: 6027–6036.

Convert P, Coutanceau C, Crouigneau P, Gloaguen F, Lamy C. 2001. Electrodes modified by electrodeposition of CoTAA complexes as selective cathodes in a direct methanol fuel cell. J Appl Electrochem 31: 945–952.

Coutanceau C, Crouigneau P, Léger JM, Lamy C. 1994. Mechanism of oxygen electroreduction at polypyrrole electrodes modified by cobalt phthalocyanine. J Electroanal Chem 379: 389–397.

Coutanceau C, El Hourch A, Crouigneau P, Léger JM, Lamy C. 1995. Conducting polymer electrodes modified by metal tetrasulfonated phthalocyanines: Preparation and electrocatalytic behaviour towards dioxygen reduction in acid medium. Electrochim Acta 40: 2739–2748.

Coutanceau C, Rakotondrainibe AF, Lima A, Garnier E, Pronier S, Léger JM, Lamy C. 2004. Preparation of Pt-Ru bimetallic anodes by galvanostatic pulse electrodeposition: characterization and application to the direct methanol fuel cell. J Appl Electrochem 34: 61–66.

Coutanceau C, Koffi RK, Léger JM, Marestin C, Mercier R, Nayoze C, Capron P. 2006. Development of materials for mini DMFC working at room temperature for portable applications. J Power Sources 160: 334.

Damjanovic A, Bockris JO'M. 1966. The rate constants for oxygen dissolution on bare and oxide-covered platinum. Electrochim Acta 11: 376–377.

Damjanovic A, Genshaw MA, Bockris JO'M. 1967. The mechanism of oxygen reduction at platinum in alkaline solutions with special reference to H_2O_2. J Electrochem Soc 114: 1107–1112.

Delime F, Léger JM, Lamy C. 1999. Enhancement of the elctrooxidation of ethanol on a Pt-PEM electrode modified by tin. Part I: Half cell study. J Appl Electrochem 29: 1249–1254.

Demarconnay L, Coutanceau C, Léger JM. 2004. Electroreduction of dioxygen (ORR) in alkaline medium on Ag/C and Pt/C nanostructured catalysts-effect of the presence of methanol. Electrochim Acta 49: 4513–4521.

Demarconnay L, Brimaud S, Coutanceau C, Léger JM. 2007. Ethylene Glycol electrooxidation in alkaline medium at pluri-metallic Pt based catalysts. J Electroanal Chem 601: 169–180.

Dinh HN, Ren X, Garzon FH, Zelenay P, Gottesfeld S. 2000. Electrocatalysis in direct methanol fuel cells: in-situ probing of PtRu anode catalyst surfaces. J Electroanal Chem 491: 222–233.

Dubau L. 2002. Electrocatalyseurs platine-ruthénium nanodispersés pour une pile à combustion directe de methanol. PhD Thesis, University of Poitiers, France.

Dubau L, Coutanceau C, Garnier E, Léger JM, Lamy C. 2003a. Electrooxidation of methanol at platinum–ruthenium catalysts prepared from colloidal precursors: Atomic composition and temperature effects. J Appl Electrochem 33: 419–429.

Dubau L, Hahn F, Coutanceau C, Léger JM, Lamy C. 2003b. On the structure effects of bimetallic PtRu electrocatalysts towards methanol oxidation. J Electroanal Chem 554/555: 407–415.

Gasteiger HA, Markovic N, Ross PN, Cairns EJ. 1993. Methanol electrooxidation on well-characterized platinum–ruthenium bulk alloys. J Phys Chem 97: 12020–12029.

Gasteiger HA, Markovic N, Ross PN, Cairns EJ. 1994. Temperature-dependent methanol electro-oxidation on well-characterized Pt-Ru alloys. J Electrochem Soc 141: 1795–1803.

Gnanamuthu DS, Petrcelli JV. 1967. A generalized expression for the Tafel slopes and the kinetics of oxygen reduction on noble metal and alloys. J Electrochem Soc 114: 1036–1041.

Gosselink JW. 2002. Pathways to a more sustainable production of energy: Sustainable hydrogen—a research objective for Shell. Int J Hydrogen Energy 27: 1125–1129.

Hamnett A. 1999. Mechanism of methanol oxidation. In: Wieckowski A, ed. Interfacial Electrochemistry: Theory, Experiments and Applications. New York: Marcel Dekker. pp. 843–883.

Hogarth MP, Ralph TR. 2002. Catalysis for low temperature fuel cells. Part III: Challenges for the direct methanol fuel cell. Platinum Metals Rev 46: 146–164.

Ianniello R, Schmidt VM, Stimming U, Stumper J, Wallau A. 1994. CO adsorption and oxidation on Pt and Pt-Ru alloys: Dependence on substrate composition. Electrochim Acta 39: 1863–1869.

Iwasita T, Nart FC. 1991. Identification of methanol adsorbates on platinum: An in situ FT-IR investigation. J Electroanal Chem 317: 291–298.

Iwasita T, Pastor E. 1994a. A DEMS and FTIR spectroscopic investigation of adsorbed ethanol on polycrystalline platinum. Electrochim Acta 39: 531–537.

Iwasita T, Pastor E. 1994b. D/H exchange of ethanol at platinum electrodes. Electrochim Acta 39: 547–551.

Iwasita T., Rasch B., Cattaneo E., Vielstich W. 1989. A SNIFTIRS study of ethanol oxidation on platinum. Electrochim Acta 34: 1073–1079.

Iwasita T, Hoster H, John-Anaker A, Lin WF, Vielstich W. 2000. Methanol oxidation on PtRu electrodes. Influence of surface structure and Pt-Ru atom distribution. Langmuir 16: 522–529.

Jiang R, Chu D. 2000. Remarkably active catalysts for the electroreduction of O_2 to H_2O for use in acidic electrolyte containing concentrated methanol. J Electrochem Soc 147: 4605–4609.

Jusys Z., Schmidt TJ, Dubau L, Lasch K, Jörissen L, Garche J, Behm RJ. 2002. Activity of $PtRuMeO_x$ (Me = W, Mo or V) catalysts towards methanol oxidation and their characterization. J Power Sources 105: 297–304.

Kabbabi A, Faure R, Durand R, Beden B, Hahn F, Léger JM, Lamy C. 1998. In situ FTIRS study of the electrocatalytic oxidation of carbon monoxide and methanol at platinum–ruthenium bulk alloy electrodes. J Electroanal Chem 444: 41–53.

Kinoshita K. 1992. Electrochemical Oxygen Technology. New York: Wiley. pp. 46–48.

Kirner JF, Dow W, Scheidt WR. 1976. Molecular stereochemistry of two intermediate-spin complexes. Iron(II) phthalocyanine and manganese(II) phthalocyanine. Inorg Chem 15: 1685–1690.

Koffi RC, Coutanceau C, Garnier E, Léger JM, Lamy C. 2005. Synthesis, characterization and electrocatalytic behaviour of non-alloyed PtCr methanol tolerant nanoelectrocatalysts for the oxygen reduction reaction (ORR). Electrochim Acta 50: 4117–4127.

Lalande G, Côté R, Tamizhmani G, Guay D, Dodelet JP. 1995. Physical, chemical and electrochemical characterization of heat-treated tetracarboxylic cobalt phthalocyanine adsorbed on carbon black as electrocatalyst for oxygen reduction in polymer electrolyte fuel cells. Electrochim Acta 40: 2635–2646.

Lalande G, Faubert G, Côté R, Guay D, Dodelet JP, Weng LT, Bertrand P. 1996. Catalytic activity and stability of heat-treated iron phthalocyanines for the electroreduction of oxygen in polymer electrolyte fuel cells. J Power Sources 61: 227–237.

Lamy C, Léger JM. 1994. Les piles à combustible: application au véhicule électrique. J Phys IV C1: 253–281.

Lamy C, Rousseau S, Belgsir EM, Coutanceau C, Léger JM. 2004. Recent progress in the direct ethanol fuel cell: Development of new platinum–tin electrocatalysts. Electrochim Acta 49: 3901–3908.

Lamy C, Demarconnay L, Coutanceau C, Leger JM. 2006. Development of electrocatalysts for the solid alkaline membrane fuel cell (SAMFC). ECS Trans 3(1): 1351–1360.

Léger JM, Rousseau S, Coutanceau C, Hahn F, Lamy C. 2005. How bimetallic electrocatalysts does work for reactions involved in fuel cells? Example of ethanol oxidation and comparison to methanol. Electrochim Acta 50: 5118–5125.

Lei HW, Suh S, Gurau B, Workie B, Liu R, Smotkin ES. 2002. Deuterium isotope analysis of methanol oxidation on mixed metal anode catalysts. Electrochim Acta 47: 2913–2919.

Lemire C, Meyer R, Shaikhutdinov S, Freund HJ. 2004. Do quantum size effects control CO adsorption on gold nanoparticles? Angew Chem Int Ed 43: 118–121.

Lima A, Coutanceau C, Léger JM, Lamy C. 2001. Investigation of ternary catalysts for methanol electrooxidation. J Appl Electrochem 31: 379–386.

Lu C, Rice C, Masel MI, Babu PK, Waszczuk P, Kim HS, Oldfield E, Wieckowski A. 2002. UHV, electrochemical NMR, and electrochemical studies of platinum/ruthenium fuel cell catalysts. J Phys Chem B 106: 9581–9589.

Lu GQ, Wang CY, Yen TJ, Zhang X. 2004. Development and characterization of a silicon-based micro direct methanol fuel cell. Electrochim Acta 49: 821–828.

Matsuoka K, Iriyama Y, Abe T, Matsuoka M, Ogumi Z. 2005. Alkaline direct alcohol fuel cells using an anion exchange membrane. J Power Sources 150: 27–31.

Mukerjee S, Srinivasan S, Soriaga MP, McBreen J. 1995. Role of structural and electronic properties of Pt and Pt alloys on electrocatalysis of oxygen reduction. J Electrochem Soc 142: 1409–1422.

Neergat M, Shukla AK, Gandhi KS. 2001. Platinum-based alloys as oxygen–reduction catalysts for solid–polymer–electrolyte direct methanol fuel cells. J Appl Electrochem 31: 373–378.

Oedegaard A, Hebling C, Schmitz A, Møller-Holst S, Tunold R. 2004. Influence of diffusion layer properties on low temperature DMFC. J Power Sources 127: 187–196.

Paffett MT, Beery JG, Gottesfeld S. 1988. Oxygen reduction at $Pt_{0.65}Cr_{0.35}$, $Pt_{0.2}Cr_{0.8}$ and roughened platinum. J Electrochem Soc 135: 1431–1436.

Perez JM, Beden B, Hahn F, Aldaz A, Lamy C. 1989. "In situ" infrared reflectance spectroscopic study of the early stages of ethanol adsorption at a platinum electrode in acid medium. J Electroanal Chem 262: 251–261.

Podlovchenko BI, Petrii OA, Frumkin AN, Lal H. 1966. The behaviour of a platinized-platinum electrode in solutions of alcohols containing more than one carbon atom, aldehydes and formic acid. J Electroanal Chem 11: 12–25.

Rightmire RA, Rowland RL, Boos DL, Beals DL. 1964. Ethyl alcohol oxidation at platinum electrodes. J Electrochem Soc 111: 242–247.

Rousseau S. 2004. Développement d'électrocatalyseurs pour l'oxydation de l'éthanol dans une pile à combustion direct (DEFC). PhD Thesis, University of Poitiers, France.

Rousseau S, Coutanceau C, Lamy C, Léger JM. 2006. Direct ethanol fuel cell (DEFC): Electrical performances and reaction products distribution under operating conditions with different platinum-based anodes. J Power Sources 158: 18–24.

Schmidt TJ, Gasteiger HA, Behm RJ. 1999. Methanol electrooxidation on a colloidal PtRu-alloy fuel-cell catalyst. Electrochem Commun 1: 1–4.

Sepa DB, Vojnovic MV, Vracar Lj. M, Damjanovic A. 1986a. Apparent enthalpies of activation of electrodic oxygen reduction at platinum in different current density regions—I. Acid solution. Electrochim Acta 31: 91–96.

Sepa DB, Vojnovic MV, Vracar Lj M, Damjanovic A. 1986b. Invariance with pH of enthalpies of activation for O_2 reduction at Pt electrodes in acid solutions. Electrochim Acta 31: 1105–1111.

Shen M, Roy S, Kuhlmann JW, Scott K, Lovell K, Horsfall JA. 2004. Grafted polymer electrolyte membrane for direct methanol fuel cells. J Memb Sci 251: 121–130.

Shropshire JA. 1965. The catalysis of the electrochemical oxidation of formaldehyde and methanol by molybdates. J Electrochem Soc 112: 465–469.

Shukla A, Neergat M, Parthasarathi B, Jayaram V, Hegde MS. 2001. An XPS study on binary and ternary alloys of transition metals with platinized carbon and its bearing upon oxygen electroreduction in direct methanol fuel cells. J Electroanal Chem 504: 111–119.

Sidik R, Anderson A. 2006. Co_9S_8 as a catalyst for electroreduction of O_2: Quantum chemistry predictions. J Phys Chem B 110: 936–941.

Smirnova NW, Petrii OA, Grzejdziak A. 1988. Effect of ad-atoms on the electro-oxidation of ethylene glycol and oxalic acid on platinized platinum. J Electroanal Chem 251: 73–87.

Ströbel R, Oszcipok M, Fasil M, Rohland B, Jörissen L, Garche J. 2002. The compression of hydrogen in an electrochemical cell based on a PE fuel cell design. J Power Sources 105: 208–215.

Toda T, Igarashi H, Watanabe M. 1999a. Enhancement of the electrocatalytic O_2 reduction on Pt-Fe alloys. J Electroanal Chem 460: 258–262.

Toda T, Igarashi H, Ushida H, Watanabe M. 1999b. Enhancement of the electroreduction of oxygen on Pt alloys with Fe, Ni and Co. J Electrochem Soc 146: 3750–3756.

Tong YY, Kim HS, Babu PK, Waszczuk P, Wieckowski A., Olfield E. 2002. An NMR investigation of CO tolerance in a Pt/Ru fuel cell catalyst. J Am Chem Soc 124: 468–473.

Varcoe JR, Slade RCT. 2006. An electron-beam-grafted ETFE alkaline anion-exchange membrane in metal-cation-free solid-state alkaline fuel cells. Electrochem Commun 8: 839–843.

Vigier F, Coutanceau C, Hahn F, Belgsir EM, Lamy C. 2004a. On the mechanism of ethanol electro-oxidation on Pt and PtSn catalysts: Electrochemical and in situ IR reflectance spectroscopy studies. J Electroanal Chem 563: 81–89.

Vigier F, Coutanceau C, Perrard A, Belgsir EM, Lamy C. 2004b. Development of anode catalysts for a direct ethanol fuel cell. J Appl Electrochem 34: 439–446.

Wang Y, Li L, Hu L, Zhuang L, Lu J, Xu B. 2003. A feasibility analysis for alkaline membrane direct methanol fuel cell: Thermodynamic disadvantages versus kinetic advantages. Electrochem Commun 5: 662.

Wasmus S, Küver A. 1999. Methanol oxidation and direct methanol fuel cells: A selective review. J Electroanal Chem 461: 14–31.

Waszczuk P, Solla-Gullón J, Kim HS, Tong YY, Montiel V, Aldaz A, Wieckowski A. 2001a. Methanol electrooxidation on platinum/ruthenium nanoparticle catalysts. J Catal 203: 1–6.

Waszczuk P, Wieckowski A, Zelenay P, Gottesfeld S, Coutanceau C, Léger JM, Lamy C. 2001b. Adsorption of CO poison on fuel cell nanoparticle electrodes from methanol solutions: A radioactive labeling study. J Electroanal Chem 511: 55–64.

Waszczuk P, Lu GU, Wieckowski A, Lu C, Rice C, Masel MI. 2002. UHV and electrochemical studies of CO and methanol adsorbed at platinum/ruthenium surfaces, and reference to fuel cell catalysis. Electrochim Acta 47: 3637–3652.

Watanabe M, Motoo S. 1975a. Electrocatalysis by ad-atoms: Part II. Enhancement of the oxidation of methanol on platinum by ruthenium ad-atoms. J Electroanal Chem 60: 267–273.

Watanabe M, Motoo S. 1975b. Electrocatalysis by ad-atoms: Part III. Enhancement of the oxidation of carbon monoxide on platinum by ruthenium ad-atoms. J Electroanal Chem 60: 275–283.

Watanabe M, Uchida M, Motoo S. 1987. Preparation of highly dispersed Pt+Ru alloy clusters and the activity for the electrooxidation of methanol. J Electroanal Chem 229: 395–406.

Xia XH, Liess HD, Iwasita T. 1997. Early stages in the oxidation of ethanol at low index single crystal platinum electrodes. J Electroanal Chem 437: 233–240.

Yamazaki Y. 2004. Application of MEMS technology to micro fuel cells. Electrochim Acta 50: 663–666.

Yang CC. 2004. Preparation and characterization of electrochemical properties of air cathode electrode. Int J Hydrogen Energy 29: 135–143.

Yang H, Alonso-Vante N, Léger JM, Lamy C. 2004. Tailoring, structure, and activity of carbon-supported nanosized Pt-Cr alloy electrocatalysts for oxygen reduction in pure and methanol-containing electrolytes. J Phys Chem 108: 1938–1947.

Yu EH, Scott K. 2004. Development of direct methanol alkaline fuel cells using anion exchange membranes. J Power Sources 137: 248–256.

Zagal J, Bindra P, Yeager E. 1980. A mechanistic study of O_2 reduction on water soluble phthalocyanines adsorbed on graphite electrodes. J Electrochem Soc 127: 1506–1517.

Zhou WJ, Li WZ, Song SQ, Zhou ZH, Jiang LH, Sun GQ, Xin Q, Poulianitis K, Kontou S, Tsiakaras P. 2004a. Bi- and tri-metallic Pt-based anode catalysts for direct ethanol fuel cells. J Power Sources 131: 217–223.

Zhou WJ, Song SQ, Li WZ, Sun GQ, Xin Q, Kontou S, Poulianitis K, Tsiakaras P. 2004b. Pt-based anode catalysts for direct ethanol fuel cells. Solid State Ionics 175: 797–803.

Zhou WJ, Song SQ, Li WZ, Zhou ZH, Sun GQ, Xin Q, Douvartzides S, Tsiakaras P. 2005. Direct ethanol fuel cells based on PtSn anodes: The effect of Sn content on the fuel cell performance. J Power Sources 140: 50–58.

CHAPTER 12

Broadband Sum Frequency Generation Studies of Surface Intermediates Involved in Fuel Cell Electrocatalysis

G. Q. LU, A. LAGUTCHEV, T. TAKESHITA, R. L. BEHRENS, DANA D. DLOTT, and A. WIECKOWSKI

Department of Chemistry, University of Illinois at Urbana-Champaign, Urbana, IL 61801, USA

12.1 INTRODUCTION

In this chapter, we discuss the use of broadband multiplex vibrational sum frequency generation (BB-SFG) spectroscopy to study electrochemical interfaces [Lu et al., 2005; Lagutchev et al., 2006]. In all vibrational SFG techniques, visible and vibrational infrared (IR) pulses are incident simultaneously on the sample of interest, and a coherent sum frequency signal in the visible range is produced (see Figs. 12.1–12.3) (Plate 12.1). In BB-SFG, as opposed to conventional (first-generation) SFG, where spectra are obtained point-by-point, the entire spectral range spanned by the IR pulse spectrum is probed simultaneously. In SFG, attenuation of the infrared beam by the electrolyte is less of a concern than in IR reflection absorption spectroscopy (IRAS), since the infrared beam need not double-pass the electrolyte in SFG. SFG is a well-known spectroscopic technique [Shen, 1989; Guyot-Sionnest and Tadjeddine, 1990; Bonn et al., 2001] and the subject of several reviews [Tadjeddine, 2000; Shen, 1994; Eisenthal, 1992; Richmond, 2002; Chen et al., 2002]. There have been a few works prior to our studies where first-generation SFG has been applied to electrochemical interfaces [Lu et al., 2004, 2005; Lagutchev et al., 2006; Guyot-Sionnest and Tadjeddine, 1990; Guyot-Sionnest, 2005; Guyot-Sionnest et al., 1987; Chou et al., 2003a, b; Hoffer et al., 2002]. This chapter is focused on the use of BB-SFG (second- and third-generation SFG) to study electrochemical interfaces.

An SFG signal exhibits characteristic features at frequencies of molecular vibrational resonances similar to the vibrational fingerprints of conventional IR

Fuel Cell Catalysis. Edited by Marc T. M. Koper
Copyright © 2009 John Wiley & Sons, Inc.

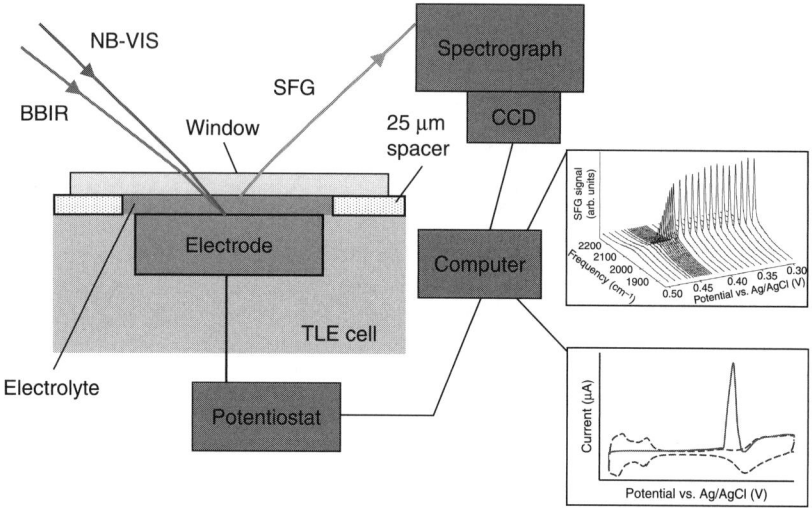

Figure 12.1 Schematic of the spectroelectrochemistry apparatus at the University of Illinois. The thin-layer spectroelectrochemical cell (TLE cell) has a 25 μm thick spacer between the electrode and window to control the electrolyte layer thickness and allow for reproducible refilling of the gap. The broadband infrared (BBIR) and narrowband visible (NBVIS) pulses used for BB-SFG spectroscopy are generated by a femtosecond laser (see Fig. 12.3). Voltammetric and spectrometric data are acquired simultaneously.

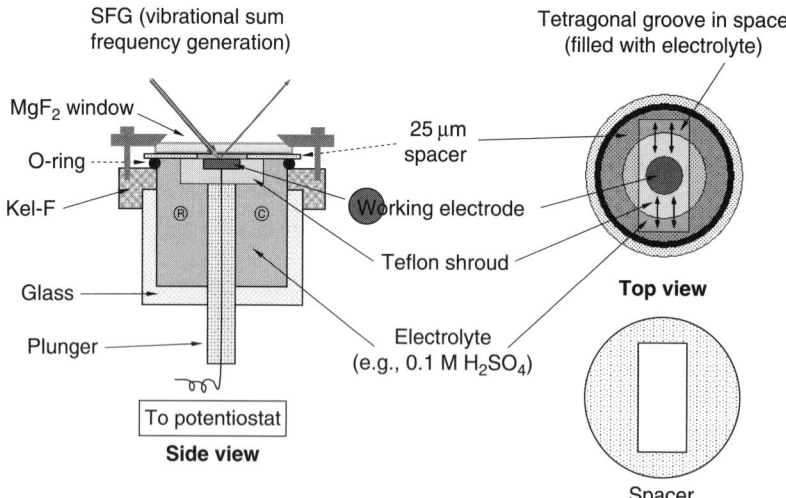

Figure 12.2 The electrochemical cell has a 25 μm Teflon spacer sandwiched between the electrode and a window (CaF_2 or MgF_2) to provide an electrolyte layer of known and controlled thickness. Working, reference, and auxiliary electrodes are indicated. Construction materials are glass and Teflon.

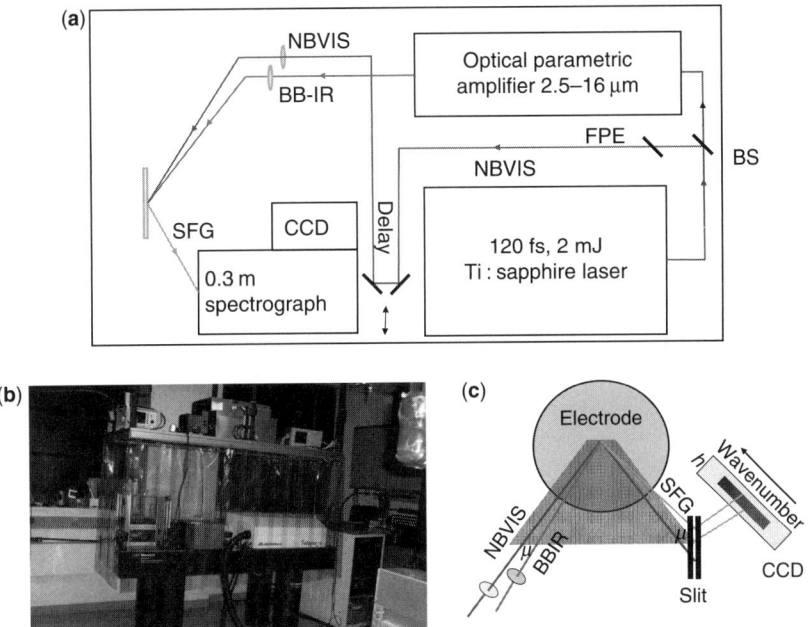

Figure 12.3 (a) Block diagram of the third-generation SFG spectrometer. A beam splitter (BS) directs $\frac{1}{4}$ of the output through a Fabry–Perot etalon (FPE) to produce the narrowband visible (NBVIS) pulses. The remaining output pumps an optical parametric amplifier to generate the broadband IR pulses (BBIR). (b) Photograph of the SFG apparatus on a 2 m × 0.8 m optical table. The laser, spectrograph, and sample chamber are shown in the foreground. (c) Schematic of the arrangement for SFG phase-matching. Different BBIR frequencies result in slightly different SFG output angles, but all SFG signals pass through the spectrograph slit and illuminate slightly different height (h) regions of the CCD. Output spectra are obtained by integrating along the CCD h-direction. (See color insert.)

spectroscopy [Shen, 1989]. SFG helps solve the two biggest problems of surface molecular spectroscopy: *sensitivity* and *selectivity*. Sensitivity results from the nonlinear coherent nature. The signal is proportional to the product of IR and visible intensities at the surface, which is huge with focused femtosecond pulses [Patterson et al., 2005]. The signal is emitted coherently, so almost all SFG photons are collected and detected by high sensitivity detectors. Selectivity for surfaces or interfaces results from the symmetry properties of the $\chi^{(2)}$ tensor, which in the dipole approximation vanishes in centrosymmetric media [Shen, 1989; Bonn et al., 2000]. Despite the low number density of interfacial molecules, the combination of intense femtosecond pulses (typically 10 GW/cm^2), coherence and the high sensitivity of CCD detectors to the visible sum-frequency photons generate excellent signals from many interfacial species.

Figures 12.1 and 12.2 show that the spectroelectrochemical cell is basically a thin-layer electrochemistry cell (TLE) with a solution gap of 25 μm [Hubbard, 1973]. The metal working electrode may be polycrystalline or a single crystal. Emptying the gap out of the adsorbate molecules due to molecules' oxidation, and refilling via molecular diffusion to the cell interior (via the channels; Fig. 12.2), can occur without changing the electrode/optical window configuration. As the working electrode is supported by a spacer, the solution gap is uniform. Therefore, the electrode may be cycled at a scan rate (up to 5 mV/s) producing state-of-the-art, ohmic-drop-free voltammetric data characteristic of the average properties of the full surface. During the scan, the surface is investigated by BB-SFG, providing BB-SFG (vibrational)–electrochemistry characteristics, as already reported [Lu et al., 2005; Lagutchev et al., 2006].

SFG also has clear advantages compared with IRAS [Kunimatsu et al., 1985a, b]. In IRAS, the signal is proportional to the IR intensity exiting the optical cell, after it has double-passed the electrolyte layer, whereas in SFG, the signal is proportional to the product of IR times visible at the electrode. The IR has only to single-pass the electrolyte layer, and the visible is barely attenuated by the electrolyte (Figs. 12.1 and 12.2). Although the effects of electrolyte IR absorption can be mitigated in both techniques by increasing the IR intensity, SFG has advantages in this regard as well. IRAS uses a thermal IR source whose brightness is limited by the meltdown temperature of the blackbody. Since the source is continuous, high intensities will result in high temperatures in the electrolyte and on the electrode surface. In SFG, the IR is generated by a laser, and although the optical physics are rather complicated, the IR pulse energy is approximately linear in the pump laser energy and is thereby limited mostly by how much one chooses to spend on the pump laser. Since the IR pulses are a low duty cycle pulsed source, where the laser is typically "on" for a short period every 1 ms, even with high intensity pulses the steady-state rises in electrolyte and surface temperatures are minimal.

In the earliest SFG experiments [Tadjeddine, 2000; Guyot-Sionnest et al., 1987; Hunt et al., 1987; Zhu et al., 1987], a first-generation data acquisition method was used, and, because of the limited signal-to-noise ratios, IR attenuation by the electrolyte solution was a substantial handicap. So, in earlier SFG studies, as in IRAS studies, measurements were performed with the electrode pressed directly against the optical window [Baldelli et al., 1999; Dederichs et al., 2000]. With the in-contact geometry, the electrolyte was a thin film of uncertain and variable depth, probably of the order of 1 μm. However, the thin nonuniform electrolyte layers can strongly distort the potential/coverage relationship and hinder the ability to study fast kinetics.

In our first BB-SFG experiments, we used a second-generation SFG spectrometer based on the scheme developed by Richter et al. [1998]. More recently, we have developed a compact new apparatus that features greatly enhanced user functionality and improved signal-to-noise ratios (the third-generation SFG). Since the original apparatus has been described in detail previously [Lu et al., 2005; Lagutchev et al., 2005, 2006], Section 12.2 below will focus on the new apparatus. The combination of BB-SFG with a TLE cell having a known controlled electrolyte gap might seem, at first glance, to be a mere technical improvement, but in actuality it represents a major advance [Guyot-Sionnest, 2005], as demonstrated in our recent publications [Lu et al., 2005; Lagutchev et al., 2006]. The SFG apparatus can obtain spectra in 200 ms or less, with a signal-to-noise ratio $> 100:1$. The thin electrolyte film and rapid spectral acquisition rate minimize the possibilities for chemical contamination. They also enable a variety of investigations involving measurements of chemical kinetics on millisecond time scales commensurate with rates of surface poisoning in fuel cell electrocatalysis.

12.2 EXPERIMENTAL

All experiments were carried out at ambient temperature (21 ± 2 °C). Potentials were measured against a commercial Ag/AgCl electrode (BAS), and are reported versus this reference scale. To date, we have investigated the following samples: a smooth polycrystalline Pt disk, a Pt single crystal 6 mm in diameter cut along (111) and (100) orientations, and a smooth polycrystalline Au disk with immobilized Pt nanoparticles (approximately 6 nm). The single crystal was prepared by annealing in a hydrogen/air flame, and cooling in an Ar/H_2 atmosphere and in ultrapure water [Clavilier, 1999]. The spontaneous deposition of Ru on Pt(111) was carried out from solutions of 1 mM of $RuCl_3$ in 0.1 M of $HClO_4$ in a preparative electrochemical cell [Crown and Wieckowski, 2001; Chrzanowski and Wieckowski, 1997; Crown et al., 2001, 2002]. After Ru deposition, the electrode was removed from the cell, rinsed with Millipore water, and "stabilized" by three voltammetric cycles between 0.01 and 0.60 V in 0.1 M H_2SO_4 electrolyte (single spontaneous deposition [Crown et al., 2002]). Cyclic voltammetric and chronoamperometric measurements were carried out using a PAR 263A potentiostat through Corrware software. The chemicals used were sulfuric, perchloric, formic, and acetic acids (GFS Chemicals), methanol (Fisher), $RuCl_3 \cdot xH_2O$ (Aldrich), CO (Matheson, research purity), and Millipore water.

The third-generation SFG apparatus is shown in Fig. 12.3. The heart of the system is a compact femtosecond laser (Quantronix Integra-C 2.0) having a diode-pumped fiber oscillator and a chirped-pulse Ti:sapphire amplifier. The output pulses at 800 nm are 120 fs in duration with 2 mJ energy at a repetition rate of 1 kHz. Three-fourths of the output energy is used to pump an optical parametric amplifier (OPA, Light Conversion TOPAS-C 800 fs DFG) equipped with both $AgGaS_2$ and GaSe difference-frequency output crystals. The OPA generates the broadband infrared (BBIR) pulses, which are tunable from 2.5 to 16 μm. The remainder of the laser output is sent through a Fabry–Perot air-spaced etalon (TecOptics) to create narrowband picosecond visible pulses (NBVIS) at 800.0 nm. These NBVIS pulses are asymmetric in time, having a steep (about 120 fs) rising edge and a gradual (a few picoseconds) falling edge, which makes them useful for suppressing the nonresonant contribution [Bain, 1995] from the metal surface [Shen, 1989]. The NBVIS and BBIR pulses are made time-coincident using an optical delay line, and both are made spatially coincident on the electrode surface. Typical pulse energies are 5 μJ in the IR and 20 μJ in the visible. In previous work, following [Patterson et al., 2005; Lagutchev et al., 2005], we made the BBIR and NBVIS pulses collinear with a dichroic beam splitter and focused them with a common CaF_2 lens. This set-up proved lossy and cumbersome, with suboptimal aberrent beam focusing, and it did not support IR wavelength tuning below 10 μm owing to CaF_2 absorption. The new apparatus uses a more efficient design shown in Fig. 12.3c, where the pulses are independently focused by individual lenses: a glass lens for the NBVIS and an antireflection-coated ZnSe lens for the BBIR.

The focused spot sizes ($1/e^2$ beam diameter) are 200 μm, and as the IR wavelength is varied in the 2.5–16 μm range, the IR spot size can be maintained by refocusing the ZnSe lens or by switching to a different ZnSe focal length. The SFG output pulses are filtered using a short wavelength pass filter and detected using a 0.3 m spectrograph and CCD (Andor Corp.).

In BB-SFG, all vibrations lying within the spectral region defined by the BBIR pulse spectrum are probed simultaneously. With our laser, this region is about 250–300 cm^{-1} in width. The spectral resolution is set by the spectral width of the NBVIS pulses, which in turn is set by the spectral bandpass of the etalon. We have several etalons available, providing a range of spectral widths. It is generally desirable to use the largest bandpass possible, because the intensity of the SFG signal is, all things being equal, proportional to the square of the etalon bandpass. Doubling the bandpass doubles the NBVIS pulse energy and reduces the NBVIS pulse duration by a factor of two. Since the vibrational transitions observed in our electrochemical measurements are typically 15 cm^{-1} or more in width, we ordinarily use an etalon that transmits a portion of the femtosecond visible pulse that is 11 cm^{-1} wide. When higher resolution is needed, we can add a second etalon in series to improve the resolution at the expense of signal.

The arrangement used to phase-match the BB-SFG signal is depicted in Fig. 12.3c. With a series of polarization rotators and analyzers, we can obtain spectra in all eight possible polarization conditions, but on metal electrodes we use only the *ppp*-polarization condition, where the two incident beams and the output beam are polarized in the plane of incidence of the NBVIS pulses. The BBIR pulses are incident upon the electrode from a few millimeters out of this plane. The spectrograph slit is perpendicular to this plane. Because there is a range of IR frequencies, the BB-SFG output has a range of output wavevectors. Since the CCD detector has a height of several millimeters, all these different wavevectors fall onto the CCD, albeit at slightly different heights h, and output spectra are obtained by integrating the signals along the h-axis. To align the system, the NBVIS beam is directed onto the slit, falling near the edge of the CCD detector. The short wavelength pass filter is then inserted. With this arrangement, tuning the BBIR pulse wavelength simply moves the SFG signal parallel to the spectrograph slit onto a slightly different region of the CCD detector, so the IR wavelength can be easily changed with minimal height realignment.

The TLE SFG electrochemical cell depicted in Fig. 12.2 was made of Kel-F and glass, with an optical window (CaF_2 or MgF_2) as the input window [Lu et al., 2005; Lagutchev et al., 2006; Vidal et al., 2002; Biggin and Gewirth, 2001]. A 25 μm thick Teflon spacer was placed between the electrode and the optical window. Originally [Lu et al., 2005] the Teflon holder was used as a plunger to compress the spacer between the working electrode and the optical window to create a uniform electrolyte layer of known thickness. In more recent work, single-crystal electrodes were used that were attached to the plunger without protecting the walls against the electrolyte contact. Voltammetry associated with spectral acquisition for the single crystals is obtained in the spectroelectrochemical cell in a meniscus configuration [Lagutchev et al., 2006].

The CCD detector and the potentiostat were synchronized in an open-loop configuration by starting the experiment with a common trigger. Careful measurements were used to show that both devices remained well synchronized throughout the duration of each SFG-electrochemistry run. The potentiostat scanned the electrode potential at 1–5 mV/s, while the CCD was usually read out at 5 spectra/s.

We have performed a detailed analysis of laser heating in Pt–CO experiments in the TLE configuration, using methods described in [Lu et al., 2005; Hare et al., 1998]. We first determined the adiabatic temperature jump ΔT after a single pulse, and then a steady-state ΔT at 1 kHz. At 2050 cm^{-1}, the electrolyte absorption coefficient is 400 cm^{-1} [Bertie and Lan, 1996]. For an IR fluence of 20 mJ/cm^2, in the electrolyte, ΔT never exceeds 1.2K. The pulses are also absorbed by the electrode; the femtosecond IR pulse being most important owing to its higher intensity. At the electrode surface, electrons, phonons, and adsorbates are not in thermal equilibrium on the femtosecond time scale. Based on ultrafast laser heating and pump-probe measurements, we were able to estimate the heat jump at the Pt surface [Venkatakrishnan et al., 2002]: $\Delta T = 30$K. However, it takes time for a hot Pt surface to heat the CO adsorbates. During the time interval of order 1 ps when the SFG signal is emitted [Persson and Ryberg, 1981], the temperature jump of CO [Bonn et al., 2000] is about 10K. These estimates were tested by varying the laser pulse energies, and we observed no changes in the spectra when the pulse intensities were reduced by up to a factor of four.

12.3 ANALYSIS OF SFG SPECTRA

For adsorbates on a metal surface, an SFG spectrum is a combination of resonant molecular transitions plus a nonresonant background from the metal. (There may also be a contribution from the water–CaF$_2$ interface that can be factored out by following electrode potential effects; see below.) The SFG signal intensities are proportional to the square of the second-order nonlinear susceptibility [Shen, 1984]:

$$I_{SFG} \propto |\chi^{(2)}|^2 \tag{12.1}$$

For simplicity, we assume nonresonant visible pulses and ignore the tensor nature of $\chi^{(2)}$ which describes the well-known polarization conditions. For adsorbates on a metal surface [Backus and Bonn, 2005; Cho et al., 2002],

$$\chi^{(2)}_{SFG} = \frac{\beta_{av}(\omega_{IR}, \omega_{vis})}{[1 + \alpha_{av}(\omega_{IR})U(0)][1 + \alpha_e U(0)]^2} \tag{12.2}$$

α_{av} and β_{av} are the linear molecular polarizability and the molecular hyperpolarizability respectively, α_e is the electronic polarizability, and $U(0)$ is the local field in the long-wavelength ($k = 0$) limit. The linear polarizability is given by Persson and Ryberg [1981] as

$$\alpha_{av} = \sum_{\mu=1}^{M} \frac{c_\mu \alpha_\mu}{1 + [\alpha_\mu(\omega_{IR}) - \alpha_{av}(\omega_{IR})]Q} \tag{12.3}$$

where Q is a transcendental function that depends on the adsorbate geometry, μ denotes the type of binding site (e.g., atop, bridge, 3-fold for CO), c_μ is the coverage, and α_μ is the polarizability of a single molecule with singleton vibrational frequency ω_0 at a μ-type site [Backus and Bonn, 2005]:

$$\alpha = \alpha_e + \frac{\alpha_v}{1 - (\omega_{IR}/\omega_0^2)(\omega_{IR} + 2i\Gamma)} \qquad (12.4)$$

where Γ is the damping constant (the inverse IR linewidth). It is usually assumed that the single-molecule hyperpolarizability β_{av} can be written in a form similar to (12.4), so that the SFG spectrum consists of one or more Lorentzian transitions plus a frequency-independent nonresonant background term [Backus and Bonn, 2005; Cho et al., 2002]. For BB-SFG, such a spectrum takes the form [Lu et al., 2005]

$$I_{SFG}(\omega) \propto \exp\left[-4\ln(2)\frac{(\omega-\Omega)^2}{\delta^2}\right]\left|A_{NR}\exp(-i\phi) + \sum_v \frac{NA_v}{\omega - \omega_v + i\Gamma_v}\right|^2 \qquad (12.5)$$

where Ω and δ are the frequency and width of the BBIR pulses, A_{NR} and ϕ are the nonresonant amplitude and phase, and, for each transition, NA_v is the product of the adsorbate surface density and the vibrational amplitude, ω_v is the central frequency, and Γ_v is the inverse linewidth. On a metal surface, the nonresonant polarization dephases rapidly, within the approximately 100 fs duration of the BBIR pulse, whereas the resonant polarization dephases more slowly, on the 1–5 ps duration of the vibrational T_2 time constants. It has recently been shown that, by appropriate time delay of a time-asymmetric NBVIS pulse, it is possible to suppress the nonresonant signal entirely, with little loss of the resonant contribution [Lagutchev et al., 2007].

12.4 BB-SFG OF ELECTROCATALYSIS AND ELECTRODE ADSORPTION ON SINGLE METAL ELECTRODES

12.4.1 SFG Spectra of Electrochemical CO Oxidation

CO electrochemistry at a Pt electrode is a model surface reaction in fuel cell catalysis, and has been intensively studied (see, e.g., Guyot-Sionnest and Tadjeddine [1990]; Lu et al. [2004]; Chou et al. [2003a, b]; Corrigan and Weaver [1988]; Petukhov et al. [1998]; Markovic and Ross [2002]; Koper et al. [2002]; Vidal et al. [2003]. Although the loss of SFG signal from electrosorbed CO on Pt can be reasonably associated with the loss of CO coverage resulting from CO oxidation, there are a number of reasons why this association need not be precise, and a number of reasons why experimental determination of the extent of correspondence has proven elusive. In previous work using IRAS or SFG, as discussed in Section 12.1, the need to minimize IR absorption in the aqueous electrolyte led to the use of cells where the electrode was pressed directly against the optical window. With such a thin electrolyte film

12.4 BB-SFG OF ELECTROCATALYSIS AND ELECTRODE ADSORPTION

(of the order of 1 μm), the scan rate must be kept below a fraction of a millivolt per second to avoid a significant ohmic drop deformation of the electric current measured as a function of time. Until recently [Lu et al., 2005; Lagutchev et al., 2006], no data on the broadband multiplex SFG apparatus [Patterson et al., 2005; Richter et al., 1998] combined with a TLE cell were reported.

Figures 12.4–12.7 show published data from our BB-SFG work on the Pt/CO system [Lu et al., 2005]. We acquired a time stream of SFG spectra of chemisorbed CO on a polycrystalline Pt electrode with a 25 μm thick Teflon spacer (Fig. 12.2) in a 0.1 M H_2SO_4 electrolyte free of CO (Figs. 12.4–12.6). The potential was scanned at 5 mV/s as spectra were obtained at 5 Hz. The voltammogram is a typical CO stripping voltammogram and is shown in Fig. 12.5. The hydrogen peaks are completely suppressed, and the CO adlayer is stable from the onset of hydrogen evolution potential to about 0.4 V (the hydrogen peaks recover on the second scan). The voltammetric CO oxidation peak begins at 0.42 V and ends at 0.51 V. Using the data in Figs. 12.4–12.6 was constructed, and shows that the potential areas of CO stability and oxidation show excellent fits. The SFG CO amplitude in Fig. 12.6 increases slightly until oxidation begins, whereupon a precipitous drop is seen; the slight deviation between the charge and SFG amplitude near the end of the oxidation process is attributed to experimental error. The CO frequency increases with a Stark shift of 28 cm^{-1} V^{-1}, in good agreement with previous reports [Chou et al., 2003a, b; Dederichs et al.,

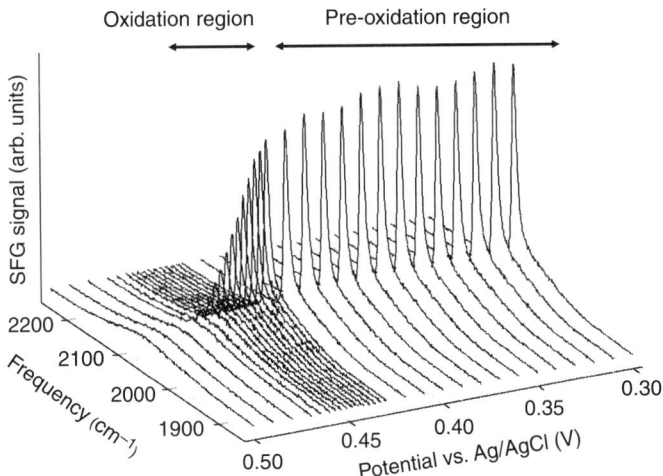

Figure 12.4 A series of SFG spectra in the CO stretch region of chemisorbed CO on polycrystalline Pt in a CO-free 0.1 M H_2SO_4 electrolyte. The atop spectra were fit to (12.5) (see text) to extract the amplitude, frequency, and width [Lu et al., 2005; Lagutchev et al., 2006] (each displayed data point is the average of three or five spectra). The electrode potential was swept at a rate of 5 mV/s, and SFG spectra were obtained every 200 ms. Spectra were obtained at 1 mV intervals, but, to avoid congestion in the plot, averaged spectra are displayed at 10 mV intervals in the pre-oxidation region (V < 0.43 V) and at 3.3 mV intervals in the oxidation region (V > 0.43 V) [Lu et al., 2005].

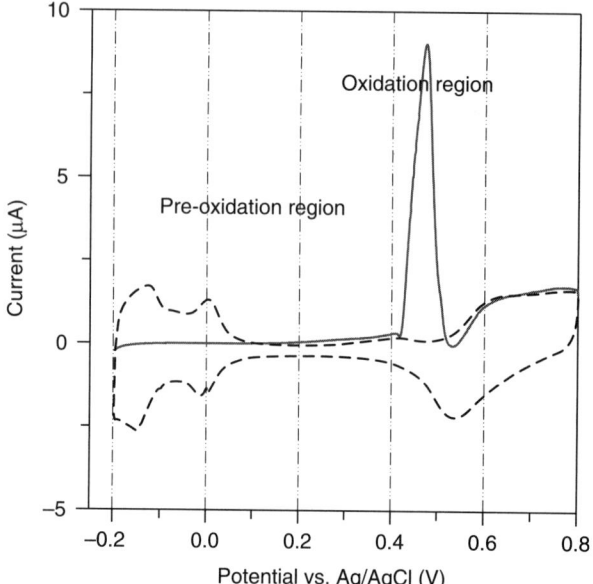

Figure 12.5 CO stripping voltammogram with a CO-free 0.1 M H_2SO_4 electrolyte. Compare the data in Fig. 12.4: the CO oxidation region begins at $V = 0.43$ V. After CO stripping, hydrogen adsorption/desorption peaks and the beginning of the Pt oxidation range are shown.

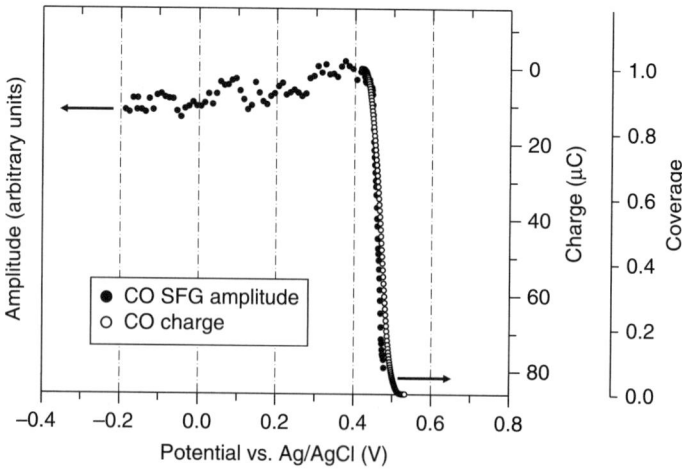

Figure 12.6 Comparison (using a 0.1 M H_2SO_4 CO-free electrolyte) of the SFG amplitude of the CO stretch (filled circles) with electric charge (open circles). The charge is computed by numerical integration of the current shown in Fig. 12.5.

2000; Corrigan and Weaver, 1998; Kunimatsu et al., 1985a, b; Korzeniewski et al., 1986; Tian et al., 1997]. At the onset of chemisorbed CO oxidation, the CO frequency evidences the anomalous Stark tuning behavior noted by Ross and Markovic and co-workers (Fig. 12.8), who associated it with compression of CO islands by adsorbed anions [Stamenkovic et al., 2005].

As mentioned above, careful comparison of the oxidation current and the SFG signal with a CO-free electrolyte (Fig. 12.6) showed a precise correspondence, so that SFG was shown for the first time to be capable of quantitative measurements of CO coverage [Lu et al., 2005]. In particular, we did not see the remarkable features reported in previous work on Pt electrodes. Baldelli et al. [1999] reported that the SFG signal vanished 200 mV below the oxidation threshold, which was interpreted as resulting from the electrochemical generation of an SFG-invisible CO adsorption state. Dederichs et al. [2000] also reported CO signal disappearance prior to the CO oxidation threshold, which was attributed to CO depletion in the quite thin electrolyte layer. Chou et al. [2003a, b] reported that the SFG signal intensity vanished immediately after the oxidation current reached its maximum. Instead, Fig. 12.6 shows that the SFG intensity perfectly tracks the oxidation current. However, when the electrolyte was CO-saturated, the charge and the SFG signal were no longer in correspondence. (The charge builds up to a considerable extent before the SFG signal starts to drop at 0.44 V; Fig. 12.7.) The current and charge are sensitive to the total number of CO equivalents oxidized in the electrolyte and on the Pt surface, whereas the SFG signal is sensitive to the CO stripping process, which reduces the CO surface coverage. Consequently, with the CO-saturated electrolyte, SFG and voltammetry measure different things.

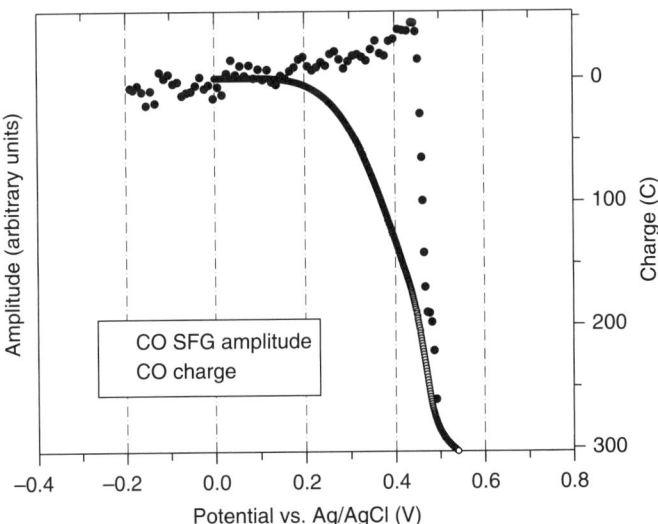

Figure 12.7 Comparison (using a 0.1 M H_2SO_4 CO-saturated electrolyte) of the SFG amplitude of the CO stretch (filled circles) with electric charge (open circles). The charge is computed by numerical integration of the current shown in Fig. 12.8.

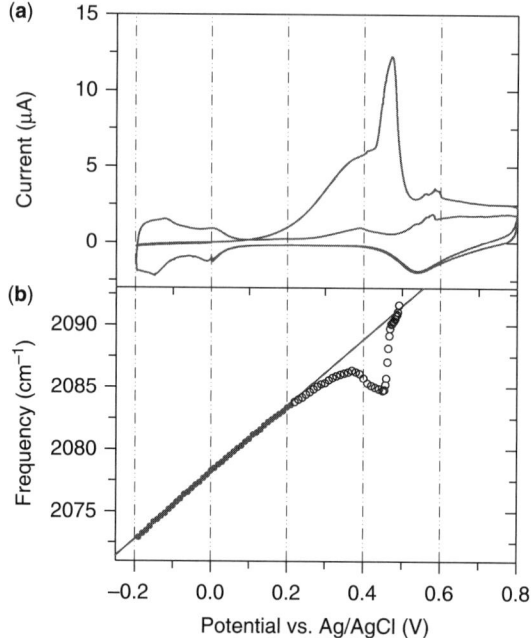

Figure 12.8 CO-saturated electrolyte in the thin cell of Fig. 12.2. (a) CO oxidation; the first and second scans are shown. (b) Comparison with the CO stretch frequency shift. (Filled circles denote linear Stark tuning behavior while open circles correspond to deviations from linear behavior during oxidation.)

12.4.2 SFG Spectra of Adsorbate Phase Transition Kinetics

CO on single-crystal Pt(111) [Lagutchev et al., 2006] evidences new phase behavior [Villegas and Weaver, 1994; Tolmachev et al., 2004; Wang et al., 2005; Lucas et al., 1999] not seen on poly Pt. We have studied the adlayer phase transition between a $(2 \times 2) - 3CO$ structure and a slightly less dense $(\sqrt{19} \times \sqrt{19})R23.4° - 13CO$ lattice [Lagutchev et al., 2006], depicted in Fig. 12.9. This adlayer phase transition is an excellent model system to understand dynamical transformations in electrochemical surface science. A representative SFG amplitude, frequency and width are plotted in Fig. 12.10 with a 0.1 M H_2SO_4 electrolyte and a potential sweep rate of 5 mV/s. The phase transition highlights a sudden atop intensity jump and is accompanied by the transformation of 3-fold sites into bridge sites (Fig. 12.9) [Villegas and Weaver, 1994; Tolmachev et al., 2004; Wang et al., 2005; Lucas et al., 1999]. At lower potentials (<0.53 V), atop and 3-fold signals are observed (Fig. 12.11a), representing the $(2 \times 2) - 3CO$ lattice consisting of atop CO sites (25% coverage) surrounded by 3-fold sites (50% coverage). At higher potentials (>0.53 V), the atop intensity jumps, and the 3-fold sites are converted to bridge sites, resulting in a $(\sqrt{19} \times \sqrt{19})R23.4° - 13CO$ structure of the atop CO (the atop CO coverage is 0.368) surrounded by bridge-bonded dimers (the bridge-bonded CO coverage is 0.316) (Fig. 12.11b).

12.4 BB-SFG OF ELECTROCATALYSIS AND ELECTRODE ADSORPTION

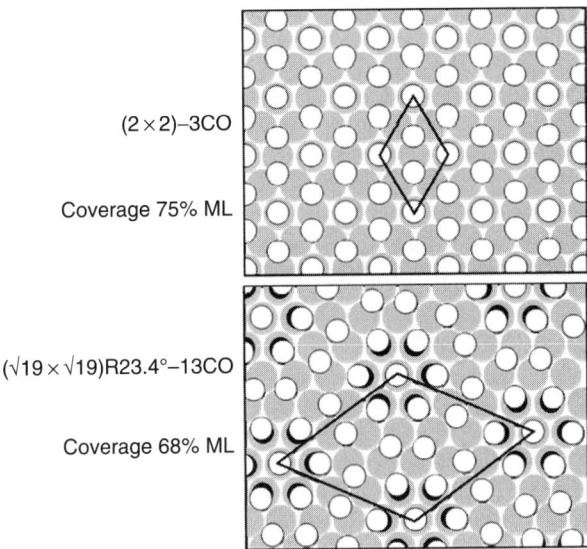

Figure 12.9 Structures of the $(2 \times 2)-3\text{CO}$ and $(\sqrt{19} \times \sqrt{19})\text{R}23.4°-13\text{CO}$ adlayers [Villegas and Weaver, 1994; Tolmachev et al., 2004; Wang et al., 2005; Lucas et al., 1999]. In the $(2 \times 2)-3\text{CO}$ lattice, every atop site is surrounded by six 3-fold sites at a distance of $1.15d$. All atop (coverage 0.25) and 3-fold sites (coverage 0.50) are equivalent. In the slightly (10%) lower packing density, in the $(\sqrt{19} \times \sqrt{19})\text{R}23.4°-13\text{CO}$ lattice, a single atop site is surrounded by six near atop neighbors and six pairs of near-bridge dimers. Each near-atop site has three nearby near-bridge sites at $1.32d$ and $1.5d$. The atop coverage is $7/19$ (0.368) and the bridge coverage is $6/19$ (0.316).

We were also able to measure how the adlayer transition depended on electrolyte-dissolved CO concentration, on adsorbing (sulfate) versus nonadsorbing (perchlorate) supporting electrolyte, and on potential scan rate. Measurements of phase transition kinetics are summarized in Fig. 12.12, where the potential was switched back and forth across the phase boundary repeatedly as atop site SFG spectra were obtained at 5 Hz. Such kinetic measurements are meaningful only if the electrolyte in the TLE cell is thick enough that its composition remains constant during surface transformations. Figure 12.12 shows only minimal differences in the kinetics between the first and second cycles of potential switching, verifying that this condition was met. As shown in Fig. 12.12, the atop site Stark shift (cf. Fig. 12.10) is observed to occur instantaneously, implying that a new potential is established on the electrode in <0.1 s. With a CO-saturated electrolyte, the forward $(2 \times 2) \rightarrow (\sqrt{19} \times \sqrt{19})$ transformation occurs with a time constant of a few seconds, whereas the back $(\sqrt{19} \times \sqrt{19}) \rightarrow (2 \times 2)$ transformation is much faster, occurring in <0.5 s (Fig. 12.12a). With CO-unsaturated electrolyte, the forward transformation becomes a bit faster and the back transformation is a bit slower (Fig. 12.12d) than with a CO-saturated electrolyte.

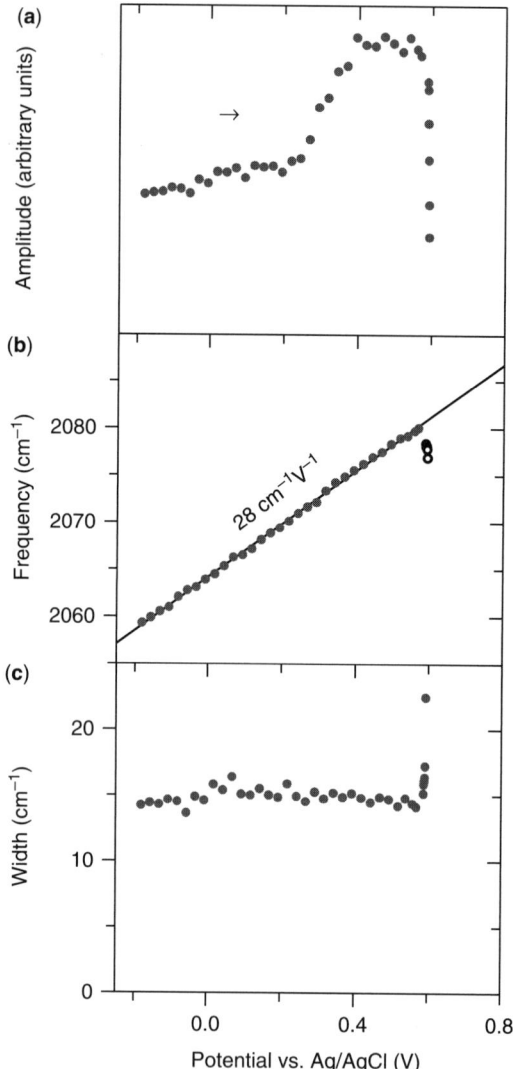

Figure 12.10 Analysis of SFG spectra from atop CO on Pt(111) using a CO-saturated 0.1 M H_2SO_4 electrolyte and a scan rate of 5 mV/s. The $(2 \times 2) - 3CO \rightarrow (\sqrt{19} \times \sqrt{19})R23.4° - 13CO$ phase transition resulted in a jump in atop amplitude (a). Stark tuning (b) and peak width versus electrode potential data (c) are also shown. (Filled circles denote linear Stark tuning behavior while open circles correspond to deviations from linear behavior during oxidation.)

Overall [Lagutchev et al., 2006], we investigated the CO phase transition where the relative proportions of atop, bridge and 3-fold sites change significantly across the phase boundary [Villegas and Weaver, 1994; Severson et al., 1995]. We found that the SFG intensity changes for these sites were different from what was expected solely on the basis of surface coverage densities. The explanation is that these sites have quite different back-bonding, so the adlayer transition changes not only the

12.4 BB-SFG OF ELECTROCATALYSIS AND ELECTRODE ADSORPTION

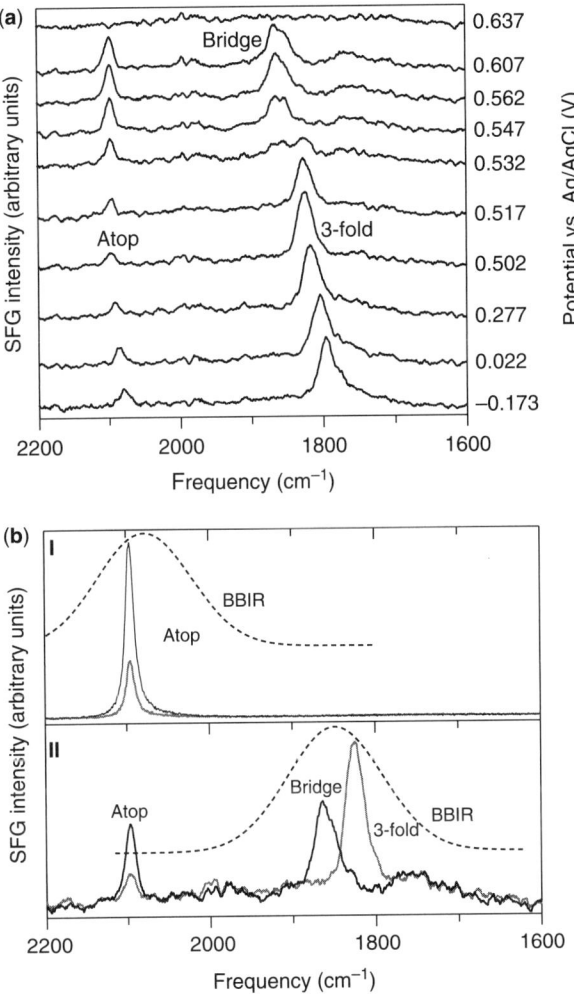

Figure 12.11 (a) SFG spectra of atop CO on a Pt(111) electrode, with a 25 μm thick CO-saturated 0.1 M H_2SO_4 electrolyte and at a scan rate of 1 mV/s. Each displayed spectrum resulted from a 15 s acquisition. Above 0.532 V, the atop intensity jumped and 3-fold sites were converted to bridge sites. At 0.547 V, bridge and 3-fold CO coexist. (b) Spectrum of atop CO in the $(2 \times 2) - 3CO$ (**I**) and $(\sqrt{19} \times \sqrt{19})R23.4° - 13CO$ (**II**) adlayers. The broadband IR BBIR pulses (dashed lines) were tuned to maximize atop (I) and the multiply bonded (II) intensities (solid lines). In case (II), the much more intense atop transitions could also be seen, despite being in the wings of the BBIR pulse spectrum. Data at ≥0.607 V show simultaneous disappearance of atop and bridge-bonded CO.

site occupancies but also the electronic polarizability of the Pt surface. The change in surface properties is clearly evident in second-harmonic generation (SHG) measurements, which are much more sensitive to the Pt surface than to the adsorbate [Severson et al., 1995]. As we reported [Lagutchev et al., 2006], the problem using SFG to determine coverage changes during a surface transformation is that the SFG

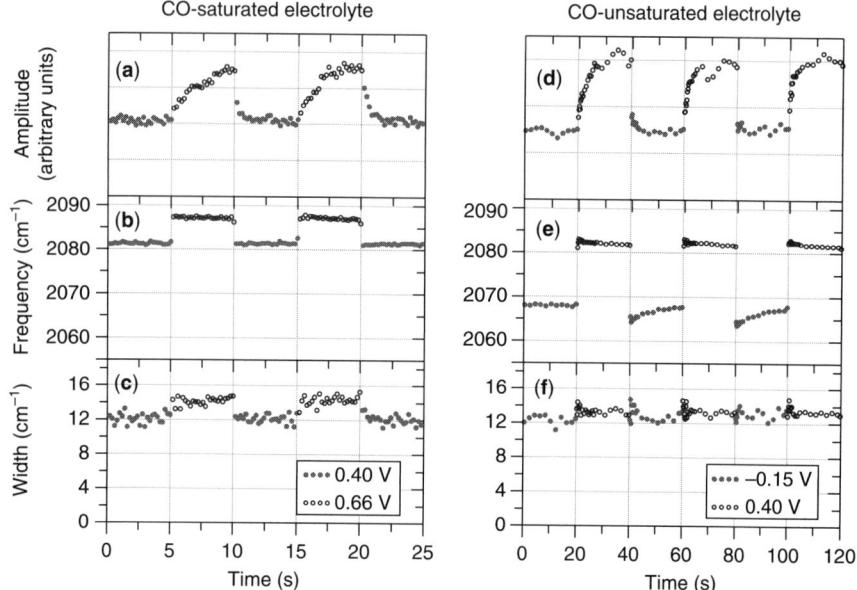

Figure 12.12 Kinetics of the $(2 \times 2) - 3\text{CO} \rightarrow (\sqrt{19} \times \sqrt{19})\text{R}23.4° - 13\text{CO}$ phase transition on a Pt(111) electrode in a CO-saturated 0.1 M H_2SO_4 electrolyte, observed via SFG of atop CO. The frequency shift data in (b) and (e) indicate that a new potential is established on the electrode within 0.2 s. The forward transformation is much slower than the reverse. There are minimal differences between the first and second cycles, indicating minimal change in electrolyte composition during kinetic measurements.

signal results from the coverage change and also from the changing $[1 + \alpha_{av}U(0)]$ and $[1 + \alpha_e U(0)]$ terms in (12.2). It can be shown that IRAS measurements depend on the $[1 + \alpha_{av}U(0)]$ term alone, and SHG on the $[1 + \alpha_e U(0)]$ term alone. Thus, it should be possible to determine these terms independently using IRAS and SHG measurements and correct the SFG measurements to extract the changing surface coverage in a quantitative manner. We derived [Lagutchev et al., 2006] a useful relation for a transformation between phases I and II, which might involve an electrochemical transformation, a phase transformation, a change in adsorbed ions, etc.:

$$\frac{C_\text{I}}{C_\text{II}} = \frac{I_\text{SFG}(\text{phase I})}{I_\text{SFG}(\text{phase II})} \frac{I_\text{IRAS}(\text{phase II})}{I_\text{IRAS}(\text{phase I})} \left[\frac{I_\text{SHG}(\text{phase II})}{I_\text{SHG}(\text{phase I})}\right]^{2/3} \quad (12.6)$$

where c is the coverage from (12.3). Using (12.6) greatly improves the quantitative accuracy of SFG surface coverage measurements. We used literature measurements of SHG [Akemann et al., 1998, 2000; Pozniak et al., 2001; Pozniak and Scherson, 2003] and IRAS [Villegas and Weaver, 1994; Kitamura et al., 1989; Yoshimi et al., 1996] intensities. Thus, this phase transition [Dederichs et al., 2000; Villegas and Weaver, 1994; Tolmachev et al., 2004; Wang et al., 2005; Lucas et al., 1999; Akemann et al., 1998, 2000; Pozniak et al., 2001; Pozniak and Scherson, 2003; Kitamura et al., 1989; Yoshimi et al., 1996] is a critical test of SFG as a quantitative probe of surface coverage. As described above, the transition causes both the vibrational and the electronic polarizability to change, so none of SFG, SHG, or IRAS alone can quantitatively determine coverage changes across this phase boundary. Combining these measurements with (12.6) allowed accurate measurements of surface coverage [Lagutchev et al., 2006]. A small (about 20%) discrepancy in the SFG determination of atop coverage was attributed to either a small amount of surface disorder or uncertainties in the SHG and IRAS measurements.

12.4.3 Decomposition of Methanol on a Pt Electrode

Tadjeddine and co-workers have used SFG [Guyot-Sionnest and Tadjeddine, 1990; Eisenthal, 1992; Richmond, 2002; Vidal et al., 2002, 2004, 2005] to study the adsorbed CO produced from a variety of solution species, including methanol [Vidal et al., 2002, 2005]. With BB-SFG, we studied the *electrochemical kinetics* of methanol chemisorption as surface CO, as shown in Fig. 12.13. We used a polycrystalline Pt electrode and 0.1 M H_2SO_4 electrolyte with 0.1 M methanol. Figure 12.13a–d characterize the potential-dependent SFG spectra obtained under the voltammetric

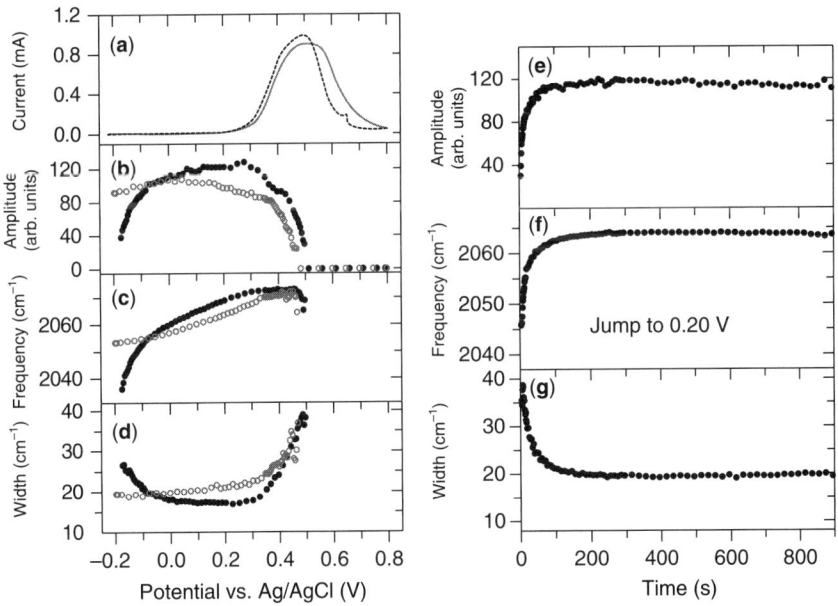

Figure 12.13 Electrochemistry and kinetics of CO resulting from methanol decomposition on polycrystalline Pt with 0.1 M H_2SO_4 electrolyte and 0.1 M methanol. (a–d) Current, SFG amplitude, frequency, and width of adsorbed CO, scanning the potential in both directions as indicated with the solid line and filled circles denoting the forward (anodic) scan and the dashed line and unfilled circles denoting the back (cathodic) scan. (e–g) Starting at 0.6 V, where the adsorbed CO is rapidly electro-oxidized, the potential is suddenly jumped to 0.2 V. The reformation of the CO layer (CO chemisorption) due to methanol decomposition occurs in about 20 s. The adsorbed CO molecules are redshifted, and have a broader spectrum at shorter times, when the adlayer coverage is low.

conditions of the CO adsorbate resulting from the methanol decomposition process. In the forward direction, starting at negative potential, the CO SFG amplitude increases, reaching a maximum at about 0.3 V, and then disappearing at about 0.5 V. In the back scan, CO reappears at about 0.47 V and reaches a maximum at about 0 V. The CO frequency shift reflects Stark tuning and intermolecular interactions among the coadsorbates. For kinetic studies (Fig. 12.13e–g), the potential was first fixed at 0.75 V, where CO was removed rapidly while the bulk methanol oxidation activity is low to minimize depletion of methanol in the solution gap (Fig. 12.13g). Jumping to a lower potential (0.20 V) where the CO adlayer is stable, SFG monitored the CO build-up from methanol. Figure 12.13e shows that the build-up occurs on the 20 s time scale, and reaches its maximum at steady state. The CO frequency shift tracks the amplitude (Fig. 12.13f). As CO is deposited on the bare electrode and coverage grows, the CO frequency blueshifts as a result of interactions among coadsorbates [Severson et al., 1995]. The CO frequency closely reflects the coverage dependence, suggesting that CO is formed randomly and is distributed uniformly instead of being arranged in dense patches.

Surface reactions that give rise to the electric current measured depend on the electrode potential range and the type of measurements reported. The dominating reactions contributing to the voltammetric data in Fig. 13a–d are chemisorbed CO oxidation, oxidation of methanol to CO_2 (on both scans), and chemisorbed CO formation on the reverse run:

$$CO + H_2O \longrightarrow CO_2 + 2H^+ + 2e^- \tag{12.7}$$

$$CH_3OH + H_2O \longrightarrow CO_2 + 6H^+ + 6e^- \tag{12.8}$$

$$CH_3OH \longrightarrow CO + 4H^+ + 4e^- \tag{12.9}$$

The drop of the voltammetric current is associated with Pt surface oxidation, and the drop on the negative-going run is due to Reaction (12.9) (surface poisoning by CO) and the Tafelian kinetics of Reaction (12.8). Further, the shift between curves in Fig. 12.13a and b indicates that in the potential range between 0.5 and 0.6 V, methanol oxidation occurs with zero or low level atop CO surface intermediate. The amplitudes on Fig. 12.13 on both scans nearly equal to each other indicate a high level of preferential (111) crystallographic orientation of the polycrystalline Pt surface used for this work, as inferred from data in [Adzic et al., 1982].

12.4.4 Decomposition of Formic Acid on Pt(111)

While very limited data are presented here, the kinetics of adsorption/decomposition of formic acid molecules [Rice et al., 2002] have been measured by BB-SFG, as shown in Fig. 12.14. A Pt(111) electrode and a 0.1 M H_2SO_4 electrolyte containing 0.1 M formic acid were used. The families of spectra at −0.200, −0.025, and 0.225 V vs. Ag/AgCl were collected as functions of decomposition time. The electrode potential was initially held at 0.75 V to produce a clean Pt(111) surface, and was next switched to monitor the CO uptake. Starting at 0s, where CO adsorption (from HCOOH decomposition) had not yet begun, the potentiostatic experiment lasted until about 500 s of the progress of reaction. The spectral position is typical of the

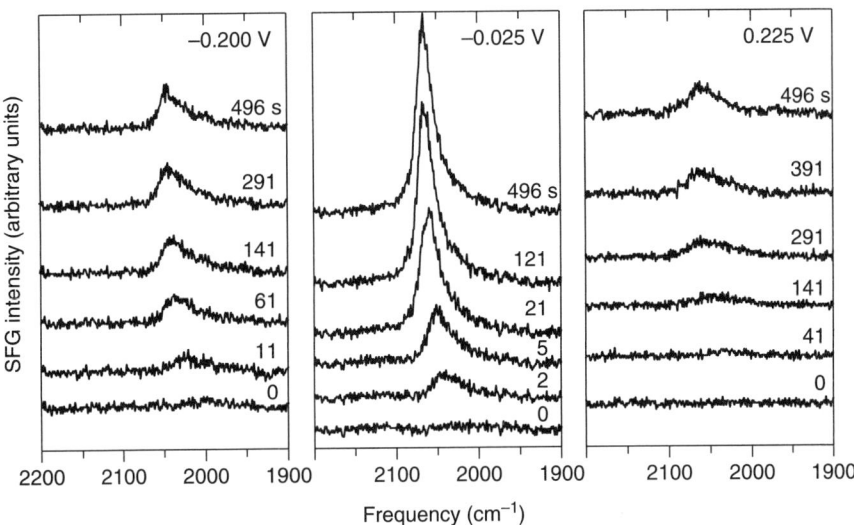

Figure 12.14 SFG spectra of the carbonyls formed during formic acid decomposition on a Pt(111) electrode in 0.1 M H_2SO_4 electrolyte containing 0.1 M formic acid. The spectral position is typical of atop CO on the Pt(111) surface. Times at which the spectra have been recorded are from 2 to 496 s, yielding HCOOH decomposition kinetics at three electrode potentials, -0.200, -0.025, and 0.225 V vs. Ag/AgCl.

atop CO on the Pt(111) surface. (Data with multiply bonded CO will be presented elsewhere [Behrens et al., submitted].) Times at which the spectra were recorded are shown. At -0.025 V, data show that maximum CO coverage is obtained after about 120 s of the decomposition process, but at the remaining two potentials the kinetics are slower.

Overall, we demonstrated electrode potential- and time-dependent properties of the atop CO adsorbate generated from the formic acid decomposition process at three potentials, and addressed the issues of formic acid reactivity and poisoning [Samjeske and Osawa, 2005; Chen et al., 2003, 2006]. There is also a consistency with the previous kinetic data obtained by electrochemical methods; the maximum in formic acid decomposition rates was obtained at -0.025 V vs. Ag/AgCl or 0.25 V vs. RHE (cf. Fig. 12.7 in [Lu et al., 1999]). However, the exact path towards the CO formation is not clear, as the main reaction is the oxidation of the HCOOH molecule:

$$HCOOH \longrightarrow CO_2 + 2H^+ + 2e^- \quad (12.10)$$

If the CO is formed from formic acid dehydration,

$$HCOOH \longrightarrow CO + H_2O \quad (12.11)$$

the rate of reaction should be electrode-potential-neutral, which is not the case (see Fig. 12.14 and data in [Lu et al., 1999]). The presence of a formate immediate on rough polycrystalline Pt was found at higher potentials [Chen et al., 2003; Samjeske et al., 2005; Samjeske and Osawa, 2005], which has not been confirmed in this report.

12.4.5 Adsorption of Acetic Acid

Acetic acid is one of the unwanted deposition products of ethanol in both acidic and basic media [Colmenares et al., 2006; Rousseau et al., 2006]. Preliminary data on adsorption of acetic acid in acidic media are shown in Fig. 12.15. The potential was swept from 0 V to 1.0 V and then back to 0 V at 1 mV/s, while the SFG spectra were recorded. Between about 0.33 and 0.81 V (Fig. 12.15a), a transition was found near 1420 cm^{-1} that is due to symmetric OCO stretching of adsorbed acetate ions [Corrigan et al., 1988; Rodes et al., 1994; Delgado et al., 2005]. The blueshift in frequency with the electrode potential produces a low Stark tuning slope of 10 cm^{-1}/V (Fig. 12.15c). On the positive run, the maximum of acetic acid adsorption occurs when the electrode potential reaches 0.60 V. Further increases reduce acetic acid adsorption owing to Pt surface oxide formation, which is reflected in the voltammogram in Fig. 12.13b. At 0.87 V and above, acetic acid adsorption is no longer observed. On the reverse scan, acetic acid adsorption occurs only after the Pt surface oxide is significantly reduced. The adsorption then increases with increasing potential, and reaches a maximum at around 0.40 V.

In this study, adsorption of acetic acid under voltammetric conditions was observed by a vibrational technique for the first time. The first work in the field was carried out using FTIR (potential difference infrared spectroscopy, PDIRS) and by radioactive labeling [Corrigan et al., 1988]. Both techniques

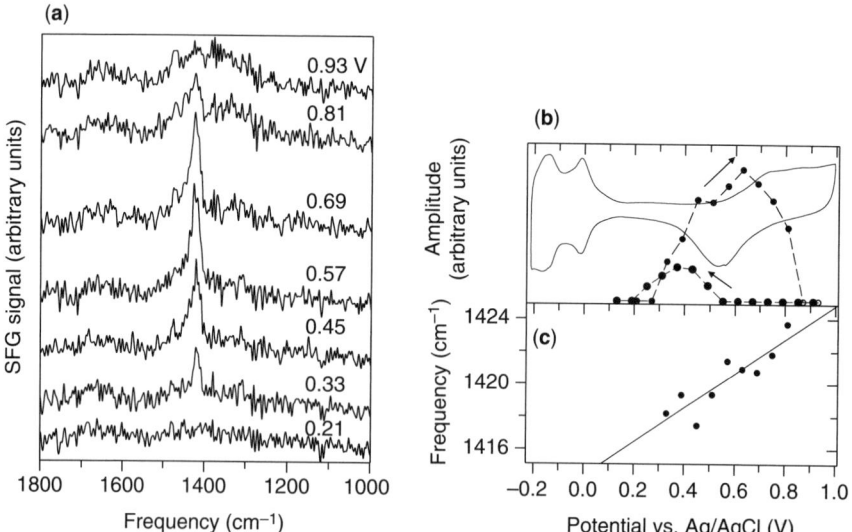

Figure 12.15 (a) Potential-dependent SFG spectra of the symmetric OCO stretch from acetic acid adsorbed on a polycrystalline Pt electrode in 0.1 M H$_2$SO$_4$ electrolyte with 0.01 M acetic acid. Each spectrum was acquired for 60 s (forward scan data only are shown). (b) A cyclic voltammetric curve for 0.1 M H$_2$SO$_4$ electrolyte with 0.01 M acetic acid solution between 0 and 1.0 V at 1 mV/s with the superimposed amplitude–electrode potential plots (vs. RHE) in forward and backward directions. (c) Stark tuning data for acetic acid adsorption on platinum.

show extensive adsorption of acetic acid, which increases at more positive potentials and maximizes at the onset of anodic oxide formation. This is what we confirmed here (Fig. 12.15). A single band near 1420 cm^{-1} was found [Corrigan et al., 1988] that is due to adsorption of acetic acid on Pt, exactly as in Fig. 12.15. The 1420 cm^{-1} feature was consistent with a symmetric carboxylate stretching mode, with both carboxylate oxygens oriented towards the metal surface. We then concluded [Corrigan et al., 1988] that the most likely modes of acetic acid adsorption involve hydrogen bonding between the carboxylate oxygens and inner layer water molecules or that the adsorption occurs by self-association to form a dimer or chain structure. While in this brief study we cannot confirm the nature of surface coordination and orientation of the adsorbed acetic acid species, it is tempting to conclude from the coincidence between the spectral data in [Corrigan et al., 1988] and in Fig. 12.15 that a similar surface structure is found, including the role of Pt oxides in guiding surface concentration of the reversibly adsorbed acetic acid adsorbate.

12.4.6 Chemisorption of CO on Nanoparticle Pt Electrodes

Pt nanoparticles, made of either pure or alloyed Pt, are well known as anodes and cathodes in hydrogen and methanol fuel cells [Vielstich et al., 2003]. Using

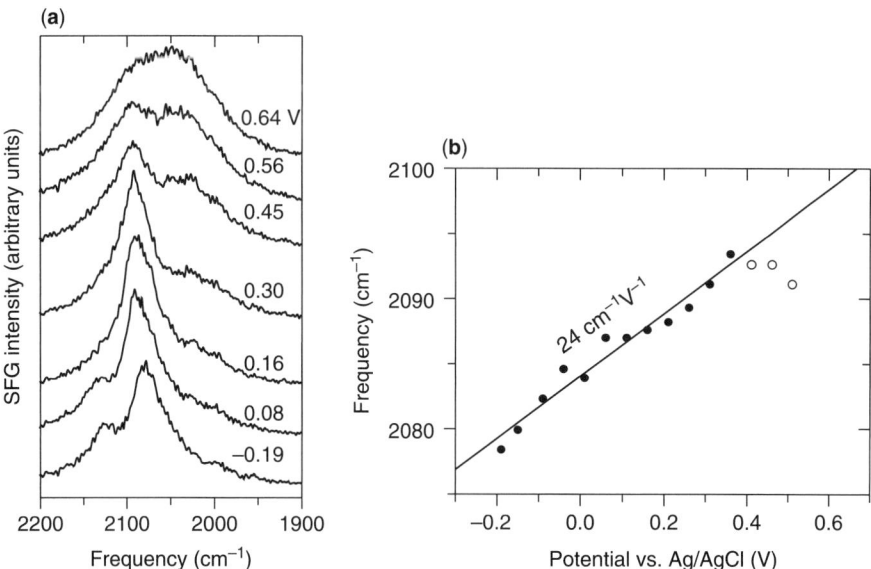

Figure 12.16 Potential dependent SFG spectra (a) and the Stark tuning plot (b) from chemisorbed CO on Pt nanoparticles in a CO-saturated 0.1 M H_2SO_4 electrolyte. Each spectrum was acquired for 10 s (forward scan data only are shown). The potential was scanned from -0.20 to 0.70 V (vs. Ag/AgCl) at 1 mV/s. Pt nanoparticles were of approximately 7 nm size, and were immobilized on an Au disk.

BB-SFG, we have investigated CO adsorption on smooth polycrystalline and single-crystal electrodes that could be considered model surfaces to those applied in fuel cell research and development. Representative data are shown in Fig. 12.16: the Pt nanoparticles were about 7 nm of Pt black, and were immobilized on a smooth Au disk. The electrolyte was CO-saturated 0.1 M H_2SO_4, and the potential was scanned from -0.19 V up to 0.64 V at 1 mV/s. The BB-SFG spectra (Fig. 12.16a) at about 2085 cm^{-1} at -0.19 V correspond to atop CO [Arenz et al., 2005], with a Stark tuning slope of about 24 cm^{-1}/V (Fig. 12.16b). Note that the Stark slope is lower than that obtained with Pt(111) (Fig. 12.9), for reasons to be further investigated. The shoulder near 2120 cm^{-1} is associated with CO adsorbed on the Au sites [Blizanac et al., 2004], and the broad background (seen clearly at 0.64 V) is from nonresonant SFG. The data shown in Figs. 12.4, 12.11a, and 12.16 represent a link between smooth and nanostructure catalyst surfaces, and will be of use in our further studies of fuel cell catalysts in the BB-SFG IR perspective.

12.5 SFG OF BIMETALLIC ELECTRODES: ADSORPTION AND OXIDATION OF CO ON Pt(111)/Ru

12.5.1 Correlation between Infrared Spectra and Reactivity

By employing spontaneous deposition to produce Ru-covered, Pt single-crystal electrodes [Crown et al., 2002], we obtained Pt(111)/Ru surfaces with a Ru coverage of approximately 0.2 monolayers (ML) (Fig. 12.17). Next, using CO-saturated, 0.1 M H_2SO_4 solutions and Pt(111)/Ru surfaces such as that shown in Fig. 12.17, we produced CO chemisorption layers in which CO was chemisorbed on both Ru and Pt sites at full coverage [Lu et al., 2002]. After flushing the cell with clean 0.1 M H_2SO_4 electrolyte, we conducted voltammetric oxidation of the CO under voltammetric conditions, into the pure H_2SO_4 electrolyte (Fig. 12.18). The current versus electrode potential measurements yielded two well-resolved current–potential peaks at 0.14 and 0.24 V vs. Ag/AgCl, at 1 mV/s [Lu et al., 2002; Tong et al., 2002]. Among other things (see below), the voltammogram in Fig. 12.18 demonstrates the high quality of the electrode surface, and confirms the cleanliness of our electrochemical systems.

Others [Massong et al., 2000] as well as us [Tong et al., 2002] have reported on the split in the voltammogram of the type shown in Fig. 12.18, and it is now fully acknowledged that the more negative cyclic voltammogram peak originates from the oxidation of CO chemisorbed on Ru sites, while the more positive peak comes from the oxidation on pure (not Ru-covered) Pt sites (see [Lu et al., 2002; Tong et al., 2002] and references therein). We may now hypothesize that the vibrational spectra [Lu et al., 2004; Friedrich et al., 1996] obtained from such CO-covered Pt(111)/Ru surfaces should follow the voltammetric distribution of the CO stripping peaks of Fig. 12.18. That is, on the positive-going voltammetric scan, the spectral amplitudes for CO adsorbed on Ru sites should disappear before the amplitudes from CO on the Pt sites can be reduced (before the ultimate disappearance at higher anodic potentials).

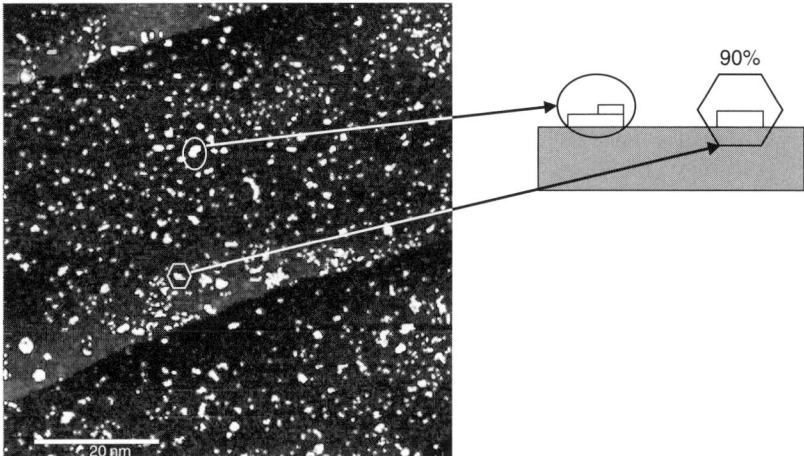

75 nm × 75 nm

Figure 12.17 Representation of a scanning tunneling microscope image of Ru spontaneously deposited on a Pt(111) surface [Crown et al., 2002]. The Ru coverage was about 0.2 ML. (Ru is in white on the figure and inset.) The inset indicates that monoatomic Ru deposition predominates; diatomic deposition is found only with 10% of the Ru deposit.

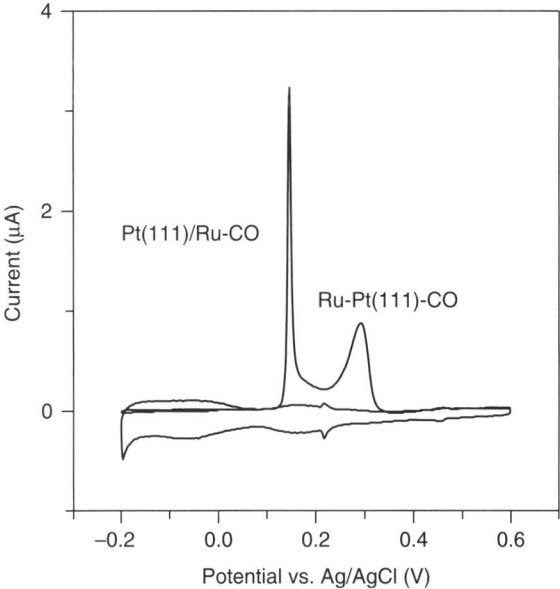

Figure 12.18 Voltammetric CO stripping from a Pt(111)/Ru electrode (Ru coverage approximately 0.2 ML) to a CO-free 0.1 M H_2SO_4 solution. The baseline voltammogram from Pt(111)/Ru after CO stripping is shown. The scan rate was 1 mV/s. CO was adsorbed at 0 V for 5 minutes, and was purged from solution by Ar bubbling for 25 minutes.

398 BROADBAND SUM FREQUENCY GENERATION STUDIES

The BB-SFG thin-layer cell (Fig. 12.2) has the ideal configuration to test the above prediction, as both vibrational amplitudes and voltammetric currents can be measured simultaneously (with no IR drop penalty). The matching (to the voltammetric data shown in Fig. 12.18) SFG spectra are shown in Figs. 12.19 and 12.20 (for details of spectral assignments of Pt-CO and Ru-CO, see [Friedrich et al., 1996]; as well as [Villegas and Weaver, 1994; Leung et al., 1988] and [Gutierrez et al., 1991; Oda et al., 1993], respectively). The data demonstrate that the correct correlation between spectra and voltammetry has been reached. From the data in Figs. 12.18 and 12.19 alone, one may conclude that the electrochemical and spectroscopic data for Ru-CO match ideally; that is, there is complete disappearance of the CO amplitude from Ru at about 0.24 V (Fig. 12.19), correlating with the saddle point of the cyclic voltammogram in Fig. 12.18 (with the common sense deconvolution of the CO stripping peaks shown in Fig. 12.18). However, there is still a discrepancy between the IR amplitudes (Fig. 12.19) and the voltammetric stripping (Fig. 12.18) for atop CO on the Pt sites: instead of the amplitude disappearing near 0.30 V, as expected from Fig. 12.18, it disappears at a more negative potential, nearly at 0.26 V (Fig. 12.19).

In order to reconcile the inconsistency, we analyzed the spectral disappearance of all CO surface forms—atop, bridge, and 3-fold—as a function of electrode potential (Fig. 12.20). While the spectra are noisy, the bridge-bonded CO survives at the Pt

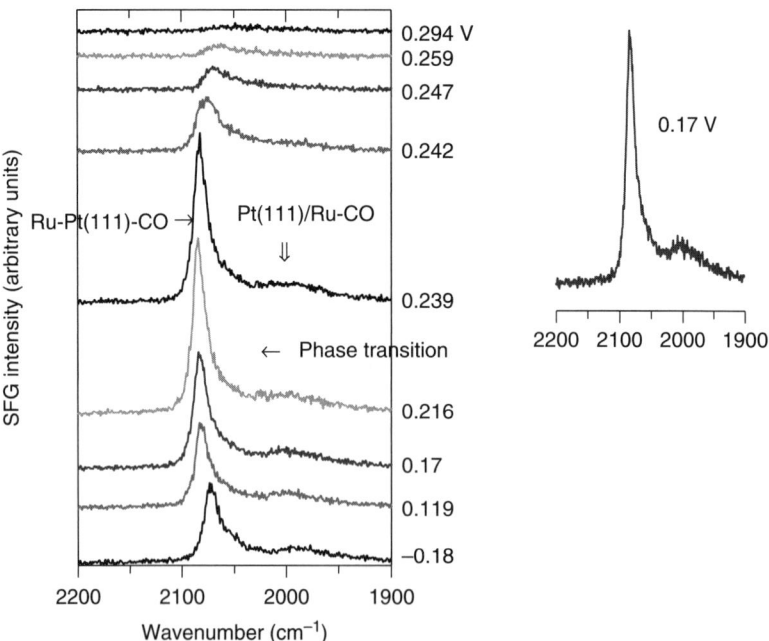

Figure 12.19 Potential-dependent SFG spectra from atop CO on a Pt(111)/Ru electrode in 0.1 M H_2SO_4 at 1 mV/s in 0.1 M H_2SO_4 (see Fig. 12.18). The scan potential for each spectrum is shown on the right. Spectra from Ru-Pt(111)-CO and Pt(111)/Ru-CO are shown, including the CO phase transition at ≥ 0.17 V (vs. Ag/AgCl). The inset shows a blow-up of the SFG spectrum from Pt(111)/Ru-CO at 0.17 V.

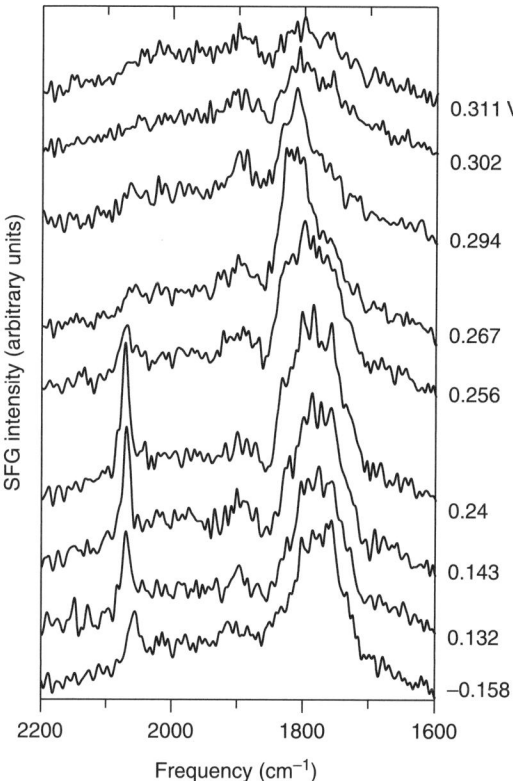

Figure 12.20 Potential-dependent SFG spectra from atop and multiply bonded CO on a Pt(111)/Ru electrode in 0.1 M H_2SO_4 at 1 mV/s (see Fig. 12.18). The scan potential for each spectrum is shown on the right. Data show disappearance of atop CO at lower potentials than multiply bonded CO.

sites at higher potential than the atop CO—namely, to nearly 0.30 V, in agreement with the voltammetric stripping behavior of Fig. 12.18. This shows that the bridge-bonded CO is more stable with respect to the electro-oxidation reaction than the atop CO (and the 3-fold CO [Lagutchev et al., 2006]). In contrast, both bridge-bonded and atop CO disappear simultaneously from pure Pt(111) (without Ru), as shown in Fig. 12.11a and in support of the data reported before [Friedrich et al., 2002].

12.5.2 Vibrational, Electrochemical, and Electronic Properties of the Pt/Ru Catalyst

The Pt/Ru catalyst is the material of choice for the direct methanol fuel cell (DMFC) (and hydrogen reformate) fuel cell anodes, and its catalytic function needs to be completely understood. In the first approximation, as is now widely acknowledged, methanol decomposes on Pt sites of the Pt/Ru surface, producing chemisorbed CO that is transferred via surface motions to the active Pt/Ru sites to become oxidized to CO_2

[Watanabe and Motoo, 1975a, b; Maillard et al., 2005]. However, a complete picture of Pt/Ru catalytic function, including the CO removal process, is only now emerging. In particular, the correlation between CO stripping voltammetry (reactivity) and spectra—IR (see above) and/or nuclear magnetic resonance (NMR)—is key to understanding the methanol (CO) oxidation process. It should be mentioned that the spectroscopic and voltammetric measurements are basically different: IR spectroscopy and/or NMR [Tong et al., 2002] probe two different CO populations on the heterogeneous Ru-island decorated Pt electrodes, but voltammetry is a probe of reactivity of CO oxidation: the cyclic voltammogram split can only be shown when the oxidizing CO is confined within the island borders [Tong et al., 2002].

The data in Figs. 12.11a, 12.19, and 12.20 indicate that while the chemisorbed CO disappears from both Ru and Pt sites of Pt(111) and Pt(111)/Ru surfaces at sufficiently positive potentials, the stability of bridge-bonded compared with atop CO is greater on P(111)/Ru (Fig. 12.20) than on pure Pt(111). The degree of stability of bridge-bonded CO (Fig. 12.20) is unexpected when compared with previous findings with clean Pt(111) [Friedrich et al., 1966, 1996, 2002], and, as noticed above, brings the electrochemical and infrared data to coherence. However, the voltammetric split can only be observed in a narrow Ru coverage range of Ru on Pt(111), not higher than 0.25 ML [Lu et al., 2002], while IR and NMR [Tong et al., 2002] mapping of independent Pt(111)-CO and Pt(111)/Ru-CO populations continues (Fig. 12.21; notice the background occurring from nonresonant SFG [Shen 1989]). At the same time, at an Ru coverage of ≥ 0.25 ML, the voltammetric split of the CO stripping from Pt(111)/Ru disappears [Lu et al., 2002].

The intriguing observation that voltammetric CO stripping disappears at a certain Ru coverage, but the spectra from Ru-CO and Pt-CO patches are still measured at high Ru coverage, will now be considered. It can be seen that three conditions must be met simultaneously to obtain avoltammetric split of the type in Fig. 12.18:

- Ru atoms need to be physically present as distinctive Ru (nano)islands on the Pt surface rather than as an intermixture with Pt surface atoms of the Pt/Ru alloy [Ianniello et al., 1994; Crown et al., 2000].
- The Ru (and surrounding Pt) nanoislands need to develop independent CO populations—a condition realized by solution CO dosing to high CO coverage on both Pt and Ru of the Pt/Ru surface.
- The island Pt/Ru edge needs to have some unique chemical properties compared with the surrounding two-dimensional Ru and Pt phases [Lu et al., 2002].

Apparently, the two first conditions are always or easily met on heterogeneous [Koper et al., 1999] Ru-island covered Pt surfaces. In contrast, the third condition is fulfilled rarely, and, on Pt(111)/Ru, is lost at Ru coverages higher than about 0.25 ML [Lu et al., 2002]. The physical property of the Ru deposit above 0.25 ML on Pt(111) that changes with Ru coverage is the transformation of the Ru islands from essentially monoatomic to multilayer adlattices [Crown et al., 2002]. At the same time, with increasing Ru coverage, both Ru and Pt surface atoms become less

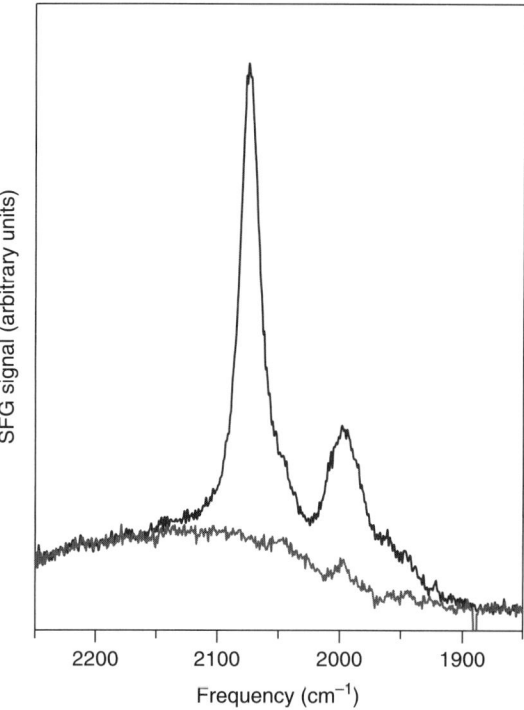

Figure 12.21 BB-SFG spectra of CO on a mixed metal electrode (in a CO-saturated 0.1 M H_2SO_4 solution) made by depositing Ru nanoislands on Pt(111) at a coverage of 0.35 ML. The electrolyte is a 25 μm thick solution layer (see Fig. 12.1). The acquisition time was 40 s; data were obtained at $E = 0.1$ V. As in Fig. 12.11b(II), the BBIR pulses are tuned to optimize the multiply bonded spectra, suppressing the atop intensity.

"metallic," and the added Ru amounts are functionally causing the suppression of back-donation of CO bonding on all surface sites [Tong et al., 2002]. (A sequence of structural properties is then activated: the reduction in back-donation weakens the metal–carbon bond, leading to a strengthening of the C–O bond, which, in turn, gives rise to a blueshift in the CO stretching frequency, particularly well seen on Ru-CO [Friedrich et al., 2002].) As the CO in the NMR measurements is just a surface probe of the state of the electrode surface, the likely surface additive that accepts a density of states from both Pt and Ru on the ruthenated Pt surfaces [Tong et al., 2002] is oxygen [Lewera et al., 2007] trapped between subsequent Ru layers deposited on Pt. The surface science literature [Stampfl et al., 1996; Bottcher et al., 1997; Todorova et al., 2002] documents thermally activated subsurface O present on Ru and/or Ru surface oxide formation, especially at high oxygen pressure. The oxygen incorporated into Pt(111)/Ru could easily function in a manner similar to Se adatoms deposited on Ru particles (as a sink of the electronic charge (DOS) [Babu et al., 2007]), and modify the properties of the Pt/Ru edge. At higher Ru coverage, the edge apparently no longer serves as an efficient barrier against CO entering the Ru domains [Lu et al.,

2002]: as soon as the barrier is crossed, CO oxidation occurs at a high rate, that is, at low potentials (Fig. 12.18). In support of these concepts, one should notice that the top of the Ru deposit on Pt is always metallic, and is available for CO chemisorption on all Pt and Ru sites.

The following are now expected: first, in the CO oxidation electrode potential range on Pt/Ru surfaces (i.e., at low potentials), entrapped oxygen should be found; second, the entrapped (subsurface) oxygen concentration between the Ru layers should increase with increasing multilayer character (and coverage) of the Ru deposit on Pt. Research focusing on these two issues is planned.

12.6 CONCLUSIONS

We have shown that BB-SFG provides a unique ability to probe the detailed behavior and kinetics of electrochemical transformations without compromising the conditions needed for high quality electrochemistry. The high sensitivity and interface selectivity of SFG allows for the use of an electrolyte layer of 25 μm up to 5 mV/s, and permits fast kinetic measurements by improving the response time to applied potential transients into the millisecond range, and by allowing rapid flow of reagents. Since the penalties for propagating IR pulses through such a thin layer are small, transparent electrodes are not needed, and well-defined single-crystal electrodes have been used, based upon low index Pt(hkl) substrates, including nanostructured mixed metal electrodes.

Information obtained by this method and reviewed above concerns:

- the coincidence between the IR BB-SFG spectra and the electrochemical behavior of CO chemisorbed on polycrystalline Pt when electro-oxidized to pure supporting electrolyte;
- evidence that the SFG observables are sensitive to the phase transition of CO adsorbed on a Pt(111) single-crystal electrode;
- coincidence between vibrational spectra and voltammetry obtained from CO-covered Pt (111)/Ru surfaces, including the split voltammetric behavior and all CO chemisorbed forms.

We have also investigated adsorption of acetate on a polycrystalline Pt electrode, which is of interest in the context of catalysts for ethanol oxidation, where acetate is an unwanted oxidation intermediate, and have demonstrated the capacity of the BB-SFG method to examine Pt nanoparticles.

This chapter has provided a quantitative link between the split voltammogram obtained from CO-covered Pt(111)/Ru surfaces and the IR spectra obtained from such surfaces. In particular, an interpretation has been given of the disappearance of the voltammetric split from stripping CO when the Ru coverage becomes higher than 0.25 ML. We have pointed out that the Ru deposits on Pt and the Pt surface sites are always available for CO adsorption, showing that neither the Ru sites nor the Pt sites are covered by oxygen, disregarding the Ru coverage. The physical property of the Ru deposit above 0.25 ML of Ru that changes with coverage is the growth of the island height from monoatomic to multilayered. Citing previous works [Lu et al., 2002; Tong et al., 2002], we note that both Ru and Pt surface atoms become less metallic with increasing Ru coverage, and both Pt and Ru atoms need to donate DOS to a surface species. This species is defined as entrapped oxygen between subsequent Ru layers deposited on Pt. The entrapped oxygen affects the properties of the Pt/Ru edge, which, at high Ru coverage, is no longer a barrier against CO entrance into the Ru domains to become oxidized. For purely monoatomic adlayers of Ru on Pt, we conclude that a voltammetric resolution of the type depicted in Fig. 12.18 should be found at high Ru coverage.

ACKNOWLEDGMENTS

This material is based upon work supported by the Army Research Office through a MURI grant (DAAD19-03-1-0169) for fuel cell research to the Case Western Reserve University by the Army Research Office under award Army W911NF-08-1-0309, by the National Science Foundation under awards DMR 0504038 and NSF CHE06-51083, and by the Air Force Office of Scientific Research under award FA9550-06-1-0235.

REFERENCES

Adzic RR, Tripkovic AV, O'Grady WE. 1982. Structural effects in electrocatalysis. Nature 296: 137–138.

Akemann W, Friedrich KA, Linke U, Stimming U. 1998. The catalytic oxidation of carbon monoxide at the platinum/electrolyte interface investigated by optical second harmonic generation (SHG): Comparison of Pt(111) and Pt(997) electrode surfaces. Surf Sci 404: 571–575.

Akemann W, Friedrich KA, Stimming U. 2000. Potential-dependence of CO adlayer structures on Pt(111) electrodes in acid solution: Evidence for a site selective charge transfer. J Chem Phys 113: 6864–6874.

Arenz M, Mayrhofer KJJ, Stamenkovic V, Blizanac BB, Tomoyuki T, Ross PN, Markovic NM. 2005. The effect of the particle size on the kinetics of CO electrooxidation on high surface area Pt catalysts. J Am Chem Soc 127: 6819–6829.

Babu PK, Lewera A, Chung J-H, Hunger R, Jaegermann W, Alonso-Vante N, Wieckowski A, Oldfield E. 2007. Selenium becomes metallic in Ru-Se fuel cell catalysts; An EC-NMR and XPS investigation. J Am Chem Soc 129: 15140–15141.

Backus EHG, Bonn M. 2005. A quantitative comparison between reflection absorption infrared and sum-frequency generation spectroscopy. Chem Phys Lett 412: 152–157.

Bain CD. 1995. Sum-frequency vibrational spectroscopy of the solid liquid interface. J Chem Soc Faraday Trans 91: 1281–1296.

Baldelli S, Markovic NM, Ross PN, Shen YR, Somorjai GA. 1999. Sum frequency generation of CO on (111) and polycrystalline platinum electrode surfaces; Evidence for SFG invisible surface CO. J Phys Chem B 103: 8920–8925.

Behrens RL, Lagutchev A, Dlott D, Wieckowski A. To be submitted.

Bertie JE, Lan Z. 1996. Infrared intensities of liquids. XX: The intensity of the OH stretching band of liquid water revisited, and the best current values of the optical constants of $H_2O(I)$ at 25 °C between 15,000 and 1 cm^{-1}. Appl Spectrosc 50: 1047–1057.

Biggin ME, Gewirth AA. 2001. Infrared studies of benzotriazole on copper electrode surfaces—Role of chloride in promoting reversibility. J Electrochem Soc 148: C339–C347.

Blizanac BB, Arenz M, Ross PN, Markovic NM. 2004. Surface electrochemistry of CO on reconstructed gold single crystal surfaces studied by infrared reflection absorption spectroscopy and rotating disk electrode. J Am Chem Soc 126: 10130–10141.

Bonn M, Denzler DN, Funk S, Wolf M. 2000. Ultrafast electron dynamics at metal surfaces; Competition between electron-phonon coupling and hot-electron transport. Phys Rev B 61: 1101–1105.

Bonn M, Hess C, Miners JH, Heinz TF, Bakker HJ, Cho M. 2001. Novel surface vibrational spectroscopy; Infrared–infrared–visible sum-frequency generation. Phys Rev Lett 86: 1566–1569.

Bottcher A, Niehus H, Schwegman S, Over H, Ertl G. 1997. CO oxidation reaction over oxygen-rich Ru(0001) Surfaces. J Phys Chem B 101: 11185–11191.

Chen YX, Miki A, Ye S, Sakai H, Osawa M. 2003. Formate, an active intermediate for direct oxidation of methanol on Pt Electrode. J Am Chem Soc 125: 3680–3681.

Chen YX, Heinen M, Jusys Z, Behm RB. 2006. Kinetics and mechanism of the electrooxidation of formic acid—Spectroelectrochemical studies in a flow cell. Angew Chem Int Ed 45: 981–985.

Chen Z, Shen YR, Somorjai GA. 2002. Studies of polymer surfaces by sum frequency generation vibrational spectroscopy. Annu Rev Phys Chem 53: 437–465.

Cho M, Hess C, Bonn M. 2002. Lateral interactions between adsorbed molecules: Investigations of CO on Ru(001) using nonlinear surface vibrational spectroscopies. Phys Rev B 65: 205423.

Chou KC, Kim J, Baldelli S, Somorjai GA. 2003a. Vibrational spectroscopy of carbon monoxide, acetonitrile, and phenylalanine adsorbed on liquid vertical bar electrode interfaces by sum frequency generation. J Electroanal Chem 554: 253–263.

Chou KC, Markovic NM, Kim JP, Ross PN, Somorjai GA. 2003b. An in situ time-dependent study of CO oxidation on Pt(111) in aqueous solution by voltammetry and sum frequency generation. J Phys Chem B 107: 1840–1844.

Chrzanowski W, Wieckowski A. 1997. Ultrathin films of ruthenium on low index platinum single crystal surfaces; An electrochemical study. Langmuir 13: 5974–5978.

Clavilier J. 1999. Flame-annealing and cleaning technique. In: Wieckowski A, ed. Interfacial Electrochemistry: Theory, Experiment, and Applications. New York: Marcel Dekker.

Colmenares L, Wang H, Jusys Z, Jiang L, Yan S, Sun GQ, Behm RJ. 2006. Ethanol oxidation on novel, carbon supported Pt alloy catalysts—Model studies under defined diffusion conditions. Electrochim Acta 52: 221–233.

Corrigan DS, Krauskopf EK, Rice LM, Wieckowski A, Weaver MJ. 1988. Adsorption of acetic acid at platinum and gold electrodes—A combined infrared spectroscopic and radiotracer study. J Phys Chem 92: 1596–1601.

Corrigan DS, Weaver MJ. 1988. Mechanisms of formic-acid, methanol, and carbon-monoxide electrooxidation at platinum as examined by single potential alteration infrared-spectroscopy. J Electroanal Chem 241: 143–162.

Crown A, Johnston C, Wieckowski A. 2002. Growth of ruthenium islands on Pt(*hkl*) electrodes obtained via repetitive spontaneous deposition. Surf Sci 506: L268–L274.

Crown A, Kin H, Lu GQ, de Moraes IR, Rice C, Wieckowski A. 2000. Research toward designing high activity catalysts for fuel cells; Structure and Reactivity. J New Mater Electrochem Syst 3: 275.

Crown A, Moraes IR, Wieckowski A. 2001. Examination of Pt(111)/Ru and Pt(111)/Os surfaces; STM imaging and methanol oxidation activity. J Electroanal Chem 500: 333–343.

Crown A, Wieckowski A. 2001. Scanning tunneling microscopy investigations of ruthenium- and osmium-modified Pt(100) and Pt(110) single crystal substrates. Phys Chem Chem Phys 3: 3290–3296.

Dederichs F, Friedrich KA, Daum W. 2000. Sum-frequency vibrational spectroscopy of CO adsorption on Pt(111) and Pt(110) electrode surfaces in perchloric acid solution; Effects of thin-layer electrolytes in spectroelectrochemistry. J Phys Chem B 104: 6626–6632.

Delgado JM, Orts JM, Rodes A. 2005. ATR–SEIRAS study of the adsorption of acetate anions at chemically deposited silver thin film electrodes. Langmuir 21: 8809–8816.

Eisenthal KB. 1992. Equilibrium and Dynamic processes at interfaces by 2nd harmonic and sum frequency generation. Annu Rev Phys Chem 43: 627–661.

Friedrich KA, Geyzers KP, Dickinson AJ, Stimming U. 2002. Fundamental aspects in electrocatalysis: From the reactivity of single-crystals to fuel cell electrocatalysis. J Electroanal Chem 524/525: 261–272.

Friedrich KA, Geyzers K-P, Henglein F, Marman A, Stimmming U, Unkauf W, Vogel R. 1966. CO Adsorption and oxidation on nanostructured electrode surfaces studied by STM and IR spectroscopy. In: Proceedings of 189th ECS Meeting; Symposium on Electrode Processes VI. Los Angeles: The Electrochemical Society.

Friedrich KA, Geyzers KP, Linke U, Stimming U, Stumper J. 1996. CO adsorption and oxidation on a Pt(111) electrode modified by ruthenium deposition: An IR spectroscopic study. J Electroanal Chem 402: 123–128.

Gutierrez C, Caram JA, Beden B. 1991. J Electroanal Chem 305: 289.

Guyot-Sionnest P. 2005. The mature years of sum-frequency generation are ahead. Surf Sci 585: 1–2.

Guyot-Sionnest P, Hunt JH, Shen YR. 1987. Sum-frequency vibrational spectroscopy of a Langmuir film: Study of molecular orientation of a two-dimensional system. Phys Rev Lett 59: 1597–1600.

Guyot-Sionnest P, Tadjeddine A. 1990. Spectroscopic investigations of adsorbates at the metal electrolyte interface using sum frequency generation. Chem Phys Lett 172: 341–345.

Hare DE, Rhea ST, Dlott DD, D'Amato RJ, Lewis TE. 1998. Pulse duration dependence of lithographic printing plate imaging by near-infrared lasers. J Imag Soc Technol 42: 90–97.

Hoffer S, Baldelli S, Chou K, Ross P, Somorjai GA. 2002. CO oxidation on electrified platinum surfaces in acetonitrile/water solutions studied by sum frequency generation and cyclic voltammetry. J Phys Chem B 106: 6473–6478.

Hubbard AT. 1973. Electrochemistry of thin layers of solution. CRC Crit Rev Anal Chem 2: 201–300.

Hunt JH, Guyot-Sionnest P, Shen YR. 1987. Observation of C–H stretch vibrations of monolayers of molecules. Optical-sum frequency generation. Chem Phys Lett 133: 189–192.

Ianniello R, Schmidt VM, Stimming U, Stumper J, Wallau A. 1994. CO adsorption and oxidation on Pt and Pt-Ru alloys, dependence on substrate composition. Electrochim Acta 39: 1863–1869.

Kitamura F, Takahashi M, Ito M. 1989. Carbon monoxide adsorption on platinum (111) single-crystal electrode surface studied by infrared reflection–absorption spectroscoy. Surf Sci 223: 493–508.

Koper MTM, Lebedeva NP, Hermse CGM. 2002. Dynamics of CO at the solid/liquid interface studied by modeling and simulation of CO electro-oxidation on Pt and PtRu electrodes. Faraday Discuss 121: 301–311.

Koper MTM, Lukkien JJ, Jansen APJ, van Santen RA. 1999. Lattice gas model for CO electrooxidation on Pt-Ru bimetallic surfaces. J Phys Chem B 103: 5522–5529.

Korzeniewski C, Pons S, Schmidt PP, Severson MW. 1986. A theoretical analysis of the vibrational spectrum of carbon monoxide on platinum metal electrodes. J Chem Phys 85: 4153–4160.

Kunimatsu K, Golden WG, Seki H, Philpott MR. 1985a. Carbon monoxide adsorption on a platinum electrode studied by polarization modulated FT-IRRAS. 1. Co Adsorbed in the double-layer potential region and its oxidation in acids. Langmuir 1: 245–250.

Kunimatsu K, Seki H, Golden WG, Gordon JG II, Philpott MR. 1985b. Electrode/electrolyte interface study using polarization modulated FTIR reflection–adsorption spectroscopy. Surf Sci 158: 596–608.

Lagutchev A, Hambir SA, Dlott DD. 2007. Nonresonant background suppression in broadband vibrational sum-frequency generation spectroscopy. J Phys Chem C 111: 13645–13647.

Lagutchev A, Patterson JE, Huang W, Dlott DD. 2005. Ultrafast dynamics of self-assembled monolayers under shock compression: Effects of molecular and substrate structure. J Phys Chem B 109: 5033–5044.

Lagutchev A, Lu GQ, Takeshita T, Dlott DD, Wieckowski A. 2006. Vibrational sum frequency generation studies of the $(2 \times 2) \rightarrow (\sqrt{19} \times \sqrt{19})$ phase transition of CO on Pt(111) electrodes. J Chem Phys 125: 154705.

Leung L-WH, Wieckowski A, Weaver MJ. 1988. In situ infrared spectroscopy of well-defined single-crystal electrodes: Adsorption and electrooxidation of carbon monoxide on platinuk(111). J Phys Chem 92: 6985–6990.

Lewera A, Inukai J, Zhou WPD, Cao D, Duong HT, Alonso-Vante N, Wieckowski A. 2007. Chalcogenide oxygen reduction reaction catalysis: X-ray photoelectron spectroscopy with Ru, Ru/Se and Ru/S samples emersed from aqueous media. Electrochim Acta 52: 5759–5765.

Lu G-Q, Crown A, Wieckowski A. 1999. Formic acid decomposition on polycrystalline platinum and palladium electrodes. J Phys Chem B 103: 9700–9711.

Lu GQ, Lagutchev A, Dlott DD, Wieckowski A. 2005. Quantitative vibrational sum-frequency generation spectroscopy of thin layer electrochemistry: CO on a Pt electrode. Surf Sci 585: 3–16.

Lu GQ, Waszczuk P, Wieckowski A. 2002. Oxidation of CO adsorbed from CO saturated solutions on the Pt(111)/Ru electrode. J Electroanal Chem 532: 49–55.

Lu G-Q, White JO, Wieckowski A. 2004. Vibrational analysis of chemisorbed CO on the Pt(111)/Ru bimetallic electrode. Surf Sci 564: 131–140.

Lucas CA, Markovic NM, Ross NM. 1999. The adsorption and oxidation of carbon monoxide at the Pt(111)/electrolyte interface; atomic structure and surface relaxation. Surf Sci 425: L381–L386.

Maillard F, Lu GQ, Wieckowski A, Stimming U. 2005. Ru-decorated Pt surfaces as model fuel cell electrocatalysts for CO electrooxidation. J Phys Chem B 109: 16230–16243.

Markovic NM, Ross PN. 2002. Surface science studies of model fuel cell electrocatalysts. Surf Sci Rep 45: 121–229.

Massong H, Wang H, Samjeske G, Baltruschat H. 2000. The co-catalytic effect of Sn, Ru and Mo decorating steps of Pt(111) vicinal electrode surfaces on the oxidation of CO. Electrochim Acta 46: 701–707.

Oda I, Inukai J, Ito M. 1993. Compression structures of carbon monoxide on a Pt(111) electrode surface studied by in situ scanning tunneling microscopy. Chem Phys Lett 203: 99–103.

Patterson JE, Lagutchev AS, Huang W, Dlott DD. 2005. Ultrafast dynamics of shock compression of molecular monolayers. Phys Rev Lett 94: 015501.

Persson BNJ, Ryberg R. 1981. Vibrational interaction between molecules adsorbed on a metal surface; The dipole–dipole interaction. Phys Rev B 24: 6954–6970.

Petukhov AV, Akemann W, Friedrich KA, Stimming U. 1998. Kinetics of electrooxidation of a co monolayer at the platinum/electrolyte interface. Surf Sci 404: 182–186.

Pozniak B, Mo YB, Stefan IC, Mantey K, Hartmann M, Scherson DA. 2001. In situ time-resolved second-harmonic generation from Pt(111) microfacetted single-crystal platinum microspheres. J Phys Chem B 105: 7874–7877.

Pozniak B, Scherson DA. 2003. Dynamics of a surface phase transition as monitored by in situ second harmonic generation. J Am Chem Soc 125: 7488–7489.

Rice C, Ha RI, Masel RI, Waszczuk P, Wieckowski A, Barnard T. 2002. Direct formic acid fuel cells. J Power Sources 111: 83–89.

Richmond GL. 2002. Molecular bonding and interactions at aqueous surfaces as probed by vibrational sum frequency spectroscopy. Chem Rev 102: 2693–2724.

Richter LT, Petralli-Mallow TP, Stephenson JC. 1998. Vibrationally resolved sum-frequency generation with broad-bandwidth infrared pulses. Opt Lett 23: 1594–1596.

Rodes A, Pastor E, Iwasita T. 1994. An FTIR study on the adsorption of acetate at the basal planes of platinum single-crystal electrodes. J Electroanal Chem 376: 109–118.

Rousseau S, Coutanceau S, Lamy C, Leger J-M. 2006. Direct ethanol fuel cell (DEFC): Electrical performances and reaction products distribution under operating conditions with different platinum-based anodes. J Power Sources 1: 18–24.

Samjeske G, Miki A, Ye S, Yamakata A, Mukouyama Y, Okamoto H, Osawa M. 2005. Potential oscillations in galvanostatic electrooxidation of formic acid on platinum: A time-resolved surface-enhanced infrared study. J Phys Chem B 109: 23509–23516.

Samjeske G, Osawa M. 2005. Current oscillations during formic acid oxidation on a Pt electrode: Insight into the mechanism by time-resolved IR spectroscopy. Angew Chem Int Ed 44: 5694–5698.

Severson MW, Stuhlmann C, Villegas I, Weaver MJ. 1995. Dipole–dipole coupling effects upon infrared spectroscopy of compressed electrochemical adlayers; Application to the Pt(111)/CO system. J Chem Phys 103: 9832–9843.

Shen YR. 1984. The Principles of Nonlinear Optics. New York: Wiley.

Shen YR. 1989. Surface properties probed by second-harmonic and sum-frequency generation. Nature 337: 519–525.

Shen YR. 1994. Surfaces probed by nonlinear optics. Surf Sci 300: 551–562.

Stamenkovic V, Chou KC, Somorjai GA, Ross PN, Markovic NM. 2005. Vibrational properties of CO at the Pt(111)–Solution interface: The anomalous Stark–Tuning slope. J Phys Chem B 109: 678–680.

Stampfl C, Schwegmann S, Over H, Scheffler M, Ertl G. 1996. Structure and stability of a high coverage (1×1) oxygen phase on Ru(0001). Phys Rev Lett 77: 3371–3374.

Tadjeddine A. 2000. Spectroscopic investigation of surfaces and interfaces by using infrared–visible sum and difference generation. Surf Rev Lett 7: 423–436.

Tian ZQ, Ren B, Mao BW. 1997. Extending surface Raman spectroscopy to transition metal surfaces for practical applications. 1. Vibrational properties of thiocyanate and carbon monoxide adsorbed on electrochemically activated platinum surfaces. J Phys Chem B 101: 1338–1346.

Todorova M, Li WX, Ganduglia-Pirovano MV, Stampfl C, Reuter, Scheffler M. 2002. Role of subsurface oxygen in oxide formation at transition metal surfaces. Phys Rev Lett 8909: 6103.

Tolmachev YV, Menzel A, Tkachuk AV, Chu YS, You H. 2004. In situ surface X-ray scattering observation of long-range ordered $(\sqrt{19} \times \sqrt{19})$R23.4°–13CO structure on Pt(111) in aqueous electrolytes. Electrochem Solid-State Lett 7: E23–E26.

Tong YY, Kim HS, Babu PK, Waszczuk P, Wieckowski A, Oldfield E. 2002. An NMR investigation of CO tolerance in a Pt/Ru fuel cell catalyst. J Am Chem Soc 124: 468–473.

Venkatakrishnan K, Tan B, Sivakumar NR. 2002. Sub-micron ablation of metallic thin film by femtosecond pulse laser. Opt Laser Tech 34: 575–578.

Vidal F, Busson B, Six C, Pluchery O, Tadjeddine A. 2002. SFG study of methanol dissociative adsorption at Pt(100), Pt(110) and Pt(111) electrodes surfaces. Surf Sci 502/503: 485–489.

Vidal F, Busson B, Six C, Tadjeddine A, Dreesen L, Humbert C, Peremans A, Thiry P. 2004. Methanol dissociative adsorption on Pt(100) as studied by nonlinear vibrational spectroscopy. J Electroanal Chem 563: 9–14.

Vidal F, Busson B, Tadjeddine A, Peremans A. 2003. Effect of a static electric field on the vibrational and electronic properties of a compressed CO adlayer on Pt(110) in nonaqueous electrolyte as probed by infrared reflection–absorption spectroscopy and infrared–visible sum-frequency generation spectroscopy. J Chem Phys 119: 12492–12498.

Vidal F, Busson B, Tadjeddine A. 2005. Probing electronic and vibrational properties at the electrochemical interface using SFG spectroscopy: Methanol electro-oxidation on Pt(110). Chem Phys Lett 403: 324–328.

Vielstich W, Lamm A, Gasteiger H, eds. 2003. Fuel Cell Handbook: Fuel Cell Electrocatalysis. Volume 2. New York; Wiley-VCH.

Villegas I, Weaver MJ. 1994. Carbon monoxide adlyaer structures on platinum (111) electrodes; A synergy between in-situ scanning tunneling microscoy and infrared spectroscopy. J Chem Phys 101: 1648–1660.

Wang JX, Robinson IK, Ocko BM, Adzic RR. 2005. Adsorbate-geometry specific subsurface relaxation in the CO/Pt(111) system. J Phys Chem B 109: 24–26.

Watanabe M, Motoo S. 1975a. Electrocatalysis by Ad-Atoms. II. Enhancement of the oxidation of methanol on platinum by ruthenium ad-atoms. J Electroanal Chem 60: 267–273.

Watanabe M, Motoo S. 1975b. Electrocatalysis by ad-atoms. III. Enhancement of the oxidation of carbon monoxide on platinum by ruthenium ad-atoms. J Electroanal Chem 60: 275–283.

Yoshimi K, Song MB, Ito M. 1996. Carbon monoxide oxidation on a Pt(111) electrode studied by in-situ IRAS and STM: Coadsorption of CO with water on Pt(111). Surf Sci 368: 389–395.

Zhu XD, Suhr H, Shen YR. 1987. Surface vibrational spectroscopy by infrared–visible sum frequency generation. Phys Rev B 35: 3047–3050.

CHAPTER 13

Methanol, Formaldehyde, and Formic Acid Adsorption/Oxidation on a Carbon-Supported Pt Nanoparticle Fuel Cell Catalyst: A Comparative Quantitative DEMS Study

Z. JUSYS and R.J. BEHM

Institute of Surface Chemistry and Catalysis, Ulm University, D-89069 Ulm, Germany

13.1 INTRODUCTION

The electrocatalytic oxidation of small organic C_1 molecules has attracted considerable interest in the last two decades, both from fundamental science aspects and because of their potential application as energy carriers in fuel cells, such as direct methanol fuel cell (DMFCs) and direct formic acid fuel cells (for reviews, see Lamy et al. [1983]; Parsons and VanderNoot [1988]; Sun [1998]; Jarvi and Stuve [1998]; Feliu and Herrero [2003]). The electro-oxidations of formaldehyde and formic acid are of direct interest also for the methanol oxidation reaction (MOR) in DMFCs, since it has repeatedly been demonstrated experimentally, both in studies on model systems [Gromyko et al., 1976; Petukhova et al., 1977; Ota et al., 1984; Shibata and Motoo, 1986; Shibata et al., 1987; Iwasita and Vielstich, 1986; Korzeniewski and Childers, 1998; Childers et al., 1999; Wang et al., 2001a, b; Jusys and Behm, 2001; Jusys et al., 2002a, b, 2003; Batista et al., 2003; Gao et al., 2004; Jambunathan et al., 2004] and in measurements at the exhaust of an operating DMFC [Wasmus et al., 1995; Lin et al., 1997; Sanicharane et al., 2002; Tkach et al., 2004; Seiler et al., 2004; Neergat et al., 2006], that these incomplete methanol oxidation products are produced as reaction intermediates or side products during methanol oxidation. A close connection between methanol oxidation and formaldehyde and formic acid oxidation had already been predicted in the classical reaction scheme of Bagotzky

---

Fuel Cell Catalysis. Edited by Marc T. M. Koper
Copyright © 2009 John Wiley & Sons, Inc.

et al. (1977), where the process of methanol dehydrogenation and electro-oxidation to CO_2 was described schematically by the following sequence of reaction steps:

$$\begin{array}{ccccc}
CH_3OH & \rightarrow & CH_2OH & \rightarrow & CHOH & \rightarrow & COH \\
& & \downarrow & & \downarrow & & \downarrow \\
& & CH_2O & \rightarrow & CHO & \rightarrow & CO \\
& & & & \downarrow & & \downarrow \\
& & & & HCOOH & \rightarrow & COOH \\
& & & & & & \downarrow \\
& & & & & & CO_2
\end{array}$$

In this reaction scheme, each step includes the release of one electron on the one hand and the release of a proton or the uptake of a hydroxyl anion—via water uptake and proton release—on the other. The educt (methanol), final product (CO_2), and stable incomplete oxidation products (formaldehyde and formic acid) are indicated here in bold to distinguish them from other possible intermediate species adsorbed on the electrode surface. Since both formaldehyde and formic acid can undergo further oxidation, the product distribution resulting from methanol oxidation (and in a similar way also for formaldehyde oxidation) will depend not only on the respective reaction rates for the formation and further oxidation of these species (formaldehyde and formic acid), but also on transport characteristics of the reaction cell and the catalyst layer/electrode surface [Jusys et al., 2003]. Desorption and rapid removal of stable, volatile reaction intermediates (incomplete oxidation products) will increase their contribution to the product distribution, while the slow, diffusion-controlled removal of these species (e.g., due to slow electrolyte flow) or an increased tendency for re-adsorption (e.g., by an increased electrode surface/porosity) will decrease their contribution and favor complete oxidation to CO_2 [Jusys et al., 2003]. These predictions are supported by a number of observations, for instance, by comparing product distributions measured in model studies with those determined at the exhaust of DMFCs or direct ethanol fuel cells (DEFCs) [Wasmus et al., 1995; Sanicharane et al., 2002; Tkach et al., 2004; Seiler et al., 2004; Neergat et al., 2006; Rao et al., 2007], or by comparison of product yields for different electrode roughnesses [Ota et al., 1984], catalyst loadings [Childers et al., 1999; Jusys et al., 2003; Bergamaski et al., 2006; Gavrilov et al., 2007; Islam et al., 2007], reactant concentrations [Wang et al., 2001a; Camara and Iwasita, 2005; Wang et al., 2004], or electrolyte flow rates [Wang et al., 2001a]. Further mechanistic information has come from studies investigating the influence of crystallographic orientation [Wang et al., 2001a; Sun and Clavilier, 1987; Shin et al., 1996; Tarnowski and Korzeniewski, 1997; Sriramulu et al., 1998; Jarvi et al., 1998; Cuesta, 2006; Housmans et al., 2006; Nakamura et al., 2007; Spendelow et al., 2007], adsorbed anions [Batista et al., 2003, 2004], or composition of bimetallic catalysts [Shibata et al., 1987; Jusys et al., 2002a, b; Gao et al., 2004; Islam et al., 2007; Entina et al., 1967; Watanabe and Motoo, 1975; Sun et al., 1988; Gasteiger et al.,

1993, 1994; Iwasita et al., 1994; Schmidt et al., 1999a; Lima et al., 2001; Jusys et al., 2002a, b; Ge et al., 2001; Basnayake et al., 2006].

More quantitative predictions of the reaction characteristics require a detailed understanding of the oxidation kinetics of the different C_1 molecules, and in particular of the product distribution in the respective oxidation reactions under defined reaction conditions and, most importantly, under defined transport conditions. The latter affect both the delivery of educts and the removal of products and side products (incomplete oxidation products). This is the topic of the present study, where we have investigated the reaction kinetics and product distribution for the oxidation of the C_1 molecules methanol, formaldehyde, and formic acid by comparative differential electrochemical mass spectrometry (DEMS) measurements under well-defined, but nevertheless close-to-realistic conditions and materials (supported catalyst, and continuous and controlled reactant transport). In contrast to purely electrochemical measurements, DEMS allows one to directly detect the volatile reaction products CO_2 and methyl formate. The latter is produced in proportion to the formic acid concentration and results from reaction between the reactant methanol and formic acid (see the next section). Formaldehyde formation is determined as the difference between the total faradaic current and the partial currents for CO_2 and formic acid formation, assuming that these are the only products contributing to the faradaic current. Defined transport conditions were achieved by using (i) a thin-layer flow cell DEMS setup [Jusys et al., 1999, 2001] and (ii) thin-film electrodes with negligible internal diffusion resistance [Schmidt et al., 1998]. The first results of this study, focusing on the adsorption and oxidation of methanol on a carbon-supported Pt (Pt/C) fuel cell catalyst [Jusys and Bchm, 2001] and on the effect of Pt catalyst loading on the MOR product yields [Jusys et al., 2003] were reported previously; additional measurements on unsupported Pt and PtRu nanoparticle catalysts were published in [Jusys et al., 2002a, b].

There are a number of excellent reviews of the numerous studies on methanol oxidation on Pt electrodes and catalysts [Lamy et al., 1983; Parsons and VanderNoot, 1988; Sun, 1998; Jarvi and Stuve, 1998; Cohen et al., 2008; Iwasita, 2002, 2003]. Most importantly, these studies showed the formation of formaldehyde and formic acid as incomplete oxidation products, in addition to the complete oxidation product CO_2. These species were detected by chemical analysis [Petukhova et al., 1977; Ota et al., 1984; Korzeniewski and Childers, 1998; Childers et al., 1999; Batista et al., 2003, 2004], transmission infrared (IR) spectroscopy [Gao et al., 2004; Islam et al., 2007], and online mass spectrometric analysis [Iwasita and Vielstich, 1986; Wang et al., 2001a, b; Jusys et al., 2002a, b; Housmans et al., 2006; Iwasita, 2002, 2003; Wonders et al., 2006; Wang and Baltruschat, 2007], while adsorbed reaction intermediates (mainly CO_{ad}) were detected by in situ techniques such as IR spectroscopy [Sanicharane et al., 2002; Tkach et al., 2004; Sun and Clavilier, 1987; Nakamura et al., 2007; Christensen et al., 1988; Perez et al., 1994; Fan et al., 1996; Xia et al., 1997; Waszczuk et al., 2001; Zhu et al., 2001; Park et al., 2002a; Coutanceau et al., 2002; Yajima et al., 2004], nuclear magnetic resonance (NMR) [Rice et al., 2000], or radiotracer [Waszczuk et al., 2001; Coutanceau et al., 2002; Sobkowski and Wieckowski, 1972; Kazarinov et al., 1975] methods. In addition to these experimental studies, the dehydrogenative adsorption/oxidation of methanol on a Pt(111) surface was investigated theoretically [Desai et al., 2002; Greeley and Mavrikakis, 2002, 2004; Okamoto et al., 2003; Cao et al., 2005; Taylor and Neurock, 2005; Hartnig and Spohr, 2005; Hartnig et al., 2007b]. It had already been proposed in the early 1970s [Capon and Parsons, 1973a, b, c] that formic acid oxidation can proceed via different pathways (the "dual-pathway mechanism"), where one pathway, the "indirect pathway," includes formation and subsequent oxidation of CO_{ad}, while the other, the "direct pathway," does not involve CO_{ad} formation, but proceeds in another, often unspecified, way directly to CO_2 (see also Fig. 13.8a). A similar dual-pathway mechanism was proposed later also for methanol oxidation [Parsons and VanderNoot, 1988; Jarvi and Stuve, 1998; Leung and Weaver, 1990; Lopes et al., 1991; Herrero et al., 1994, 1995]. It should be noted that in these earlier studies, both pathways were assumed to lead to CO_2 as final product, whereas in some later studies, the formation and desorption of the incomplete oxidation products formaldehyde and formic acid was considered as the second pathway [Cao et al., 2005] (see also the discussion in Section 13.4.1). Based on in situ IR and DEMS observations, COH [Xia et al., 1997; Iwasita et al., 1987, 1992; Iwasita and Nart, 1997], CHO [Willsau and Heitbaum, 1986; Wilhelm et al., 1987], COOH [Zhu et al., 2001], and HCOO [Chen et al., 2003] have been proposed as possible adsorbed reaction intermediates. The role of the last of these as a reactive intermediate [Chen et al., 2003] is still under debate [Nakamura et al., 2007].

Formic acid oxidation on Pt catalysts and metal electrodes has equally been intensively studied, for example, by Parsons and co-workers and by other groups [Parsons and VanderNoot, 1988; Capon and Parsons, 1973; Anastasijevic et al., 1989; Wolter et al., 1985; Sun et al., 1988, 1994; Iwasita et al., 1994; Pastor et al., 1996; Xia, 1999; Sun and Yang, 1999; Schmidt et al., 2000; Yang and Sun, 2002; Jiang and Kucernak, 2002; Okamoto et al., 2004]. Detailed reviews are given in [Jarvi and Stuve, 1998; Markovic and Ross, 2002]. As mentioned abovee, a dual-pathway mechanism was proposed, involving a direct oxidation path via a reactive intermediate (the direct pathway) in addition to oxidation via decomposition to CO_{ad} and its subsequent oxidation (the indirect pathway) (see also Fig. 13.8a) [Parsons and VanderNoot, 1988; Sun et al., 1988; Iwasita et al., 1996; Lu et al., 1999; Miki et al., 2002; Samjeské and Osawa, 2005; Chen et al., 2006a, b, c]. An adsorbed bridge-bonded formate was detected during the reaction by in situ IR spectroscopy studies in an attenuated total reflection (ATR) configuration [Miki et al., 2002; Samjeské and Osawa, 2005; Chen et al., 2006a, b, c; Samjeské et al., 2005, 2006]. This species was proposed to act as active reaction intermediate [Miki et al., 2002; Samjeské and Osawa, 2005; Samjeské et al., 2005, 2006]. Based on a quantitative evaluation of spectro-electrochemical data for formic acid oxidation over a Pt film electrode, which were measured under continuous mass transport conditions as a function of formic acid concentration [Chen et al., 2006a, b] and reaction temperature [Chen et al., 2006c], we favored the direct dehydrogenation of formic acid to CO_2 as the major pathway, rather than a pathway via the adsorbed formate species detected by IR spectroscopy. This interpretation, where the IR-detected adsorbed formates act as "spectator species" rather than as reactive intermediates, was also supported by recent density functional theory (DFT) calculations [Hartnig et al., 2007a].

Formaldehyde electro-oxidation has been studied much less extensively compared with methanol and formic acid oxidation [Breiter, 1967; Loucka and Weber, 1968; Sidheswaran and Lal, 1971; Spasojevic et al., 1980; Beltowska-Brzezinska and Heitbaum, 1985; Napporn et al., 1995; Nakabayashi, 1998; Nakabayashi et al., 1998; Mishina, et al., 2002; Okamoto et al., 2005; Mai et al., 2005; Batista and Iwasita, 2006; Samjeské et al., 2007; de Lima et al., 2007]. This may partly be related to experimental problems: formaldehyde disproportionates to methanol and formic acid in the absence of methanol (the Canizzarro reaction). Furthermore, a technical application of formaldehyde as a fuel had not been anticipated, because of its toxicity. The latter point is also important for DMFC applications, where formaldehyde emissions should be avoided. In potentiodynamic measurements, formaldehyde oxidation was found to start at significantly more positive potentials than methanol oxidation, and adsorption of formaldehyde equally results in a reaction-inhibiting CO adlayer [Sun, 1998; Olivi et al., 1994; Miki et al., 2004; Chen et al., to be published]. Formic acid formation during formaldehyde oxidation was reported and quantified [Batista and Iwasita, 2006]. Formate species were detected upon formaldehyde oxidation on Group 1b metals in alkaline solution [Anastasijevic et al., 1993; Jusys, 1994; ten Kortenaar et al., 1999, 2001; Stadler et al., 2002] and as adsorbed species on Pt electrodes in acidic solution [Miki et al., 2004; Chen et al., to be published]. On the latter substrates [Samjeské et al., 2007], they were proposed to act as reaction intermediates [Miki et al., 2004], similar to methanol oxidation [Chen et al., 2003] and formic acid oxidation [Miki et al., 2002; Samjeské et al., 2007].

In the following, after a brief description of the experimental setup and procedures (Section 13.2), we will first focus on the adsorption and on the coverage and composition of the adlayer resulting from adsorption of the respective C_1 molecules at a potential in the H_{upd} range as determined by adsorbate stripping experiments (Section 13.3.1). Section 13.3.2 deals with bulk oxidation of the respective reactants and the contribution of the different reaction products to the total reaction current under continuous electrolyte flow, first in potentiodynamic experiments and then in potentiostatic reaction transients, after stepping the potential from 0.16 to 0.6 V, which was chosen as a typical reaction potential. The results are discussed in terms of a mechanism in which, for methanol and formaldehyde oxidation, the commonly used dual-pathway mechanism is extended by the possibility that reaction intermediates can desorb as incomplete oxidation products and also re-adsorb for further oxidation (for the formic acid oxidation mechanism, see [Samjeské and Osawa, 2005; Chen et al., 2006a, b; Miki et al., 2004]).

13.2 EXPERIMENTAL

The DEMS setup and experimental procedures used in this study were the same as described in more detail elsewhere [Jusys et al., 2001]. Briefly, the DEMS setup consisted of two differentially pumped chambers, a Balzers QMS 112 quadrupole mass spectrometer (MS), a Pine Instruments potentiostat, and a computerized data acquisition system.

The electrochemical measurements were performed in parallel to online mass spectrometry measurements using a dual thin-layer flow cell [Jusys et al., 1999]. The second thin-layer compartment was interfaced to the MS inlet through a porous membrane (Scimat, 60 μm thick, 50% porosity), and interconnected via four capillaries with the first thin-layer compartment. The latter was designed to accommodate a glassy carbon disk (9 mm in diameter, Sigradur G from Hochtemperatur Werkstoffe GmbH), on which a thin layer of the carbon-supported Pt catalyst (E-TEK Inc., Pt particles supported on carbon Vulcan XC 72, 20 wt% Pt, mean Pt particle size 3.0 nm [Jusys et al., 2003], Pt loading 7 μg cm^{-2}, catalyst film diameter 6 mm) was deposited by successive pipetting–drying cycles of aqueous catalyst suspension and Nafion solution, respectively [Jusys et al., 2003; Schmidt et al., 1998]. Two Pt wires were used as counter-electrodes in the thin-layer cell. A saturated calomel electrode (SCE) connected to the outlet of the DEMS cell through the Teflon capillary served as a reference electrode. All potentials are quoted, however, with respect to that of the reversible hydrogen electrode (RHE).

The CO_2 signal was calibrated by comparing against the well-known amount of CO_2 formation produced during oxidation of a saturated CO adlayer ("CO stripping") [Jusys et al., 2001], determining the calibration constant K^* between the two signals via the relation

$$K^* = zQ_{MS}/Q_F \qquad (13.1)$$

where Q_{MS} and Q_F are the mass spectrometric and the faradaic charge, respectively, and $z = 2$ is the number of electrons per CO_{ad} molecule oxidized to CO_2. Alternatively, for calibration of the CO_2 production in bulk reactions, the potentiostatic oxidation of formic acid was used for comparison. In this reaction, CO_2 formation is the only reaction pathway (two-electron reaction). The methyl formate signal was calibrated by evaluating the product distribution (CO_2 and formic acid yields) during methanol oxidation at high catalyst loadings, where formaldehyde production is negligible [Ota et al., 1984; Childers et al., 1999; Jusys et al., 2003]. The partial current for formaldehyde formation was calculated from the difference between the calculated faradaic currents (integrated charges in potentiodynamic measurements) for CO_2 and formic acid formation on the one hand and the measured faradaic current (charge) on the other, assuming that these three are the only reaction products [Wang et al., 2001a; Jusys et al., 2003].

The supporting electrolyte (0.5 M sulfuric acid) was prepared using Millipore Milli Q water and ultrapure sulfuric acid (Merck, suprapur), and deaerated by high purity Ar (MTI Gase, N 6.0). For adsorbate stripping experiments, the respective reactant was pre-adsorbed at a constant electrode potential for about 10 minutes, after inserting 2 mL of 0.5 M H_2SO_4 solution saturated either with CO (CO_{ad} saturation) or with a 10% CO in Ar mixture (CO_{ad} submonolayer coverage) (CO: Messer-Griesheim, N 4.7), or a similar amount of electrolyte containing 0.1 M of the corresponding C_1 species (methanol, formaldehyde, or formic acid) by an all-glass syringe through a separate port. Subsequently, the cell was carefully flushed (10 minutes) by the supporting electrolyte at the adsorption potential, and then the adsorbate stripping experiment was performed. The potentiodynamic bulk oxidation data represent cycles taken during continuous cycling. For potentiostatic bulk oxidation, the potential was first stepped to 1.16 V for 1 s to oxidize adsorbed species, then lowered to 0.16 V for 3 minutes to accumulate an adlayer comparable to that before the adsorbate stripping experiments, and finally stepped to the reaction potential of 0.6 V. These experiments were carried out in 0.5 M H_2SO_4 solution containing 0.1 M of methanol (Merck, p.a.), formaldehyde (Alfa Aesar, 16 wt% aqueous formaldehyde solution, methanol-free) or formic acid (Merck, p.a.). All solutions were deaerated by high purity Ar (MTI Gase, N 6.0) prior to the measurements.

13.3 RESULTS AND DISCUSSION

13.3.1 Pre-Adsorbed Residue Stripping

In this section, we will present and discuss cyclic voltammetry and potential-step DEMS data on the electro-oxidation ("stripping") of pre-adsorbed residues formed upon adsorption of formic acid, formaldehyde, and methanol, and compare these data with the oxidative stripping of a CO adlayer formed upon exposure of a Pt/Vulcan catalyst to a CO-containing (either CO- or CO/Ar-saturated) electrolyte as reference. We will identify adsorbed species from the ratio of the mass spectrometric and faradaic stripping charge, determine the adsorbate coverage relative to a saturated CO adlayer, and discuss mass spectrometric and faradaic current transients after adsorption at 0.16 V and a subsequent potential step to 0.6 V.

13.3.1.1 Potentiodynamic Electro-Oxidation of Adsorbed Species

Figure 13.1 shows representative faradaic (a) and $m/z = 44$ (b) current responses of a Pt/Vulcan electrode during oxidative stripping of the adsorbate formed upon exposure to solutions saturated with CO (solid lines), or containing 0.1 M HCHO (dashed lines), HCOOH (dash–dotted lines) or CH_3OH (dash–double-dotted lines) (adsorption parameters: 0.11 V, 10 minutes). In addition to the positive-going scan of the electrode potential in the same electrolyte, a subsequent base cyclic voltammogram (CV) in the supporting electrolyte is shown as a dotted line. The faradaic current peak for pre-adsorbed CO_{ad} monolayer stripping (adsorbed from CO-saturated solution: solid line in Fig. 13.1a) is centered at about 0.77 V. The mass spectrometric signal (solid line in Fig. 13.1b) largely reproduces the faradaic current signal, except for the contributions from double-layer charging and Pt oxidation. (Note that the corresponding mass spectrometric current peak is shifted by about 0.02 V to more positive potential owing to the time delay between product formation and detection of about 2 s in the dual thin-layer flow cell.) Very small amounts of CO_2 formed at potentials positive of 1.0 V (Fig. 13.1b) are attributed to oxidation of the carbon support [Roen et al., 2004; Willsau and Heitbaum, 1984]. This assignment, which contrasts with the interpretation in a recent IR spectroscopic study [Maillard et al., 2004b], is based on our observation of similar amounts of CO_2 formation also during a base CV. From the CO_{ad} stripping charge (two-electron reaction), which is about twice as

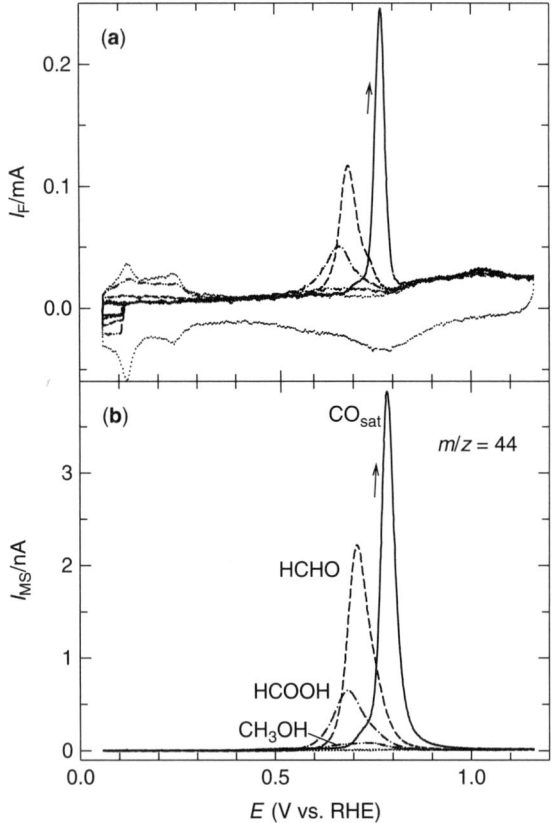

Figure 13.1 Electrooxidation of CO_{ad} and C_1 adsorbate layers pre-adsorbed on a Pt/Vulcan thin-film electrode (7 μg_{Pt} cm^{-2}, geometric area 0.28 cm^2) in 0.5 M H_2SO_4 solution during a first positive-going potential scan, and subsequent response of the faradaic (a) and $m/z = 44$ ion current (b) to the electrode potential in the thin-layer DEMS flow cell. The potential scan rate was 10 mV s^{-1} and the electrolyte flow rate was 5 μL s^{-1}, at room temperature. The respective adsorbates were adsorbed at 0.11 V for 10 minutes from CO-saturated solution (solid line), 0.1 M HCHO solution (dashed line), 0.1 M HCOOH solution (dash–dotted line), and 0.1 M CH_3OH solution (dash–double-dotted line).

much as that for hydrogen monolayer adsorption (one-electron reaction) in the subsequent base CV (Fig. 13.1a, dotted line), the coverage of the saturated CO adlayer on Pt/Vulcan is about 0.75 monolayers (ML) [Jusys and Behm, 2001], assuming an H_{upd} coverage of 0.77 ML at the onset of bulk hydrogen evolution and an H_{upd} monolayer charge of 0.21 mC cm^{-2} for polycrystalline Pt [Biegler et al., 1971]. This value agrees well with results reported previously [Jusys et al., 2001, 2003].

Oxidation of the adsorbed species resulting from interaction with formaldehyde, formic acid, and methanol, respectively, leads to stripping peaks that are downshifted to more negative potentials. Furthermore, the adsorbate coverage is significantly lower

TABLE 13.1 Number of Electrons per Resulting CO_2 Molecule Required for Oxidizing the Stable Adsorbed Decomposition Product from Adsorption of C_1 Molecules to CO_2, and C_1 Adsorbate Coverage Relative to that of a Saturated CO Adlayer after Adsorption of C_1 Reactants[a]

CO_{ad} from Molecules[b]	C_1 $n_e \pm 0.3$	$\theta_{CO,rel} \pm 0.02$	$\theta_{CO,rel} \pm 0.05$
HCHO	1.9	0.87	0.70
HCOOH	2.4	0.36	0.47
CH_3OH	2.1	0.10	0.22

[a] As determined from potentiodynamic (data from Fig. 13.1, adsorption potential 0.11 V) and potentiostatic (data from Fig. 13.2c, adsorption potential 0.16 V) stripping experiments.
[b] $c = 0.1$ M, $E_{ad} = 0.11$ V, 10 minutes.

compared with a saturated CO adlayer (Fig. 13.1a). Correspondingly, the oxidation of adsorbed hydrogen in the potential range 0.06–0.3 V during the positive-going scan is less suppressed than for saturated CO adlayer stripping. The numbers of electrons formed per CO_2 molecule are summarized in Table 13.1. Based on the number of about 2 electrons per CO_2 product molecule, the stable, adsorbed residues developed upon dehydrogenative adsorption of the three C_1 molecules are predominantly CO_{ad} species. The somewhat higher value (about $2.4\,e^-$ molecule^{-1}) determined for formic acid adsorbate oxidation, which was obtained in a number of experiments, might point to contributions from a non-CO_{ad} adsorbate requiring 3 electrons per CO_2 molecule formation (e.g., COH), as had been proposed previously [Willsau and Heitbaum, 1986]. Considering the error margin in these experiments of about ± 0.3 electron per CO_2 molecule, however, the deviation is still close to the precision of these measurements, and thus is not considered as proof for any other type of adsorbate than CO_{ad}.

The conclusion that CO_{ad} represents the only stable adsorbate after interaction with these three C_1 molecules is supported also by the findings in previous in situ IR studies, where adsorbed CO was detected as the only stable adsorbate on various Pt surfaces, including single-crystal surfaces, after adsorption of methanol, formic acid, or formaldehyde over a wide potential range (≤ 0.6 V) [Xia et al., 1997; Waszczuk et al., 2001; Rice et al., 2000; Iwasita and Nart, 1997; Beden et al., 1981; Nichols and Bewick, 1988; Hamnett et al., 1990; Park et al., 2002]. Similar results were also obtained very recently from highly sensitive in situ IR measurements on a Pt film electrode in an ATR-IRS configuration [Chen Y-X, et al., to be published]. Despite their high surface sensitivity, these measurements did not show any indication of other stable adsorbates after removal of the reactant solution. It should be noted that in early DEMS studies on porous Pt electrodes, values of about 3 electrons per CO_2 product molecule were found for both methanol and formic acid adsorbate oxidation [Willsau and Heitbaum, 1986; Wolter et al., 1985; Willsau et al., 1985]. Most plausibly, the higher values can be explained by an incomplete removal of the respective reactants from the porous layer after electrolyte exchange [Jusys and Behm, 2001], and/or variations in the collection efficiency [Wolter and Heitbaum, 1984], which can occur for very high CO_2 formation rates on these porous electrodes.

The CO_{ad} coverages resulting from 10-minute interaction with formaldehyde, formic acid and methanol containing solution at 0.11 V were determined by comparing with the mass spectrometric charges for CO_2 obtained for saturated CO adlayer stripping. Under present adsorption conditions, the CO_{ad} coverages correspond to about 87% (formaldehyde), 36% (formic acid), and 10% (methanol), respectively, of that of a saturated CO adlayer (see Table 13.1), i.e., they are significantly lower than after CO adsorption. It is important to note that the shift of the adsorbate stripping peak to lower potentials compared with CO_{ad} stripping cannot be explained by the reduced adsorbate coverage, since in a previous study on similar Pt/C catalysts, the coverage dependence of the CO_{ad} stripping peak after exposure to different amounts of CO was found to be marginal (<40 mV at 15% of the CO_{ad} saturation coverage) [Behm and Jusys, 2006]. Other effects, originating, for example, from a different structure and/or lateral distribution of the adlayer, must contribute as well, and are dominant.

Similar trends to much lower CO_{ad} coverages upon adsorption of methanol, compared with formaldehyde and formic acid adsorption, at potentials in the H_{upd} region were found in recent in situ IR spectroscopy [Park et al., 2002] and DEMS [Lanova et al., 2006] measurements on Pt/Vulcan (adsorption potentials 0.06 V [Park et al., 2002] and 0.3 V [Lanova et al., 2006], respectively) and on Pt films [Chen et al., unpublished results] (adsorption potential 0.06 V). Other studies, in contrast, reported significantly higher adsorbate coverages, up to 60% of CO_{ad} saturation, for methanol adsorption on a Pt/Vulcan catalyst in the H_{upd} region [Rice et al., 2000]. The discrepancy between the latter study on the one hand and our as well as other results on the other may be due to the much longer adsorption time in the experiments in [Rice et al., 2000]. While slow adsorption of additional methanol cannot be ruled out, such long-term adsorption experiments can also be affected by adsorption of trace impurities present in the reactant. Similarly, even very slow oxidation of methanol to formaldehyde at low potentials [Korzeniewski and Childers, 1998], and subsequent adsorption of the resulting formaldehyde, could contribute to long-term adsorption experiments in a stagnant electrolyte.

The pronounced difference in CO_{ad} saturation coverage obtained upon interaction with methanol-, formaldehyde-, or formic acid-containing solution under present reaction conditions (see Table 13.1) is attributed to a different influence of adsorbed H_{upd} species on the probability of dehydrogenation of the three C_1 reactants, which are most likely related to different spatial requirements for the dehydrogenation step [Desai et al., 2002; Greeley and Mavrikakis, 2002, 2004; Okamoto et al., 2003; Cao et al., 2005; Taylor and Neurock, 2005; Hartnig and Spohr, 2005]. Potential-dependent measurements for C_1 adsorption on Pt/C [Jusys and Behm, 2001] indeed showed a pronounced increase in CO_{ad} coverage with decreasing H_{upd} coverage for methanol adsorption, while, for formaldehyde adsorption, the resulting CO_{ad} coverage is practically independent of the adsorption potential and thus of the H_{upd} coverage [Chen et al., unpublished results]. Hence, formaldehyde adsorption/decomposition is not inhibited by a H_{upd} adlayer, and can lead to a displacement of the H_{upd} adlayer by the resulting CO_{ad}. Formic acid dehydrogenation to CO_{ad} is between the two cases of methanol dehydrogenation and formaldehyde dehydrogenation (for a quantitative evaluation and discussion, see [Chen et al., 2006a, b, c]).

These trends agree well with recent theoretical studies, which found that several neighboring (three) Pt atoms are required for methanol dehydrogenation [Desai et al., 2002; Greeley and Mavrikakis, 2002, 2004; Okamoto et al., 2003; Cao et al., 2005], while these requirements are less stringent for formaldehyde dehydrogenation [Desai et al., 2002; Greeley and Mavrikakis, 2002].

13.3.1.2 Potential-Step Electro-Oxidation of Adsorbed Species

Additional information on the adsorbate layer resulting from adsorption of the different C_1 reactants is obtained from potential step oxidation transients, recorded after adsorption of the respective reactants at a potential in the H_{upd} regime (10 minutes at 0.16 V), subsequent electrolyte exchange, and finally a potential step to 0.6 V. Figure 13.2 shows faradaic (Fig. 13.2a) and $m/z = 44$ mass spectrometric (Fig. 13.2b) current transients measured during oxidation of the adlayer at 0.6 V. For comparison, we also include similar transients recorded after CO_{ad} adsorption from a CO-containing supporting electrolyte and a control experiment performed with a clean Pt/C catalyst in pure supporting electrolyte. The insets show the corresponding current transients on an expanded time scale, during the first 2–3 minutes. (For better separation, the current transients in the insets are progressively shifted by 30 s against each other.)

The control experiment in pure supporting electrolyte (dotted lines in Fig. 13.2) shows a sharp faradaic current spike, which is mainly due to pseudocapacitive contributions (adsorption of (bi)sulfate and rearrangement of the double layer) plus oxidation of adsorbed H_{upd} (dotted lines in Fig. 13.2a), but no measurable increase in the CO_2 partial pressure ($m/z = 44$ current) above the background level (dotted lines in Fig. 13.2b). Therefore, a measurable adsorption of trace impurities from the base electrolyte can be ruled out on the time scale of our experiments. Moreover, this experiment also demonstrates the advantage of mass spectrometric transient measurements compared with faradaic current measurements, since the initial reaction signal is not obscured by pseudocapacitive effects and the related faradaic current spike.

The initial current spike appears as a dominant feature also in the faradaic current transients obtained upon oxidation of pre-adsorbed CO and of the adsorbates resulting from adsorption of C_1 molecules. For oxidation of a saturated CO adlayer (CO adsorption from a CO-saturated electrolyte), the faradaic current decays within about 30 s to 0.5 µA and remains approximately constant for about 3 minutes (solid lines in Fig. 13.2a). Later on, the faradaic current increases slowly, passes through a maximum value of about 1.2 µA about 10 minutes after the potential step, and then decreases again to zero. The characteristic shape of the CO oxidation transient recorded during oxidation of a saturated CO adlayer (CO adsorption from a CO-saturated electrolyte), with a pronounced induction period after the initial current spike and a distinct oxidation peak, closely resembles previous observations on supported Pt/C catalysts [Lanova et al., 2006; Friedrich et al., 2000; Gustavsson et al., 2004; Maillard et al., 2004a, 2005; Arenz et al., 2005] and solid Pt electrodes [Santos et al., 1991; Petukhov et al., 1998; Koper et al., 1998; Bergelin et al., 1999; Korzeniewski and Kardash, 2001; Lebedeva et al., 2002; Heinen et al., 2007]. For oxidation of formaldehyde adsorbate, the current transient still shows a distinct

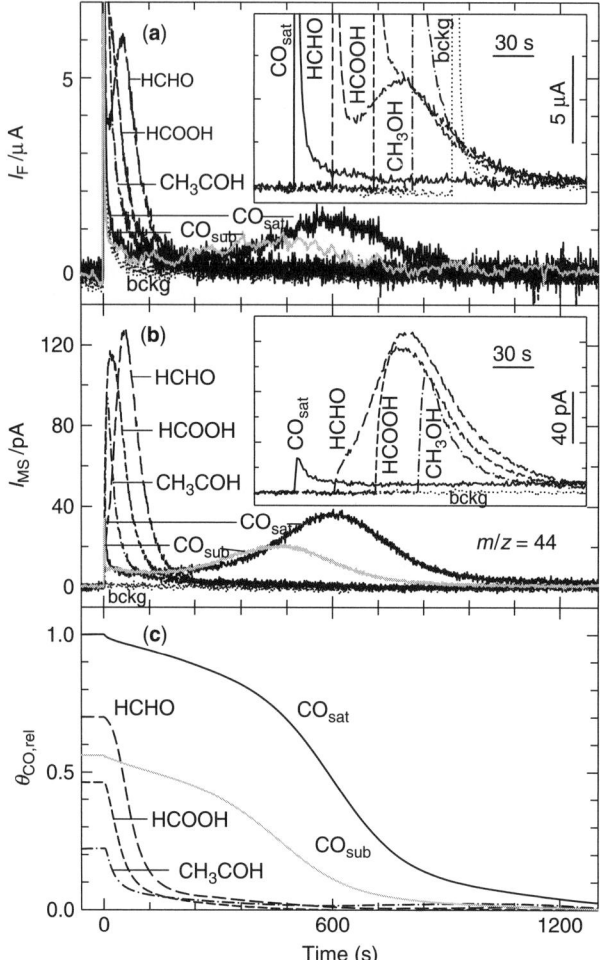

Figure 13.2 Potential-step electrooxidation of pre-adsorbed CO_{ad} and C_1 adsorbate adlayers on a Pt/Vulcan thin-film electrode (7 $\mu g_{Pt}\,cm^{-2}$, geometric area 0.28 cm^2) in 0.5 M H_2SO_4 solution: faradaic (a), $m/z = 44$ ion current (b), and CO_{ad} coverage (c) transients stepping the potential from 0.16 to 0.6 V in a thin-layer DEMS flow cell. The electrolyte flow rate was 5 $\mu L\,s^{-1}$, at room temperature. CO was pre-adsorbed at 0.16 V for 10 minutes from CO-saturated solution (solid lines), 0.1 M HCHO solution (dashed lines), 0.1 M HCOOH solution (short-dashed lines), and CH_3OH (dash–dotted lines). Insets in (a) and (b) show the corresponding initial faradaic (a) and $m/z = 44$ ion (b) current transients progressively shifted by 30 s. Dotted lines in (a) and (b) show background (bckg) signals recorded on the adsorbate-free Pt/C catalyst upon a potential step from 0.16 to 0.6 V in 0.5 M H_2SO_4 solution.

oxidation peak after the initial current spike, which appears, however, already at much shorter time, after about 50 s. For oxidation of the formic acid adsorbate, the distinct main peak is replaced by a broader tail at the shoulder of the initial current spike, extending over about 30 s, and for methanol adsorbate oxidation, the main oxidation peak fully overlaps with the initial current spike and can hardly be separated. Furthermore, the methanol adsorbate oxidation peak is much narrower compared with the other adsorbate oxidation peaks.

The identification of the actual adlayer oxidation process is much simpler in the mass spectrometric signals (Fig. 13.2b) than from the faradaic current transients. In all cases, the sharp initial spike is absent, confirming that this is due to pseudocapacitive contributions (see above). For constant potential oxidation of a saturated CO adlayer (CO pre-adsorption), the transient shows a small initial peak, which, however, is much smaller and also wider than the sharp spike in the faradaic current peak. Therefore, the onset of CO_{ad} oxidation upon the potential step cannot be followed via the faradaic current, because of the above-mentioned capacitive contributions, which are often overlooked [Maillard et al., 2004a, 2005; Arenz et al., 2005; Santos et al., 1991; Petukhov et al., 1998; Koper et al., 1998; Lebedeva et al., 2002; Andreaus et al., 2006; Andreaus and Eikerling, 2007]. The subsequent induction period and the bell-shaped CO_{ad} oxidation peak resemble the features observed in the faradaic current. The observation of small amounts of CO_{ad} oxidation right upon the potential step (instantaneous CO_{ad} oxidation) agrees well with previous findings for CO_{ad} oxidation on nanostructured Pt/C model catalysts [Gustavsson et al., 2004] and on Pt film electrodes [Heinen et al., 2007]. Also, the other transients (C_1 adsorbate oxidation transients) largely resemble the faradaic current transients, with the exception of the initial current spike, which is absent in the mass spectrometric transients. Because of the increasing overlap of the initial current spike with the actual adsorbate oxidation peak in the faradaic current signal, the mass spectrometric current transients allow a much better identification and quantification of the adsorbate oxidation current. Furthermore, exact evaluation of CO_{ad} coverage from the following transient is hardly possible, owing to the capacitive contribution to the measured faradaic current from anion re-adsorption upon removal of CO_{ad} [Heinen et al., 2007]. By integrating the mass spectrometric CO_2 signal, we determined the initial CO_{ad} coverages and their decay during the transient measurements [Gustavsson et al., 2004; Heinen et al., 2007]. The CO_{ad} coverage at a given moment is calculated as the (integrated) saturation coverage minus the ratio of integrated ion current to the total mass spectrometric charge. (This evaluation is based on the assumption that CO_{ad} is the only stable adsorbed species resulting from C_1 molecule adsorption; see Section 13.3.1.1). The resulting coverage transients are plotted in Fig. 13.2c. In agreement with the potentiodynamic experiments, the CO_{ad} coverage decreases in the order $CO_{sat} >$ formaldehyde $>$ formic acid $>$ methanol. The exact values are collected in Table 13.1. The difference with respect to the values determined in the potentiodynamic stripping experiments is explained by the different adsorption potential in the potentiostatic measurements (0.16 V vs. 0.11 V). The increase in adsorbate coverage is most pronounced for methanol adsorption. This observation agrees closely with previous findings for potentiodynamic and potentiostatic methanol adsorbate stripping experiments performed after adsorption at different adsorption potentials [Jusys and Behm, 2001; Lanova et al., 2006]. Recent in situ ATR-IR experiments equally showed that the potential dependence is most pronounced for methanol adsorption, less for formic acid adsorption, and essentially absent for formaldehyde adsorption [Chen et al., unpublished results].

For saturated CO adlayer oxidation (solid line in Fig. 13.2c), the CO_{sat} coverage transient shows only a small initial decay of 1–2% of the adlayer coverage (a small decrease in the CO_{ad} coverage at zero time). Then the CO_{sat} coverage decreases nearly linearly for about 3 minutes down to about 0.8 of its initial value, changes to a steeper slope in the coverage range from about 0.7 to 0.3 (400–700 s after the potential step), and finally changes slope again to an almost linear decay from about 0.2 down to nearly zero. A subsequent stripping experiment confirmed that the initial, saturated CO adlayer was almost completely removed after 25–30 minutes' oxidation at 0.6 V (remaining $CO_{ad} \approx 3\%$). The CO_{ad} coverage transients obtained after adsorption of the three C_1 molecules differ significantly from these characteristics. They exhibit a pronounced coverage decay in the initial phase (Fig. 13.2b), where almost all of the adsorbate is removed.

The different characteristics of the adsorbate stripping transients can be qualitatively explained by using the concepts derived for describing the oxidation behavior of a saturated CO adlayer, after adsorption from CO-containing solution [Gilman, 1963, 1964]. It is generally agreed (for carbon-supported Pt/C catalysts, see [Friedrich et al., 2000; Maillard et al., 2004a, 2005; Arenz et al., 2005; Andreaus et al., 2006; Andreaus and Eikerling, 2007]; for smooth Pt electrodes, see [Santos et al., 1991; Petukhov et al., 1998; Koper et al., 1998; Bergelin et al., 1999; Korzeniewski and Kardash, 2001; Lebedeva et al., 2002; Housmans et al., 2007]) that electro-oxidation of adsorbed CO on Pt electrodes proceeds via a Langmuir–Hinshelwood mechanism by reaction between CO_{ad} and neighboring, electrosorbed OH_{ad} species. In the initial peak, this includes initial (instantaneous) nucleation of OH_{ad}, possibly on defect sites, reaction with weakly bound CO_{ad} species, and slow relaxation of the CO adlayer into a more strongly bound state. Progressive nucleation of OH_{ad} on free Pt sites and reaction with neighboring "strongly bound" CO_{ad} leads to the main oxidation peak. For all adsorbate stripping transients, the mass spectrometric measurements clearly reveal a fast onset of CO_{ad} electro-oxidation immediately after stepping the electrode potential, which supports the proposal of instantaneous nucleation of OH_{ad} species and reaction on a limited number of defect sites within the CO adlayer [Korzeniewski and Kardash, 2001]. For the saturated adlayer, the amount of initial, instantaneous OH_{ad} nucleation is small, as expected for a surface fully covered by a stable saturated adlayer. It is somewhat higher after formaldehyde adsorption, in agreement with the slightly lower CO_{ad} coverage. For the other two reactants, OH_{ad} nucleation is no longer rate-limiting. An increasingly faster onset of the main oxidation peak for transient C_1 adsorbate oxidation with decreasing CO_{ad} coverage would be consistent with a Langmuir–Hinshelwood mechanism, since nucleation of OH_{ad} should become more facile at lower CO_{ad} coverages, which in turn increases the adlayer oxidation rate and shifts the main oxidation peak towards the initial spike. The asymmetric shape of the initial peak with a significant tailing at longer times can be explained by the increasing overlap of the initial "instantaneous" OH_{ad} nucleation and reaction step [Korzeniewski and Kardash, 2001], and CO_2 formation via progressive OH_{ad} nucleation and reaction with CO_{ad} [Lebedeva et al., 2000, 2002; Andreaus and Eikerling, 2007; Housmans et al., 2007].

The different CO_{ad} coverages obtained after CO and C_1 molecule adsorption are sufficient to qualitatively account for the observed change in the overall shape of the respective stripping transients. For a more quantitative comparison, we included a CO_{ad} stripping transient recorded upon oxidation of a medium coverage CO adlayer obtained by adsorption from CO containing solution (gray lines in Fig. 13.2, CO_{sub}). Although the CO_{ad} submonolayer coverage is between those obtained after formic acid and after formaldehyde adsorption, respectively, the general shape of the transient is much closer to that obtained upon saturated CO adlayer oxidation than to those for C_1 adsorbate oxidation. In agreement with the potentiodynamic stripping experiments, C_1 adsorbate stripping is significantly faster at the same potential as that of a similar coverage CO adlayer produced by CO adsorption.

In total, the differences between C_1 adsorbate stripping and CO_{ad} stripping, in both potentiodynamic measurements and potentiostatic transients, can be qualitatively explained by the lower CO_{ad} coverage after C_1 adsorption. A quantitative comparison with a lower coverage CO adlayer reveals, however, that these coverage effects are not sufficient and that contributions from other effects, most likely related to the structure of the CO adlayer, are important for the different oxidation behavior of the respective adsorbates as well.

13.3.2 Bulk Oxidation of Formic Acid, Formaldehyde, and Methanol: Potentiodynamic Measurements

In this section, we present results of potentiodynamic DEMS measurements on the continuous (bulk) oxidation of formic acid, formaldehyde and methanol on a Pt/Vulcan catalyst, and compare these results with the adsorbate stripping data in Section 13.3.1. We quantitatively evaluate the partial oxidation currents, product yields, and current efficiencies for the respective products (CO_2 and the incomplete oxidation products). In the presentation, the order of the reactants follows the increasing complexity of the oxidation reaction, with formic acid oxidation discussed first (one reaction product, CO_2), followed by formaldehyde oxidation (two reaction products) and methanol oxidation (three reaction products).

13.3.2.1 General Characteristics DEMS data recorded during the potentiodynamic electro-oxidation of the three C_1 molecules on a Pt/Vulcan electrode are shown in Fig. 13.3a (formic acid oxidation), 13.3b (formaldehyde oxidation), and 13.3c (methanol oxidation). They include the faradaic current (solid lines in the top panels), the CO_2^+-related $m/z = 44$ ion current [solid lines in the lower panels in (a) and (b) and the middle panel in (c)], and, for methanol oxidation [bottom panel in (c)], also the $m/z = 60$ signal of methyl formate. The latter is commonly used to detect formic acid in methanol-containing solutions [Iwasita and Vielstich, 1986; Iwasita, 2002, 2003]. The faradaic current signals closely resemble those reported earlier for methanol oxidation [Iwasita and Vielstich, 1986; Jusys and Behm, 2001; Jusys et al., 2003; Iwasita, 2002, 2003], formaldehyde oxidation [Loucka and Weber, 1968; Beltowska-Brzezinska and Heitbaum, 1985; Nakabayashi, 1998; Nakabayashi et al., 1998; Batista and Iwasita, 2006; Samjeské et al., 2007;

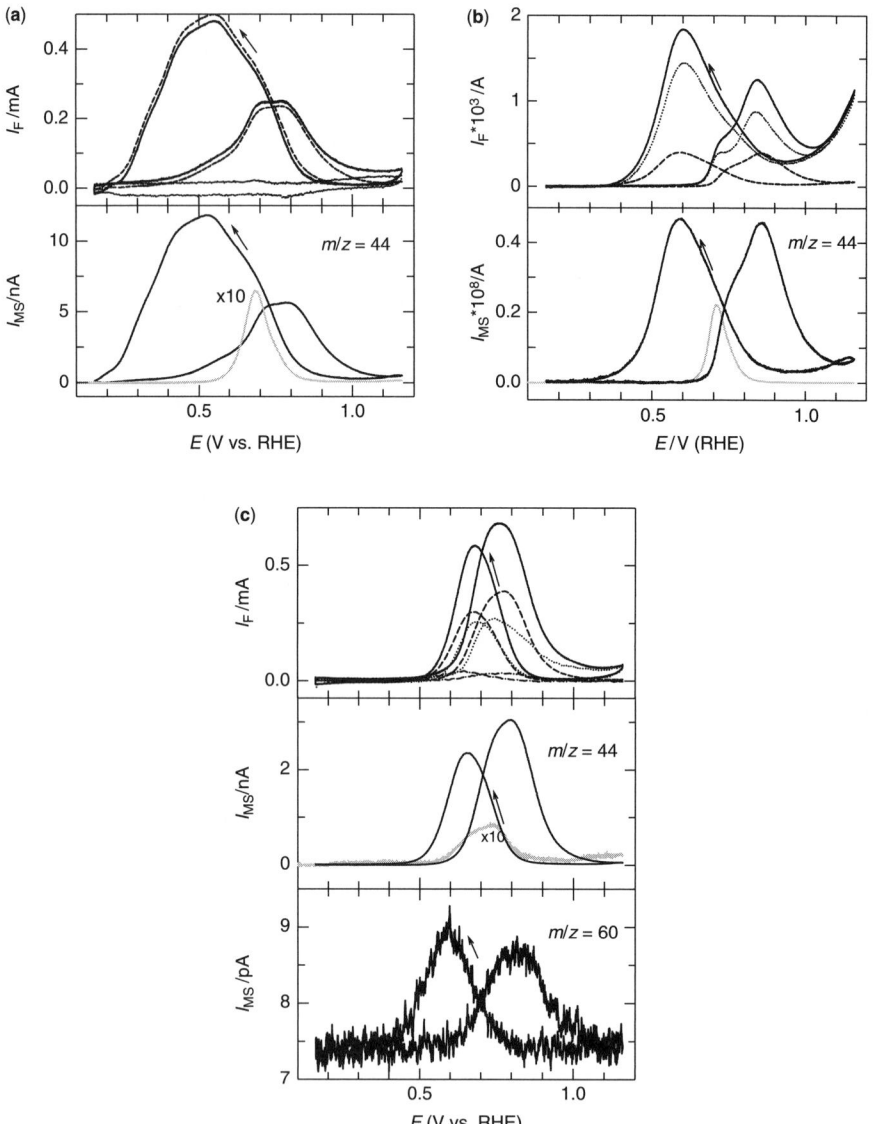

Figure 13.3 Potentiodynamic electrooxidation of (a) formic acid, (b) formaldehyde, and (c) methanol on a Pt/Vulcan thin-film electrode (7 $\mu g_{Pt}\,cm^{-2}$, geometric area $0.28\,cm^2$) in $0.5\,M\,H_2SO_4$ solution containing $0.1\,M$ HCOOH (a), HCHO (b), or CH_3OH (c). The potential scan rate was $10\,mV\,s^{-1}$ and the electrolyte flow rate was $5\,\mu L\,s^{-1}$, at room temperature). The top panels show the faradaic current (solid lines), the partial currents for C_1 oxidation to CO_2 (dashed lines) and for formic acid formation (dash-dotted line), calculated from the respective ion currents, and the difference between the measured faradaic current and the partial current for CO_2 oxidation (formic acid oxidation (a), formaldehyde oxidation (b)), or the difference between faradaic current and the sum of the partial currents for CO_2 formation and formic acid oxidation (methanol oxidation, (c)) (dotted line). The solid lines in the lower panels in

de Lima et al., 2007; Miki et al., 2004], and formic acid oxidation [Okamoto et al., 2004; Samjeské and Osawa, 2005; Breiter, 1967; Loucka and Weber, 1968; Spasojevic et al., 1980; Napporn et al., 1995; Okamoto et al., 2005; Park et al., 2002b] on polycrystalline Pt and Pt nanoparticle electrodes. For all three reactants, they exhibit a distinct hysteresis between positive-going and negative-going scan, which had been associated with the different states of the catalyst prior to the onset of the reaction in the two scan directions, with the Pt surface being CO_{ad}-covered at low potentials, before the positive-going scan, and OH_{ad} covered/partly oxidized but CO_{ad}-free at high potentials, before the negative-going scan [Parsons and VanderNoot, 1988; Capon and Parsons, 1973]. For methanol and formaldehyde oxidation, the reaction is largely hindered by adsorbed CO at potentials negative of 0.5 V and 0.6 V, respectively, while formic acid oxidation starts already at much lower potentials, at around 0.2 V in the positive-going scan. The current decay at more positive potentials (>0.8–0.9 V) is attributed to an increasing hindrance of the C_1 electrooxidation reaction by PtO formation [Angerstein-Kozlowska et al., 1973]. In the reverse, negative-going scan, the oxidation current starts simultaneously with PtO reduction on an initially CO_{ad}-free (CO_{ad} can be efficiently oxidized at these potentials [Jusys et al., 2001; Gilman, 1964; Stonehart and Kohlmayr, 1972; Gasteiger et al., 1995; Schmidt et al., 1999b]) and increases steadily. It then passes through a maximum and decreases to zero at lower potentials, where OH electrosorption and therefore also oxidation of the CO_{ad} species are inhibited by the CO adlayer. For all three reactants, the reaction extends to potentials negative of the onset of CO_{ad} stripping in the positive-going scan, which is attributed to the finite time required for the build-up of a reaction inhibiting CO adlayer in the negative-going scan. Interestingly, the ratio of the peak heights in the positive- and negative-going scans changes significantly from methanol to formic acid oxidation, becoming increasingly higher in the negative-going scan. The oxidation of formic acid even occurs on a CO_{ad}-blocked surface at potentials lower than 0.4 V, where OH_{ad} generation on the Pt catalyst electrode is negligible. This can be rationalized by the presence of two oxygen atoms in the carboxylic acid group, which allows CO_2 formation without requiring the formation of and reaction with adsorbed oxygen species. The same is true also for formaldehyde oxidation, considering that formaldehyde is largely hydrated to methylene glycol in aqueous formaldehyde solutions, with two OH groups attached to the carbon atom (see Batista and Iwasita [2006]).

The mass spectrometric currents follow largely, but not completely the faradaic current signals. The contributions to the respective faradaic currents resulting from complete oxidation to CO_2, which are calculated using the calibration constant K^* (see Section 13.2), are plotted as dashed lines in the top panels in Fig. 13.3. For the calculations of the partial reaction currents, we assumed six electrons per CO_2 molecule formation and considered the shift in the potential scale caused by the time

---

Figure 13.3 (*Continued*) (a) and (b) and in the middle panel in (c) show the $m/z = 44$ ion current response to the electrode potential, the gray lines illustrate the oxidation of preformed CO, derived upon C_1 adsorption at 0.11 V, in reactant-free H_2SO_4 solution. The bottom panel in (c) shows the $m/z = 60$ ion current response to the electrode potential.

delay (about 2 s) between faradaic current and mass spectrometric current detection. For methanol oxidation (Fig. 13.3c), the partial current for CO_2 formation (complete methanol oxidation) constitutes only about half of the measured faradaic current; for formaldehyde oxidation (Fig. 13.3b), it is even less than 30%. As expected, for formic acid oxidation (Fig. 13.3a), CO_2 formation accounts for almost all of the faradaic current, except for the small contributions from double-layer charging and from PtO formation/reduction, which also shows the precision of the partial current determination.

Similarly, the $m/z = 60$ ion current signal was converted into the partial current for methanol oxidation to formic acid in a four-electron reaction (dash–dotted line in Fig. 13.3c; for calibration, see Section 13.2). The resulting partial current of methanol oxidation to formic acid does not exceed about 10% of the methanol oxidation current. Obviously, the sum of both partial currents of methanol oxidation to CO_2 and formic acid also does not reach the measured faradaic current. Their difference is plotted in Fig. 13.3c as a dotted line, after the PtO formation/reduction currents and pseudocapacitive contributions, as evident in the base CV of a Pt/Vulcan electrode (dotted line in Fig. 13.1a), were subtracted as well. Apparently, a significant fraction of the faradaic current is used for the formation of another methanol oxidation product, other than CO_2 and formic acid. Since formaldehyde formation has been shown in methanol oxidation at ambient temperatures as well, parallel to CO_2 and formic acid formation [Ota et al., 1984; Iwasita and Vielstich, 1986; Korzeniewski and Childers, 1998; Childers et al., 1999], we attribute this current difference to the partial current of methanol oxidation to formaldehyde. (Note that direct detection of formaldehyde by DEMS is not possible under these conditions, owing to its low volatility and interference with methanol-related mass peaks, as discussed previously [Jusys et al., 2003]). Assuming that formaldehyde is the only other methanol oxidation product in addition to CO_2 and formic acid, we can quantitatively determine the partial currents of all three major products during methanol oxidation, which are otherwise not accessible. Similarly, subtraction of the partial current for formaldehyde oxidation to CO_2 from the measured faradaic current for formaldehyde oxidation yields an additional current, which corresponds to the partial oxidation of formaldehyde to formic acid. The characteristics of the different C_1 oxidation reactions are presented in more detail in the following sections.

13.3.2.2 Formic Acid Oxidation

Despite the relatively simple oxidation reaction (two-electron oxidation of formic acid leads to CO_2 as the only reaction product), the potential sweep results in a rather complicated faradaic and mass spectrometric current response (Fig. 13.3a). It closely resembles that reported previously for polycrystalline Pt [Feliu et al., 2003; Capon and Parsons, 1973b, c; Willsau and Heitbaum, 1986; Wolter et al., 1985; Okamoto et al., 2004; Lu et al., 1999; Chen et al., 2006a, b, c; Samjeské et al., 2006] and Pt nanoparticle electrodes [Park et al., 2002], with the faradaic current starting to increase slowly at 0.15 V and then more strongly at $E > 0.35$ V and again at $E > 0.6$ V. After passing through two maxima at about 0.72 and 0.80 V, it then decays again. The current decay at more positive potential is attributed to increasing OH_{ad} accumulation and PtO formation, where the latter is not active for formic acid oxidation in this potential regime. In the negative-going scan, the current starts to increase steeply at 0.9 V, after reduction of the Pt oxides. It passes through a wide peak with several ill-resolved maxima, and starts to decay again at $E < 0.58$ V, reaching zero current at the negative potential limit of 0.06 V. In the low potential region, the electrocatalytic activity of Pt is suppressed owing to surface blocking by reaction-inhibiting side products/reaction intermediates, which in in situ IR spectroscopy studies were identified as CO_{ad} species [Sun et al., 1988; Sun and Yang, 1999; Miki et al., 2002; Samjeské and Osawa, 2005; Chen et al., 2006a, b, c; Samjeské et al., 2005; Park et al., 2002] (see also Section 13.3.1). The latter species are responsible for the complex current response on the electrode potential and the pronounced hysteresis, with a low initial coverage of poisoning adsorbate species immediately after PtO reduction in the negative-going potential scan, but a high coverage at the negative potential limit. Further mechanistic implications will be discussed in more detail in Section 13.4.

In Fig. 13.3a (lower panel), we show the $m/z = 44$ mass signal for continuous formic acid oxidation in 0.1 M HCOOH solution (solid line) and, in addition, the 10-fold magnified $m/z = 44$ ion current for the oxidation of formic acid adsorption residues developed after adsorption at 0.11 V (gray line, data taken from Fig. 13.1b). Obviously, the steep increase in the formic acid oxidation rate at about 0.7 V in the positive-going scan (Fig. 13.3a) coincides with the oxidation of adsorbed CO, and is therefore attributed to the appearance of CO_{ad}-free, reactive Pt sites. Similar conclusions were drawn by Willsau and Heitbaum from DEMS experiments, where they investigated the oxidation of pre-adsorbed labeled formic acid adsorbate in non-labeled formic acid solution [Wilhelm and Heitbaum, 1986]. On the other hand, there is hardly any intensity in the $m/z = 44$ signal for oxidation of formic acid adsorbate (CO_{ad} generated by "chemical" dehydration/adsorption of formic acid) at potentials negative of 0.55 V, while CO_2 formation during continuous HCOOH oxidation starts already at potentials as negative as 0.2 V in the positive-going scan. In agreement with the interpretation in [Willsau and Heitbaum, 1986], this clearly demonstrates the presence of an additional reaction pathway that does not proceed via formation and oxidation of CO_{ad} (the "dual-pathway mechanism"; for further discussion, see Section 13.4 and [Samjeské and Osawa, 2005; Chen et al., 2006a, b; Samjeské et al., 2005, 2006]).

The dashed line in Fig. 13.3a (top panel) shows the faradaic current resulting from this reaction, calculated by assuming that CO_2 is the only reaction product during formic acid oxidation, independent of the reaction mechanism, and corrected by the time constant of the DEMS setup (2 s). The calculated faradaic current obtained this way is free from contributions arising from PtO formation/reduction, which appear in the measured faradaic current at potentials positive of 0.8 V, or from pseudocapacitive double-layer charging effects in the potential region 0.2–0.8 V. These contributions are determined by subtracting the calculated faradaic current from the measured current (dotted line in the top panel of Fig. 13.3a). This difference signal closely resembles the base CV of the same Pt/Vulcan electrode (Fig. 13.1a), except for the H_{upd}-related features due to CO_{ad} blocking.

13.3.2.3 Formaldehyde Oxidation

The general characteristics of the faradaic current dependence on the potential (solid line in the top panel of Fig. 13.3b) are similar to those for formic acid oxidation, with a few distinct differences. First, formaldehyde oxidation is more suppressed at low potentials than formic acid oxidation. Second, the faradaic current peak for formaldehyde oxidation is about four to five times higher than the respective formic acid oxidation faradaic current maxima. Third, at potentials positive of 1.0 V, the formaldehyde oxidation current increases again, up to the upper potential limit of 1.16 V, and decreases reversibly in the negative-going scan to a current minimum at about 0.95 V. The re-start of formaldehyde oxidation in the negative-going scan coincides with the reduction of the PtO cover layer at potentials lower than 0.9 V; it leads to a broad peak centered at about 0.6 V. The subsequent suppression of the formaldehyde oxidation reaction at lower potentials is explained by the increasing coverage of poisoning CO_{ad} species (see Section 13.3.2.2). The earlier decay of the reaction current in the negative-going scan compared with formic acid oxidation can be explained by the higher tendency of formaldehyde for CO_{ad} formation, which was observed also in the formaldehyde adsorption experiments (see Section 3.1.2 and Table 13.1) and in recent in situ IR spectroscopy experiments [Miki et al., 2004; Chen et al., to be published; Park et al., 2002]. The increasing formaldehyde oxidation current at potentials >1.0 V indicates a significant activity of the oxidized Pt surface for formaldehyde oxidation at these potentials, in contrast to the negligible activity for formic acid oxidation under similar conditions (Fig. 13.3a).

In the mass spectrometric trace (solid line in the lower panel of Fig. 13.3b), CO_2 formation starts at 0.6 V, almost simultaneously with the onset of formaldehyde adsorbate oxidation (gray line in the lower panel of Fig. 13.3b, data from Fig. 13.1b). The main CO_2 formation peak coincides with the main faradaic current peak. Its intensity, however, exhibits a rather different behavior compared with that for formic acid oxidation, with rather similar peak currents in the positive- and negative-going scans. While the peak current is about similar to that for formic acid oxidation in the positive-going scan, it is almost three times lower in the negative-going scan (solid lines in the lower panels of Fig. 13.3a, b). Compared with the formaldehyde adsorbate stripping peak (gray line in the lower panel of Fig. 13.3b, panel, data from Fig. 13.1b), the main onset of formaldehyde bulk oxidation in the positive-going scan is shifted to slightly more anodic potentials. Most likely, this results from re-formation of CO_{ad} species by decomposition of formaldehyde during the potential scan. Finally, the $m/z = 44$ ion current (solid line in the lower panel of Fig. 13.3b) does not show the significant increase in the high potential region observed in the faradaic current. Under these conditions, incomplete oxidation of formaldehyde to formic acid prevails (two-electron reaction). The same is true down to about 0.85 V in the negative-going scan (solid line in the lower panel of Fig. 13.3b).

The partial faradaic current for formaldehyde oxidation to CO_2, calculated from the $m/z = 44$ ion current, is plotted as a dashed line in Fig. 13.3b (upper panel). Complete oxidation of formaldehyde to CO_2 contributes only one-third (positive-going scan) or one-quarter (negative-going scan) of the corresponding faradaic current peaks (solid line in the upper panel of Fig. 13.3b). The difference between the measured net current and the calculated faradaic current, which is plotted as a dotted line in Fig. 13.3b (upper panel), reflects the partial current for incomplete formaldehyde oxidation to formic acid. Hence, under the present reaction conditions, incomplete oxidation of formaldehyde to formic acid prevails over complete oxidation to CO_2.

The current efficiencies for CO_2 formation and formic acid formation during potentiodynamic formaldehyde oxidation, calculated from the data in Fig. 13.3b as the ratio of the partial currents to the total faradaic current (in %), are plotted in Fig. 13.4a.

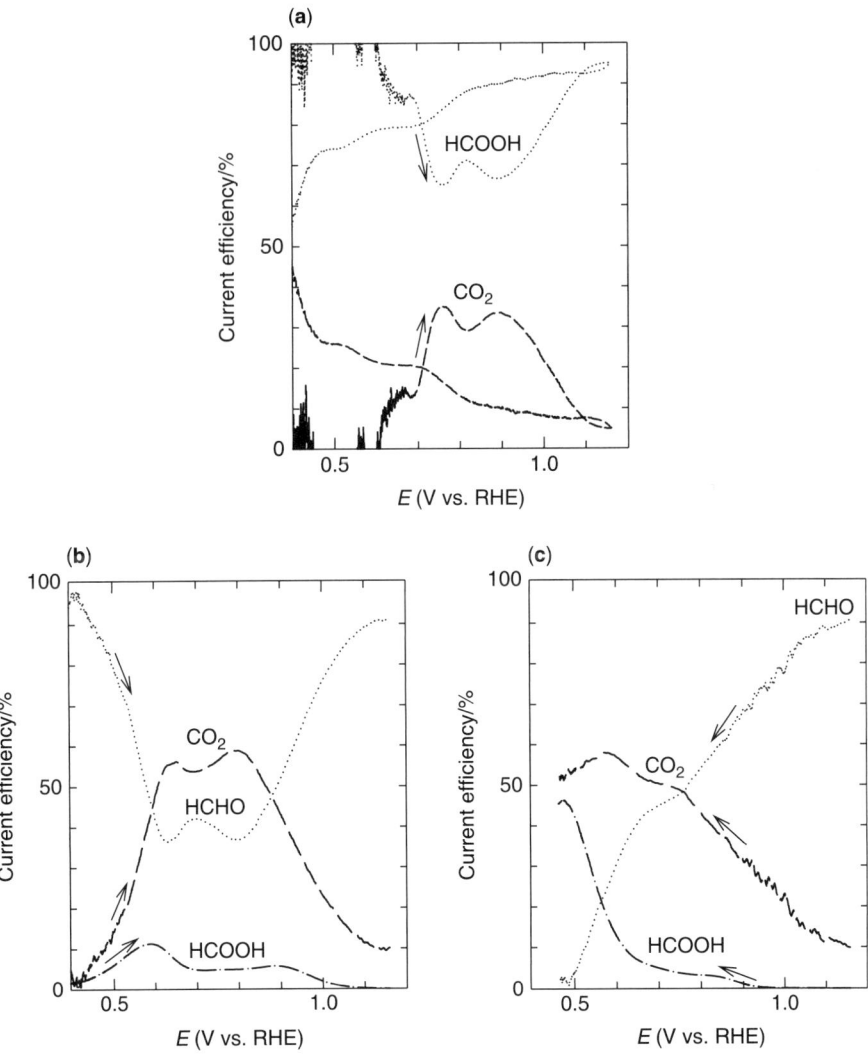

Figure 13.4 Current efficiency plots for the potentiodynamic electro-oxidation of formaldehyde (a) and methanol (b: positive-going scan; c: negative-going scan) on a Pt/Vulcan thin-film electrode (data from Fig. 13.3a, b): dashed lines, current efficiency for CO_2 formation; dash–dotted lines, current efficiency for HCOOH formation; dotted lines, current efficiency for HCHO formation.

(For formaldehyde oxidation with only two reaction products, the current efficiency for formic acid formation simply mirrors that for CO_2 formation.) Because of the very low currents in the low potential region, we have considered only potentials anodic of 0.6 V in the positive-going scan and anodic of 0.4 V in the negative-going scan, respectively. In the potential region from 0.6 to 0.7 V in the positive-going scan, the current efficiency exhibits constant values of about 15% for CO_2 and, correspondingly, about 85% for formic acid formation. With further potential increase, the current efficiency for CO_2 formation increases and passes through a double maximum (about 33%) with peaks at 0.75 and 0.9 V, and finally decreases to about 5% at the positive potential limit of 1.16 V. On the oxidized Pt surface, the selectivity for formic acid formation is close to 100%. In the negative-going scan, the current efficiency for CO_2 formation first increases rather slowly, down to potentials around 0.8 V, followed by a faster increase to 25% until 0.45 V, and finally 50% at 0.4 V. The current efficiency for formic acid formation changes correspondingly with the electrode potential.

For the understanding of the complex shape of the current efficiency for CO_2 formation during formaldehyde oxidation in the positive-going scan, one has to consider contributions from the oxidation of CO_{ad}, which was formed during the potential excursion to more negative potentials, as well. Comparing to the potential of the formaldehyde adsorbate stripping peak (dashed line in Fig. 13.1b), we attribute the first peak in the current efficiency for CO_2 formation (dashed line in Fig. 13.4a) to the oxidation of adsorbed CO_{ad} pre-formed at lower potentials. This occurs in addition to the oxidation of CO_{ad} produced by the ongoing formaldehyde decomposition. Therefore, the first peak can be considered as an additional feature, superimposed on a broad peak in CO_2 efficiency with its maximum at about 0.8 V. Accordingly, the second peak in the CO_2 current efficiency at 0.8 V must represent an inherent maximum in CO_2 selectivity, which is most likely related to the fact that, under these conditions, surface blocking, either by CO_{ad} or by OH_{ad}, is minimized. Apparently, formaldehyde oxidation to CO_2, either directly or via methylene glycol oxidation [Batista and Iwasita, 2006], is more site-demanding than oxidation to formic acid. This interpretation is supported by recent observations showing transient CO_2 formation upon formaldehyde adsorption/oxidation on a CO_{ad}-free Pt surface even at potentials around 0.3 V, before the reaction is inhibited by accumulation of a CO adlayer [Heinen et al., 2009]. Finally, at potentials positive of 1.0 V, formaldehyde oxidation is dominated by selective oxidation to formic acid. The ability of the oxidized Pt surface to catalyze formaldehyde oxidation to formic acid under these conditions, but not methanol oxidation, cannot be explained by a more facile decomposition of formaldehyde to CO_{ad}, compared with methanol, since, in that case, CO_2 should be the dominant oxidation product in formaldehyde oxidation, rather than formic acid. Hence, oxidation of the Pt surface prevents formaldehyde decomposition. Similar observations were made also for acetaldehyde, where oxidation at high potentials is facile and results exclusively in acetic acid formation [Wang et al., 2006]. Apparently, the oxidative dissociation of the aldehyde C–H bond is possible on an oxidized Pt catalyst at potentials $E > 0.9$ V. In the reverse, negative-going scan, accumulation of CO_{ad} is inhibited at potentials anodic of the onset of Pt oxidation, and therefore the structure of the CO_2 efficiency is rather simple, with a more or less continuous increase to more cathodic potentials.

13.3.2.4 Methanol Oxidation Methanol oxidation on a Pt/Vulcan catalyst was studied and discussed extensively in our previous papers [Jusys and Behm, 2001; Jusys et al., 2003]. Therefore, we will focus her on the most important features and refer to those previous studies for more detailed discussions.

At potentials negative of 0.5 V, methanol oxidation is largely hindered by a reaction inhibiting CO adlayer, which results from dehydrogenative methanol adsorption in the preceding potential scan. Comparison with the 10-fold magnified methanol adsorbate stripping peak, both for the faradaic current (dotted line in the top panel of Fig. 13.3c) and for the CO_2 mass spectrometric signal (gray line in the middle panel of Fig. 13.3c), shows that both methanol adsorbate and methanol bulk oxidation start at the same potential in the positive-going scan. At potentials positive of 1.0 V, methanol oxidation is suppressed by PtO formation. It is re-activated again by PtO reduction in the negative-going scan. Overall, these effects result in the well-known bell-shaped polarization curves. The corresponding $m/z = 44$ and $m/z = 60$ mass spectrometric currents (solid lines in the middle and bottom panels of Fig. 13.3c) generally follow the faradaic current, but exhibit some differences in the ratio of the peak heights and in the peak shape, which point to a potential-dependent variation of the product distribution for the two scan directions (for details, see [Jusys and Behm, 2001; Jusys et al., 2003]).

The calibrated $m/z = 44$ and $m/z = 60$ ion currents were converted into the respective partial reaction faradaic currents as described above, and are plotted in Fig. 13.3c as dashed ($m/z = 44$) and dash–dotted ($m/z = 60$) lines, using electron numbers of 6 electrons per CO_2 molecule and 4 electrons per formic acid molecule formation. The calculated partial current for complete methanol oxidation to CO_2 contributes only about one-half of the measured faradaic current. The partial current of methanol oxidation to formic acid is in the range of a few percent of the total methanol oxidation current. The remaining difference, after subtracting the PtO formation/reduction currents and pseudocapacitive contributions as described above, is plotted in Fig. 13.3c (top panel) as a dotted line. As mentioned above (see the beginning of Section 13.3.2), we attribute this current difference to the partial current of methanol oxidation to formaldehyde. This way, we were able to extract the partial currents of all three major products during methanol oxidation reaction, which are otherwise not accessible.

The current efficiencies for the different products during the positive-going scan in potentiodynamic methanol oxidation are plotted in Fig. 13.4b. (To reduce the noise level of the formic acid current efficiency, which results from the poor signal-to-noise ratio of the $m/z = 60$ mass signal, this was fitted by a Gaussian.) At potentials of 0.4–0.5 V, where the Pt surface is largely blocked by adsorbed CO, the (very slow) reaction is dominated by incomplete oxidation of methanol to mainly formaldehyde, with a current efficiencies of about 80–100%. With increasing potential, the current efficiency for CO_2 formation increases, reaching a double peak with maxima of about 60% at 0.62 and 0.8 V. At the same time, the current efficiency for formaldehyde formation decreases gradually. In the potential range $E > 0.5$ V, it mirrors that for

CO_2, at least qualitatively, with two minima coincident with the maxima in the CO_2 current efficiency. The pronounced decrease in current efficiency for CO_2 formation at more positive potentials, anodic of 0.8 V, is accompanied by a comparable increase in formaldehyde formation, while that for formic acid formation remains about constant. Simultaneously, the current efficiency for formic acid formation increases to about 10% at about 0.6 V, then decreases to about 5% and remains at this value up to 0.9 V, finally decaying to zero at 1.1 V. It should be mentioned, however, that the absolute currents are rather small for potentials cathodic of 0.5 V or anodic of 1.0 V, respectively. This is true, of course, also for the partial current for formaldehyde formation, which results in larger uncertainties in the calculated current efficiencies in these potential regions. In the negative-going scan (Fig. 13.4c), formaldehyde formation decreases continuously and reaches zero at 0.5 V, while the current efficiency for CO_2 formation increases steadily from about 10% at 1.16 V to about 50% at 0.7 V. It then remains at that level. The formic acid current efficiency behaves similarly as in the positive-going scan, starting to increase at about 0.7 V and reaching a value of 50% at 0.5 V.

The distinct double-peak structure in the CO_2 current efficiency (Fig. 13.4b) is in clear contrast to the slightly asymmetric peak for the partial current for CO_2 formation (dashed line in top panel of Fig 13.3c) and the rather symmetric faradaic current peak (solid line in the top panel of Fig. 13.3c) in the positive-going scan. Similar to formaldehyde oxidation, the additional peak in the current efficiency for CO_2 formation at more negative potentials is attributed to the oxidation of CO_{ad} that was formed upon adsorption of methanol at more negative potentials (for comparison, see the methanol adsorbate stripping peak in Fig. 13.3c: the gray line in the middle panel) and which is superimposed on the signal for bulk oxidation of methanol to CO_2. As a result of the difference in electron number for the oxidation of CO_{ad} and of methanol to CO_2, with a two-electron reaction for CO_{ad} oxidation and a six-electron reaction for methanol oxidation, the current efficiency for CO_2 formation is overestimated in the more cathodic current efficiency peak, since for conversion of the $m/z = 44$ ion current to faradaic current, a six-electron reaction was assumed. This is different for the (almost simultaneous) CO_{ad} formation and oxidation during the potential scan, since in that case the four electrons released during CO_{ad} formation would be included in the current signal as well. (Note that such problems are not encountered in the evaluation of the steady-state current efficiencies during potentiostatic bulk oxidation.)

13.3.3 C_1 Molecule Bulk Oxidation: Potential-Step Transients

In order to distinguish more clearly between effects induced by the varying potential and kinetic contributions, the continuous oxidation of the three C_1 molecules was followed at a constant potential after the potential step. The corresponding faradaic and mass spectrometric ($m/z = 44$) current transients recorded after 3 minutes' adsorption at 0.16 V and a subsequent potential step to 0.6 V (see Section 13.2) are reproduced in Figs. 13.5–13.7. In all cases, the faradaic current exhibits a small initial spike, which is associated with double-layer charging when stepping the electrode potential to 0.6 V.

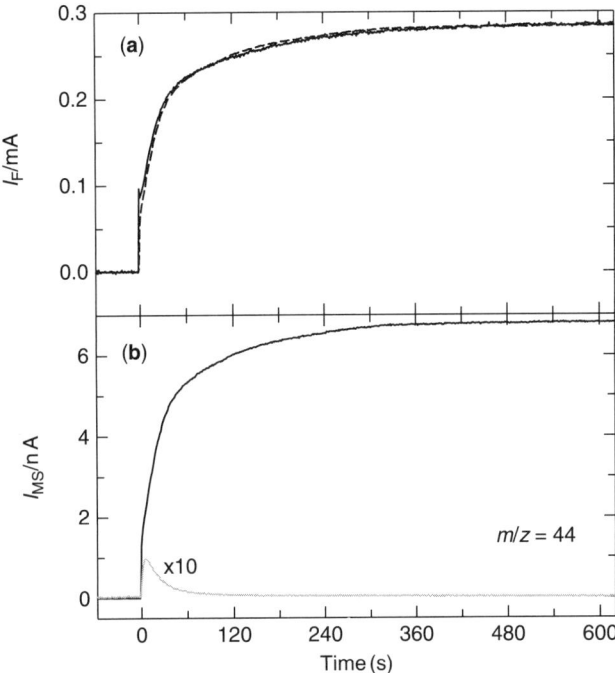

Figure 13.5 Potential-step electro-oxidation of formic acid on a Pt/Vulcan thin-film electrode (7 $\mu g_{Pt}\,cm^{-2}$, geometric area 0.28 cm^2) in 0.5 M H$_2$SO$_4$ solution containing 0.1 M HCOOH upon stepping the potential from 0.16 to 0.6 V (electrolyte flow rate 5 μL s^{-1}, at room temperature). (a) Solid line, faradaic current transients; dashed line, partial current for HCOOH oxidation to CO$_2$. (b) Solid line, $m/z = 44$ ion current transients; gray line, potential-step oxidation of pre-adsorbed CO derived upon HCOOH adsorption at 0.16 V, in HCOOH-free H$_2$SO$_4$ solution.

13.3.3.1 Formic Acid Oxidation

After passing through the initial spike, the faradaic current continues to increase, and finally reaches saturation after 7–8 minutes. Similar types of current transients were also obtained in potential-step rotating disk electrode (RDE) experiments for formic acid oxidation on bare Pt(111) electrodes and on Pt(111) electrodes irreversibly modified by adsorbed Bi [Schmidt et al., 2000; Yang and Sun, 2002], as well as on Pt film electrodes [Chen et al., 2006a, b, c]. The $m/z = 44$ mass signal closely follows the faradaic current transients, except for the initial current spike, which is absent in this case, confirming that this feature is caused by pseudocapacitive effects (see also Fig. 13.2 and the discussion in Section 13.3.1.2). For comparison, we also include the CO$_2$ signal (multiplied by 10) obtained for formic acid adsorbate oxidation (gray solid line in Fig. 13.5b, data from Fig. 13.2b). Conversion of the $m/z = 44$ ion current into a faradaic current for formic acid oxidation to CO$_2$ (dashed line in Fig. 13.5a) shows excellent agreement with the measured faradaic current (solid line in Fig. 13.5a) between the two signals, similar to our findings for potentiodynamic formic acid oxidation. A 100% current

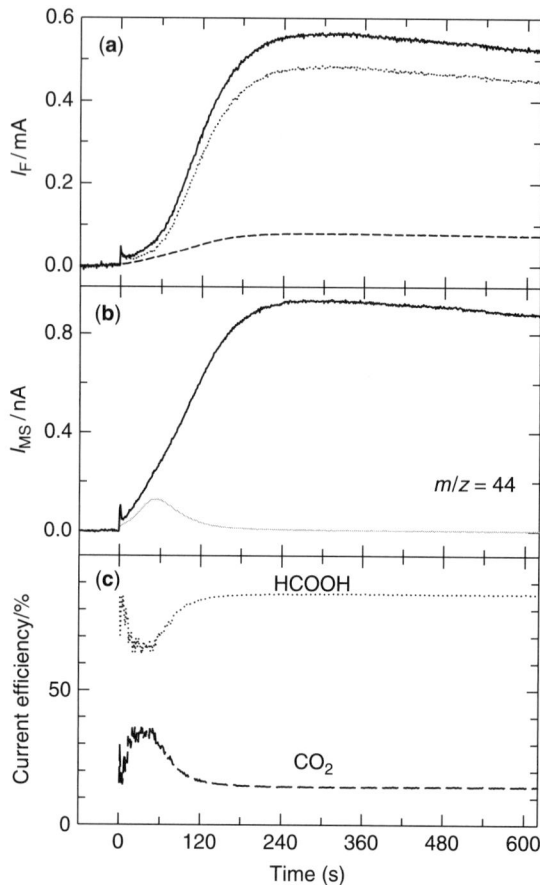

Figure 13.6 Potential-step electro-oxidation of formaldehyde on a Pt/Vulcan thin-film electrode (7 μg_{Pt} cm^{-2}, geometric area 0.28 cm^2) in 0.5 M H_2SO_4 solution containing 0.1 M HCHO upon stepping the potential from 0.16 to 0.6 V (electrolyte flow rate 5 $\mu L\ s^{-1}$, at room temperature). (a) Solid line, faradaic current transients; dashed line, partial current for HCHO oxidation to CO_2; dotted line, difference between the net faradaic current and that for CO_2 formation. (b) Solid line, $m/z = 44$ ion current transients; gray line: potential-step oxidation of pre-adsorbed CO derived upon HCHO adsorption at 0.16 V, in HCHO-free sulfuric acid solution. (c) Current efficiency transients for CO_2 formation (dashed line) and formic acid formation (dotted line).

Figure 13.7 (*Continued*) (b) Solid line, $m/z = 44$ ion current transients; gray line, potential-step oxidation of pre-adsorbed CO derived upon CH_3OH adsorption at 0.16 V, in CH_3OH-free H_2SO_4 solution. (c) $m/z = 60$ ion current transients. (d) Current efficiency transients for CO_2 formation (dashed line), formic acid formation (dash–dotted line), and formaldehyde formation (dotted line).

13.3 RESULTS AND DISCUSSION 437

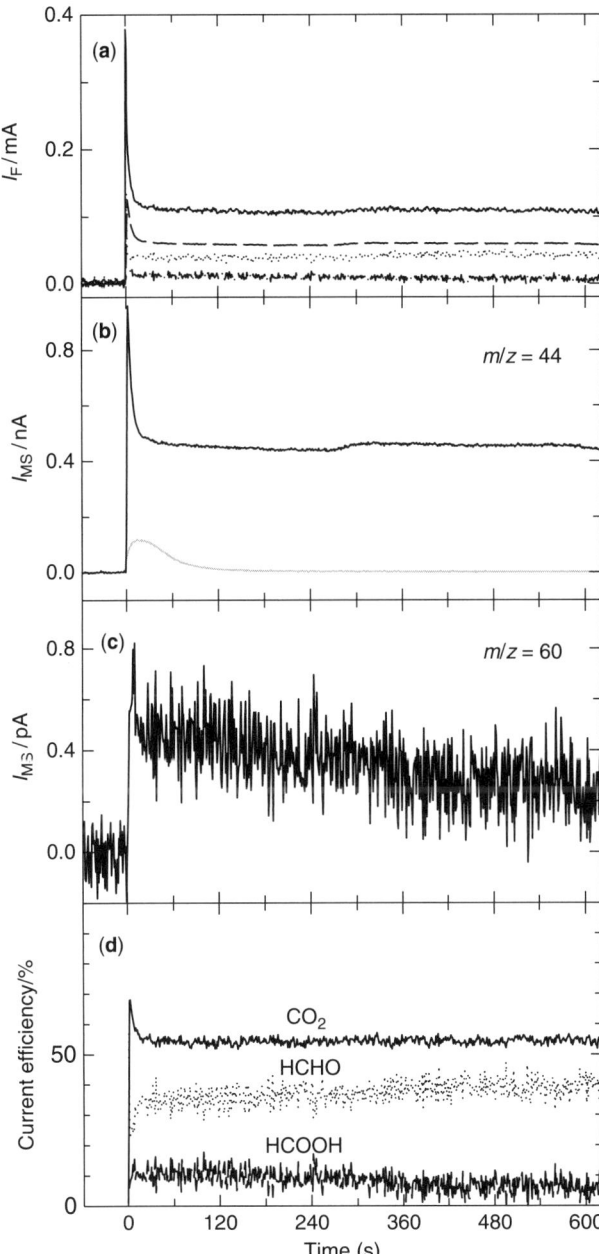

Figure 13.7 Potential-step electrooxidation of methanol on a Pt/Vulcan thin-film electrode (7 $\mu g_{Pt}\,cm^{-2}$, geometric area 0.28 cm^2) in 0.5 M H$_2$SO$_4$ solution containing 0.1 M CH$_3$OH upon stepping the potential from 0.16 to 0.6 V (electrolyte flow rate 5 $\mu L\,s^{-1}$, at room temperature). (a) Solid line, faradaic current transients; dashed line, partial current for CH$_3$OH oxidation to CO$_2$; dotted line, difference between the net faradaic current and that for CO$_2$ formation.

efficiency for formic acid oxidation to CO_2 has been reported also for potentiostatic formic acid oxidation on a polycrystalline Pt electrode based on transmission IR spectroscopy measurements in a micro flow cell, where the intensities were calibrated against solutions with known CO_2 concentrations [Huang et al., 2002].

The saturation values for both faradaic and mass spectrometric currents obtained in the potentiostatic experiment (Fig. 13.5a, b) are about three times higher than the potentiodynamic formic acid oxidation currents at the same potential in the positive-going scan, but lower than those in the negative-going scan (Fig. 13.3a). These differences can be explained by the higher CO_{ad} coverage in the positive-going scan and the lower CO_{ad} coverage in the negative-going scan compared with the steady-state situation. Therefore, the increase in the formic acid oxidation rate after the potential step is caused by the increasing removal of reaction-inhibiting CO_{ad} species, which allows "direct" formic acid oxidation (see the discussion in Section 13.4) on an increasing area of the Pt surface. Because of the steady supply of CO_{ad}, due to continuous adsorption and dehydrogenation of formic acid, the faradaic current and the CO_2 formation rate decrease with time, and finally reach (potential- and concentration-dependent) steady-state values once the rates for CO_{ad} formation and CO_{ad} oxidation are equivalent. The lower steady-state formic acid oxidation rate compared with the peak rate in the negative-going scan in the potentiodynamic measurements can be explained by the lack of time to build up a CO adlayer of comparable coverage in the latter case. In the same way, the higher steady-state formic acid oxidation rate compared with that at 0.6 V in the positive-going scan is attributed to the higher CO_{ad} coverage in the potentiodynamic measurement, resulting from the low CO_{ad} oxidation rate at potentials $E < 0.6$ V and the CO_{ad} coverage of about 50% CO_{ad} of the saturation value in the positive-going scan before the onset of CO_{ad} oxidation [Chen et al., 2006a, b, c]. Finally, the much smaller faradaic current and CO_2 formation rate upon formic acid adsorbate oxidation compared with the steady-state bulk oxidation current/rate indicates that CO_{ad} oxidation makes only a minor contribution to formic acid oxidation under these conditions and that most of the formic acid oxidation current results from direct oxidation of formic acid. This result agrees well with recent conclusions based on quantitative data from in situ IR spectroscopy measurements [Chen et al., 2006a, b, c], where, depending on the reaction conditions, the indirect pathway was found to contribute only in the range of 1% or less (for details, see Samjeské et al. [2005, 2006]).

13.3.3.2 Formaldehyde Oxidation The faradaic current for formaldehyde oxidation (solid line in Fig. 13.6a) is largely suppressed during the first minute after stepping to 0.6 V, and then increases with time, resulting in an S-shaped transient. It reaches its maximum current about 4–5 minutes after the potential step, followed by a slow and nearly linear decay of the faradaic current with time (solid line in Fig. 13.6a). In comparison with formic acid oxidation (solid line in Fig. 13.5a), three major differences can be noted:

(i) Formaldehyde oxidation is more suppressed during the initial stages of the oxidation reaction than formic acid oxidation.

(ii) The maximum faradaic current for formaldehyde oxidation is about twice that for formic acid oxidation.
(iii) The formaldehyde oxidation current shows a slow decay with time, while the formic acid oxidation current is fairly stable.

On the other hand, during potentiodynamic formaldehyde oxidation (solid line in the upper panel of Fig. 13.3b), there is only a small faradaic current at 0.6 V in the positive-going scan, in contrast to the much higher steady-state value (about 0.55 mA) attained in the potentiostatic experiment.

The CO_2 formation transient (solid line in Fig. 13.6b) exhibits a small $m/z = 44$ ion current spike upon stepping the electrode potential, which is absent for the corresponding adsorbate stripping trace (see also Fig. 13.2b) and also for formic acid oxidation. Apparently, formaldehyde oxidation to CO_2 starts instantaneously, similar to oxidation of a saturated CO adlayer (Fig. 13.2b). Subsequently, the current trace for CO_2 formation continues to grow, until reaching a maximum after about 4–5 minutes. The current efficiency for CO_2 formation (dashed line in Fig. 13.6c) passes through an initial maximum of about 33% after about 1 minute, and then decays to a constant value of about 14%. The maximum CO_2 current efficiency has about the same value as that reached in the positive-going scan at potentials of 0.65–0.7 V (Fig. 13.4b), indicating that the adlayer composition is similar in both cases. The time evolution of the current efficiency for formaldehyde oxidation to formic acid (solid line in Fig. 13.6c), simply mirrors that of the current efficiency for CO_2 formation.

Comparison with the formaldehyde adsorbate stripping trace (gray line in Fig. 13.6b, data from Fig. 13.2b) implies that within the first minute after the potential step, oxidation of preformed CO_{ad}, which was generated upon interaction of formaldehyde with the Pt electrocatalyst at 0.06 V, contributes significantly to the total faradaic current and to the total amount of CO_2 formed during formaldehyde oxidation at 0.6 V. Accordingly, the initial increase and later decay in selectivity for CO_2 production during the first 2 minutes can be explained by oxidation of CO_{ad} that was pre-adsorbed before the potential step. Since the CO_{ad} species initially present on the surface are oxidized to CO_2 in a two-electron reaction, whereas a four-electron reaction was assumed when calculating the CO_2 current efficiency, the CO_2 efficiency calculated during this initial phase is higher than the actual value. Finally, the slight decreases in both faradaic and $m/z = 44$ ion currents observed also after 10 minutes of formaldehyde oxidation at 0.6 V are most likely due to a slight increase in the steady-state CO_{ad} coverage with time and/or a structural transformation in the resulting CO adlayer.

Conversion of the $m/z = 44$ ion current into a partial faradaic reaction current for formaldehyde oxidation to CO_2 (four-electron reaction) shows that, under these experimental conditions, formaldehyde oxidation to CO_2 is only a minority reaction pathway (dashed line in Fig. 13.6a). Assuming CO_2 and formic acid to be the only stable reaction products, most of the oxidation current results from the incomplete oxidation to formic acid (dotted line in Fig. 13.6a). The partial reaction current for CO_2 formation on Pt/Vulcan at 0.6 V is only about 30% of that during formic acid oxidation under similar conditions, although the faradaic current for formaldehyde oxidation (solid line in Fig. 13.6a) is about twice as high as that for formic acid oxidation (solid line in Fig. 13.5a). In total, this results in a steady-state current efficiency for CO_2 formation of 14% for formaldehyde oxidation. This is significantly less than the current efficiency for CO_2 formation of about 55% (product yield of 37%) reported from IR spectroscopy measurements on Pt(111) at a similar potential [Batista and Iwasita, 2006]. While we cannot rule out differences in the electrochemical properties of the two electrodes, we mainly attribute the much lower CO_2 efficiency in our measurements to the different mass transport conditions in the two experiments. In the IR spectroscopy measurements, which were performed in a thin-layer configuration with very slow diffusion of reactants/products to and from the electrode surface, the probability of further oxidation of formic acid via re-adsorption is much higher than in our flow cell measurements, where the off-transport of reaction products is much more rapid.

We speculate that the high selectivity for formic acid formation during formaldehyde oxidation under steady-state conditions is related to a relatively high CO adlayer coverage under these conditions compared with the other C_1 reactants (see Fig. 13.1). This statement is based on the higher tendency of formaldehyde towards oxidative CO_{ad} formation compared with the other reactants. Assuming that the CO_{ad} oxidation rate does not depend on the nature of the initial C_1 species, this should result in a higher steady-state CO_{ad} coverage during formaldehyde oxidation than during formic acid or methanol oxidation under the present conditions (for comparison, see also [Chen et al., to be published]). Considering further that formaldehyde oxidation to CO_2 is more site-demanding than oxidation to formic acid (see above), this directly explains the high current efficiency for formic acid formation. The nevertheless rather high faradaic current, which is significantly higher than the steady-state reaction currents for methanol or formic acid oxidation, could be explained by considering that in aqueous solution formaldehyde is largely hydrated to methylene glycol [Batista and Iwasita, 2006], which does not require OH_{ad} for formic acid or CO_2 formation.

13.3.3.3 Methanol Oxidation The faradaic current transient (solid line in Fig. 13.7a) exhibits the most pronounced initial current spike among the C_1 bulk oxidation transients, which may be due to the higher H_{upd} coverage (itself due to the lower initial CO_{ad} coverage). Subsequently, the faradaic current drops within about 30 s to 0.12 mA and then stays about constant for the remaining time of the experiment (10 minutes). In that respect, the MOR transient is distinctly different to the initial faradaic current increase during formic acid oxidation (Fig. 13.5a) or the delayed current increase observed during formaldehyde oxidation. Because of the significant pseudocapacitive double-layer charging current contributions, it is hardly possible to interpret the magnitude of the initial spike [Franaszczuk et al., 1992; Housmans and Koper, 2003]. It is clear, however, from the similar feature in the mass spectrometric $m/z = 44$ ion current transient that this feature is not due solely to pseudocapacitive charging effects, but contains reactive contributions as well [Jusys et al., 2003], similar to the observation for formaldehyde oxidation (see Section 13.3.3.2). (Note that because of the limited time resolution (1 s^{-1}) of the

MS measurements, the actual height of the initial spike may be significantly greater than indicated in Fig. 13.7b.) The steady-state MOR current of 0.12 mA at the end of the measurement is significantly lower than that for formic acid oxidation and much lower than that for formaldehyde oxidation under similar conditions.

Comparison with the methanol adsorbate stripping trace (gray solid line in Fig. 13.7b) shows that also for methanol oxidation the continuous CO_2 formation rate under steady-state conditions is much higher than the maximum oxidation rate of preformed CO_{ad}, which resembles our observations for formic acid (solid line in Fig. 13.5b) and formaldehyde (solid line in Fig. 13.6b) oxidation to CO_2. However, in contrast to the initial increase in the $m/z = 44$ ion current with time for formic acid and formaldehyde oxidation, the CO_2 formation rate for methanol oxidation drops within half a minute to a value that is at most half of that attained initially (in the spike). Apparently, the higher CO_2 efficiency during this initial phase not only results from pre-adsorbed CO_{ad} oxidation (for comparison, see the methanol adsorbate stripping curve with its much lower current during the initial spike, which is included as gray line in Fig. 13.7b), but also reflects a higher selectivity for CO_2 formation during initial methanol bulk oxidation.

The $m/z = 60$ ion current (Fig. 13.7c), which results from methyl formate and hence is indicative of formic acid formation, equally increases in a pronounced step after raising the potential, and then decays slowly with time. (The higher noise in these data is due to the very low concentration of methyl formate species.)

The current efficiencies for the different reaction products CO_2, formaldehyde, and formic acid obtained upon potential-step methanol oxidation are plotted in Fig. 13.7d. The CO_2 current efficiency (solid line) is characterized by an initial spike of up to about 70% directly after the potential step, followed by a rapid decay to about 54%, where it remains for the rest of the measurement. The initial spike appearing in the calculated current efficiency for CO_2 formation can be at least partly explained by a similar artifact as discussed for formaldehyde oxidation before, caused by the fact that oxidation of the pre-formed CO_{ad} contributes only two electrons per CO_2 molecule rather than six electrons, as assumed in the calculation of the current efficiency. The current efficiency for formic acid oxidation steps to a value of about 10% at the initial period of the measurement, and then decreases gradually to about 5% at the end of the measurement. Finally, the current efficiency for formaldehyde formation, which was not measured directly, but calculated from the difference between total faradaic current and partial reaction currents for CO_2 and formic acid formation, shows an apparently slower increase during the initial phase and then remains about constant (final value about 40%). The imitial increase is at least partly caused by the same artifact as discussed above for CO_2 formation, only in the opposite sense.

13.4 DISCUSSION

13.4.1 Mechanistic Aspects

The data presented in the preceding sections provide clear proof that the oxidation of both methanol and formaldehyde over carbon-supported Pt/C catalysts can lead to very significant amount of incompletely oxidized reaction products in addition to the complete oxidation product CO_2. This is in perfect agreement with the formal reaction scheme introduced by Bagotzky et al. (1977) and also later work [Petukhova et al., 1977; Ota et al., 1984; Iwasita and Vielstich, 1986; Korzeniewski and Childers, 1998; Childers et al., 1999; Wang et al., 2001a; Jusys and Behm, 2001; Jusys et al., 2003; Housmans et al., 2006]. Formic acid is produced in both reactions; methanol oxidation also results in formaldehyde formation. For formic acid oxidation, CO_2 is the only reaction product. Under the present reaction conditions, CO_2 formation contributes about 50% of the faradaic current (or about 27% of the reaction products) during methanol oxidation at 0.6 V; for formaldehyde oxidation under similar reaction conditions, CO_2 is only a minority reaction product.

While the data provide clear evidence for the formation of incomplete oxidation products, and help to identify the nature of the stable adsorbate(s) formed upon interaction with the respective C_1 molecules, the molecular-scale information on the actual reaction mechanism and the main reaction intermediates is very indirect. Also, the reaction step(s) at which branching into the different reaction pathways occurs (e.g., "direct" versus "indirect" pathway, or complete oxidation versus incomplete oxidation) cannot be identified directly from these data. Nevertheless, by combining these and the many previous experimental data, as well as theoretical results, conclusions on the molecular-scale mechanism are possible, and are substantiated by a solid data base.

Before going into more detail, we will briefly summarize the different mechanistic proposals and the experimental information available on possible reaction intermediates.

1. For both methanol oxidation and formic acid oxidation, a "dual-pathway mechanism" has been proposed (for methanol oxidation, see Lamy et al. [1983]; Jarvi and Stuve [1998]; Cuesta [2006]; Housmans et al. [2006]; Iwasita [2003]; for formic acid oxidation, see Parsons and VanderNoot [1988]; Sun et al. [1988]; Willsau and Heitbaum [1986]; Miki et al. [2002]; Samjeské and Osawa [2005]; Chen et al. [2006a, b, c]; Samjeské et al. [2005, 2006]; Miki et al. [2004], Chang et al. [1989]), in which one reaction pathway proceeds via formation and subsequent oxidation of CO_{ad} (P, "indirect pathway"), while the other leads, via one or more reaction intermediates RI, "directly" to CO_2 ("direct pathway") (Fig. 13.8a).

2. A similar reaction scheme is also likely for formaldehyde oxidation [Loucka and Weber, 1968].

3. Whereas in the indirect pathway, CO_{ad} is clearly identified as a reaction intermediate, the specific nature of the intermediate(s) in the direct pathway is under debate. For methanol oxidation, species such as COH [Xia et al., 1997; Iwasita et al., 1987, 1992; Iwasita and Nart, 1997], CHO [Zhu et al., 2001; Willsau and Heitbaum, 1986; Wilhelm et al., 1987], COOH [Zhu et al., 2001], and adsorbed formate species [Chen et al., 2003] have been proposed. Adsorbed formate species were identified during formaldehyde oxidation [Samjeské et al., 2007], methanol oxidation [Nakamura et al., 2007; Chen et al., 2003, unpublished], and formic acid oxidation [Miki et al., 2002, 2004; Samjeské and Osawa, 2005; Chen et al., 2006a, b, c; Samjeské et al., 2005, 2006].

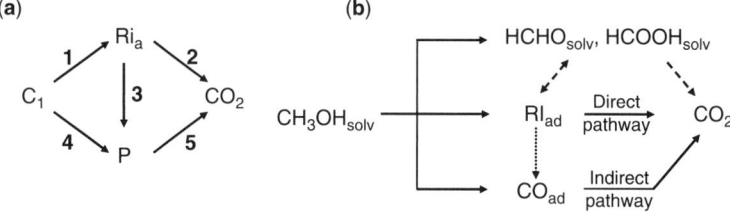

Figure 13.8 (a) Schematic representation of the original dual-pathway mechanism proposed for the oxidation of small organic molecules. C_1, reaction educt (e.g., formic acid or methanol); RI, reaction intermediate; P, poisoning species (CO_{ad}); CO_2, complete oxidation product. The numbers indicate partial reaction pathways: **1**, C_1 oxidation to the reactive intermediate; **2**, further oxidation of the reactive intermediate to CO_2; **3**, decomposition of the reactive intermediate to CO_{ad}; **4**, C_1 decomposition to CO_{ad}; **5**, CO_{ad} oxidation to CO_2. (b) Schematic representation of the expanded dual-pathway mechanism, including the formation of incomplete oxidation products (formaldehyde and formic acid). (Reproduced with permission from Jarvi and Stuve [1998].)

4. The adsorbed formate species identified by IR were proposed as reactive intermediates for C_1 oxidation [Chen et al., 2003; Miki et al., 2002, 2004; Samjeské and Osawa, 2005; Samjeské et al., 2005, 2006]. For methanol [Nakamura et al., 2007] and formic acid [Chen et al., 2006a, b, c] oxidation, this is, however, under debate. Based on a quantitative evaluation of the correlation between formate intensities and reaction rate at different concentrations [Chen et al., 2006a, b, c], we favored a mechanism where adsorbed formic acid is "directly" dehydrogenated to CO_2 as the dominant reaction pathway under those reaction conditions, while the IR-detected adsorbed formate species acts as a spectator rather than as a reaction intermediate. The latter proposal is also supported by recent DFT calculations [Hartnig et al., 2007a]. Final agreement, however, has not yet been reached.

In the original proposal of the dual-pathway mechanism (for formic acid oxidation, see [Capon and Parsons, 1973a, b, c]; for methanol oxidation, see [Parsons and VanderNoot, 1988; Jarvi and Stuve, 1998; Leung and Weaver, 1990; Lopes et al., 1991; Herrero et al., 1994, 1995]), both pathways lead to CO_2 as the final product, as illustrated in the reaction scheme depicted in Fig. 13.8a [Jarvi and Stuve, 1998]. In this mechanism, desorption of incomplete oxidation products was not included. The existence of a direct reaction pathway for methanol oxidation, following the dual-pathway mechanism, was justified by the observation of a methanol oxidation current at potentials where CO_{ad} oxidation is not yet active [Sriramulu et al., 1998, 1999; Herrero et al., 1994, 1995]. The validity of this interpretation was questioned, however, by Vielstich and Xia (1995), who claimed that CO_2 formation is observed only with the onset of CO_{ad} oxidation and that the faradaic current measured at lower potentials is due to the formation of the incomplete oxidation products formaldehyde and formic acid. The latter findings were later confirmed by Wang et al. [2001], Korzeniewski and Childers [1998], and Jusys et al. [2001, 2003]. In more recent studies, the definition of the dual-pathway mechanism was not always consistent, and in some cases, the reaction pathway leading to incomplete oxidation products was described as second reaction pathway in the dual-pathway mechanism [Cao et al., 2005]. For clarity, we suggest that the formation of incomplete oxidation products be included in the reaction scheme, in addition to the direct and the indirect pathway, which results in the expanded reaction scheme presented in Fig. 13.8b. A more detailed reaction scheme proposed recently by Housmans et al. [2006] agrees in most aspects with that simple scheme. The dashed arrows indicate the possible formation of the incomplete oxidation products formaldehyde and formic acid from the reaction intermediate in the direct pathway and their possible readsorption (two-headed arrow), and the (irreversible) formation of CO_2 from these two reaction products. The dotted arrow accounts for CO_{ad} formation upon re-adsorption and dehydrogenation of the incomplete oxidation products.

Furthermore, it is important to distinguish between adsorbed side products on the one hand, which may be formed reversibly or irreversibly, but are not involved in the actual reaction pathway ("spectator species"), and real reaction intermediates on the other hand, where the latter are part of the reaction pathway to the final product. In a reaction network including different parallel pathways to the final product, such reaction intermediates may occur in each of these pathways, for example, in the dominant "majority" reaction pathway and in minority pathways. As one example, it was shown by quantitative ATR-IRS measurements that for formic acid oxidation on a Pt film electrode at potentials around 0.6 V, the indirect pathway is a minority pathway [Chen et al., 2006b, c; Samjeské et al., 2005, 2006], contributing less than 1% to the total faradaic current [Chen et al., 2006b, c]. In that case, CO_{ad} can be considered as reaction intermediate in a minority pathway. Often reaction intermediates are short-lived species, which, because of their very low steady-state coverage, are not detectable spectroscopically. It is important to note that both spectator species and reaction intermediates in a minority pathway can significantly affect the dominant reaction pathway as well, for example by blocking the reactive surface. In the above case, the reaction intermediate in the minority pathway, CO_{ad}, affects the main reaction path by blocking the surface with a reaction-inhibiting CO adlayer.

In the following, we will discuss four points that are relevant for the mechanistic understanding of the C_1 oxidation reaction and where the present data can contribute. Since formic acid oxidation leads to a single reaction product only and the mechanism for formic acid oxidation has been discussed in detail in recent publications [Miki et al., 2002, 2004; Samjeské and Osawa, 2005; Chen et al., 2006a, b, c; Samjeské et al., 2005, 2006], we will concentrate on methanol and formaldehyde oxidation. For these reactions, one may formulate the following questions:

1. Does methanol (or formaldehyde) oxidation along the direct pathway lead "directly" to CO_2 or are incomplete oxidation products formed first, which are then oxidized further to finally result in CO_2 formation (dashed arrows in Fig. 13.8b)?
2. A similar question may also be asked for the indirect pathway: Is CO_{ad} directly formed by methanol decomposition, or does it result from a follow-up reaction,

such as methanol oxidation to formaldehyde in a first step, and subsequent formaldehyde decomposition to CO_{ad} (dotted arrow in Fig. 13.8b).

3. What are the probabilities of incomplete oxidation product formation in a single adsorption/reaction event, without considering effects related to re-adsorption and subsequent further reaction.
4. At which stage of the reaction process does the branching into the different parallel pathways occur?

The first question asks essentially the same question as has been raised previously by Vielstich and Xia for methanol oxidation [Vielstich and Xia, 1995], namely, whether it is possible to "directly" produce CO_2 along the "direct" pathway (not via formation and oxidation of CO_{ad}), or whether this occurs via formation and subsequent re-adsorption and further oxidation of the incomplete oxidation products formaldehyde and formic acid. We have shown previously for bulk oxidation of methanol over Pt/Vulcan thin-film electrodes that the product distribution depends sensitively on catalyst loading [Jusys et al., 2003]. The fraction of CO_2 increased with increasing catalyst loading, mainly on the expense of formaldehyde formation, while the formic acid content in the product distribution changed little. This was explained by a "desorption–re-adsorption" mechanism, where volatile incomplete oxidation products can desorb into the diffusion layer and subsequently either leave the diffusion layer into the flowing bulk electrolyte or adsorb again. In the latter case, the adsorbed species can either undergo further oxidation or desorb again, to start the same cycle. This mechanism leads to an increasing fraction of incomplete oxidation products in the exit stream with decreasing catalyst loading [Childers et al., 1999; Wang et al., 2001b; Jusys et al., 2003] or increasing electrolyte flow [Wang et al., 2001a]. Similar trends were reported also for ethanol oxidation over polycrystalline Pt electrodes and a Pt/C thin-film catalyst electrode, where the amount of acetic acid increased at the expense of acetaldehyde for the higher surface area Pt/C catalyst [Wang et al., 2004]. These findings are in perfect agreement with concepts developed in heterogeneous catalysis, where it is well known that with increasing space velocity (lower catalyst loading at constant reactant flow, or higher reactant flow at constant catalyst loading) the system and therefore also the product distribution move further away from the equilibrium composition [Thomas and Thomas, 1997]. Hence, the present findings clearly indicate that desorption and re-adsorption plus further reaction of the volatile incomplete oxidation products play an important role in the reaction process and contribute significantly to the observed CO_2 formation. Based on the present data, we cannot exclude, however, the possibility that CO_2 can be formed "directly," without involving the formation of adsorbed or dissolved formaldehyde and/or formic acid (at zero probability for re-adsorption). Hence, reaction along the direct reaction pathway in Fig. 13.8b could not be confirmed, but also could not be ruled out.

With regard to the second question, while CO_{ad} formation from methanol is slow at potentials in the H_{upd} region, formaldehyde adsorption transients on a Pt film electrode showed rapid CO_{ad} formation under these condition [Chen et al., to be published]. Furthermore, Korzeniewski and Childers [1998] reported increasingly higher formaldehyde yields with decreasing potential, at potentials below 0.2 V (up to 30–40% at about 0.05 V) for methanol oxidation over a polycrystalline Pt electrode. On the other hand, despite the high formaldehyde yields, the absolute rates and partial currents for formaldehyde formation are very low under these conditions, i.e., the high formaldehyde yields during methanol oxidation result solely from the fact that at potentials cathodic of 0.5 V, the CO_2 formation rate is even lower than the formaldehyde formation rate. Considering that the amount of oxidized methanol, and hence also of formaldehyde formation, during potentiostatic methanol oxidation is of the order of one nanomole per second at 0.6 V, and even much lower in the H_{upd} potential range, this sequential pathway for CO_{ad} formation does not seem to be sufficient to explain the CO_{ad} formation rate upon interaction with methanol. CO_{ad} formation by direct decomposition of methanol has to be possible, although contributions from a pathway via initial oxidation to (free) formaldehyde and subsequent re-adsorption plus dehydrogenation to CO_{ad} will be present [Olivi et al., 1994].

The reaction sequence of formaldehyde formation and subsequent CO_{ad} formation can proceed either as sequential reactions of adsorbed species, or it can involve formation and desorption of formaldehyde into the electrolyte and subsequent re-adsorption and further decomposition of formaldehyde to CO_{ad}. Considering the significant transport and catalyst loading effects discussed above, it is clear that desorption and subsequent re-adsorption plus dehydrogenation of formaldehyde will play an important role also for CO_{ad} formation, although a direct reaction of adsorbed RI_{ad} species can not be ruled out.

I the context of the third question, for comparison with theory, it would be interesting to know the selectivities for direct CO_2 formation and for desorption of the incomplete oxidation products formaldehyde and formic acid during methanol oxidation in a single adsorption/reaction event. Under steady-state conditions and at 0.6 V, the measured (overall) selectivity for CO_2 formation (CO_2 current efficiency) reached values as low as 30% on a Pt/C catalyst for low catalyst loadings (7 $\mu g_{Pt}\,cm^{-2}$) [Jusys et al., 2003]. For smooth Pt electrodes, this value was even lower [Wang et al., 2001a]. On the other hand, even on a smooth surface and under enforced electrolyte transport, a molecule desorbing from the surface is likely to undergo several collisions with the surface, before it finally leaves the diffusion layer into the flowing electrolyte and is transported away. Each of these collisions can be considered as one attempt for re-adsorption. For a catalyst layer, this number will be much higher. Although it is clear from the present data that the selectivity for "direct" CO_2 formation in a single reaction event, compared with desorption of an incompletely oxidized reaction product, must be low (of the order of 10% or less), it is not possible at the moment to deduce quantitative numbers, because of the unknown number of re-adsorption events.

The much lower amount of formic acid formation during methanol oxidation compared with formaldehyde oxidation agrees with expectations if we assume that formic acid is predominantly formed by further oxidation of (free) molecular formaldehyde produced in a first step of methanol oxidation. Under the present reaction conditions, only a very small fraction, about 1 part per thousand, of the total reactant passing through the cell reacts to give formaldehyde, formic acid, or CO_2. The rest leaves the cell without reaction. Even accounting for the possibility that the probability for re-adsorption and further oxidation to formic acid is much higher for formaldehyde formed within the catalyst layer than for formaldehyde entering the reaction cell, the relative amount of formic acid formed this way must be smaller than during bulk oxidation of a 0.1 M formaldehyde solution.

The answer to the fourth question—the reaction step responsible for the observed branching into the different pathways—can only be speculated upon. If during methanol oxidation this is decided already in the first reaction step, by either C–H bond breaking (–CH_2OH formation and subsequent decomposition to CO_{ad}) or O–H bond breaking (–OCH_3 formation and subsequent dehydrogenation to adsorbed HCHO), as was proposed in theoretical studies [Cao et al., 2005; Taylor and Neurock, 2005], this would mean that at 0.6 V and under the present reaction conditions, O–H bond breaking is the dominant initial step (40% current efficiency or 65% product selectivity for formaldehyde formation). This would be very surprising in view of previous discussions, where C–H bond breaking was considered to be much more effective at not too high potentials [Greeley and Mavrikakis, 2004; Cao et al., 2005; Taylor and Neurock, 2005]. On the other hand, if, adsorbed formaldehyde is formed in a concerted step of C–metal bond formation (C–H bond breaking) and transfer of the –OH hydrogen atom to an adjacent, hydrogen-bridge-bonded water molecule, as it was proposed by Hartnig et al. [2005, 2007], then methanol oxidation/dehydrogenation would start by C–H bond breaking for all reaction pathways. In that case, branching would occur upon formation of adsorbed formaldehyde, for example by desorption of adsorbed formaldehyde competing with its further dehydrogenation to CO_{ad}.

Similar ideas can be applied to formaldehyde oxidation. For bulk formaldehyde oxidation, we found predominant formic acid formation under current reaction conditions rather than CO_2 formation. Hence, it cannot be ruled out, and may even be realistic, that formaldehyde is first oxidized to formic acid, which can subsequently be oxidized to CO_2. The steady-state product distribution at 0.6 V is much more favorable for such a mechanism as in the case of methanol oxidation. On the other hand, because of the high efficiency of CO_{ad} formation from formaldehyde, this process is likely to proceed directly from formaldehyde adsorption rather than via formation and re-adsorption of formic acid. Alternatively, the second oxygen can be introduced via formaldehyde hydration to methylene glycol, which could be further oxidized to formic acid and finally to CO_2 (see the next paragraph).

The present and previous findings discussed in the preceding paragraphs are summarized in the more detailed reaction scheme depicted in Fig. 13.9. There, we explicitly include also the reversible exchange between adsorbed species, dissolved educts, and products in the catalyst layer and the same species in the bulk electrolyte. Starting with methanol adsorption and dehydrogenation, there seems to be general agreement among theoretical studies that adsorption via the carbon atom (C–H activation) and stepwise dehydrogenation finally results in CO_{ad} [Greeley and Mavrikakis, 2004], which at sufficiently anodic potentials can be oxidized to CO_2 ("indirect pathway"). Furthermore, there is general agreement in these studies also about the formation of an adsorbed formaldehyde species [Greeley and Mavrikakis,

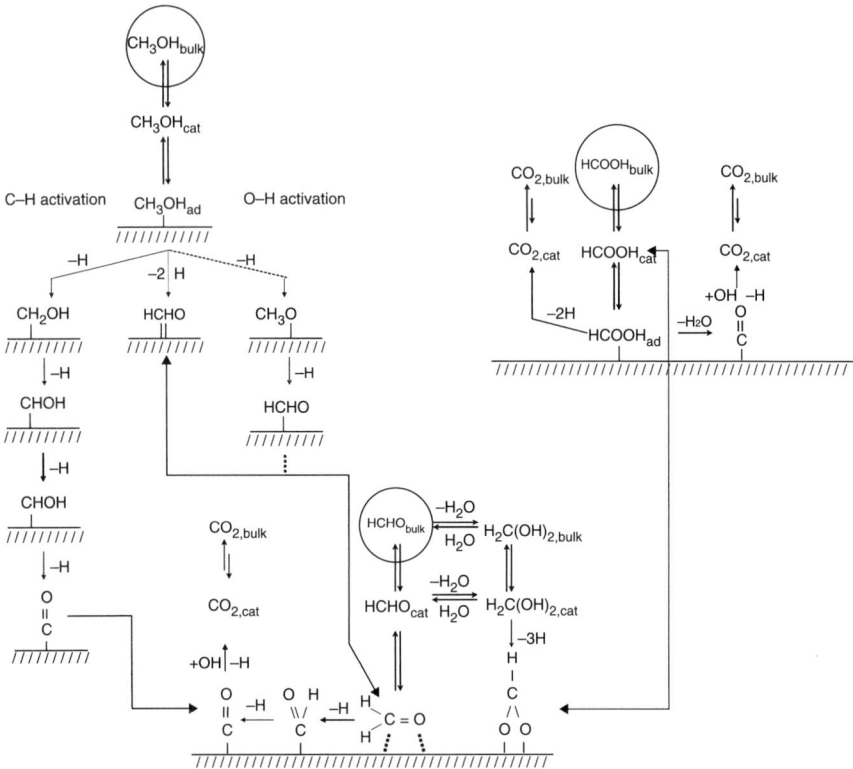

Figure 13.9 Reaction scheme for C_1 molecule oxidation on a Pt/C catalyst electrode, including reversible diffusion from the bulk electrolyte into the catalyst layer, (reversible) adsorption/desorption of the reactants/products, and the actual surface reactions. The different original reactants (educts) and products are circled. For removal/addition of H, we do not distinguish between species adsorbed on the Pt surface and species transferred directly to neighboring water molecule (H_{ad}, H^+); therefore, no charges are included (H^+, e^-). For a description of the individual reaction steps, see the text.

2004; Hartnig and Spohr, 2005], although the proposals for the reaction pathways differ. In several studies, this was proposed to result from O–H bond breaking and subsequent transfer of a C-bonded H atom to the Pt substrate [Greeley and Mavrikakis, 2004]. Hartnig et al. predicted a mechanism in which the CH_3OH molecule is polarized by formation of a hydrogen bond between the OH hydrogen and a neighboring water molecule. Upon adsorption (via the C atom), one H atom is transferred to the Pt substrate, and the OH hydrogen is fully transferred to the neighboring water molecule, resulting in an adsorbed $HCHO_{ad}$ molecule [Hartnig and Spohr, 2005; Hartnig et al., 2007b]. Desorption of this species into the catalyst layer will result in formaldehyde as an incomplete oxidation product, which can either re-adsorb or finally diffuse into the bulk electrolyte. Based on the above theoretical studies, there would be no "direct" pathway to CO_2 in addition to that via formation, possibly desorption and re-adsorption, and further oxidation of the incomplete oxidation product formaldehyde.

For formaldehyde oxidation, the formaldehyde reactant molecule diffuses reversibly from the bulk electrolyte ($HCHO_{bulk}$, circled in Fig. 13.9) into the catalyst layer ($HCHO_{cat}$) and finally adsorbs (reversibly). Independent of its origin, from methanol oxidation or formaldehyde adsorption, adsorbed formaldehyde is in equilibrium with formaldehyde dissolved in the catalyst layer ($HCHO_{cat}$) and in the bulk ($HCHO_{bulk}$). On the other hand, further dehydrogenation results in adsorbed formyl species and finally in adsorbed CO. Both reaction steps are strongly exothermic and therefore irreversible (see the references cited above). In aqueous solution, formaldehyde is in equilibrium with methylene glycol, which is proposed to form adsorbed formate species upon adsorption (see also Housmans et al. [2006]). These adsorbed species in turn can desorb as formic acid (upon protonation). Formic acid was shown in previous studies to dehydrate to CO_{ad} or to form CO_2, either "directly" [Chen et al., 2006a, b, c] or via the IR spectroscopically detected adsorbed formate [Miki et al., 2002] (in addition to CO_{ad} formation and oxidation via the indirect pathway). Owing to the additional supply of oxygen via formaldehyde hydration, CO_2 formation during formaldehyde oxidation is possible without the requirement for OH_{ad} formation.

Although the present data do not allow us to decide on the reaction steps leading to adsorbed formaldehyde during methanol oxidation, the large fraction of formaldehyde formation at 0.6 V (about 65%) leads us to favor the reaction pathway via C–H activation rather than via O–H activation (see above), as proposed by Hartnig et al. [2005, 2007a, b], but in contrast to other theoretical studies [Greeley and Mavrikakis, 2004; Cao et al., 2005; Taylor and Neurock, 2005]. This is also a major difference from the very detailed reaction scheme proposed recently by Housmans et al. [2006]. Since the formation of CO_{ad} and CO_2 is irreversible under present reaction conditions, while most other steps are reversible (see Fig. 13.9), an increasing probability of re-adsorption and further oxidation of the incomplete oxidation products formaldehyde and formic acid will drive the system more towards these stable products, as discussed above. The reaction scheme in Fig. 13.9 underlines that it is indeed possible to produce both CO_2 and CO_{ad} via re-adsorption of the incomplete methanol oxidation products formaldehyde and formic acid. It is still not possible, however, to provide unambiguous proof of the existence of a "direct" pathway as defined in Fig. 13.8b for methanol oxidation (and similarly for formaldehyde oxidation) that does not proceed via formation of the incomplete oxidation products formaldehyde and/or formic acid.

In summary, this discussion illustrates the general importance of transport processes in many (electro)catalytic reactions. These have to be addressed properly for a detailed (and quantitative) understanding of the molecular-scale mechanism. Because of the problems associated with the direct identification of the reaction intermediates (see above), experiments on nanostructured model electrodes with a well-defined distribution of reaction sites of controlled, variable distance and under equally well-defined transport conditions (first attempts in this direction are described in [Lindström et al., submitted; Schneider et al., 2008]), in combination with detailed simulations of the ongoing transport processes and theoretical calculations of the electrode surface reactions, seem to be a promising alternative for elucidating the mechanism of the complex reaction process for methanol oxidation on Pt.

13.4.2 Role of Incomplete Oxidation Products for Fuel Cell Applications

The finding of very substantial amounts of incomplete oxidation products for methanol and formaldehyde oxidation can have considerable consequences for technical applications, such as in DMFCs. In that case, the release of formaldehyde at the fuel cell exhaust has to be avoided not only from efficiency and energetic reasons, but in particular because of the toxicity of formaldehyde. While in standard DMFC applications the catalyst loading is sufficiently high that this is not a problem, i.e., only CO_2 is detected [Aricò et al., 1998], the trend to reducing the catalyst loading or applications in micro fuel cells may lead to situations where the formation of incomplete oxidation products could indeed become problematic (see also Wasmus et al. [1995]). For such purposes, one could define a maximum space velocity above which formation of incomplete oxidation products may become critical.

Finally, it should also be noted that the product distribution will depend on the reaction temperature, with an increasing fraction of CO_2 being expected for higher temperatures. Considering typical reaction temperatures in a DMFC (80–120 °C), this would further shift the critical limit for catalyst loading to lower values (or the critical space velocity to higher values). Accordingly, this effect would be more limiting for applications requiring lower operating temperatures (e.g., at or slightly above room temperature), as met, for example, in portable applications.

In summary, the formation of incomplete oxidation products seems to be unproblematic for standard DMFC applications, at least at the present catalyst loadings. Nevertheless, the general problematics have to be kept in mind, not only for DMFC applications, but also for other fuels (and fuel cells), such as ethanol oxidation, where incomplete oxidation products are equally possible. Properly performed model studies, using appropriate reaction conditions, can give valuable information on the product distribution under certain reaction conditions, and on the critical values of catalyst loading or space velocity for the appearance of certain incomplete oxidation products, and are often more informative for elucidating these basic reaction characteristics than measurements at the exhaust of a fuel cell, which largely represent the specific design (flow field geometry, catalyst loading and utilization, etc.), rather than intrinsic reaction and catalyst properties [Rao et al., 2007]. Nevertheless, in the end, proposals based on model studies like the present one have to be verified in fuel cell test measurements, by evaluating the composition of the exhaust by mass spectrometry [Wasmus et al., 1995; Lin et al., 1997; Seiler et al., 2004; Neergat et al., 2006; Rao et al., 2007].

13.4.3 Implications for Reaction Modeling

Kinetic results such as those presented in the previous sections, which could be further extended by varying the reaction parameters (reactant concentration, electrode potential, catalyst loading, electrolyte flow rate, and reaction temperature), can serve as basis for a systematic modeling of the reaction process in a DMFC or direct formic acid fuel cell. In contrast to standard modeling procedures, where the oxidation of the reactant is described by an overall kinetic rate law (most simply some kind of a power law), data of the present type would allow the reaction process to be described in a much more realistic way, including also the formation and subsequent further oxidation of the incompletely oxidized reaction side products. Measurements such as those described here can provide rate constants for all different partial reactions; methanol oxidation to formaldehyde, formic acid, and CO_2; further oxidation of formaldehyde to formic acid and CO_2; and finally oxidation of formic acid to CO_2.

For a simple description, one could use a segmented plug flow reactor model, where the reaction-induced changes in reactant and product concentrations are calculated for each segment and where the outlet concentrations of the nth segment are identical to the input concentrations of the $(n + 1)$th segment [Levenspiel, 1972]. Considering the measured steady-state reaction currents of 100 µA, which are equivalent to a reaction rate of about 0.5 nmol s^{-1}, and an electrolyte flow rate of 5 µL s^{-1}, the reaction-induced decrease in reactant concentration in the present setup and under present reaction conditions is negligible (about 1 part per thousand conversion). Hence, the DEMS cell is operated under differential reaction conditions (constant reactant concentration throughout the reactor). Therefore, the reaction rates and the product distributions measured under steady-state conditions for the different C_1 molecules are characteristic for reaction of 0.1 M reactant solution under the present reaction conditions (space velocity and temperature). For modeling the reaction under technical conditions, with significant conversion in the cell (integral reaction conditions), a number of measurements of a similar type would be required, covering the full range of reactant concentrations present in the reaction cell.

The situation becomes more complicated by the fact that the incompletely oxidized side products are not only reactive, but may also affect the activity of the catalyst for the main reaction, for example by modifying the composition of the steady-state adlayer. This may be particularly important in the later part of the reactor, where the concentration of these side products will be higher than at the reactor inlet. As a result, the reactions of the different reactive components in the electrolyte can no longer be considered as independent. Therefore, in order to properly describe the accumulation and further oxidation of the incompletely oxidized reaction products along the reaction cell, measurements using representative mixtures of the three C_1 molecules are required.

Finally, for relevant fuel cell modeling, the kinetic model studies should be performed under similar temperature and pressure conditions, i.e., at temperatures in the range of 80–120 °C and pressures up to 3 bar, and at comparable space velocities. The first flow-cell and DEMS measurements under such temperature and pressure conditions are currently underway in our laboratory.

13.5 SUMMARY

The adsorption and oxidation of the C_1 molecules methanol, formaldehyde, and formic acid over a carbon-supported Pt/C fuel cell catalyst under continuous electrolyte flow have been investigated in a quantitative, comparative online DEMS study.

Based on the number of electrons required for their oxidation, the stable adsorbates remaining on the surface after adsorption in the H_{upd} potential region and subsequent electrolyte exchange were identified as CO_{ad}. The saturation coverages reached after 10 minutes' adsorption from 0.1 M C_1 solution decrease in the order formaldehyde > formic acid > methanol, assuming coverages between 87% and 10% of the coverage of a saturated CO adlayer obtained by exposure to a CO-containing electrolyte. The C_1 adsorbates are easier to oxidize than a CO adlayer of similar coverage prepared by CO adsorption. This is reflected by a downshift of the oxidation peak, both of the onset and of the maximum, in potentiodynamic stripping experiments and in a faster oxidation with no induction period in potentiostatic stripping experiments. The differences in oxidation behavior are tentatively assigned to a different distribution of the CO_{ad} molecules after C_1 and after CO adsorption, respectively.

Potentiodynamic oxidation of methanol and formaldehyde reveals considerable formation of incomplete oxidation species (formaldehyde and formic acid, respectively) under the present reaction conditions, with potential-averaged CO_2 current efficiencies of about 50% for methanol oxidation and 20% for formaldehyde oxidation. In the positive-going scan, the onset of the oxidation reaction differs significantly for the three reactants, with onset potentials of 0.1 V (formic acid), 0.4 V (methanol), and 0.55 V (formaldehyde), respectively. The higher onset potential for formaldehyde oxidation compared with methanol oxidation is attributed to the much higher CO_{ad} coverage, which inhibits the reaction start. For formic acid oxidation, the reaction is possible already in the presence of a CO adlayer and at potentials far below the onset of OH_{ad} formation, which was the first evidence for a direct pathway for formic acid oxidation. In the negative-going scan, the similar onset potential for the main oxidation peak for all three reactants is correlated to the reduction of the oxidized/OH_{ad}-covered Pt surface, underlining also the much higher activity of the bare metallic Pt surface compared with the oxidized surface. In both scan directions, the maximum formaldehyde oxidation rate is at least twice as high as those for methanol and formic acid oxidation.

The current efficiency for formic acid formation during methanol oxidation is below 10%, leaving formaldehyde as the main incomplete oxidation product under the present reaction conditions. The dominant formation of incomplete oxidation products at the onset of the reaction in the positive-going scan points to a lower site requirement for incomplete oxidation compared with complete oxidation to CO_2. For both methanol and formaldehyde oxidation, the CO_2 current efficiency exhibits a characteristic double-peak structure in the positive-going scan, with maxima at about 0.62/0.7 V and 0.8/0.9 V for methanol and formaldehyde oxidation, respectively. The first maximum, which coincides with the CO_{ad} stripping peak, is attributed to the oxidation of CO_{ad} that was formed in the low potential region of the preceding negative-going scan and the present positive-going scan. The second maximum reflects an inherently higher activity for CO_2 formation on the Pt surface in the absence of reaction-hindering adlayers. It results mainly from CO_2 formation by direct oxidation of the respective reactant. At increasing potential, both the overall reaction rate and the CO_2 efficiency decrease owing to the increasing coverage of adsorbed

OH/oxide species. At potentials anodic of 1 V, incomplete oxidation of formaldehyde to formic acid is activated, while methanol oxidation is almost completely hindered. This reflects an easier oxidation of the C–H group in the aldehyde than in the alcohol. For the negative-going scan, where the CO_{ad} coverage at the onset of oxidation is negligible, there is no such double-peak structure in the current efficiency.

The oxidation transients, recorded after stepping the potential from 0.16 to 0.6 V, show similar characteristics. At 0.6 V, the steady-state faradaic current decreases in the order formaldehyde > formic acid > methanol. A higher CO_2 efficiency in the initial stage of the transient is attributed to the oxidation of CO_{ad} that was formed while holding the electrode at 0.16 V. The current efficiencies and product distributions follow the trends observed for potentiodynamic oxidation, with a significantly higher steady-state current at 0.6 V for formaldehyde oxidation than for formic acid and methanol oxidation. The steady-state current efficiency for CO_2 formation is much lower during formaldehyde formation (8%) than for methanol oxidation (50% CO_2 current efficiency and 10% formic acid current efficiency). For all three reactants, the steady-state currents at 0.6 V are between those obtained for potentiodynamic oxidation in the positive-going and negative-going scans at the same potential, and the same is true also for the CO_2 current efficiencies during methanol and formaldehyde oxidation. This was explained by a reaction-dominating role of the adlayer during the reaction, with higher adlayer coverage during the positive-going scan and lower coverage during the negative-going scan compared with the steady-state situation.

The results have been compared with the earlier proposal of a dual-pathway mechanism for C_1 oxidation, and, together with previous experimental and theoretical results, summarized in a comprehensive reaction scheme that explicitly includes also the (reversible) exchange between adsorbed species, dissolved product species in the catalyst layer, and similar species in the bulk electrolyte. The traditional dual-pathway mechanism, where both the direct and indirect pathways lead to CO_2 formation, has been extended by adding a third pathway that accounts for formation and desorption of incomplete oxidation products. In the mechanistic discussion, we have focused on the role in and contribution to the C_1 oxidation process of the formation/desorption and re-adsorption plus further oxidation of incomplete oxidation products. This not only leads to faradaic currents exceeding that for CO_2 formation, but may result in additional CO_{ad} and CO_2 formation, via adsorption and oxidation of the incomplete oxidation products.

Finally, we have discussed the effect of incomplete C_1 oxidation product formation for fuel cell applications and the implications of these processes for reaction modeling. While for standard DMFC applications, formaldehyde and formic acid formation will be negligible, they may become important for low temperature applications and for microstructured cells with high space velocities. For reaction modeling, we have particularly stressed the need for an improved kinetic data base, including kinetic data under defined reaction and transport conditions and kinetic measurements on the oxidation of C_1 mixtures with defined amounts of formaldehyde and formic acid, for a better understanding of cross effects between the different reactants at an operating fuel cell anode.

ACKNOWLEDGMENTS

This work was supported by the Deutsche Forschungsgemeinschaft (Project Be 1201/12-2). We gratefully acknowledge discussions with C. Hartnig and E. Spohr.

REFERENCES

Anastasijevic NA, Baltruschat H, Heitbaum J. 1989. DEMS as a tool for the investigation of dynamic processes: Galvanostatic formic acid oxidation on a Pt electrode. J Electroanal Chem 272: 89–100.

Anastasijevic NA, Baltruschat H, Heitbaum J. 1993. On the hydrogen evolution during the electrochemical oxidation of aldehydes at Ib metals. Electrochim Acta 38: 1067–1072.

Andreaus B, Eikerling M. 2007. Active site model for CO adlayer electrooxidation on nanoparticle catalysts. J Electroanal Chem 607: 121–132.

Andreaus B, Maillard F, Kocylo J, Savinova ER, Eikerling M. 2006. Kinetic modeling of CO_{ad} monolayer oxidation on carbon-supported platinum nanoparticles. J Phys Chem B 110: 21028–21040.

Angerstein-Kozlowska H, Conway BE, Sharp WBA. 1973. The real conditions of electrochemically oxidized platinum surfaces. Part 1. Resolution of component processes. J Electroanal Chem 43: 9–36.

Arenz M, Mayrhofer KJ, Stamenkovic V, Blizanac BB, Tomoyuki T, Ross PN, Markovic NM. 2005. The effect of the particle size on the kinetics of CO electrooxidation on high surface area Pt catalysts. J Am Chem Soc 127: 6819–6829.

Aricò AS, Creti P, Antonucci PL, Antonucci V. 1998. Comparison of ethanol and methanol oxidation in a liquid-feed solid polymer electrolyte fuel cell at high temperature. Electrochem Sol Lett 1: 66–68.

Bagotzky VS, Vassiliev YB, Khazova OA. 1977. Generalized scheme of chemisorption, electrooxidation and electroreduction of simple organic compounds on platinum group metals. J Electroanal Chem 81: 229–238.

Basnayake R, Li Z, Katar S, Zhou W, Rivera H, Smotkin ES, Casadonte DJ Jr, Korzeniewski C. 2006. PtRu nanoparticle electrocatalyst with bulk alloy properties prepared through a sonochemical method Langmuir 22: 10446–10450.

Batista EA, Iwasita T. 2006. Adsorbed intermediates of formaldehyde oxidation and their role in the reaction mechanism. Langmuir 22: 7912–7916.

Batista EA, Malpass GRP, Motheo AJ, Iwasita T. 2003. New insight into the pathways of methanol oxidation. Electrochem Commun 5: 843–846.

Batista EA, Malpass GRP, Motheo AJ, Iwasita T. 2004. New mechanistic aspects of methanol oxidation. J Electroanal Chem 571: 273–282.

Beden B, Lamy C, Bewick A, Kunimatsu K. 1981. Electrosorption of methanol on a platinum electrode. IR spectroscopic evidence for adsorbed species. J Electroanal Chem 121: 343–347.

Behm RJ, Jusys Z. 2006. The potential of model studies for the understanding of catalyst poisoning and temperature effects in polymer electrolyte fuel cell reaction. J Power Sources 154: 327–342.

Beltowska-Brzezinska M, Heitbaum J. 1985. On the anodic oxidation of formaldehyde on Pt, Au and Pt/Au-alloy electrodes in alkaline solution. J Electroanal Chem 183: 167–181.

Bergamaski K, Pinheiro ALN, Teixeira-Neto E, Nart FC. 2006. Nanoparticle size effects on methanol electrochemical oxidation on carbon supported platinum catalysts. J Phys Chem B 110: 19271–19279.

Bergelin M, Herrero E, Feliu JM, Wasberg M. 1999. Oxidation of CO adlayers on Pt(111) at low potentials: An impinging jet study in H_2SO_4 electrolyte with mathematical modeling of the current transients. J Electroanal Chem 467: 74–84.

Biegler T, Rand DAJ, Woods R. 1971. Limiting oxygen coverage on platinized platinum: Relevance to determination of real platinum area by hydrogen adsorption. J Electroanal Chem 29: 269–277.

Breiter MW. 1967. A study of the intermediates adsorbed on platinized platinum during the steady-state oxidation of methanol, formic acid and formaldehyde. J Electroanal Chem 14: 407–413.

Camara GA, Iwasita T. 2005. Parallel pathways of ethanol oxidation: The effect of ethanol concentration. J Electroanal Chem 578: 315–321.

Cao D, Lu G-Q, Wieckowski A, Wasileski SA, Neurock M. 2005. Mechanisms of methanol decomposition on platinum: A combined experimental and ab-initio approach. J Phys Chem B 109: 11622–11633.

Capon A, Parsons R. 1973a. The oxidation of formic acid at noble metal electrodes. Part I. Review of previous work. J Electroanal Chem 44: 1–7.

Capon A, Parsons R. 1973b. The oxidation of formic acid on noble metal electrodes. Part II. A comparison of the behavior of pure electrodes. J Electroanal Chem 44: 239–254.

Capon A, Parsons R. 1973c. The oxidation of formic acid at noble metal electrodes. Part III. Intermediates and mechanisms on platinum electrodes. J Electroanal Chem 45: 205–231.

Chang S-C, Leung L-WH, Weaver MJ. 1989. Metal crystallinity effects in electrocatalysis as probed by real-time FTIR spectroscopy: Electrooxidation of formic acid, methanol and ethanol on ordered low-index platinum surfaces. J Phys Chem 94: 6013–6021.

Chen Y-X, Miki A, Ye S, Sakai H, Osawa M. 2003. Formate, an active intermediate for direct oxidation of methanol on Pt electrode. J Am Chem Soc 125: 3680–3681.

Chen Y-X, Heinen M, Jusys Z, Behm RJ. 2006a. Bridge-bonded formate: Active intermediate or spectator in formic acid oxidation on a Pt film electrode? Langmuir 22: 10399–10408.

Chen Y-X, Heinen M, Jusys Z, Behm RJ. 2006b. Kinetics and mechanism of formic acid electrooxidation—Spectro-electrochemical studies in a novel flow cell configuration. Angew Chem Int Ed 45: 981–985.

Chen Y-X, Ye S, Heinen M, Jusys Z, Osawa M, Behm RJ. 2006c. Application of in-situ ATR–FTIR spectroscopy for the understanding of complex reaction mechanism and kinetics: Formic acid oxidation on a Pt electrode at elevated temperatures. J Phys Chem B 110: 9534–9544.

Chen Y-X, Heinen M, Jusys Z, Behm RJ. Dissociative adsorption and oxidation of formaldehyde on a Pt film electrode under controlled mass-transport conditions, an in-situ spectro-electrochemical flow-cell study. To be published.

Chen Y-X, Heinen M, Jusys Z, Behm RJ. Unpublished results.

Childers CL, Huang H, Korzeniewski C. 1999. Formaldehyde yields from methanol electrochemical oxidation on carbon-supported platinum catalysts. Langmuir 15: 786–789.

Christensen PA, Hamnett A, Weeks SA. 1988. In-situ FTIR study of adsorption and oxidation of methanol on platinum and platinized glassy carbon electrodes in sulphuric acid solution. J Electroanal Chem 250: 127–142.

Cohen JL, Volpe DJ, Abruña, HD. 2008. Electrochemical determination of activation energies for methanol oxidation on polycrystalline platinum in acidic and alkaline electrolytes. Phys Chem Chem Phys 9: 49–77.

Coutanceau C, Hahn F, Waszczuk P, Wieckowski A, Lamy C, Léger J-M. 2002. Radioactive labeling study and FTIR measurements of methanol adsorption and oxidation on fuel cell catalysts. Fuel Cells 2: 153–158.

Cuesta A. 2006. At least three contiguous atoms are necessary for CO formation during methanol electrooxidation on platinum. J Am Chem Soc 128: 13332–13333.

de Lima RB, Massafera MP, Batista EA, Iwasita T. 2007. Catalysis of formaldehyde oxidation by electrodeposits of PtRu. J Electroanal Chem 603: 142–148.

Desai SK, Neurock M, Kourtakis K. 2002. A periodic density functional theory analysis of the dehydrogenation of methanol over Pd(111). J Phys Chem B 106: 2559–2568.

Entina VS, Petrii OA, Rysikova VT. 1967. On the nature of products of methanol chemisorption on Pt + Ru electrode surface. Elektrokhimiya 3: 758–761.

Fan Q, Pu C, Ley KL, Smotkin ES. 1996. In situ FTIR–diffuse reflectance spectroscopy of the anode surface in a direct methanol fuel cell. J Electrochem Soc 143: L21–L23.

Feliu JM, Herrero E. 2003. Formic acid oxidation. In: Vielstich W, Gasteiger HA, Lamm A, eds. Handbook of Fuel Cells. Volume 2: Electrocatalysis. Chichester: Wiley. pp. 625–634.

Franaszczuk K, Herrero E, Zelenay P, Wieckowski A, Wang J, Masel RI. 1992. A comparison of electrochemical and gas-phase decomposition of methanol on platinum surfaces. J Phys Chem 96: 8509–8516.

Friedrich KA, Henglein F, Stimming U, Unkauf W. 2000. Size dependence of the CO monolayer oxidation on nanosized Pt particles supported on gold. Electrochim Acta 45: 3283–3293.

Gao L, Huang H, Korzeniewski C. 2004. The efficiency of methanol conversion to CO_2 on thin films of Pt and PtRu fuel cell catalysts. Electrochim Acta 49: 1281–1287.

Gasteiger HA, Markovic NM, Ross PN, Cairns EJ. 1993. Methanol electrooxidation on well-characterized Pt-Ru alloys. J Phys Chem 97: 12020–12029.

Gasteiger HA, Markovic NM, Ross PN, Cairns EJ. 1994. Electro-oxidation of small organic molecules on well-characterized Pt-Ru alloys. Electrochim Acta 39: 1825–1832.

Gasteiger HA, Markovic NM, Ross PN. 1995. H_2 and CO electrooxidation on well-characterized Pt, Ru, and Pt-Ru. 1. Rotating disk electrode studies of the pure gases including temperature effects. J Phys Chem 99: 8290–8301.

Gavrilov AN, Savinova ER, Simonov PA, Zaikovskii VI, Cherepanova SV, Tsirlina GA, Parmon VN. 2007. On the influence of the metal loading on the structure of carbon-supported PtRu catalysts and their electrocatalytic activities in CO and methanol electrooxidation. Phys Chem Chem Phys 9: 5476–5489.

Ge Q, Desai S, Neurock M, Kourtakis K. 2001. CO adsorption on Pt-Ru surface alloys and on the surface of Pt-Ru bulk alloy. J Phys Chem B 106: 9533–9536.

Gilman S. 1963. The mechanism of electrochemical oxidation of carbon monoxide and methanol on platinum. I. Carbon monoxide adsorption and desorption and simultaneous oxidation of the platinum surface at constant potential. J Phys Chem 67: 1989–1905.

Gilman S. 1964. The mechanism of electrochemical oxidation of carbon monoxide and methanol on platinum. II. The "reactant pair" mechanism for the electrochemical oxidation of carbon monoxide and methanol. J Phys Chem 68: 70–80.

Greeley J, Mavrikakis M. 2002. A first-principles study of methanol decomposition on Pt(111). J Am Chem Soc 124: 7193–7201.

Greeley J, Mavrikakis M. 2004. Competitive paths for methanol decomposition on Pt(111). J Am Chem Soc 126: 3910–3909.

Gromyko VA, Khazova OA, Vassiliev YB. 1976. The differences in adsorption and electrocatalytic properties of smooth and platinized platinum. Elektrokhimiya 12: 1352–1357.

Gustavsson M, Fredriksson H, Kasemo B, Jusys Z, Jun C, Behm RJ. 2004. Nanostructured platinum-on-carbon model electrocatalysts prepared by colloidal lithography. J Electroanal Chem 568: 371–377.

Hamnett A, Weeks SA, Kennedy BJ, Troughton G, Christensen PA. 1990. Long-term poisoning of methanol anodes. Ber Bunsenges Phys Chem 94: 1014–1020.

Hartnig C, Spohr E. 2005. The role of water in the initial steps of methanol oxidation on Pt(111). Chem Phys Lett 319: 185–191.

Hartnig C, Grimminger J, Spohr E. 2007a. Adsorption of formic acid on Pt(111) in the presence of water. J Electroanal Chem 607: 133–139.

Hartnig C, Grimminger J, Spohr E. 2007b. The role of water in the initial steps of methanol oxidation on Pt(2 1 1). Electrochim Acta 52. 2236–2243.

Heinen M, Chen YX, Jusys Z, Behm RJ. 2007. In situ ATR–FTIRS coupled with on-line DEMS under controlled mass transport conditions—A novel tool for electrocatalytic reaction studies. Electrochim Acta 52: 5634–5643.

Heinen M, Jusys Z, Behm RJ. 2009. Reaction pathways analysis and reaction intermediate detection via simultaneous differential electrochemical mass spectrometry (DEMS) and attenuated total reflection Fourier transform infrared spectroscopy (ATR-FTIRS). In: Vielstich W, Gasteiger HA, Yokokawa H, eds. Handbook of Fuel Cells. Volume 5: Advances in Electrocatalysis. Chichester: John Wiley & Sons, Ltd., in press.

Herrero E, Franaszczuk K, Wieckowski A. 1994. Electrochemistry of methanol at low index crystal planes of platinum: An integrated voltammetric and chronoamperometric study. J Phys Chem 98: 5074–5083.

Herrero E, Chrzanowski W, Wieckowski A. 1995. Dual path mechanism in methanol oxidation on a platinum electrode. J Phys Chem 99: 10423–10424.

Housmans THM, Koper MTM. 2003. Methanol oxidation on stepped Pt[n(111) (110)] electrodes: A chronoamperometric study. J Phys Chem B 107: 8557–8567.

Housmans THM, Wonders AH, Koper MTM. 2006. Structure sensitivity of methanol electrooxidation pathways on platinum: An on-line electrochemical mass spectrometry study. J Phys Chem B 110: 10021–10031.

Housmans THM, Hermse CGM, Koper MTM. 2007. CO oxidation on stepped single crystal electrodes: A dynamic Monte Carlo study. J Electroanal Chem 607: 69–82.

Huang H, Korzeniewski C, Vijayaraghavan G. 2002. Quantification of CO_2 from electrochemical oxidation reactions with a strategy based on transmission infrared spectroscopy. Electrochim Acta 47: 3675–3679.

Islam M, Basnayake R, Korzeniewski C. 2007. A study of formaldehyde formation during methanol oxidation over PtRu bulk alloys and nanometer scale catalyst. J Electroanal Chem 599: 31–40.

Iwasita T. 2002. Electrocatalysis of methanol oxidation. Electrochim Acta 47: 3663–3674.

Iwasita T. 2003. Methanol and CO electrooxidation. In: Vielstich W, Lamm A, Gasteiger HA, eds. Handbook of Fuel Cells. Volume 2: Electrocatalysis. Chichester: Wiley. pp. 603–624.

Iwasita T, Nart FC. 1997. In-situ infrared spectroscopy at electrochemical interfaces. Prog Surf Sci 55: 271–340.

Iwasita T, Vielstich W. 1986. On-line mass spectroscopy of volatile products during methanol oxidation at platinum in acid solutions. J Electroanal Chem 201: 403–408.

Iwasita T, Vielstich W, Santos E. 1987. Identification of the adsorbate during methanol oxidation. J Electroanal Chem 229: 367–376.

Iwasita T, Dalbeck R, Pastor E, Xia X. 1994. Progress in the study of electrocatalytic reactions of organic species. Electrochim Acta 39: 1817–1823.

Iwasita T, Nart FC, Lopez B, Vielstich W. 1992. On the study of adsorbed species at platinum from methanol, formic acid and reduced carbon dioxide via in situ FTIR spectroscopy. Electrochim Acta 37: 2361–2367.

Iwasita T, Xia X, Herrero E, Liess H-D. 1996. Early stages during the oxidation of HCOOH on single-crystal Pt electrodes as characterized by infrared spectroscopy. Langmuir 12: 4260–4265.

Jambunathan K, Jayaraman S, Hillier AC. 2004. A multielectrode electrochemical and scanning differential electrochemical mass spectrometry study of methanol oxidation on electrodeposited Pt_xRu_y. Langmuir 20: 1856–1863.

Jarvi TD, Sriramulu S, Stuve EM. 1998. Reactivity and extent of poisoning during methanol electrooxidation on platinum (100) and (111): A comparative study. Colloids Surf A 134: 145–153.

Jarvi TD, Stuve EM. 1998. Fundamental aspects of the vacuum und electrocatalytic reactions of methanol and formic acid on platinum surfaces. In: Lipkowski J, Ross PN, eds. Electrocatalysis. New York: Wiley-VCH. pp. 75–153.

Jiang J, Kucernak A. 2002. Nanostructured platinum as an electrocatalyst for the electrooxidation of formic acid. J Electroanal Chem 520: 64–70.

Jusys Z. 1994. H/D substitution effect on formaldehyde oxidation rate at a copper anode in alkaline medium studied by differential electrochemical mass spectrometry. J Electroanal Chem 375: 257–262.

Jusys Z, Behm RJ. 2001. Methanol oxidation on a carbon supported Pt fuel cell catalyst—A kinetic and mechanistic study by differential electrochemical mass spectrometry. J Phys Chem B 105: 10874–10883.

Jusys Z, Massong H, Baltruschat H. 1999. A new approach for simultaneous DEMS and EQCM: Electro-oxidation of adsorbed CO on Pt and Pt-Ru. J Electrochem Soc 146: 1093–1098.

Jusys Z, Kaiser J, Behm RJ. 2001. Electrooxidation of CO and H_2/CO mixtures on a carbon supported Pt catalyst—A kinetic and mechanistic study by differential electrochemical mass spectrometry. Phys Chem Chem Phys 3: 4650–4660.

Jusys Z, Kaiser J, Behm RJ. 2002a. Composition and activity of high surface area PtRu catalysts towards adsorbed CO and methanol electrooxidation. A DEMS study. Electrochim Acta 47: 3693–3706.

Jusys Z, Schmidt TJ, Dubau L, Lasch K, Jörissen L, Garche J, Behm RJ. 2002b. Activity of $PtRuMeO_x$ (Me = W, Mo or V) catalysts towards methanol oxidation and their characterization. J Power Sources 105: 297–304.

Jusys Z, Kaiser J, Behm RJ. 2003. Methanol electrooxidation over Pt/C fuel cell catalysts—Dependence of product yields on catalyst loading. Langmuir 19: 6759–6769.

Kazarinov VE, Tysyachnaya GY, Andreev VN. 1975. On the reasons for the discrepancies in the data on methanol adsorption on platinum. J Electroanal Chem 65: 391–400.

Koper MTM, Jansen APJ, van Santen RA. 1998. Monte Carlo simulations of a simple model for the electrocatalytic CO oxidation on platinum. J Chem Phys 109: 6051–6062.

Korzeniewski C, Childers CL. 1998. Formaldehyde yields from methanol electrochemical oxidation on platinum. J Phys Chem B 102: 489–492.

Korzeniewski C, Kardash D. 2001. Use of a dynamic Monte Carlo simulation in the study of nucleation-and-growth models for CO electrochemical oxidation. J Phys Chem B 105: 8663–8671.

Lamy C, Léger J-M, Clavilier J, Parsons R. 1983. Structural effects in electrocatalysis: A comparative study of the oxidation of CO, HCOOH and CH_3OH on single crystal Pt electrodes. J Electroanal Chem 150: 71–77.

Lanova B, Wang H, Baltruschat H. 2006. Methanol oxidation on carbon supported Pt and Ru-modified Pt nanoparticles: A comparison with single crystal and polycrystalline electrodes. Fuel Cells 3/4: 214–224.

Lebedeva NP, Koper MTM, Feliu JM, van Santen RA. 2000. Role of crystalline defects in electrocatalysis: Mechanism and kinetics of CO adlayer oxidation on stepped polatinum electrodes. J Phys Chem B 106: 12938–12947.

Lebedeva NP, Koper MTM, Feliu JM, van Santen RA. 2002. Mechanism and kinetics of the electrochemical CO adlayer oxidation on Pt(111). J Electroanal Chem 524–525: 242–251.

Leung L-WH, Weaver MJ. 1990. Influence of adsorbed carbon monoxide on the electrocatalytic oxidation of simple organic molecules at platinum and palladium electrodes in acidic solution: A survey using real-time FTIR spectroscopy. Langmuir 6: 323–333.

Levenspiel O. 1972. Chemical Reaction Engineering. New York: Wiley.

Lima A, Coutanceau C, Léger J-M, Lamy C. 2001. Investigation of ternary catalysts for methanol electrooxidation. J Appl Electrochem 31: 379–386.

Lin W-F, Wang J-T, Savinell RF. 1997. On-line FTIR spectroscopic investigations of methanol oxidation in a direct methanol fuel cell. J Electrochem Soc 144: 1917–1922.

Lindström RW, Seidel YE, Jusys Z, Gustavsson M, Kasemo B, Behm RJ. Submitted for publication.

Lopes MIS, Beden B, Hahn F, Léger J-M, Lamy C. 1991. On the nature of the adsorbates resulting from the chemisorption of methanol at a platinum electrode in acid medium: An EMIRS study. J Electroanal Chem 313: 323–339.

Loucka T, Weber J. 1968. Adsorption and oxidation of formaldehyde at the platinum electrode in acid solutions. J Electroanal Chem 21: 329–344.

Lu GQ, Crown A, Wieckowski A. 1999. Formic acid decomposition on polycrystalline platinum and palladized platinum electrodes. J Phys Chem B 103: 9700–9711.

Mai C-F, Shue C-H, Yang Y-C, Yang L-YO, Yau S-L, Itaya K. 2005. Adsorption of formaldehyde on Pt(111) and Pt(100) electrodes: Cyclic voltammetry and scanning tunneling microscopy. Langmuir 21: 4964–4970.

Maillard F, Eikerling M, Cherstiouk OV, Schreier S, Savinova ER, Stimming U. 2004a. Size effects on reactivity of Pt nanoparticles in CO monolayer oxidation: The role of surface mobility. Faraday Discuss 125: 357–377.

Maillard F, Savinova ER, Simonov PA, Zaikovskii VI, Stimming U. 2004b. Infrared spectroscopic study of CO adsorption and electro-oxidation on carbon-supported Pt nanoparticles: Interparticle versus intraparticle heterogeneity. J Phys Chem B 108: 17893–17904.

Maillard F, Schreier S, Hanzlik M, Savinova ER, Weinkauf S, Stimming U. 2005. Influence of particle agglomeration on the catalytic activity of carbon-supported Pt nanoparticles in CO monolayer oxidation. Phys Chem Chem Phys 7: 385–393.

Markovic NM, Ross PN, Jr. 2002. Surface science studies of model fuel cell electrocatalysts. Surf Sci Rep 45: 117–229.

Miki A, Ye S, Osawa M. 2002. Surface-enhanced IR absorption on platinum nanoparticles: an application to real-time monitoring of electrocatalytic reactions. Chem Commun (14): 1500–1501.

Miki A, Ye S, Sensaki T, Osawa M. 2004. Surface-enhanced infrared study of catalytic electrooxidation of formaldehyde, methyl formate, and dimethoxymethane on platinum electrodes in acidic solution. J Electroanal Chem 563: 23–31.

Mishina E, Karantonis A, Yu Q-K, Nakabayashi S. 2002. Optical second harmonic generation during the electrocatalytic oxidation of formaldehyde on Pt(111): Potentiostatic regime versus galvanostatic potential oscillations. J Phys Chem B 106: 10199–10204.

Nakabayashi S, Tamura M, Uosaki K. 1998. Kinetic coupling of formaldehyde oxidation and oxide formation on a platinum electrode. Bull Chem Soc Jpn 71: 67–71.

Nakabayashi S. 1998. Reaction pathway of four-electron oxidation of formaldehyde on platinum electrode as observed by in situ optical spectroscopy. Surf Sci 386: 82–88.

Nakamura M, Shibutani K, Hoshi N. 2007. In-situ flow-cell IRAS observation of intermediates during methanol oxidation on low-index platinum surfaces. ChemPhysChem 8: 1846–1849.

Napporn WT, Laborde H, Léger J-M, Lamy C. 1995. Electro-oxidation of C_1 molecules at Pt-based catalysts highly dispersed into a polymer matrix: Effect of the method of preparation. J Electroanal Chem 404: 153–159.

Neergat M, Seiler T, Savinova ER, Stimming U. 2006. Improvement of the performance of a direct methanol fuel cell using a pulse technique. J Electrochem Soc 153: A997–A1003.

Nichols RJ, Bewick A. 1988. SNIFTIRS with a flow cell: The identification of the reaction intermediates in methanol oxidation at Pt anodes. Electrochim Acta 33: 1691–1694.

Okamoto Y, Sugino O, Mochizuki Y, Ikeshoji T, Morikawa Y. 2003. Comparative study of dehydrogenation of methanol at Pt(111)/water and Pt(111)/vacuum interfaces. Chem Phys Lett 377: 236–242.

Okamoto H, Kon W, Mukouyama Y. 2004. Stationary voltammogram for oxidation of formic acid on polycrystalline platinum. J Phys Chem B 108: 4432–4438.

Okamoto H, Kon W, Mukouyama Y. 2005. Five current peaks in voltammograms for oxidation of formic acid, formaldehyde, and methanol on platinum. J Phys Chem B 109: 15659–15666.

Olivi P, Bulhoes LOS, Léger J-M, Hahn F, Beden B, Lamy C. 1994. New results in the electrooxidation of formaldehyde on a platinum electrode in an acid medium. J Electroanal Chem 370: 241–249.

Ota K-I, Nakagawa Y, Takahashi M. 1984. Reaction products of anodic oxidation of methanol in sulfuric acid solution. J Electroanal Chem 179: 179–186.

Park S, Tong YY, Wieckowski A, Weaver MJ. 2002a. Infrared spectral comparison of electrochemical carbon monoxide adlayers formed by direct chemisorption and methanol dissociation on carbon-supported platinum nanoparticles. Langmuir 18: 3233–3240.

Park S, Xie Y, Weaver MJ. 2002b. Electrocatalytic pathways on carbon supported platinum nanoparticles: Comparison of particle-size-dependent rates of methanol, formic acid and formaldehyde electrooxidation. Langmuir 18: 5792–5798.

Parsons R, VanderNoot T. 1988. The oxidation of small organic molecules: A survey of recent fuel cell related research. J Electroanal Chem 257: 9–45.

Pastor E, Castro CM, Rodriguez JL, Gonzalez S. 1996. On-line mass spectrometric studies on the interaction between organic adlayers on platinum. Part 1. Consecutive adsorption of formic acid and propargyl alcohol. J Electroanal Chem 404: 77–88.

Perez JM, Munoz E, Morallón E, Cases F, Vazquez JL, Aldaz A. 1994. Formation of CO during adsorption on platinum electrodes of methanol, formaldehyde, ethanol and acetaldehyde in carbonate medium. J Electroanal Chem 368: 285–291.

Petukhov AV, Akemann W, Friedrich KA, Stimming U. 1998. Kinetics of electrooxidation of a CO monolayer at the platinum/electrolyte interface. Surf Sci 402–404: 182–186.

Petukhova RP, Stenin VF, Podlovchenko BI. 1977. About the composition of the products of methanol electrooxidation on smooth platinum. Elektrokhimiya 14: 755–756.

Radmilovic V, Gasteiger HA, Ross PN. 1995. Structure and chemical composition of a supported Pt-Ru electrocatalyst for methanol oxidation. J Catal 154: 98–106.

Rao V, Cremers C, Stimming U, Cao L, Sun S, Yan S, Sun G, Xin Q. 2007. Electrooxidation of ethanol at gas diffusion electrodes a DEMS study. J Electrochem Soc 154: B1138–B1147.

Rice C, Tong YY, Oldfield E, Wieckowski A, Hahn F, Gloaguen F, Léger J-M, Lamy C. 2000. In situ infrared study of carbon monoxide adsorbed onto commercial fuel-cell-grade carbon-supported platinum nanoparticles: correlation with ^{13}C NMR results. J Phys Chem B 104: 5803–5807.

Roen LM, Paik CH, Jarvi TD. 2004. Electrocatalytic corrosion of carbon support in PEMFC cathodes. Electrochem Sol Lett 7: 19–22.

Samjeské G, Osawa M. 2005. Current oscillations during formic acid oxidation on a Pt electrode: Insight into the mechanism by time-resolved IR spectroscopy. Angew Chem Int Ed 44: 5694.

Samjeské G, Miki A, Ye S, Yamakata A, Mukouyama Y, Okamoto H, Osawa M. 2005. Potential oscillations in galvanostatic electrooxidation of formic acid on platinum: A time-resolved surface-enhanced infrared study. J Phys Chem B 109: 23509–23516.

Samjeské G, Miki A, Ye S, Osawa M. 2006. Mechanistic study of electrocatalytic oxidation of formic acid at platinum in acidic solution by time-resolved surface-enhanced infrared absorption spectroscopy. J Phys Chem B 110: 16559–16566.

Samjeské G, Miki A, Osawa M. 2007. Electrocatalytic oxidation of formaldehyde on platinum under galvanostatic and potential sweep conditions studied by time-resolved surface-enhanced infrared spectroscopy. J Phys Chem 111: 15074–15083.

Sanicharane S, Bo A, Sompalli B, Gurau B, Smotkin ES. 2002. In-situ 50 °C FTIR spectroscopy of Pt and PtRu direct methanol fuel cell membrane electrode assembly anodes. J Electrochem Soc 149: A554–A557.

Santos E, Leiva EPM, Vielstich W. 1991. CO adsorbate on Pt(111) single crystal surfaces. Electrochim Acta 36: 555–561.

Schmidt TJ, Gasteiger HA, Stäb GD, Urban PM, Kolb DM, Behm RJ. 1998. Characterization of high-surface area electrocatalysts using a rotating disk electrode configuration. J Electrochem Soc 145: 2354–2358.

Schmidt TJ, Gasteiger HA, Behm RJ. 1999a. Methanol electrooxidation on a colloidal PtRu alloy fuel cell catalyst. Electrochem Commun 1: 1–4.

Schmidt TJ, Gasteiger HA, Behm RJ. 1999b. Rotating disk electrode measurements on a high-surface area Pt/Vulcan carbon fuel cell catalyst. J Electrochem Soc 146: 1296–1304.

Schmidt TJ, Behm RJ, Grgur BN, Markovic NM, Ross PN. 2000. Formic acid oxidation on pure and bi-modified Pt(111): Temperature effects. Langmuir 16: 8159–8166.

Schneider A, Colmenares L, Seidel YE, Jusys Z, Wickman B, Kasemo B, Behm RJ. 2008. Transport effects in the oxygen reduction reaction on nanostructured, planar glassy carbon supported Pt/GC model electrodes. Phys Chem Chem Phys 10: 1931–1943.

Seiler T, Savinova ER, Friedrich KA, Stimming U. 2004. Poisoning of PtRu/C catalysts in the anode of a direct methanol fuel cell: A DEMS study. Electrochim Acta 49: 3927–3936.

Shibata M, Furuya N, Watanabe M. 1987. Electrocatalysis by ad-atoms. Part XXI. Catalytic effects on the elementary steps in methanol oxidation by non-oxygen-adsorbing ad-atoms. J Electroanal Chem 229: 385–394.

Shibata M, Motoo S. 1986. Electrocatalysis by ad-atoms. Part XX. Rate determining step in methanol oxidation enhanced by oxygen-adsorbing ad-atoms. J Electroanal Chem 209: 151–158.

Shin J, Tornquist WJ, Korzeniewski C, Hoaglund CS. 1996. Elementary steps in the oxidation and dissociative chemisorption of ethanol on smooth and stepped surface planes of platinum electrodes. Surf Sci 364: 122–130.

Sidheswaran P, Lal H. 1971. A study of intermediates adsorbed on platinized platinum during the anodic oxidation of formaldehyde. J Electroanal Chem 34: 173–183.

Sobkowski J, Wieckowski A. 1972. A new approach to the radiometric study of methanol adsorption on platinum. J Electroanal Chem 34: 185–189.

Spasojevic MD, Adzic RR, Despic AR. 1980. Electrocatalysis on surfaces modified by foreign metal adatoms: Oxidation of formaldehyde on platinum. J Electroanal Chem 109: 261–269.

Spendelow JS, Xu Q, Goodpaster JD, Kenis PJA, Wieckowski A. 2007. The role of surface defects in CO oxidation, methanol oxidation, and oxygen reduction on Pt(111). J Electrochem Soc 154: F238–F242.

Sriramulu S, Jarvi TD, Stuve EM. 1998. A kinetic analysis of distinct reaction pathways in methanol electrocatalysis on Pt(111). Electrochim Acta 44: 1127–1134.

Sriramulu S, Jarvi TD, Stuve EM. 1999. Reaction mechanism and dynamics of methanol electrooxidation on platinum (111). J Electroanal Chem 467: 132–142.

Stadler R, Jusys Z, Baltruschat H. 2002. Hydrogen evolution during the oxidation of formaldehyde on Au: The influence of single crystal structure and Tl-upd. Electrochim Acta 47: 4485–4500.

Stonehart P, Kohlmayr G. 1972. Effect of poisons on the kinetic parameters for platinum electrocatalyst sites. Electrochim Acta 17: 369–382.

Sun S-G. 1998. Studying electrocatalytic oxidation of small organic molecules with in-situ infrared spectroscopy. In: Lipkowski J, Ross PN, editors. Electrocatalysis. New York: Wiley-VCH. pp. 243–290.

Sun S-G, Clavilier J. 1987. Electrochemical study on the poisoning intermediate formed from methanol dissociation at low-index and stepped platinum surfaces. J Electroanal Chem 236: 95–112.

Sun S-G, Yang Y-Y. 1999. Studies of kinetics of HCOOH oxidation on Pt(100), Pt(110), Pt(111), Pt(510) and Pt(911) single crystal electrodes. J Electroanal Chem 467: 121–131.

Sun S-G, Clavilier J, Bewick A. 1988. The mechanism of electrocatalytic oxidation of formic acid on Pt(100) and Pt(111) in sulphuric acid solution: An EMIRS study. J Electroanal Chem 240: 147–159.

Sun S-G, Lin Y, Li N-H, Mu J-Q. 1994. Kinetics of dissociative adsorption of formic acid on Pt(100), Pt(610), Pt(210) and Pt(110) single crystal electrodes in perchloric acid solutions. J Electroanal Chem 370: 273–280.

Tarnowski DJ, Korzeniewski C. 1997. Effects of surface step density on the electrochemical oxidation of ethanol to acetic acid. J Phys Chem B 101: 253–258.

Taylor CD, Neurock M. 2005. Theoretical insights into the structure and reactivity of the aqueous/metal interface. Curr Opin Solid State Mater Sci 9: 49–65.

ten Kortenaar MV, Tessont C, Kolar ZI, van der Weijde H. 1999. Anodic oxidation of formaldehyde on gold studied by electrochemical impedance spectroscopy: An equivalent circuit approach. J Electrochem Soc 146: 2146–2155.

ten Kortenaar MV, Kolar ZI, de Goeij JJM, Frens G. 2001. Electrocatalytic oxidation of formaldehyde on gold studied by differential electrochemical mass spectrometry and voltammetry. J Electrochem Soc 148: E327–E335.

Thomas JM, Thomas WJ. 1997. Principles and Practice of Heterogeneous Catalysis. Weinheim: VCH.

Tkach I, Panchenko A, Kaz T, Gogel V, Friedrich KA, Roduner E. 2004. In situ study of methanol oxidation on Pt and Pt/Ru-mixed with Nafion anodes in a direct methanol fuel cell by means of FTIR spectroscopy. Phys Chem Chem Phys 6: 5419–5426.

Vielstich W, Xia XH. 1995. Comments on "Electrochemistry of methanol at low index crystal planes of platinum: An integrated voltammetric and chronoamperometric study." J Phys Chem 99: 10421–10422.

Wang H, Baltruschat H. 2007. DEMS study on methanol oxidation at poly- and monocrystalline platinum electrodes: The effect of anion, temperature, surface structure, Ru adatom, and potential. J Phys Chem C 111: 7038–7048.

Wang H, Löffler T, Baltruschat H. 2001a. Formation of intermediates during methanol oxidation: A quantitative DEMS study. J Appl Electrochem 31: 759–765.

Wang H, Wingender Ch, Baltruschat H, Lopez M, Reetz MT. 2001b. Methanol oxidation on Pt, PtRu, and colloidal Pt electrocatalysts: A DEMS study of product formation. J Electroanal Chem 509: 163–169.

Wang H, Jusys Z, Behm RJ. 2004. Ethanol electrooxidation on a carbon-supported Pt catalyst: Reaction kinetics and product yields. J Phys Chem B 108: 19413–19424.

Wang H, Jusys Z, Behm RJ. 2006. Electrooxidation of acetaldehyde on carbon-supported Pt, PtRu and Pt$_3$Sn and unsupported PtRu$_{0.2}$ catalysts: A quantitative DEMS study. J Appl Electrochem 36: 1187–1198.

Wasmus S, Wang J-T, Savinell RF. 1995. Real-time mass spectrometric investigation of the methanol oxidation in a direct methanol fuel cell. J Electrochem Soc 142: 3825–3833.

Waszczuk P, Wieckowski A, Zelenay P, Gottesfeld S, Coutanceau C, Léger J-M, Lamy C. 2001. Adsorption of CO poison on fuel cell nanoparticle electrodes from methanol solutions: A radioactive labeling study. J Electroanal Chem 511: 55–64.

Watanabe M, Motoo S. 1975. Electrocatalysis by ad-atoms. Part II. Enhancement of the oxidation of methanol on platinum by ruthenium adatoms. J Electroanal Chem 60: 267–273.

Wilhelm S, Vielstich W, Buschmann HW, Iwasita T. 1987. Direct proof of the hydrogen in the methanol adsorbate at platinum—An ECTDMS study. J Electroanal Chem 229: 377–384.

Willsau J, Heitbaum J. 1984. The influence of Pt-activation on the corrosion of carbon in gas diffusion electrodes: A DEMS study. J Electroanal Chem 161: 93–101.

Willsau J, Heitbaum J. 1986. Analysis of adsorbed intermediates and determination of surface potential shifts by DEMS. Electrochim Acta 31: 943–948.

Willsau J, Wolter O, Heitbaum J. 1985. On the nature of the adsorbate during methanol oxidation at platinum: A DEMS study. J Electroanal Chem 185: 163–170.

Wolter O, Heitbaum J. 1984. The adsorption of CO on a porous Pt-electrode in sulfuric acid studied by DEMS. Ber Bunsenges Phys Chem 88: 6–10.

Wolter O, Willsau J, Heitbaum J. 1985. Reaction pathways of the anodic oxidation of formic acid on Pt evidenced by ^{18}O labeling—A DEMS study. J Electrochem Soc 132: 1635–1638.

Wonders AH, Housmans THM, Rosca V, Koper MTM. 2006. On-line mass spectrometry systems for measurements at single-crystal electrodes in hanging meniscus configuration. J Appl Electrochem 36: 1215–1221.

Xia XH. 1999. New insights into the influence of upd Sn on the oxidation of formic acid on platinum in acidic solution. Electrochim Acta 45: 1057–1066.

Xia XH, Iwasita T, Ge F, Vielstich W. 1997. Structural effects and reactivity in methanol solution on polycrystalline and single-crystal platinum. Electrochim Acta 41: 711–718.

Yajima T, Uchida H, Watanabe M. 2004. In-situ ATR-FTIR spectroscopic study of electo-oxidation of methanol and adsorbed CO at Pt-Ru alloy. J Phys Chem B 108: 2654–2659.

Yang Y-Y, Sun S-G. 2002. Effects of Sb adatoms on kinetics of electrocatalytic oxidation of HCOOH at Sb-modified Pt(100), Pt(111), Pt(110), Pt(320), and Pt(331) surfaces: An energetic modeling and quantitative analysis. J Phys Chem B 106: 12499–12507.

Zhu Y, Uchida H, Yajima T, Watanabe M. 2001. Attenuated total reflection–Fourier transform infrared study of methanol oxidation on sputtered Pt film electrode. Langmuir 17: 146–154.

CHAPTER 14

The Effect of Structurally Well-Defined Pt Modification on the Electrochemical and Electrocatalytic Properties of Ru(0001) Electrodes

H.E. HOSTER and R.J. BEHM

Institute of Surface Chemistry and Catalysis, Ulm University, D-89069 Ulm, Germany

This chapter is dedicated to Professor Teresa Iwasita on the occasion of her 65th birthday and in recognition of her contributions to electrocatalysis

14.1 INTRODUCTION

Carbon-supported PtRu catalysts have become the state-of-the-art anode catalysts in polymer electrolyte fuel cells (PEFCs) operated by CO-containing fuel gases [Petry et al., 1965] resulting, for example, from steam reforming of hydrocarbons or alcohols [Peppley et al., 2003], and in direct methanol fuel cells (DMFCs) [Hamnett, 2003]. Their higher CO tolerance and activity with respect to methanol oxidation compared with Pt catalysts has been attributed to a bifunctional effect, where facile formation of adsorbed oxygen species on Ru sites and their reaction with CO_{ad} adsorbed on neighboring Pt sites results in a reduction of the steady-state CO_{ad} coverage at significantly lower potentials than on Pt, and hence in improved CO tolerance [Watanabe and Motoo, 1975; Gasteiger et al., 1994a]. More recently, a reduced CO adsorption energy, and hence an increased tendency for CO_{ad} desorption, was proposed as a second effect contributing to the higher CO tolerance of these bimetallic catalysts [Buatier de Mongeot et al., 1998; Koper, 2004; Diemant et al., 2003; Liu et al., 2003]. The mechanistic aspects underlying the improved CO tolerance and methanol oxidation activity of PtRu catalysts have been investigated in a number of experimental and theoretical studies on bimetallic PtRu model systems [Gasteiger et al., 1994a, b; Binder et al., 1972; Stimming and

---

Fuel Cell Catalysis. Edited by Marc T. M. Koper
Copyright © 2009 John Wiley & Sons, Inc.

Vogel, 1998; Iwasita et al., 2000; Hoster et al., 2001a, b; Crown et al., 2001, 2002; Herrero et al., 1999; Brankovic et al., 2001a, 2002b; Markovic and Ross, 2002]. Most of the experimental model studies have been performed on bulk alloy substrates [Gasteiger et al., 1994a, b; Binder et al., 1972; Hoster et al., 2001a, b; Markovic and Ross, 2002] or on bimetallic electrode surfaces prepared by electrochemical or electroless deposition of Ru on Pt(111) [Stimming and Vogel, 1998; Iwasita et al., 2000; Crown et al., 2001, 2002; Herrero et al., 1999] or of Pt on Ru(0001) [Brankovic et al., 2001a, b, 2002a, b] substrates, respectively. These bimetallic electrodes are characterized by a large number of small deposit islands, which, depending on the amount of the respective material deposited, are mostly several layers high (multilayer islands) [Stimming and Vogel, 1998; Herrero et al., 1999; Brankovic et al., 2001a]. This morphology is very different from that of bimetallic (bulk) PtRu nanoparticles, where, depending on the pretreatment, either Pt and Ru atoms are expected to coexist on the surface or the bimetallic core of the nanoparticle is covered by a Pt skin layer [Nashner et al., 1997; Watanabe et al., 1999; Christoffersen et al., 2001; Han et al., 2006]. Such morphologies can be approximated in planar model systems by depositing a thin (sub)mono- or multilayer Pt film on a Ru substrate (Pt skin layer) [Brankovic et al., 2001b; Hoster et al., 2004] or by deposition of controlled amounts of Pt on a Ru substrate and subsequent controlled annealing to form a surface alloy, where the Pt is confined to the outermost layer [Hoster et al., 2008].

Aiming at a microscopic, atomic-scale understanding of the chemical properties of planar, bimetallic PtRu model systems, we recently investigated the interaction of CO, H_2 and H_2/CO with such structurally well-defined bimetallic Pt/Ru(0001) and PtRu/Ru(0001) surfaces under ultrahigh vacuum (UHV) conditions [Buatier de Mongeot et al., 1998; Rauscher et al., 2007]. The structure of each type of surface was quantitatively characterized by scanning tunneling microscopy (STM) [Hoster et al., 2008]. Results on the interaction of the above adsorbates with Ru(0001) substrates modified by different amounts of Pt monolayer islands or a Pt monolayer film were reported in [Buatier de Mongeot et al., 1998; Diemant et al., 2003, in preparation] (see also Schlapka et al. [2002, 2003]; Käsberger et al. [2003]; Jakob and Schlapka [2007]). Similar studies were also performed on the adsorption of CO and deuterium on PtRu monolayer surface alloys [Diemant et al., 2003, 2008, in preparation; Rauscher et al., 2007]. Based on atomic-resolution STM images with chemical contrast [Buatier de Mongeot et al., 1998; Schmid et al., 1993; Nielsen et al., 1995], the Pt surface atoms are almost randomly distributed in the surface layer of these samples [Hoster et al., 2008]. From these spectroscopic and structural data and their combination with results of theoretical studies [Liu et al., 2003; Christoffersen et al., 2001; Schlapka et al., 2003; Mavrikakis et al., 1998; Ge et al., 2001; Koper et al., 2002; Davies et al., 2005; Greeley and Mavrikakis, 2005; Groß, 2006; Lischka et al., 2007], we have been able to derive clear trends for the chemical properties of individual local PtRu nanostructures. The Pt-induced changes in the local chemical properties of these surfaces have been described and discussed in terms of (i) the geometric ensemble effect, which includes modifications in the adsorption/reaction properties due to variations in size and composition of the local adsorption ensemble, (ii) the electronic ligand effect, which describes modifications in the electronic structure of the adsorption ensemble due to different neighbors, and

(iii) electronic strain effects reflecting changes in the chemical properties due to modifications in the surface lattice geometry imposed by the substrate [Mavrikakis et al., 1998; Sachtler, 1973, 1997; Soma-Noto and Sachtler, 1974; Sachtler and Somorjai, 1983; Hammer et al., 1996; Gsell et al., 1998; Liu and Nørskov, 2001b].

In this chapter, we present and discuss results of a similar type of study on the fundamental electrochemical and electrocatalytic properties of structurally well-defined bimetallic PtRu surfaces, focusing on the effect of Pt modification of Ru(0001) substrates. Surface preparation and STM imaging were performed under UHV conditions. The electrochemical and electrocatalytic characterization occurred under controlled mass transport conditions in an electrochemical flow cell, which was attached to the UHV–STM chamber via a sample transfer system. In analogy to the UHV experiments, this study includes (i) Ru(0001) surfaces that are covered by pseudomorphic Pt monolayer islands or a closed Pt monolayer [Hoster et al., 2004] or (ii) PtRu/Ru(0001) monolayer surface alloys, where the Ru(0001) substrate is covered by a two-dimensional (2D) Pt_xRu_{1-x} alloy layer, with different Pt coverages or different amounts of Pt in the surface layer, respectively. For comparison, we include similar measurements on an unmodified Ru(0001) electrode. In all three cases, the surface layer rests on a Ru(0001) substrate, i.e., the neighbors in vertical direction are exclusively Ru atoms, while the lateral neighbors can be either Ru or Pt, depending on the respective type of model surface.

The chapter is organized as follows. After a brief description of the experimental setup and procedures, we will first focus on the electrochemical properties of the respective electrodes with regard to the adsorption/desorption of hydrogen and oxygenated species (OH_{ad}, O_{ad}) and the characteristic modifications introduced by the different types of Pt modification (Section 14.3.1). This includes a summary of the structural characteristics of the bimetallic electrodes and of their structure-dependent adsorption properties under UHV conditions determined previously. From a quantitative evaluation of the Pt coverage-dependent H_{upd} and OH_{ad} adsorption capacities of the bimetallic electrodes, we determine the nature of the stable adsorption sites on these surfaces. In Section 14.3.2, we describe the electrochemical bulk oxidation of CO on these model surfaces, focusing again on the effect of Pt modification on the activity of Ru(0001) electrodes. These will be discussed in comparison with data obtained under UHV conditions and with results from theoretical studies. We will finally discuss the consequences of the structure-dependent surface properties for the atomic-scale understanding of the activity of PtRu electrodes in general.

14.2 EXPERIMENTAL

14.2.1 Experimental Setup

The experiments were performed in a combined system for UHV and electrochemical measurements. It consists of a UHV system equipped with standard facilities for surface preparation and characterization and a pocket-size scanning tunneling microscope (STM) [Kopatzki, 1994], a pre-chamber containing a flow cell for electrochemical measurements, which was attached to the main UHV system via a gate valve, and

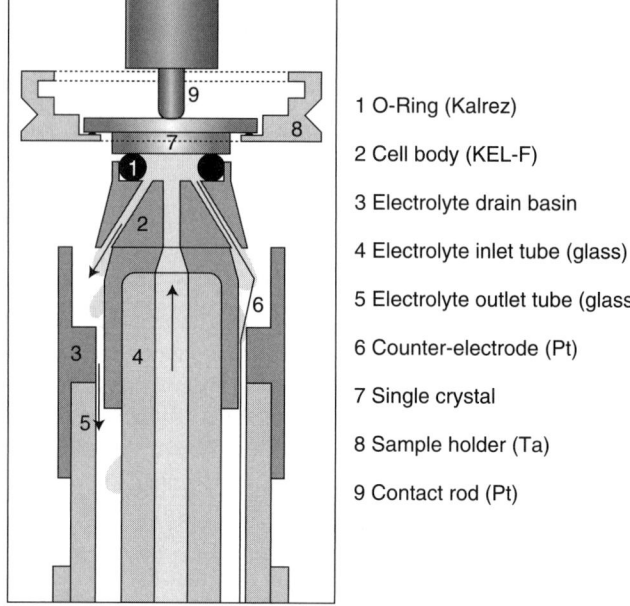

Figure 14.1 Scheme of the electrochemical flow cell used in the UHV-STM/EC transfer system.

facilities for the reversible sample transfer between UHV and electrochemical cell. The UHV system contained two metal evaporators for Pt and Ru evaporation, a cylindrical mirror analyzer (Physical Instruments) for Auger electron spectroscopy (AES), and a quadrupole mass spectrometer (Balzers QMS 112) for residual gas analysis. The single-crystal sample, which was shaped as a flat "hat," was mounted onto a tantalum or molybdenum sample holder with an outer diameter of 18 mm (see Fig. 14.1), which in turn was held in a precision sample manipulator. Sample heating was possible by radiation heating from a filament behind the sample or by electron bombardment, accelerating electrons from the filament to the sample. The latter allowed to flash anneal the sample to temperatures of up to 1900K (heating rate $\approx 100 \text{ K s}^{-1}$). The temperature was measured by an infrared pyrometer (Impac Infratherm IGA 140). For STM measurements, the sample was transferred from the manipulator to the STM by two wobble sticks.

14.2.2 Sample Preparation

The Ru(0001) surface was prepared by cycles of Ar^+ ion bombardment (sputtering) and annealing to temperatures >1100K, followed by oxygen adsorption/desorption cycles and final annealing to about 1750K to remove residual adsorbed oxygen [Buatier de Mongeot et al., 1998; Diemant et al., 2003]. After this treatment, the surface is characterized by atomically smooth terraces of 100–200 nm width separated by monolayer steps (see also Fig. 14.5a). Surface contamination was below the detection limit of AES (0.01 monolayers (ML)). Pt was evaporated from an electron beam evaporator (Omicron FOCUS EFM 3), with the substrate at 300–350K and typical deposition rates of about 0.25 ML min^{-1}. The resulting Pt coverages were analyzed by quantitative evaluation of STM images and by AES. In some cases, the Ru(0001) surface was covered with oxygen prior to Pt deposition to attain a higher Pt island density. Surface alloys were formed by flash annealing to 1350K.

14.2.3 Electrochemical Measurements

For the electrochemical measurements, sample and sample holder were transferred into a small pre-chamber via a magnetically coupled transfer rod. The pre-chamber was continuously pumped by a turbomolecular pump (Pfeiffer, TMU 71) to minimize possible contaminations of the clean surface or of the main chamber (typical working pressure 10^{-9} mbar). After the sample transfer, the pre-chamber was separated from the main chamber and the turbo pump by two gate valves and filled with clean nitrogen to ambient pressure. Subsequently, a miniaturized electrochemical flow cell, made of KEL-F and mounted on top of a glass tube, was moved up into the pre-chamber through an opened gate valve and brought into contact with the single crystal from below (Fig. 14.1b). The model electrode was then pressed to an O-ring gasket (made of Kalrez) on top of the flow cell by a small wobble stick, which also provided a direct electrical contact to the sample. A continuous electrolyte flow (about 0.5 mL s^{-1}) could be maintained for measurements requiring enhanced mass transport. Cyclic base voltammograms (base CVs) were recorded either in resting or in flowing electrolyte. The mass-transport-limited currents attained under electrolyte flow conditions are in the range of those reached by a rotating disk electrode at 900 rev min^{-1}, for example around 1.5 mA cm^{-2} for the electro-oxidation of CO [Markovic et al., 1999]. A reversible hydrogen electrode (RHE), which was coupled to the cell via the electrolyte inlet, was used as reference electrode. All potentials in this chapter are given with respect to that of the RHE electrode. Potential values from the work of other groups were converted into RHE potentials for easier comparison. In all measurements at Ru(0001) model electrodes, a total charge of one electron per surface atom equals 0.253 mC cm^{-2}. All integrated charges are corrected for a double layer contribution of 20 μF cm^{-2} as previously used for bare Ru(0001) [El-Aziz and Kibler, 2002].

14.3 RESULTS

14.3.1 Structure and Electrochemical Properties of Bare and Pt Film-Modified Ru(0001) Electrodes

14.3.1.1 Unmodified Ru(0001) Electrodes

Related UHV Studies The interaction of molecules relevant for the present purpose, such as H_2, O_2, and H_2O, with Ru(0001) surfaces under UHV conditions has been studied in much detail over the last decades [Diemant et al., 2003; Liu et al., 2003; Danielson et al., 1978; Schwarz, 1979; Shimizu et al., 1980; Feulner and Menzel, 1985; Christmann, 1988; Sun and Weinberg, 1985; Jachimowski et al., 1995; Lindroos et al., 1987; Held et al., 1992]. Hydrogen adsorbs dissociatively on Ru(0001) with an initial sticking coefficient of $s_0 = 0.21$ and a maximum coverage of 1 ML, with the H_{ad} atoms adsorbed on the fcc-type threefold-hollow sites [Feulner and Menzel, 1985; Lindroos et al., 1987; Held et al., 1992]. Comparing hydrogen adsorption on Ru(0001) and on Pt(111), the hydrogen adsorption energy is significantly higher on Ru(0001) than on Pt(111). Temperature programmed desorption (TPD) experiments and density functional theory (DFT) calculations yielded Ru–H bond energies of 2.86 eV [Feulner and Menzel, 1985; Christmann, 1988] and 2.70–2.97 eV [Liu et al., 2003; Greeley and Mavrikakis, 2005], respectively. For Pt(111), the experimental values are 2.59 eV [Gdowski et al., 1983] and 2.64 eV [Christmann, 1988], whereas DFT calculations lead to values of 2.5–2.56 eV [Liu et al., 2003; Greeley and Mavrikakis, 2005; Liu and Nørskov, 2001; Hammer and Nørskov, 1995]. Hence, both theory and experiment predict a higher stability of H_{ad} on Ru(0001) than on Pt(111), with a difference in the range between 150 and 300 meV.

Compared with hydrogen, the adsorption of oxygen on Ru(0001) is significantly stronger. Increasing exposures to O_2 lead to the sequential formation of $(2 \times 2)O$ [Madey and Engelhardt, 1975], $(2 \times 1)O$ [Madey and Engelhardt, 1975], $(2 \times 2)3O$ [Kostov et al., 1997], and $(1 \times 1)O$ [Stampfl et al., 1996] phases with coverages of 0.25, 0.5, 0.75, and 1.0 ML (or larger). Even at $\theta_O = 1$, desorption of oxygen was found to set in only at $T > 800K$ [Böttcher et al., 1997]. Ru–O bond energies range from 5.2 eV at $\theta_O = 0.25$ to 4.5 eV at $\theta_O = 1$ [Stampfl et al., 1999]. Under ambient conditions, Ru(0001) will be covered by a $(1 \times 1)O$ phase, since the thermodynamically more favorable oxide formation (RuO_2) is kinetically hindered [Assmann et al., 2003].

Water (D_2O) formation was shown to occur in UHV even at 80–90K upon exposure of (i) a Ru(0001)-$(2 \times 1)O$ phase to atomic hydrogen [Schick et al., 1996] or (ii) a Ru(0001)-$(1 \times 1)D$ phase to atomic oxygen [Weiss et al., 1998]. In contrast, the reaction of a Ru(0001)-$(2 \times 1)O$ phase with *molecular* H_2 (at 400–900K) was very slow (with a time scale of 10–100 minutes), which was attributed to a hindered dissociative H_2 adsorption [Shi et al., 1981].

H_2O adsorbed on Ru(0001) forms hydrogen-bonded adlayers [Thiel and Madey, 1987]. According to a recent DFT study, such adlayers should be metastable against partial dissociation [Feibelman, 2002; Weissenrieder et al., 2004]. Owing to the presence of kinetic barriers, however, dissociation sets in only at $T > 180K$, in competition with H_2O desorption [Andersson et al., 2004]. The formation of O_{ad} via $H_2O_{ad} \leftrightarrow OH_{ad} + H_{ad} \leftrightarrow O_{ad} + 2H_{ad}$ is kinetically limited by the simultaneous formation of site-blocking H_{ad}, which, because of the strong Ru–H bond, does not desorb at low temperatures [Weissenrieder et al., 2004]. Even in the presence of coadsorbed oxygen, H_2 desorption ranges from 130K up to 240K [Schiffer et al., 2000]. Thermodynamically, however, H_{ad} is metastable against displacement by O_{ad} because of the higher Ru–O bond energy. Similar effects are expected for the formation of OH_{ad} and O_{ad} in an acidic electrolyte via H_2O dissociation, which will likewise be kinetically hindered by strongly bound, coadsorbed H_{ad} (H_{upd}). At sufficiently high potentials, however, H_{upd} can be removed oxidatively as H^+, which allows also higher OH_{ad} and O_{ad} coverages to be formed.

Electrochemical Properties Typical cyclic voltammograms (CVs) of Ru(0001) in an $HClO_4$ electrolyte are shown in Fig. 14.2. The figure also includes information on (i) the ex situ low energy electron diffraction/reflection high energy electron diffraction (LEED/RHEED) patterns reported for emersion at three different potentials by Zei and Ertl [Zei and Ertl, 2000] and (ii) the anodic charges, relative to a surface free of adsorbed species that were determined by cyclic voltammetry in combination with CO displacement experiments by El-Aziz and Kibler [El-Aziz and Kibler, 2002]. A positive charge (H_{upd} displacement) was only observed when CO was offered at $E < 0.1$ V, indicating that only under these conditions is the adlayer dominated by

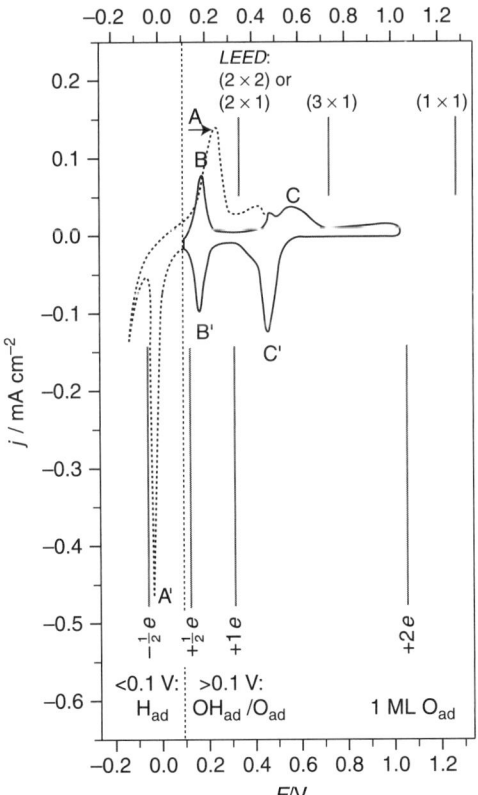

Figure 14.2 Cyclic base voltammograms of Ru(0001) in 0.1 M $HClO_4$, 50 mV s^{-1}: scan range 0.1–1.05 V (solid line) and −0.12–1.05 V (dotted line). Also indicated are LEED patterns reported in [Zei and Ertl, 2000] after emersion at 0.35, 0.75, and 1.2 V and anodic charges per surface atom transferred at different potentials according to El-Aziz and Kibler [2002].

adsorbed hydrogen (note that a potential sweep to $E < 0.1$ V was necessary to form this adlayer) [El-Aziz and Kibler, 2002]. In this way, a H_{upd} coverage of $\theta_H = 0.5-0.6$ ML was determined for $E = 0.08$ V [El-Aziz and Kibler, 2002]. For $E \geq 0.11$ V, negative charge transients were observed, which were attributed to the displacement of roughly 0.5–0.6 ML of OH_{ad} [Hoster et al., 2004; El-Aziz and Kibler, 2002]. Considering the additional positive charge accumulated in the positive-going scan up to a positive potential limit of 1.05 V, which results from H_2O dissociation, the Ru(0001) surface must be covered by approximately 1 ML O_{ad} at that potential [El-Aziz and Kibler, 2002]. This also fits to the (1×1) LEED/RHEED pattern found after emersing the electrode in this potential region [Zei and Ertl, 2000]. (It should be noted, however, that a (1×1) pattern would be observed also for a disordered adlayer, independent of the coverage.)

In the cathodic potential scan, the charge exchanged between 1.05 V and 0.3 V corresponds to about $1\,e^-$ per surface atom [El-Aziz and Kibler, 2002], equivalent to the reduction of 1 ML O_{ad} to 1 ML OH_{ad} or to 0.5 ML O_{ad}. Comparing with the (2×2) LEED/RHEED diffraction pattern, which was observed after emersion at 0.3 V [Zei and Ertl, 2000], we assign this to a $(2 \times 1)O$ phase with $\theta = 0.5$ ML O_{ad}, similar to that formed on Ru(0001) under UHV conditions [Pfnür et al., 1989]. In this case, the cathodic peak B' in the range 0.1–0.3 V with an integrated charge of 0.133 mC cm^{-2} (about $0.5\,e^-$ per Ru atom) corresponds to the reduction of 0.5 ML O_{ad} to a 0.5 ML OH_{ad} adlayer. Extending the potential scan to a lower limit in the hydrogen evolution region, a large cathodic peak A' (solid line in Fig. 14.2b) develops, which, after subtraction of the hydrogen evolution current, contains a charge of approximately $1.1\,e^-$ per surface atom [El-Aziz and Kibler, 2002]. Based on the CO displacement experiments described above, this peak must reflect the replacement of an OH adlayer by a more stable H adlayer [Hoster et al., 2004; El-Aziz and Kibler, 2002].

In the anodic scan, the oxidation of the H adlayer formed below 0.1 V and the re-formation of OH_{ad}/O_{ad} (both in peak A) are shifted to markedly higher potentials compared with the O_{ad}/OH_{ad} removal and H_{upd} formation (peak A') in the cathodic scan (Fig. 14.2b). Furthermore, it overlaps with the peak B (OH_{ad} oxidation) observed for a cathodic scan limit of 0.1 V. At low scan rates, peak A starts at 0.1–0.15 V and reaches up to 0.48 V. Hence, compared with a scan with a cathodic limit of $E > 0.1$ V, the equilibration of the O_{ad}/OH_{ad} adlayer is shifted from 0.28 to 0.48 V. The charge in peak A integrated in the range 0.1–0.48 V corresponds to $1.5\,e^-$ per surface atom, which is equal to the sum of the charges in peaks B' and A' in the negative-going scan.

The above discussion clearly illustrates that the electrochemistry of Ru(0001) is largely influenced by its high affinity to adsorbed H, OH, and O species. Because of the close relation between the metal–hydrogen bond strength and the potential of electrochemical hydrogen adsorption [Jerkiewicz, 1998; Karlberg et al., 2007], the anodic limit of the H_{upd} potential range on Ru(0001) should be about 0.15–0.22 V higher than on Pt(111). This means that H_{upd} should be stable up to about 0.5 V with respect to desorption as H^+. On the other hand, the strong bonding of O_{ad} and OH_{ad} to Ru(0001) stabilizes adsorbed oxygen-containing species down to potentials around 0 V. This leads to a substantial overlap of OH_{ad} adsorption and hydrogen adsorption on Ru(0001) [Hoster et al., 2004; El-Aziz and Kibler, 2002], in contrast to the electrochemical behavior of Pt(111), where the potential regions of reversibly adsorbed H_{upd} and OH_{ad} are separated by the "double-layer region" (see, e.g., Fig. 14.4b, where a Pt(111) CV is included as reference) [Garcia-Araez et al., 2006]. The overlap of the H_{upd} and OH_{ad} potential regions on Ru(0001) agrees with results of a recent first-principles study that predicted the equilibrium potentials for the formation of H_{upd} and OH_{ad} adlayers on that surface to be so close to each other that the double-layer region should virtually vanish in acidic electrolytes [Taylor et al., 2007].

The overlap of the stability regions of the two adsorbates has two major consequences. First, the potentials for H_{upd} and OH_{ad} formation are determined by the equilibrium potential for $OH_{ad} \leftrightarrow H_{upd}$ exchange rather than by the equilibria between H_{upd} and H^+ and between OH_{ad} and H_2O, respectively. This is completely different from Pt(111), where H_{upd} and OH_{ad} adsorption are determined independently by the equilibria with H^+ and H_2O, respectively. Second, this exchange process is held responsible for the pronounced hysteresis of about 0.2 V between $OH_{ad} \rightarrow H_{upd}$ replacement in the cathodic scan (peak A′) and the reverse reaction in the anodic scan (peak A). As will be discussed in more detail in the following sections, the equilibrium potential for $OH_{ad} \leftrightarrow H_{ad}$ exchange on Ru(0001) in 0.1 M $HClO_4$ can be estimated to be around 0.1 V. The onset of $OH_{ad} \rightarrow H_{upd}$ exchange on this surface at significantly lower potentials, around 0.05 V (onset of peak A′ with a maximum at -0.02 V), can be explained by the stability of the (metastable) OH_{ad} layer, which will measurably react with H^+ only at potentials significantly below the equilibrium potential for $OH_{ad} \leftrightarrow H_{upd}$ exchange. Hence, the OH_{ad} layer is kinetically stabilized. In a microscopic picture, this can be tentatively explained by a mechanism where $OH_{ad} \leftrightarrow H_{upd}$ replacement in peak A′ proceeds via a two-step process:

$$OH_{ad} + H^+ + e^- \longrightarrow H_2O_{ad} \longrightarrow H_2O + \square \qquad (14.1)$$

$$\square + H^+ + e^- \longrightarrow H_{upd} \qquad (14.2)$$

In this reaction scheme, hydrogen adsorption (on an empty site $\square$) can occur only after OH removal. Since the adsorption energy of hydrogen is gained only in the second step, this may shift the onset of the reaction to more cathodic potentials. We will see later that on a surface covered by Pt monolayer islands that allow an easy formation of H_{ad} (see below), a homolytic reaction according to

$$OH_{ad} + H_{upd} \longrightarrow H_2O_{ad} + \square \longrightarrow H_2O + 2\square \qquad (14.1a)$$

is more probable than the heterolytic Reaction (14.1). For a pure Ru surface, however, our data do not allow us to rule out one or the other possibility. Recently, Taylor et al. (2007) suggested that a homolytic pathway is dominant for surfaces where the H_{upd} and OH_{ad} stability regions overlap.

In the reverse anodic scan, similar arguments hold true for the removal of the H_{upd} adlayer in peak A, which is likely to proceed in a comparable two-step process:

$$H_{upd} + H_2O \longrightarrow H_3O^+ + e^- + \Box \quad (14.3)$$

$$2\Box + H_2O \longrightarrow OH_{ad} + H_{upd} \quad (14.4)$$

Also in this case, a heterolytic reaction according to

$$\Box + 2H_2O \longrightarrow OH_{ad} + H_3O^+ + e^- \quad (14.4a)$$

would be possible as well, and cannot, on the basis of our data, be excluded for reaction on a pure Ru(0001) electrode.

Because of the high stability of the H_{upd} adlayer, the first step (14.3), which is required for nucleation of a vacancy in the closed H_{upd} adlayer, starts only at potentials considerably above the equilibrium potential for $H_{upd} \rightarrow OH_{ad}$ exchange, at about 0.15 V. With increasing potential, H_{upd} is replaced by OH_{ad} [Reactions (14.3) and (14.4) or (14.4a)] and finally OH_{ad} by O_{ad} according to

$$OH_{ad} + H_2O \longleftrightarrow O_{ad} + H_3O^+ + e^- \quad (14.5)$$

Only at $E = 0.45$–0.5 V, do the CVs obtained for low (<0.1 V) and higher (0.1 V) cathodic scan limits match again, indicating that at this point the respective surface states are identical, independent of the previous treatment. Most simply, at this point, O_{ad} formation via Reactions (14.3), (14.4), and (14.5) is completed, and similar amounts of O_{ad} (0.75 ML; see the next paragraph) are deposited. (About 1.5 ML OH_{ad} would be possible as well based on the charge, but this is not consistent with other observations discussed above.)

In this picture, the kinetic barriers hindering the exchange between the two adlayers are related to the presence of metastable, but rather strongly bound, adsorbed species (H_{upd} and OH_{ad}), which cannot be removed easily, and which block the surface for adsorption of the respective other species. The nonequilibrium situation is also reflected in the shape of the corresponding peaks A' and A, where the anodic one (A) is less sharp and extends over a larger potential range.

The symmetric pair of voltammetric peaks in the Ru(0001) base CV in the range 0.1–0.3 V (peaks B and B'), which is best seen for a lower potential limit of $E_{min} = 0.1$ V, is tentatively assigned to (14.5), which can run reversibly in both directions. This assignment is based on the assumption that the surface is covered by 0.5 ML O_{ad} at 0.3 V. Only for more negative potential limits, when OH_{ad} is further reduced to H_2O and replaced by H_{upd} according to (14.1) and (14.2), does the re-oxidation of the adlayer require H_2O dissociation according to (14.3) and (14.4). This provides a simple explanation why the pronounced hysteresis between OH_{ad} removal (peak A') and reformation of OH_{ad}/O_{ad} (peak A) is only observed when the potential is scanned to $E < 0.1$ V.

14.3.1.2 Pt Film-Modified Ru(0001) Electrodes

Surface Structure and Adsorption Properties at the Solid/Vacuum Interface For deposition at temperatures around room temperature, Pt growth on Ru(0001) proceeds via nucleation and lateral growth of monolayer islands with a distinct triangular shape, which at higher coverages start to coalesce (Fig. 14.3a) [Buatier de Mongeot et al., 1998; Diemant et al., 2003; Käsberger et al., 2003]. The island shape is indicative of a dendritic growth process [Hwang et al., 1991; Brune et al., 1996]. At Pt coverages around 1 ML, essentially perfect monolayer films can be produced by Pt deposition at 300–350K and subsequent annealing to 800K (Fig. 14.4a) [Hoster et al., 2008; Käsberger et al., 2003]. Up to film thicknesses of at least four atomic layers, Pt grows pseudomorphically on Ru(0001), i.e., for films in this thickness range, the Pt lattice is laterally compressed by 2.5% with respect to Pt(111) to adapt to the lattice constant of the Ru(0001) surface [Buatier de Mongeot et al., 1998; Schlapka et al., 2003].

The chemical properties of these bimetallic surfaces were characterized previously by adsorption of CO and deuterium. The interaction of a Pt monolayer-covered Ru(0001) surface with these adsorbates is very weak—significantly weaker than for adsorption on the respective bulk substrates Ru(0001) or Pt(111) [Buatier de Mongeot et al., 1998; Schlapka et al., 2002, 2003; Jakob and Schlapka, 2007; Davies et al., 2005; Greeley and Mavrikakis, 2005; Groß, 2006]. With increasing Pt film thickness, the binding energies of CO_{ad} and O_{ad} increase again and finally approach constant values [Schlapka et al., 2003; Lischka et al., 2007]. The differences between these latter values and the adsorption energies on Pt(111) were attributed to the compressive strain in the pseudomorphic layers, while the additional destabilization observed for very thin films was associated with an electronic modification of the Pt

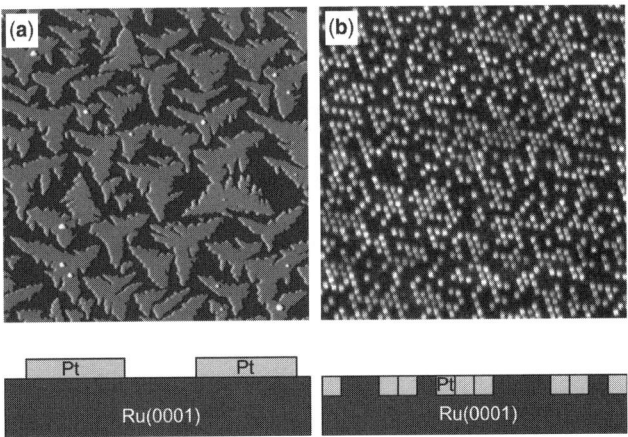

Figure 14.3 (a) Deposition of 0.5 ML Pt vapor onto Ru(0001) at 40 °C (217 nm × 217 nm, $I_t = 0.56$ nA, $U_t = 1$ V). (b) The same surface after formation of a $Pt_{0.5}Ru_{0.5}/Ru(0001)$ surface alloy by flash annealing to 1350K; Pt and Ru atoms appear dark and bright, respectively (11 nm × 11 nm, $I_t = 28$ nA, $U_t = 52$ mV).

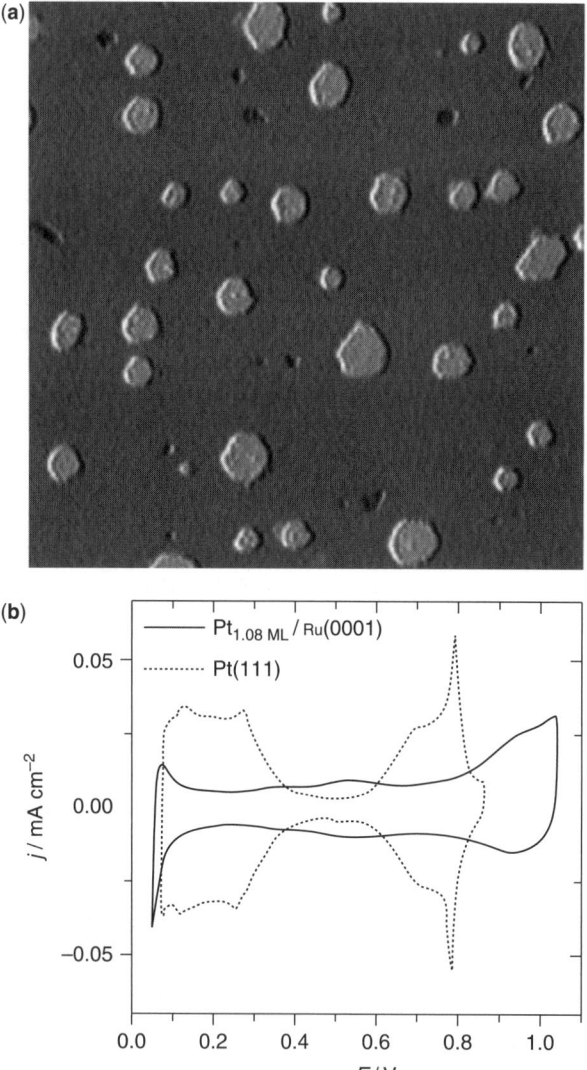

Figure 14.4 (a) STM image (150 nm × 150 nm², $U_t = 0.9$ V, $I_t = 0.56$ nA) and (b) cyclic voltammogram (0.1 M $HClO_4$, 50 mV s^{-1}) of a smooth 1.08 ML Pt film on Ru(0001) and of Pt(111). (Reprinted with permission from [Hoster et al., 2004]. Copyright 2004 American Chemical Society.)

surface layer due to the interaction with the Ru(0001) substrate ("vertical ligand effect") [Schlapka et al., 2003; Lischka et al., 2007].

Electrochemical Properties Figures 14.4a and 14.4b show an STM image of the surface morphology and a cyclic base voltammogram, respectively, of an atomically smooth 1.08 ML Pt film on Ru(0001), which was prepared by Pt deposition at 300K and subsequent annealing to 870K [Hoster et al., 2004]. Apart from a few vacancy islands in the monolayer Pt film or second-layer Pt islands, the surface is covered by a pseudomorphic Pt monolayer, and hence the base CV is characteristic of a Pt monolayer-covered Ru(0001) surface. The absence of any distinct adsorption/desorption features in the base CV over a wide potential range (from about 0.15 V to about 0.80 V) points to rather low/high onset potentials for the adsorption of H_{upd}/OH_{ad} species, respectively, compared with Pt(111) (dotted line in Fig. 14.4b). This is indicative of a rather weak interaction with these species [Karlberg, 2007; Climent et al., 2006], which agrees with the results of recent theoretical studies [Koper et al., 2002; Hoster et al., 2009a] and fits to the data obtained for hydrogen [Diemant et al., 2003, in preparation] and oxygen adsorption on $Pt_{ML}/Ru(0001)$ [Lischka et al., 2007] under UHV conditions (see above). Consequently, base CVs of Ru(0001) surfaces partly covered by pseudomorphic Pt monolayer islands are largely governed by features associated with adsorption on Ru sites [Hoster et al., 2004; Zhou et al., 2007]. This is demonstrated in the base CVs in Fig. 14.5 g–j for Ru(0001) electrodes covered by 0.03–0.8 ML Pt. STM images of the respective surface morphologies are shown in Fig. 14.5b–e [Hoster et al., 2004]. In the anodic potential region ($E > 0.2$ V), the CV basically resembles that obtained on bare Ru(0001) (Fig. 14.5a, f). However, the distinct signals A and A′ that reflect oxidation of H_{upd} and formation of OH_{ad} from H_2O, and vice versa, now set in at 0.1 V in either scan direction, respectively, compared with 0.05 V (negative-going scan) and 0.15 V (positive-going scan) on a Pt-free Ru(0001) surface. The charges in peaks A and A′ decrease with increasing Pt coverage, while their general shape and position change only slightly. The charge decay of the cathodic peak A′ with increasing Pt content follows a linear relation (Fig. 14.6) and approaches zero for $x_{Pt} = 1$. These characteristics in combination indicate that the peaks result from Pt-catalyzed adsorption/removal of OH_{ad} and H_{upd} species on Ru sites (most likely Ru_3 adsorption ensembles; see below), whose number decreases linearly with increasing Pt island coverage [Hoster et al., 2004]. The maximum charge of 0.29 mC cm^{-2} (equivalent to $1.16\,e^-$ per surface atom), which includes contributions from H_{upd} adsorption and OH_{ad} removal in the cathodic scan, points to coverages in the range of 0.5–0.6 ML for either adsorbate at potentials negative (0.08 V) and positive (0.12 V) of the A′ peak, respectively [Hoster et al., 2004]. This is identical to the respective coverage values at potentials directly above and below the corresponding peak around 0 V reported for bare Ru(0001) [El-Aziz and Kibler, 2002]. On the pseudomorphic Pt islands, the H_{upd} coverage is very low under these conditions, which can be rationalized by the weak bonding power of these monolayer islands discussed above. By charge integration, we estimated a H_{upd} coverage of at most 0.12 ± 0.1 ML at the onset of H_2 evolution on these islands [Hoster et al., 2004].

The onset of $H_{upd} \leftrightarrow OH_{ad}$ exchange at 0.1 V on the Pt monolayer island-modified Ru(0001) surface in either scan direction is the basis for our estimate that the thermodynamic equilibrium potential for $H_{upd} \leftrightarrow OH_{ad}$ exchange on Ru(0001) is at about this value (see the preceding section). Therefore, at a potential of 0.1 V, H_{upd} and OH_{ad} are about equally stable. With increasing potential, they become more weakly

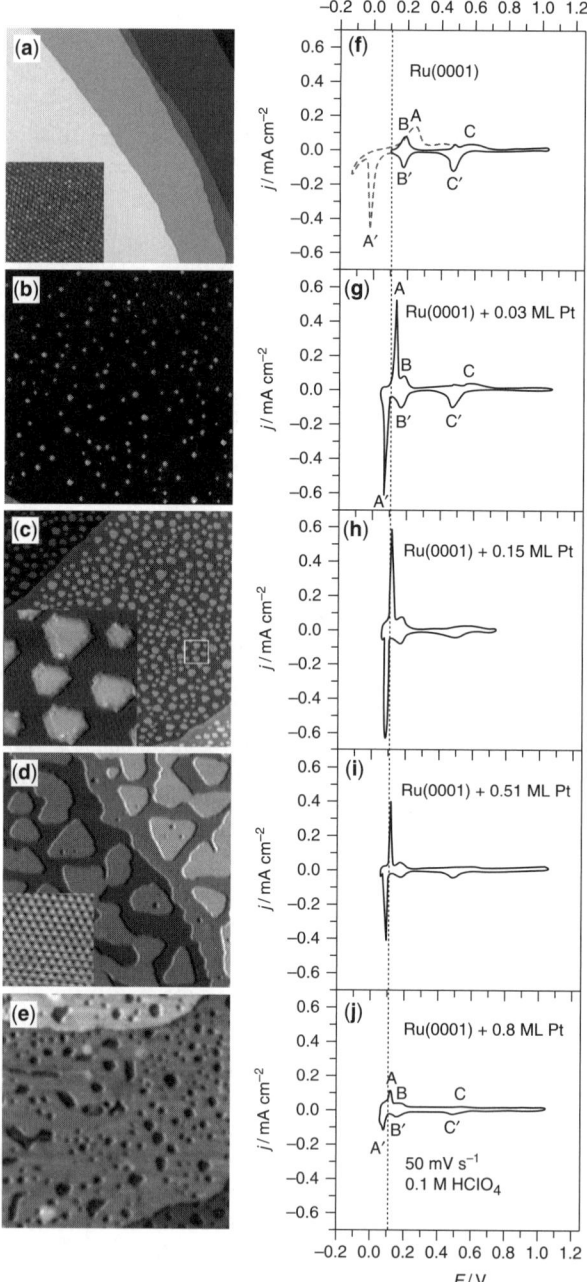

Figure 14.5 STM data (140 nm × 140 nm) (a–e) and corresponding cyclic voltammograms (f–j) of Ru(0001) and Ru(0001) covered by increasing amounts of pseudomorphic Pt submonolayers. (Reprinted with permission from [Hoster et al., 2004]. Copyright 2004 American Chemical Society.)

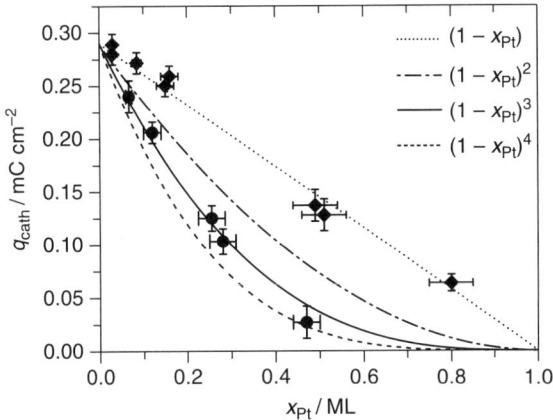

Figure 14.6 Charge in the cathodic peak between 0.11 and 0.06 V as a function of Pt surface content: diamonds, Pt_x submonolayers on Ru(0001); circles, Pt_xRu_{1-x}/Ru(0001) surface alloys; the lines are predicted trends for linear or polynomial correlations between charge and Pt surface content.

(H_{ad}) or more strongly (O_{ad}/OH_{ad}) bound. In Fig. 14.7, we illustrate the resulting potential-dependent adlayer formation and replacement processes for anodic (upper part) and cathodic (lower part) scan directions. In the negative-going scan, H_{upd} formed on the Pt islands can react with OH_{ad} on neighboring Ru sites and desorb as H_2O [equivalent to Reaction (14.1a)]. Spillover of further H_{upd} from the Pt islands to the Ru terraces or direct adsorption of H_{upd} on the Ru areas results in further OH_{ad} removal and subsequent replacement by H_{ad}. The pronounced shift of peak A′ from

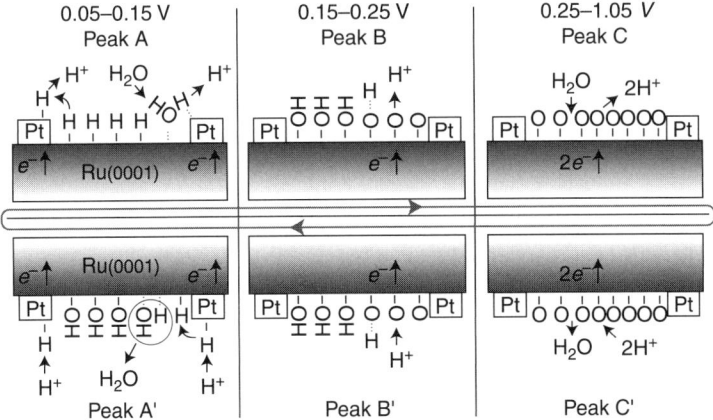

Figure 14.7 Illustration of the formation, removal, and exchange of adlayers on Ru(0001) in the presence of Pt islands/sites as observed in the peaks A/A′, B/B′, and C/C′ (see also Figs. 14.2 and 14.8). Processes in the anodic/cathodic potential scan direction are shown in the upper/lower part; for simplicity, H^+ is used instead of H_3O^+.

-0.02 V on bare Ru(0001) (Figs. 14.2 and 14.5f) to $+0.05$ V on Pt-modified Ru(0001) (Fig. 14.5 g–i) shows that the Pt monolayer islands act as catalyst for the replacement process, as described above.

In the positive-going scan, the Pt monolayer islands also catalyze the $H_{upd} \rightarrow OH_{ad}$ exchange reaction, reducing the onset potential of peak A from 0.15 V on Pt-free Ru(0001) to 0.1 V in the presence of the Pt islands. At these potentials positive of the equilibrium potential for $H_{upd} \leftrightarrow OH_{ad}$ exchange, H_{upd} removal is driven by the energy gain due to OH_{ad} adsorption. On the Pt island-modified surface, the mechanism for exchange of H_{upd} by OH_{ad} must be such that the Pt islands can act as catalysts. This makes a heterolytic process as described by (14.3) and (14.4) unlikely, since (14.3), which controls the onset of the exchange reaction, should not be affected by the presence of the Pt monolayer islands. A catalytic acceleration of the exchange process is possible, however, if the exchange is initiated by spillover of (small amounts of) H_{upd} from the Ru(0001) areas to Pt islands, (14.6), where it becomes oxidized, (14.7), in combination with dissociative H_2O adsorption/OH_{ad} formation at the interface between the Ru substrate and the Pt islands, (8):

$$H_{upd}(Ru) + \square(Pt) \rightarrow H_{upd}(Pt) + \square(Ru) \quad (14.6)$$

$$H_{upd}(Pt) + H_2O \rightarrow \square(Pt) + H_3O^+ + e^- \quad (14.7)$$

$$\square(Ru) + \square(Pt) + H_2O \rightarrow OH_{ad}(Ru) + H_{upd}(Pt) \quad (14.8)$$

Neither reaction step/sequence is hindered by the dense H adlayer on the Ru(0001) areas, and can therefore proceed faster than without Pt. The dissociation of water is supported by the higher binding energy of the H_{upd} species at the edge of the Pt monolayer islands compared with H_{upd} adsorption on the Pt islands [Hammer et al., 1997]. Furthermore, the spillover of H_{upd} from the Ru(0001) terraces to the Pt islands, (14.5), which on first view is in contrast to the much higher adsorption energy of H_{upd} on the Ru(0001) substrate than on the Pt monolayer islands, is facilitated by the high (local) H_{upd} coverage (the adlayer on the Ru areas is essentially saturated throughout the exchange process, until complete replacement of H_{upd} by OH_{ad} is reached), which reduces the H_{upd} bond strength on the Ru areas owing to repulsive interactions. Finally, it is important to note that the ongoing reaction requires mobility of the OH_{ad} and H_{upd} species in the adlayer on the Ru areas.

At higher potentials, positive of the $H_{upd} \leftrightarrow OH_{ad}$ exchange, the CVs of the Pt island-modified Ru(0001) surface closely resemble those of the Pt-free Ru(0001) electrode, except for the lower currents/charges in the characteristic features. This simply reflects the fact that at these potentials, the surface reactivity is dominated by the electrochemical properties of the remaining exposed Ru surface. As already mentioned, the Pt monolayer islands themselves contribute only little to the voltammetric behavior, which is due to the weak bonding and hence low adsorbate coverages on these islands.

14.3.1.3 Pt_xRu_{1-x} Surface Alloys on Ru(0001)

Surface Structure and Adsorption Properties at the Solid/Vacuum Interface Heating Ru(0001) surfaces covered by submonolayer amounts of Pt to temperatures $T \geq 850K$ leads to Pt–Ru intermixing [Hoster et al., 2008, 2009b], and brief annealing to 1300–1350K results in the formation of atomically dispersed monolayer surface alloys with an essentially random distribution of the two components in the surface layer (see the atomic-resolution STM images in Fig. 14.3b and [Buatier de Mongeot et al., 1998; Diemant et al., 2003; Hoster et al., 2008]). Owing to a highly negative surface segregation energy of Pt impurities in Ru [Christensen et al., 1997; Ruban et al., 1999] and a barrier for Pt diffusion into deeper Ru regions, which must be significantly higher than that for Pt exchange with Ru surface atoms [Hoster et al., 2008], Pt–Ru intermixing is largely confined to the outermost layer under these conditions. Based on quantitative STM measurements, the loss of Pt into deeper layers is negligible up to Pt surface contents of 0.8 under these conditions, and less than 10% up to 1 ML Pt pre-deposition [Hoster et al., 2008]. Representative high-resolution STM images illustrating the atom distribution in $Pt_xRu_{1-x}/Ru(0001)$ surface alloys with five different Pt contents and the base CVs obtained on them are presented in Fig. 14.8. (Note that the surface in Fig. 14.8e is the same as that used for recording the CV in Fig. 14.4.)

TPD measurements characterizing the desorption of deuterium from $Pt_xRu_{1-x}/Ru(0001)$ monolayer surface alloys revealed a continuous decrease in the deuterium desorption barrier with increasing Pt coverage [Diemant et al., 2003, 2008]. This was attributed to an increasingly weaker binding of D adatoms to Pt_nRu_{3-n} threefold sites (adsorption ensembles)—or to Pt_nRu_{5-n} pentamers, when considering the recombinative desorption of adjacent deuterium adatoms—with increasing number of Pt surface atoms in the respective adsorption ensembles [Diemant et al., 2008]. CO desorption experiments on the same type of surfaces, on the other hand, showed two dominating adsorption states at high and low temperatures, which were assigned to atop CO adsorption on Ru and Pt surface atoms, respectively [Rauscher et al., 2007]. Both states were destabilized with increasing Pt content, which was rationalized by electronic ligand and strain effects [Rauscher et al., 2007]. This Pt concentration-dependent destabilization is, however, less pronounced than the modification introduced by mixed adsorption ensembles for deuterium adsorption. Hence, for both probe molecules, the surface alloys offer adsorption sites with intermediate stability that exist neither on Ru(0001) nor on pseudomorphic Pt/Ru(0001) monolayers and that affect the CO and deuterium adsorption behavior in a distinct way. As will be shown in the following, similar effects are also observed in electrochemical experiments.

Electrochemical Properties All CVs are presented on two different scales to show both the larger and smaller peaks in sufficient detail. At low Pt surface concentrations, the base CVs are very similar to those of the Pt island-modified Ru(0001) surfaces (see Fig. 14.5). With increasing Pt surface content, however, the charge in the $H_{upd} \leftrightarrow OH_{ad}$ exchange peaks A and A' at about 0.1 V decreases much faster than linearly. A faster than linear decay of the charge in this peak would be expected if Ru ensembles with more than one Ru atom were required for OH_{ad} and/or H_{upd} adsorption. Since the atom distribution in $Pt_xRu_{1-x}/Ru(0001)$ surface alloys is very close to a random distribution [Hoster et al., 2008], the number of Ru_n sites is proportional to x_{Ru}^n or $(1 - x_{Pt})^n$. As is evident from the plot in Fig. 14.6, the experimental data agree very

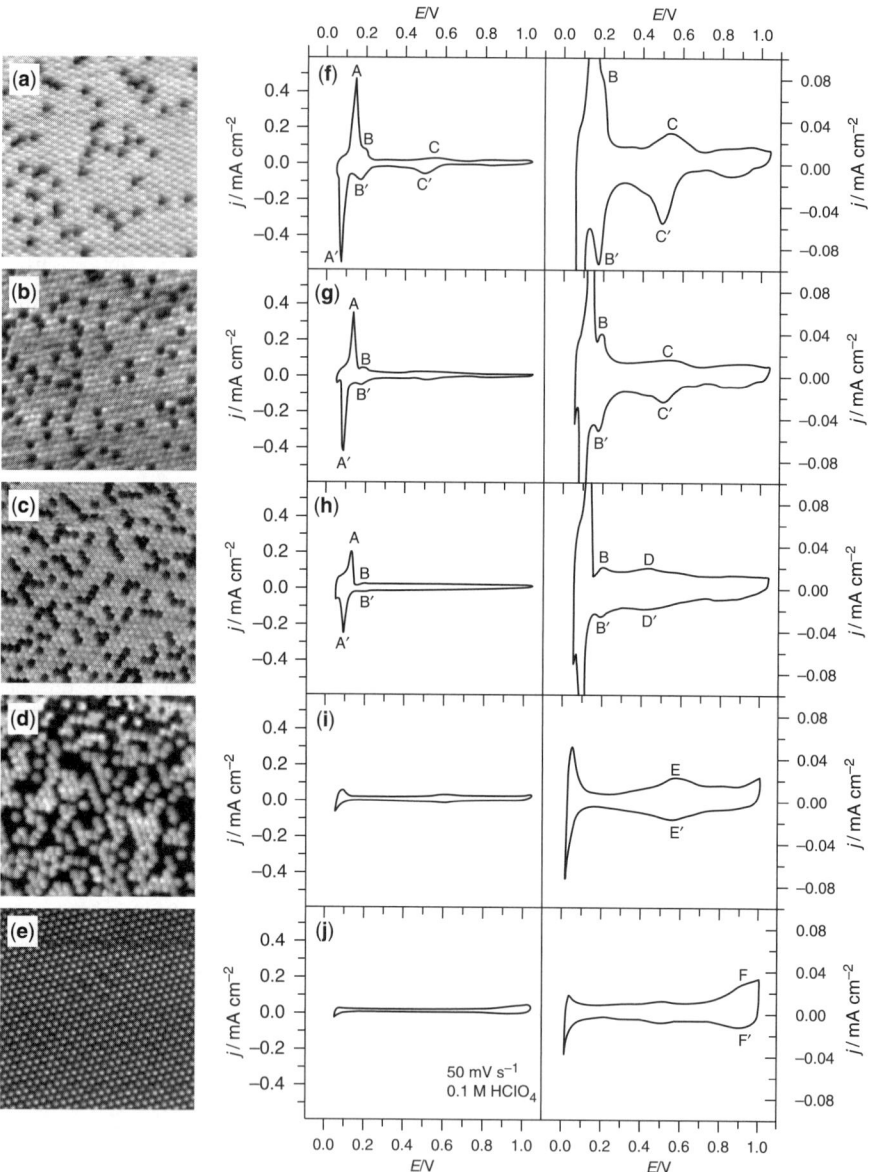

Figure 14.8 STM images (7 nm × 7 nm) (a–e) and corresponding cyclic voltammograms (f–j) of different $Pt_xRu_{1-x}/Ru(0001)$ surface alloys. The voltammograms on the right-hand side have an expanded current scale. (a, f) $x_{Pt} = 0.07$; (b, g) 0.12; (c, h) 0.25; (d, i) 0.53; (e, j) 1.05. [Hoster et al., 2008]—Reproduced by permission of the PCCP Owner Societies.

well with a $(1 - x_{Pt})^3$ relation between Pt content and the experimentally determined peak charges on the $Pt_xRu_{1-x}/Ru(0001)$ surface alloy. Therefore, we conclude that H_{upd} and OH_{ad} adsorption take place only on Ru_3 sites in this potential region, and that therefore the abundance of these sites dominates the charge in the narrow and sharp replacement signals.

Although the sharp signals for H_{upd}/OH_{ad} exchange dominate the CVs, the charges outside this potential region are not negligible. Similar to the observations at Pt island-modified surfaces (see Fig. 14.5), the peak pairs B/B' and C/C' decrease in amplitude with increasing x_{Pt}. They disappear, however, already long before approaching $x_{Pt} = 1$. They are essentially invisible already for $x_{Pt} = 0.25$ (Fig. 14.5g) and $x_{Pt} = 0.53$ (Fig. 14.5h), respectively. Simultaneously, additional broad features develop at 0.4 and 0.6 V, which are labeled as D/D' and E/E', respectively. For the full Pt monolayer, the peaks F/F' at 1.0 V, which we assign to reversible OH_{ad} formation, mark the most pronounced feature. Although an unambiguous assignment of the changes for $E > 0.3$ V is less straightforward than for the peaks A/A', we will discuss some basic trends at this point.

As stated above, our UHV data [Diemant et al., 2003, 2008] indicate that both CO_{ad} and H_{upd} bind more weakly to $Pt_xRu_{1-x}/Ru(0001)$ surface alloys than to Ru(0001); according to DFT calculations, a similar trend can be assumed also for OH and O [Liu et al., 2003; Lischka et al., 2007; Hoster et al., 2009c]. In cyclic voltammograms, weaker H_{upd} adsorption gives rise to a negative shift of the corresponding peaks [Karlberg et al., 2007], whereas OH_{ad} and O_{ad} formation will require higher potentials [Climent et al., 2006]. Consequently, the new peaks D/D', E/E', and F/F' reflect OH_{ad} adsorption/replacement and H_{upd} adsorption/replacement on these less strongly binding adsorption sites. The binding energies of H_{upd} and OH_{ad} on mixed Pt_mRu_n will be between the values of the Ru_3 site and the (pseudomorphic) Pt_3 sites (see also [Diemant et al., 2008]). Recent DFT calculations yielded values of -2.89 eV (-3.28 eV), -2.74 eV (-2.73 eV), -2.56 eV (-2.04 eV), and -2.44 eV (-1.46 eV) for the binding energies of H_{upd} (OH_{ad}) on Ru_3, Ru_2Pt, $RuPt_2$, and Pt_3 sites in $Ru_{3-n}Pt_n/Ru(0001)$ ($n = 0-3$) surfaces, respectively [Hoster et al., 2009c]. Hence, the charge decrease in peak A', which we showed to be proportional to $(1 - x_{Pt})^3$, will be accompanied by the appearance of new features at lower or higher potentials. Furthermore, on the mixed sites, the stability regions of H_{upd} and OH_{ad} are likely to no longer overlap, in contrast to the Ru_3 sites. The peaks reflecting H_{upd} adsorption/desorption on/from Pt_3 sites are expected to be shifted to below 0.1 V, where they are masked by H_2 evolution [Hoster et al., 2004]. On the other hand, for surface alloys with lower Pt content, the region between peaks B and C also exhibits larger currents than for bare or Pt island-modified Ru(0001) surfaces. This is tentatively associated with desorption of H_{upd}/adsorption of OH_{ad} on mixed Pt_xRu_{3-x} sites. They disappear only when approaching $x_{Pt} > 0.5$. Therefore, we suggest that on $Ru_{3-n}Pt_n$ sites with $n \geq 2$, H_{upd} and OH_{ad} are formed only at $E < 0.2$ V and $E > 0.3$ V, respectively.

The peak pairs B/B' and C/C' are proposed to reflect the reversible deprotonation of 0.5 ML OH_{ad} in peak B/B' (Reaction (14.5)) and completion of the O_{ad} coverage up to 1 ML in peak C/C' [Reactions (14.4) and (14.5)], respectively. Both peak pairs become smaller with increasing Pt content, accompanied by the appearance and growth of new voltammetric features at higher potentials that we have labeled as D/D′, E/E′, and F/F′. Integration from 0.3 to 1.05 V yields about 0.5 and 0.4 electrons per surface atom for $x_{Pt} = 0.57$ and the full Pt monolayer, respectively, equivalent to at most 0.25 and 0.2 ML O_{ad}, respectively, at 1.05 V. Again, electrochemically this cannot be distinguished from an OH_{ad} or mixed OH_{ad}/O_{ad} adlayer with accordingly higher coverages, and, for the mixed sites, the nature of the adsorbate at 1.05 V is in fact not clear. The lower coverage of OH_{ad} and/or O_{ad} at the anodic potential limit and the higher potentials necessary to stabilize these species agree well with the expected smaller binding energies of OH_{ad} on the mixed sites of the Pt_xRu_{1-x} surface alloys. One may speculate that the increasing symmetry between positive-going and negative-going scan in the new features at $E > 0.3$ V, with increasing Pt surface content, is due to an additional catalytic promotion of homolytic H_2O formation and dissociation by the mixed Pt_nRu_{3-n} sites, similar to the mechanism discussed above for OH_{ad} formation and reduction at $E = 0.1$ V [Reactions (14.6)–(14.8)].

Overall, the comparative cyclic voltammetry analysis of bare Ru(0001), Pt adlayer-modified Ru(0001), and Pt_xRu_{1-x}/Ru(0001) surface alloys has demonstrated that the electrochemical adsorption properties of these surfaces can be largely described by a local picture, including mainly electronic ligand and strain effects in the pseudomorphic surface layer, and ensemble effects induced by the composition of the threefold adsorption ensembles. Long-range effects, mediated, for example, by elastic distortions of the surface layer, do not seem to be important. Pseudomorphic Pt monolayers on Ru(0001) adsorb H_{upd}, OH_{ad}, and O_{ad} much more weakly than Pt(111) or Ru(0001), in agreement with the findings in recent adsorption studies under UHV conditions and theoretical predictions. Mixed M_3 adsorption ensembles containing fewer than three Ru atoms or even Pt_3 ensembles provide new adsorption states that interact less strongly with H_{ad}, OH_{ad}, and O_{ad} than Ru_3 sites, and thus give rise to additional, more reversible voltammetric features that first appear in addition to those of larger Ru ensembles and replace them at increasing Pt content. The addition of Pt, either as Pt monolayer islands or in a PtRu surface alloy, has a pronounced catalytic effect on the kinetics of the H_{upd}/OH_{ad} exchange process, by providing an additional channel for H_{upd} formation (cathodic scan) and its oxidative replacement by OH_{ad} (anodic scan) on the OH_{ad}/H_{upd} blocked Ru(0001) surface areas or Ru_3 ensembles.

14.3.2 CO Electro-oxidation on Bare and Pt-Modified Ru(0001) Electrodes

In this section, we focus on the Pt-induced modifications of the electrocatalytic properties of Ru(0001), using the electro-oxidation of CO (CO bulk oxidation) as example.

14.3.2.1 Pt-Free Ru(0001) CO adsorption on Ru(0001) under UHV conditions has been extensively studied [Buatier de Mongeot et al., 1998; Diemant et al., 2003; Thomas and Weinberg, 1979; Williams and Weinberg, 1979; Pfnür et al., 1980, 1983; Pfnür and Menzel, 1983; Michalk et al., 1983; Kostov et al., 1992]. A comprehensive description of the coverage-dependent adsorption behavior and an overview of previous work can be found in [McEwen and Eichler, 2007]. The interaction between CO and Ru(0001) is characterized by an initial adsorption energy of 165–175 kJ mol^{-1} up to saturation of a 0.33 ML ($\sqrt{3} \times \sqrt{3}$)R30° CO adlayer [Pfnür et al., 1983]. The latter adlayer desorbs in a peak at 450–480K [Buatier de Mongeot et al., 1998; Pfnür et al., 1983; Kostov et al., 1992]. When the CO coverage is increased to 0.67 ML, a low temperature shoulder develops in the TPD spectra at 400–410K [Buatier de Mongeot et al., 1998; Pfnür et al., 1983; Kostov et al., 1992], reflecting a reduced adsorption energy of 120 kJ mol^{-1} at these higher coverages [Pfnür et al., 1983]. Coadsorption of CO and hydrogen or CO and oxygen leads to the formation of mixed adlayers:

(i) a $CO_{ad} + H_{ad}$ adlayer (stable at $T < 400K$ [Peebles et al., 1982])
(ii) a $2CO_{ad} + O_{ad}$ adlayer ($T < 220K$ [Schiffer et al., 1997])
(iii) a $CO_{ad} + O_{ad}$ adlayer ($T <$ room temperature [Kostov et al., 1992])
(iv) a $2O_{ad} + CO_{ad}$ adlayer ($T < 330K$ [Kostov et al., 1992; Narloch et al., 1994])

Adlayers containing CO_{ad} and O_{ad} were found to be ordered and intermixed, whereas CO_{ad} and H_{ad} segregate into islands [Peebles et al., 1982; Ciobica et al., 2003; Riedmüller et al., 2002]. H_{ad} is destabilized by coadsorbed CO_{ad} (via compression of the H adlayer) [Diemant et al., 2003; Peebles et al., 1982], and the same is true for CO_{ad} in the presence of O_{ad} (via repulsive CO_{ad}–O_{ad} interactions) [Schiffer et al., 1997]. Oxygen desorption sets in only at $T > 800K$ [Böttcher et al., 1997], where CO_{ad} has already left the surface. The strong adsorption bond of O_{ad} makes the Ru(0001) surface essentially inactive for CO oxidation under UHV conditions [Kostov et al., 1992]. CO oxidation was observed only at high O_2 pressures and elevated temperatures [Peden and Goodman, 1986], where active, oxygen-rich surface phases can be formed [Over and Muhler, 2003; Blume et al., 2006; Goodman et al., 2007; Over et al., 2007].

A similar inhibition was found also for electrochemical CO oxidation. In CO_{ad} stripping experiments, numerous potential cycles up to 1 V were necessary to remove all CO_{ad} from a smooth Ru(0001) surface [Zei and Ertl, 2000; Lin et al., 2000; Wang et al., 2001]. CO bulk oxidation experiments under enforced mass transport conditions on polycrystalline Ru [Gasteiger et al., 1995] and on carbon-supported Ru nanoparticle catalysts [Jusys et al., 2002] led to similar results. Hence, CO_{ad} can coexist with nonreactive OH_{ad} or O_{ad} species on Ru(0001) at lower potentials ($E < 0.55$ V) [El-Aziz and Kibler, 2002].

Based on electrochemical experiments combined with ex situ analysis by AES, LEED, and RHEED, Wang et al. (2001) suggested the formation of a (2×2) $(2CO + O)$ adlayer on Ru(0001) at $E = 0.2$ V in CO-saturated $HClO_4$, similar to the phase formed in UHV after CO adsorption on a (2×2)O-covered surface [Schiffer et al., 1997]. From the total charge density transferred after a potential step to 1.05 V in a CO-free electrolyte, they concluded that only 60% of the CO content in such an adlayer can be oxidized under these conditions [Wang et al., 2001].

This would be consistent with a transformation of a CO_{ad}-rich $(2 \times 2)(2CO + O)$ adsorbate phase [Schiffer et al., 1997] into an oxygen-rich, but still CO_{ad}-containing phase, for example the $(2 \times 2)(CO + 2O)$ phase known from UHV experiments [Narloch et al., 1994]. CO adsorption on the Ru(0001) surface at 0.7 V is essentially inhibited [Wang et al., 2001], most likely as a result of surface blocking by OH_{ad}/O_{ad} species.

Potentiodynamic CO bulk oxidation on a nonmodified Ru(0001) electrode is shown in Fig. 14.9 (Plate 14.1) (solid lines), including cycles with increasing anodic limit. For comparison, a base CV (dotted line) is included as well (as in Fig. 14.2, but at a scan rate of $10 \, \text{mV s}^{-1}$). Only for $E > 0.55$ V are anodic currents observed in a CO-containing electrolyte in the positive-going and (less pronounced) also in the negative-going potential scan, indicative of continuous CO oxidation. In the negative-going scan, we find reduction charges at potentials $0.55 \, \text{V} > E > 0.1$ V. They grow for higher anodic potential limits, and finally approach a value of about $0.5 \, \text{mC cm}^{-2}$ at an anodic limit ≥ 1.2 V. This is the charge expected for the reduction of 1 ML O_{ad} to H_2O, in good agreement with our assignment of a $(1 \times 1)O$ adlayer on Ru(0001) at $E > 1.05$ V in a CO-free electrolyte (see above). In CO-free electrolyte, this reduction takes place in the peaks C′, B′, and part of A′, whereas in the presence of CO, the voltammetric pattern is totally different, even for $E < 0.55$ V, where CO oxidation does not occur. We explain this difference by the gain in energy due to adsorption of strongly adsorbing CO in a CO-containing solution, compared with O_{ad} reduction and the $OH_{ad} \rightarrow H_{upd}$ exchange in the absence of

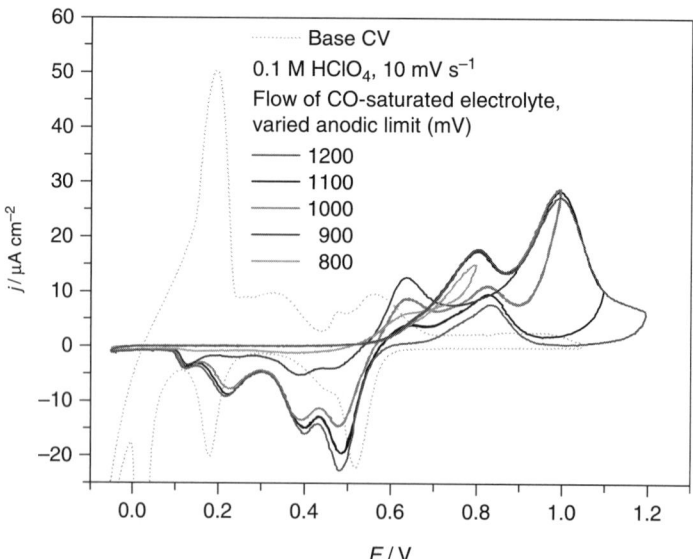

Figure 14.9 CO bulk electro-oxidation at bare Ru(0001) in flow cell; dotted line, CO free electrolyte; solid lines: flow of CO saturated electrolyte, with varied upper scan limits (see key on figure). (See color insert.)

CO. At $E < 0.1$ V, the resulting CO adlayer blocks the formation of H_{upd} that would take place in a CO-free electrolyte. The absence of anodic charges for $E < 0.55$ V in the positive-going scan, in contrast to the significant OH_{ad} uptake in this potential range in a CO-free electrolyte, indicates that CO_{ad} also blocks the re-formation of the OH/O adlayer. This only sets in at 0.55 V. In a potentiodynamic measurement, however, about 90% of the OH_{ad}/O_{ad} produced in the potential range up to 0.8 V is directly consumed for CO_{ad} oxidation. This can be concluded from the very small cathodic reduction charge visible when the scan is reversed at $E = 0.8$ V. Only for higher anodic potential limits ($E_{lim} > 0.8$ V), does the simultaneous increase of both reduction and oxidation charge for $E < 0.55$ V (reduction)/$E > 0.55$ V (oxidation) point towards an increasing enrichment of the high potential adlayer in OH_{ad}/O_{ad}. The potentials required to attain certain OH_{ad}/O_{ad} coverage are much higher than in the absence of CO. This shift is mainly attributed to kinetic effects, arising from the slow removal of CO_{ad}. From these data, we conclude that only on scanning the potential to $E < 0.1$ V or $E > 1.1$ V do the respective adlayers become dominated either by CO_{ad} or by OH_{ad}/O_{ad}, respectively. When the scan direction is reversed, mixed adlayers start to form at $E = 0.55$ V in either direction. As a consequence of the strong interaction of their constituents with Ru(0001), these mixed adlayers show little reactivity for CO_2 formation. This agrees well with results of the CO_{ad} stripping experiments in base electrolyte discussed above [Wang et al., 2001]. Based on electrochemical data alone, however, an unambiguous interpretation of the complex oxidation and reduction peak patterns to CO_2 and/or OH_{ad}/O_{ad} formation is not possible. For this purpose, DEMS measurements under continuous electrolyte flow are planned for the future.

In conclusion, cyclovoltammetric results in CO-saturated electrolyte demonstrate that CO oxidation takes place only for $E > 0.55$ V, and this with very low rates [Wang et al., 2001; Brankovic et al., 2002a]. Assuming a similar reaction mechanism for CO oxidation on Ru(0001) as for the extensively studied reaction on Pt surfaces [Santos et al., 1991; Lebedeva et al., 2000; Shubina et al., 2004], the overall oxidation of adsorbed CO proceeds via

$$2H_2O + \square(Ru) \longrightarrow OH_{ad}(Ru) + H_3O^+ + e^- \quad [\text{see } (14.4a)]$$

$$CO_{ad}(Ru) + OH_{ad}(Ru) \xrightarrow{rds} COOH_{ad}(Ru) + \square(Ru) \quad (14.9)$$

$$COOH_{ad}(Ru) + H_2O \xrightarrow{fast} CO_2 + \square(Ru) + H_3O^+ + e^- \quad (14.10)$$

$$\square(Ru) + CO \longrightarrow CO_{ad}(Ru) \quad (14.11)$$

where $COOH_{ad}$ may be a stable reaction intermediate or a transition state. The empty Ru site, $\square(Ru)$, must result from fluctuations in the CO adlayer and/or from displacement of CO_{ad} at the high coverage (low adsorption energy), similar to exchange of CO_{ad} in a CO adlayer at room temperature [Heinen et al., 2006].

If O_{ad} instead of OH_{ad} is the reactive oxygen species for CO oxidation, (14.4a) would be followed by (14.5), and CO_2 would then be formed directly via

$$CO_{ad} + O_{ad} \xrightarrow{rds} CO_2 \quad (14.12)$$

Owing to their strong bond on Ru(0001), mixed $CO_{ad} + OH_{ad}/O_{ad}$ layers are very stable against CO_2 formation, similar to the coadsorption behavior at the solid/gas interface discussed above. We assume that, for $E > 0.55$ V, the shift of the equilibrium between water and adsorbed OH_{ad}/O_{ad} towards the latter increases the density of the respective species in the intermixed adlayer, which increases the repulsions between the adsorbed species and hence leads to more weakly bound OH_{ad}/O_{ad} and CO_{ad} species. These latter species are less stable against $COOH_{ad}$ or CO_2 formation, because of the reduced reaction barrier ("Brønsted–Polanyi–Evans" relation [Bronstedt, 1928]), and can support a reaction via (14.9) or (14.12), respectively, at low rates. (Note that the total density of the adlayer does not need to remain constant, although also this is possible.)

14.3.2.2 Pt Monolayer Island-Modified Ru(0001)

The presence of Pt monolayer islands results in a dramatic increase in CO oxidation activity compared with a Pt-free Ru(0001) electrode. This is illustrated by potentiodynamic CO oxidation scans recorded for Ru(0001) surfaces modified by 0.05, 0.23, and 0.9 ML of Pt monolayer islands (Fig. 14.10). Already, 0.05 ML Pt (dashed line in Fig. 14.10) is sufficient to reach at least half of the mass-transport-limited current obtained for the electrodes with higher Pt coverages, which is more than one order of magnitude higher than the maximum oxidation current obtained on a Pt-free surface under similar conditions. The general shape of the scans is comparable for the three different Pt coverages. Similar to the behavior of the Pt-free surface, CO oxidation starts at about 0.55 V in the positive-going scan. The current first rises continuously up to a bending point (G, G', G''), and then increases steeply until reaching its maximum value. In the negative-going scan, the current decreases slightly right after the scan reversal for the low and medium Pt coverage (0.05 and 0.23 ML Pt) electrodes. For the high Pt coverage sample (0.9 ML Pt), the current remains constant at the mass-transport-limited value (0.9 ML Pt, dash–dotted curve in Fig 14.10). Starting at $E \approx 0.9$ V (0.05 and 0.23 ML Pt) or $E \approx 0.82$ V (0.9 ML Pt), it then decreases continuously, until reaching zero at about 0.55 V. With increasing amount of Pt, the maximum current grows continuously, and appears at lower potentials. For 0.9 ML Pt, the mass-transport-limited current is reached at $E \approx 0.9$ V. Also, the increase in slope (points G, G', G'') shifts to lower potentials with higher Pt coverage. In the negative-going scan, the activity is at least equal or even higher than in the positive-going scan at the same potential.

Therefore, we arrive at the same conclusion for the mechanism of CO_{ad} oxidation in the lower potential regime as for Pt-free Ru(0001), postulating that at potentials $E < 0.55$ V, only strongly bound OH_{ad}/O_{ad} species are present in the mixed $CO_{ad} + OH_{ad}/O_{ad}$ adlayer, which are not reactive towards CO_2 formation, while for $E \geq 0.55$ V, additional, weakly adsorbed OH_{ad}/O_{ad} species are formed, which can react with the (likewise destabilized) CO_{ad}. Similar to CO_{ad} oxidation on a Ru(0001) surface, the reaction starts by dissociative adsorption of H_2O on the Ru(0001) surface (no shift in the onset potential). In this case, however, the Pt islands can accelerate the reaction by accepting the H_{upd} resulting from a homolytic dissociation process. Thus, we tentatively propose a mechanism for CO oxidation at potentials between the reaction onset up to the bending point (see also Lin et al. [1999]), which is

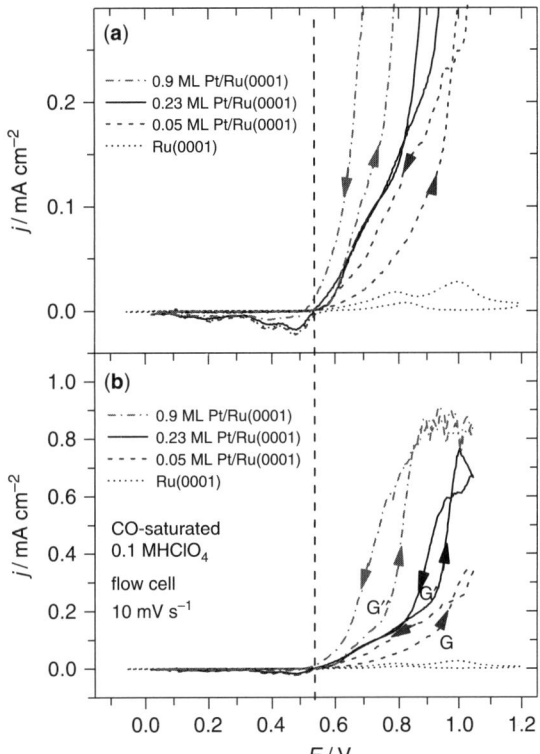

Figure 14.10 CO bulk electro-oxidation at Ru(0001) and Ru(0001) modified by 0.05, 0.23, and 0.9 ML Pt, measured at 10 mV s^{-1} in a flow cell with a CO-saturated electrolyte (0.1 M HClO$_4$): (a) expanded current scale to visualize the onset behavior; (b) entire current region.

described by the following scheme and also illustrated in Fig. 14.11:

$$\Box(\text{Ru}) + \Box(\text{Pt edge}) + \text{H}_2\text{O} \longrightarrow \text{OH}_{ad}(\text{Ru}) + \text{H}_{ad}(\text{Pt edge}) \quad [\text{see } (14.8)]$$

$$\text{H}_{ad}(\text{Pt, edge}) + \text{H}_2\text{O} \longrightarrow \Box(\text{Pt}) + \text{H}_3\text{O}^+ + e^- \quad [\text{see } (14.7)]$$

$$\text{CO}_{ad}(\text{Ru}) + \text{OH}_{ad}(\text{Ru}) \xrightarrow{\text{rds}} \text{COOH}_{ad}(\text{Ru}) + \Box(\text{Ru}) \quad [\text{see } (14.9)]$$

$$\text{COOH}_{ad}(\text{Ru}) + \text{H}_2\text{O} \xrightarrow{\text{fast}} \text{CO}_2 + \Box(\text{Ru}) + \text{H}_3\text{O}^+ + e^- \quad [\text{see } (14.10)]$$

$$\Box(\text{Ru}) + \text{CO} \longrightarrow \text{CO}_{ad}(\text{Ru}) \quad [\text{see } (14.11)]$$

It is important to note that reaction of CO$_{ad}$ occurs only at sufficiently high coverages, equivalent to a reduced reaction barrier (see the discussion of CO oxidation on Ru(0001) above). The high coverage is maintained by continuous OH$_{ad}$ formation, in competition with re-adsorption of CO. The Pt islands help in maintaining the high coverage via (14.8). Finally, additional CO adsorption on the Pt monolayer islands and reaction with OH$_{ad}$ on the Ru(0001) areas may be possible as well, and this would further increase the overall reaction rate. At these potentials, however,

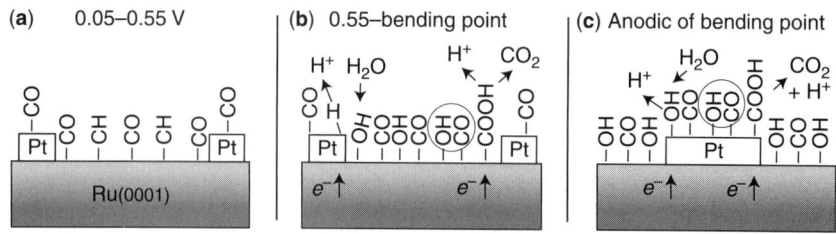

Figure 14.11 Illustration of CO electro-oxidation at Pt-modified Ru(0001): (a) mixed, non-reactive adlayer; (b) Pt-assisted formation of OH_{ad} at high local adsorbate coverages on the Ru areas; (c) CO oxidation at the Pt islands. For simplicity, H^+ is used instead of H_3O^+.

no OH_{ad} or O_{ad} will be formed on the Pt islands themselves (see also the base voltammograms).

The sudden increase in slope of the j–E curves at the bending points in the positive-going scans is most simply explained by a change in the dominant reaction pathway, which may either be directly induced by the potential or by potential-induced modifications of the adlayer composition. In the present case, we assume that, for potentials negative of the bending points, both Pt and Ru sites take part in the CO oxidation process, while at more positive potentials, CO oxidation on the Pt monolayer islands becomes dominant (Fig. 14.11c). This tentative assignment agrees with conclusions based on the Tafel slopes b of the j–E curves, which we estimate to be $b > 300$ meV/decade and $b \approx 120$ meV/decade for the regions cathodic and anodic of the bending points, respectively. Values of $b \gg 120$ meV/decade are expected for pathways involving adsorbed intermediates, whose coverage are very high and thus vary only weakly with potential [Trasatti, 2003], as expected for the densely packed, mixed CO + OH_{ad}/O_{ad} adlayers on the Pt-free Ru(0001) areas. The lower value of the Tafel slope at potentials positive of the bending points is much closer to those reported for CO electro-oxidation on Pt electrodes [Santos et al., 1991; Lebedeva et al., 2000; Shubina et al., 2004], in agreement with our assumption that in this potential region, CO oxidation takes place predominantly at Pt sites.

Similar to bulk Pt electrodes, we correlate the onset potential for CO oxidation on the Pt islands, at potentials close to the bending points, with the formation of OH_{ad}/O_{ad} species on these areas (see the base CV in Fig. 14.4), possibly at island edge sites, in competition with CO adsorption. The subsequent reaction between CO_{ad} and OH_{ad} is apparently facile under these conditions. Since the binding energies of CO_{ad} [Schlapka et al., 2003] and OH_{ad}/O_{ad} [Lischka et al., 2007] to pseudomorphic Pt layers on Ru(0001) change with the thickness of the Pt film, the onset potential for CO oxidation should also change with Pt film thickness. This explains why for the 0.9 ML Pt electrode the bending points in the positive-going and negative-going scans, and the maximum in the positive-going scan, are shifted to lower potentials compared with the surfaces with lower Pt contents: At 0.9 ML Pt, about 10% of the surface is covered by second-layer islands on top of the monolayer Pt film. Considering preliminary measurements on a 1.5 ML Pt-covered sample, which show j–E curves comparable to those for 0.9 ML Pt, the presence of second-layer islands seems to be decisive for the downshift of the bending point potential for CO oxidation. Both O_{ad}/OH_{ad} [Lischka et al., 2007] and CO_{ad} [Schlapka, 2003] bind more strongly to $Pt_{2ML}/Ru(0001)$ than to $Pt_{1ML}/Ru(0001)$, but less strongly than to Pt(111). Therefore, OH_{ad} formation is expected at lower potentials than on $Pt_{1ML}/Ru(0001)$, but at higher potentials than on Pt(111) [Climent et al., 2006]. Similar effects occur also for CO adsorption, resulting in an increasingly pronounced surface blocking by CO_{ad} in the order on $Pt_{1ML}/Ru(0001) <$ on $Pt_{2ML}/Ru(0001) <$ on Pt(111) $<$ on Ru(0001). Obviously, the stabilization of OH_{ad} overcompensates the increased tendency for CO_{ad} surface blocking for the $Pt_{2ML}/Ru(0001)$ electrode, while for the even stronger adsorbing Pt(111) electrode, the onset for CO bulk oxidation shifts upwards again. For the electrodes with low and medium Pt contents (0.05 ML Pt and 0.23 ML Pt), the potentials of the bending points and the maxima are about equal.

These arguments can be summarized in the following proposed reaction mechanism for CO oxidation at potentials anodic of the bending point:

$$\Box(Pt) + CO \longrightarrow CO_{ad}(Pt) \quad (14.13)$$

$$\Box(Pt) + H_2O \longrightarrow \{OH_{ad}(Pt) + H_{ad}(Pt)\} \longrightarrow OH_{ad}(Pt) + H^+ + e^-$$

[see (14.8)]

$$CO_{ad}(Pt) + OH_{ad}(Pt) \xrightarrow{rds} COOH_{ad}(Pt) + \Box(Pt) \quad (14.14)$$

$$COOH_{ad}(Pt) + H_2O \xrightarrow{fast} CO_2 + \Box(Pt) + H_3O^+ + e^- \quad (14.15)$$

Because of the lower adsorption energies on the Pt monolayer islands, the steady-state coverage on the islands is relatively low, and surface blocking plays no role.

14.3.2.3 PtRu/Ru(0001) Surface Alloys

Potentiodynamic CO oxidation voltammograms recorded on three $Pt_xRu_{1-x}/Ru(0001)$ surface alloys with different Pt contents are presented in Fig. 14.12a, b, together with a voltammogram for the $Pt_{0.23ML}/Ru(0001)$ surface that was already shown in Fig. 14.10. At the lowest Pt content ($x = 0.07$), the general shape of the $j–E$ trace resembles that observed for the Pt monolayer island-modified electrodes, but with a smaller slope around the onset potential at 0.55 V, which only becomes steeper at 0.8 V. After passing through a maximum at 0.9 V, it steadily decreases up to the anodic potential limit at 1.05 V. For medium ($x = 0.25$) and high ($x = 0.47$) Pt contents, the shape of the $j–E$ curves is rather similar. For the medium ($x = 0.25$) Pt content electrode, CO oxidation starts at the same potential as for the low Pt content surface ($x = 0.07$), increases slowly, and then rises steeply at $E \approx 0.8$ V. For the high Pt content ($x = 0.47$) surface, CO oxidation starts already at 0.4 V, and the transition to a steep slope occurs at $E \approx 0.75$ V. Compared with the low Pt content surface, the steep current increase of the $j–E$ curve is downshifted by about 0.06 V for $x = 0.25$ and 0.47. In both cases, the current reaches a plateau at about 0.85 V, where it slowly decreases again with further increasing potential. At the anodic limit, these electrodes still show significantly higher CO oxidation currents than are found for the $Pt_{0.07}Ru_{0.93}/Ru(0001)$ electrode. The current in the plateau, which varied with the electrolyte flow rate, reflects the transport-limited continuous CO oxidation current.

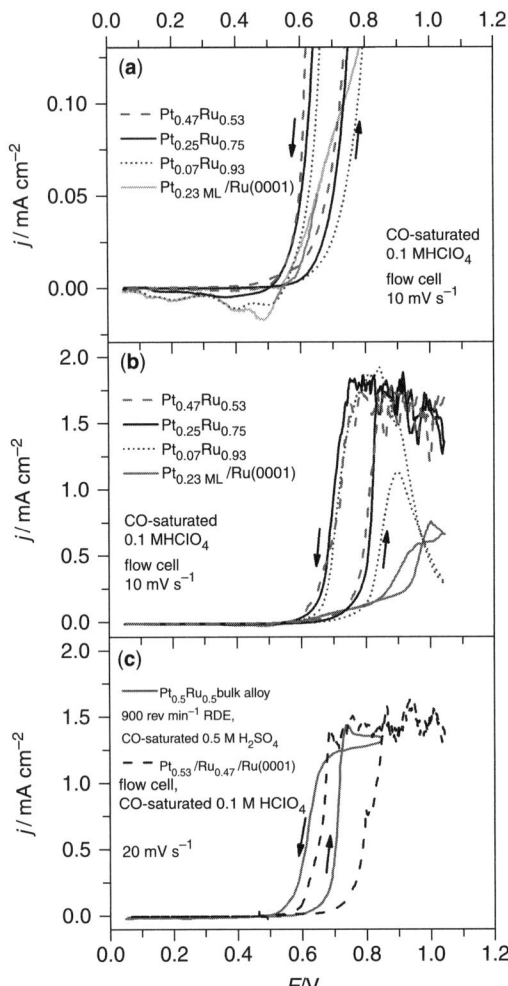

Figure 14.12 CO bulk electro-oxidation at PtRu alloys. (a, b) $Pt_xRu_{1-x}/Ru(0001)$ ($x = 0.07$, 0.25, 0.47) surface alloys measured in a flow cell with a CO-saturated electrolyte. (c) Freshly sputtered $Pt_{0.5}Ru_{0.5}$ bulk alloy in a rotating disk electrode setup (data from Gasteiger et al. [1995]), compared with a $Pt_{0.53}Ru_{0.47}/Ru(0001)$ surface alloy.

In the negative-going scan, the current remains at the same constant value for the medium and high Pt content samples ($x = 0.25$ and 0.47), until it decays steeply at about 0.75 V, and vanishes completely at 0.51 V. For the low Pt content sample, the situation is somewhat different, since the current increases first, essentially following the current trace of the positive-going scan, down to about 0.8 V, where it bends off and decays steeply at 0.69 V. For all surface alloys, we observe a negative current below 0.55 V, which for the low Pt content electrode ($x = 0.07$) has a voltammetric profile very similar to that of the $Pt_{0.23ML}/Ru(0001)$ surface. This points to a similar mixed but nonreactive $CO_{ad} + OH_{ad}/O_{ad}$ adlayer on the surface alloy and on the Pt monolayer island-covered surface at lower potentials and for low Pt content. On the PtRu/Ru(0001) surface alloys, however, the reduction current decreases much more rapidly with increasing Pt content than on the Pt monolayer island-modified surface. This agrees well with the lower OH_{ad}/O_{ad} coverage expected from the base voltammetry at $E = 0.55$ V on mixed Pt_nRu_m sites.

The CO oxidation behavior in the potential range around the reaction onset (see Fig. 14.12a) is very similar to that of the Pt monolayer island-modified Ru(0001) electrode for the low Pt content sample ($x = 0.07$). For the higher Pt surface contents, the onset of the CO oxidation reaction is shifted to lower potentials and the increase in oxidation current with potential is much steeper than for the adlayer-modified surfaces. For the $Pt_{0.47}Ru_{0.53}$/Ru(0001) surface, the onset potential for CO oxidation in the positive-going scan and the collapse of the CO oxidation current in the negative-going scan are about 0.05 V more cathodic than for surfaces dominated by truly "Ru-like" sites, as they appear on nonmodified Ru(0001) electrodes, Pt monolayer-modified Ru(0001) electrodes, or PtRu (surface) alloys with low Pt content. On the other hand, the onset potential and also the decay potential are still more anodic than that on a sputtered, polycrystalline $Pt_{0.5}Ru_{0.5}$ bulk alloy (see Fig. 14.12c) [Gasteiger et al., 1995]. It should be noted, however, that measurements on the latter sample were performed in a different electrolyte, in 0.5 M H_2SO_4 [Gasteiger et al., 1995] rather than 0.1 M $HClO_4$ (our data), which can give rise to slight differences in the potential-dependent behavior.

The significant downshift in the onset potential for CO oxidation on the PtRu/Ru(0001) monolayer surface alloys with high Pt content compared with the Pt island-modified Ru(0001) surface is explained by modifications of the local adsorption properties. Assuming that the onset of the CO oxidation Reaction (14.9) or (14.12) is determined by the activity of the O_{ad}/OH_{ad} species, and considering that the barrier for $CO_{ad} + OH_{ad}$ reaction is correlated with the stability of the initial and/or final state (Brønsted–Polanyi–Evans relation [Brønsted, 1928; Evans and Polanyi, 1938; Bondzie et al., 1999; Logadottir et al., 2001]), the downshift of the reaction onset implies a destabilization of these adspecies with increasing Pt content. As mentioned above, our studies of CO and deuterium adsorption on similar-type PtRu/Ru(0001) surface alloys under UHV conditions revealed that these adsorbates are increasingly destabilized with increasing Pt surface content owing to electronic effects (lateral ligand and strain effects), i.e., with increasing number of Pt neighbors [Rauscher et al., 2007, Diemant et al., 2008]. Similar trends can be expected also for OH_{ad}/O_{ad}, and were in fact also determined in recent DFT calculations [Hoster et al., 2009c]. These effects were made responsible also for the upshift of the peaks associated with OH_{ad}/O_{ad} formation in the base CVs with increasing Pt content (Fig. 14.8).

The much steeper increase of the current with potential in Fig. 14.12 after the initial slow increase, as compared with bare or Pt monolayer island-modified Ru(0001), is attributed to an increased abundance of OH_{ad}/O_{ad} species with increasing potential. In agreement with our previous arguments, this can be rationalized by the weaker adsorption of these species on the mixed $Ru_{3-n}Pt_n$ sites, while on Ru(0001) the "active" adsorbates required for reaction can be generated only by strongly repulsive adsorbate–adsorbate interactions. (The CO_{ad} species are also less strongly bound on the mixed surface than on Ru(0001), but this does not change directly with potential.) Owing to the lower steady-state adsorbate coverage on the (less strongly bonding) surface alloy, the formation of OH_{ad} and therefore also the following reactive removal of CO_{ad} increase more strongly with potential than on Ru(0001). This is equivalent to the general concepts based on the Tafel slope [Trasatti, 2003] described above for reaction on a Pt monolayer island-modified Ru(0001).

14.3.2.4 CO Oxidation on Bimetallic PtRu Electrodes: General Aspects

The general characteristics of the j–E curves for CO bulk oxidation on bimetallic PtRu electrodes in the region $E > 0.55$ V largely resemble those reported for CO bulk oxidation on Pt electrodes [Markovic et al., 1999]/supported Pt catalysts [Schmidt et al., 1999; Jusys et al., 2001] and on PtRu bulk alloys [Gasteiger et al., 1995] (see also Fig. 14.12c). In all cases, there is a pronounced hysteresis between the onset of CO oxidation in the positive-going scan and the current decay in the negative-going scan. Such hysteresis is generally attributed to a bistable behavior of the surface, which "switches" between two states [Ertl et al., 1982; Cox et al., 1983; Behm et al., 1983; Imbihl et al., 1984]. For CO oxidation at the gas/solid interface, the two regimes are characterized by a CO_{ad}-covered surface ("low rate branch" [Engel and Ertl, 1982]) and an essentially clean surface ("high rate branch"). For CO electro-oxidation, similar ideas apply, but with some differences. At lower potentials, the barrier for CO_2 formation rather than surface blocking by CO_{ad} is rate-limiting [Santos et al., 1991; Shubina et al., 2004; Levia and Sánchez, 2003]. At higher potentials, the surface is covered by an OH_{ad}/O_{ad} adlayer rather than being essentially adsorbate-free as for CO oxidation in the high rate branch in the gas phase (see also the simulations of CO electro-oxidation on a Pt(111) electrode [Markovic et al., 1999; Koper et al., 2001; Saravanan et al., 2002]). These assignments and explanations are largely applicable also for the Ru(0001)-based model systems. The main difference between Pt(111) and Ru(0001) is that in the latter case the removal of the strongly bound OH_{ad}/O_{ad} species is rather slow at all potentials above the onset of the reaction. This results in a high steady-state OH_{ad}/O_{ad} coverage at these potentials. In this respect, the reaction behavior of the surface alloys is much closer to that of Pt electrodes, since the dominance of strongly bound OH_{ad}/O_{ad} decreases rapidly with increasing Pt content (see also the base voltammetry data in Fig. 14.8).

Our results clearly demonstrate that the physical origin of the enhanced CO oxidation activity of bimetallic PtRu surfaces or nanoparticles depends markedly on the respective surface structure and composition. Adding Pt atoms to the Ru(0001) surface layer generally promotes the dissociation of H_2O by providing a fast pathway for adsorption and desorption of H^+. This increases the reaction rate considerably at potentials $E > 0.55$ V, where the OH_{ad}/O_{ad} species become sufficiently reactive. In the potential range $E < 0.55$ V, in contrast, the reaction of coadsorbed CO_{ad} and OH_{ad} is inhibited by the high stability of the OH_{ad}/O_{ad} species on the Ru(0001) surface areas, and this is not changed by the presence of Pt monolayer islands. Therefore, the onset potential of the CO oxidation reaction is identical for the unmodified

Ru(0001) surface and the Pt monolayer island-modified Ru(0001) surface. On the Pt_xRu_{1-x}/Ru(0001) surface alloys, the onset potential can be modified by the reduced bond energy of the OH_{ad}/O_{ad} and CO_{ad} species on the mixed Pt_xRu_{3-x} sites compared with the Ru_3 sites. This increases their reactivity and thus allows the CO oxidation reaction to start at lower potentials than on a bare or Pt island-modified Ru(0001) surface. Compared with nonmodified Pt surfaces, on the other hand, the OH_{ad}/O_{ad} bonding is strong enough on the mixed Pt_xRu_{1-x} ($x = 1, 2$) sites that H_2O splitting can occur at potentials significantly below that on nonmodified Pt surfaces (see Figs. 14.4 and 14.8). This results in the observed downshift of the onset potential to lower values compared with CO oxidation on these Pt surfaces.

Comparing PtRu monolayer surface alloys on the one hand and PtRu bulk alloy surfaces on the other, these two differ in the slightly different electronic properties of their respective surface atoms, owing to their interactions with different metal atoms in the second layer (vertical ligand effects) and the different lattice constants (strain effects) of these alloys. Both of these factors can result in further slight modifications of the electronic surface properties, and thus can affect, for example, the onset potential for CO oxidation. It is not expected, however, that this will change the major result of the present measurements, and therefore we explain the higher CO oxidation reactivity of PtRu bulk alloys compared with pure Pt or Ru electrodes in the same way as for Pt_xRu_{1-x}/Ru(0001) surface alloys (see the previous paragraph).

For comparison with bimetallic electrode surfaces prepared by electrochemical or electroless deposition of Ru on Pt(111) [Stimming and Vogel, 1998; Iwasita et al., 2000; Crown et al., 2001, 2002; Herrero et al., 1999] or Pt on Ru(0001) [Brankovic et al., 2001a, b, 2002a, b] substrates, respectively, one has to consider the rather different morphology of these deposits. These bimetallic electrodes are characterized by a large number of small deposit islands, which, depending on the amount of the respective material deposited, are mostly several layers high (multilayer islands) [Stimming and Vogel, 1998; Herrero et al., 1999; Brankovic et al., 2001a]. Because of their different electronic and geometric properties, comparison of Ru/Pt(111) surfaces with the surfaces studied here is possible only on a rather qualitative scale. They will, of course, also provide bifunctional sites at the Ru island edges, but the electronic and geometric properties of these surfaces differ considerably from those of the present system, owing to the different bulk substrate and to the different lattice constant of the deposit. For Pt on Ru(0001) [Brankovic et al., 2001a, b; Zhou et al., 2007] and Pt/Ru($10\bar{1}0$) [Brankovic et al., 2002a, b; Pinheiro et al., 2005], the electronic properties should be comparable to those of the present systems. Owing to the different morphology of the electrochemically deposited islands (multilayer island formation; see above), however, we expect considerable differences in the electrochemical and electrocatalytic properties. Base CVs of the Pt/Ru(0001) electrodes prepared by electrodeposition resemble our data [Zhou et al., 2007]; CO bulk oxidation data are not available so far.

Finally, we want to compare the main mechanistic findings of our study with the classic bifunctional mechanism, which is generally used to explain the improved CO oxidation reactivity of PtRu surfaces and catalyst particles [Watanabe and Motoo, 1975]. According to that mechanism, Ru acts as a promotor for the formation of oxygenated adspecies on bimetallic PtRu surfaces, which can then react with CO adsorbed on Pt sites. In this way, the reaction-limiting lack of oxygenated adspecies characteristic of Pt surfaces at lower potentials is counteracted by the enhanced formation of these species on the Ru sites, and the subsequent reaction between CO_{ad} (on Pt sites) and OH_{ad} (on Ru sites) is considered to be facile (for Pt, see [Lebedeva et al., 2000; Shubina et al., 2004; van Santen, 1991]). Based on the present data, we propose, in contrast, that on mixed Pt- and Ru-containing surfaces (PtRu surface alloys and PtRu bulk alloys), mixed Pt_2Ru and $PtRu_2$ sites act as active centers for the formation of OH_{ad}/O_{ad} species of optimized stability and reactivity for the CO oxidation reaction. The reaction-enhancing effect of the mixed sites with their intermediate OH_{ad}/O_{ad} stability, which is between those on pure Ru or Pt electrodes, is a classic example of the well-known Sabatier principle [Sabatier, 1913], which predicts that there is an optimum stability of the reactants for a given catalytic reaction, with lower activities resulting for a too high or too low stability of the adsorbed reactants, or the variation of the catalytic activity with the position of transition metals in the periodic table described by the volcano curves as first described by Balandin [1969]. Although closely related in its physical origin, the concept of a reaction-enhancing effect of the mixed Pt_2Ru and $PtRu_2$ sites for the CO oxidation reaction on PtRu (surface) alloys is distinctly different from the classic bifunctional mechanism.

14.4 CONCLUSIONS

The influence of Pt modifications on the electrochemical and electrocatalytic properties of Ru(0001) electrodes has been investigated on structurally well-defined bimetallic PtRu surfaces. Two types of bimetallic surfaces were considered: Ru(0001) electrodes covered by monolayer Pt islands and monolayer PtRu/Ru(0001) surface alloys with a highly dispersed and almost random distribution of the respective surface atoms, with different Pt surface contents for both types of structures. The morphology of these surfaces differs significantly from that of bimetallic PtRu surfaces prepared by electrochemical deposition of Pt on Ru(0001), where Pt predominantly exists in small multilayer islands. The electrochemical and electrocatalytic measurements, base CVs, and CO bulk oxidation under continuous electrolyte flow, led to the following conclusions:

1. Owing to their high affinity to H_{upd}, OH_{ad}, and O_{ad}, Ru(0001) surfaces are covered by strongly bound adlayers in the entire potential region between 0 and 1 V, with overlapping stability ranges of the respective adsorbates. These adlayers inhibit catalytic reactions and thus make this surface and Ru in general a rather poor electrocatalyst, for example for CO oxidation.
2. Pseudomorphic Pt monolayers on Ru(0001) interact very weakly with H_{upd}, OH_{ad}, or O_{ad}, because of electronic ligand (vertical ligand effects) and strain effects (tensile strain), in agreement with results obtained under UHV conditions and in DFT calculations. Therefore, base CVs on these surfaces do not show pronounced voltammetric features.
3. Ru(0001) surfaces partly covered by Pt monolayer islands exhibit very interesting catalytic effects. In reactions such as $H_{upd} \leftrightarrow O_{ad}/OH_{ad}$ exchange or CO

oxidation, they provide local adsorption channels for the respective second reactant, whose adsorption is inhibited on local Ru(0001) areas, and can thus catalyze the reactions. This gives rise to an accelerated exchange between OH_{ad} and H_{upd} layers with onset potentials around 0.1 V, while on an unmodified Ru(0001) electrode, the exchange reaction exhibits a much more pronounced hysteresis, with much broader peaks and less well-defined onset potentials of 0–0.05 V in the negative-going and 0.15 V in the positive-going scan.

4. For the same reason, Ru(0001) modification by Pt monolayer islands results in a pronounced promotion of the CO oxidation reaction at potentials above 0.55 V, which on unmodified Ru(0001) electrodes proceeds only with very low reaction rates. The onset potential for the CO oxidation reaction, however, is not measurably affected by the presence of the Pt islands, indicating that they do not modify the inherent reactivity of the O/OH adlayer on the Ru sites adjacent to the Pt islands. At potentials between the onset potential and a bending point in the j–E curves, CO_{ad} oxidation proceeds mainly by dissociative H_2O formation/OH_{ad} formation at the interface between the Ru(0001) substrate and Pt islands, and subsequent reaction between OH_{ad} and CO_{ad}. The Pt islands promote homolytic H_2O dissociation, and thus accelerate the reaction. At potentials anodic of the bending point, where the current increases steeply, H_2O adsorption/OH_{ad} formation and CO_{ad} oxidation are proposed to proceed on the Pt monolayer islands. The lower onset potential for CO oxidation in the presence of second-layer Pt islands compared with monolayer island-modified Ru(0001) is assigned to the stronger bonding of a double-layer Pt film (more facile OH_{ad} formation).

5. Pt_xRu_{1-x} monolayer surface alloys with an atomically disperse distribution of the two surface species offer adsorption sites ("mixed adsorption ensembles"), which are not present on pure or Pt monolayer island-modified Ru(0001) surfaces. The adsorption strength of these mixed sites, for example for H, O, or OH adsorption, decreases with increasing Pt content of the adsorption ensemble, in this case from Ru_3 via $PtRu_2$ and Pt_2Ru to Pt_3. The lower binding power of the Pt surface atoms in the mixed adsorption ensembles results from the same effects as described for Pt monolayer islands. It is further reduced by Ru neighbors, owing to the stronger interactions between Pt and Ru surface atoms compared with Pt surface atoms ("lateral electronic ligand effects"). On the other hand, compact Ru_3 ensembles maintain an adsorption behavior rather similar to that of Pt-free Ru(0001). This is illustrated by their ability to adsorb hydrogen in a sharp peak starting at 0.1 V in the negative-going scan, via reactive replacement of OH_{ad}. This proceeds in a similar mechanism as for Pt monolayer island-modified Ru(0001). Accordingly, the charge in the replacement peak scales with $(1 - \theta_{Pt})^3$, as expected for a random distribution of the respective surface atoms and OH/H adsorption on threefold sites. For mixed sites, OH adsorption and H adsorption are shifted to higher and lower potentials, respectively.

6. The same energetic modifications also affect the CO bulk oxidation. Because of the lower binding energy of the adsorbed reactants (CO_{ad} and OH_{ad}) on the intermixed surface, the barrier for reaction of these species and hence for CO_2 formation is significantly reduced compared with reaction on Ru(0001)

and also Pt(111) (energetic modification). In addition, the lower overall coverage on these surfaces affords a more facile H_2O dissociation/OH_{ad} formation process (kinetic modification). In total, these two effects result in a further downshift of the onset potential for CO bulk oxidation at higher Pt contents, and in a significantly steeper slope of the $j-E$ curves compared with Ru(0001), Pt(111), and Pt monolayer-modified Ru(0001).

From a methodological point of view, the results presented and discussed in this chapter demonstrate the potential of using structurally well-defined bimetallic electrode surfaces, whose local surface structure and composition is quantitatively known on a atomic scale, as model systems for systematic studies of the local adsorption properties, and thus of the modifications brought about by the bimetallic nature of the surface. In contrast to surfaces of bulk alloys, the bulk composition and therefore also the neighborhoods of the surface atoms, are known. Hence, maps of the atomic distribution in the outermost layer contain all the information needed to quantitatively link structure and composition on the one hand to adsorption and electrocatalytic properties on the other. Because of their well-defined structure and composition, these planar model systems allow a distinction to be drawn between contributions from different effects such as the electronic ligand, the geometric ensemble, or the lattice strain effect, and even to quantify the order of magnitude of their different contributions. Furthermore, these surfaces are particularly suited for quantitative comparison with theory, in particular with results of DFT calculations and Monte Carlo simulations. The combination of experiment and theory leads to an unprecedented understanding of the electrochemical and electrocatalytic properties of bimetallic electrodes and catalysts on a microscopic scale, and in an atomistic picture. The good agreement between experiment and theory achieved so far indicates that the description of the above properties in a local picture can, at least for the present scale of experimental and theoretical accuracy, provide an appropriate model of electrochemical/electrocatalytic reactions as sums of elementary processes.

ACKNOWLEDGMENTS

We are indebted to O. B. Alves for providing data prior to publication. Furthermore, we gratefully acknowledge fruitful discussions with O. B. Alves, A. Bergbreiter, J. Bansmann, T. Diemant, A. Groß, L. A. Kibler (all Ulm University), M. Janik (Pennsylvania State University, USA), M. T. M. Koper (Leiden University, Leiden, The Netherlands), and J. K. Nørskov (Danish Technical University, Lyngby, Denmark).

REFERENCES

Andersson K, Nikitin A, Pettersson LGM, Nilsson A, Ogasawara H. 2004. Water dissociation on Ru(001): An activated process. Phys Rev Lett 93: 196101.

Assmann J, Narkhede V, Breuer A, Muhler M, Seitsonen AP, Knapp M, Crihan D, Farkas A, Mellau G, Over H. 2003. Heterogeneous oxidation catalysis on ruthenium: Bridging the pressure and materials gaps and beyond. J Phys Cond Matt 20: 184017.

Balandin A. 1969. Modern state of the multiplet theory of heterogeneous catalysis. Adv Catal 19: 1.

Behm RJ, Thiel PA, Norton PR, Ertl G. 1983. The interaction of CO and Pt(100). I. Mechanism of adsorption and Pt phase transition, J Chem Phys 78: 7437.

Binder H, Köhling A, Sandstede G. 1972. Effect of alloying components on the catalytic activity of platinum in the case of carbonaceous fuels. In: Sandstede G, ed. From Electrocatalysis to Fuel Cells. Seattle University of Washington Press. p. 43.

Blume R, Hävecker M, Zafeiratos S, Teschner D, Kleimenov E, Knop-Gericke A, Schlögl R, Barinov A, Dudin P, Kiskinova M. 2006. Catalytically active states of Ru(0001) catalyst in CO oxidation reaction. J Catal 239: 354.

Bondzie VA, Parker SC, Campbell CT. 1999. The kinetics of CO oxidation by adsorbed oxygen on well-defined gold particles on TiO_2(110). Catal Lett 63: 143.

Böttcher A, Niehus H, Schwegmann S, Over H, Ertl G. 1997. CO oxidation reaction over oxygen-rich Ru(0001) surfaces. J Phys Chem B 101: 11185.

Brønsted JN. 1928. Acid and basic catalysis. Chem Rev 5: 231.

Brankovic SR, McBreen J, Adzic RR. 2001a. Spontaneous deposition of Pt on the Ru(0001) surface. J Electroanal Chem 503: 99.

Brankovic SR, Wang JX, Adzic RR. 2001b. Metal monolayer deposition by replacement of metal adlayers on electrode surfaces. Surf Sci 474: L173.

Brankovic SR, Marinkovic NS, Wang JX, Adzic RR. 2002a. Carbon monoxide oxidation on bare and Pt-modified Ru(10-10) and Ru(0001) single crystal electrodes. J Electroanal Chem 532: 57.

Brankovic SR, Wang JX, Zhu Y, Sabatini Y, McBreen J, Adzic RR. 2002b. Electrosorption and catalytic properties of bare and Pt modified single crystal and nanostructured Ru surfaces. J Electroanal Chem 524/525: 231.

Brune H, Röder H, Bromann K, Kern K, Jacobsen J, Stoltze P, Jacobsen K, Nørskov JK. 1996. Anisotropic corner diffusion as origin for dendritic growth on hexagonal substrates. Surf Sci 349: L115.

Buatier de Mongeot F, Scherer M, Gleich B, Kopatzki E, Behm RJ. 1998. CO adsorption and oxidation on bimetallic Pt/Ru(0001) surfaces—A combined STM and TPD/TPR study. Surf Sci 411: 249.

Christensen A, Ruban AV, Stoltze P, Jacobsen KW, Skriver HL, Nørskov JK, Besenbacher F. 1997. Phase diagram for surface alloys. Phys Rev B 56: 5822.

Christmann K. 1988. Interaction of hydrogen with solid surfaces. Surf Sci Rep 9: 1.

Christoffersen E, Liu P, Ruban A, Skriver HL, Nørskov JK. 2001. Anode materials for low temperature fuel cells—A density functional theory study. J Catal 199: 123.

Ciobica IM, Kleyn AW, van Santen RA. 2003. Adsorption and coadsorption of CO and H on ruthenium surfaces. J Phys Chem B 107: 164.

Climent V, Gómez R, Orts JM, Feliu JM. 2006. Thermodynamic analysis of the temperature dependence of OH adsorption on Pt(111) and Pt(100) electrodes in acidic media in the absence of specific anion adsorption. J Phys Chem B 110: 11344.

Cox MP, Ertl G, Imbihl R, Rüstig J. 1983. Non-equilibrium surface phase transitions during the catalytic oxidation of CO on Pt(100). Surf Sci 134: L517.

Crown A, Moraes IR, Wieckowski A. 2001. Examination of Pt(111)/Ru and Pt(111)/Os surfaces: STM imaging and methanol oxidation activity. J Electroanal Chem 500: 333.

Crown A, Johnston C, Wieckowski A. 2002. Growth of ruthenium islands on Pt(*hkl*) electrodes obtained via repetitive spontaneous deposition. Surf Sci 506: L268.

Danielson LR, Dresser MJ, Donaldson EE, Dickinson JT. 1978. Adsorption and desorption of ammonia, hydrogen and nitrogen on Ru(0001). Surf Sci 71: 599.

Davies JC, Bonde J, Logadottir A, Nørskov JK, Chorkendorff I. 2005. The ligand effect; CO desorption from PtRu catalysts. Fuel Cells 4: 429.

Diemant T, Hager T, Hoster HE, Rauscher H, Behm RJ. 2003. Hydrogen adsorption and coadsorption with CO on well-defined bimetallic PtRu surfaces—A model study on the CO tolerance of bimetallic PtRu anode catalysts in low temperature polymer electrolyte fuel cells, Surf Sci 541: 137.

Diemant T, Hager T, Rauscher H, Behm RJ. In preparation.

Diemant T, Rauscher H, Behm RJ. 2008. Interaction of deuterium with well-defined Pt_xRu_{1-x}/Ru(0001) surface alloys. J Phys Chem C 112: 8381.

El-Aziz AM, Kibler LA. 2002. New information about the electrochemical behaviour of Ru(0001) in perchloric acid solutions. Electrochem Commun 4: 866.

Engel T, Ertl G. 1982. Oxidation of carbon monoxide. In: King DA, Woodruff DP, eds. The Chemical Physics of Solid Surfaces and Heterogeneous Catalysis. Volume 4. Amsterdam: Elsevier.

Ertl G, Norton PR, Rüstig J. 1982. Kinetic oscillations in the platinum-catalyzed oxidation of CO. Phys Rev Lett 49: 177.

Evans MG, Polanyi M. 1938. Inertia and driving force of chemical reactions. Trans Faraday Soc 34: 11.

Feibelman PJ. 2002. Partial dissociation of water on Ru(0001). Science 295: 99.

Feulner P, Menzel D. 1985. The adsorption of hydrogen on Ru(0001): Adsorption states, dipole moments and kinetics of adsorption and desorption. Surf Sci 154: 465.

Garcia-Araez N, Climent V, Herrero E, Feliu J, Lipkowski J. 2006. Thermodynamic approach to the double layer capacity of a Pt(111) electrode in perchloric acid solutions. Electrochim Acta 51: 3787.

Gasteiger HA, Markovic N, Ross PN, Cairns EJ. 1994a. CO electrooxidation on well-characterized Pt-Ru alloys. J Phys Chem 98: 617.

Gasteiger HA, Markovic NM, Ross PN, Cairns EJ. 1994b. Electro-oxidation of small organic molecules on well-characterized Pt-Ru alloys, Electrochim Acta 39: 1825.

Gasteiger HA, Markovic NM, Ross PN. 1995. H_2 and CO electrooxidation on well-characterized Pt, Ru, and Pt-Ru. 1. Rotating disk electrode studies of the pure gases including temperature effects. J Phys Chem 99: 8290.

Gdowski GE, Fair JA, Madix RJ. 1983. Reactive scattering of small molecules from platinum crystal surfaces: D_2CO, CH_3OH, HCOOH and the nonanomalous kinetics of hydrogen atom recombination. Surf Sci 127: 541.

Ge Q, Desai S, Neurock M, Kourtakis K. 2001. CO adsorption on Pt-Ru surface alloys and on the surface of Pt-Ru bulk alloy. J Phys Chem B 106: 9533.

Goodman DW, Peden CHF, Chen MS. 2007. CO oxidation on ruthenium: The nature of the active catalytic surface. Surf Sci 601: L124.

Greeley J, Mavrikakis M. 2005. Surface and subsurface hydrogen: Adsorption properties on transition metals and near-surface alloys. J Phys Chem B 109: 3460.

Groß A. 2006. Reactivity of bimetallic systems studied from first principles. Top Catal 37: 29.

Gsell M, Jakob P, Menzel D. 1998. Effect of substrate strain on adsorption. Science 280: 717.

Hammer B, Nørskov JK. 1995. Why gold is the noblest of all the metals. Nature 376: 238.

Hammer B, Morikawa Y, Nørskov JK. 1996. CO chemisorption at metal surfaces and overlayers. Phys Rev Lett 76: 2141.

Hammer B, Nielsen OH, Nørskov JK. 1997. Structure sensitivity in adsorption: CO interaction with stepped and reconstructed Pt surfaces. Catal Lett 46: 31.

Hamnett A. 2003. In: Vielstich W, Gasteiger HA, Lamm A, eds. Electrocatalysis. Volume 2. Chichester: Wiley.

Han BC, Van der Ven A, Ceder G, Hwang BJ. 2006. Surface segregation and ordering of alloy surfaces in the presence of adsorbates. Phys Rev B 72: 205409.

Heinen M, Chen Y-X, Jusys Z, Behm RJ. 2006. Room temperature CO_{ad} desorption/exchange kinetics on Pt electrodes—A combined in-situ IR and mass spectrometry study. ChemPhysChem 8: 2484.

Held G, Pfnür H, Menzel D. 1992. A LEED-IV investigation of the Ru(001)-p(2×1)-H structure. Surf Sci 271: 21.

Herrero E, Feliu JM, Wieckowski A. 1999. Scanning tunneling microscopy images of ruthenium submonolayers spontaneously deposited on a Pt(111) electrodes. Langmuir 15: 4944.

Hoster H, Bergbreiter A, Erne P, Hager T, Rauscher H, Behm RJ. 2008. Atomic distribution in well-defined Pt_xRu_{1-x}/Ru(0001) monolayer surface alloys. Phys Chem Chem Phys 10: 3812.

Hoster H, Iwasita T, Baumgärtner H, Vielstich W. 2001a. Current–time behavior of smooth and porous PtRu surfaces for methanol oxidation. J Electrochem Soc 148: A496.

Hoster H, Iwasita T, Baumgärtner H, Vielstich W. 2001b. Pt-Ru model catalysts for anodic methanol oxdiation: Influence of structure and composition on the reactivity. Phys Chem Chem Phys 3: 337.

Hoster H, Richter B, Behm RJ. 2004. Catalytic influence of Pt monolayer islands on the hydrogen electrochemistry of Ru(0001) studied by ultrahigh vacuum scanning tunneling microscopy and cyclic voltammetry. J Phys Chem B 108: 14780.

Hoster HE, Alves OB, Koper MTM, Behm RJ. 2009a. In preparation.

Hoster HE, Bergbreiter A, Erne P, Behm RJ. 2009b. In preparation.

Hoster HE, Janik MJ, Neurock M, Behm RJ. 2009c. In preparation.

Hwang RQ, Schröder J, Günther C, Behm RJ. 1991. Fractal growth of two-dimensional islands: Au on Ru(0001). Phys Rev Lett 67: 3279.

Imbihl R, Cox MP, Ertl G, Müller H, Brenig W. 1984. Kinetic oscillations in the catalytic CO oxidation on Pt(100): Theory. J Chem Phys 83: 1578.

Iwasita T, Hoster H, John-Anacker A, Lin W-F, Vielstich W. 2000. Methanol oxidation on PtRu-electrodes: Influence of surface structure and Pt-Ru distribution. Langmuir 16: 522.

Jachimowski TA, Meng B, Johnson DF, Weinberg WH. 1995. Thermal desorption studies of high-coverage hydrogen overlayers on Ru(001) created with gas-phase atomic hydrogen. J Vac Sci Technol A 13: 1564.

Jakob P, Schlapka A. 2007. CO adsorption on epitaxially grown Pt layers on Ru(0001), Surf Sci 601: 3556.

Jerkiewicz G. 1998. Hydrogen sorption at/in electrodes. Prog Surf Sci 57: 137.

Jusys Z, Kaiser J, Behm RJ. 2001. Electrooxidation of CO and H_2/CO mixtures on a carbon supported Pt catalyst—A kinetic and mechanistic study by differential electrochemical mass spectrometry. Phys Chem Chem Phys 3: 4650.

Jusys Z, Kaiser J, Behm RJ. 2002. Composition and activity of high surface area PtRu catalysts towards adsorbed CO and methanol electrooxidation. A DEMS study. Electrochim Acta 47: 3693.

Käsberger U, Jakob P. 2003. Growth and thermal evolution of submonolayer Pt films on Ru(0001) studied by STM. Surf Sci 540: 76.

Karlberg GS, Jaramillo DM, Skúlason E, Rossmeisl J, Bligaard T, Nørskov JK. 2007. Cyclic Voltammograms for H on Pt(111) and Pt(100) from first principles. Phys Rev Lett 99: 126101.

Kopatzki KE. 1994. Sauerstoffadsorption, Oxidbildung und Homoepitaxie auf Ni(100) Oberflächen—Eine Untersuchung mit dem Rastertunnelmikroskop. PhD Dissertation, Ludwig-Maximilians-Universität, München.

Koper MTM. 2004. Electrocatalysis on bimetallic and alloy surfaces. Surf Sci 548: 1.

Koper MTM, Schmidt TJ, Markovic NM, Ross PN. 2001. Potential oscillations and S-shaped polarization curve in the continuous electro-oxidation of CO on platinum single-crystal electrodes. J Phys Chem B 105: 8381.

Koper MTM, Shubina TE, van Santen RA. 2002. Periodic density functional study of CO and OH adsorption on Pt-Ru alloy surfaces: Implications for CO-tolerant fuel cell catalysts. J Phys Chem 106: 686.

Kostov KL, Rauscher H, Menzel D. 1992. Adsorption of CO on oxygen-precovered Ru(0001). Surf Sci 278: 62.

Kostov KL, Gsell M, Jakob P, Moritz T, Widdra W, Menzel D. 1997. Observation of a novel high density $3O(2 \times 2)$ structure on Ru(001). Surf Sci 394: L138.

Lebedeva NP, Koper MTM, Feliu JM, van Santen RA. 2000. Role of crystalline defects in electrocatalysis: Mechanism and kinetics of CO adlayer oxidation on stepped polatinum electrodes. J Phys Chem B 106: 12938.

Leiva EPM, Sánchez C. 2003. Theoretical aspects of some prototypical fuel cell reactions. In: Vielstich W, Gasteiger HA, Lamm A, eds. Electrocatalysis. Volume 2. Chichester: Wiley.

Lin W-F, Zei MS, Eiswirth M, Ertl G, Iwasita T, Vielstich W. 1999. Electrocatalytic activity of Ru modified Pt(111) electrodes towards CO oxidation. J Phys Chem B 103: 6968.

Lin W-F, Christensen PA, Hamnett A, Zei MS, Ertl G. 2000. The electro-oxidation of CO at the Ru(0001) single-crystal electrode surface. J Phys Chem B 104: 6642.

Lindroos M, Pfnür H, Menzel D. 1987. Investigation of a disordered adsorption system by electron reflection: H/Ru(0001) at intermediate coverages. Surf Sci 192: 421.

Lischka M, Mosch C, Gross A. 2007. Tuning catalytic properties of bimetallic surfaces: Oxygen adsorption on pseudomorphic Pt/Ru overlayers, Electrochim Acta 52: 2219.

Liu P, Nørskov JK. 2001a. Kinetics of the anode processes in PEM fuel cells—The promoting effect of Ru in PtRu anodes. Fuel Cells 1: 192.

Liu P, Nørskov JK. 2001b. Ligand and ensemble effects in adsorption on alloy surfaces. Phys Chem Chem Phys 3: 3814.

Liu P, Logadóttir, Nørskov JK. 2003. Modeling the electro-oxidation of CO and H_2/CO on Pt, Ru, PtRu and Pt_3Sn. Electrochim Acta 48: 3731.

Logadottir A, Rod TH, Nørskov JK, Hammer B, Dahl S, Jacobsen CJH. 2001. The Brønsted–Evans–Polanyi relation and the volcano plot for ammonia synthesis over transition metal catalysts. J Catal 197: 229.

Madey TE, Engelhardt HA, Menzel D. 1975. Adsorption of oxygen and oxidation of CO on the ruthenium (001) surface. Surf Sci 48: 304.

Markovic NM, Ross PN Jr.. 2002. Surface science studies of model fuel cell electrocatalysts. Surf Sci Rep 45: 117.

Markovic NM, Schmidt TJ, Grgur BN, Gasteiger HA, Behm RJ, Ross PN. 1999. The effect of temperature on the surface process at the Pt(111)–liquid interface: Hydrogen adsorption, oxide formation and CO oxidation. J Phys Chem B 103: 8568.

Mavrikakis M, Hammer B, Nørskov JK. 1998. Effect of strain on the reactivity of metal surfaces. Phys Rev Lett 81: 2819.

McEwen JS, Eichler A. 2007. Phase diagram and adsorption-desorption kinetics of CO on Ru(0001) from first principles. J Chem Phys 126: 094701.

Michalk G, Moritz W, Pfnür H, Menzel D. 1983. A LEED determination of the structures of Ru(001) and of CO/Ru(001)-$\sqrt{3} \times \sqrt{3}\,R30°$. Surf Sci 129: 92.

Narloch B, Held G, Menzel D. 1994. Structural rearrangement by coadsorption: a LEED IV determination of the Ru(001)-p(2×2)($2O + CO$) structure. Surf Sci 317: 131.

Nashner MS, Frenkel AI, Adler DL, Shapley JR, Nuzzo RG. 1997. Structural characterization of carbon supported platinum-ruthenium nanoparticles from the molecular cluster precursor PtRu$_5$(CO)$_{16}$. J Am Chem Soc 119: 7760.

Nielsen LP, Besenbacher F, Stensgaard I, Laegsgaard E, Engdahl C, Stoltze P, Jacobsen KW, Nørskov JK. 1995. Initial growth of Au on Ni(110): Surface alloying of non-miscible metals. Phys Rev Lett 71: 754.

Over H, Muhler M. 2003. Catalytic CO oxidation over ruthenium—Bridging the pressure gap. Prog Surf Sci 72: 3.

Over H, Muhler M, Seitsonen AP. 2007. Comment on "CO oxidation on ruthenium: The nature of the active catalytic surface" by Goodman DW, Peden CHF, Chen MS. Surf Sci 601: 5659.

Peden CHF, Goodman DW. 1986. Kinetics of CO oxidation over Ru(0001). J Phys Chem 1360.

Peebles DE, Schreifels JA, White JM. 1982. The interaction of coadsorbed hydrogen and carbon monoxide on Ru(0001). Surf Sci 116: 117.

Peppley BA, Amphlett JC, Mann RF. 2003. In: Vielstich W, Lamm A, Gasteiger HA, eds. Handbook of Fuel Cells—Fundamentals Technology and Applications. Volume 3: Fuel Cells and Technology. Chichester: Wiley.

Petry OA, Podlovchenko BI, Frumkin AN, Lal H. 1965. The behaviour of platinized-platinum and platinum–ruthenium electrodes in methanol solutions. J Electroanal Chem 10: 253.

Pfnür H, Menzel D. 1983. The influence of adsorbate interactions on kinetics and equilibrium for CO on Ru(001). I. Adsorption kinetics. J Chem Phys 79: 2400.

Pfnür H, Menzel D, Hoffmann FM, Ortega A, Bradshaw AM. 1980. High resolution vibrational spectroscopy of CO on Ru(001): The importance of lateral interactions. Surf Sci 93: 431.

Pfnür H, Feulner P, Menzel D. 1983. The influence of adsorpate interactions on kinetics and equilibrium for CO on Ru(001)—II. Desorption kinetics and equilibrium. J Chem Phys 79: 4613.

Pfnür H, Held G, Lindroos M, Menzel D. 1989. Oxygen induced reconstruction of a close-packed surface: A LEED IV study on Ru(001)-p(2×1)O. Surf Sci 220: 43.

Pinheiro ALN, Zei MS, Ertl G. 2005. Electro-oxidation of carbon monoxide and methanol on bare and Pt-modified Ru(10−10) electrodes. Phys Chem Chem Phys 7: 1300.

Rauscher H, Hager T, Diemant T, Hoster H, Buatier de Mongeot F, Behm RJ. 2007. Interaction of CO with atomically well-defined Pt_xRu_y/Ru(0001) surface alloys. Surf Sci 601: 4608.

Riedmüller B, Papageorgopoulos DC, Berenbak B, van Santen RA, Kleyn AW. 2002. "Magic" island formation of CO coadsorbed with H on Ru(0001). Surf Sci 515: 323.

Ruban AV, Skriver HL, Nørskov JK. 1999. Surface segregation energies in transition-metal alloys. Phys Rev B 59: 15990.

Sabatier P. 1913. La Catalyse en Chimie Organique. Paris: Librairie Polytechnique Béranger.

Sachtler WMH. 1973. Surface composition of alloys in equilibrium, Le Vide 164: 67.

Sachtler WMH, Somorjai GA. 1983. Influence of ensemble size on CO chemisorption and catalytic n-hexane conversion by Au-Pt(111) bimetallic single-crystal surfaces. J Catal 81: 77.

Sachtler WMH. 1997. Factors influencing catalytic action—Ensemble and ligand effects in metal catalysis. In: Ertl G, Knözinger H, Weitkamp J, eds. Handbook of Heterogeneous Catalysis. Volume 3. Weinheim: VCH-Wiley.

Santos E, Leiva EPM, Vielstich W. 1991. CO adsorbate on Pt(111) single crystal surfaces. Electrochim Acta 36: 555.

Saravanan C, Koper MTM, Markovic NM, Head-Gordon M, Ross PN. 2002. Modeling base voltammetry and CO electrooxidation at the Pt(111)–electrolyte interface: Monte Carlo simulations including anion adsorption. Phys Chem Chem Phys 4: 2660.

Schick M, Xie J, Mitchell WJ, Weinberg WH. 1996. Interaction of gas-phase atomic deuterium with the Ru(001)-p(1 × 2)-O surface: Kinetics of hydroxyl and water formation, J Chem Phys 104: 7713.

Schiffer A, Jakob P, Menzel D. 1997. The (2CO + O)(2 × 2)/Ru(001) layer: Preparation, characterization, and analysis of interaction effects in vibrational spectra. Surf Sci 389: 116.

Schiffer A, Jakob P, Menzel D. 2000. Structure and lateral interactions in binary and ternary coadsorbate layers of O, H and CO on Ru(001). Surf Sci 465: 198.

Schlapka A, Käsberger U, Menzel D, Jakob P. 2002. Vibrational spectroscopy of CO used as a local probe to study the surface morphology of Pt on Ru(0001) in the submonolayer regime. Surf Sci 502/503: 129.

Schlapka A, Lischka M, Groß A, Käsberger U, Jakob P. 2003. Surface strain versus substrate interaction in heteroepitaxial metal layers: Pt on Ru(0001). Phys Rev Lett 91: 016101.

Schmid M, Stadler H, Varga P. 1993. Direct observation of surface chemical order by scanning tunneling microscopy. Phys Rev Lett 70: 1441.

Schmidt TJ, Gasteiger HA, Behm RJ. 1999. Rotating disk electrode measurements on a high-surface area Pt/Vulcan carbon fuel cell catalyst. J Electrochem Soc 146: 1296.

Schwarz JA. 1979. Adsorption-desorption kinetics of H_2 from clean and sulfur covered Ru(0001). Surf Sci 87: 525.

Shi S-K, Schreifels JA, White JM. 1981. Titration of chemisorbed oxygen by hydrogen on Ru(001). Surf Sci 105: 1.

Shimizu H, Christmann K, Ertl G. 1980. Model studies on bimetallic Cu/Ru catalysts II. Adsorption of hydrogen. J Catal 61: 412.

Shubina TE, Hartnig C, Koper MTM. 2004. Density functional theory study of the oxidation of CO by OH on Au(110) and Pt(111) surface. Phys Chem Chem Phys 6: 4215.

Soma-Noto Y, Sachtler WMH. 1974. Infrared spectra of carbon monoxide adsorbed on supported palladium and palladium-silver alloys, J Catal 32: 315.

Stampfl C, Schwegmann S, Over H, Scheffler M, Ertl G. 1996. Structure and stability of a high-coverage (1×1) oxygen phase on Ru(0001). Phys Rev Lett 77: 3371.

Stampfl C, Kreuzer HJ, Payne SH, Pfnür H, Scheffler M. 1999. First-principles theory of surface thermodynamics and kinetics. Phys Rev Lett 83: 2993.

Stimming U, Vogel R. 1998. In-situ local probe techniques at electrochemical interfaces. In: Lorenz WJ, Plieth W, eds. Electrochemical Nanotechnology. Weinheim: Wiley-VCH.

Sun YK, Weinberg WH. 1985. Determination of the absolute saturation coverage of hydrogen on Ru(0001). Surf Sci 214: L246.

Taylor CD, Kelly RG, Neurock M. 2007. First-principles prediction of equilibrium potentials for water activation by a series of metals. J Electrochem Soc 154: F217.

Thiel PA, Madey TE. 1987. The interaction of water with solid surfaces: Fundamental aspects. Surf Sci Rep 7: 211.

Thomas GE, Weinberg WH. 1979. The vibrational spectrum and adsorption site of CO on the Ru(0001) surface. J Chem Phys 70: 1437.

Trasatti S. 2003. Reaction mechanism and rate determining steps. In: Vielstich W, Gasteiger HA, Lamm A, eds. Electrocatalysis. Volume 2. Chichester: Wiley.

van Santen RA. 1991. Chemical basis of metal catalyst promotion. Surf Sci 251/252: 6.

Wang WB, Zei MS, Ertl G. 2001. Electrosorption and electrooxidation of CO on Ru(0001). Phys Chem Chem Phys 3: 3307.

Watanabe M, Motoo S. 1975. Electro catalysis by ad-atoms. Part III. Enhancement of the oxidation of carbon monoxide on platinum by ruthenium adatoms. J Electroanal Chem 60: 275.

Watanabe M, Igarashi H, Fujino T. 1999. Design of CO tolerant anode catalysts for polymer electrolyte fuel cells. Electrochemistry 67: 1194.

Weiss MJ, Hagedorn CJ, Weinberg WH. 1998. Observation of the reaction of gas phase atomic oxygen with Ru(0001)-p(1×1)D at 80 K. J Vac Sci Technol A 16: 3521.

Weissenrieder J, Mikkelsen A, Andersen JN, Feibelman PJ, Held G. 2004. Experimental evidence for a partially dissociated water bilayer on Ru{0001}. Phys Rev Lett 93: 196102.

Williams ED, Weinberg WH. 1979. The geometric structure of carbon monoxide chemisorbed on the ruthenium (001) surface at low temperatures. Surf Sci 82: 93.

Zei MS, Ertl G. 2000. Structural changes of a Ru(0001) surface under the influence of electrochemical reactions. Phys Chem Chem Phys 2: 3855.

Zhou W-P, Lewera A, Bagus PS, Wieckowski A. 2007. Electrochemical and electronic properties of platinum deposits on Ru(0001): Combined XPS and cyclic voltammetric study. J Phys Chem C 111: 13490.

CHAPTER 15

Size Effects in Electrocatalysis of Fuel Cell Reactions on Supported Metal Nanoparticles

FREDERIC MAILLARD

Laboratoire d'Electrochimie et de Physico-chimie des Matériaux et des Interfaces, LEPMI UMR 5631 CNRS/UJF/G-INP, BP75, 38402 Saint Martin d'Hères, France

SERGEY PRONKIN and ELENA R. SAVINOVA*

LMSPC UMR-7515, ECPM, Université de Strasbourg, 67087 Strasbourg, France

15.1 INTRODUCTION

For several decades, fuel cell research and development have grown in line with the worldwide demand for energy, particularly with global warming concerns and the consequent requirements to reduce consumption of fossil fuels (petrol, coal, and gas), which produce greenhouse gases. Recently, considerable effort has been directed to the development of polymer electrolyte membrane fuel cells (PEMFCs), which are considered as a promising solution to provide electricity for small stationary, mobile and portable applications [Carrette et al., 2000]. Various metal-rich electrocatalysts, usually based on carbon-supported Pt and/or its alloys, are used to accelerate fuel cell reactions in PEMFCs: the oxygen (electro)reduction reaction (ORR) at the cathode and the hydrogen (electro)oxidation reaction (HOR), formic acid (electro)oxidation reaction (FOR), or methanol (electro)oxidation reaction (MOR) at the anode. Decreasing the particle size to nanometer dimensions allows an increase in the ratio between the number of atoms on the surface N_s and the overall number of metal atoms N_{total} (the so-called dispersion $D = N_s/N_{total}$), thus saving expensive precious metals. If, for a given reaction, the *specific electrocatalytic activity* (i.e., the activity

---

*Former address: Boreskov Institute of Catalysis, Russian Academy of Sciences, Pr. Akademika Lavrentieva 5, 630090 Novosibirsk, Russia.

Fuel Cell Catalysis. Edited by Marc T. M. Koper
Copyright © 2009 John Wiley & Sons, Inc.

per unit surface area of the active component) does not depend on the size of the particles, the reaction is classified as "facile" or "structure-insensitive" by analogy with the definition given by Boudart for heterogeneous catalysis [Boudart, 1969]. For this kind of reaction, dispersion may be increased effectively until the metal–insulator transition is reached, with the *mass activity* (i.e., the activity per gram of the active component) increasing in inverse proportion to the particle size. On the other hand, if the specific electrocatalytic activity of nanoparticles depends on the particle size, the reaction is classified as "demanding" or "surface-sensitive" [Boudart, 1969]. This holds especially true for multielectron reactions occurring in PEMFCs at temperatures below about 130 °C. Indeed, it is now well established that the electrocatalytic activity of Pt for the ORR [Kabbabi et al., 1994; Takasu et al., 1996; Gamez et al., 1996; Genies et al., 1998], the ORR in methanol-containing electrolytes [Maillard et al., 2002], the MOR [Yahikozawa et al., 1991; Kabbabi et al., 1994; Frelink et al., 1995; Takasu et al., 2000; Cherstiouk et al., 2003a] and CO electrooxidation [Takasu et al., 1997, 2000; Cherstiouk et al., 2003a, b; Maillard et al., 2004a, b, 2005, 2007b; Mayrhofer et al., 2005b] is strongly dependent on the particle size. In this case, the following questions arise:

- How do particle size and structure influence catalytic activity?
- What are the physical reasons for this dependence?
- Ultimately, can catalytic activity and stability be purposely tuned via modification of particle size and nanostructure?

These questions are of both applied and fundamental importance, since answering them will further our understanding of electrocatalysis.

This chapter is organized as follows. In Section 15.2, we will briefly discuss structural and electronic properties characteristic of supported metal nanoparticles, and will provide the reader with references where more information can be found. In Section 15.3, we will analyze various model systems currently utilized for studying particle size effects (PSEs) in electrocatalysis, along with their advantages and limitations. Sections 15.4 and 15.5 will be devoted to the adsorption and electrocatalytic properties of supported metal nanoparticles. Owing to obvious space constraints, it will not be possible to give a comprehensive treatment of all the published data, so we will rather present a selective review essentially pinpointing studies in which we have been directly involved. In Section 15.6, we will attempt to summarize the current state of understanding of PSEs in electrocatalysis, and will present our vision of further developments in the vitally important area of electrocatalysis concerned with size and structural effects.

15.2 STRUCTURE AND PROPERTIES OF METAL NANOPARTICLES

An inherent property of nanoparticulate materials is their high dispersion D, which for spheres scales with the inverse diameter $1/d$ and with $N_{\text{total}}^{-1/3}$. For example, for a spherical metal (e.g., Pt) particle with a size of 1 nm, $D \approx 0.9$, i.e., 90% of the atoms are exposed to the surface, and this fact may strongly affect the physical and chemical properties of the particle. In the so-called scalable interval of particle sizes, physical properties of materials that are determined by the surface-to-volume ratio scale with the inverse diameter [Roduner, 2006]. The so-called nonscalable interval provides the most spectacular manifestations of size effects on material properties, and this has been widely documented in solid state physics. However, until now, all published investigations of the electrochemical and electrocatalytic properties of materials have fallen in the scalable size interval. As discussed below, a variety of physical properties of nanoparticulate materials in the scalable size interval may be explained within the thermodynamic approach. After a short introduction to the thermodynamics of disperse systems, we will briefly discuss structural and electronic properties of metal nanoparticles, which will be essential for further analysis of PSEs in electrocatalysis.

15.2.1 Thermodynamics of Small Particles

The foundations of classical thermodynamics of disperse materials were laid by Gibbs and are described in nearly every undergraduate course of chemical thermodynamics. To learn about more recent developments, the reader is referred to Hill's treatise [Hill, 1963, 1964]. The essential quantity in the thermodynamics of disperse materials is the surface tension γ, which for liquids equals the surface free energy σ. The excess surface energy results in the chemical potential of disperse materials $\mu(d)$ being higher than that for the corresponding bulk. For example, for spherical liquid droplets of diameter d, the chemical potential is given by the Gibbs–Thompson relation:

$$\mu(d) = \mu(d = \infty) + \frac{4v_m \gamma}{d} \quad (15.1)$$

where v_m is the molar volume. An important consequence is that the pressure inside a droplet exceeds the outer pressure by the so-called Laplace pressure, which for a sphere is equal to

$$\Delta p_L = \frac{4\gamma}{d} \quad (15.2)$$

Although (15.1) and (15.2) hold strictly only for a liquid in equilibrium with its vapor, they have been commonly applied also to solid materials, in particular for describing nanoparticle sintering (see the discussion in [Campbell et al., 2002]). However, a number of complications must be considered for solid materials. First of all, $\gamma \neq \sigma$, since for a solid a change in the surface area A can be realized either by increasing the number of surface atoms without changing the interatomic distances between them (this is related to the first term in (15.3)) or by introducing a strain (this is related to the second term in (15.3)):

$$\gamma = \sigma + A \frac{\partial \sigma}{\partial A} \quad (15.3)$$

Equation (15.3) is valid for isotropic solids. For anisotropic solid materials, γ is a tensor. For a crystal comprising terraces, edges, and vertices, the surface energy may be obtained as a sum over all structural elements. Depending on the strain resistance of a material, γ may be either larger or smaller than σ. For various metals, γ has been calculated using pseudopotentials [Payne et al., 1989]. For (111) surfaces of Al, Ir, Pt, and Au, the ratio γ/σ was calculated as 1.3, 1.6, 2.5, and 2.2, respectively. Thus, for particles of these metals, the chemical potential is expected to increase with decreasing particle size even more strongly than predicted by the Gibbs–Thompson relation.

As pointed out by Nagaev, a further complication that has to be considered for solid particles that are not in equilibrium with their melt or vapor is that material exchange between the particles and the environment is lacking [Nagaev, 1991, 1992]. He has demonstrated that in this case of "one-phase thermodynamics," the Laplace excess pressure is not a real physical force, but may be considered a formal quantity describing the surface renormalization of the chemical potential. Nevertheless, the chemical potential does depend on the size, and, strictly speaking, this dependence is not determined by the Laplace pressure, but rather by the *high contribution of surface atoms* whose energy is different from that of the bulk. Thus, alterations in the chemical potential are predicted also for thin films, although their curvature is zero. For the particular case of isotropic spherical solid particles, the predicted dependence of μ on the radius is similar to that given by the Gibbs–Thompson relation [Nagaev, 1992]. This is a very important conclusion, which lends validity to attempts to utilize thermodynamic notions to predict (at least qualitatively) the properties of particulate materials.

An essential issue concerns the size down to which the laws of classical thermodynamics apply. A simplified answer is that macroscopic thermodynamics is applicable as long as the splitting δ between the electronic energy levels is less than the thermal energy (see Section 15.2.2):

$$\delta \ll k_B T \qquad (15.4)$$

It should be pointed out, however, that even when the inequality (15.4) holds, quantitative deviations from the laws of classical thermodynamics may be substantial if the numbers of the surface and volume atoms become comparable. Still, in many cases, classical thermodynamics provides a reasonable account of the influence of size on physical properties (e.g., the melting temperature) down to the nanometer range. For example, using a scanning electron diffraction technique, Buffat and Borel performed an elegant study of the influence of the size of Au nanocrystals supported on carbon on their melting temperature [Buffat and Borel, 1976]. Their phenomenological model, based essentially on the equations of classical thermodynamics, is in excellent agreement with the experimental results in the particle interval from 25 nm down to about 2.5 nm. At smaller sizes, deviations may, however, be significant. A number of refinements have been proposed in order to apply thermodynamics to nano-objects [Hill, 1963, 1964; Rusanov, 2005]. Campbell and co-workers performed microcalorimetric measurements of the heat of adsorption of Pb onto MgO(100), and showed that the energy of a metal atom in a nanoparticle increases with decreasing size below about 2.5 nm much more dramatically than predicted by the Gibbs–Thompson relation [Campbell et al., 2002]. They proposed a simple pairwise bond-additivity model, which reproduces the dramatic dependence of the energy on the cluster size very well in the size interval from 1 to 4 nm. Sun developed a bond order–length–strength correlation model based on the following principles: (i) relaxation of bond length in nanoparticles because of the reduced atomic coordination numbers (CN) at the surface; (ii) increased binding energy of the relaxed bonds; and (iii) concomitant variation of cohesive energy of the surface atoms from the bulk value [Sun, 2007]. The model is capable of explaining a variety of nanoscale phenomena, including phase transitions, sintering, chemical reactivity, and thermal stability.

Although, for obvious reasons, classical thermodynamics cannot provide a quantitative account of the vast variety of phenomena occurring on the nanoscale, it does make some useful semiquantitative predictions in the scalable size interval. For example, based on the Gibbs–Thompson relation, Plieth predicted that for a metal composed of small particles, the redox potential $E(d)$ of the M^{z+}/M^0 transition must scale with the inverse particle diameter and shift negatively with respect to the value characteristic of the bulk metal [Plieth, 1982]:

$$E(d) = E(d = \infty) - \frac{4v_m \gamma}{zFd} \tag{15.5}$$

Despite the limitations of (15.5) [Horanyi, 1985], it explains reasonably well the Ostwald ripening of Ag particles by the dissolution of small particles and the growth of large particles [Redmond et al., 2005; Schroeder et al., 2006]. The negative shift of the Cu^{2+}/Cu^0 redox potential for 3 nm colloidal Cu particles with respect to the value for bulk Cu has been confirmed, for example, in Savinova et al. [1988].

It has recently been suggested that the thermodynamic approach may also be helpful in understanding PSEs in catalysis and adsorption [Parmon, 2007; Savinova, 2006]. The idea is essentially based on the assumption that the activation energy for adsorption (desorption) occurring on the surface of heterogeneous catalysts must depend on the chemical potential of the latter through the Brønsted–Evans–Polanyi rule establishing a linear relationship between the activation energy and the heat of reaction. Given that the chemical potential of a catalyst in a dispersed state depends on the particle size, this simple reasoning predicts a particle size dependence of adsorption/desorption and heterogeneous catalytic reactions. In other words, the adsorption/desorption energetics must be influenced by particle size. An influence of the chemical potential of metal nanoparticles on the rates of electrochemical reactions on their surfaces has been proposed [Nagaev, 1992]. It would, therefore, be very interesting to use more rigorous computational approaches to estimate chemical potentials of metal nanoparticles relevant to fuel cell electrocatalysis and consider their possible influence on the activation barriers for electrocatalytic reactions. Although one may hardly expect quantitative agreement with experiment, since the surface energy must depend on the chemical environment and on the electrode potential, the thermodynamic approach may help in understanding and predicting some essential trends.

15.2.2 Structural Properties

Structure and morphology of supported metal nanoparticles may differ drastically, depending on (i) their size, (ii) their interaction with support, (iii) the (electro)chemical environment, and, (iv) since very often particles do not attain equilibrium shapes, also on the preparation conditions and sample prehistory.

15.2.2.1 Equilibrium Shape The equilibrium shape of crystals is given by the Wulff construction, according to which $\sigma_i/h_i = $ const. [Frenkel et al., 2001; Henry, 1998]. Here h_i is the distance from the center to the ith crystal face. A theoretical study by Romanowski using the concept of localized bonds supposed that only the nearest neighbor interactions are important to determine the minimum surface free energy for nanoparticles [Romanowski, 1969]. For free fcc, bcc, and hcp metal crystals, the structures of minimum surface energy are the cubo-octahedron, the rhombo-dodecahedron, and the truncated hexagonal bipyramid, respectively [Romanowski, 1969]. In this chapter, we will mainly focus on fcc metals of interest in electrocatalysis.

In the case of supported metallic particles, the construction is modified by introducing the adhesion energy (Wulff–Kaishew construction) [Henry, 1998]. The equilibrium shape is a Wulff polyhedron, which is truncated at the interface by an amount Δh_s, according to the relation $\Delta h_s/h_i = \beta/\sigma_i$, where β is the adhesion energy of the crystal on the substrate.

For an ideal particle of an equilibrium shape, the contribution of various structural elements to the surface is a function of the particle size and can easily be computed. As the particle size decreases, the surface fraction of vertices (CN = 6) and edges (CN = 7) increases, while that of the atoms associated with (100) and (111) facets (CN = 8 and 9, respectively) decreases [Van Hardeveld and Hartog, 1969; Kinoshita, 1988]. Therefore, the average first-shell coordination number scales with the inverse diameter and decreases well below the theoretical value of CN = 12 to about 6 for a 1 nm particle [Frenkel et al., 2001]. In order to minimize their surface energy and maximize the bonding between metal atoms, nanometer-sized particles may adopt a structure different from that characteristic of bulk metals. For example, Burton suggested that very small crystallites can possess different thermodynamically stable shapes than expected from Romanowski's calculations [Burton, 1974]. He reported small crystallites having five-fold symmetry and icosahedral shape to be more stable than the expected close-packed structures. The surface of such a crystallite consists of triangular (111) facets. A low symmetry icosahedral shape has indeed been observed experimentally, for example, for giant Pd_{561} clusters [Vargaftik et al., 1991]. In some cases, the tendency of decreasing surface energy results in the formation of multiple twins [Nagaev, 1992]. On the other hand, it seems unrealistic to assume that metal nanoparticles always comprise an exact number of atoms to fit one of the geometric models described above. As a result, nanoparticles in "real" supported metal catalysts may not have the shape of the Wulff polyhedron, but adopt the form corresponding to the minimum surface free energy. In some cases, decreasing the particle size may result in their amorphization. Calculations have produced a phase diagram that predicts that small Au clusters at room temperature may be in a state of quasi-melt [Ajayan and Marks, 1988], and this has been indeed confirmed experimentally [Henry, 2003].

Relaxation of a metal nanoparticle to its equilibrium shape requires substantial activation energy, and thus may not be reached on the time scale of catalytic reactions. This is especially the case for catalysts obtained by low temperature preparation approaches (e.g., colloidal or micellar) and used in model electrochemical cells in the temperature range between 20 and 80 °C or in low temperature fuel cells operating below about 130 °C. Various approaches to obtain preferentially oriented cubic, pyramidal, elongated, etc. particles have been developed [Burda et al., 2005; El-Deab et al., 2005], and exploration of their electrocatalytic properties [Solla-Gullon et al., 2006; Hernandez et al., 2005] is a fascinating area of research that adds to the understanding of the role of sites of different geometry in (electro)catalysis.

So far, we have discussed the shapes of nanocrystals in contact with vacuum. Immersion of the particles in gas or liquid environments may result in substantial changes. Indeed, it is known that adsorption decreases the surface energy [Herring, 1952]. Since the adsorption energy usually depends on the surface crystallography, the equilibrium particle shape may be altered upon adsorption. The influence of adsorption on the shapes and lattice parameters of metal nanoparticles has been predicted theoretically and confirmed experimentally (see the references cited in Henry [2003]). Adsorption/desorption occurring during heterogeneous catalytic reactions may thus result in dynamic changes of particle shapes, as predicted by Monte Carlo simulations [Zhdanov, 2002; Kovalyov et al., 2003, 2008]. In electrochemical systems, a potential drop develops at the interface between the catalytic nanoparticles and the electrolyte. This may also induce changes in particle shape. Experimental data concerning the particle shapes determined in situ during electrochemical operation are rare. Mukerjee and McBreen performed in situ extended X-ray absorption fine structure (EXAFS) studies of carbon-supported Pt particles with the average diameters ranging from 3 to 9 nm in 1 M $HClO_4$ [Mukerjee and McBreen, 1998]. Pt–Pt apparent CNs at 0.0 V (with respect to a reversible hydrogen electrode, RHE) were in excellent agreement with those calculated for a cubo-octahedral shape. However, an increase of the electrode potential to 0.54 V vs. RHE resulted in a substantial drop in Pt–Pt CNs, which led Mukerjee and McBreen to conclude that there was a change in particle morphology from a cubo-octahedral to a plane raft-like configuration. Further in situ studies would be desirable for establishing the influence of the interfacial potential drop and the adsorbates on the shapes of metal nanoparticles.

In fuel cell electrocatalysis, high metal loadings per unit area of support are normally utilized in order to minimize the thickness of the catalytic layers [Vielstich et al., 2003]. This leads to yet another complication: particle coalescence and formation of complex structures where nanometer-sized particles contact each other through grain boundaries. This is illustrated in Fig. 15.1, which shows Pd (a) and Pt-Ru (b) particles supported on model carbon supports of the Sibunit family. One can see that the Pd catalyst comprises isolated 2–5 nm particles consisting of single grains, while in high loading Pt-Ru catalysts, individual grains merge together to form complex structures with a high density of grain boundary regions.

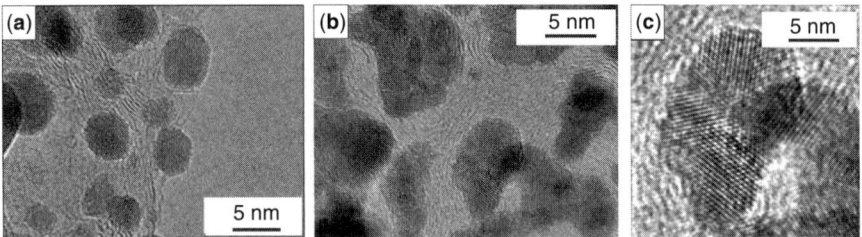

Figure 15.1 High resolution transmission electron microscopy images (HR-TEM) of 5 wt% Pd (a) and 50 wt% Pt-Ru (b) particles supported on carbon supports of the Sibunit family with surface areas of about 6 m^2 g^{-1} (a) and 72 m^2 g^{-1} (b). (c) Fourier-transformed image of (b). ((a) Reprinted from Pronkin et al. [2007], Copyright 2007, with permission from Elsevier. (b) and (c) reprinted from Gavrilov et al. [2007]—Reproduced by permission of the PCCP Owner Societies.)

The Fourier-transformed image in Fig. 15.1c shows that individual grains are randomly oriented, breaking the continuity of lattice planes in the crystallites. Unless high temperature annealing procedures are employed (which often is not the case), these grain boundary regions may have very long lifetimes. The described materials are usually called "nanostructured" or "nanocrystalline" [Gleiter, 1992]. The high volume density of disordered grain boundary regions leads to unique physical properties, differentiating these materials from single crystals, coarsely grained polycrystalline materials, and nanometer-sized supported metal particles. As we will show in Section 15.5, the presence of grain boundary regions may strongly influence the electrocatalytic activities of heterogeneous (electro)catalysts.

15.2.2.2 Lattice Parameter Along with the particle shape, another quantity of relevance to catalysis is the lattice parameter. In the early studies of size effects (see, e.g., the discussion in the review [Uvarov and Boldyrev, 2001]), changes in the lattice parameter were explained by the influence of the Laplace excess pressure compressing crystallites of high curvature. Indeed, lattice constant contraction was reported for nanoparticles of some fcc metals—Au [Mays et al., 1968], Pt [Wasserman and Vermaak, 1972], and Ag [Mays et al., 1968]—relative to the values characteristic of bulk materials. However, the simplified approach resting upon the notion of Laplace excess pressure neglects the strain energy. When the latter is considered, the excess pressure may exceed the Laplace pressure and thus lead to lattice compression, but may also become negative, resulting in lattice dilatation [Mays et al., 1968; Vermaak et al., 1968; Vermaak and Kuhlmann-Wilsdorf, 1968; Nagaev, 1992]. Thus, for Cu particles [Wasserman and Vermaak, 1972], the lattice constant appeared to remain unchanged or to increase slightly with decreasing particle size, while for diamond and silicon particles, dilatation was reported [Gamarnik, 1990]. All-electron density functional (DFT) calculations by Krueger and co-workers, performed for high symmetry Au and Pd crystals, have shown a linear correlation between bond length and average coordination numbers (and thus with inverse particle size) [Krueger et al., 1997]. For further details on the influence of size on the lattice parameter, the reader is referred to Nagaev [1992]; Vermaak et al. [1968]; Vermaak and Kuhlmann-Wilsdorf [1968]; Gamarnik [1993]; and Henry [1998].

Along with size, interaction with the support may induce considerable changes in the lattice parameter. For example, employing transmission electron microscopy (TEM), Nepijko and co-workers found that the lattice constant of tantalum clusters on a thin Al_2O_3 film epitaxially grown on NiAl(110) decreased with decreasing cluster size, with the greatest observed reduction being 4.5% for a cluster with a diameter of 1.25 nm [Nepijko et al., 1998], while for 1 nm Pt on NiAl(110) [Klimenkov et al., 1997], a contraction of up to 10% was detected. Such strong deviations of the lattice parameter from the bulk value may induce considerable changes in the reactivity of the clusters.

15.2.3 Electronic Properties

The influence of cluster size on electronic properties has recently been the subject of numerous studies by both theorists and experimentalists, so we will only touch briefly upon this issue. More information can be found in the reviews [Nagaev, 1992; Henry, 1998, 2003; Binns, 2001; Pacchioni and Illas, 2003; Roduner, 2006]. From the standpoint of electronic structure, metal clusters fall between isolated atoms with localized atomic orbitals and the bulk metallic phase, where a continuous conduction band is formed. For an isolated cluster, according to Kubo, splitting between electron levels scales with the inverse number of atoms, so that at room temperature, for low symmetry systems, the inequality (15.4) should normally hold for metal particles comprising more than 50–100 atoms [Kubo, 1962]. For smaller clusters, one may expect quantization of electron levels and, upon reaching a critical size, a metal–insulator transition. However, for high symmetry systems, quantum effects may be observed at rather large characteristic sizes owing to degeneracy of the levels [Nagaev, 1992]. Until now, rigorous quantum chemical description of the electronic structure of transition metals has been an open problem. An opinion generally held is that the cluster size strongly influences the electronic properties of metal particles comprising less than 20–50 atoms, while larger clusters have the electronic properties of bulk metals. Indeed, remarkable size effects have been demonstrated in the low dimension, nonscalable interval both experimentally (e.g., by photoemission spectroscopy and scanning tunneling spectroscopy) and computationally, and are widely discussed in the literature [Henry, 1998, 2003; Binns, 2001; Pacchioni and Illas, 2003]. However, quantum chemical calculations performed by Nagaev strongly suggest that size may influence the electronic properties of much larger particles in the so-called weak quantization limit, when $\delta \ll k_B T$, resulting in a shift of the Fermi level, variation of the density of states (DOS) at the Fermi level, and, in some cases, the appearance of surface electronic states [Nagaev, 1992]. Various reasons have been put forward to account for the size dependence of the Fermi level E_F [Grigorieva et al., 1987], including spatial quantization of the kinetic energy of electrons stipulated by the size confinement—with a concomitant upshift of the Fermi energy—and creation of a surface conduction band, lying below the volume band—with a concomitant downshift of the Fermi level. Overall, both upshift and downshift may be expected, depending on the metal, the size of the particles, etc. The magnitude of the shift ΔE_F of the Fermi level for a metal particle (or film) versus the corresponding position for the bulk material may be very roughly estimated as [Nagaev, 1992; Grigorieva et al., 1987]

$$\Delta E_F \sim \frac{E_F}{k_F L} \tag{15.6}$$

where k_F is the Fermi momentum and L is the characteristic dimension, which for a spherical particle is equal to its diameter and for a film to its thickness. Estimation of ΔE_F for a silver cluster with $L = 5$ nm, $k_F^{-1} \approx 0.5$ nm and $E_F \approx 5$ eV gives $\Delta E_F \approx 0.5$ eV [Grigorieva et al., 1987]. Although this is a rough estimate, it shows that Fermi level shifts may be expected even for relatively "large" particles of a few nanometers in size. In the case of formation of surface electronic levels, yet stronger shifts of E_F may be expected. It should be stressed that the above-mentioned size dependence of the Fermi energy is often neglected in the literature, and the size dependence of the work function, which has been documented in numerous publications (see, e.g., Knickelbein et al. [1990]; Ekardt [1984]), is entirely attributed to the variation of the surface potential (due to an effective surface dipole) with particle size. The reader is referred to the comprehensive review article by Nagaev for an in-depth analysis of the influence of the size confinement on the Fermi energy [Nagaev, 1992]. Experimentally, the electronic structures of supported clusters are generally studied with photoemission spectroscopy: X-ray photoelectron spectroscopy (XPS) and ultraviolet photoelectron spectroscopy (UPS). The broadening of the valence band, and the shift of the core levels and the valence band toward the Fermi level, have been observed with increasing cluster size [Henry, 1998, 2003]. The difficulties in interpreting the changes observed in photoemission spectra arise from various contributions associated with initial and final state effects. The former are related to the shift of the electronic levels, i.e., they are "true" size effects, while the latter correspond to an increase in the binding energy that is due to the nonperfect screening of the hole created during the photoemission process [Henry, 1998, 2003]. For example, for a free cluster, this additional energy w_c may be estimated as the Coulomb energy associated with the charge left on the particle upon photoemission:

$$w_c = \frac{e^2}{4\pi \varepsilon \varepsilon_0 d} \tag{15.7}$$

where w_c scales with the inverse particle diameter and often outweighs the changes occurring because of μ_e^M (initial state effect). Unfortunately, some researches neglect final state effects in XPS and UPS, leading them to erroneous conclusions. Differentiating between initial and final state effects is difficult, but not impossible. For further details, the reader is referred to the review by Henry [1998] and references therein. Overall, it should be mentioned that the understanding of the size dependence of the electronic properties of metals is still in its infancy, and concerted efforts of theorists and experimentalists are necessary in order to further it.

15.3 MODEL SYSTEMS TO STUDY SIZE EFFECTS IN ELECTROCATALYSIS

In PEMFCs, the membrane electrode assembly (MEA, Fig. 15.2a) is a multilayer sandwich composed of catalytic layers (CLs) where electrochemical reactions take place, gas-diffusion media providing access of gases to the CLs, and a proton exchange membrane (PEM) such as Nafion®. The CL is a multiphase multicomponent medium comprising:

- a catalyst, usually nanometer-sized noble metal particles supported on porous carbon materials, accelerating the rates of electrochemical reactions
- an ionomer, allowing flow of protons through the MEA
- gas- and liquid-filled pores, providing access of the reagents to and the products from the catalyst surface

In practice, the catalytic layers are prepared by brushing or spraying "catalyst ink" (a suspension of the catalyst particles in water and/or an organic solvent with addition of ionomer) either onto diffusion media (carbon paper or carbon cloth, also referred to as substrates), resulting in so-called catalyst-coated substrates (CCS), or directly onto

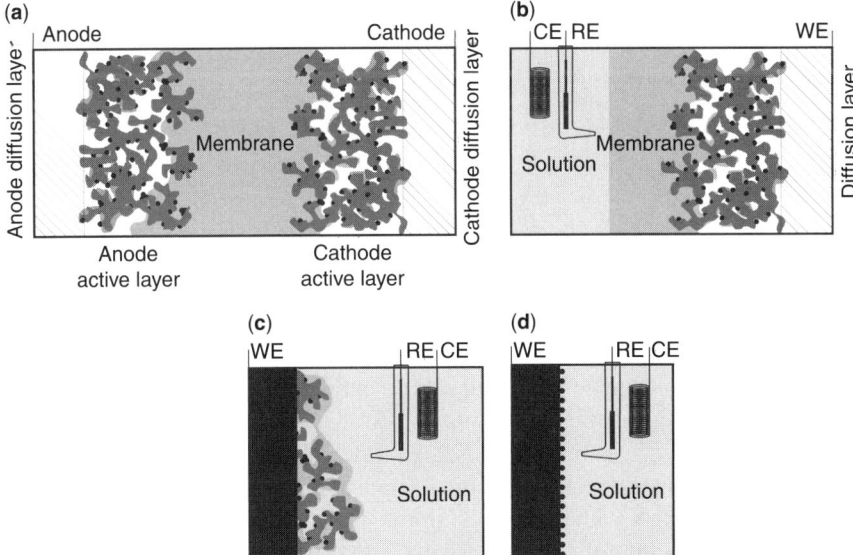

Figure 15.2 Schematic representation of different electrochemical cell types used in studies of electrocatalytic reactions: (a) proton exchange membrane single cell, comprising a membrane electrode assembly; (b) electrochemical cell with a gas diffusion electrode; (c) electrochemical cell with a thin-layer working electrode; (d) electrochemical cell with a model nonporous electrode. CE, counter-electrode; RE, reference electrode; WE, working electrode.

the PEM, resulting in a so-called catalyst-coated membrane (CCM) [Gasteiger et al., 2005].

It would certainly be desirable to evaluate catalyst performance and understand size and structural effects directly under the conditions of fuel cell operation. However, determination of kinetic parameters in a single-cell fuel cell is associated with a number of limitations. Let us consider some of them.

The first limitation is related to interference of the anode and the cathode. The finite permeability of the Nafion® membrane to fuel and oxygen results in crossover of fuel from the anode to the cathode, and oxygen crossover in the opposite direction. This may have a significant influence on electrode kinetics.

The second limitation is concerned with the high time constant τ_{cell} determined by the high capacitance of the CLs and by substantial cell resistance. τ_{cell} for a thick MEA of large geometric area may exceed several minutes, thus strongly limiting the application of transient methods and methods based on the concept of impedance.

The third limitation is concerned with the numerous contributions to the cell voltage V_{cell}, which, along with the difference in the electrode reversible potentials ΔE_{eq}, comprises overpotentials at the cathode, η_C, and the anode, η_A, as well as the ohmic drop ΔE_{ohmic}:

$$V_{cell} = \Delta E_{eq} - |\eta_C| - |\eta_A| - \Delta E_{ohmic} \tag{15.8}$$

For isolating the overpotential of the working electrode, it is common practice to admit hydrogen to the counter-electrode (the anode in a PEMFC; the cathode in a direct methanol fuel cell, DMFC) and create a so-called dynamic reference electrode. Furthermore, the overpotential comprises losses associated with sluggish electrochemical kinetics, $\eta_{C,A}^k$, as well as a concentration polarization $\eta_{C,A}^{conc}$, related to hindered mass transport:

$$\eta_{C,A} = \eta_{C,A}^k + \eta_{C,A}^{conc} \tag{15.9}$$

Ohmic losses ΔE_{ohmic} originate from (i) membrane resistance, (ii) resistance of CLs and diffusion layers, and (iii) contact resistance between the flow field plates. Although it is common practice to split current–voltage characteristics of an MEA into three regions—"kinetic" (low currents), "ohmic" (intermediate currents), and "mass transport" (high currents) [Winter and Brodd, 2004]—implicit separation of η_i^k, η_i^{conc}, and ΔE_{ohmic} is not always straightforward, and thus studies of size and structural effects under conditions of non-negligible mass transport and ohmic contributions may be fraught with significant errors.

In practice, reduction of the size of metal particles supported on porous carbon materials is usually implemented by decreasing the metal loading, which can be done either by decreasing the metal percentage at constant carbon surface area or by keeping the metal percentage constant but increasing the specific surface area of the support. These changes may affect not only kinetic but also mass transport and ohmic losses [Kaiser et al., 2007]. Evaluation of the influence of particle size on electrochemical kinetics can be performed by applying appropriate mathematical models, allowing separation of the various contributions. Since mathematical models include many unknown parameters (agglomerate structure, diffusivities in ionomer films, etc.), quantification of intrinsic catalyst activity may be associated with large uncertainties. Hence, unambiguous analyses require that kinetic measurements be performed in the low overpotential interval. As we will see from our further analysis of the literature, this is not always the case, and conclusions concerning PSEs are sometimes made in the potential interval of hindered mass transport.

The fourth limitation is related to the fact that the catalyst utilization factor u may fall well below 1. This factor is commonly defined as the ratio of the metal surface atoms participating in an electrochemical reaction to the overall number of metal surface atoms. There is currently no agreement in the literature regarding the utilization factor achieved in state-of-the-art PEMFCs. Indeed, Gasteiger and co-workers suggest that u may be close to 1 [Gasteiger et al., 2005], while, according to Eikerling and co-workers, the maximum achievable u in a random CL cannot exceed about 0.4 [Eikerling et al., 2005]. This apparent discrepancy is likely to be due to somewhat different meanings assigned to this parameter by different authors. It was formerly widely accepted that only surface sites located at the three-phase boundary, i.e., at the interface between catalytic particles, proton-conducting electrolyte, and gas-filled pores, are electrochemically active. However, a number of recent fuel cell as well as model studies suggest that proton transfer may rather efficiently occur in a thin aqueous film in the absence of a proton-conducting ionomer [Paulus et al., 2003; Eikerling et al., 2005]. Paulus and co-workers have shown that metal utilization is high for electrodes where Pt particles form a contiguous network (e.g., Pt black) and much lower if metal particles are distributed on a carbon surface [Paulus et al., 2003]. Thus, proton conduction may occur in a network of hydrogen-bonded water molecules adsorbed on the Pt surface. Rao and co-workers have demonstrated that u may significantly vary depending on the porous structure of carbon materials [Rao et al., 2005]. This adds to the uncertainty in studying size and structural effects with the single-cell approach.

The fifth limitation is related to the heterogeneity of the catalyst. The point is that metal nanoparticles are usually deposited on high surface area carbon supports, and, in order to provide high currents at low ohmic losses, often contain the active component (e.g., Pt) in a high percentage. This gives rise to polydispersed catalysts containing particle agglomerates. The presence and structure of the latter may have dramatic influence on catalytic activity, as discussed in Cherstiouk et al. [2000, 2003a]; Maillard et al. [2004b, 2005, 2007a]; Gavrilov et al. [2007]. In the investigation of PSEs, both commercial and "home-made" catalysts are used. The latter are prepared through various synthetic approaches, including colloidal, microemulsion, incipient wetness impregnation, adsorption, electrodeposition, and chemical and physical vapor deposition. The reader is referred to review articles for more information about various preparative approaches to supported nanoparticle catalysts [Aiken and Finke, 1999; Lewis, 1993; Simonov and Likholobov, 2003; Liu et al., 2006; Ralph and Hogarth, 2002a, b; Hogarth and Ralph, 2002]. A detailed discussion is beyond the scope of this chapter.

Despite the above limitations, electrocatalytic activities are often evaluated from polarization measurements in a single-cell configuration (for references, see the review [Gasteiger et al., 2005]). Various methods are applicable for the investigation of electrocatalytic materials, including conventional electrochemical [Gasteiger et al., 2005; Dinh et al., 2000], impedance spectroscopy [Diard, 1998; Mazurek et al., 2006], and recently also in situ Fourier transform infrared (FTIR) spectroscopy [Tkach et al., 2004] and X-ray absorption (XAS) spectroscopy (see the review [Russell and Rose, 2004] and references therein), provided that the characteristic time of the measurements exceeds the time constant of the cell. The high selectivity of XAS allows detection of particular elements in multicomponent fuel cell systems. XAS measurements in transmission form, however, require either removal of the catalyst from the side of the MEA not under investigation [Roth et al., 2002] or exclusion of the absorbing element from this electrode [Viswanathan et al., 2002]. In situ XAS allows the monitoring of transformations of particle size, morphology, and oxidation state of individual components, and, for particles with sufficiently high dispersion, also detection of adsorbate molecules [Russell and Rose, 2004]. Nevertheless, studies of size effects in fuel cell configurations are rare, because of the complexity of the MEA and its nonuniform structure and composition.

Another approach to measuring the activity of fuel cell catalysts is based on gas diffusion electrodes (GDEs) in the so-called "half-cell" configuration (Fig. 15.2b). The catalytic layer is prepared as outlined above and sandwiched between a gas diffusion medium and a proton-conducting membrane, the other side of the latter being in contact with an aqueous electrolyte of a conventional electrochemical cell. The working electrode potential is measured against a reference electrode within the accuracy of the ohmic drop in the PEM, and processes at the counter-electrode do not affect the catalyst under study [Antoine et al., 1998]. Hence, the first of the above-mentioned limitations is lifted, while the second and third are partially relaxed, allowing the application of a wider range of experimental techniques with somewhat shorter time constants. Compared with the single-cell approach, this allows more detailed studies of electrochemical kinetics and size effects. Antoine and co-workers applied the GDE approach to the investigation of size effects in the HOR and ORR [Antoine et al., 1998]. While the results of this study will be discussed in more detail in Section 15.5, here we would like to mention that Antoine et al. concluded that ORR kinetic currents could be measured quite accurately, whereas those for the HOR were noticeably influenced by hydrogen mass transport. The latter currents were therefore subjected to correction using a mathematical model taking into account mass transport losses in the GDE and the discrete character of the nanoparticle distribution, resulting in the overlap of diffusion zones at adjacent catalytic particles.

At the next level of abstraction are measurements performed at a thin film of fuel cell catalyst immobilized on the surface of an inert substrate, such as glassy carbon (GC) or gold (Fig. 15.2c). Essentially, three versions of this approach have been described in the literature. In the first case (a "porous electrode"), an ink containing catalyst and Nafion® ionomer is spread onto an inert nonporous substrate [Gloaguen et al., 1994; Gamez et al., 1996; Kabbabi et al., 1994]. In the second case (a "thin-film electrode"), the ink does not contain Nafion®, but the latter is deposited as a thin film on the surface of the dried catalyst layer to stabilize it [Schmidt et al., 1998a, 1999]. In the third case, there is no binder, and the catalyst is fixed on the surface by adhesion [Park et al., 2001; Maillard et al., 2004b]. The electrodes thus prepared may then be used for performing model studies in conventional three-electrode cells with liquid electrolytes using various techniques: electrochemical, impedance spectroscopy, FTIR, differential electrochemical mass spectroscopy (DEMS), and others. This approach has been widely applied for the investigation of size effects in fuel cell-related processes, including hydrogen electro-oxidation [Markovic and Ross, 2000; Schmidt et al., 2001] and oxygen electroreduction [Gloaguen et al., 1994; Kabbabi et al., 1994; Schmidt et al., 1998b; Paulus et al., 2001; Maillard et al., 2002; Gasteiger et al., 2005] reactions, CO monolayer [Friedrich et al., 2001; Park et al., 2002b; Arenz et al., 2005; Mayrhofer et al., 2005a; Maillard et al., 2004b] and CO bulk [Arenz et al., 2005; Mayrhofer et al., 2005a; Schmidt et al., 1999] electro-oxidation, and methanol electro-oxidation [Jusys and Behm, 2001; Park et al., 2002b].

Thin catalyst layers on a GC rotating disk electrode (RDE) or a rotating ring–disk electrode (RRDE) serve for studies of ORR kinetics. In order to separate the kinetic current j_k from the measured current j, Schmidt and co-workers [Schmidt et al., 1998b] corrected the latter for the influence of oxygen diffusion in the aqueous electrolyte and in the polymer film using the following equation:

$$\frac{1}{j} = \frac{1}{j_k} + \frac{1}{j_d} + \frac{1}{j_f} = \frac{1}{j_k} + \frac{1}{BC_O\omega^{1/2}} + \frac{L}{nFC_fD_f} \quad (15.10)$$

where B is the Levich constant, C_O is the oxygen concentration in the electrolyte solution, C_f is the oxygen concentration in the Nafion® film, L is the film thickness, D_f is the diffusion coefficient of oxygen molecules in the film, ω is the electrode rotation rate, j_d is the mass-transport-limiting current in the liquid electrolyte, and j_f is the limiting current in the polymer film, the other terms having their usual meaning. Paulus and co-workers [Paulus et al., 2001] have shown that if the polymer film thickness is reduced below about 0.1 μm, j_f becomes significantly larger than j_k and j_d; the influence of the oxygen diffusion in the film on the measured current density is then negligible, and the last term in (15.10) may be ignored.

Gloaguen and co-workers underlined the fact that it is also necessary to correct the measured current for mass transport and ohmic losses *within the catalytic layer* [Gloaguen et al., 1994]. They proposed a mathematical model for a fully flooded catalytic layer comprising ionic and electronic conductors (no voids), taking into account gas diffusion, ohmic drops, and interfacial kinetics within the CL. The so-called "effectiveness factor" f, i.e., the ratio of the actual reaction rate to the rate expected in the absence of mass and ion transport limitations, is a function of the roughness factor γ (the ratio of total catalyst area to geometric area), the exchange current density j_0, the diffusion coefficient D, the concentration of electroactive species, and the overpotential. Figure 15.3 shows that f drops sharply with increasing exchange current density and with decreasing product DC_O. For oxygen in an aqueous electrolyte at room temperature, $DC_O \approx 10^{-11}$ mol cm^{-1} s^{-1}. Thus, the effectiveness factor of

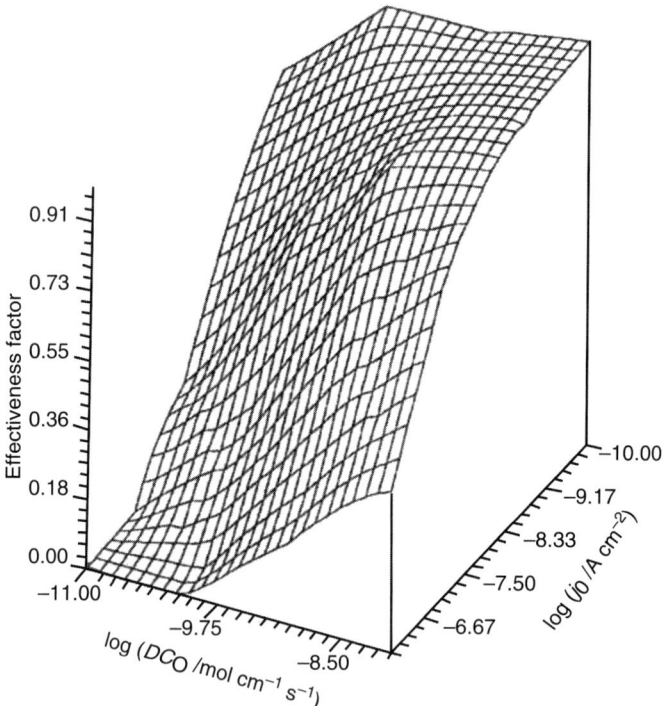

Figure 15.3 Simulated effectiveness factor for porous carbon electrode as a function of the exchange current density j_0 and DC_O for $|\eta| = 0.4$ V for a 10 wt% Pt/C catalyst layer with $\gamma = 10$, $A = 140 \, m^2 \, g^{-1}$, $\rho = 2 \, g \, cm^{-3}$, Nafion® volume fraction 0.6, thickness μm, and ionic conductivity 0.05 S cm^{-1}. See the text for details. (Reproduced from Gloaguen et al. [1994], with kind permission from Springer Science and Business Media.)

the flooded layer in the aqueous electrolyte is significantly lower compared with the GDE owing to slow diffusion and low solubility of gases in the liquid phase [Gloaguen et al., 1998]. As far as the exchange current density and the specific electrocatalytic activity for the ORR are concerned, the values published in the literature show significant scatter. For example, according to Wang and co-workers, the intrinsic exchange current density for the ORR on Pt(111) is in the range 7 μA cm^{-2} < j_0 < 25 μA cm^{-2} at room temperature in HClO$_4$, with an intrinsic Tafel slope from -118 to -130 mV dec^{-1} [Wang et al., 2004]. Gasteiger and co-workers compared activities for polycrystalline Pt, Pt black, and Pt/C catalysts in HClO$_4$ at 0.9 V and 60 °C [Gasteiger et al., 2005]. Assuming a Tafel slope of 60 mV dec^{-1} in the high potential region, the reported values for polycrystalline Pt translate into $j_0 \approx 20$ nA cm^{-2}. Meanwhile, the values of the apparent rate constant for the ORR at 0.8 V and 60 °C reported by Yano and co-workers [Yano et al., 2006] for polycrystalline Pt and carbon-supported Pt nanoparticles after recalculation give $j_0 \approx 1$ nA cm^{-2}.

Inspection of Fig. 15.3 reveals that while for $j_0 \approx 0.1$ nA cm^{-2}, the effectiveness factor is expected to be close to 1, for a faster reaction with $j_0 \approx 1$ μA cm^{-2}, it will drop to about 0.2. This is the case of internal diffusion limitation, well known in heterogeneous catalysis, when the reagent concentration at the outer surface of the catalyst grains is equal to its volume concentration, but drops sharply inside the pores of the catalyst. In this context, it should be pointed out that when the pore size is decreased below about 50 nm, the predominant mechanism of mass transport is Knudsen diffusion [Malek and Coppens, 2003], with the diffusion coefficient being less than the Fick diffusion coefficient and dependent on the porosity and pore structure. Moreover, the discrete distribution of the catalytic particles in the CL may also affect the measured current owing to overlap of diffusion zones around closely positioned particles [Antoine et al., 1998].

The above brief analysis underlines that the porous structure of the carbon substrate and the presence of an ionomer impose limitations on the application of porous and thin-layer RDEs to studies of the size effect. Unless measurements are carried out *at very low currents*, corrections for mass transport and ohmic limitations *within the CL* [Gloaguen et al., 1998; Antoine et al., 1998] must be performed, otherwise evaluation of kinetic parameters may be erroneous. This is relevant for the ORR, and even more so for the much faster HOR, especially if the measurements are performed at high overpotentials and with relatively thick CLs. Impurities, which are often present in technical carbons, must also be considered, given the high purity requirements in electrocatalytic measurements in aqueous electrolytes at room temperature and for samples with small surface area.

In order to overcome problems associated with the use of porous carbon materials, a model approach was proposed based on the application of metal nanoparticles supported on flat nonporous substrates (Fig. 15.2d), such as highly oriented pyrolytic graphite (HOPG) [Zoval et al., 1998; Savinova et al., 2000], GC [Tateishi et al., 1991; Takasu et al., 1996; Cherstiouk et al., 2000, 2003a; Maillard et al., 2004a, 2005, 2007b], Au [Friedrich et al., 2000; Pronkin et al., 2001], or TiO$_{2-x}$ [Guerin et al., 2006a, b; Hayden et al., 2007]. Pt, Pd, Pt-Ru, or Au nanoparticles with relatively narrow size distributions were immobilized from colloidal solutions [Friedrich et al., 2000; Pronkin et al., 2001], electrodeposited [Gloaguen et al., 1997; Zoval et al., 1998; Liu and Penner, 2000; Maillard et al., 2007b], chemically deposited [Cherstiouk et al., 2003a], vacuum-deposited [Takasu et al., 1989; Guerin et al., 2006b], etc. The relatively simple structure of these model electrodes and the short time constants of model electrochemical cells allow detailed in situ characterization and direct observation of structural and size effects, for example in CO monolayer [Friedrich et al., 2000; Cherstiouk et al., 2003a; Maillard et al., 2007b] and CO bulk [Guerin et al., 2006b] electro-oxidation, and in the ORR [Takasu et al., 1996; Guerin et al., 2006a], the MOR [Yahikozawa et al., 1991; Takasu et al., 2000], and the FOR [Zhang et al., 1995]. Despite numerous advantages of model electrodes, it should be mentioned that there is a disadvantage concerned with particle coalescence, which is especially relevant to atomically flat substrates weakly interacting with metal nanoparticles such as highly oriented pyrolytic graphite (HOPG) [Savinova et al., 2000]. Utilization of metal oxides (e.g., TiO_2) as supports for model and fuel cell electrocatalysts is a very interesting novel direction [Guerin et al., 2006a, b; Hayden et al., 2007]. TiO_2 interacts with metal particles much more strongly than do carbon materials, which on the one hand allows tuning of electrochemical properties of metal particles [Hayden et al., 2007] and on the other stabilizes them against coalescence. Application of metal electrodes as supports for particle deposition may result in alloy formation. Model nanoparticles on flat nonporous substrates offer better control over the particle distribution and allow internal diffusion effects inside the pores to be avoided. A further improvement in creating model systems has been suggested in a number of publications [Kasemo et al., 2000; Gustavsson et al., 2004; Kumar and Zou, 2006], and is based on the use of ordered nanoparticle arrays. This emerging approach seems very promising, and in the future will hopefully allow one to separate effects arising from the influence of particle size on intrinsic catalytic activities from effects of interparticle separation on the overlap of diffusion zones. When considering various model approaches towards understanding properties of nanosized particles, one must not leave out the investigation of single-crystal electrodes. The motivation for this approach is the analogy between low coordinated sites on high index (stepped) crystals and particle edges and vertices. The influence of steps of specific geometry was demonstrated in studies of the ORR on Au [Strbac et al., 1994] and Pt [Macia et al., 2004; Kuzume et al., 2007], CO adsorption [Kim et al., 1993] and electro-oxidation [Lebedeva et al., 2000, 2002a], and electro-oxidation of organic molecules on Pt [Housmans and Koper, 2003; Sun and Yang, 1999; Tarnowski and Korzeniewski, 1997]. Feliu and co-workers have documented negative shifts of the potential of total and free zero charge with increasing step density [Attard et al., 2004; Climent et al., 2006]. Application of single-crystalline surfaces resulted in significant progress in the understanding of the roles of different types of sites in various electrocatalytic processes. However, it should be kept in mind that some essential characteristics of supported metal nanoparticles cannot be simulated with single-crystal surfaces. The first concerns the influence of support, which may affect geometric (particle shape, lattice parameter, etc.) and electronic (DOS, shift of the d-band, etc.) properties of metal nanoparticles, and, if reagents adsorb on the support, supply them via surface diffusion [Zhdanov and Kasemo, 2000]. The second characteristic is size confinement. Indeed, as outlined in Section 15.2, metal nanoparticles may have different chemical potentials, Fermi energies, and DOS owing to their small dimensions. Zhdanov and Kasemo have pointed out that different electric field distributions on nanoparticles compared with flat surfaces may add to the differences in the rates of electrochemical reactions [Zhdanov and Kasemo, 2002]. The third factor is related to the interplay of reaction kinetics on crystal facets of different configurations [Zhdanov and Kasemo, 2000]. Komanicky and co-workers investigated the ORR at "nanofaceted" Pt electrodes with alternating (111)–(100) plane edges [Komanicky et al., 2005]. They demonstrated that the catalytic activity of "nanofaceted" electrodes is higher than the activities of individual (111) and (100) planes, which was attributed to the kinetic interplay between them.

We would also like to point out some essential differences in site geometry between particle edges and monoatomic steps, which may be crucial for some catalytic

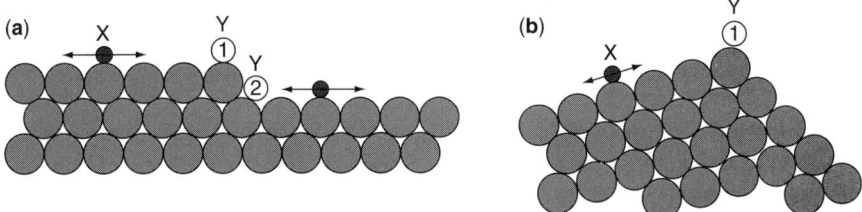

Figure 15.4 Schematic representation of the Langmuir–Hinshelwood reaction between two adsorbates: mobile X (black) and immobile Y (white) on a stepped single-crystalline surface (a) and a facetted nanoparticle (b).

reactions. Let us consider a Langmuir–Hinshelwood (L–H) reaction between a mobile (X) and an immobile (Y) adsorbate, occurring either on a stepped single crystal (Fig. 15.4a) or on a nanoparticle (Fig. 15.4b). Let us assume that Y is adsorbed irreversibly either at a low coordinated site on the step, labeled (1), or at a high coordinated site in the step (2). In the first case, X diffuses over the upper terrace and reacts with Y, and, from the geometric standpoint, the active site configurations on a stepped single crystal and at the nanoparticle edge will be similar. On the contrary, if Y is preferentially adsorbed in the step, and the most favorable catalytic site is realized when X approaches Y from the lower terrace, it is obvious that from the geometric standpoint stepped single-crystal surface will offer a grossly different catalytic site geometry compared with a facetted nanoparticle.

For example, Yates [1995] performed a detailed investigation of heterogeneous catalytic CO oxidation at stepped Pt single crystals, and proposed that the reaction occurs between an O atom in the step site with a neighboring CO molecule on the lower terrace. This configuration was assumed to provide a favorable O$\cdots$C—O approach distance of about 0.25 nm. As discussed below, the difference in the site geometries between steps and edges may to some extent be responsible for the different behavior of stepped Pt single crystals and supported metal nanoparticles in electrochemical CO oxidation.

Last, but not least, one must not forget that steps and kinks appear as structural defects on atomically flat surfaces of single crystals, while edges and vertices are inherent structural elements of metal nanoparticles.

The general problem of model studies in aqueous electrolytes is related to anion-specific adsorption. Kucernak and co-workers have compared the kinetics of the HOR [Jiang and Kucernak, 2004] and the MOR [Jiang and Kucernak, 2005] on Pt electrodes in H_2SO_4 and in a solid state model cell of their own design, in which Nafion® was used as the electrolyte. Differences were observed in the MOR kinetics, and were related to suppression of OH_{ads} formation in the aqueous electrolyte due to the presence of sulfate/bisulfate anions. The anion effect can be reduced, but, unfortunately, not fully eliminated, by using weakly adsorbing electrolytes, such as $HClO_4$. It should also be mentioned that the concentration and activity of water may differ substantially between model electrochemical cells with aqueous electrolytes and PEMFCs. Meanwhile, recent spectroscopic studies have demonstrated the important role that water plays in electrocatalytic processes. For example, water molecules are co-adsorbed with CO on noble metal surfaces surfaces, and act as the source of surface oxygen-containing species in the electrochemical oxidation of CO. Moreover, it has been proposed that co-adsorbed water screens lateral repulsions within the CO adlayer and stabilizes it [Chang et al., 1990; Zou et al., 1999; Pronkin and Wandlowski, 2004].

Another type of model electrode uses multilayer electrolytic deposits, which attracted the interest of electrochemists long before physical methods for their structural characterization were introduced. These electrodes were usually characterized by their roughness factors rather than particle size, the former being of the order of $10^2 - 10^3$ (for original references, see the review [Petrii and Tsirlina, 2001]). Multilayer electrolytic deposits have very complex structures [Plyasova et al., 2006] consisting of nanometer-sized crystallites joined together via grain boundaries, and hence have very peculiar electrocatalytic properties [Cherstiouk et al., 2008]; they will not be considered further in this chapter.

Summing up this section, we would like to note that understanding size effects in electrocatalysis requires the application of appropriate model systems that on the one hand represent the intrinsic properties of supported metal nanoparticles, such as small size and interaction with their support, and on the other allow straightforward separation between kinetic, ohmic, and mass transport (internal and external) losses and control of readsorption effects. This requirement is met, for example, by *metal particles and nanoparticle arrays on flat nonporous supports*. Their investigation allows unambiguous access to reaction kinetics and control of catalyst structure. However, in order to understand how catalysts will behave in the fuel cell environment, these studies must be complemented with GDE and MEA tests to account for the presence of aqueous electrolyte in model experiments.

15.4 DEPENDENCE OF ADSORPTION PROPERTIES ON PARTICLE SIZE

The structural sensitivity of chemisorption at the interface between metal electrodes and liquid electrolytes has been widely documented [Markovic and Ross, 2002]. Among the most thoroughly investigated processes are hydrogen, oxygen, CO and anion adsorption/desorption on low and high index Pt single-crystalline electrodes. Owing to the high structural sensitivity of hydrogen (H_{UPD}) and coupled hydrogen/anion adsorption/desorption, it has become common practice to use the potentiodynamic responses of Pt electrodes in liquid electrolytes to identify their surface crystallography. The validity of this approach has been recently verified by combined electrochemical (in situ) and transmission electron microscopy (ex situ) characterization of preferentially oriented Pt nanoparticles [Solla-Gullon et al., 2006] of size $d \geq 4$ nm. For smaller nanoparticles, the applicability of this approach may be questioned, since decreasing the particle size below about 3–4 nm may not only result in variations in the contributions of facets, edges, and vertices to the surface, but also lead to changes in adsorption energies.

15.4 DEPENDENCE OF ADSORPTION PROPERTIES ON PARTICLE SIZE

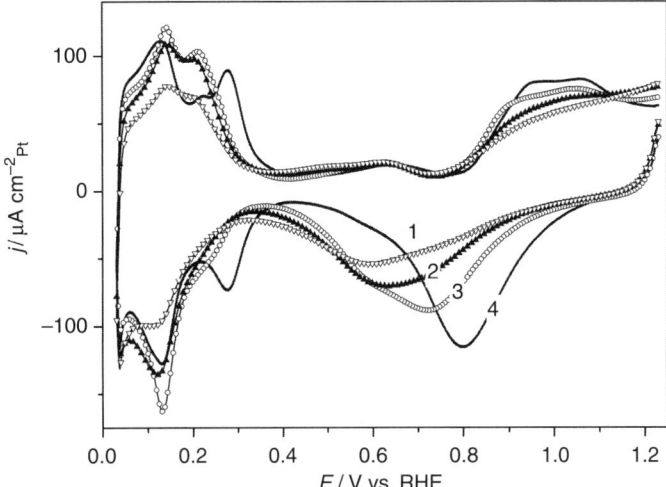

Figure 15.5 Cyclic voltammograms obtained in 0.1 M H_2SO_4 at a sweep rate of 0.1 V s^{-1} for Pt/GC electrodes with mean particle sizes $\bar{d}_N$ = 1.8 nm (1), 2.4 nm (2), and 2.8 nm (3), and Pt electrodeposited on GC (4). See the text for details. (Curves have been replotted from Maillard et al. [2004a, 2005]—Reproduced by permission of The Royal Society of Chemistry and the PCCP Owner Societies.)

In Fig. 15.5, we plot cyclic voltammograms (CVs) for model Pt nanoparticles supported on GC with different mean particle sizes $\bar{d}_N$ = 1.8, 2.4, and 2.8 nm after subtraction of the current originating from the GC support. As one may see, subtraction does not completely eliminate currents originating from electro-oxidation/reduction of surface groups on GC. We explain this by alterations of GC interfacial properties upon Pt deposition. An essential step in making a fair comparison for particles of different average diameters is normalization of the currents to the real Pt surface area. A common approach is to determine the latter from the charge of H_{UPD}. This approach is valid only if neither the stoichiometry of adsorption nor the coverage of the surface with H_{UPD} is altered by the PSEs. As we will show below, however, this does not seem to be the case.

Table 15.1 compares surface areas of Pt calculated using coulometry of H_{UPD} (A_H), CO_{ad} stripping (A_{CO}), and TEM (A_{TEM}) for Pt nanoparticles supported on GC and for comparison for Pt foil. A_{TEM} was calculated assuming a (hemi)spherical particle shape. One may see that for Pt foil, $A_{CO} < A_H$, in agreement with the fact that CO coverage on low index Pt single crystals and on polycrystalline Pt is below 1. For electrodeposited Pt(ed) and for particles about 3 nm in diameter, A_{CO} and A_H are in excellent agreement with TEM data. This may be explained by the fact that CO preferably adsorbs in an atop configuration on Pt atoms with low average coordination numbers. For example, Climent and co-workers have documented an increase in CO coverage along with an increase in step density [Climent et al., 1999].

The most remarkable fact to emerge from Table 15.1 is that for particles of average size below about 3 nm, H_{UPD} leads to considerable underestimation (up to a factor

TABLE 15.1 Comparison of Surface Areas A_i and Charges Required for Electrochemical Desorption of H_{UPD} (Q_H), CO_{ads} (Q_{CO}), and Surface Oxidation (Q_{ox})[a]

$\bar{d}_N$/ nm[b]	$\bar{d}_S$/ nm[b]	A_H/ m^{2}g^{-1}	A_{CO}/ m^{2}g^{-1}	A_{TEM}/ m^{2}g^{-1}	$2Q_H$/ Q_{CO}	Q_H/ Q_{ox}	Q_{CO}/ Q_{ox}
Pt foil	—	5.5[c]	4.6[c]	—	1.2	0.83	2.0
Pt(ed)	—	—	—	—	0.96	0.91	1.9
3.2	4.3	80	87	88	0.92	0.87	1.9
3.1	3.8	64	84	90	0.76	0.74	1.9
2.4	2.8	98	127	117	0.77	0.76	2.0
1.8	2.3	77	144	156	0.53	0.54	2.0

[a]Data have been recalculated from voltammograms in Maillard et al. [2004a, 2005]—Reproduced with permission from The Royal Society of Chemistry and the PCCP Owner Societies. The scan rate was 0.1 V s^{-1} and the positive limit was 1.23 V. See the text for details.
[b]The average particle size $\bar{d}_N$ and surface average particle size $\bar{d}_S$ are given by $\bar{d}_N = \frac{\sum_i N_i d_i}{\sum_i N_i}$, $\bar{d}_S = \frac{\sum_i N_i d_i^3}{\sum_i N_i d_i^2}$.
[c]Total surface area in cm^2.

of two) of the surface areas, while CO stripping data are in very good agreement with electron microscopy. We rule out contamination as a possible reason for H_{UPD} suppression on small Pt particles on the grounds of the excellent agreement between surface areas measured by CO adsorption and calculated from TEM data. Note that CVs were obtained after potential cycling and were stable throughout the measurements.

Thus, we conclude that *the traditional method of surface evaluation based on H_{UPD} coulometry is inappropriate for measurement of the surface areas of particles smaller than 3 nm*. This conclusion may have a major influence on the comparison of the activities of Pt nanoparticles. Instead, we suggest calculating specific surface areas of Pt particles in the size range 1–4 nm from the CO stripping charge. It should be stressed that our conclusion is based on measurements on *model Pt nanoparticles* supported on a clean nonporous GC substrate. These samples comprise particles that have relatively narrow size distributions, with very low fractions of agglomerated particles at moderate Pt loadings, and are fully accessible to electrolyte. The last of these may not always be the case for porous electrodes. For example, Bergamaski and co-workers reported $A_{CO}/A_{TEM} = 0.35$ and 0.52 for 10 and 30 wt% Pt/Vulcan XC-72 catalysts from E-Tek, respectively [Bergamaski et al., 2006]. They attributed these discrepancies to factors such as carbon particle wetting and pore blockage that depend on the electrode preparation. Meanwhile, for a 20 wt% Pt/Vulcan XC-72 containing Pt particles with average size 3.7 nm and purchased from the same manufacturer, Jusys and co-workers observed excellent agreement between the surface areas calculated from H_{UPD}, CO stripping, and TEM [Jusys and Behm, 2001]. For saturation CO coverage, obtained by adsorption in the H_{UPD} region, they found $2Q_H/Q_{CO} = 1.05$, which is in very good agreement with the data of Table 15.1 for "large" particles. These data underline the importance of control of the electrode preparation procedure.

Considering the above discussion, the currents in Fig. 15.5 are normalized to the surface areas estimated from CO stripping. Similar to massive electrodes, CVs for Pt nanoparticles show three characteristic potential regions (all potentials vs. RHE): (i) an H_{UPD} region 0.05 V $< E <$ 0.40 V, followed by (ii) the so-called "double-layer region" 0.40 V $< E <$ 0.60 V and (iii) for $E >$ 0.7 V, the oxygen adsorption/desorption region.

Two adsorption peaks and three desorption peaks can be distinguished in the H_{UPD} region. Based on the data obtained for basal [Clavilier et al., 1980] and stepped [Attard et al., 2004; Climent et al., 2006] Pt single-crystal electrodes, the so-called "strongly bound" (about 0.25–0.28 V vs. RHE) hydrogen may be attributed to processes on Pt(111) and Pt(100) facets [Clavilier et al., 1980, 1981a, b; Markovic and Ross 2002], while "weakly bound" hydrogen (about 0.13–0.15 V vs. RHE) may be tentatively assigned to Pt(110) sites [Clavilier et al., 1980; Inaba et al., 2006; Komanicky et al., 2005]. We have observed that decreasing the size of the crystallites results in suppression of H_{UPD} charge, which is especially remarkable in the potential interval of "strongly bound" hydrogen. Note that suppression of hydrogen adsorption has also been claimed by Zoval and co-workers for Pt nanoparticles with $d \leq 4$ nm supported on HOPG [Zoval et al., 1998].

There is no consensus in the literature regarding H_{UPD} adsorption sites on Pt metals. It is usually believed that H_{UPD} occupies multicoordinated sites (three-fold sites on (111) and four-fold sites on (100) surfaces) (see Jerkiewicz [1998] and references therein). However, somewhat conflicting evidence comes from infrared–visible sum frequency generation (IR-VIS SFG) investigations by Tadjeddine and co-workers, which favor monocoordinated H_{UPD} [Peremans and Tadjeddine, 1994; Tadjeddine and Peremans, 1996]. According to DFT calculations by Watson and co-workers, on Pt(111), the sites with maximum adsorption energy are the atop and three-fold hollow sites [Watson et al., 2002]. The study by Teliska and co-workers, performed for highly dispersed 1.5–2.0 nm Pt particles supported on carbon in an $HClO_4$ electrolyte with Pt $L_{2,3}$ XAS combined with real-space full-multiple-scattering calculations on model clusters [Teliska et al., 2004], lead the authors to the following conclusions: (i) at low coverage, a chemisorbed hydrogen atom is highly mobile and possibly delocalized on the surface; (ii) at higher coverage, it localizes into fcc sites; and (iii) at very high coverage, H is also found in atop sites, presumably at or near edges. Note that a study of bimetallic Pd_xAu_y electrodes of different compositions has shown that hydrogen adsorption coverage correlates with the fraction of Pd_3 sites, while CO adsorption correlates with Pd_1 sites [Maroun et al., 2001].

Given the above discussion, the observed decrease in the fraction of "strongly adsorbed" hydrogen may be tentatively ascribed to the "ensemble effect" related to the decrease of the fraction of the most energetically favourable three-fold hollow sites with decreasing particle size. On the other hand, one cannot exclude a possible "electronic effect" resulting in a change in the adsorption energy with decreasing size of Pt crystallites. Yet another factor that may have an impact on H_{UPD} on Pt nanoparticles is the negative shift of the potential of total zero charge with decreasing particle size documented by Mayrhofer and co-workers [Mayrhofer et al., 2005b]. In any case, the observation of reduced H_{UPD} coverage for nanometer-size Pt particles may have major consequences for electrocatalysis, since the charge required for electro-oxidation of H_{UPD} will be a function of particle size, and not equal to 210 μC per cm^2 of Pt, as assumed up to now.

At positive potentials, water splitting occurs on the Pt surface, manifested by an increase of the current density in the CVs positive of about 0.75 V. According to Anderson, the equilibrium potential of OH formation on a Pt surface, $E^{eq}_{H_2O/OH}$ occurs around 0.62 V on bulk Pt [Anderson, 2002]. The onset of surface oxidation, as well as the charge corresponding to OH/O formation in H_2SO_4, do not seem to depend on the particle size, as confirmed by Table 15.1 and Fig. 15.5 (for further discussion, see Maillard et al. [2005]). This is in line with the modeling work by Andreaus and co-workers, which established $E^{eq}_{H_2O/OH} \approx 0.67-0.69$ V vs. RHE regardless of particle size [Andreaus et al., 2006]. Meanwhile, Arenz and co-workers performed measurements in $HClO_4$, and reported a negative shift in the OH_{ads} formation potential with decreasing Pt particle size, although no clear trend was observed, because of different metal loadings and thus a contribution from the pseudocapacitance of the carbon support [Arenz et al., 2005]. The reason for this discrepancy may be associated with the different electrolytes utilized, but also with the different approaches to determining Pt surface areas: H_{UPD} [Arenz et al., 2005] and CO stripping [Maillard et al., 2004a, 2005].

As can be clearly seen from Fig. 15.5, the potential of the oxide reduction peak systematically shifts negative with decreasing Pt particle size, in agreement with the previous observations of other authors made both in $HClO_4$ and H_2SO_4 electrolytes [Gloaguen et al., 1994; Kabbabi et al., 1994; Frelink et al., 1995; Takasu et al., 1996; Genies et al., 1998; Maillard et al., 2004a, 2005]. This, along with the marginal dependence of the onset of Pt oxidation on particle size, may be attributed to an increased electrochemical irreversibility of the oxidation of the surface of Pt nanoparticles compared with bulk Pt. Conway attributed the irreversibility in the anodic region of voltammograms of noble metal electrodes to the so-called "place exchange" process, resulting in penetration of oxygen in subsurface sites (see the review [Conway, 1995] and references therein). Recently, application of state-of-the-art surface science characterization methods has shown that penetration of oxygen in the metal lattice (e.g., Pd) is facilitated by decreasing particle size [Schalow et al., 2007]. It would be very interesting to perform in situ spectroscopic investigations of model supported metal nanoparticles with controlled size and shape in an electrochemical environment in order to better understand how their size affects interaction with oxygen, hydrogen, and other atoms.

CO adsorption on Pt particles supported on Vulcan XC-72 [Park et al., 2001, 2002a; Maillard et al., 2004b] and on low surface area (about 6 $m^2 g^{-1}$) Sibunit carbon [Park et al., 2001, 2002a; Maillard et al., 2004b] has been investigated with in situ FTIR spectroscopy. The following conclusions were drawn. On small (1–2 nm) Pt particles, CO is preferentially adsorbed in an atop configuration, but with increasing particle size, the fraction of bridge-bonded CO grows [Park et al., 2001, 2002a, 2004b]. The vibrational frequency of the C—O bond of atop CO_{ads} redshifts with decreasing particle size, giving *indirect* support for stronger CO bonding to small nanoparticles. At low surface coverages, CO preferentially adsorbs to the particle edges, but with increasing coverage, also occupies particle facets [Maillard et al., 2004b]. Formation of dense CO_{ads} islands is favored on nanoparticles compared with extended surfaces [Maillard et al., 2004b].

15.5 INFLUENCE OF PARTICLE SIZE ON ELECTROCATALYTIC ACTIVITIES: CASE STUDIES

15.5.1 Hydrogen Oxidation Reaction (HOR)

Reactions occurring at the hydrogen electrode were the first electrochemical reactions to be investigated, and ever since they have been studied as model reactions in electrochemistry;

$$2H^+ + 2e^- \longleftrightarrow H_2 \quad (15.11)$$

The hydrogen evolution reaction (HER) is important for water electrolysis, while the hydrogen oxidation reaction (HOR) occurs at the anode of a PEMFC. The HER and the HOR as well as related processes of hydrogen adsorption have been reviewed comprehensively (see, e.g., Frumkin [1961, 1963]; Breiter [2003]; Markovic and Ross [2002]; Conway [1999]; Conway and Tilak [2002]; Jerkiewicz [1998]; Vetter [1961]; Petrii and Tsirlina [1994]; Kita [2003] and references therein), and they are, perhaps, the most studied electrocatalytic reactions, owing to their relatively simple mechanisms and their practical importance. Investigation of the HOR, especially with very active Pt catalysts, is hampered by mass transport limitations. Several experimental approaches have been used for Pt electrodes in order to extract kinetic information: RDE [Frumkin and Aykazjian, 1955; Breiter and Clamroth, 1954], electrochemical impedance spectroscopy (EIS) [Conway and Tilak, 2002; Safonov et al., 1975], S-shaped RDEs [Bagotsky and Osetrova, 1973], and reaction on a surface partially blocked with adsorbed CO [Breiter, 1975]. The HOR/HER is believed to occur in a sequence of three elementary steps: the Tafel (T) (15.12), Volmer (V) (15.13), and Heyrovski (H) (15.14) reactions:

$$H_2 + 2^* \longleftrightarrow 2H_{ads} \quad (15.12)$$

$$H_{ads} \longleftrightarrow {}^* + H^+ + e^- \quad (15.13)$$

$$H_2 + {}^* \longleftrightarrow H_{ads} + H^+ + e^- \quad (15.14)$$

where * stands for a free site on the electrode surface. The overall process occurs via Tafel–Volmer (T–V) and/or Heyrovski–Volmer (H–V) mechanisms [Vetter, 1961; Conway and Tilak, 2002; Breiter, 2003].

Although the HOR on Pt has been studied very extensively [Markovic and Ross, 2002; Kita, 2003; Safonov et al., 1975; Bagotsky and Osetrova, 1973; Breiter, 1975; Chen and Kucernak, 2004a, b; Markovic et al., 1997b; Seto et al., 1987; Kita et al., 1992; Croissant et al., 1998], there are several disagreements in the literature concerning the values of the exchange current density, the detailed reaction mechanism (T–V or H–V), and the nature of the hydrogen intermediate involved, as well as the influence of the structure on the reaction kinetics and the mechanism. Chen and Kucernak have proposed a novel approach for the investigation of electrochemical reactions using single submicron Pt particles with size down to about 40 nm grown on the tip of a carbon fiber as ultrasmall electrodes [Chen and Kucernak, 2004a, b]. Diffusion to the surface of such an ultramicroelectrode is hemispherical, with the mass transport coefficient being inversely proportional to the particle radius. Thus, for a 36 nm particle, the apparent mass transport coefficient was estimated as about 10 cm s^{-1}. In order to achieve such a mass transport rate at an RDE, the rotation speed would have to be 4.6×10^8 rev min^{-1} [Chen and Kucernak, 2004a, b], which is obviously not feasible. The approach described allows electrochemical reactions to be studied in the absence of mass transport limitations. The work by Chen and Kucernak puts into question a number of previous studies on HOR kinetics and mechanism, where separation of kinetic and mass transport losses might have been inadequate, and has led to renewed interest to the mechanism of this important reaction [Chialvo and Chialvo, 2006; Quaino et al., 2006; Wang et al., 2006]. These recent studies also challenge previous conclusions on the structural sensitivity of the HOR. For example, according to Markovic and co-workers, the exchange current density j_0 for Pt(*hkl*) basal planes falls in the range $0.45-0.98 \text{ mA cm}^{-2}$ and the reaction mechanisms differ between crystal planes, with Pt(111) being the least active [Markovic et al., 1997b]. In contrast, Seto and co-workers suggested that the (111) surface has the highest HOR activity, j_0 for Pt(*hkl*) basal planes being in the range $1.7-3.0 \text{ mA cm}^{-2}$, while the mechanism (T–V) is the same for all basal planes [Seto et al., 1987]. Meanwhile, according to the single-particle study by Chen and Kucernak, j_0 is much higher, amounting to about 20 mA cm^{-2} [Chen and Kucernak, 2004a, b]. This value is in agreement with $j_0 \approx 27 \text{ mA cm}^{-2}$ estimated by Gasteiger and co-workers from MEA studies [Gasteiger et al., 2004].

Fast HOR kinetics coupled with slow mass transport of H_2 hinder investigation of the PSEs. The reaction rates reported for small Pt particles by Takasu and co-workers [Takasu et al., 1989] are several orders of magnitude lower than those deduced from the work of Chen and Kucernak, and therefore may be questioned. A rare example of a direct observation of the size dependence of HOR kinetics is the study by Antoine and co-workers, who investigated the reaction using a GDE approach in the wide Pt particle size interval from 2.5 to 28 nm by increasing the Pt loading on a Vulcan XC-72 carbon support from 10 wt% to 80 wt% [Antoine et al., 1998]. The authors determined specific activity (SA) per unit surface area of Pt, and have shown that it increases by about a factor of ten when the particle size is decreased from about 10 down to 2.8 nm. The authors applied a mathematical model in order to clarify whether this remarkable effect is driven by electrochemical kinetics or is related to mass transport, namely transition from linear to spherical diffusion along with a decrease in metal loading and a concomitant increase in interparticle separation. After correction for mass transport effects, Antoine et al. concluded that intrinsic catalytic activity does indeed increase with decreasing particle size.

15.5 INFLUENCE OF PARTICLE SIZE ON ELECTROCATALYTIC ACTIVITIES: CASE STUDIES

Given that the HOR is one of the rare examples of "positive" size effects, it would be desirable to verify this conclusion with planar model electrodes of simpler structure, where overlap of diffusion zones can be modeled in a straightforward manner. If this is a true kinetic rather than a diffusion-related effect, it is important to understand its origin and find out whether it is related to the decreased particle coverage with H_{UPD} as discussed in Section 15.4. It would also be interesting to know whether it is related to changes in the metal nanostructure, namely transition from single-grain particles at low metal loadings to multigrain structures at high loadings.

15.5.2 Oxygen Reduction Reaction (ORR)

The ORR is the most sluggish reaction in a PEMFC, and thus its understanding is very important from both fundamental and application points of view:

$$O_2 + 4e^- + 4H^+ \longrightarrow 2H_2O$$
$$E^\circ = 1.23 \text{ V vs. SHE at 298K (in acidic electrolytes)}$$
(15.15)

A simplified scheme showing different pathways of the ORR, including the intermediate formation of H_2O_2, was proposed by Wroblowa, Bagotskii, and co-workers [Wroblowa et al., 1976; Bagotskii et al., 1969] and is represented in Fig. 15.6. The $4e^-$ path (k_1) is usually called the *direct* or *parallel* pathway, while the one involving intermediate H_2O_2 formation (k_2, k_3) is called the *indirect* or *series* pathway. For a discussion of the ORR mechanism in greater detail, see Chapter 9. Despite numerous investigations (see the reviews Tarasevich et al. [1983]; Damjanovic [1992]; Kinoshita [1992]; Adzic [1998]; Gattrell and MacDougall [2003]; Durand et al. [1996] and references therein), the mechanism of the ORR is still not fully understood. Depending on the electrode material and the reaction conditions, various reaction intermediates may be anticipated, in particular $O_{2,ad}$, $O^-_{2,ad}$, $HO_{2,ad}$, O_{ad}, OH_{ad}, and $H_2O_{2,ad}$. The last of these is the only stable intermediate; it can desorb from the electrode surface and can be detected in the electrolyte. Determination of whether the series or the

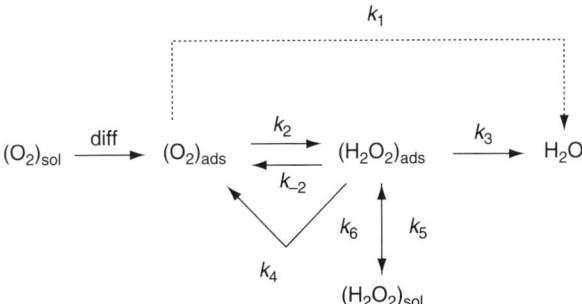

Figure 15.6 Simplified scheme of the ORR [Wroblowa et al., 1976; Bagotskii et al., 1969].

direct pathway is dominant is usually made on the basis of RRDE studies. Low amounts of H_2O_2 detected at the ring of a Pt-disk RRDE is often taken as support for a direct four-electron mechanism of the ORR on Pt.

The ORR has been most thoroughly investigated on Pt electrodes of various structures and morphologies. The reaction rate is first order with respect to the oxygen partial pressure P_{O_2}, which implies that the O_2 bond break occurs after or during the ORR rate-determining step (RDS). The reaction order with respect to H^+ concentration is equal to $1\frac{1}{2}$ at low and 1 at high current densities [Zinola et al., 1994]. The Tafel slopes for Pt electrodes in acidic solutions have been found to change from about -120 mV dec^{-1} in the potential interval below about 0.8 V vs. RHE, where the Pt surface is free from oxide, to about -60 mV dec^{-1} at higher electrode potentials, where the surface is oxide-covered. Based on the experimental observations, the RDS on Pt is usually believed to be the first electron transfer coupled with the proton transfer:

$$(O_2)_{ads} + H^+ + e^- \longrightarrow (O_2H)_{ads} \quad (15.16)$$

The reaction rate can therefore be expressed as [Markovic and Ross, 2002; Chen and Kucernak, 2004a, b]:

$$j_k = -k_0 nFP_{O_2}[H^+](1-\theta)\exp\left(-\frac{\Delta G_\theta}{RT}\right)\exp\left(-\frac{\alpha_c FE}{RT}\right) \quad (15.17)$$

where k_0 is the heterogeneous rate constant of the RDS, $[H^+]$ is the concentration of protons near the electrode surface, $1-\theta$ is the free surface site coverage, and ΔG_θ is the coverage-dependent free energy of oxygen adsorption, the other terms having their usual meanings.

ORR kinetics is usually investigated using an RDE or RRDE, and the kinetic current density is isolated from the measured current via Levich–Koutecky analysis. Since the exchange current density is very low, the activity is usually characterized by the kinetic current density at specified electrode potential in the kinetic region (usually at 0.9 V) either as specific activity, SA (in A cm^{-2}), or, for dispersed materials, as mass activity, MA (in A g^{-1}). Chen and Kucernak [2004a, b] have studied the ORR on Pt under enhanced mass transport conditions using the single-particle approach described above. Particles with a radius of several micrometers demonstrated behavior similar to that of polycrystalline Pt, with the effective number of electrons transferred in the overall reaction, n_{eff}, being close to 4. However, decrease of the particle radii below about 1 μm resulted in a systematic decrease of n_{eff} to 3.5, which suggests that only about 75% of the reactant oxygen molecules are reduced to water, with the rest being only reduced as far as hydrogen peroxide. This means that the small amount of H_2O_2 usually detected on Pt electrodes in a conventional RRDE or thin-layer cell configuration is not necessarily due to the four-electron direct pathway, but should rather be attributed to the slow diffusion of H_2O_2 from the electrode surface, its readsorption, and further reaction via the series pathway.

15.5 INFLUENCE OF PARTICLE SIZE ON ELECTROCATALYTIC ACTIVITIES: CASE STUDIES

Thus, this work challenges the validity of some earlier concepts on the mechanism of the ORR, and suggests that the serial pathway may be equally important, if not dominant. This shows how little, in fact, we understand about this practically important reaction.

Note that for metal nanoparticles supported on porous carbon materials, it is even more difficult to establish the mechanism of the ORR. Indeed, for the above-described thin layer or porous RRDE (Section 15.3), H_2O_2 has very little chance to escape from the CL and be detected at the ring. H_2O_2 can readsorb either on Pt particles or on the carbon support, and undergo chemical decomposition or further electrochemical reduction, while diffusing out of the CL. This implies great difficulties in establishing the detailed ORR mechanism on nanometer-sized metal nanoparticles.

The effect of Pt particle size on electrocatalytic activity in the ORR has been the subject of many controversial studies. The so-called "negative" PSEs, with SA decreasing with decreasing particle size, were first discovered in H_3PO_4 in connection with the development of cathode catalysts for phosphoric acid fuel cells (see Kinoshita [1992] and references therein). Similar dependences were then found in H_2SO_4, as well as in electrolytes containing weakly adsorbing anions, such as $HClO_4$, with SA declining when the particle size decreased below about 4–5 nm [Bregoli, 1978; Sattler and Ross, 1986; Kinoshita, 1988, 1990; Mukerjee, 1990; Giordano et al., 1991; Gloaguen et al., 1994; Kabbabi et al., 1994; Gamez et al., 1996; Takasu et al., 1996; Genies et al., 1998; Maillard et al., 2002; Guerin et al., 2004; Gasteiger et al., 2005]. The MA decreases below about 3–4 nm, where it reaches its maximum, and decreases again at larger particle sizes owing to the decrease in specific surface area [Bregoli, 1978; Sattler and Ross, 1986; Kinoshita, 1990; Gloaguen et al., 1994; Gamez et al., 1996; Genies et al., 1998; Antoine et al., 2001; Maillard et al., 2002; Gasteiger et al., 2005]. This means that it is not advantageous to use very small particles, because the increase in specific surface area is counterbalanced by the loss of SA. PSEs are weakest in $HClO_4$ [Maillard et al., 2002], much stronger in NaOH [Genies et al., 1998], and very strong in H_2SO_4 and H_3PO_4 [Gamez et al., 1996; Gloaguen et al., 1994; Kabbabi et al., 1994; Mukerjee, 1990]. This trend is illustrated in Fig. 15.7, which shows SA from Maillard et al. [2002]; Gasteiger et al. [2005]; Guerin et al. [2004] as a function of the inverse Pt particle diameter.

The greater influence of particle size in the presence of strongly adsorbing anions is in agreement with the stronger structural sensitivity for Pt single crystals in H_2SO_4 and H_3PO_4 as compared with $HClO_4$. For example, El-Kadiri and co-workers, and later Markovic and co-workers, reported that ORR kinetic currents in H_2SO_4 decrease strongly in the order (110) > (100) > (111) [El-Kadiri et al., 1991; Markovic et al., 1994, 1997a]. The same order of catalytic activity was observed in H_3PO_4 [El-Kadiri et al., 1991; Adzic, 1998]. Meanwhile, in $HClO_4$, the differences are relatively small, with SA increasing in the order (100) < (110) ~ (111) [Markovic et al., 1997a]. Similar structure sensitivity was observed in KOH, with SA increasing in the order (100) < (110) < (111), but with larger differences [Markovic et al., 1997a]. These effects may be rationalized by considering the very strong adsorption of

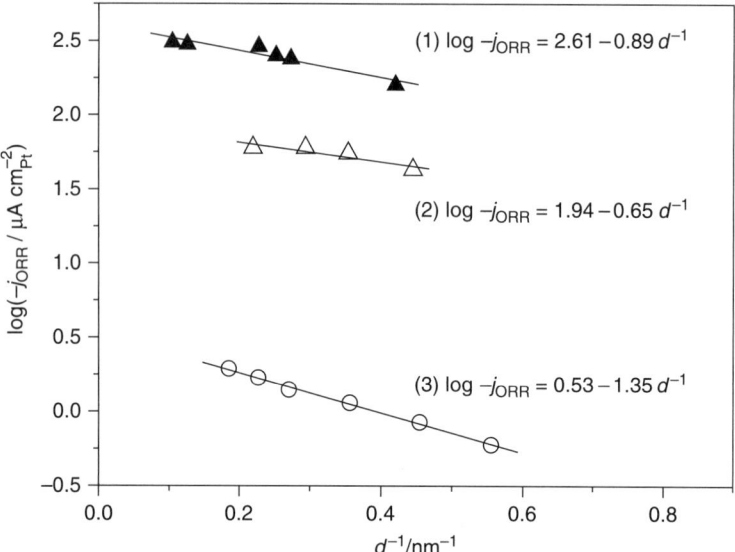

Figure 15.7 Logarithm of the kinetic current for the ORR in oxygen-saturated liquid electrolytes versus inverse diameter for Pt particles supported on Vulcan XC-72: (1) 0.9 V vs. RHE at 60 °C [Gasteiger et al., 2005]; (2) 0.85 V vs. RHE at room temperature [Maillard et al., 2002]; (3) 0.85 V vs. SHE at room temperature [Guerin et al., 2004]. For curves 1 and 2, measurements were performed with the "thin-layer" RDE in 0.1 M $HClO_4$; for curve 3, they were performed with stationary voltammetry in 0.5 M H_2SO_4. (Curves have been replotted from Maillard et al. [2002]; Gasteiger et al. [2005], Copyright 2002 and 2005, with permission from Elsevier; and from Guerin et al. [2004], Copyright 2004 American Chemical Society.)

PO_4^{3-} and HSO_4^-/SO_4^{2-} anions on Pt(111) planes, whereas adsorption is much weaker on (110) and (100) planes. Recently, Kuzume and co-workers investigated the ORR on [1$\bar{1}$1] stepped Pt single crystals [Kuzume et al., 2007]. In both H_2SO_4 and $HClO_4$ solutions, the Pt(111) surface showed the lowest catalytic activity among those studied. In both H_2SO_4 and $HClO_4$, the plot of electrocatalytic activity versus step density was nonlinear. In H_2SO_4, the activity increased strongly from Pt(111) to Pt(221), but did not change for (331), (551), (771), and (110) planes. In $HClO_4$, the overall influence of crystalline structure on ORR electrocatalysis was much less pronounced and the plot of activity versus step density exhibited a maximum. The experimental results were rationalized in terms of the site blocking effect of sulfate/bisulfate but the site blocking and energetic effect of adsorbed OH and surface oxide. Kinoshita and co-workers have noticed that, qualitatively, the dependence of MA on particle size followed the same trend as the mass averaged distribution (where MAD(hkl) = number of atoms in the (hkl) orientation/total number of atoms) of (111) and (100) facets [Giordano et al., 1991; Kinoshita, 1990]. Indeed, MAD(111) and MAD(100) reach maxima at particle sizes around 3.5 nm. If these

15.5 INFLUENCE OF PARTICLE SIZE ON ELECTROCATALYTIC ACTIVITIES: CASE STUDIES

facets were the most active in the ORR, then the observed correlation would explain why the maximum of MA was attained between 3 and 4 nm. Although this idea has been invoked for explaining PSEs in numerous original and review articles, it does not seem very convincing in view of the fact that (111) facets are the least active in the ORR in strongly adsorbing H_2SO_4 and H_3PO_4 electrolytes.

An alternative standpoint is to explain the PSEs on the ORR kinetics by electronic effects concomitant to the decrease of particle size or arising owing to metal–support interactions. This can be correlated with the increasing concentration of low coordination atoms, which have low d-band occupancy and are thus prone to strong adsorption of atoms and molecules [Hanson and Boudart, 1978; Parmigiani et al., 1990; Kinoshita, 1990, 1992; Giordano et al., 1991; Hwang and Chung, 1993; Mukerjee and McBreen, 1998]. Stronger adsorption of oxygenated species on nanoparticles was first reported by Hanson and Boudart in gas-phase catalysis over Pt/SiO_2 catalysts while studying the $H_2 + O_2$ reaction in the interval 273–373K [Hanson and Boudart, 1978]. They reported strong chemisorption of O_2 on Pt nanoparticles with a particle size below 3 nm. In liquid electrolyte, stronger adsorption of OH_{ads} on smaller Pt/C particles was derived by Mukerjee and McBreen from analysis of the white line in the in situ X-ray absorption near edge structure (XANES) spectra [Mukerjee and McBreen, 1998]. They found that the calculated Pt $5d$-band vacancy normalized to the number of surface atoms increased slightly with decreasing particle size both at 0.0 V and at 0.84 V vs. RHE.

According to Markovic and co-workers and Chen and Kucernak, anions and OH_{ads}, along with the site-blocking effect (the $1 - \theta$ term in (15.17)), may also affect the free energy of oxygen adsorption through lateral surface interactions [Wang et al., 2004; Chen and Kucernak, 2004a, b]. Following the discussion in Section 15.4, it is likely that for smaller Pt particles, aside from oxygen adsorption on the surface, its penetration into the subsurface is also possible, strongly affecting the electrocatalytic properties of the metal. It is well known that the formation of oxide films on metal electrodes significantly attenuates their activity in the ORR [Damjanovic and Hudson, 1988; Damjanovic, 1992].

Aside from the studies advocating the influence of particle size on ORR kinetics, some authors have claimed that the intrinsic catalytic activity of Pt is independent of particle size and that the observed changes in reaction rate are related to diffusion [Blurton et al., 1972; Bett et al., 1973; Vogel and Baris, 1977; Watanabe et al., 1988, 1989; Yano et al., 2006; Stonehart, 1990]. In view of the fact that understanding structure–function relationships is of paramount importance for a fundamental understanding of electrocatalysis, we will analyze this point of view in some detail. The idea originally proposed by Stonehart and by Watanabe is that the apparent correlation between the catalytic activity of nanoparticles and particle size stems from diffusive interference between Pt crystallites [Stoneheart, 1990; Watanabe et al., 1989]. If two Pt particles are located in close proximity on the carbon support, their O_2 diffusion spheres overlap, leading to a decrease in the measured current density, which translates into decreased apparent kinetic current and SA. These results were contradicted by Giordano and co-workers in H_3PO_4 and by Gamez and co-workers in H_2SO_4 [Giordano et al., 1991; Gamez et al., 1996]. According to Gamez et al. [1996], the local current density at a Pt/C particle is much smaller than the local limiting current density expected for spherical diffusion, and suggests that in the interval of low currents/low overpotentials, diffusion limitations occur only when the interparticle distance is less than five times the size of the Pt particles. Recently, a diffusion-related effect has been reported for model Pt nanoparticle arrays supported on flat indium–tin oxide (ITO)-coated glass substrates [Kumar and Zou, 2006]. O_2 electro-reduction on Pt particle arrays with the same particle size of 4 nm shows a negative shift in the peak potential and a significant increase in current density with increasing interparticle spacing from 40 to 80 nm. These changes have been explained by the change of the extent of O_2 diffusion field overlap.

Yano and co-workers have recently undertaken an RDE study of the ORR on carbon-supported Pt nanoparticles combined with ^{195}Pt electrochemical nuclear magnetic resonance (NMR) spectroscopy [Yano et al., 2006]. They conclude that there is a negligible difference in the surface electronic properties of these Pt/CB (carbon black) catalysts owing to size variations and therefore that ORR activities are not affected by differences in particle size. It is unfortunate that the authors of this otherwise very careful study have chosen to determine the kinetic currents at rather low electrode potentials of 0.70, 0.76, and 0.80 V vs. RHE. Under these conditions, the measured current is approaching the mass-transport-limiting value, and proper evaluation of the intrinsic catalytic activity is hardly feasible. Note that according to Chen and Kucernak, even at 0.9 V the measured current is strongly influenced by mass transport, not to speak of more negative electrode potentials [Chen and Kucernak, 2004a, b]. Higuchi and co-workers put special effort into preparing a very thin layer (<0.1 μm) film of carbon particles on a solid support and dispersing the catalyst very evenly [Higuchi et al., 2005]. Unfortunately, they did not take into consideration the fact that even for these very thin CLs, mass transport limitations are overwhelming and the effectiveness factor may be well below 1 at the high overpotentials utilized in their work. Thus, the absence of the PSE claimed by Yano and co-workers is likely related to severely underestimated SAs due to mass transport limitations [Yano et al., 2006]. Our concerns are confirmed by the values of the kinetic currents furnished by these authors. For example, for bulk Pt, the apparent rate constant $k_{app} = j_k/(4F[H^+][O_2])$ plotted in Fig. 15.6 of [Yano et al., 2006] amounts to about 10^2 cm^4 mol^{-1} s^{-1} at about 0.8 V vs. RHE and 60 °C. Assuming a Tafel slope of 60 mV dec^{-1} this translates into about 10^2 μA cm^{-2} at 0.9 V vs. RHE. This is an order of magnitude lower than the value reported by Gasteiger and co-workers for polycrystalline Pt under relevant conditions [Gasteiger et al., 2005]. Note that for Pt/CB samples, along with the external mass transport losses, internal losses in the pores of carbon materials cannot be ruled out. Unfortunately, Yano et al. did not supply details of the specific surface areas of the carbons utilized. Low values for kinetic currents are reported also in Higuchi et al. [2005].

In concluding this discussion, we would like to point out that understanding PSEs in the ORR is a very important issue, and special care should be taken to determine appropriate experimental conditions.

15.5.3 Oxidation of C_1 Molecules: CO, HCOOH, and CH_3OH

15.5.3.1 Oxidation of Adsorbed CO The electro-oxidation of CO has been extensively studied given its importance as a model electrochemical reaction and its relevance to the development of CO-tolerant anodes for PEMFCs and efficient anodes for DMFCs. In this section, we focus on the oxidation of a CO_{ads} monolayer and do not cover continuous oxidation of CO dissolved in electrolyte. An invaluable advantage of CO_{ads} electro-oxidation as a model reaction is that it does not involve diffusion in the electrolyte bulk, and thus is not subject to the problems associated with mass transport corrections and desorption/readsorption processes.

It is now widely accepted that CO_{ads} electro-oxidation proceeds via the L–H mechanism, which includes the reaction steps of CO adsorption, water splitting (15.18), and $CO_{ads} + OH_{ads}$ recombination (15.19) on two adjacent sites to yield CO_2:

$$H_2O + Pt \longleftrightarrow Pt - OH_{ads} + H^+ + e^- \qquad (15.18)$$

$$CO_{ads} + OH_{ads} \longrightarrow COOH_{ads} \longrightarrow CO_2 + H^+ + e^- \qquad (15.19)$$

Owing to the high adsorption energy of CO on Pt, high surface coverages of CO_{ads} are formed at adsorption potentials in the H_{UPD} region [Cuesta et al., 2006].

The $COOH_{ads}$ species was proposed by Gilman [1964] as an intermediate in $CO_{ads} + OH_{ads}$ recombination (15.19), and the feasibility of this reaction scheme has recently been validated by DFT calculations [Anderson and Neshev, 2002; Saravanan et al., 2003; Shubina et al., 2004; Filhol and Neurock, 2006]. Formation of $COOH_{ads}$ species was suggested by Zhu and co-workers based on the results of attenuated total reflection (ATR)-FTIR spectroscopy [Zhu et al., 1999]. $CO_{ads} + OH_{ads}$ recombination has been suggested to be controlled by the number of active sites that are able to dissociate water molecules [Maillard et al., 2004a; Andreaus et al., 2006]. Lebedeva and co-workers have shown that on Pt single crystals, OH_{ads} species are preferentially formed at monoatomic steps [Lebedeva et al., 2002a, b]. An increase in step density strongly enhances the electrocatalytic activity of Pt, and the reaction rate increases linearly with defect density [Petukhov et al., 1998; Lebedeva et al., 2002b], confirming that steps of (110) configuration are indeed the active sites for CO_{ads} electro-oxidation.

Typical CO_{ads} stripping voltammograms on Pt/GC electrodes with different mean particle sizes are presented in Fig. 15.8. The position of the main CO_{ads} stripping peak for $\bar{d}_N = 3.2$ nm ($E = 0.86$ V vs. RHE) is in good agreement with that for commercial 30 wt% Pt/C (E-Tek) [Maillard et al., 2002; Guerin et al., 2004]. As the particle size decreases, both the onset and the peak are shifted positively, and the peak becomes broader, with tailing on the descending slope [Friedrich et al., 1998, 2000; Cherstiouk et al., 2003b; Maillard et al., 2004a, b, 2005, 2007b; Mayrhofer et al., 2005b]. The peak shift is very systematic, scaling with $1/d_N$ and providing clear evidence that particle edges cannot be the active sites for CO monolayer electro-oxidation. Indeed, if this were the case, then the overpotential should have diminished with decreasing particle size, since the fraction of edges increases as the particle size

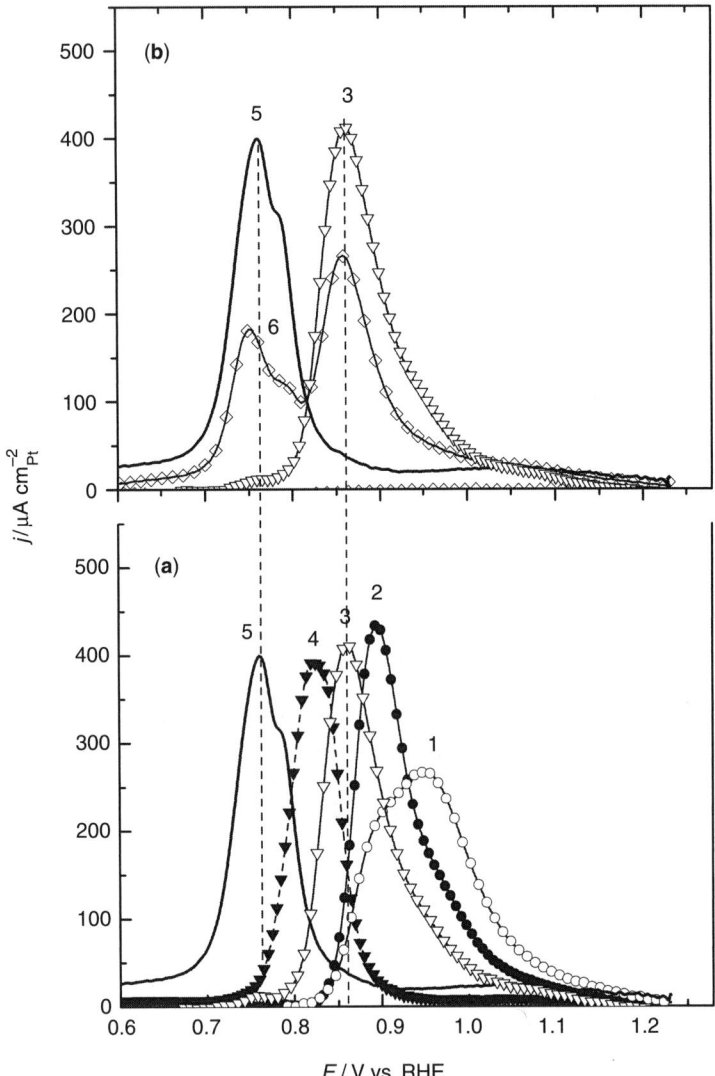

Figure 15.8 Background-subtracted CO_{ads} stripping voltammograms in 0.1 M H_2SO_4 at a sweep rate of 0.1 V s^{-1} for model Pt/GC electrodes with Pt mean particle sizes $\bar{d}_N = 1.8$ nm (1), 2.8 nm (2), and 3.2 nm (3), for polycrystalline Pt foil (4), and for Pt electrodeposited on GC comprising multigrained Pt with a grain size of about 5 nm (5). Curve 6 corresponds to a Pt/GC sample containing single-grained Pt particles with average size 3.5 nm along with multigrained Pt structures. (Curves have been replotted from Maillard et al. [2004a, 2005]—Reproduced by permission of The Royal Society of Chemistry and the PCCP Owner Societies.)

goes down. Hence, the peak potential should have shifted negatively rather than positively. The fact that particle edges do not provide active sites for CO electro-oxidation may be explained by their different site geometry, compared with steps, as discussed in Section 15.3.

Mayrhofer and co-workers suggested that the active sites are located at the defect sites on nanoparticle facets [Mayrhofer et al., 2005a]. Based on ex situ high resolution TEM (HRTEM) images, they speculated that large particles bear a larger fraction of defects, and hence show better activity in CO_{ads} electro-oxidation. Although we believe that this is a reasonable assumption, it should be noted that making conclusions on the basis of ex situ characterization is not unambiguous. The low potential peak in CO stripping voltammograms, which appears in curve 5 of Fig. 15.8a, was first documented by Cherstiouk and co-workers at high Pt loadings and attributed to Pt agglomeration [Cherstiouk et al., 2003a]. Maillard and co-workers confirmed that this low potential peak correlates with the number of Pt agglomerates detected with TEM [Maillard et al., 2005]. They examined Pt/GC agglomerates by HRTEM, and found complex irregular structures composed of individual Pt nanoparticles interconnected via grain boundaries (so-called "multigrained" structures). These nanocrystalline materials feature a high density of surface and bulk crystalline defects (steps and intergrain boundaries), which are believed to be the likely reason for their high electrocatalytic activity [Cherstiouk et al., 2000, 2003a; Maillard et al., 2005]. This peak assignment is proven by CO stripping from electrodeposited Pt with multigrained structure (curve 5 in Fig. 15.8), which exhibits exclusively the low potential peak. It has recently been demonstrated that the catalytic activity of bimetallic Pt-Ru catalysts in CO and methanol electro-oxidation also correlates with the concentration of intergrain boundaries [Gavrilov et al., 2007; Maillard et al., 2007a]. Metal atoms in the vicinity of grain boundaries usually have a decreased number of neighbors in their first coordination shell [Gleiter, 1992], and thus are expected to show modified adsorption properties compared with metal atoms of densely packed terraces. For further discussion of the electrocatalytic properties of these promising materials, the reader is referred to Maillard et al. [2005, 2007b]; Gavrilov et al. [2007]; Cherstiouk et al. [2000, 2003a, 2008].

Peak multiplicity in CO_{ads} stripping voltammograms and chronoamperograms has given rise to some controversy in the literature. Guerin and co-workers ascribed the low potential peak to electro-oxidation of CO_{ads} on Pt terraces, whereas they attributed the high potential peak to CO electro-oxidation on edges and corner sites [Guerin et al., 2004]. Solla-Gullon and co-workers attributed the peak multiplicity in CO stripping voltammograms to electro-oxidation of ordered and disordered CO_{ads} domains on facets of different crystallographic orientation [Solla-Gullon et al., 2005, 2006]. Arenz and co-workers reported double-peaked versus single-peaked chronoamperograms for 2 and 1 nm Pt particles, respectively [Arenz et al., 2005], and explained this finding by the increasing fraction of surface defects on 2 versus 1 nm Pt/C particles [Arenz et al., 2005; Mayrhofer et al., 2005a, b]. In our opinion, a more likely reason for the peak multiplicity in chronoamperograms and in stripping voltammograms of supported metal nanoparticles is so-called "interparticle heterogeneity," i.e., the presence of particles of different sizes and structures, although other factors may also be relevant. For example, multiple-peaked stripping voltammograms for polycrystalline Pt are likely to originate from the contribution of different crystallographic planes to the electrode surface. For further discussion, the reader is referred to Maillard et al. [2004b, 2005, 2007b].

CO monolayer electro-oxidation on Pt/C has also been investigated with chronoamperometry [Cherstiouk et al., 2003b; Maillard et al., 2004a, 2007b]. Peaked current transients (Fig. 15.9) typical of CO electro-oxidation on polycrystalline

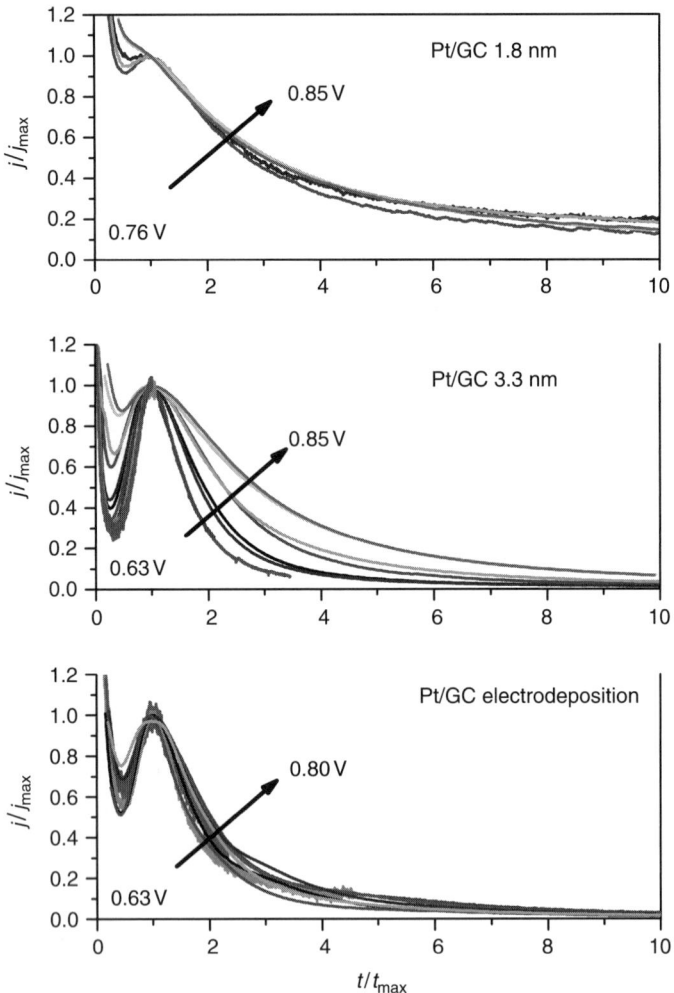

Figure 15.9 Normalized current transients for CO_{ads} monolayer electro-oxidation in 0.1 M H_2SO_4 after a potential step from 0.1 V vs. RHE to various electro-oxidation potentials for Pt/GC electrodes with different particle sizes. The CO adsorption potential was 0.1 V vs. RHE, $T = 298$K. (Reprinted from Maillard et al. [2007b], Copyright 2007, with permission from Elsevier.)

15.5 INFLUENCE OF PARTICLE SIZE ON ELECTROCATALYTIC ACTIVITIES: CASE STUDIES

[McCallum and Pletcher, 1976] and single-crystalline [Love and Lipkowski, 1988; Koper et al., 1998; Lebedeva et al., 2002b] Pt were observed. For extended surfaces, the transient shape was explained by the L–H mechanism [Koper et al., 1998] or the L–H mechanism complicated with nucleation and growth of OH_{ads} islands [McCallum and Pletcher, 1976; Love and Lipkowski, 1988].

Upon decreasing the particle size from above 5 nm to 1.8 nm, CO_{ads} electro-oxidation kinetics slows down, as manifested by the increase in the time of the current maxima and the decrease in their amplitude, concomitant with a pronounced current tailing on the descending slope of the transient, strongly deviating from the transients reported for extended surfaces [Andreaus et al., 2006; Arenz et al., 2005; Kobayashi et al., 2005; Maillard et al., 2004a, 2007b]. Most remarkable are the size-dependent changes in the shape of the transient (Fig. 15.9), which suggest an interplay of electrochemical and nonelectrochemical steps. Detailed experimental investigations of the current transients and stripping voltammograms as functions of particle size and experimental conditions (electrode potential, sweep rate, CO_{ads} coverage, etc.), together with kinetic modeling, have provided important information concerning the mechanism of CO monolayer electro-oxidation on carbon-supported Pt nanoparticles, which can be summarized as follows [Andreaus et al., 2006]:

- Water splitting (15.18) is restricted to active sites, whose proportion on the surface does not correlate with the fraction of edge sites and vertices. Its potential dependence follows a Tafel law with transfer coefficient $\alpha \approx 0.5$ and a particle-size-independent equilibrium potential $E_{H_2O/OH}^{eq} \approx 0.67 - 0.69$ V vs. RHE.
- CO_{ads} surface mobility is essential for the electro-oxidation reaction, with a diffusion coefficient that depends strongly on particle size. For "large" (>5 nm) particles with multigrained structure, diffusion is fast compared with OH_{ads} formation and $CO_{ads} + OH_{ads}$ recombination, and does not impose limitations on the reaction kinetics. The lower limit for D_{CO} for "large" particles was estimated as over 10^{-14} cm^2 s^{-1}. Hence, relatively symmetric current transients are observed. As the particle size is decreased to 1.8 nm, CO_{ads} diffusivity drops by at least two orders of magnitude to about 10^{-16} cm^2 s^{-1}. Thus, the slow current decay for small particles in the whole potential interval studied and for intermediate size (3.3 nm) particles at high potentials (above about 0.8 V vs. RHE) can be identified with slow CO_{ads} diffusion to the active sites.
- $CO_{ads} + OH_{ads}$ recombination occurs in an electrochemical step and follows a Tafel law with $\alpha \approx 0.5$, and is particle-size-dependent: when the size decreases to 1.8 nm, k_{ox} falls by about an order of magnitude.

An important result of this study is the conclusion of a particle-size-dependent CO_{ads} surface mobility. The value obtained for "large" Pt particles is significantly smaller than D_{CO} at a solid/gas interface. However, Kobayashi and co-workers, using solid state NMR, performed measurements of the tracer diffusion coefficient D_{CO}^* at the solid/electrolyte interface and for Pt-black particles (about 5 nm grain size), obtained $D_{CO}^* = 3.6 \times 10^{-13}$ cm^2 s^{-1}, which is considerably smaller than the value previously found for single-crystalline surfaces at a solid/gas interface [Kobayashi et al., 2005]. The reasons for the lower diffusion coefficient at the solid/electrolyte compared with the solid/gas interface may be manifold, and include anion effects, influence of the solvent, and influence of the surface charge density. Therefore, it would be of interest to explore the influence of anion-specific adsorption on CO_{ads} electro-oxidation kinetics [Maillard et al., 2004a; Arenz et al., 2005].

The inferred influence of particle size on D_{CO} is in agreement with the ^{13}C NMR study by Becerra and co-workers, who reported a considerable increase in the activation energy for CO_{ads} diffusion with decreasing size of Pt particles supported on Al_2O_3 [Becerra et al., 1993]. They explained their data by a model in which surface diffusion requires the formation of bridge-bonded CO molecules, and by a decrease in the fraction of the latter with decreasing particle size. A reduced contribution of bridge-bonded and multiply bonded CO with decreasing size of Pt particles is supported by FTIR studies [Park et al., 2001]. It should be mentioned, however, that quantitative determination of the surface coverages of bridge- and atop-bonded CO molecules from FTIR spectra obtained at high surface coverages is obstructed by dipole–dipole coupling of the adsorbates.

While size effects on the kinetics of CO_{ads} monolayer electro-oxidation are unequivocal, their origin is still not fully understood. Takasu and co-workers discussed the changes induced by particle size on particles' electronic properties and on CO bonding by exploring the UPS spectra of CO over Pd/graphite model catalysts [Takasu et al., 1984]. Unfortunately, they did not make a distinction between the initial and final state effects, and thus the validity of their conclusions may be questioned. Mukerjee and McBreen discussed the influence of d-band vacancies on the strength of adsorbate bonding [Mukerjee and McBreen, 1998]. It is likely that both geometric (the proportion of low coordinated sites and the types of surface sites) and electronic properties of Pt nanoparticles change as their size drops below about 4–5 nm resulting in a stronger CO interaction with the surface, increased irreversibility of surface oxidation, a decrease in the $OH_{ads} + CO_{ads}$ recombination rate constant, and a strong decrease in CO_{ads} surface mobility. An increased propensity of small Pt particles to stabilize dense CO_{ads} islands may also have a negative effect on electro-oxidation kinetics.

An influence of particle size on the kinetics of CO_{ads} electro-oxidation has been shown by Maillard and co-workers with FTIR spectroscopy. It has been suggested that the reaction starts on the terraces of "large" (≥ 3 nm) particles, and then propagates to the particle edges. Electro-oxidation of CO_{ads} on small (<2 nm) particles commences at more positive potentials, when CO_{ads} on large particles is oxidized.

15.5.3.2 Formic Acid Oxidation (FOR)
Formic acid electro-oxidation is a two-electron electrochemical reaction,

$$HCOOH \longrightarrow CO_2 + 2H^+ + 2e^- \quad (15.20)$$

which on Pt electrodes is known to occur via a so-called "dual pathway" mechanism involving (i) "dehydrogenation" and (ii) dehydration pathways [Markovic and Ross, 2002]. The former involves the formation of adsorbed CO, which acts as a catalytic poison, while the latter does not. Formation of CO_{ads} in the course of formic acid electro-oxidation results in strong self-inhibition effects. Takasu and co-workers, and later Park and co-workers, demonstrated that SA for the electro-oxidation of formic acid on carbon-supported Pt nanoparticles *increases* as the particle size decreases, with no significant difference in electro-oxidation kinetics between $HClO_4$ and H_2SO_4 [Yahikozawa et al., 1991; Park et al., 2002b].

Strong PSEs in the FOR have been attributed to the structural sensitivity of the reaction, previously confirmed for Pt(*hkl*) single crystals by Clavilier, Adzic, and co-workers [Clavilier et al., 1981b; Adzic et al., 1983]. The reaction kinetics has been studied with cyclic voltammetry and chronoamperometry. The peak current densities for Pt(*hkl*) in CVs follow the order (110) > (111) > (100) in the positive-going sweep and (110) > (100) > (111) in the negative-going sweep [Adzic et al., 1983]. Pt(111) has the lowest activity but the highest tolerance to self-poisoning. Sun and Yang obtained similar results, with Pt(110) > Pt(111) > Pt(100), in a chronoamperometric study [Sun and Yang, 1999]. Not only the reaction rate, but also the apparent activation energy demonstrates remarkable structure sensitivity [Sun and Yang, 1999].

Park and co-workers attributed the PSE in the FOR to the "ensemble effect," where reactant dehydrogenation to form poisoning CO_{ads} species is impeded by the sharply decreasing availability of contiguous Pt terrace sites for particles with $d < 4$ nm [Park et al., 2002b]. This structural model is consistent with infrared measurements, which for small particles below about 4 nm show a decreased rate of surface poisoning with CO_{ads} formed upon particle contact with formic acid [Park et al., 2002b]. Since the oxidation of formic acid to CO_2 does not necessarily require its dehydrogenation, the decreased contribution of this pathway results in the enhancement of the overall reaction rate.

It is interesting that for carbon-supported Pd nanoparticles, PSEs are manifested in a different manner [Zhang et al., 1995; Zhou et al., 2006]. SA gradually increases as the particle size decreases to about 4 nm, but for smaller particles it falls steeply. Zhang et al. and Zhou et al. attributed the observed changes to the variation of the electronic structure of Pd clusters, in particular the DOS at the Fermi level, as accessed by XPS and UPS. Unfortunately, they did not attempt to separate initial and final state effects in the photoelectron spectra, and therefore one may challenge their conclusions. Dissimilarities in the behavior of Pd and Pt nanoparticles may be tentatively ascribed to differences in the mechanism of the FOR for these metals. Indeed, it is known from single-crystal studies that for Pd electrodes the contribution of the dehydrogenation pathway is marginal, resulting in their insignificant poisoning.

15.5.3.3 Methanol Oxidation (MOR) The MOR on noble metal electrodes has been studied for decades (see, e.g., the reviews [Petrii et al., 1965; Léger, 2001; Wasmus and Kuver, 1999; Iwasita, 2003; Waszczuk et al., 2002; Gasteiger et al., 1993] and references therein):

$$CH_3OH + H_2O \longrightarrow CO_2 + 6H^+ + 6e^- \qquad (15.21)$$

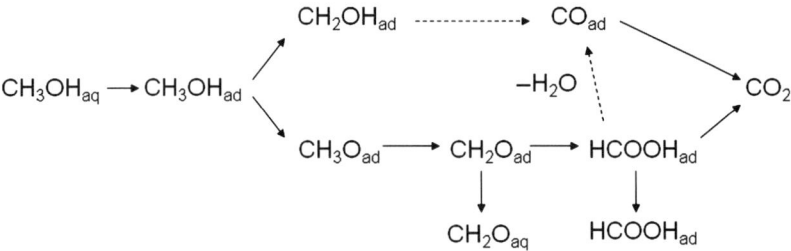

Figure 15.10 Schematic representation of the "dual pathway" mechanism of the MOR.

An in-depth review of the mechanism of the MOR is beyond the scope of this chapter and will be discussed only inasmuch as it is necessary for understanding PSEs. For a more detailed discussion, the reader is referred to Chapters 6, 11, and 13 and to the original papers and reviews cited in this section.

A simplified scheme of the "dual pathway" electrochemical methanol oxidation on Pt resulting from recent advances in the understanding of the reaction mechanism [Cao et al., 2005; Housmans et al., 2006] is shown in Fig. 15.10. The term "dual pathway" encompasses two reaction routes: one ("indirect") occurring via the intermediate formation of CO_{ads}, and the other ("direct") proceeding through partial oxidation products such as formaldehyde.

Adsorption via dehydrogenation was proposed many years ago by Petrii and co-workers [Petrii et al., 1965]. This is a surface-sensitive reaction, as underlined, for example, in Papoutsis et al. [1987], and requires three adjacent sites [Beden et al., 1981; Léger, 2001], as confirmed recently in an elegant study [Cuesta, 2006]. Methanol dehydrogenation appears as an anodic current in the positive-going scan of the CV starting from the H_{UPD} region. The simultaneously registered mass spectrometric current shows zero at $m/z = 44$, suggesting that CO_2 is not formed in this potential interval [Krausa and Vielstich, 1994]. Stepwise dehydrogenation via C—H bond breaking results in a hydroxymethyl (CH_2OH) intermediate, which is then further dehydrogenated to CO_{ads}, the latter acting as a reaction intermediate but also as a poison. Formation of CO_{ads} has been confirmed in numerous electrochemical and spectroscopic investigations (see Biegler and Koch [1967]; Beden et al. [1981] and the reviews [Petrii et al., 1965; Léger, 2001; Wasmus and Kuver, 1999; Iwasita, 2003; Waszczuk et al., 2002; Gasteiger et al., 1993]). As outlined above, CO_{ads} electro-oxidation proceeds in a surface reaction with OH_{ads} and thus does not occur below the threshold potential of water splitting. CO_{ads} electro-oxidation in the potential interval relevant to DMFCs (<0.35 V vs. RHE) is very slow, and thus CO_{ads} accumulates on the surface, inhibiting further reaction. Thus, some authors have identified water dissociation on Pt electrodes as the rate-determining step for the MOR [Redondo et al., 1989; Léger, 2001]. It should be noted, however, that the problem is not necessarily the slow OH_{ads} formation, but rather its low coverage in the potential interval relevant to the MOR.

Recent experimental and theoretical studies [Cao et al., 2005; Housmans et al., 2006] suggest that, depending on the surface structure of the electrode and on the

15.5 INFLUENCE OF PARTICLE SIZE ON ELECTROCATALYTIC ACTIVITIES: CASE STUDIES

experimental conditions (in particular, the nature of the supporting electrolyte and the electrode potential), methanol dissociative adsorption may follow an O—H bond cleavage pathway, resulting in a reactive methoxy adsorbate and avoiding surface poisoning with CO_{ads}. There is also spectroscopic evidence for reaction intermediates such as CHO_{ads} and $COOH_{ads}$, which demonstrates that some methanol molecules are partially dehydrogenated and yield formaldehyde or formic acid [Iwasita and Vielstich, 1986; Beden et al., 1987a, b; Wasmus et al., 1995; Zhu et al., 1999].

The MOR is a surface-sensitive reaction [Clavilier et al., 1981a]. The influence of surface crystallography on the partitioning between the "direct" and the "indirect" pathways was investigated by Housmans and co-workers using electrochemistry, mass spectrometry, and modeling [Housmans et al., 2006; Housmans and Koper, 2003].

The first observation of PSEs for the MOR on Pt was reported by Attwood and co-workers in 1980. They found a bell-shaped curve for the dependence of SA on particle size, with the maximum at about 2 nm [Attwood et al., 1980]. Takasu and co-workers reported a decrease in SA with decreasing particle size for Pt particles deposited on GC by vacuum deposition, as depicted in Fig. 15.11 [Yahikozawa et al., 1991]. This PSE was detected in both $HClO_4$ and H_2SO_4 and at various electrolyte concentrations [Yahikozawa et al., 1991]. The PSE was confirmed by Kabbabi

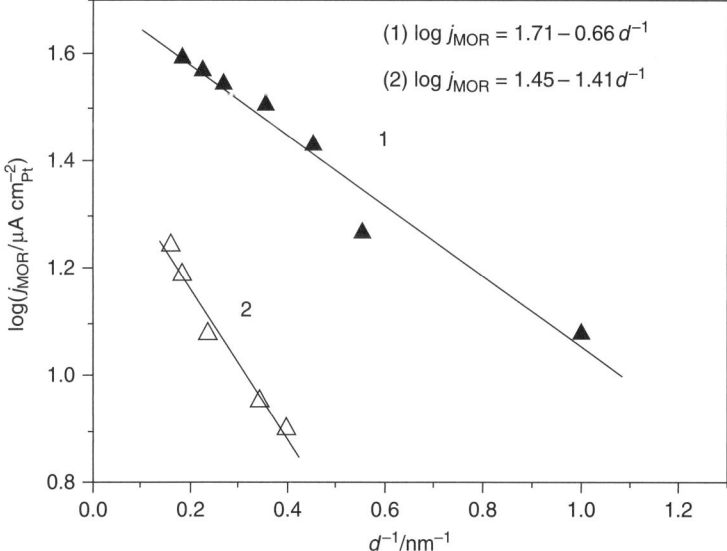

Figure 15.11 Specific activities for the MOR on Pt/Vulcan XC-72 (1) [Guerin et al., 2004] and Pt/GC (2) [Takasu et al., 2000] in aqueous electrolytes as functions of particle size. For curve 1, currents were measured after 1 minute at 0.6 V vs. RHE in 1 M CH_3OH + 0.5 M H_2SO_4 at room temperature; for curve 2, they were measured after 10 minutes at 0.6 V vs. RHE in 0.1 M CH_3OH + 0.02 M $HClO_4$ at 298K. (Curves have been replotted from Guerin et al. [2004], Copyright 2004 American Chemical Society; and from Takasu et al. [2000], Copyright 2000, with permission from Elsevier.)

and co-workers for Pt/Vulcan nanoparticles of different sizes provided by E-Tek [Kabbabi et al., 1994], and later by other authors for Pt particles supported on various carbon materials [Frelink et al., 1995; Takasu et al., 2000; Gloaguen et al., 1997; Gojkovic and Vidakovic, 2001; Maillard et al., 2002; Cherstiouk et al., 2003a]. Frelink and co-workers prepared Pt/C electrocatalysts with particle sizes ranging between 1.2 and 10 nm by different elaboration techniques (ion exchange, impregnation, impregnation followed by sintering, and aging of colloids) [Frelink et al., 1995]. The plot of specific activity versus particle size was bell-shaped, with a maximum between 4.5 and 10 nm [Frelink et al., 1995].

Most of these authors ascribed the decrease in SA to the increased coverage of oxygenated species on the surface of small Pt/C particles and concomitant blockage of the methanol dissociative adsorption. Takasu and co-workers [Yahikozawa et al., 1991] related the observed decrease in SA for the MOR to the decrease in the number of Pt(100) planes, which are the most active among the three low index planes for this reaction [Housmans et al., 2006]. Park and co-workers investigated the MOR using cyclic voltammetry and FTIR spectroscopy, and attributed the decrease in SA to an "ensemble effect," whereby methanol dehydrogenation to form CO_{ads} is impeded by the sharply decreasing availability of contiguous Pt terrace sites for $d < 4$ nm [Park et al., 2002b]. Indeed, using FTIR spectroscopy, they observed slower accumulation of CO_{ads} on the surface of small Pt particles. This explanation is based on considering CO_{ads} as the reactive intermediate, while it acts also as a poison. Hence, Cherstiouk and co-workers suggested that the lower activity of small Pt particles in the MOR may be related to their lower activity in CO_{ads} electro-oxidation [Cherstiouk et al., 2003a].

Recently, Bergamaski and co-workers investigated the MOR using differential electrochemical mass spectrometry (DEMS) over commercial Pt/Vulcan XC-72 catalysts with Pt weight percentage ranging from 10% to 80% [Bergamaski et al., 2006]. It was found that the partitioning among the end products depends strongly on the metal percentage in the catalysts and thus on mean particle size and particle size distribution. CO_2 efficiency was minimal for 10% and 80% catalysts, but increased greatly for the intermediate metal percentages of 30% and 60%. In order to account for this remarkable observation, Bergamaski et al. proposed a model of diffusional coupling between large and small particles on the nanometer scale, where formaldehyde desorbs from large particles and is readsorbed and further oxidized to CO_2 on smaller particles. We find this idea very interesting. It should be noted, however, that the total Pt surface area in porous electrodes employed by Bergamaski et al. varied strongly with metal percentage. Comparison of the data on CO_2 efficiency with total Pt area shows a one-to-one correlation between them (see Fig. 15.10 and Table 15.1 in Bergamaski et al. [2006]), suggesting that the effect observed by Bergamaski et al. may be explained within the readsorption–reoxidation model proposed earlier by Jusys and co-workers when studying the Pt/C loading effect [Jusys et al., 2003]. Note that Jusys et al. did not change the type of electrocatalyst but rather changed its amount in the thin layer, and observed that with increasing Pt loading, the current efficiency for partial electro-oxidation to formaldehyde decayed significantly (from 40% to almost 0%), while that for complete electro-oxidation to

CO_2 increased from 50% to 80%. It would be highly desirable to investigate MOR on model nonporous substrates with DEMS in order to determine the influence of particle size on various reaction steps and the partitioning between the "direct" and "indirect" pathways.

Finally, it should be pointed out that physical and chemical properties of carbon supports and catalyst preparation methods may influence methanol electro-oxidation kinetics [Attwood et al., 1980; Gloaguen et al., 1997; Gojkovic and Vidakovic, 2001]. For example, Shukla and co-workers established a correlation between electrocatalytic activity for the MOR and the concentration of surface functional groups [Shukla et al., 1994]. Pt nanoparticles were supported on Ketjenblack carbons with different concentrations of acid/base functional groups. Shukla et al. claimed higher electrocatalytic activities for electrodes with low concentrations of surface functional groups (pH_{zpc} values between 6 and 7). On the other hand, Frelink, Gloaguen, and co-workers arrived at the opposite conclusion, namely, that the presence of acidic functionalities on carbon slightly enhances electrocatalytic activity for the MOR [Frelink et al., 1995; Gloaguen et al., 1997]. It should also be pointed out that Rao and co-workers demonstrated a pronounced influence of carbon porosity on electrocatalytic activity for the MOR in MEAs [Rao et al., 2005]. They prepared Pt-Ru/Sibunit particles with constant size of metal particles but different specific surface areas of carbon supports, S_{BET}. Both MA and SA increased with decreasing S_{BET}. This observation highlights the importance of the catalyst support, and suggests that the latter may obscure PSEs. For example, catalyst utilization and mass transport losses may depend on the porosity of the support.

The above discussion emphasizes the limitations imposed by the use of metal particles on porous substrates, and calls for further efforts in designing model systems for better understanding of PSEs in complex multistep electrochemical reactions.

15.5.3.4 ORR in the Presence of Methanol
We have seen in the preceding sections that Pt particle size has significant effects on the kinetics of both the ORR and the MOR. On the other hand, "crossover" of methanol from the anode to the cathode is a serious practical problem limiting the performance of DMFCs. Although some methanol-tolerant electrocatalysts have been developed during the last decade, Pt-based electrocatalysts are still the materials of choice, because of their relatively high stability in acidic and oxidizing media. Studies of oxygen reduction in the presence of methanol have confirmed that the ORR and the MOR occur simultaneously at a Pt electrode [Bittins-Cattaneo et al., 1993; Chu and Gilman, 1994]. For carbon-supported Pt nanoparticles, the potential loss ΔE with respect to the ORR in a methanol-free electrolyte increases significantly with increasing size of Pt particles [Maillard et al., 2002]. This occurs simultaneously with the positive shift of the methanol electro-oxidation peak. SA for oxygen electro-reduction in a methanol-containing electrolyte at 0.85 V vs. RHE was reported to decrease with increasing Pt particle size [Maillard et al., 2002]. Since for both the ORR and the MOR, SA decreases with decreasing particle size (i.e., it exhibits "negative" particle size effects), the enhanced tolerance of smaller Pt particles to methanol may be explained by a stronger decrease in SA for the MOR with particle size.

15.6 SUMMARY AND OUTLOOK

We have presented a selected survey of the literature devoted to the investigation of PSEs in fuel cell-related electrochemical reactions. This review has been almost entirely devoted to the discussion of Pt/C, with some occasional allusions to other materials. This largely reflects today's situation in electrocatalysis at supported metal nanoparticles, and makes generalizations difficult, if not impossible. In order to further understanding of PSEs, it would be desirable to extend the range of metals and supports investigated. Studies are now underway in Hayden's group, utilizing Au and other metals supported on oxide-based (e.g., TiO_2) supports Guerin et al. [2006a, b]; Hayden et al. [2007] (see Chapter 16). Such investigations are imperative for better understanding of PSEs and particle–support interactions.

Despite some controversies, most workers in this field admit the existence of PSEs in the particle interval from 1 to 5 nm. Most often, so-called "negative" PSEs are observed, with SA decreasing with decreasing particle size. This has been documented for the ORR and the MOR and for CO monolayer electro-oxidation. Meanwhile, for the FOR and the HOR, the activity increases with decreasing particle size, exhibiting a so-called "positive" PSE. As underlined in this chapter, up to now PSEs have been investigated in the scalable size range. They scale with inverse particle diameter, suggesting that they are related to properties dependent on the surface-to-volume ratio rather than to nonscalable quantum size effects. In order to account for PSEs, the following reasons have been invoked:

- An "ensemble" effect related to the variation of the number of sites of specific configuration with particle size. Ensemble effects have been, for example, suggested to account for the decrease in H_{UPD} coverage, and the SA in the MOR.
- A surface crystallography effect, which is, in fact, closely related to the "ensemble" effect.
- Lower average CNs related to the increase in the contribution of edge and corner sites, concomitant with the d-band shift and an increase in the proportion of d-band vacancies, resulting in changes in adsorbate bond strength (stronger OH/O and CO bonding to Pt nanoparticles), and also changes in the type of adsorption sites (cf. atop CO on Pt nanoparticles versus partitioning among atop, bridge, and three-fold hollow site bonded CO on Pt(111)).
- Changes in electronic properties, such as Fermi level shifts and changes in the DOS, which still have to be confirmed experimentally. For example, Yano and co-workers used ^{195}Pt NMR to probe possible changes in electronic properties induced by particle size [Yano et al., 2006]. They concluded that the surface Knight shifts as well as E_F-LDOS of surface Pt atoms showed no noticeable size dependence in the particle size range from 4.8 down to 1.6 nm; hence, electronic properties are independent of size in this interval.
- Changes in the scaling of chemical potential with $1/d$ [Nagaev, 1991, 1992; Parmon, 2007; Savinova, 2006]. This could account for the similar slopes of the SA versus $1/d$ plots for different electrochemical reactions.

- Dependence of the electric field distribution in the double layer on particle size [Zhdanov and Kasemo, 2002; Chen and Kucernak, 2004a, b], which, according to Zhdanov and Kasemo, should result in an increase in the rates of electrochemical reactions on nanometer-sized metal particles.

Further studies are needed in order to better understand the origins of the PSEs in electrocatalysis.

The strong sensitivity of reaction kinetics to point and extended defects requires that studies of size and structural effects be performed with materials whose size and structure are characterized using a variety of complementary imaging and spectroscopic techniques, preferably in situ. This would greatly aid in reducing controversies and misinterpretations in the electrocatalysis of nanoparticulate systems. Although size and structural effects are closely related, it is worth separating effects inherent to small systems from structural effects that are common also to extended surfaces. A noteworthy example concerns monoatomic steps, which may be present on single-crystalline surfaces and likely also on nanoparticle facets. Another example concerns intergrain boundaries—extended defects whose presence strongly affects electrocatalytic activity, as confirmed for CO and methanol electro-oxidation on Pt and Pt-Ru materials. Multigrained materials with a high density of intergrain boundaries are generic for fuel cell electrocatalysts containing metals in high weight percentage. Unfortunately, their formation is often ignored, and this may lead to misinterpretation of PSEs.

Understanding of PSEs is complicated by a lack of understanding of the intimate mechanisms of electrocatalytic reactions, for example for the ORR and the MOR. Further insights into the reaction mechanisms and identification of the rate determining steps will help to elucidate PSEs.

Interpretation of published data is often complicated by the fact that rather complex catalytic materials are utilized, namely, polydisperse nonuniform metal particles, highly porous supports, etc., where various secondary effects may influence or even submerge PSEs. These include mass transport and discrete particle distribution effects in porous layers, as confirmed by Gloaguen, Antoine, and co-workers [Gloaguen et al., 1994, 1998; Antoine et al., 1998], and diffusion–readsorption effects, as shown by Jusys and co-workers for the MOR and by Chen and Kucernak for the ORR [Jusys et al., 2003; Chen and Kucernak, 2004a, b]. Novel approaches to the design of ordered nanoparticle arrays where nanoparticle size and interparticle distances can be varied independently are expected to shed light on PSEs in complex multistep multielectron processes such as the MOR and the ORR.

As the reader might have noticed, many conclusions in electrocatalysis are based on results obtained with electrochemical techniques. In situ characterization of nanoparticles with imaging and spectroscopic methods, which is performed in a number of laboratories, is invaluable for the understanding of PSEs. Identification of the types of adsorption sites on supported metal nanoparticles, as well as determination of the influence of particle size on the adsorption isotherms for oxygen, hydrogen, and anions, are required for further understanding of the fundamentals of electrocatalysis.

Understanding the physical origins of PSEs requires a coordinated effort from experimentalists and theoreticians, and will proceed together with the development of novel approaches toward modeling supported metal nanoparticles using semiempirical and ultimately ab initio approaches. We will not be able to claim that we understand PSEs until we are able to predict them. This is important both for the formulation of the fundamentals of electrocatalysis as scientific discipline and for the development of low cost energy-efficient fuel cells.

ACKNOWLEDGMENTS

We would like to express our gratitude to all the collaborators who have contributed to the work described in this chapter. We are also indebted to M. T. M. Koper for a thorough reading of the manuscript and for suggesting a number of improvements. The authors gratefully acknowledge the financial support received from the ANR (in the framework of "Chaire d'Excellence" (E.S.) and the Carbocell and MDM projects (F.M.)), RFBR, DFG, and BMBF.

REFERENCES

Adzic R. 1998. Recent advances in the kinetics of oxygen reduction. In: Lipkowski J, Ross PN. editors. Electrocatalysis. New York: Wiley-VCH.

Adzic RR, Tripkovic AV, Markovic NM. 1983. Structural effects in electrocatalysis: oxidation of formic acid and oxygen reduction on single-crystal electrodes and the effects of foreign metal adatoms. J Electroanal Chem 150: 79–88.

Aiken JDI, Finke RG. 1999. A review of modern transition-metal nanoclusters: their synthesis, characterization, and applications in catalysis. J Mol Catal A 145: 1–44.

Ajayan PM, Marks LD. 1988. Quasimelting and phases of small particles. Phys Rev Lett 60: 585–587.

Anderson AB. 2002. O_2 reduction and CO oxidation at the Pt–electrolyte interface. The role of H_2O and OH adsorption bond strengths. Electrochim Acta 47: 3759–3763.

Anderson AB, Neshev NM. 2002. Mechanism for the electro-oxidation of carbon monoxide on platinum, including electrode potential dependence—Theoretical determination. J Electrochem Soc 149: E383–E388.

Andreaus B, Maillard F, Kocylo J, Savinova ER, Eikerling M. 2006. Kinetic modeling of CO monolayer oxidation on carbon-supported platinum catalyst nanoparticles. J Phys Chem B 110: 21028–21040.

Antoine O, Bultel Y, Durand R, Ozil P. 1998. Electrocatalysis, diffusion and ohmic drop in PEMFC: particle size and spatial discrete distribution effects. Electrochim Acta 43: 3681–3691.

Antoine O, Bultel Y, Durand R. 2001. Oxygen reduction reaction kinetics and mechanism on platinum nanoparticles inside Nafion®. J Electroanal Chem 499: 85–94.

Arenz M, Mayrhofer KJJ, Stamenkovic V, Blizanac BB, Tomoyuki T, Ross PN, Markovic NM. 2005. The effect of the particle size on the kinetics of CO electrooxidation on high surface area Pt catalysts. J Am Chem Soc 127: 6819–6829.

Attard GA, Hazzazi O, Wells PB, Climent V, Herrero E, Feliu JM. 2004. On the global and local values of the potential of zero total charge at well-defined platinum surfaces: Stepped and adatom modified surfaces. J Electroanal Chem 568: 329–342.

Attwood PA, McNicol BD, Short RT. 1980. Electrocatalytic oxidation of methanol in acid electrolyte—Preparation and characterization of noble-metal electrocatalysts supported on pretreated carbon-fiber papers. J Appl Electrochem 10: 213–222.

Bagotskii VS, Tarasevich MR, Filinovskii VY. 1969. Calculation of the kinetic parameters of conjugated reactions of oxygen and hydrogen peroxide. Elektrokhimiya 5: 1218.

Bagotsky VS, Osetrova NV. 1973. Investigation of hydrogen ionization on platinum with the help of micro-electrodes. J Electroanal Chem 43: 233–249.

Becerra LR, Klug CA, Slichter CP, Sinfelt JH. 1993. NMR-study of diffusion of CO on alumina-supported Pt clusters. J Phys Chem 97: 12014–12019.

Beden B, Lamy C, Bewick A, Kunimatsu K. 1981. Electrosorption of methanol on a platinum-electrode—IR spectroscopic evidence for adsorbed CO species. J Electroanal Chem 121: 343–347.

Beden B, Hahn F, Juanto S, Lamy C, Léger JM. 1987a. Infrared spectroscopic study of the methanol adsorbates at a platinum electrode: Part I. Influence of the bulk concentration of methanol upon the nature of the adsorbates. J Electroanal Chem 225: 215–225.

Beden B, Juanto S, Léger JM, Lamy C. 1987b. Infrared spectroscopic study of the methanol adsorbates at a platinum electrode: Part III Structural effects and behaviour of a polycrystalline surface. J Electroanal Chem 238: 323–331.

Bergamaski K, Pinheiro ALN, Teixeira-Neto E, Nart FC. 2006. Nanoparticle size effects on methanol electrochemical oxidation on carbon supported platinum catalysts. J Phys Chem B 110: 19271–19279.

Bett JAS, Lundquist J, Washington E, Stonehart P. 1973. Platinum crystallite size considerations for electrocatalytic oxygen reduction—I. Electrochim Acta 18: 343–348.

Biegler T, Koch DFA. 1967. Adsorption and oxidation of methanol on a platinum electrode. J Electrochem Soc 114: 904–909.

Binns C. 2001. Nanoclusters deposited on surfaces. Surf Sci Rep 44: 1–49.

Bittins-Cattaneo B, Wasmus S, Lopez-Mishima B, Vielstich W. 1993. Reduction of oxygen in an acidic methanol oxygen (air) fuel-cell—An online MS study. J Appl Electrochem 23: 625–630.

Blurton KF, Greenberg P, Oswin HG, Rutt DR. 1972. The electrochemical activity of dispersed platinum. J Electrochem Soc 119: 559–564.

Boudart M. 1969. Catalysis by supported metals. Adv Catal 20: 153–165.

Bregoli LJ. 1978. Influence of platinum crystallite size on electrochemical reduction of oxygen in phosphoric-acid. Electrochim Acta 23: 489–492.

Breiter MW. 1975. Influence of chemisorbed carbon monoxide on the oxidation of molecular hydrogen at smooth platinum in sulfuric acid solution. J Electroanal Chem 65: 623–634.

Breiter MW. 2003. Reaction mechanisms of the H_2 oxidation/evolution reaction. In: Vielstich W, Gasteiger HA, Lamm A editors. Handbook of Fuel Cells—Fundamentals, Technology and Applications. Volume 2. Chichester: Wiley.

Breiter MW, Clamroth R. 1954. Ueber die wasserstoffueberspannung an elektroden mit zwei hintereinander ablaufenden geschwindigkeitbestimmenden vorgaengen. Z. Elektrochem 58: 493–505.

Buffat P, Borel J-P. 1976. Size effect on the melting temperature of gold particles. Phys Rev A 13: 2287–2298.

Burda C, Chen X, Narayanan R, El-Sayed MA. 2005. Chemistry and properties of nanocrystals of different shapes. Chem Rev 105: 1025–1102.

Burton JJ. 1974. Structure and properties of microcrystalline catalysts. Catal Rev 9: 209–222.

Campbell CT, Parker SC, Starr DE. 2002. The effect of size-dependent nanoparticle energetics on catalyst sintering. Science 298: 811–814.

Cao D, Lu G-Q, Wieckowski A, Wasileski SA, Neurock M. 2005. Mechanisms of methanol decomposition on platinum: A combined experimental and ab initio approach. J Phys Chem B 109 (23): 11622–11633.

Carrette L, Friedrich KA, Stimming U. 2000. Fuel cells: Principles, types, fuels, and applications. Chem Phys Chem 1: 162–193.

Chang SC, Roth JD, Ho YH, Weaver MJ. 1990. New developments in electrochemical infrared-spectroscopy—Adlayer structures of carbon-monoxide on monocrystalline metal-electrodes. J Electron Spectrosc Relat Phenom 54: 1185–1203.

Chen SL, Kucernak A. 2004a. Electrocatalysis under conditions of high mass transport rate: oxygen reduction on single submicrometer-sized Pt particles supported on carbon. J Phys Chem B 108: 3262–3276.

Chen SL, Kucernak A. 2004b. Electrocatalysis under conditions of high mass transport: investigation of hydrogen oxidation on single submicron Pt particles supported on carbon. J Phys Chem B 108: 13984–13994.

Cherstiouk OV, Simonov PA, Chuvilin AL, Savinova ER. 2000. The global climate change: A coordinated response by electrochemistry and solid-state science and technology. Paper read at ECS, October 22–27, Phoenix, AZ, USA.

Cherstiouk OV, Simonov PA, Savinova ER. 2003a. Model approach to evaluate particle size effects in electrocatalysis: Preparation and properties of Pt nanoparticles supported on GC and HOPG. Electrochim Acta 48: 3851–3860.

Cherstiouk OV, Simonov PA, Zaikovskii VI, Savinova ER. 2003b. CO monolayer oxidation at Pt nanoparticles supported on glassy carbon electrodes. J Electroanal Chem 554: 241–251.

Cherstiouk OV, Gavrilov AN, Plyasova LM, Molina IY, Tsirlina GA, Savinova ER. 2008. Influence of structural defects on the electrocatalytic activity of platinum. J Solid State Electrochem 12: 497–509.

Chu D, Gilman S. 1994. The influence of methanol on O_2 electroreduction at a rotating Pt disk electrode in acid electrolyte. J Electrochem Soc 141: 1770–1773.

Clavilier J, Faure R, Guinet G, Durand R. 1980. Preparation of monocrystalline Pt microelectrodes and electrochemical study of the plane surfaces cut in the direction of the 111 and 110 planes. J Electroanal Chem 107: 205–209.

Clavilier J, Lamy C, Léger JM. 1981a. Electrocatalytic oxidation of methanol on single-crystal platinum-electrodes—Comparison with polycrystalline platinum. J Electroanal Chem 125 (1): 249–254.

Clavilier J, Parsons R, Durand R, Lamy C, Léger JM. 1981b. Formic acid oxidation on single crystal platinum electrodes—Comparison with polycrystalline platinum. J Electroanal Chem 124: 321–326.

Climent V, Gomez R, Feliu JM. 1999. Effect of increasing amount of steps on the potential of zero total charge of Pt(111) electrodes. Electrochim Acta 45: 629–637.

Climent V, Garcia-Araez N, Herrero E, Feliu JM. 2006. Potential of zero total charge of platinum single crystals: A local approach to stepped surfaces vicinal to Pt(111). Russ J Electrochem 42: 1145–1160.

Conway BE. 1995. Electrochemical oxide film formation at noble metals as a surface-chemical process. Prog Surf Sci 49: 331–452.

Conway BE. 1999. Electrochemical processes involving H adsorbed at metal electrode surfaces. In: Wieckowski A. editor. Interfacial electrochemistry. Theory, experiment, and applications. New York: Marcel Dekker.

Conway BE, Tilak BV. 2002. Interfacial processes involving electrocatalytic evolution and oxidation of H_2, and the role of chemisorbed H. Electrochim Acta 47: 3571–3594.

Costamagna P, Srinivasan S. 2001. Quantum jumps in the PEMFC science and technology from 1960s to the year 2000. J Power Sources 102: 242–252.

Croissant MJ, Napporn T, Léger J-M, Lamy C. 1998. Electrocatalytic oxidation of hydrogen at platinum-modified polyaniline electrodes. Electrochim Acta 43: 2447–2457.

Cuesta A. 2006. At least three contiguous atoms are necessary for CO formation during methanol electrooxidation on platinum. J Am Chem Soc 128: 13332–13333.

Cuesta A, Couto A, Rincon A, Perez MC, Lopez-Cudero A, Gutierrez C. 2006. Potential dependence of the saturation CO coverage of Pt electrodes: the origin of the pre-peak in CO-stripping voltammograms. Part 3: Pt(poly). J Electroanal Chem 586: 184–195.

Damjanovic A. 1992. Progress in the studies of oxygen reduction during the last thirty years. In: Murphy OJ, Srinivasan S, Conway BE, editors. Electrochemistry in Transition. New York: Plenum Press.

Damjanovic A, Hudson PG. 1988. On the kinetics and mechanism of O_2 reduction at oxide film covered Pt electrodes. I. Effect of oxide film thickness on kinetics. J Electrochem Soc 135: 2269–2273.

de Chialvo MRG, Chialvo AC. 2006. Hydrogen diffusion effects on the kinetics of the hydrogen electrode reaction. Part I. Theoretical aspects. Phys Chem Chem Phys 6: 4009–4017.

Diard J-P. 1998. Impedance measurements of polymer electrolyte membrane fuel cells running on constant load. J Power Sources 74: 244–245.

Dinh HN, Ren X, Garzon FH, Zelenay P, Gottesfeld S. 2000. Electrocatalysis in direct methanol fuel cells: In-situ probing of PtRu anode catalyst surfaces. J Electroanal Chem 491: 222–233.

Durand R, Faure R, Gloaguen F, Aberdam D. 1996. Oxygen reduction reaction on platinum in acidic medium: From bulk material to nanoparticles. In: Adzic RR, Anson FC, Kinoshita K. editors. Proceedings of the Symposium on Oxygen Electrochemistry. Pennington, NJ: The Electrochemical Society.

Eikerling M, Kornyshev A, Kulikovsky A. 2005. Can theory help to improve fuel cells? Fuel Cell Rev 1: 15–25.

Ekardt W. 1984. Work function of metal particles: Self consistent spherical jellium-background model. Phys Rev B 29: 1558–1564.

El-Deab MS, Sotomura T, Ohsaka T. 2005. Morphological selection of gold nanoparticles electrodeposited on various substrates. J Electrochem Soc 152: C730–C737.

El Kadiri F, Faure R, Durand R. 1991. Electrochemical reduction of molecular-oxygen on platinum single-crystals. J Electroanal Chem 301: 177–188.

Filhol JS, Neurock M. 2006. Elucidation of the electrochemical activation of water over Pd by first principles. Angew Chem Int Ed 45: 402–406.

Frelink T, Visscher W, vanVeen JAR. 1995. Particle-size effect of carbon-supported platinum catalysts for the electrooxidation of methanol. J Electroanal Chem 382: 65–72.

Frenkel AI, Hills CW, Nuzzo RG. 2001. A view from the inside: complexity in the atomic scale ordering of supported metal nanoparticles. J Phys Chem B 105: 12689–12703.

Friedrich KA, Henglein F, Stimming U, Unkauf W. 1998. Investigation of Pt particles on gold substrates by IR spectroscopy—Particle structure and catalytic activity. Colloids Surf A 134: 193–206.

Friedrich KA, Henglein F, Stimming U, Unkauf W. 2000. Size dependence of the CO monolayer oxidation on nanosized Pt particles supported on gold. Electrochim Acta 45: 3283–3293.

Friedrich KA, Henglein F, Stimming U, Unkauf W. 2001. In-situ vibrational spectroscopy on Pt electrocatalysts. Electrochim Acta 47: 689–694.

Frumkin AN. 1961. Hydrogen overvoltage and adsorption phenomena. Part I. In: Delahay P, Tobias CW. editors. Advances in Electrochemistry and Electrochemical Engineering. Volume 1. New York: Interscience.

Frumkin AN. 1963. Hydrogen overvoltage and adsorption phenomena. Part II. In: Delahay P, Tobias CW. editors. Advances in Electrochemistry and Electrochemical Engineering. Volume 3. New York: Interscience.

Frumkin AN, Aykazjian EA. 1955. The kinetics of the ionization of molecular hydrogen and the role of anions. Dokl Akad Nauk SSSR 100: 315–318.

Gamarnik MY. 1990. Size changes of lattice parameters in ultradisperse diamond and silicon. Phys Status Solidi B 161: 457–462.

Gamarnik MY. 1993. The physical nature of changes of lattice parameters in small particles. Phys Status Solidi B 178: 59–69.

Gamez A, Richard D, Gallezot P, Gloaguen F, Faure R, Durand R. 1996. Oxygen reduction on well-defined platinum nanoparticles inside recast ionomer. Electrochim Acta 41: 307–314.

Gasteiger HA, Markovic N, Ross PN, Cairns EJ. 1993. Methanol electrooxidation on well-characterized Pt-Ru alloys. J Phys Chem 97: 12020–12029.

Gasteiger HA, Panels JE, Yan SG. 2004. Dependence of PEM fuel cell performance on catalyst loading. J Power Sources 127: 162–171.

Gasteiger HA, Kocha SS, Sompalli B, Wagner FT. 2005. Activity benchmark and requirements for Pt, Pt-alloy, and non-Pt oxygen reduction catalysts for PEMFCs. Appl Catal B 56: 9–35.

Gattrell G, MacDougall B. 2003. Reaction mechanisms of the O_2 reduction/evolution reaction. In: Vielstich W, Lamm A, Gasteiger HA. editors. Handbook of Fuel Cells—Fundamentals, Technology and Applications. Volume 2. Chichester: Wiley.

Gavrilov AN, Savinova ER, Simonov PA, Zaikovskii VI, Cherepanova SV, Tsirlina GA, Parmon VN. 2007. On the influence of the metal loading on the structure of carbon-supported PtRu catalysts and their electrocatalytic activities in CO and methanol electrooxidation. Phys Chem Chem Phys 40: 5476–5489.

Genies L, Faure R, Durand R. 1998. Electrochemical reduction of oxygen on platinum nanoparticles in alkaline media. Electrochim Acta 44: 1317–1327.

Gilman S. 1964. The mechanism of electrochemical oxidation of carbon monoxide and methanol on platinum. II. The "reactant-pair" mechanism for electrochemical oxidation of carbon monoxide and methanol. J Phys Chem 68: 70–80.

Giordano N, Passalacqua E, Pino L, Arico AS, Antonucci V, Vivaldi M, Kinoshita K. 1991. Analysis of platinum particle-size and oxygen reduction in phosphoric-acid. Electrochim Acta 36: 1979–1984.

Gleiter H. 1992. Materials with ultrafine microstructures: retrospectives and perspectives. Nanostruct Mater 1: 1–19.

Gloaguen F, Andolfatto F, Durand R, Ozil P. 1994. Kinetic-study of electrochemical reactions at catalyst-recast ionomer interfaces from thin active layer modeling. J Appl Electrochem 24: 863–869.

Gloaguen F, Léger JM, Lamy C. 1997. Electrocatalytic oxidation of methanol on platinum nanoparticles electrodeposited onto porous carbon substrates. J Appl Electrochem 27: 1052–1060.

Gloaguen F, Convert P, Gamburzev S, Velev OA, Srinivasan S. 1998. An evaluation of the macro-homogeneous and agglomerate model for oxygen reduction in PEMFCs. Electrochim Acta 43: 3767–3772.

Gojkovic SL, Vidakovic TR. 2001. Methanol oxidation on an ink type electrode using Pt supported on high area carbons. Electrochim Acta 47: 633–642.

Grigorieva LK, Lidorenko NS, Nagaev EL, Chizhik SP. 1987. [Effect of sizes of metal small particles and thin plates on the rate of the chemical reactions on their surfaces]. Poverkhnost' 8: 131–137 (in Russian).

Guerin S, Hayden BE, Lee CE, Mormiche C, Owen JR, Russell AE, Theobald B, Thompsett D. 2004. Combinatorial electrochemical screening of fuel cell electrocatalysts. J Comb Chem 6: 149–158.

Guerin S, Hayden BE, Pletcher D, Rendall ME, Suchsland J-P. 2006a. A combinatorial approach to the study of particle size effects on supported electrocatalysts: oxygen reduction on gold. J Comb Chem 8: 679–686.

Guerin S, Hayden BE, Pletcher D, Rendall ME, Suchsland J-P, Williams LJ. 2006b. Combinatorial approach to the study of particle size effects in electrocatalysis: synthesis of supported gold nanoparticles. J Comb Chem 8: 791–798.

Gustavsson M, Fredriksson H, Kasemo B, Jusys Z, Kaiser J, Jun C, Behm RJ. 2004. Nanostructured platinum-on-carbon model electrocatalysts prepared by colloidal lithography. J Electroanal Chem 568: 371–377.

Hanson FV, Boudart M. 1978. The reaction between H_2 and O_2 over supported platinum catalysts. J Catal 53: 56–67.

Hayden BE, Pletcher D, Suchsland J-P. 2007. Enhanced activity for electrocatalytic oxidation of carbon monoxide on titania-supported gold nanoparticles. Angew Chem Int Ed 46: 3530–3532.

Henry CR. 1998. Surface studies of supported model catalysts. Surf Sci Rep 31: 235–325.

Henry C. 2003. Adsorption and reaction at supported model catalysts. In: Wieckowski A, Savinova ER, Vayenas CG. editors. Catalysis and Electrocatalysis at Nanoparticle Surfaces. New York: Marcel Dekker.

Hernandez J, Solla-Gullon J, Herrero E, Aldaz A, Feliu JM. 2005. Characterization of the surface structure of gold nanoparticles and nanorods using structure sensitive reactions. J Phys Chem B 109: 12651–12654.

Herring C. 1952. The use of classical macroscopic concepts in surface-energy problems. In: Gomer R, Smith CS. editors. Structure and Properties of Solid Surfaces. Chicago: Chicago University Press.

Higuchi E, Uchida H, Watanabe M. 2005. Effect of loading level in platinum-dispersed carbon black electrocatalysts on oxygen reduction activity evaluated by rotating disk electrode. J Electroanal Chem 583: 69–76.

Hill TL. 1963. Thermodynamics of Small Systems. Part I. New York: Benjamin.

Hill TL. 1964. Thermodynamics of Small Systems. Part II. New York: Benjamin.

Hogarth MP, Ralph TR. 2002. Catalysis for low temperature fuel cells. Part III: Challenges for the direct methanol fuel cell. Platinum Metals Rev 46: 146–164.

Horanyi G. 1985. Comments on the electrochemical behavior of small metal particles. J Phys Chem 89: 2967–2968.

Housmans THM, Koper MTM. 2003. Methanol oxidation on stepped Pt[n(111) × (110)] electrodes: A chronoamperometric study. J Phys Chem B 107 (33): 8557–8567.

Housmans THM, Wonders AH, Koper MTM. 2006. Structure sensitivity of methanol electrooxidation pathways on platinum: An on-line electrochemical mass spectrometry study. J Phys Chem B 110: 10021–10031.

Hwang JT, Chung JS. 1993. The morphological and surface properties and their relationship with oxygen reduction activity for platinum–iron electrocatalysts. Electrochim Acta 38: 2715–2723.

Inaba M, Ando M, Hatanaka A, Nomoto A, Matsuzawa K, Tasaka A, Kinumoto T, Iriyama Y, Ogumi Z. 2006. Controlled growth and shape formation of platinum nanoparticles and their electrochemical properties. Electrochim Acta 52: 1632–1638.

Iwasita T. 2003. Methanol and CO electrooxidation. In: Vielstich W, Gasteiger HA, Lamm A. editors. Handbook of Fuel Cells—Fundamentals, Technology and Applications. Volume 2. Chichester: Wiley.

Iwasita T, Vielstich W. 1986. Online mass-spectroscopy of volatile products during methanol oxidation at platinum in acid-solutions. J Electroanal Chem 201: 403–408.

Jerkiewicz G. 1998. Hydrogen sorption at/in electrodes. Prog Surf Sci 57: 137–186.

Jiang J, Kucernak A. 2004. Investigations of fuel cell reactions at the composite microelectrode | solid polymer electrolyte interface. I. Hydrogen oxidation at the nanostructured Pt | Nafion® membrane interface. J Electroanal Chem 567: 123–137.

Jiang J, Kucernak A. 2005. Solid polymer electrolyte membrane composite microelectrode investigations of fuel cell reactions. II: Voltammetric study of methanol oxidation at the nanostructured platinum microelectrode | Nafion® membrane interface. J Electroanal Chem 576: 223–236.

Jusys Z, Behm RJ. 2001. Methanol oxidation on a carbon-supported Pt fuel cell catalyst—A kinetic and mechanistic study by differential electrochemical mass spectrometry. J Phys Chem B 105: 10874–10883.

Jusys Z, Kaiser J, Behm RJ. 2003. Methanol electrooxidation over Pt/C fuel cell catalysts: Dependence of product yields on catalyst loading. Langmuir 19: 6759–6769.

Kabbabi A, Gloaguen F, Andolfatto F, Durand R. 1994. Particle-size effect for oxygen reduction and methanol oxidation on Pt/C inside a proton-exchange membrane. J Electroanal Chem 373: 251–254.

Kaiser J, Simonov PA, Zaikovskii VI, Hartnig C, Joerissen L, Savinova ER. 2007. Influence of the carbon support on the performance of the platinum based oxygen reduction catalysts in a polymer electrolyte fuel cell. J Appl Electrochem 37: 1429–1437.

Kasemo B, Johansson S, Persson H, Thornmahlen P, Zhdanov VP. 2000. Catalysis in the nm-regime: manufacturing of supported model catalysts and theoretical studies of the reaction kinetics. Top Catal 13: 43–53.

Kim CS, Tornquist WJ, Korzeniewski C. 1993. Infrared-spectroscopy as a probe of CO adsorption at Pt(335) under aqueous electrochemical conditions. J Phys Chem 97: 6484–6491.

Kinoshita K. 1988. Carbon: Electrochemical and Physicochemical Properties. New York: Wiley.

Kinoshita K. 1990. Particle-size effects for oxygen reduction on highly dispersed platinum in acid electrolytes. J Electrochem Soc 137: 845–848.

Kinoshita K. 1992. Electrochemical Oxygen Technology. New York: Wiley.

Kita H. 2003. Horiuti's generalized rate expression and hydrogen electrode reaction. J Mol Catal A 199: 161–174.

Kita H, Ye S, Gao Y. 1992. Mass transfer effect in hydrogen evolution reaction on Pt single-crystal electrodes in acid solution. J Electroanal Chem 334: 351–357.

Klimenkov M, Nepijko S, Kuhlenbeck H, Baümer M, Schlögl R, Freund H-J. 1997. The structure of Pt-aggregates on a supported thin aluminum oxide film in comparison with unsupported alumina: a transmission electron microscopy study. Surf Sci 391: 27–36.

Knickelbein MB, Yang S, Riley SJ. 1990. Near-threshold photoionization of nickel clusters: Ionization potentials for Ni_3 to Ni_{90}. J Chem Phys 93: 94–104.

Kobayashi T, Babu PK, Gancs L, Chung JH, Oldfield E, Wieckowski A. 2005. An NMR determination of CO diffusion on platinum electrocatalysts. J Am Chem Soc 127: 14164–14165.

Komanicky V, Menzel A, Chang K-C, You H. 2005. Nanofaceted platinum surfaces: a new model system for nanoparticle catalysts. J Phys Chem B 109: 23543–23549.

Koper MTM, Jansen APJ, van Santen RA, Lukkien JJ, Hilbers PAJ. 1998. Monte Carlo simulations of a simple model for the electrocatalytic CO oxidation on platinum. J Chem Phys 109: 6051–6062.

Kovalyov EV, Resnyanskii ED, Elokhin VI, Bal'zhinimaev BS, Myshlyavtsev AV. 2003. Novel statistical lattice model for the supported nanoparticle. Features of the reaction performance influenced by the dynamically changed shape and surface morphology of the supported active particle. Phys Chem Chem Phys 5: 784–790.

Kovalyov EV, Elokhin VI, Myshlyavtsev AV. 2008. Stochastic simulation of physicochemical processes performance over supported metal nanoparticles. J Comput Chem 29: 79–86.

Krausa M, Vielstich W. 1994. Study of the electrocatalytic influence of Pt/Ru and Ru on the oxidation of residues of small organic-molecules. J Electroanal Chem 379: 307–314.

Krueger S, Vent S, Roesch N. 1997. Size dependence of bond length and binding energy in palladium and gold clusters. Ber Bunsenges Phys Chem 101: 1640–1643.

Kubo R. 1962. Electronic properties of metallic fine particles. I. Phys Soc Jpn 17: 975–986.

Kumar S, Zou S. 2006. Electroreduction of O_2 on uniform arrays of Pt and PtCo nanoparticles. Electrochem Commun 8: 1151–1157.

Kuzume A, Herrero E, Feliu JM. 2007. Oxygen reduction on stepped platinum surfaces in acidic media. J Electroanal Chem 599: 333–343.

Lebedeva NP, Koper MTM, van Santen RA, Herrero E, Feliu JM. 2000. CO oxidation on stepped Pt[n(111) × (111)] electrodes. J Electroanal Chem 487: 37–44.

Lebedeva NP, Rodes A, Feliu JM, Koper MTM, van Santen RA. 2002a. Role of crystalline defects in electrocatalysis: CO adsorption and oxidation on stepped platinum electrodes as studied by in situ infrared spectroscopy. J Phys Chem B 106: 9863–9872.

Lebedeva NP, Koper MTM, Feliu JM, van Santen RA. 2002b. Role of crystalline defects in electrocatalysis: Mechanism and kinetics of CO adlayer oxidation on stepped platinum electrodes. J Phys Chem B 106: 12938–12947.

Léger JM. 2001. Mechanistic aspects of methanol oxidation on platinum-based electrocatalysts. J Appl Electrochem 31: 767–771.

Lewis LN. 1993. Chemical catalysis by colloids and clusters. Chem Rev 93: 2693–2730.

Liu H, Penner RM. 2000. Size-selective electrodeposition of mesoscale metal particles in the uncoupled limit. J Phys Chem B 104: 9131–9139.

Liu H, Songa C, Zhang L, Zhang J, Wang H, Wilkinson DP. 2006. A review of anode catalysis in the direct methanol fuel cell. J Power Sources 155: 95–110.

Love B, Lipkowski J. 1988. Effect of surface crystallography on electrocatalytic oxidation of carbon-monoxide on platinum-electrodes. ACS Symp Ser 378: 484–496.

McCallum C, Pletcher D. 1976. An investigation of mechanism of oxidation of carbon monoxide adsorbed onto a smooth Pt electrode in aqueous acid. J Electroanal Chem 70: 277–290.

Macia MD, Campina JM, Herrero E, Feliu JM. 2004. On the kinetics of oxygen reduction on platinum stepped surfaces in acidic media. J Electroanal Chem 564: 141–150.

Maillard F, Martin M, Gloaguen F, Léger JM. 2002. Oxygen electroreduction on carbon-supported platinum catalysts. Particle-size effect on the tolerance to methanol competition. Electrochim Acta 47: 3431–3440.

Maillard F, Eikerling M, Cherstiouk OV, Schreier S, Savinova E, Stimming U. 2004a. Size effects on reactivity of Pt nanoparticles in CO monolayer oxidation: The role of surface mobility. Faraday Discuss 125: 357–377.

Maillard F, Savinova E, Simonov PA, Zaikovskii VI, Stimming U. 2004b. Infrared spectroscopic study of CO adsorption and electrooxidation on carbon-supported Pt nanoparticles: Inter-particle versus intra-particle heterogeneity. J Phys Chem B 108: 17893–17904.

Maillard F, Schreier S, Hanzlik M, Savinova ER, Weinkauf S, Stimming U. 2005. Influence of particle agglomeration on the catalytic activity of carbon-supported Pt nanoparticles in CO monolayer oxidation. Phys Chem Chem Phys 7: 385–393.

Maillard F, Bonnefont A, Chatenet M, Guétaz L, Doisneau-Cottignies B, Roussel H, Stimming U. 2007a. Effect of the structure of Pt-Ru/C particles on CO_{ad} monolayer vibrational properties and electrooxidation kinetics. Electrochim Acta 53: 811–822.

Maillard F, Savinova ER, Stimming U. 2007b. CO monolayer oxidation on Pt nanoparticles: Further insights into the particle size effects. J Electroanal Chem 599: 221–232.

Malek K, Coppens MO. 2003. Knudsen self- and Fickian diffusion in rough nanoporous media. J Chem Phys 119: 2801–2811.

Markovic NM, Ross PN. 2000. Electrocatalysts by design: From the tailored surface to a commercial catalyst. Electrochim Acta 45: 4101–4115.

Markovic NM, Ross PN. 2002. Surface science studies of model fuel cell electrocatalysts. Surf Sci Rep 45: 121–229.

Markovic NM, Adzic RR, Cahan BD, Yeager EB. 1994. Structural effects in electrocatalysis—Oxygen reduction on platinum low-index single-crystal surfaces in perchloric-acid solutions. J Electroanal Chem 377: 249–259.

Markovic N, Gasteiger H, Ross PN. 1997a. Kinetics of oxygen reduction on Pt(*hkl*) electrodes: Implications for the crystallite size effect with supported Pt electrocatalysts. J Electrochem Soc 144: 1591–1597.

Markovic NM, Grgur BN, Ross PN. 1997b. Temperature-dependent hydrogen electrochemistry on platinum low-index single-crystal surfaces in acid solutions. J Phys Chem B 101: 5405–5413.

Maroun F, Ozanam F, Magnussen OM, Behm RJ. 2001. The role of atomic ensembles in the reactivity of bimetallic electrocatalysts. Science 293: 1811–1814.

Mayrhofer KJJ, Arenz M, Blizanac BB, Stamenkovic VR, Ross PN, Markovic NM. 2005a. CO surface electrochemistry on Pt-nanoparticles: A selective review. Electrochim Acta 50: 5144–5154.

Mayrhofer KJJ, Blizanac BB, Arenz M, Stamenkovic VR, Ross PN, Markovic NM. 2005b. The impact of geometric and surface electronic properties of Pt-catalysts on the particle size effect in electrocatalysis. J Phys Chem B 109: 14433–14440.

Mays CW, Vermaak JS, Kuhlmann-Wilsdorf D. 1968. On surface stress and surface tension: II. Determination of the surface stress of gold. Surf Sci 12: 134–140.

Mazurek M, Benker N, Roth C, Buhrmester T, Fuess H. 2006. Electrochemical impedance and X-ray absorption spectroscopy (EXAFS) as in situ methods to study PEMFC anode. Fuel Cells 6: 16–20.

Mukerjee S. 1990. Particle-size and structural effects in platinum electrocatalysis. J Appl Electrochem 20: 537–548.

Mukerjee S, McBreen J. 1998. Effect of particle size on the electrocatalysis by carbon-supported Pt electrocatalysts: an in situ XAS investigation. J Electroanal Chem 448: 163–171.

Nagaev EL. 1991. Equilibrium properties of small particles at their ultralow vapour pressures Surf Sci 243: 252–260.

Nagaev EL. 1992. Equilibrium and quasiequilibrium properties of small particles. Phys Rep 222: 199–307.

Nepijko SA, Klimenkov M, Kuhlenbeck H, Zemlyanov D, Herein D, Schlögl R, Freund H-J. 1998. TEM study of tantalum clusters on Al_2O_3/NiAl(110). Surf Sci 412/413: 192–201.

Pacchioni G, Illas F. 2003. Electronic structure and chemisorption properties of supported metal clusters: model calculations. In: Wieckowski A, Savinova ER, Vayenas CG. editors. Catalysis and Electrocatalysis at Nanoparticle Surfaces. New York: Marcel Dekker.

Papoutsis A, Léger JM, Lamy C. 1987. New results for the electrosorption of methanol on polycrystalline platinum in acid medium obtained by programmed potential voltammetry. J Electroanal Chem 234: 315–327.

Park S, Wasileski SA, Weaver MJ. 2001. Electrochemical infrared characterization of carbon-supported platinum nanoparticles: A benchmark structural comparison with single-crystal electrodes and high-nuclearity carbonyl clusters. J Phys Chem B 105: 9719–9725.

Park S, Tong YT, Wieckowski A, Weaver MJ. 2002a. Infrared spectral comparison of electrochemical carbon monoxide adlayers formed by direct chemisorption and methanol dissociation on carbon-supported platinum nanoparticles. Langmuir 18: 3233–3240.

Park S, Xie Y, Weaver MJ. 2002b. Electrocatalytic pathways on carbon-supported platinum nanoparticles: Comparison of particle-size-dependent rates of methanol, formic acid, and formaldehyde electrooxidation. Langmuir 18: 5792–5798.

Parmigiani F, Kay E, Bagus PS. 1990. Anomalous oxidation of platinum clusters studied by X-ray photoelectron-spectroscopy. J Electron Spectrosc Relat Phenom 50: 39–46.

Parmon VN. 2007. Thermodynamic analysis of the effect of the nanoparticle size of the active component on the adsorption equilibrium and the rate of heterogeneous catalytic processes. Dokl Phys Chem 413: 42–48.

Paulus UA, Schmidt TJ, Gasteiger HA, Behm RJ. 2001. Oxygen reduction on a high-surface area Pt/Vulcan carbon catalyst: A thin-film rotating ring-disk electrode study. J Electroanal Chem 495: 134–145.

Paulus UA, Veziridis V, Schnyder B, Kuhnke M, Scherer GG, Wokaun A. 2003. Fundamental investigation of catalyst utilization at the electrode/solid polymer electrolyte interface. Part I. Development of a model system. J Electroanal Chem 541: 77–91.

Payne M, Roberts N, Needs R, Needels M, Joannopoulos J. 1989. Total energy and stress of metal and semiconductor surfaces. Surf Sci 211/212: 1–20.

Peremans A, Tadjeddine A. 1994. Vibrational spectroscopy of electrochemically deposited hydrogen on platinum. Phys Rev Lett 73: 3010–3013.

Petrii OA, Tsirlina GA. 1994. Electrocatalytic activity prediction: for hydrogen electrode reaction: intuition, art, sceince. Electrochim Acta 39: 1739–1747.

Petrii OA, Tsirlina GA. 2001. Size effects in electrochemistry. RussChem Rev 70: 285–298.

Petrii OA, Podlovchenko BI, Frumkin AN, Hira-Lal. 1965. The behaviour of platinized-platinum and platinum–ruthenium electrodes in methanol solutions. J Electroanal Chem 10: 253–269.

Petukhov AV, Akemann W, Friedrich KA, Stimming U. 1998. Kinetics of electrooxidation of a CO monolayer at the platinum/electrolyte interface. Surf Sci 402–404: 182–186.

Plieth WJ. 1982. Electrochemical properties of small clusters of metal atoms and their role in surface enhanced Raman scattering. J Phys Chem 86: 3166–3170.

Plyasova LM, Molina IY, Gavrilov AN, Cherepanova SV, Cherstiouk OV, Rudina NA, Savinova ER, Tsirlina GA. 2006. Electrodeposited platinum revisited: tuning nanostructure via the deposition potential. Electrochim Acta 51: 4477–4488.

Pronkin S, Wandlowski T. 2004. ATR–SEIRAS—An approach to probe the reactivity of Pd-modified quasi-single crystal gold film electrodes. Surf Sci 573: 109–127.

Pronkin SN, Tsirlina GA, Petrii OA, Vassiliev SY. 2001. Nanoparticles of Pt hydrosol immobilized on Au support: An approach to the study of structural effects in electrocatalysis. Electrochim Acta 46: 2343–2351.

Pronkin SN, Simonov PA, Zaikovskii VI, Savinova ER. 2007. Model Pd-based bimetallic supported catalysts for nitrate electroreduction. J Mol Catal A 265: 141–147.

Quaino PM, JL Fernàndez, Chialvo MRG, Chialvo AC. 2006. Hydrogen oxidation reaction on microelectrodes: Analysis of the contribution of the kinetic routes. J Mol Catal A. 252: 156–162.

Ralph TR, Hogarth MP. 2002a. Catalysis for low temperature fuel cells. Part I: The cathode challenges. Platinum Metals Rev 46: 3–14.

Ralph TR, Hogarth MP. 2002b. Catalysis for low temperature fuel cells. Part II: The anode challenges. Platinum Metals Rev 46: 117–135.

Rao V, Simonov PA, Savinova ER, Plaksin GV, Cherepanova SV, Kryukova GN, Stimming U. 2005. The influence of carbon support porosity on the activity of PtRu/Sibunit anode catalysts for methanol oxidation. J Power Sources 145: 178–187.

Redmond PL, Hallock AJ, Brus LE. 2005. Electrochemical Ostwald ripening of colloidal Ag particles on conductive substrates. Nano Lett 5: 131–135.

Redondo A, Ticianelli EA, Gottesfeld S. 1989. Ellipsometric studies of conducting polymers. Synth Met 29: E265–E270.

Roduner E. 2006. Nanoscopic Materials: Size-dependent Phenomena. Cambridge: The Royal Society of Chemistry.

Romanowski W. 1969. Equilibrium forms of very small metallic crystals. Surf Sci 18: 373–388.

Roth C, Martz N, Buhrmester T, Scherer J, Fuess H. 2002. In-situ XAFS fuel cell measurements of a carbon-supported Pt-Ru anode electrocatalyst in hydrogen and direct methanol operation. Phys Chem Chem Phys 4: 3555–3557.

Rusanov AI. 2005. Surface thermodynamics revisited. Surf Sci Rep 58: 111–239.

Russell AE, Rose A. 2004. X-ray absorption spectroscopy of low temperature fuel cell catalysts. Chem Rev 104: 4613–4635.

Safonov VA, Yazkowska K, Petrii OA. 1975. The influence of the nature of solution anions on the impedance of smooth platinum. Elektrokhimiya 11: 1628–1630.

Saravanan S, Dunietz BD, Markovic NM, Somorjai G, Ross PN, Head-Gordon M. 2003. Electro-oxidation of CO on Pt-based electrodes simulated by electronic structure calculations. J Electroanal Chem 554/555: 459–465.

Sattler PM, Ross PN. 1986. The surface structure of Pt crystallites supported on carbon black. Ultramicroscopy 20: 21–28.

Savinova ER. 2006. Size and structural effects in electrocatalysis. Doctor of Science Dissertation, Boreskov Institute of Catalysis, Novosibirsk.

Savinova ER, Chuvilin AL, Parmon VN. 1988. Copper colloids stabilized by water-soluble polymers: Part I. Preparation and Properties. J Mol Catal 48: 217–229.

Savinova ER, Lebedeva NP, Simonov PA, Kryukova GN. 2000. Electrocatalytic properties of platinum anchored to the surface of highly oriented pyrolytic graphite. Russ J Electrochem. 36 (9): 952–959.

Schalow T, Brandt B, Starr DE, Laurin M, Shaikhutdinov SK, Schauermann S, Libuda J, Freund HJ. 2007. Particle size dependent adsorption and reaction kinetics on reduced and partially oxidized Pd nanoparticles. Phys Chem Chem Phys 9: 1347–1361.

Schmidt TJ, Noeske M, Gasteiger HA, Behm RJ, Britz P, Boennemann H. 1998a. PtRu alloy colloids as precursors for fuel cell catalysts. J Electrochem Soc 145: 925–931.

Schmidt TJ, Gasteiger HA, Stab GD, Urban PM, Kolb DM, Behm RJ. 1998b. Characterization of high-surface area electrocatalysts using a rotating disk electrode configuration. J Electrochem Soc 145: 2354–2358.

Schmidt TJ, Gasteiger HA, Behm RJ. 1999. Rotating disk electrode measurements on the CO tolerance of a high-surface area Pt/Vulcan carbon fuel cell catalysts. J Electrochem Soc 146: 1296–1304.

Schmidt TJ, Jusys Z, Gasteiger HA, Behm RJ, Endruschat U, Boennemann H. 2001. On the CO tolerance of novel colloidal PdAu/carbon electrocatalysts. J Electroanal Chem 501: 132–140.

Schroeder A, Fleig J, Gryaznov D, Maier J, Sitte W. 2006. Quantitative model of electrochemical Ostwald ripening and its application to the time-dependent electrode potential of nanocrystalline metals. J Phys Chem B 110: 12274–12280.

Seto K, Iannelli A, Love B, Lipkowski J. 1987. The influence of surface crystallography on the rate of hydrogen evolution at Pt electrodes. J Electroanal Chem 226: 351–360.

Shubina TE, Hartnig C, Koper MTM. 2004. Density functional theory study of the oxidation of CO by OH on Au(110) and Pt(111) surfaces. Phys Chem Chem Phys 6: 4215–4221.

Shukla AK, Ravikumar MK, Roy A, Barman SR, Sarma DD, Arico AS, Antonucci V, Pino L, Giordano N. 1994. Electrooxidation of methanol in sulfuric-acid electrolyte on platinized-carbon electrodes with several functional-group characteristics. J Electrochem Soc 141: 1517–1522.

Simonov PA, Likholobov VA. 2003. Physicochemical aspects of preparation of carbon-supported noble metal catalysts. In: Wieckowski A, Savinova ER, Vayenas CG. editors. Catalysis and Electrocatalysis at Nanoparticle Surfaces. New York: Marcel Decker.

Solla-Gullon J, Vidal-Iglesias FJ, Rodriguez P, Herrero E, Feliu JM, Aldaz A. 2005. Shape-dependent electrocatalysis: CO monolayer oxidation at platinum nanoparticles. Paper presented at Meeting ECS, May 15–20, 2005, Quebec City.

Solla-Gullon J, Solla-Gullon J, Vidal-Iglesias FJ, Herrero E, Feliu JM, Aldaz A. 2006. CO monolayer oxidation on semi-spherical and preferentially oriented (100) and (111) platinum nanoparticles. Electrochem Commun 8: 189–194.

Stonehart P. 1990. Development of advanced noble metal-alloy electrocatalysts for phosphoric-acid fuel cells (PAFC). Ber Bunsenges Phys Chem 94: 913–921.

Strbac S, Anastasijevic NA, Adzic RR. 1994. Oxygen reduction on Au(111) and vicinal Au(332) faces: a rotating disc and disc–ring study. Electrochim Acta 39: 983–990.

Sun CQ. 2007. Size dependence of nanostructures: Impact of bond order deficiency. Prog Solid State Chem 35: 1–159

Sun SG, Yang YY. 1999. Studies of kinetics of HCOOH oxidation on Pt(100), Pt(110), Pt(111), Pt(510) and Pt(911) single crystal electrodes. J Electroanal Chem 467: 121–131.

Tadjeddine A, Peremans A. 1996. Vibrational spectroscopy of the electrochemical interface by visible infrared sum-frequency generation. Surf Sci 368: 377–383.

Takasu Y, Matsuda Y, Toyoshima I. 1984. A photoelectron spectroscopic study of the effect of particle-size on the adsorbed state of carbon-monoxide over supported palladium catalysts. Chem Phys Lett 108: 384–387.

Takasu Y, Fujii Y, Yasuda K, Iwanaga Y, Matsuda Y. 1989. Electrocatalytic properties of ultrafine platinum particles for hydrogen electrode reaction in an aqueous solution of sulfuric acid. Electrochim Acta 34: 453–458.

Takasu Y, Ohashi N, Zhang XG, Murakami Y, Minagawa H, Sato S, Yahikozawa K. 1996. Size effects of platinum particles on the electroreduction of oxygen. Electrochim Acta 41: 2595–2600.

Takasu Y, Zhang XG, Minoura S, Murakami Y. 1997. Size effects of ultrafine palladium particles on the electrocatalytic oxidation of CO. Appl Surf Sci 121: 596–600.

Takasu Y, Iwazaki T, Sugimoto W, Murakami Y. 2000. Size effects of platinum particles on the electro-oxidation of methanol in an aqueous solution of $HClO_4$. Electrochem Commun 2: 671–674.

Tarasevich MR, Sadkowski A, Yeager E. 1983. Oxygen electrochemistry. In: Conway BE, Bockris JOM, Yeager E, Khan SUM, White RE. editors. Comprehensive Treatise of Electrochemistry. New York: Plenum Press.

Tarnowski DJ, Korzeniewski C. 1997. Effects of surface step density on the electrochemical oxidation of ethanol to acetic acid. J Phys Chem 101: 253–258.

Tateishi N, Yahikozawa K, Nishimura K, Masato S, Iwanaga Y, Watanabe M, Enami E, Matsuda Y, Takasu Y. 1991. Electrochemical properties of ultra-fine palladium particles for adsorption and absorption of hydrogen in an aqueous $HClO_4$ solution. Electrochim Acta 36: 1235–1240.

Teliska M, O'Grady WE, Ramaker DE. 2004. Determination of H adsorption sites on Pt/C electrodes in $HClO_4$ from Pt L_{23} X-ray absorption spectroscopy. J Phys Chem B 108: 2333–2344.

Tkach I, Panchenko A, Dilger H, Kaz T, Gogel V, Friedrich KA, Roduner E. 2004. In situ study of methanol oxidation on Pt and Pt/Ru-mixed with Nafion® anodes in a direct methanol fuel cell by means of FTIR spectroscopy. Phys Chem Chem Phys 6: 5419–5426.

Uvarov NF, Boldyrev VV. 2001. Size effects in chemistry of heterogeneous systems. Russ Chem Rev 70: 265–284.

Van Hardeveld R, Hartog F. 1969. Statistics of surface atoms and surface sites on metal crystals. Surf Sci 15: 189–230.

Vargaftik MN, Moiseev II, Kochubey DI, Zamaraev KI. 1991. Giant palladium clusters—Synthesis and characterization. Faraday Discuss 92: 13–29.

Vermaak JS, Kuhlmann-Wilsdorf D. 1968. Measurement of the average surface stress of gold as a function of temperature in the temperature range 50–985 °C. J Phys Chem 72: 4150–4154.

Vermaak JS, Mays CW, Kuhlmann-Wilsdorf D. 1968. On surface stress and surface tension: I. Theoretical considerations. Surf Sci 12: 128–133.

Vetter KJ. 1961. Elektrochemische Kinetik. Berlin: Springer-Verlag.

Vielstich W, Lamm A, Gasteiger HA, editors. 2003. Handbook of Fuel Cells—Fundamentals, Technology and Applications. 5 volumes. Chichester: Wiley.

Viswanathan R, Hou GY, Liu RX, Bare SR, Modica F, Mickelson G, Segre CU, Leyarovska N, Smotkin ES. 2002. In-situ XANES of carbon-supported Pt-Ru anode electrocatalyst for reformate–air polymer electrolyte fuel cells. J Phys Chem B 106: 3458–3465.

Vogel WM, Baris JM. 1977. Reduction of oxygen on platinum black in acid electrolytes. Electrochim Acta 22: 1259–1263.

Wang JX, Markovic NM, Adzic RR. 2004. Kinetic analysis of oxygen reduction on Pt(111) in acid solutions: intrinsic kinetic parameters and anion adsorption effects. J Phys Chem B 108: 4127–4133.

Wang JX, Springer TE, Adzic RA. 2006. Dual-pathway kinetic equation for the hydrogen oxidation reaction on Pt electrodes. J Electrochem Soc 153: A1732–A1740.

Wasmus S, Kuver A. 1999. Methanol oxidation and direct methanol fuel cells: A selective review. J Electroanal Chem 461: 14–31.

Wasmus S, Wang JT, Savinell RF. 1995. Real-time mass-spectrometric investigation of the methanol oxidation in a direct methanol fuel-cell. J Electrochem Soc 142: 3825–3833.

Wasserman HJ, Vermaak JS. 1972. On the determination of the surface stress of copper and platinum. Surf Sci 32: 168–174.

Waszczuk P, Lu GQ, Wieckowski A, Lu C, Rice C, Masel RI. 2002. UHV and electrochemical studies of CO and methanol adsorbed at platinum/ruthenium surfaces, and reference to fuel cell catalysis. Electrochim Acta 47: 3637–3652.

Watanabe M, Saegusa S, Stonehart P. 1988. Electro-catalytic activity on supported platinum crystallites for oxygen reduction in sulfuric-acid. Chem Lett 17: 1487–1490.

Watanabe M, Sei H, Stonehart P. 1989. The influence of platinum crystallite size on the electroreduction of oxygen. J Electroanal Chem 261: 375–387.

Watson GW, Wells RPK, Willock DJ, Hutchings GJ. 2001. A comparison of the adsorption and diffusion of hydrogen on the {111} surfaces of Ni, Pd, and Pt from density functional theory calculations. J Phys Chem B 105: 4889–4894.

Winter M, Brodd RJ. 2004. What are batteries, fuel cells, and supercapacitors? Chem Rev 104: 4245–4269.

Wroblowa HS, Pan YC, Razumney G. 1976. Electroreduction of oxygen—New mechanistic criterion. J Electroanal Chem 69: 195–201.

Yahikozawa K, Fujii Y, Matsuda Y, Nishimura K, Takasu Y. 1991. Electrocatalytic properties of ultrafine platinum particles for oxidation of methanol and formic-acid in aqueous-solutions. Electrochim Acta 36: 973–978.

Yano H, Inukai J, Uchida H, Watanabe M, Babu PK, Kobayashi T, Chung JH, Oldfield E, Wieckowski A. 2006. Particle-size effect of nanoscale platinum catalysts in oxygen reduction reaction: an electrochemical and Pt^{195} EC-NMR study. Phys Chem Chem Phys 8: 4932–4939.

Yates JT. 1995. Surface-chemistry at metallic step defect sites. J Vac Sci Technol A 13: 1359–1367.

Zhang X-G, Arikawa T, Murakami Y, Yahikozawa K, Takasu Y. 1995. Electrocatalytic oxidation of formic acid on ultrafine particles supported on glassy carbon. Electrochim Acta 40: 1889–1897.

Zhdanov VP. 2002. Monte Carlo simulations of oscillations, chaos and pattern formation in heterogeneous catalytic reactions. Surf Sci Rep 45: 233–326.

Zhdanov VP, Kasemo B. 2000. Simulations of the reaction kinetics on nanometer supported catalyst particles. Surf Sci Rep 39: 25–104.

Zhdanov VP, Kasemo B. 2002. Kinetics of electrochemical reactions: from single crystals to nm-sized supported particles. Surf Sci 521: L655–L661.

Zhou WP, Lewera A, Larsen R, Masel RI, Bagus PS, Wieckowski A. 2006. Size effects in electronic and catalytic properties of unsupported palladium nanoparticles in electrooxidation of formic acid. J Phys Chem B 110: 13393–13398.

Zhu YM, Uchida H, Watanabe M. 1999. Oxidation of carbon monoxide at a platinum film electrode studied by Fourier transform infrared spectroscopy with attenuated total reflection technique. Langmuir 15: 8757–8764.

Zinola CF, AM Castro Luna, Arvia AJ. 1994. Temperature dependence of kinetic parameters related to oxygen electroreduction in acid solutions on platinum electrodes. Electrochim Acta 39: 1951–1959.

Zou S, Gomes R, Weaver MJ. 1999. Infrared spectroscopy of carbon monoxide and nitric oxide on palladium(111) in aqueous solution: unexpected adlayer structural differences between electrochemical and ultrahigh-vacuum interfaces. J Electroanal Chem 474: 155–166.

Zoval JV, Lee J, Gorer S, Penner RM. 1998. Electrochemical preparation of platinum nanocrystallites with size selectivity on basal plane oriented graphite surfaces. J Phys Chem B 102: 1166–1175.

CHAPTER 16

Support and Particle Size Effects in Electrocatalysis

BRIAN E. HAYDEN and JENS-PETER SUCHSLAND

School of Chemistry, University of Southampton, Southampton SO17 1BJ, UK

16.1 INTRODUCTION

The aim of catalysis is to lower the activation energy and/or increase the selectivity [Somorjai and Rioux, 2005] for a particular chemical reaction. For heterogeneous catalysis, at the gas/solid interface, activity is usually defined in terms of the turnover frequency (TOF) per surface area of catalyst (or per active site: specific activity) at fixed temperature. In the case of electrocatalysis, the electrochemical current per catalyst area (or active site) at a fixed potential provides a measure of activity (or specific activity). In order to produce high utilization of catalyst material (particularly for precious metals), high dispersion of the active materials on a support will also yield optimal mass activity. It is well established in heterogeneous catalysis, however, that this does not take into account the apparent structural sensitivity of many catalytic reactions [Henry et al., 1997]. A decrease in specific activity with increased dispersion, for example, will counteract the positive effect of dispersion on mass activity. An important example of a structurally sensitive electrocatalytic reaction is the apparent deactivation of the oxygen reduction reaction (ORR) in acid media on platinum with decreasing particle size [Guerin et al., 2004]. This results in an optimal mass activity of carbon-supported Pt of about 3 nm. In addition to particle size effects, the nature of the supporting substrate can strongly influence the overall catalytic activity of a supported metal. This can result from stabilization of particle sizes and morphologies, or from modification of the electronic properties of the supported metal. A well-documented example of this effect in heterogeneous catalysis is the induced activity of gold supported by a titania support in low temperature oxidation reactions. Support effects are less well documented in the case of electrocatalysis. This chapter summarizes a series of experiments carried out on model supported

---

Fuel Cell Catalysis. Edited by Marc T. M. Koper
Copyright © 2009 John Wiley & Sons, Inc.

gold catalysts designed to investigate the effect of particle size and substrate on electrocatalytic activity.

We describe the application of a high throughput synthesis and screening method to investigate the effect of substrate and particle size on electrocatalytic activity. Au supported on carbon and titania surfaces has been used to investigate these effects because of the strong influence of particle size and support on the heterogeneous counterparts. Combinatorial, or high throughput, approaches have been increasingly applied to the optimization of the composition and structure of materials for specific physical or chemical properties, not least in the fields of electrocatalyst and battery materials [Reddington et al., 1998; Chen et al., 2001; Sun et al., 2001; Morris and Mallouk, 2002; Spong et al., 2003; Guerin et al., 2004]. Recently, Guerin and Hayden described a physical vapor deposition system that has been developed for the high throughput synthesis of thin-film materials [Guerin and Hayden, 2006]. It employs source shutters to achieve controlled gradients of the depositing elements across a substrate or an array of electrodes [Guerin and Hayden, 2006]. This allows, for example, the synthesis of compositional libraries of alloys [Guerin et al., 2006a] and particle centers supported on homogeneous supports [Guerin et al., 2006b, c; Hayden et al., 2007a, c]. This combinatorial approach has been applied to the optimization of Pd/Pt/Au ORR catalysts [Guerin et al., 2006a] and the investigation of particle size and substrate effects in electrocatalysis by Au and Pt particles [Guerin et al., 2006b, c; Hayden et al., 2007a, c, 2009]. The work described here centers on how the electrocatalyzed ORR and CO oxidation reaction are influenced by particle size and support, and we emphasize the close relationship in some cases between the activities of structures supporting electrocatalytic and heterogeneous catalyzed reactions.

16.2 PARTICLE SIZE AND GEOMETRIC EFFECTS IN ELECTROCATALYSIS

16.2.1 Oxygen Reduction on Pt

The ORR has been widely studied in recent years because of its role in the cathode reaction of low temperature polymer electrolyte membrane fuel cells (PEMFCs). Even if a precious metal such as Pt is used as the cathode catalyst, which is known to promote almost exclusively the four-electron reduction of oxygen to water, relatively high overpotentials and low exchange current densities are observed. Nevertheless, Pt appears to be one of the most effective catalysts for the ORR in acidic media, but, because of the high price and scarcity of the element, it must be significantly dispersed in the catalyst to be viable economically [Gasteiger et al., 2004, 2005]. State-of-the-art fuel cell catalysts are precious metal nanoparticles supported on high surface area carbons, where the metal is commonly Pt or a Pt alloy [Gasteiger et al., 2004]. The aim of the support and the alloying is to maximize the weight activity of Pt. This approach has yielded a significant cost reduction over the last three decades of fuel cell catalyst development, although there remains a significant potential loss in PEMFCs (under load, cell potentials are commonly in the range of about 0.6–0.85 V), mainly as a result of the associated low kinetics and hence high overpotentials for oxygen reduction at the cathode. The increase in mass activity gained by dispersion of Pt appears to be limited, and there seems to be a "critical" particle size (approximately 3 nm in diameter) at which further dispersion leads to a loss of mass activity and hence an economic penalty [Kinoshita, 1990].

There has been a considerable effort to understand the structural and electrolyte dependence of the ORR through experiments on single-crystal surfaces. Significant electrolyte-dependent oxygen reduction activity on Pt was reported and explained by the effects of specific anion adsorption [Hsueh et al., 1983]. The activity was found to increase in the order $H_2SO_4 \approx CF_3SO_3H < H_3PO_4 < HClO_4$. Oxygen reduction on Pt was also found to depend on the structure of the Pt(hkl) surface [Markovic et al., 1995, 1997]. The activity in weakly adsorbing $HClO_4$ was found to increase in the order $(100) < (110) \approx (111)$, with only small differences in the rate of ORR, while in H_2SO_4, where strongly adsorbing HSO_4^- is present, the difference between highest and lowest activity was two orders of magnitude, following the order $(111) \ll (100) < (110)$. Measurements on stepped single-crystal electrodes in H_2SO_4 showed higher activity, which was explained by a lowering of (bi)sulfate adsorption, whereas in $HClO_4$, the activities of the stepped surfaces were not found to be enhanced significantly [Macía et al., 2004; Komanicky et al., 2005].

On the basis of the single-crystal observations, one may expect a structure-sensitive (particle-size-dependent) rate dependence for oxygen reduction in the presence of strongly adsorbing anions. However a nonlinear dependence of mass activity on dispersion that indicates lower specific activity at small particles has been reported not only in H_2SO_4 and H_3PO_4 environments, but also in $HClO_4$ [Kinoshita, 1990; Antoine et al., 2001; Guerin et al., 2004]. It has also recently been reported that if the interspacing of dispersed particles is taken into account, there is in fact no particle size dependence in $HClO_4$ electrolyte [Yano et al., 2006]; this is supported by Stonehart's earlier work [Stonehart, 1994], where it was found that the specific ORR activity was independent for particles separated by distances at least 10 times the particle size [Watanabe et al., 1989]. The origin of the electrolyte dependence or the intrinsic activity of the surface with regard to oxygen reduction has been described in terms of the influence of surface structure or particle size on the adsorption of OH. Generally, the OH adsorption strength seems to increase as the particle size decreases; that is, Pt oxide formation/reduction is found to become more irreversible [Takasu et al., 1996; Markovic et al., 1997; Antoine et al., 2001; Guerin et al., 2004], and this can be correlated with a progressively positive shift in binding energy of the f-states of Pt for smaller particles [Takasu et al., 1996].

16.2.2 Oxygen Reduction on Au

Au has recently received less attention than Pt as a supported catalyst because of its lower impact in PEMFC energy conversion technology, since the ORR is dominated by a two-electron reduction process, at what is a high overpotential, in acidic media. Nevertheless, it is an important oxygen reduction catalyst in alkaline media, and, in contrast to Pt, is oxide-free in the potential range where oxygen reduction occurs.

In addition, it sustains CO electro-oxidation at relatively low overpotential, and there are crystal face dependences for both the ORR and CO oxidation. Since Au is also a system that exhibits both particle size and support effects in heterogeneous catalysis, it provides an interesting model system for studying such effects in electrocatalysis.

The ORR on Au was found to be dependent on the crystallographic orientation of the low index Miller planes in alkaline [Adzic et al., 1984; Markovic et al., 1984], neutral [Prieto et al., 2003], and acidic [Alvarez-Rizatti and Jüttner, 1983] media. The activity decreased with pH in the order (100) $\gg$ (110) $>$ (111) [Alvarez-Rizatti and Jüttner, 1983; Adzic et al., 1984; Markovic et al., 1984; Schmidt et al., 2002; Prieto et al., 2003]. The (100) crystallographic orientation of Au [Adzic and Markovic, 1982; Adzic et al., 1984; Markovic et al., 1994; Strbac and Adzic, 1996a, b; Blizanac et al., 2004a] catalyzes oxygen reduction to water in a four-electron process over a pH range of 3 $>$ pH $>$ 14 [Strbac and Adzic, 1996a, b] at potentials where OH appears to be stable at the surface. The reasons for these structure and pH dependences are still not fully understood [Kim and Gewirth, 2006; Vassilev and Koper, 2007].

More recently, there have been a number of publications describing the ORR on supported Au nanoparticle electrodes [Maye et al., 2004; Yagi et al., 2004; Zhang et al., 2004; El-Deab et al., 2005a, b; Gao et al., 2005; Hernández et al., 2005; Alexeyeva et al., 2006; Baker et al., 2006]. El-Deab and co-workers studied the ORR on electrodeposited Au nanoparticles on Au [El-Deab and Ohsaka, 2002a, b] and glassy carbon [El-Deab et al., 2003, 2005a, b; Gao et al., 2005] supports in acidic, neutral, and basic solutions.

Very low coverages of electrodeposited Au on an Au electrode result in a lower overpotential for oxygen reduction in acid electrolytes [El-Deab and Ohsaka, 2002a, b], with a similar increase in activity observed on glassy carbon-supported Au nanoparticles in both acid [Gao et al., 2005] and alkaline [El-Deab et al., 2005a] media. These studies also suggest that both Au- and carbon-supported particles can catalyze the reduction of hydrogen peroxide to water, although in neutral media, the dominant reaction on glassy carbon-supported small Au particles is hydrogen peroxide formation [El-Deab et al., 2003]. The particles in these studies are relatively large (10–100 nm), and the results can be explained in terms of the different activities of the exposed facets of the particles. Smaller particle sizes (3.7–4.8 nm) of Au supported on carbon synthesized with preferential growth of the (100) facets [Hernández et al., 2004, 2005] exhibited enhanced oxygen reduction activity, with a shift towards a four-electron reduction of oxygen. It has been reported recently [Baker et al., 2006] that tin dioxide-supported Au nanoparticles of diameter about 1 nm reduce oxygen in 0.1 M $HClO_4$ to water. Baker et al. speculated that either an electronic effect (charge transfer from the support to the Au particles) or a "bifunctional" mechanism might be responsible for this behavior. The latter would involve oxygen adsorption to form an O_2^- radical at the support, and the migration of this species to the perimeter of the Au particles, where further reduction would take place.

16.2.3 CO Electro-Oxidation on Pt

Catalytic oxidation of CO heterogeneously in the gas phase and electrochemically in solution are important reactions for the removal of CO from reactant stock gases in

PEMFC applications, since CO poisons Pt-based anode catalysts. Almost a monolayer of CO is adsorbed if traces of CO (about 10 ppm) are present in the feedstock gas hydrogen at the PEMFC anode at temperatures below 100 °C [Li et al., 2003]. The alloying of a second, or even third, metal component in Pt has been the favored method of providing CO-tolerant anode catalysts, and to date the most active catalytic system for both applications is Pt/Ru alloy [Wasmus and Kuver, 1999]. The efficacy of the alloying component in providing CO tolerance has been attributed to either lowering of the overpotential for CO electro-oxidation and oxidation of CO through a "bifunctional" mechanism [Watanabe and Motoo, 1975a, b] or an electronic effect of Ru on Pt that reduces the CO binding energy on the Pt [Hammer and Nørskov, 1995; Hammer et al., 1996; Shubina and Koper, 2002; Tsuda and Kasai, 2006], concomitantly reducing the mean coverages of CO.

The overall electrocatalytic oxidation of CO can be expressed as

$$CO + H_2O \longrightarrow CO_2 + 2H^+ + 2e^- \tag{16.1}$$

with a standard potential $E° = -0.106$ V (with respect to a standard hydrogen electrode, SHE) [Galus, 1985]. A Langmuir–Hinshelwood mechanism for this reaction is widely accepted for polycrystalline as well as single-crystal Pt surfaces [McCallum and Pletcher, 1976; Markovic and Ross, 2002]; the overpotential for the surface reaction is associated with activation of water to produce the surface oxidant, and the surface reaction may take place through a hydroxycarbonyl intermediate [Herrero et al., 2000].

Both the binding energy of CO and the activation of water may be expected to be structure-sensitive, and hence the CO tolerance of Pt may also be expected to be dependent on particle size and particle morphology. There is an effect of Pt particle size on the stripping voltammetry of CO (see Guerin et al. [2004]; Maillard et al. [2004a, b, 2005]; Arenz et al. [2005]; Mayrhofer et al. [2005a, b] and references therein). Generally, a displacement of the CO stripping peak is observed towards higher potentials at particles below about 3 nm [Guerin et al., 2004; Maillard et al., 2004a, b]. This can be attributed to either an increasing overpotential for water activation at small particles or changes in the mobility of the reactants in the subsequent Langmuir–Hinshelwood reaction over particle facets.

16.2.4 CO Electro-Oxidation on Au

The mechanism of anodic oxidation of CO at polycrystalline Au remains uncertain. Several groups have reported that the voltammetry of Au in acidic electrolytes is straightforward, with a well-formed oxidation wave/peak [Stonehart, 1966; Gibbs et al., 1977; Kita et al., 1985; Sun et al., 1999]. There is, however, no voltammetric evidence for the adsorption of CO on the Au surface, and spectroscopic studies indicate only a weak interaction of CO with polycrystalline Au surfaces in acidic solutions [Kunimatsu et al., 1986; Cuesta et al., 2003]. Moreover, there is little evidence for the formation of oxidizing species at the potential where the oxidation process is observed. Certainly, the oxidation of CO occurs at a potential over 500 mV less positive than that where bulk Au oxide is formed, and, indeed, the formation of this oxide strongly inhibits CO oxidation. The literature, however, implies that the mechanism at Au is similar to that at other precious metals, but adsorbed CO and adsorbed OH species are present only at very low coverages. This may implicate specific, low coordinate sites on the Au surface as critical to the oxidation reaction, and also suggests that the intermediates are both formed and react very rapidly [Burke and Nugent, 1997, 1998; Burke, 2004]. A structure sensitivity of the CO oxidation activity is indeed evident on single-crystal Au electrodes [Chang et al., 1991; Blizanac et al., 2004b], with the activity decreasing in the order Au(110) ≥ Au(100) > Au(111). This also supports the notion that defect sites at surfaces may dominate the electro-oxidation of CO. Recent studies of structurally induced changes in activity on Au(111) single crystals have also been explained by an increased number of low coordinated sites during the lifting of the surface reconstruction [Gallagher et al., 2005]. Several groups have reported studies of CO oxidation in acid solution at single-crystal Au surfaces [Chang et al., 1990, 1991; Edens et al., 1996; Blizanac et al., 2004a, b], although the voltammetry is similar to polycrystalline Au, the conclusions are somewhat different. Blizanac and co-workers have provided evidence that oxidation of the Au surface initiates at potentials as low as $+0.3$ V vs. SHE [Blizanac et al., 2004b], and have highlighted competition for surface sites between 'OH' and anions of the electrolyte [Blizanac et al., 2004a]. Also, infrared spectroscopy of the surface confirms the adsorption of CO, although there is disagreement about the extent of adsorption [Chang et al., 1990, 1991; Edens et al., 1996; Blizanac et al., 2004a, b]. Weaver and co-workers concluded that the coverage is never above 0.1 monolayer in acid solution [Chang et al., 1990, 1991], while Blizanac and co-workers reported almost full monolayer coverage as well as well-formed CO stripping peaks on cyclic voltammograms [Blizanac et al., 2004b]. The differences between the data at polycrystalline and single-crystal surfaces are very surprising, and explanation must await further studies.

There have been even fewer studies of CO electro-oxidation on supported Au particles than on supported Pt. One may expect, however, on the basis of the studies on extended Au surfaces, that there will be a structure sensitivity to the CO electro-oxidation reaction on supported Au particles.

16.3 SYNTHESIS AND SCREENING OF MODEL SUPPORTED ELECTROCATALYSTS

16.3.1 High Throughput Synthesis of Supported Nanoparticles

Electrocatalytic activity of supported metal particles has been investigated on surfaces prepared in an ultrahigh vacuum (UHV) molecular beam epitaxy system (DCA Instruments) modified to allow high throughput (parallel) synthesis of thin-film materials [Guerin and Hayden, 2006]. The system is shown in Fig. 16.1, and consisted of two physical vapor deposition (PVD) chambers, a sputtering chamber, and a surface characterization chamber (CC), all interconnected by a transfer chamber (TC). The entire system was maintained at UHV, with a base pressure of 10^{-10} mbar. Sample access was achieved through a load lock, and samples could be transferred

16.3 SYNTHESIS AND SCREENING OF MODEL SUPPORTED ELECTROCATALYSTS

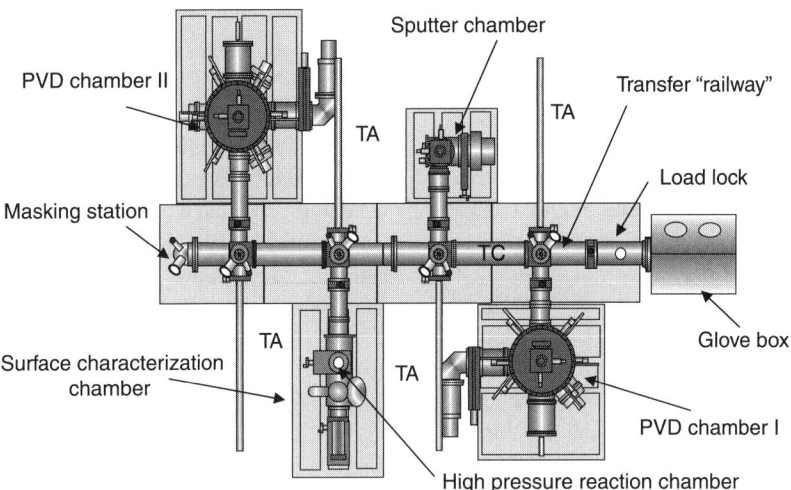

Figure 16.1 Schematic of the UHV deposition system. TA, transfer arm; TC, Transfer chamber.

between chambers by a transfer "railway" system and transfer arms. The "railway" consisted of two trolleys on rails transporting the samples along the TC. Adjacent to each chamber, a pickup mechanism and transfer arm allowed samples to be brought into the chamber for deposition. The PVD deposition chambers were cryo-pumped (Helix Tech. Corp.) and titanium sublimation-pumped (Varian), the TC and CC were ion-pumped (Varian) and titanium sublimation-pumped. The load lock was pumped by an oil-free rotary pump (Pfeiffer) and a turbo-molecular pump (Pfeiffer).

Deposition of the materials was undertaken in growth chambers incorporating effusion (Knudsen Source) and e-beam evaporation sources. Chamber I was equipped with three off-axis Knudsen cells (K-Cell, DCA HTKS) and one centered electron beam (e-gun, Temescal) evaporator. Chamber II consisted of six off-center sources, three K-Cells, and three e-gun evaporators. Titanium (granules 99.95%, Goodfellow Metals) was evaporated from either a high temperature K-Cell for operation up to 2000 °C using a pyrolitic graphite 10 cm^3 (Sintec Keramik) crucible or an e-beam evaporator. Substoichiometric TiO_x layers were deposited by oxidation of the titanium during deposition by a collimated beam of molecular oxygen (BOC Special Gases, N6 grade) directed at the sample face. Carbon layers were deposited using an e-beam evaporator source and graphite rods (Alfa Aesar, type 231-955-3). Carbon layers were also deposited by means of an arc carbon coating system (BOC Edwards) using carbon rods (BOC Edwards, type E085-19-030). Nanoparticles of Au and Pt were grown by evaporation at low rates for short times. Au (pellets 99.99%, Goodfellow Metals) were evaporated from a low temperature K-Cell for operation up to 1400 °C equipped with a 30 cm^3 Fabmate crucible (Kurt Lesker). Pt was evaporated from an e-beam evaporator.

The graduation of material across a wafer was achieved using a "wedge shutter" controlling the deposition profile of each source independently; the principle is discussed in detail elsewhere [Guerin and Hayden, 2006]. For uniform depositions such as carbon and titania support materials, the sample holder was equipped with a motor drive that allowed rotation of the substrate during deposition.

Au particles were deposited from a low temperature K-Cell with a source temperature $T_{dep} = 1548$ or 1623K. These temperatures were found to yield a highly stable Au flux, equivalent to the deposition of 0.16 ± 0.015 nm min^{-1} and 0.51 ± 0.05 nm min^{-1}. Deposition rates were calculated by the deposition of several thick layers over extended times (1800–7200 s), demonstrating that the thickness (as determined on contact masked samples by atomic force microscopy, AFM) was proportional to the deposition time. Variable coverage of Au over the substrate was achieved under the same conditions, using the fixed "wedge" shutter technique [Guerin and Hayden, 2006]. The characteristics of the "wedge" growth were established (Fig. 16.2) from AFM and ellipsometry measurements of Au deposited for times sufficient to achieve continuous films, and concomitantly lower times were used subsequently to achieve the lower equivalent coverage required for particle growth. Pt depositions have been characterized following the same procedure [Guerin and Hayden, 2006; Hayden et al., 2009].

Two examples of Au depositions are shown in Fig. 16.2, with different orientations of the source with respect to the substrate (samples deposited in chamber I show a diagonal deposition profile and those from chamber II a perpendicular profile). In Fig. 16.2a, a uniform thickness of about 18 nm would be achieved in the absence of the wedge shutter. AFM measurements show a variation of a factor of seven in coverage is achieved across the (10 × 10) array. This deposition geometry produces electrodes with 19 different coverages, and the electrodes perpendicular to the arrow should be identical. The ellipsometry measurements were automated and therefore faster than the AFM measurements. Also, the measurement error (standard deviation between measurements) seems to be smaller. In Fig. 16.2b, Au was deposited using a geometry (inset) that produced 10 rows of electrodes of different thicknesses, each row consisting of 10 electrodes with the same thickness (deposition perpendicular to the "wedge" direction is constant [Guerin and Hayden, 2006]).

Electrochemical experiments have been carried out on materials deposited by PVD on silicon microfabricated arrays of Au pad electrodes [Guerin et al., 2006a]. The substrate is made up of a square silicon wafer capped with silicon nitride (31.8 mm × 31.8 mm), which has an array of 100 individually addressable Au pad electrodes. These electrodes make up a square matrix on the wafer, which can be masked when placed in a PVD chamber, allowing deposition of thin films on the Au electrodes. Figure 16.3 is a schematic drawing of the configuration. Small electrical contact pads in Au for the individual addressing of electrodes (0.8 mm × 0.8 mm) are placed on the boundaries.

16.3.2 Electrochemical Measurements

The electrochemical responses of the 100 electrodes in the array were measured simultaneously using a potentiostat, two 64-channel current followers, and two data

16.3 SYNTHESIS AND SCREENING OF MODEL SUPPORTED ELECTROCATALYSTS 575

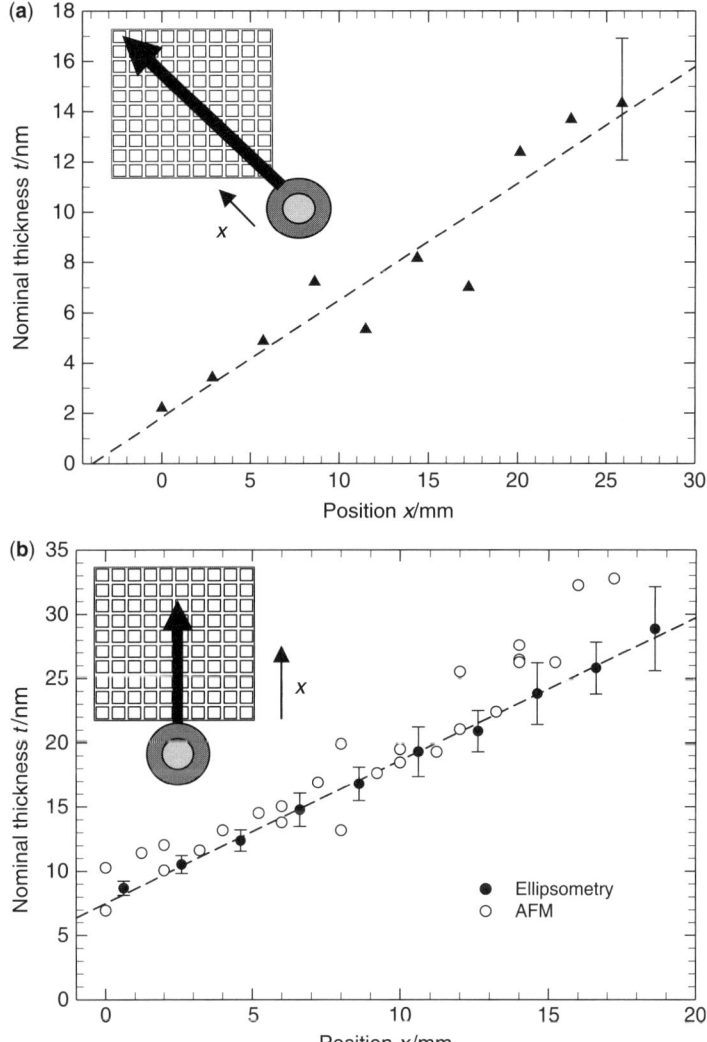

Figure 16.2 Thickness determination of Au deposition onto a bare silicon wafer using a 10 × 10 contact mask in two geometries (see insets), using (a) AFM along the diagonal of an array of 100 electrodes and (b) AFM and ellipsometry for a deposition geometry that allowed an average of 10 fields of identical thickness across the "wedge." The source temperatures and deposition times were (a) 1548K, 7200 s and (b) 1623K and 4500 s.

acquisition cards controlled/monitored by a PC as shown in Fig. 16.4. The data were then visualized with software written in the laboratory. The software allowed simultaneous monitoring of the current, or single cyclic voltammograms (CVs), or potential step experiments at each of the 100 electrodes. This instrumentation has been described previously [Guerin et al., 2004], but the sensitivities of the current followers

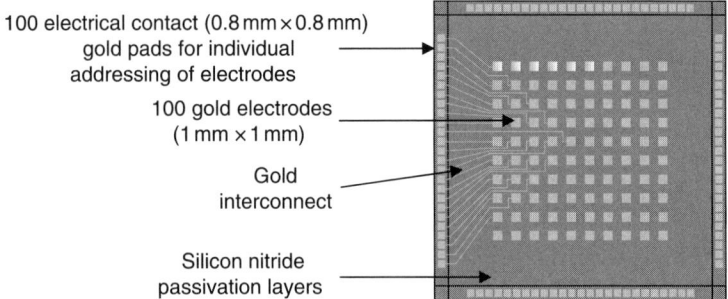

Figure 16.3 Silicon nitride pacified silicon wafer with an array of 100 individually addressable square Au electrodes.

used here were 10 $\mu A\,V^{-1}$. A similar setup was used for the single-electrode measurements, but with a bi-potentiostat and current followers that had the ability to switch between measurement resistances for the disk and ring responses and therefore the sensitivity (1, 10, 100, and 1000 $\mu A\,V^{-1}$). An adapted single-electrode version of the software was used.

The electrochemical experiments with the TiO_x/Au arrays were carried out in a new three-compartment glass cell with a water jacket (Fig. 16.5) and those with

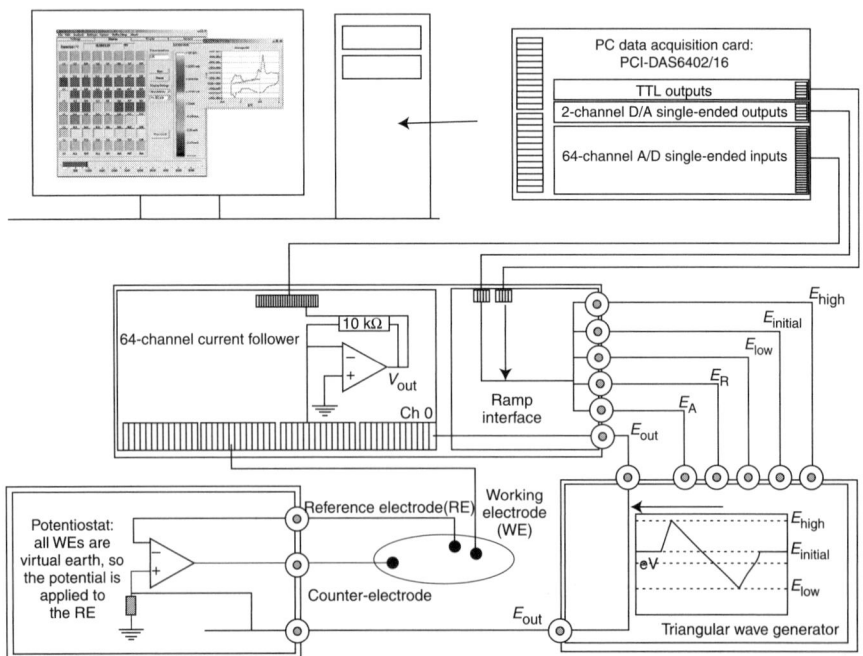

Figure 16.4 Electrochemical screening instrumentation, consisting of a wave generator, a potentiostat, a current follower, and a PC with a data acquisition card [Guerin et al., 2004].

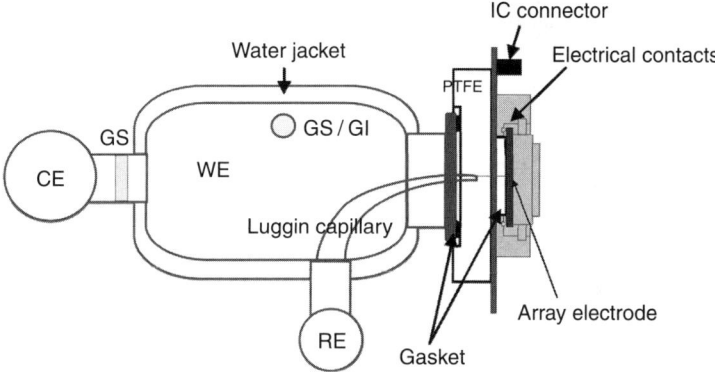

Figure 16.5 Schematic of the electrochemical array cell. CE, counter-electrode compartment; WE, working electrode compartment; RE, reference electrode compartment; GS, glass sinter; GS/GI, gas inlet with glass sinter.

the C/Au arrays in the previously described cell [Guerin et al., 2006b]. The new cell had a counter-electrode (CE) compartment (CE Au gauze, Alfa Aeser 99.99%), separated from the working electrode (WE) compartment by a glass sinter. The reference electrode (RE) was mounted in a Luggin capillary whose tip was placed a few millimeters from the array. A specifically designed socket allowed precise positioning of the array to ensure electrical contact. The array was sealed with a Viton gasket to the WE compartment of the glass cell. The glass cell was cleaned by repeated boiling in ultrapure water, followed by a concentrated $HClO_4$ rinse, and was finally washed with ultrapure water.

16.4 CHARACTERIZATION OF SUPPORTED PARTICLES

Au was evaporated and particles nucleated on TiO_x and carbon on substrates. For particle size characterization, depositions on transmission electron microscope (TEM) grids were performed under identical conditions to those used for other substrates (support materials were deposited for shorter times to ensure passage of the electron beam of the microscope). As an example of this approach, Fig. 16.6 shows TEM images for five deposition times (50, 300, 600, 900, and 1200 s), corresponding respectively to nominal thicknesses of 0.13, 0.8, 1.6, 2.3, and 3.2 nm of Au on TiO_x estimated from the known rate of Au deposition. The light field image clearly allows identification of the Au particles. The particles appear round at the base and randomly distributed at low coverages, and they coalesce at higher coverages.

From the TEM micrographs, particle sizes and the number of particles per unit area could be estimated. Figure 16.6 provides a quantitative analysis of the particle sizes as a function of deposition time. It is evident from the particle size distributions that at low nominal Au thickness (0.13 nm), mean particle diameters are about 1.4 nm and fall in a narrow range of sizes. As the nominal thickness becomes higher, the particle

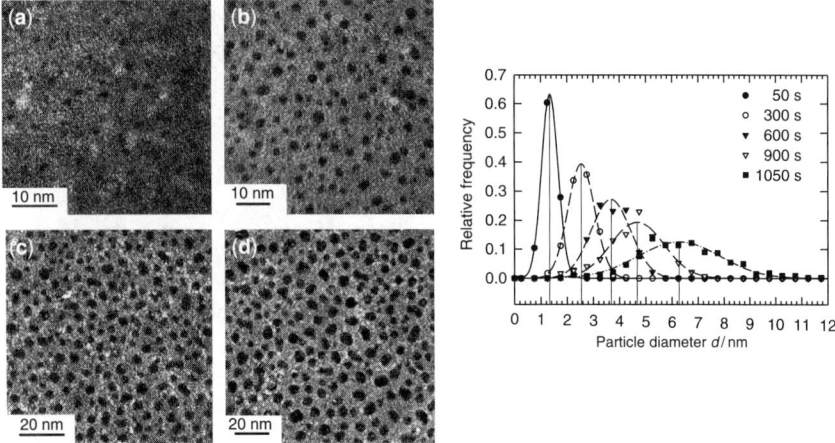

Figure 16.6 TEM micrographs of titania-supported Au particles. The nominal thickness of Au was (a) 0.13 nm; (b) 0.78 nm; (c) 1.56 nm; (d) 2.33 nm. The Au deposition rate was 2.6×10^{-3} nm s^{-1}. Particle size distributions of Au for various deposition times are shown in the plot, with the distributions fitted to a normal Gaussian function.

size increases and the distributions become broader. The characteristics of particle growth on both carbon and titania supports are described in more detail elsewhere [Guerin et al., 2006c].

16.5 OXYGEN REDUCTION ON SUPPORTED Au AND Pt NANOPARTICLES

The electrochemical reduction of oxygen has been measured using a chronoamperometric methodology in HClO$_4$ electrolyte on Au-supported nanoparticles [Guerin et al., 2006b]. The potential was stepped from 0.8 V (with respect to a reversible hydrogen electrode, RHE) to 0.5, 0.4, and 0.3 V vs. RHE, and the current was recorded during 90 s holding time. Figure 16.7 shows the TEM surface area normalized currents of the 0.4 V step (average of the last 20% of the steps) for (a) carbon- and (b) titania-supported Au particles in 0.5 M HClO$_4$. It is evident from these results that the ORR on supported Au particles is sensitive to particle size on both substrates. On both carbon and titania substrates, the reaction is strongly deactivated below a critical particle size of about 3 nm diameter [Guerin et al., 2006b]. Rotating disk electrode measurements confirmed for both supports that there was no change in reaction mechanism, and mainly an indirect reduction to form hydrogen peroxide was promoted [Guerin et al., 2006b] on all surfaces. In addition, in the case of the titania-supported Au, there appears to be a small activation of the reaction as particles reduce in size between 6 and 3 nm.

Recent studies of Pt supported on titania synthesized using the high throughput method on arrays have also revealed a particle size dependence of the ORR

16.5 OXYGEN REDUCTION ON SUPPORTED Au AND Pt NANOPARTICLES 579

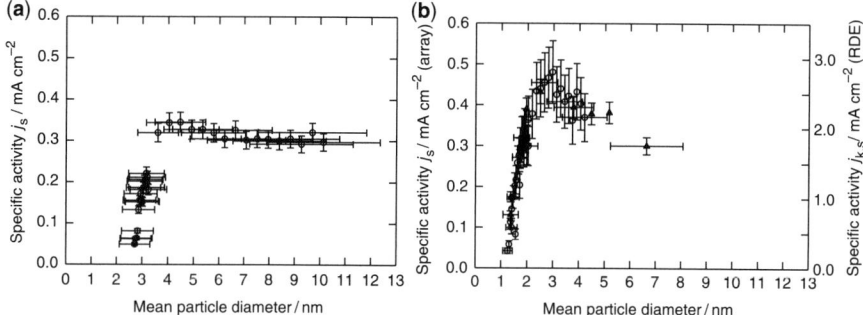

Figure 16.7 Specific activity for oxygen reduction at 0.3 V vs. RHE on carbon- and titania-supported Au nanoparticles [Guerin et al., 2006b]: (a) C/Au; (b) TiO$_x$/Au. In the case of TiO$_x$/Au (b), results are shown for data obtained on arrays of electrodes (◇) and on rotating disk electrodes (RDE) (▲).

[Hayden et al., 2009]. Figure 16.8 (Plate 16.1) shows the effect of the equivalent thickness of deposited Pt on titania on the activity of the ORR carried out in a 0.5 M HClO$_4$ electrolyte at 20 °C. In this case, the activity was assessed by determining the potential (vs. RHE) at which the specific current density reaches 0.01 mA cm^{-2}: A lower potential indicates a higher overpotential required to achieve this rate of reaction, and hence corresponds to a reduction in activity. The shaded region of the plot, below about 1 nm equivalent thickness, corresponds to coverages over which distinct particles are

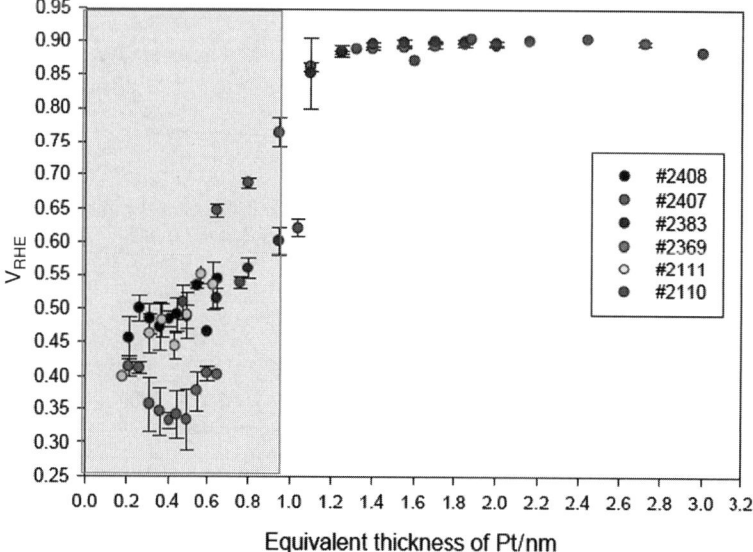

Figure 16.8 Pt/TiO$_x$-catalyzed oxygen reduction potential, where 0.01 mA cm^{-2} is reached during the negative scan in a cyclic voltammetry experiment (scan rate 20 mV s^{-1}) in oxygen-saturated 0.5 M HClO$_4$ at 25 °C. (See color insert.)

formed, with the smallest particles (at lowest equivalent thickness) corresponding to about 1 nm. These results are clearly consistent with the results obtained on high area carbon-supported Pt, where a decrease in specific activity in 0.5 M H_2SO_4 electrolyte was also observed [Guerin et al., 2004]. In both cases, the decrease in oxygen reduction activity is accompanied by a concomitant shift in the Pt surface reduction potential as particle sizes decrease [Guerin et al., 2004].

16.6 CO ELECTRO-OXIDATION ON SUPPORTED Au AND Pt NANOPARTICLES

As for the reduction of oxygen, supported Au particles have been tested for their ability to catalyze oxidation of carbon monoxide in $HClO_4$. Figure 16.9 shows the electrochemical responses of eight electrodes out of arrays of electrodes of either carbon- or titania-supported Au nanoparticles. The measurements were carried out in CO-saturated 0.5 M $HClO_4$. Each of the four pairs of voltammograms corresponds to surfaces supporting similar Au particle sizes on carbon and titania, allowing a direct comparison of the influence of the support. Results for particles between about 2.5 and 7.5 nm are shown. While CO oxidation is seen at all the surfaces studied, and the peak current

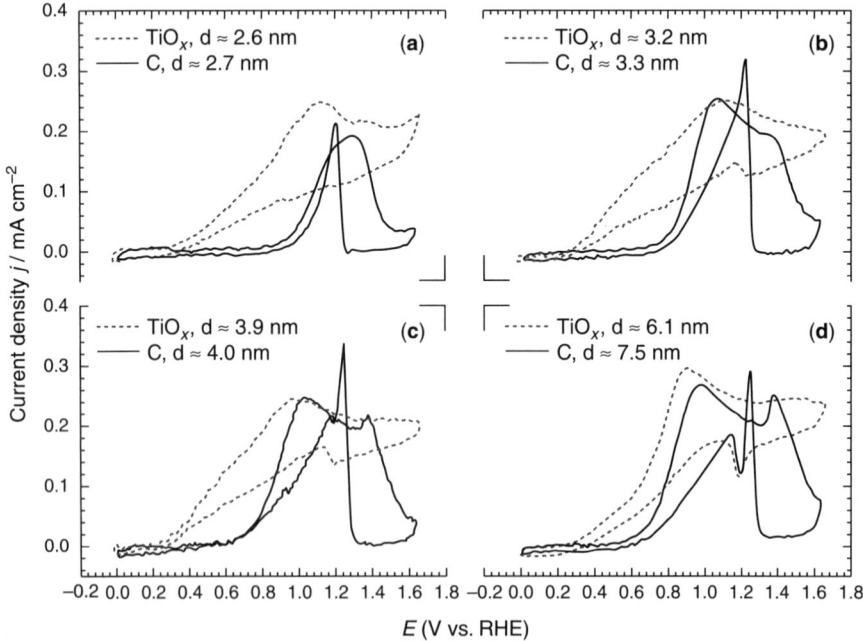

Figure 16.9 Comparison of cyclic voltammetry in a CO-saturated electrolyte (0.5 M $HClO_4$) of Au supported on carbon (solid curves) and titania (dashed curves) for four different particle sizes (indicated). The measurements were made at a temperature of 298K and a scan rate of 50 mV s^{-1}.

densities are similar to those expected for diffusion-controlled oxidation, it is immediately evident that the voltammograms on the C and TiO_x substrates are markedly different. Most interestingly, the TiO_x-supported Au nanoparticles show substantially higher catalytic activity for CO oxidation at low positive potentials. Indeed, CO oxidation commences at $+0.2$ V at the TiO_x-supported samples, and this compares with over $+0.45$ V at both of the C-supported nanoparticles. A striking difference between the two systems under investigation was found at potentials positive to the onset of Au oxidation. At the C-supported Au nanoparticles, the voltammograms show features associated with inhibition of CO oxidation by Au oxide formation and reactivation on the reverse scan at the potential, where Au oxide is reduced. In contrast, at Au on a TiO_x support, inhibition of CO oxidation does not occur at high anodic potentials. The apparent insensitivity of CO oxidation to anodic potentials can only be the result of an alternative mechanism of CO oxidation that does not rely on clean Au sites or modified redox behavior of the titania-supported Au particles (for further details, see Hayden et al. [2007a, b]).

The voltammograms in Fig. 16.9 clearly show that in both systems the reaction is strongly dependent on particle size. In order to illustrate this, in Figs. 16.10 and 16.11, specific activities are plotted against mean particle size for three systems. Figure 16.10 shows the particle size dependence of the CO oxidation activity of the C/Au system at a fixed potential of 0.8 V, taken from the voltammograms presented in Fig. 16.9. In the case of carbon-supported Au surfaces, CO electro-oxidation catalysis is similar to that observed on bulk Au for nanoparticles larger than 3 nm. The voltammetry is similar in that the onset of the oxidation takes place at the same overpotential, and oxidation of the supported Au strongly inhibits the reaction [Hayden et al.,

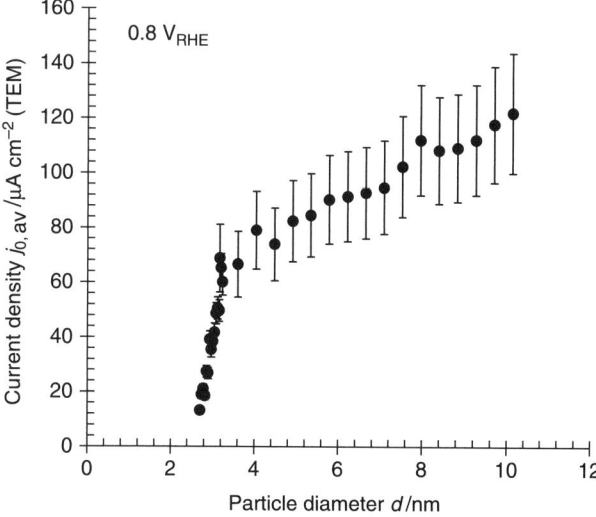

Figure 16.10 Particle-size-dependent specific activity at 0.8 V vs. RHE for Au nanoparticles supported on carbon.

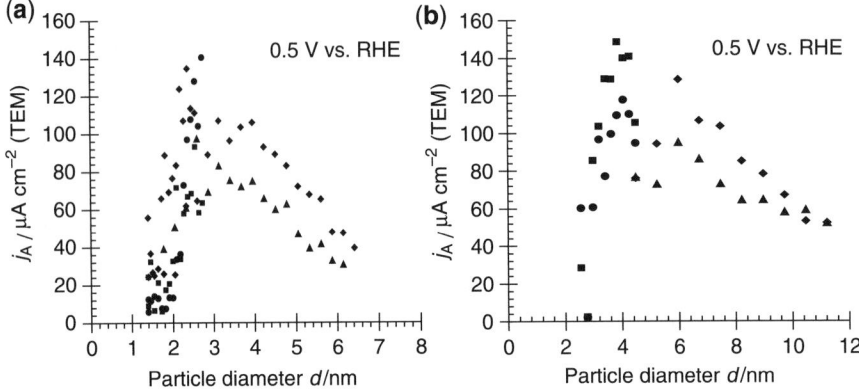

Figure 16.11 Titania (TiO_x)-supported (a) and niobium-doped titania ($Nb_{0.08}Ti_{0.92}O_x$/Au)-supported (b) Au particle-size-dependent specific activities for—CO oxidation at a potential step of 0.5 V vs. RHE. The current densities have been corrected for the TEM-determined surface areas.

2007a]. It is evident from the particle size dependence in Fig. 16.10, however, that at particle sizes below 3 nm, the activation shows a sharp fall with decreasing particle size. This result is similar to those obtained for the particle size dependence of the ORR on carbon-supported Au, which is also characterized by a sharp de-activation of the reaction for nanoparticles smaller than 3 nm at 0.4 and 0.3 V vs. RHE [Guerin et al., 2006b].

Figure 16.11a shows the results of chronoamperometric measurements on TiO_x/Au electrodes, where the potential was stepped from initially 0 V vs. RHE in 50 mV steps to 0.5 V. The results shown are for the real surface-area-corrected current densities at 0.5 V vs. RHE. A similar particle size dependence is also observed at 0.3 V vs. RHE [Hayden et al., 2007c]. In the case of the titania support, the conductivity of the oxide derives from the oxygen defects incorporated in the bulk during synthesis [Guerin et al., 2006c]. Since the stability of such substrates in oxidative conditions may be poor, with oxidation leading to the formation of a more stoichiometric material with low conductivity, experiments have been carried out on niobium-doped titania, which provides materials with a dopant-induced conductivity, which should lead to greater stability. Figure 16.11b shows the results of CO electrooxidation on niobium-doped titania-supported Au nanoparticle ($Nb_{0.08}Ti_{0.92}O_x$/Au) electrodes. The measurements were made under identical conditions to those shown for titania-supported Au in Fig. 16.11a. Both the doped and substoichiometric titania-supported Au nanoparticles exhibit a maximum in the activity for CO oxidation at a particle size of 3 nm. At a potential of 0.5 V vs. RHE, the carbon-supported nanoparticles show no activity. This particle-dependent behavior (an increase in addition to a decrease in specific activity), taken together with the strong lowering of the overpotential for the reaction, is strong evidence for a support-induced modification of Au activity.

16.7 INTRINSIC PARTICLE SIZE EFFECTS AND SUBSTRATE MODIFICATIONS

A number of observations can be made concerning the particle size and support dependences of the reactions described above:

- Oxygen reduction on both carbon- and titania-supported Pt particles is dependent on particle size. A deactivation of the catalytic activity is observed for decreasing particle size on both supports. In addition, there is no evidence of any activation of the Pt above that of bulk Pt on either support.
- Oxygen reduction on both carbon- and titania-supported Au exhibits a similar dependence on particle size to that observed for Pt, namely, a decrease in activity with decreasing particle size. This decrease occurs at particle sizes below about 3 nm. In addition to the decrease in activity, a small increase in activity is also observed for titania-supported Au nanoparticles.
- CO electro-oxidation exhibits a strong particle size dependence on both carbon- and titania-supported Au catalysts: a strong deactivation of the reaction is observed for particle sizes below about 3 nm. In the case of the titania supports, however, a distinct activation of the reaction is also evident. This manifests itself in a strong decrease in the overpotential for the reaction, and an increase in activity as the particle size decreases in the range 8–3 nm. The result is a maximum in the catalytic activity with particle size.

The results can be understood in terms of the influence of an intrinsic particle size effect (independent of the support) and a support-induced particle size effect. For both reactions and both supported metals, the intrinsic effect manifests itself in a decrease in activity with decreasing particle size below about 3 nm.

The deactivation of the CO electro-oxidation reaction [Hayden et al., 2007a, c], and the deactivation of the ORR [Guerin et al., 2006b] on both carbon- and titania-supported Au for particle sizes of 3 nm and below, have led to the suggestion that there is a common origin to the effect, and the insensitivity to the support indicates that it is an intrinsic particle size effect. For one of these reactions, namely, oxygen reduction, it appears in addition that the reaction is also deactivated on small Pt particles supported both on carbon [Guerin et al., 2004] and on titania [Hayden et al., 2009]. An explanation for the latter that has been proposed is the increased adsorption energy of oxygen or OH on small particles, resulting in an inhibition of oxygen reduction. This is qualitatively consistent with the optimal position of extended Pt in the theoretically predicted volcano curve for oxygen reduction [Nørskov et al., 2004]. In the case of Au, however, it is the weak binding of O or OH that is correlated with the low activity for oxygen reduction, and one may expect qualitatively that an increasing binding energy on small particles of Au may result in increasing oxygen reduction activity. If the adsorption energy of oxidizing surface intermediates were increased sufficiently strongly on Au that small particles were irreversibly oxidized, this could account for deactivation. There is, however, no evidence that this is the case [Hayden et al., 2007b]. An alternative explanation involving the reduced activity of the different facets exposed at small particles has also been put forward in the case of the supported Pt [Kinoshita, 1990]; there may indeed be a similar explanation for the deactivation of reactivity observed for supported Au. A possible alternative explanation for an intrinsic particle size effect proposed for supported Pt [Maillard et al., 2004a; Mayrhofer et al., 2005b] and Au [Hayden et al., 2007a, c] electrocatalysts is that there is a change in electronic properties as a result of quantum size effects below about 3 nm.

Notwithstanding the origin of the deactivation, it is the superposition of this "intrinsic" particle size effect with a particle size dependence of the titania-supported activation that has been suggested to be the primary origin of the maxima observed in the CO oxidation reaction [Hayden et al., 2007a, c] and the ORR [Guerin et al., 2006b] for these catalysts. It should be emphasized that the enhanced activity is unique to titania-supported Au catalysts. In view of the extensive literature on support-induced activity for titania-supported Au in gas-phase reactions in particular for CO oxidation, and the particle size dependence of activity with a maximum at about 3 nm, it seemed reasonable to seek a common explanation for the effects in heterogeneous catalysis and electrocatalysis [Hayden et al., 2007a, c]. In the case of heterogeneous catalysis of CO oxidation, the promoting supports are metal oxides, most notably titania [Haruta et al., 1993; Haruta, 1997; Valden et al., 1998; Bond and Thompson, 2000; Daté and Haruta, 2001; Haruta and Daté, 2001; Choudhary and Goodman, 2002; Grisel et al., 2002; Davis, 2003; Daté et al., 2004; Lopez et al., 2004a; Chen and Goodman, 2006], and the specific activity also shows a maximum using nanoparticles with a mean diameter of 3 nm [Haruta et al., 1993; Valden et al., 1998; Lopez et al., 2004a; Chen and Goodman, 2006]. This suggested that the underlying mechanisms responsible for the substrate-induced activity, as well as the deactivation at small particle sizes, of titania-supported Au in heterogeneous catalysis and electrocatalysis have common origins [Hayden et al., 2007a, c]. It is important to note, however, that in the electrochemical case, the oxidant is provided through activation of water (rather than oxygen) as the surface oxidant in a Langmuir–Hinshelwood reaction. A number of explanations for the titania-induced activity of Au in heterogeneous catalysis have been put forward, and these have been considered in the light of electrocatalytic oxidation.

16.7.1 Titania-Induced Morphology of Au

The increased activity may be the result of a higher binding energy (and hence surface concentration) of reactant molecules, or the lowering of an oxygen dissociation barrier through a stronger interaction [Grunwaldt et al., 1999; Lopez and Nørskov, 2002; Lemire et al., 2004; Lopez et al., 2004b]. For example, it was reported that the heat of adsorption of CO increased substantially at small titania-supported Au particle diameters [Meier and Goodman, 2004]; the heat of adsorption at bulk Au was reported to be 52 kJ mol^{-1}, while at particles of about 2.5 nm mean diameter, it was 76 kJ mol^{-1}. Also, oxygen seems to interact more strongly with small thin Au particles, as evidenced by an increase in desorption temperature at TiO_x/Au surfaces [Bondzie et al., 1999]. In this regard, there is indeed strong evidence of a surface structural dependence in CO electrocatalytic oxidation on Au [Blizanac et al., 2004b].

In the case of the electro-oxidation of CO, the increased activity may also be due to an enhancement in CO coverage, or a lowering of the activation energy for water in the provision of adsorbed OH. Increased CO binding energy, and hence CO coverages, would result in the possibility of observing adsorbed CO in CO stripping voltammetry, as on single-crystal surfaces [Bondzie et al., 1999], and no such stripping peaks on titania-supported Au could be observed [Hayden et al., 2007a, c]. It is also true to say that if the increased activity of CO electro-oxidation on titania-supported Au were a result of modifications in CO adsorption, this would not provide an explanation for the apparent activation in the ORR [Guerin et al., 2006b]. One cannot so easily rule out the possibility of the enhancement of water activation, particularly since this may also influence the ORR. There is no evidence for different morphologies of Au particles on carbon and titania supports—at least from TEM characterization [Guerin et al., 2006c].

16.7.2 Titania-Induced Electronic Modification of Au

The most active Au particle size for CO oxidation was observed where a band gap of 0.2–0.6 eV was measured by scanning tunneling spectroscopy (STS); these are centers about two atomic layers thick and about 3 nm in diameter [Valden et al., 1998] In a study of monolayers of Au supported on thin TiO_x films, it was shown that a "bilayer" of Au was the most active structure (and it was suggested that there is no influence of the perimeter) [Chen and Goodman, 2004]. It is also apparent that structural and electronic modifications associated with substrate-induced strain in the Au particle [Mavrikakis et al., 2000; Xu and Mavrikakis, 2003] or charge transfer may contribute to the activity of $Au^{\delta+}$ nanoparticles on titania substrates [Sanchez et al., 1999; Okazawa et al., 2006]. Further evidence for the CO oxidation activity of positively charged Au^+ derives from the apparent activity of partially oxidized Au clusters [Park and Lee, 1999; Guzman and Gates, 2004], and, in the limit of atomic dispersion, positively charged Au ions were suggested to be crucial for the activity of titania-supported Au in the water gas shift reaction, where zero-valent Au atoms were found to be inactive [Fu et al., 2003].

The activation observed in titania-supported Au electrocatalysts is unlikely to arise from electronic effects in monolayer or bilayer Au [Valden et al., 1998; Chen and Goodman, 2004], since the electrocatalytic activity was correlated with the size of three-dimensional titania-supported Au particles [Guerin et al., 2006b; Hayden et al., 2007a, c]. The possibility that titania-induced electronic modification of three-dimensional particles below 6.5 nm is responsible for the induced activity, however, could not be excluded. It was pointed out, though, that such electronic effects should dominate for the smaller particle regime (<3 nm), where deactivation of the Au is observed on all supports.

16.7.3 Reactions at the Titania–Au Perimeter

It has been suggested that spillover of oxygen from the support at the particle–support interface, or active sites at the particle perimeter itself, may be responsible for enhancement of CO oxidation [Boccuzzi et al., 2001; Liu et al., 2003; Molina et al., 2004]. The results here suggest that a mechanism based on the reaction of a substrate-accommodated oxidant diffusing and reacting at the perimeter of the nanoparticles is unlikely [Haruta et al., 1993; Boccuzzi et al., 2001], since there is no evidence that titania can activate water to produce such an oxidant (OH) at these low potentials. However, a high activity induced by the activation of water at perimeter sites cannot be ruled out [Hayden et al., 2007a, c]. A correlation between the total length of the particle perimeters and the activity in CO electro-oxidation as a function of Au nanoparticle size supported on titania has been suggested [Hayden et al., 2007c], at least in so far as the maximum in activity is predicted at about 2.5 nm. However, the decrease in activity observed experimentally is much steeper below 3 nm than the decrease in perimeter length, although this could be explained by the additional contribution of intrinsic particle size affects.

16.8 CONCLUSIONS

There are clearly two important contributions that must be considered in evaluating the influence of particle size on electrocatalytic reactivity: intrinsic particle size effects and support-induced particle effects. The model electrocatalyst studies reveal a deactivation of Au and Pt particle activity when particles are 3 nm and smaller. These studies also reveal that the support can strongly influence the activity of the catalyst, and that there are strong similarities between such effects in heterogeneous catalysis and electrocatalysis. The decrease in the activity of small particles as a result of some intrinsic (support-independent) effect clearly will act to limit the advantage gained by dispersion in optimizing mass activity. It is evident, however, that gains in specific and mass activity can be achieved by varying the support material. This provides a powerful method of optimization for supported metal electrocatalysts.

REFERENCES

Adzic RR, Markovic NM. 1982. Structural effects in electrocatalysis: Oxygen and hydrogen peroxide reduction on single crystal gold electrodes and the effects of lead ad-atoms. J Electroanal Chem 138: 443–447.

Adzic RR, Markovic NM, Vesovic, VB. 1984. Structural effects in electrocatalysis: Oxygen reduction on the Au(100) single crystal electrode. J Electroanal Chem 165: 105–120.

Alexeyeva N, Laaksonen T. 2006. Oxygen reduction on gold nanoparticle/multi-walled carbon nanotubes modified glassy carbon electrodes in acid solution. Electrochem Commun 8: 1475–1480.

Alvarez-Rizatti M, Jüttner K. 1983. Electrocatalysis of oxygen reduction by UPD of lead on gold single-crystal surfaces. J Electroanal Chem 144: 351–363.

Antoine O, Bultel Y, Durand R. 2001. Oxygen reduction reaction kinetics and mechanism on platinum nanoparticles inside Nafion®. J Electroanal Chem 499: 85–94.

Arenz M, Mayrhofer KJJ, Stamenkovic V, et al. 2005. The effect of the particle size on the kinetics of CO electrooxidation on high surface area Pt catalysts. J Am Chem Soc 127: 6819–6829.

Baker WS, Pietron JJ, Teliska ME, et al. 2006. Enhanced oxygen reduction activity in acid by tin-oxide supported Au nanoparticle catalysts. J Electrochem Soc 153: A1702–A1707.

Blizanac BB, Lucas CA, Gallagher ME, et al. 2004a. Anion adsorption, CO oxidation, and oxygen reduction reaction on a Au(100) surface: The pH effect. J Phys Chem B 108: 625–634.

Blizanac BB, Arenz M, Ross PN, Markovic NM. 2004b. Surface electrochemistry of CO on reconstructed gold single crystal surfaces studied by infrared reflection absorption spectroscopy and rotating disk electrode. J Am Chem Soc 126: 10130–10141.

Boccuzzi F, Chiorino A, Manzoli M, et al. 2001. Au/TiO_2 nanosized samples: A catalytic, TEM, and FTIR study of the effect of calcination temperature on the CO oxidation. J Catal 202: 256–267.

Bond GC, Thompson DT. 2000. Gold-catalysed oxidation of carbon monoxide. Gold Bull 33(2): 41–51.

Bondzie VA, Parker SC, Campbell CT. 1999. The kinetics of CO oxidation by adsorbed oxygen on well-defined gold particles on TiO_2(110). Catal Lett 63: 143–151.

Burke LD. 2004. Scope for new applications for gold arising from the electrocatalytic behaviour of its metastable surface states. Gold Bull 37(1–2): 125–135.

Burke LD, Nugent PF. 1997. The electrochemistry of gold: I The redox behaviour of the metal in aqueous media. Gold Bull 30(2): 43–53.

Burke LD, Nugent PF. 1998. The electrochemistry of gold: II The electrocatalytic behaviour of the metal in aqueous media. Gold Bull 31(2): 39–50.

Chang SC, Hamelin A, Weaver MJ. 1990. Reactive and inhibiting adsorbates for the catalytic electrooxidation of carbon-monoxide on gold (210) as characterized by surface infrared-spectroscopy. Surf Sci 239: L543–L547.

Chang SC, Hamelin A, Weaver MJ. 1991. Dependence of the electrooxidation rates of carbon-monoxide at gold on the surface crystallographic orientation—A combined kinetic–surface infrared spectroscopic study. J Phys Chem 95: 5560–5567.

Chen GY, Delafuente DA, Sarangapani S, Mallouk TE. 2001. Combinatorial discovery of bifunctional oxygen reduction–water oxidation electrocatalysts for regenerative fuel cells. Catal Today 67: 341–355.

Chen MS, Goodman DW. 2004. The structure of catalytically active gold on titania. Science 306: 252–255.

Chen MS, Goodman DW. 2006. Structure–activity relationships in supported Au catalysts. Catal Today 111: 22–33.

Choudhary TV, Goodman DW. 2002. Oxidation catalysis by supported gold nano-clusters. Top Catal 21: 25–34.

Cuesta A, Lopez N, Gutierrez C. 2003. Electrolyte electroreflectance study of carbon monoxide adsorption on polycrystalline silver and gold electrodes. Electrochim Acta 48: 2949–2956.

Daté M, Haruta M. 2001. Moisture effect on CO oxidation over Au/TiO_2 catalyst. J Catal 201: 221–224.

Daté M, Okumura M, Tsubota S, Haruta M. 2004. Vital role of moisture in the catalytic activity of supported gold nanoparticles. Angew Chem Int Ed 43: 2129–2132.

Davis RJ. 2003. CHEMISTRY: All That Glitters Is Not Au^0. Science 301: 926–927.

Edens GJ, Hamelin A, Weaver MJ. 1996. Mechanism of carbon monoxide electrooxidation on monocrystalline gold surfaces: Identification of a hydroxycarbonyl intermediate. J Phys Chem 100: 2322–2329.

El-Deab M, Ohsaka T. 2002a. Hydrodynamic voltammetric studies of the oxygen reduction at gold nanoparticles-electrodeposited gold. Electrochim Acta 47: 4255–4261.

El-Deab MS, Ohsaka T. 2002b. An extraordinary electrocatalytic reduction of oxygen on gold nanoparticles-electrodeposited gold electrodes. Electrochem Commun 4: 288–292.

El-Deab MS, Okajima T, Ohsaka T. 2003. Electrochemical reduction of oxygen on gold nanoparticle-electrodeposited glassy carbon electrodes. J Electrochem Soc 150: A851–A857.

El-Deab M, Sotomura T, Ohsaka T. 2005a. Oxygen reduction at electrochemically deposited crystallographically oriented Au(100)-like gold nanoparticles. Electrochem Commun 7: 29–34.

El-Deab M, Sotomura T, et al. 2005b. Size crystallographic orientation of gold nanoparticles electrodeposited on GC electrodes. J Electrochem Soc 152: C1–C6.

Fu Q, Saltsburg H, Flytzani-Stephanopoulos M. 2003. Active nonmetallic Au and Pt species on ceria-based water-gas shift catalysts. Science 301: 935–938.

Gallagher ME, Blizanac BB, Lucas CA, et al. 2005. Structure sensitivity of CO oxidation on gold single crystal surfaces in alkaline solution: Surface X-ray scattering and rotating disk measurements. Surf Sci 582: 215–226.

Galus Z. 1985. Carbon, silicon, germanium, tin, and lead. In: Bard AJ, Parsons R, Jordan J, editors. Standard Potentials in Aqueous Solution. New York: Marcel Dekker.

Gao F, El-Deab M, Okajima T, Ohsaka T. 2005. Electrochemical preparation of a Au crystal with peculiar morphology and unique growth orientation and its catalysis for oxygen reduction. J Electrochem Soc 152: A1226–A1232.

Gasteiger HA, Panels JE, Yan SG. 2004. Dependence of PEM fuel cell performance on catalyst loading. J Power Sources 127: 162–171.

Gasteiger HA, Kocha SS, Sompalli B, Wagner FT. 2005. Activity benchmarks and requirements for Pt, Pt-alloy, and non-Pt oxygen reduction catalysts for PEMFCs. Appl Catal B: Environ 56: 9–35.

Gibbs TK, McCallum C, Pletcher D. 1977. The oxidation of carbon monoxide at platinum and gold metallized membrane electrodes. Electrochim Acta 22: 525–530.

Grisel R, Weststrate KJ, Gluhoi A, Nieuwenhuys BE. 2002. Catalysis by gold nanoparticles. Gold Bull 35(2): 39–45.

Grunwaldt J-D, Maciejewski M, Becker OS, et al. 1999. Comparative Study of Au/TiO_2 and Au/ZrO_2 catalysts for low-temperature CO oxidation. J Catal 186: 458–469.

Guerin S, Hayden BE. 2006. Physical vapor deposition method for the high-throughput synthesis of solid-state material libraries. J Comb Chem 8: 66–73.

Guerin S, Hayden BE, Lee CE, et al. 2004. Combinatorial electrochemical screening of fuel cell electrocatalysts. J Comb Chem 6: 149–158.

Guerin S, Hayden BE, Lee CE, et al. 2006a. High-throughput synthesis and screening of ternary metal alloys for electrocatalysis. J Phys Chem B 110: 14355–14362.

Guerin S, Hayden BE, Pletcher D, et al. 2006b. A combinatorial approach to the study of particle size effects on supported electrocatalysts: Oxygen reduction on gold. J Comb Chem 8: 679–686.

Guerin S, Hayden BE, Pletcher D, et al. 2006c. Combinatorial approach to the study of particle size effects in electrocatalysis: Synthesis of supported gold nanoparticles. J Comb Chem 8: 791–798.

Guzman J, Gates BC. 2004. Catalysis by supported gold: Correlation between catalytic activity for CO oxidation and oxidation states of gold. J Am Chem Soc 126: 2672–2673.

Hammer B, Nørskov JK. 1995. Electronic factors determining the reactivity of metal surfaces. Surf Sci 343: 211–220.

Hammer B, Morikawa Y, Nørskov JK. 1996. CO chemisorption at metal surfaces and overlayers. Phys Rev Lett 76: 2141.

Haruta M. 1997. Size- and support-dependency in the catalysis of gold. Catal Today 36: 153–166.

Haruta M, Daté M. 2001. Advances in the catalysis of Au nanoparticles. Appl Catal A: Gen 222: 427–437.

Haruta M, Tsubota S, Kobayashi T, et al. 1993. Low-temperature oxidation of CO over gold supported on TiO_2, α-Fe_2O_3, and Co_3O_4. J Catal 144: 175–192.

Hayden BE, Pletcher D, Rendall ME, Suchsland J-P. 2007a. CO oxidation on gold in acidic environments: Particle size and substrate effects. J Phys Chem C 111: 17044–17051.

Hayden BE, Pletcher D, Rendall ME, et al. 2007b. The influence of substrate and particle size on the electrochemistry of the Au/AuO couple.

Hayden BE, Pletcher D, Suchsland J-P. 2007c. Enhanced activity for electrocatalytic oxidation of carbon monoxide on titania supported gold nanoparticles. Angew Chem Int Ed 46: 3530–3532.

Hayden BE, Pletcher D, Suchsland J-P, Williams LJ. 2009. The influence of Pt particle size on the surface oxidation of platinum—Part 1: Reduced titania support. Phys Chem Chem Phys. DOI: 10.1039/b817553e.

Henry CR, Chapon C, Giorgio S, Goyhenex C. 1997. Size effects in heterogeneous catalysis. In: Lambert RM, Pacchioni G, editors. Chemisorption and Reactivity on Supported Clusters and Thin Films. Dordrecht, Kluwer Academic. pp. 117–152.

Hernández J, Solla-Gullón J, Herrero E. 2004. Gold nanoparticles synthesized in a water-in-oil microemulsion: Electrochemical characterization and effect of the surface structure on the oxygen reduction reaction. J Electroanal Chem 574: 185–196.

Hernández J, Solla-Gullón J, Herrero E, et al. 2005. Characterization of the surface structure of gold nanoparticles and nanorods using structure sensitive reactions. J Phys Chem B 109: 12651–12654.

Herrero E, Feliu JM, Blais S, et al. 2000. Temperature dependence of CO chemisorption and its oxidative desorption on the Pt(111) electrode. Langmuir 16: 4779–4783.

Hsueh KL, Gonzalez ER, Srinivasan S. 1983. Electrolyte effects on oxygen reduction kinetics at platinum—A rotating-ring disk electrode analysis. Electrochim Acta 28: 691–697.

Kim J, Gewirth A. 2006. Mechanism of oxygen electroreduction on gold surfaces in basic media. J Phys Chem B 110: 2565–2571.

Kinoshita K. 1990. Particle-size effects for oxygen reduction on highly dispersed platinum in acid electrolytes. J Electrochem Soc 137: 845–848.

Kita H, Nakajima H, Hayashi K. 1985. Electrochemical oxidation of CO on Au in alkaline solution. J Electroanal Chem 190: 141–156.

Komanicky V, Menzel A, You H. 2005. Investigation of oxygen reduction kinetics at (111)–(100) nanofaceted platinum surfaces in acidic media. J Phys Chem B 109: 23550–23557.

Kunimatsu K, Aramata A, Nakajima N, Kita H. 1986. Infrared spectra of carbon monoxide adsorbed on a smooth gold electrode: Part II. EMIRS and polarization-modulated IRRAS study of the adsorbed CO layer in acidic and alkaline solutions. J Electroanal Chem 207: 293–307.

Lemire C, Meyer R, Shaikhutdinov S, Freund HJ. 2004. Do quantum size effects control CO adsorption on gold nanoparticles? Angew Chem Int Ed 43: 118–121.

Li Q, He R, Gao J-A, et al. 2003. The CO poisoning effect in PEMFCs operational at temperatures up to 200 °C. J Electrochem Soc 150: A1599–A1605.

Liu Z-P, Gong X-Q, Kohanoff J, et al. 2003. Catalytic role of metal oxides in gold-based catalysts: A first principles study of CO oxidation on TiO_2 supported Au. Phys Rev Lett 91: 266102.

Lopez N, Nørskov JK. 2002. Catalytic CO oxidation by a gold nanoparticle: A density functional study. J Am Chem Soc 124: 11262–11263.

Lopez N, Janssens TVW, Clausen BS, et al. 2004a. On the origin of the catalytic activity of gold nanoparticles for low-temperature CO oxidation. J Catal 223: 232–235.

Lopez N, Nørskov JK, Janssens TVW, et al. 2004b. The adhesion and shape of nanosized Au particles in a Au/TiO_2 catalyst. J Catal 225: 86–94.

McCallum C, Pletcher D. 1976. An investigation of the mechanism of the oxidation of carbon monoxide adsorbed onto a smooth Pt electrode in aqueous acid. J Electroanal Chem 70: 277–290.

Macía MD, Campiña JM, Feliu JM. 2004. On the kinetics of oxygen reduction on platinum stepped surfaces in acidic media. J Electroanal Chem 564: 141–150.

Maillard F, Eikerling M, Cherstiouk OV, et al. 2004a. Size effects on reactivity of Pt nanoparticles in CO monolayer oxidation: The role of surface mobility. Faraday Discuss 125: 357–377.

Maillard F, Savinova ER, Simonov PA, et al. 2004b. Infrared spectroscopic study of CO adsorption and electro-oxidation on carbon-supported Pt nanoparticles: Interparticle versus intraparticle heterogeneity. J Phys Chem B 108: 17893–17904.

Maillard F, Schreier S, Hanzlik M, et al. 2005. Influence of particle agglomeration on the catalytic activity of carbon-supported Pt nanoparticles in CO monolayer oxidation. Phys Chem Chem Phys 7: 385–393.

Maillard F, Savinova ER, Stimming U. 2007. CO monolayer oxidation on Pt nanoparticles: Further insights into the particle size effects. J Electroanal Chem 599: 221–232.

Markovic NM, Ross PN. 2002. Surface science studies of model fuel cell electrocatalysts. Surf Sci Rep 45: 121–229.

Markovic NM, Adzic RR, Vesovic VB. 1984. Structural effects in electrocatalysis: Oxygen reduction on the gold single crystal electrodes with (110) and (111) orientations. J Electroanal Chem 165: 121–133.

Markovic NM, Tidswell IM, Ross PN. 1994. Oxygen and hydrogen peroxide reduction on the gold(100) surface in alkaline electrolyte: the roles of surface structure and hydroxide adsorption. Langmuir 10: 1–4.

Markovic NM, Gasteiger HA, Ross PN. 1995. Oxygen reduction on platinum low-index single-crystal surfaces in sulphuric acid solution: Rotating Ring–Pt(*hkl*) Disk Studies. J Phys Chem 99: 3411–3415.

Markovic N, Gasteiger H, Ross PN. 1997. Kinetics of oxygen reduction on Pt(*hkl*) electrodes: Implications for the crystallite size effect with supported Pt electrocatalysts. J Electrochem Soc 144: 1591–1597.

Mavrikakis M, Stoltze P, Nørskov JK. 2000. Making gold less noble. Catal Lett 64: 101–106.

Maye MM, Kariuiki NN, et al. 2004. Electrocatalytic reduction of oxygen: Gold and gold–platinum nanoparticle catalysts prepared by two-phase protocol. Gold Bull 37(3–4): 217–223.

Mayrhofer KJJ, Arenz M, Blizanac BB, et al. 2005a. CO surface electrochemistry on Pt-nanoparticles: A selective review. Electrochim Acta 50: 5144–5154.

Mayrhofer KJJ, Blizanac BB, Arenz M, et al. 2005b. The impact of geometric and surface electronic properties of Pt-catalysts on the particle size effect in electrocatalysis. J Phys Chem B 109: 14433–14440.

Meier DC, Goodman DW. 2004. The influence of metal cluster size on adsorption energies: CO adsorbed on Au clusters supported on TiO_2. J Am Chem Soc 126: 1892–1899.

Molina LM, Rasmussen MD, Hammer B. 2004. Adsorption of O_2 and oxidation of CO at Au nanoparticles supported by $TiO_2(110)$. J Chem Phys 120: 7673–7680.

Morris ND, Mallouk TE. 2002. A high-throughput optical screening method for the optimization of colloidal water oxidation catalysts. J Am Chem Soc 124: 11114–11121.

Nørskov JK, Rossmeisl J, Logadottir A, et al. 2004. Origin of the overpotential for oxygen reduction at a fuel-cell cathode. J Phys Chem B 108: 17886–17892.

Okazawa T, Kohyama M, Kido Y. 2006. Electronic properties of Au nano-particles supported on stoichiometric and reduced $TiO_2(110)$ substrates. Surf Sci 600: 4430–4437.

Park ED, Lee JS. 1999. Effects of pretreatment conditions on CO oxidation over supported Au catalysts. J Catal 186: 1–11.

Prieto A, Hernández J, Herrero E, Feliu JM. 2003. The role of anions in oxygen reduction in neutral and basic media on gold single-crystal electrodes. J Solid State Electrochem 7: 599–606.

Reddington E, Sapienza A, Gurau B, et al. 1998. Combinatorial electrochemistry: A highly parallel, optical screening method for discovery of better electrocatalysts. Science 280: 1735–1737.

Sanchez A, Abbet S, Heiz U, et al. 1999. When gold is not noble: Nanoscale gold catalysts. J Phys Chem A 103: 9573–9578.

Schmidt TJ, Stamenkovic V, Arenz M, et al. 2002. Oxygen electrocatalysis in alkaline electrolyte: Pt(*hkl*), Au(*hkl*) and the effect of Pd-modification. Electrochim Acta 47: 3765–3776.

Shubina TE, Koper MTM. 2002. Quantum-chemical calculations of CO and OH interacting with bimetallic surfaces. Electrochim Acta 47: 3621–3628.

Somorjai GA, Rioux RM. 2005. High technology catalysts towards 100% selectivity: Fabrication, characterization and reaction studies. Catal Today 100: 201–215.

Spong AD, Vitins G, Guerin S, et al. 2003. Combinatorial arrays and parallel screening for positive electrode discovery. J Power Sources 119: 778–783.

Stonehart P. 1966. In: Proceedings of the Fifth International Symposium on Non-Mechanical Power Sources. Oxford: Pergamon Press, p. 509.

Stonehart P. 1994. The role of electrocatalysis in solid polymer electrolyte fuel cells. In: Drake JAG, editor. Electrochemistry and Clean Energy. Cambridge: The Royal Society of Chemistry.

Strbac S, Adzic RR. 1996. The influence of OH-chemisorption on the catalytic properties of gold single crystal surfaces for oxygen reduction in alkaline solutions. J Electroanal Chem 403: 169–181.

Strbac S, Adzic RR. 1996. The influence of pH on reaction pathways for O_2 reduction on the Au(100) face. Electrochim Acta 41: 2903–2908.

Sun SG, Cai WB, Wan LJ, Osawa M. 1999. Infrared absorption enhancement for CO adsorbed on Au films in perchloric acid solutions and effects of surface structure studied by cyclic voltammetry, scanning tunneling microscopy, and surface-enhanced IR spectroscopy. J Phys Chem B 103: 2460–2466.

Sun YP, Buck H, Mallouk TE. 2001. Combinatorial discovery of alloy electrocatalysts for amperometric glucose sensors. Anal Chem 73: 1599–1604.

Takasu Y, Ohashi N, Zhang XG, et al. 1996. Size effects of platinum particles on the electroreduction of oxygen. Electrochim Acta 41: 2595–2600.

Tsuda M, Kasai H. 2006. Ab initio study of alloying and straining effects on CO interaction with Pt. Phys Rev B 73: 155405.

Valden M, Lai X, et al. 1998. Onset of catalytic activity of gold clusters on titania with the appearance of nonmetallic properties. Science 281: 1647–1650.

Vassilev P, Koper MTM. 2007. Electrochemical reduction of oxygen on gold surfaces: A density functional theory study of intermediates reaction paths. J Phys Chem C 111: 2607–2613.

Wasmus S, Kuver A. 1999. Methanol oxidation and direct methanol fuel cells: A selective review. J Electroanal Chem 461: 14–31.

Watanabe M, Motoo S. 1975a. Electrocatalysis by ad-atoms. Part II. Enhancement of the oxidation of methanol on platinum by ruthenium ad-atoms. J Electroanal Chem 60: 267.

Watanabe M, Motoo S. 1975b. Electrocatalysis by ad-atoms. Part III. Enhancement of the oxidation of carbon monoxide on platinum by ruthenium ad-atoms. J Electroanal Chem 60: 275–283.

Watanabe M, Sei H, et al. 1989. The influence of platinum crystallite size on the electroreduction of oxygen. J Electroanal Chem 261: 375–387.

Xu Y, Mavrikakis M. 2003. Adsorption and dissociation of O_2 on gold surfaces: Effect of steps and strain. J Phys Chem B 107: 9298–9307.

Yagi I, Ishida T, et al. 2004. Electrocatalytic reduction of oxygen to water at Au nanoclusters vacuum-evaporated on boron-doped diamond in acidic solution. Electrochem Commun 6: 773–779.

Yano H, Inukai J, Uchida H, et al. 2006. Particle-size effect of nanoscale platinum catalysts in oxygen reduction reaction: An electrochemical and ^{195}Pt EC-NMR study. Phys Chem Chem Phys 8(42): 4932–4939.

Zhang Y, Suryanarayanan V, et al. 2004. Electrochemical behaviour of Au nanoparticle deposited on as-grown and O-terminated diamond electrodes for oxygen reduction in alkaline solution. Electrochim Acta 2004: 5235–5240.

CHAPTER 17

Electrocatalysis for Fuel Cells at Enzyme-Modified Electrodes

K. A. VINCENT

Inorganic Chemistry Laboratory, University of Oxford, South Parks Road, Oxford OX1 3QR, UK

S. C. BARTON

Department of Chemical Engineering and Materials Science, Michigan State University, East Lansing, MI 48824, USA

G. W. CANTERS

Leiden Institute of Chemistry, Leiden University, Einsteinweg 55, 2333 CC Leiden, The Netherlands

H. A. HEERING

Kavli Institute of NanoScience, Delft University of Technology, Lorentzweg 1, 2628, CJ Delft, The Netherlands; and Leiden Institute of Chemistry

17.1 INTRODUCTION

A number of reactions relevant to fuel cell catalysis are carried out under ambient conditions by enzymes that employ catalytic centers comprising common metals or organic cofactors (Fig. 17.1). One example is the efficient catalytic reduction of O_2 to H_2O at copper centers in the fungal enzyme laccase. Another is the rapid and reversible oxidation or production of H_2 by microbial hydrogenase enzymes that incorporate catalytic centers comprising iron, or nickel and iron. These reactions are not limited to microorganisms living in extreme environments: on the contrary, most fungi use laccases to couple O_2 reduction to the oxidation of organic molecules, and hydrogen cycling is integral to communities of bacteria in the human gut, in soil, and in waterways. Variation in amino acid residues around the active site as well as further away in the protein provides scope for subtle alteration of the properties of the catalytic centers in enzymes, and, over millions of years, microorganisms have evolved to carry out energy cycling reactions selectively in complex environments. For example, a class

---

Fuel Cell Catalysis. Edited by Marc T. M. Koper
Copyright © 2009 John Wiley & Sons, Inc.

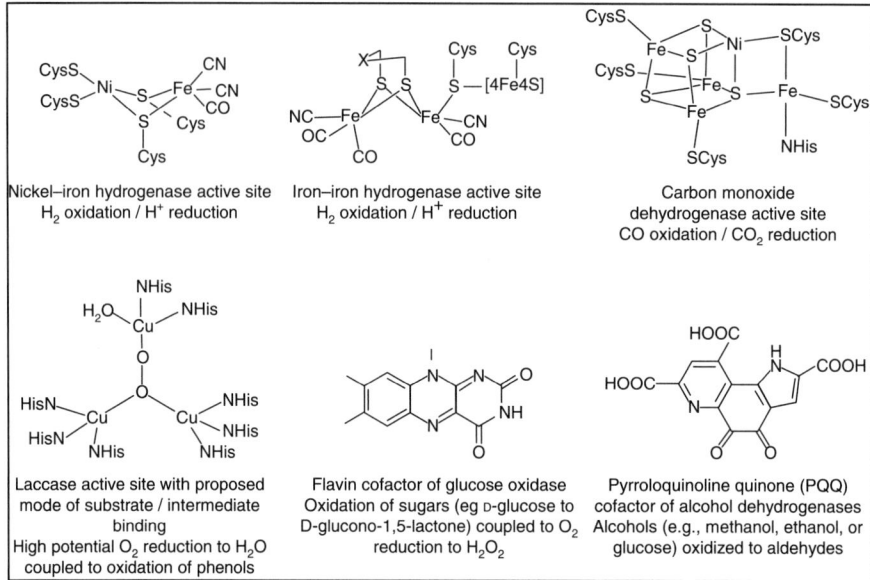

Figure 17.1 Active sites of some enzymes that carry out reactions relevant to fuel cell catalysis.

of bacteria that live on trace H_2 in aerobic conditions expresses hydrogenases that are able to oxidize low levels of H_2 in air (even at high levels of carbon monoxide), and the hydrogenases from sulfate-reducing bacteria are tolerant to sulfide over a wide potential window [Vincent et al., 2007]. The remarkable substrate specificity of many enzymes forms the basis for their application in novel fuel cell devices that would not be possible with conventional metal catalysts. Other interesting examples of enzyme catalysis do indeed arise from organisms found in unusual environments, such as the reversible interconversion of CO and CO_2 at an iron- and nickel-containing cluster in carbon monoxide dehydrogenase [Parkin et al., 2007], or H_2 oxidation at 100 °C by hydrogenases from thermophilic microorganisms.

Enzymes that catalyze redox reactions are usually large molecules (molecular mass typically in the range 30–300 kDa), and the effects of the protein environment distant from the active site are not always well understood. However, the structures and reactions occurring at their active sites can be characterized by a combination of spectroscopic methods, X-ray crystallography, transient and steady-state solution kinetics, and electrochemistry. Catalytic states of enzyme active sites are usually better defined than active sites on metal surfaces.

The large size of redox enzymes means that diffusion to an electrode surface will be prohibitively slow, and, for enzyme in solution, an electrochemical response is usually only observed if small, soluble electron transfer mediator molecules are added. In this chapter, discussion is limited to examples in which the enzyme of interest is attached to the electrode surface. Electrochemical experiments on enzymes can be very simple, involving direct adsorption of the protein onto a carbon or modified metal surface from dilute solution. Protein film voltammetry, a method in which a film of enzyme in direct electronic contact with an electrode surface is probed electrochemically under turnover or nonturnover conditions, is increasingly providing important information about enzyme reactions, and leads directly into technological applications of redox enzymes [Léger et al., 2003; Vincent et al., 2007]. In many cases, electron transfer between the adsorbed enzyme and the electrode surface is very fast, meaning that the electrocatalytic current response reports directly on the kinetics of processes within the enzyme. In other cases, strategies for "wiring" redox centers in the protein to the electrode are employed. The use of an electrode modified with a conducting polymer in which redox-active groups mediate transfer of electrons allows multiple layers of enzyme molecules to be addressed, increasing the current response. This chapter describes methods for attachment of enzymes to electrodes, examples of enzyme electrocatalysis of reactions that are relevant to fuel cells, and the ways in which enzymes have been utilized in energy cycling devices.

17.1.1 Relationship to Energy Harnessing in Biology

In the ubiquitous bacterium *Escherichia coli*, oxidation of fuels (including succinate, formate, glycerol-3-phosphate, lactate, and H_2) catalyzed by a series of enzymes facing the periplasm is coupled to reduction of a range of electron acceptors in the cytoplasm (e.g., fumarate, nitrate, nitrite, or O_2) via a pool of electron-carrier quinols that are soluble in the lipid bilayer of the cytoplasmic membrane. In *E. coli*, H_2 oxidation occurs only under anaerobic growth conditions, but in certain strictly aerobic bacteria such as *Ralstonia* species, oxidation of H_2 is coupled to O_2 reduction (similar to the situation in an H_2/O_2 fuel cell; Fig. 17.2) [Burgdorf et al., 2005]. The energy released during the biological reactions is stored in the form of a proton gradient across the membrane, generated by uptake of H^+ from the cytoplasmic side of the membrane and H^+ release on the periplasmic side. (This is equivalent to using a fuel cell to charge a capacitor.) Controlled diffusion of protons back through the large membrane-spanning enzyme ATP synthase drives the phosphorylation of adenosine diphosphate (ADP) to generate adenosine triphosphate (ATP), and hydrolysis of ATP can drive chemical reactions elsewhere in the cell [Nicholls and Ferguson, 2002].

Microbes grow on a wide range of energy sources (e.g., starch, glucose, lactate, methanol, glycerol, and H_2), meaning that bioderived catalysts for oxidation of a range of fuels (including biomass and sewerage waste) are potentially available. Here, we focus mainly on biological catalysts for oxidation of H_2, sugars, and other alcohols, and for reduction of O_2 (simply provided from air), since these reactions are directly relevant to fuel cell electrocatalysis. Enzymes usually catalyze very specific reactions; for example, laccase cleanly catalyzes the four-electron reduction of O_2 to H_2O without release of peroxide or superoxide. Methanol is oxidized only to formate by alcohol dehydrogenase, but, with a series of enzymes operating in series, it is possible to carry out a cascade of oxidation reactions to form CO_2 [Yue and Lowther, 1986; Palmore et al., 1998].

17.1.2 Enzymes in Fuel Cell Electrocatalysis

Motivations for exploring enzymes in fuel cell catalysis are both intellectual and applied. Using enzymes as electrocatalysts, there is scope for creation of fuel cells

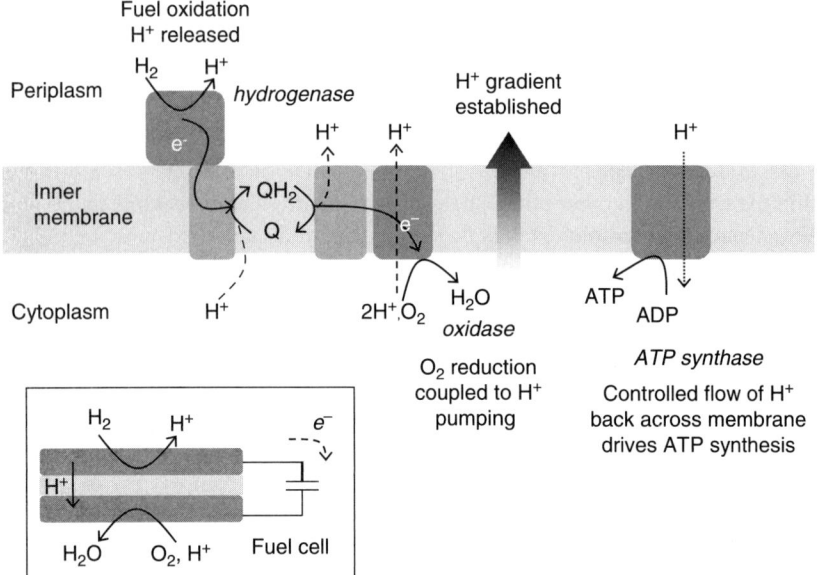

Figure 17.2 Bacterial cells harness energy from fuel oxidation coupled to oxidant reduction in the form of a proton circuit. In aerobic bacteria such as *Ralstonia* species, H_2 oxidation in the periplasmic compartment of the cell is coupled to O_2 reduction in the cytoplasm. Electrons are transferred across the membrane by lipid-soluble electron-carrier quinol/quinone molecules, and electron transfer is coupled to uptake of protons from the cytoplasm and their release into the periplasm. This leads to a proton gradient across the membrane, analogous to charging a capacitor in a fuel cell circuit (inset). A quinol oxidase makes a further contribution to the proton gradient by pumping protons across the membrane as O_2 is reduced. The "load" on this proton circuit is the transmembrane enzyme ATP synthase, which uses the proton gradient to drive phosphorylation of ADP to generate ATP; hydrolysis of ATP elsewhere in the cell then provides the energy to drive chemical reactions [Nicholls and Ferguson, 2002].

that operate under special conditions where traditional fuel cell catalysts fail. These applications rely upon the high selectivity of enzymes for a substrate or group of substrates, and the enormous variety of enzymes available in biology for specific catalysis of a wide range of different reactions. Examples include devices that operate on biologically derived fuels (e.g., sugars, glycerol, biomass, and sewage sludge), contaminated fuels, or mixtures of fuel and oxidant. Fuel cells running on glucose are particularly well developed. Research into enzyme electrocatalysis of O_2 reduction and H_2 or methanol oxidation for fuel cells has been driven by a desire for low cost, renewable alternatives to precious metal catalysts. The high turnover rates and impressive selectivity of enzyme electrocatalysis have stimulated the development of synthetic catalysts that mimic the structures of enzyme active sites (see, e.g., Liu et al. [2005]; Mahadevan et al. [2000]). Studies of enzyme electrocatalysis have provided extensive information on reactions and mechanisms of redox enzymes. Finally, application of enzymes in working fuel cells is providing information on the ways in which biological catalysts function in demanding environments and the way that pairs of redox enzymes function together as they do in biology.

As we demonstrate in this chapter, enzymes can be extremely active electrocatalysts at ambient temperatures and mild pH, and have significantly higher reaction selectivity than precious metals. The main disadvantage in applying redox enzymes for electrocatalysis arises from their large size, which means that the catalytic active site density is low. Enzymes also have a relatively short lifetime (usually not more than a few months), making them more suited to disposable applications.

Enzymes such as glucose oxidase, alcohol dehydrogenases, and laccase are commercially available in reasonably pure form. Other enzymes, such as hydrogenase, are technically difficult to isolate [Cammack et al., 2001]. On a laboratory scale, biocatalysts are still expensive (probably more so than noble metals) owing to the cost of labor for small scale production. However, microorganisms can be grown on cheap carbon/nitrogen sources, and costs decrease dramatically for large scale production. Increasingly it is becoming possible to genetically manipulate readily grown organisms such as *E. coli* or yeasts to express the enzymes of other organisms in high yield and with affinity tags genetically attached for one-step purification. Importantly, biocatalysts are indefinitely renewable.

The protein environment around the buried active site in an enzyme must provide routes for electron transfer to or from the surface and for mass transport of substrate and product. These features are highlighted on the structure of laccase from the white rot fungus, *Trametes versicolor*, in Fig. 17.3 (Plate 17.1) [Bertrand et al., 2002]. Fungal laccases are involved in pigment formation, detoxification, and lignin degradation, using O_2 as an electron acceptor in the oxidation of a range of organic substrates [Solomon et al., 1996]. The presence of a co-crystallized 2,5-xylidene molecule in a wide, hydrophobic pocket at the surface of *T. versicolor* laccase provides strong evidence that this is the binding site for organic substrates. The pocket is close to the mononuclear "blue" Cu center, and thus provides a route for sequential transfer of electrons to the site of O_2 reduction, the trinuclear Cu cluster located approximately 1.3 nm away. This distance is sufficiently short to allow fast electron tunneling, and is typical of the spacing between electron transfer relay centers in proteins [Page et al., 1999]. The redox centers are buried in the protein, ensuring that the environment of the metals is precisely tuned by the amino acids that ligate or surround them. Two solvent accessible channels leading to the trinuclear center are evident from the presence of water molecules with well-defined electron density in the crystal structure, and probably provide routes for O_2 entry [Piontek et al., 2002].

Important inherent characteristics of an enzyme that should be considered are the substrate affinity, characterized by the Michaelis constant K_M, the rate of turnover k_{cat}, providing the catalytic efficiency k_{cat}/K_M, and the catalytic potential. Several attempts to compare enzyme catalysis with that of platinum have been published. Direct comparisons are difficult, because enzyme electrodes must be operated in aqueous electrolyte containing dissolved substrate, whereas precious metal electrodes are often supplied with a humidified gaseous stream of fuel or oxidant, and produce water as steam. It is not straightforward to compare true optimal turnover rates per active site, as it is often unclear how many active sites are being engaged in a film of enzyme on an electrode.

598   ELECTROCATALYSIS FOR FUEL CELLS AT ENZYME-MODIFIED ELECTRODES

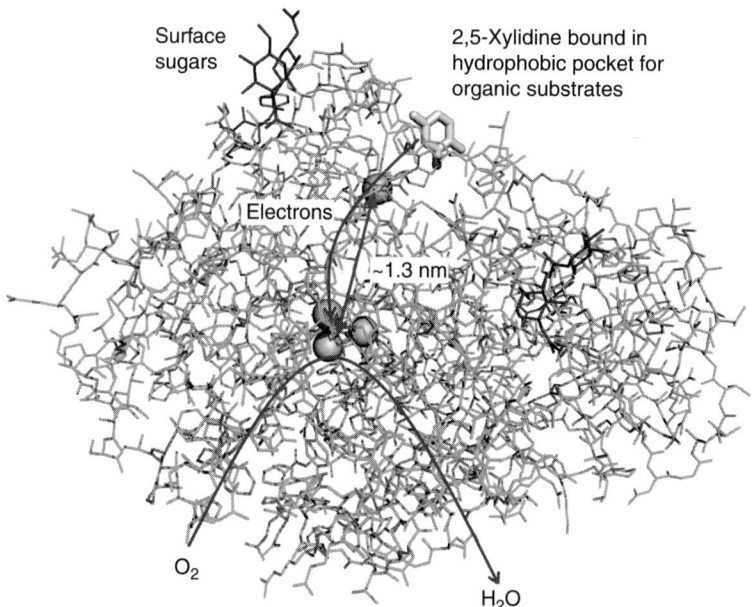

Figure 17.3 Anatomy of a redox enzyme: representation of the X-ray crystallographic structure of *Trametes versicolor* laccase III (PDB file 1KYA) [Bertrand et al., 2002]. The protein is represented in green lines and the Cu atoms are shown as gold spheres. Sugar moieties attached to the surface of the protein are shown in red. A molecule of 2,5-xylidine that co-crystallized with the protein (shown in stick form in elemental colors) is thought to occupy the broad-specificity hydrophobic binding pocket where organic substrates are oxidized by the enzyme. Electrons from substrate oxidation are passed to the mononuclear "blue" Cu center and then to the trinuclear Cu active site where O_2 is reduced to H_2O. (See color insert.)

A paper by Heller and co-workers compares the electrocatalytic activity of "blue" copper oxidases and Pt. Specifically, they have compared the overpotentials required to reduce O_2 at 37 °C at a Pt fiber electrode operating in 0.5 M H_2SO_4 and an electrode involving laccase linked into a hydrogel containing Os redox centers (see Section 17.2.1.1 for details of this assembly) and operating at pH 5 [Soukharev et al., 2004]. Exact comparison is complicated by pH effects, but the laccase electrode operates at a lower overpotential. Hydrogenases, like Pt, cycle H_2 with negligible overpotential [Vincent et al., 2007]. As with laccase, it is difficult to gauge the absolute activity, because the electroactive site density is rarely known. A rotating disk electrode (RDE) modified with a nickel–iron hydrogenase has been shown to oxidize H_2 at the same diffusion-controlled rate as a platinized electrode when both electrodes are operated at 45 °C, 1 bar H_2, pH 7 and 0.242 V versus SHE [Jones et al., 2002]. However, limiting interfacial electron transfer arising from poor coupling of some enzyme molecules to the electrode meant that the platinized electrode was superior at lower applied potentials. Other enzymes also exhibit electrocatalysis at negligible overpotentials. For example, carbon monoxide dehydrogenase provides the only known example of "reversible" electrocatalytic CO_2/CO cycling [Parkin et al., 2007], and Complex I (NADH dehydrogenase) on an electrode provides the only known example of efficient $NAD^+/NADH$ interconversion [Barker et al., 2007].

The laccase molecule represented in Fig. 17.3 has a diameter of about 5 nm, giving it a footprint of about 20 nm^2: this means that the maximum coverage supported on a flat electrode surface would be about 8 pmol cm^{-2}, i.e., the surface density of active sites is very low. A key implication is that in order to obtain high current densities (e.g., greater than 10 mA cm^{-2}) at enzyme electrodes, the effective surface area of the electrode must be large and/or the coating of enzyme must be fairly thick. Increasing thickness brings additional challenges, including the need to move electrons, substrate, and product through multiple layers. Strategies for "wiring" multiple layers of enzyme at an electrode are discussed in Sections 17.2 and 17.3.

Many of the methods discussed in this chapter are amenable to other applications besides fuel cells. For example, the effort invested in enzymatic fuel cells is dwarfed by efforts to develop enzymatic electrochemical sensors for species such as glucose, methanol, and ethanol using very similar technologies. Less well developed, but also with significant potential value, is the study of chemical conversions by enzyme electrocatalysis, for example selective oxidation of a single enantiomer of a secondary alcohol by an alcohol dehydrogenase [Kroutil et al., 2004]. Furthermore, the investigation of enzyme catalysis on electrodes has provided much insight into the potential-dependent reactions of enzymes that would be difficult to control and study for enzymes in solution.

Two key features distinguish enzymatic fuel cells from sensors and highlight the challenges of implementing enzymes in power-producing devices. First, it is generally desirable to minimize current draw in a sensor, whereas in a fuel cell it is desirable to maximize both cell current and voltage under operating conditions. The demand for high cell voltage is not difficult to meet, because enzymes are available to catalyze the oxidation of H_2, sugars, or other alcohols and the reduction of O_2 at fairly low overpotentials. Achieving high currents in enzyme fuel cells is a much more difficult problem: although enzyme turnover rates per active site can be very high, their large size means that the turnover rate per unit volume is low compared with precious metals. Trapping enzymes within a redox-active hydrogel or expanding a solid conducting support into three-dimensional space are strategies that have been exploited to address this problem.

Second, sensors are often intended for a single use, or for usage over periods of one week or less, and enzymes are capable of excellent performance over these time scales, provided that they are maintained in a mild environment at moderate temperature and with minimal physical stress. Stabilization of enzymes on conducting surfaces over longer periods of time presents a considerable challenge, since enzymes may be subject to denaturation or inactivation. In addition, the need to feed reactants to the biofuel cell means that convection and therefore viscous shear are often present in working fuel cells. Application of shear to a soft material such as a protein-based film can lead to accelerated degradation due to shear stress [Binyamin and Heller, 1999]. However, enzymes on surfaces have been demonstrated to be stable for several months (see below).

Electrocatalysis by redox enzymes has now been demonstrated at the anode and/or cathode of a variety of small scale proof-of-concept fuel cells. It remains to be seen whether redox enzymes can be produced at low cost and can be sufficiently robust to enable their widespread use in fuel cell technologies. Enzymes almost always require an aqueous environment to function, and generally will not tolerate extremes of temperature or pH, so it is unlikely that they will compete with precious metal catalysts in high power output devices, for which elevated operating temperatures are desirable. (In some cases, ambient biocatalysis of reactions that otherwise require high temperatures can be an advantage, for example reversible interconversion of CO and CO_2 at an electrode modified with carbon monoxide dehydrogenase [Parkin et al., 2007].) However, the specificity of enzyme catalysis may open up new concepts and designs for fuel cells, leading to specialized applications, such as biologically implantable power sources [Barton et al., 2004], disposable fuel cells where limited lifetime is not a problem, or fuel cells that function on contaminated fuels [Vincent et al., 2005, 2006]. Whether or not enzymes find applications in real energy cycling devices, their extremely high specificity, discussed in detail in this chapter, provides a benchmark for desirable properties of fuel cell catalysts, and should provide inspiration for the development of new catalysts for energy cycling.

17.1.3 Strategies for Exploiting Enzyme Electrocatalysis

17.1.3.1 Enzyme Immobilization and Electron Transfer For fuel cell applications, it is highly advantageous to immobilize the enzymes at the respective anode or cathode in order to facilitate efficient electron transfer and to avoid washing out the biocatalyst if the fuel is supplied in liquid form. Redox centers in the enzymes must be in close electronic contact with an electronically conducting support. This may be direct, or may involve co-immobilized electron transfer mediator molecules that are able to exchange electrons with redox centers buried in the protein or centers far from the electrode. Earlier work used soluble mediators (see, e.g., Yahiro et al. [1964]; Davis et al. [1983]; Yue and Lowther [1986]; Palmore et al. [1999]). This unnecessarily complicates cell design: a single compartment fuel cell is possible with a pair of immobilized enzymes that are selective for fuel and oxidant, but if electron transfer mediators are present in the electrolyte solution, the anode and cathode must be in separate compartments to prevent cross reaction of the mediators.

The choice of immobilization strategy obviously depends on the enzyme, electrode surface, and fuel properties, and on whether a mediator is required, and a wide range of strategies have been employed. Some general examples are represented in Fig. 17.4. Key goals are to stabilize the enzyme under fuel cell operating conditions and to optimize both electron transfer and the efficiency of fuel/oxidant mass transport. Here, we highlight a few approaches that have been particularly useful in electrocatalysis directed towards fuel cell applications.

Achieving fast electron transfer to enzyme active sites need not be complicated. As mentioned above, many redox enzymes incorporate a relay of electron transfer centers that facilitate fast electron transfer between the protein surface and the buried active site. These may be iron–sulfur clusters, heme porphyrin centers, or mononuclear

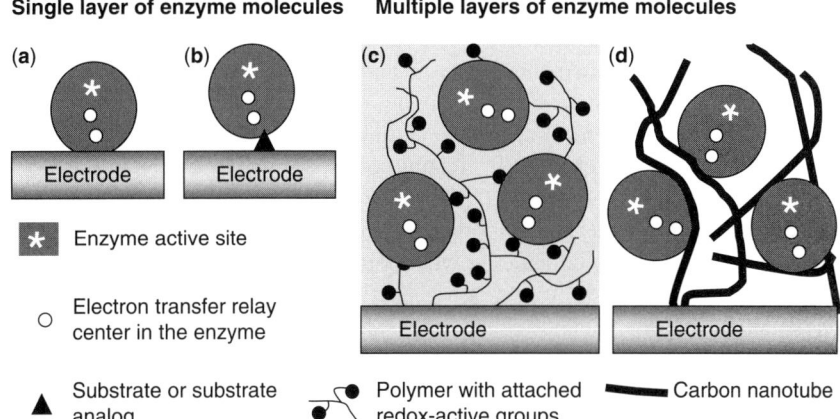

Figure 17.4 Cartoon representation of strategies for studying and exploiting enzymes on electrodes that have been used in electrocatalysis for fuel cells. (a) Attachment or physisorption of an enzyme on an electrode such that redox centers in the protein are in direct electronic contact with the surface. (b) Specific attachment of an enzyme to an electrode modified with a substrate, cofactor, or analog that contacts the protein close to a redox center. Examples include attachment of the modifier via a conductive linker. (c) Entrapment of an enzyme within a polymer containing redox mediator molecules that transfer electrons to/from centers in the protein. (d) Attachment of an enzyme onto carbon nanotubes prepared on an electrode, giving a large surface area conducting network with direct electron transfer to each enzyme molecule.

Cu centers, for example. Provided that the relay center closest to the surface of the protein makes good contact with the electrode (less than about 1.2 nm), so that interfacial electron transfer is fast, the electrocatalytic current measured will generally reflect the rate of turnover at the active site. In many cases, direct adsorption of the enzyme onto a pyrolytic graphite edge (PGE) electrode or a gold electrode modified with an alkanethiolate monolayer provides a highly electroactive film (Fig. 17.4a). Physisorption relies upon noncovalent interactions: hydrogen bonding, electrostatics, hydrophobic bonding, and van der Waals forces. Studies of enzymes in this configuration are termed protein film voltammetry [Léger et al., 2003]. This strategy has been particularly useful in studies and exploitation of H_2 oxidation and evolution by hydrogenases, which comprise an electron relay chain of iron–sulfur clusters leading from the surface to the deeply buried active site (see Section 17.3.2) [Vincent et al., 2007]. In this configuration, all enzyme molecules are exposed to the electrolyte solution, ensuring optimal accessibility of fuel or oxidant to the active site. Disadvantages of this approach are the low density of catalytic sites achieved, due to the large size of redox enzymes, and lack of control over the orientation of an enzyme, meaning that some molecules are incorrectly positioned for fast electron transfer. In some cases, addition of a charged co-adsorbate helps to stabilize an adsorbed enzyme or to favor correct orientation (for specific examples, see Sections 17.2 and 17.3).

Covalent chemical attachment of an enzyme to an electrode or a polymer coating requires mild and specific chemistry to react with amino acids in the protein in aqueous solution at near-neutral pH and ambient temperature. Examples include a direct covalent bond between cysteine thiols and metals such as gold and silver (see, e.g., Hasan et al. [2006]) and direct amide bonding between activated surface carboxylates or hydroxides on an electrode and lysine amines on the surface of an enzyme (see, e.g., Cardosi [1994]) or between amine functionalities introduced onto an electrode surface and carboxylic acid side chains (of glutamic or aspartic acid) on the surface of an enzyme [Alonso-Lomillo et al., 2007]. In addition to covalent coupling approaches, metal coordination chemistry can sometimes be exploited; for example, a protein with a genetically introduced His_6 tag will coordinate Ni^{2+} introduced onto an electrode surface by chelation to nitrilotriacetic acid (NTA) [Mayer et al., 2005]. Reconstitution of the ligand sphere at a metal site within a protein by a coordinating head group of a linker provides another strategy for protein attachment. For example, Canters and co-workers generated a genetic mutant of azurin, a copper electron transfer protein, that lacked a histidine ligand to the metal and were able to attach an imidazole-capped linker to this site [de Jongh et al., 2006]. Specific examples of coupling strategies used in enzyme fuel cell electrocatalysis are discussed in Sections 17.2 and 17.3.

Covalent attachment of enzymes to surfaces is often intuitively perceived as being more reliable than direct adsorption, but multisite physical interactions can in fact yield a comparably strong and stable union, as demonstrated by several biological examples. The biotin/streptavidin interaction requires a force of about 0.3 nN to be severed [Lee et al., 2007], and protein/protein interactions typically require 0.1 nN to break, but values over 1 nN have also been reported [Weisel et al., 2003]. These forces are comparable to those required to rupture weaker chemical bonds such as the gold–thiolate bond (1 nN for an alkanethiol, and even only 0.3 nN for a 1,3-alkanedithiol [Langry et al., 2005]) and the poly(His)–Ni(NTA) bond (0.24 nN, [Lévy and Maaloum, 2005]).

Immobilized substrates, cofactors, or their analogs have been used to orientate, anchor, and electronically couple enzymes to conducting surfaces in an approach that has been developed extensively by Katz, Willner, and co-workers. Examples include attachment of apo-glucose oxidase (lacking the flavin adenine dinucleotide (FAD) cofactor) to a gold surface modified with FAD attached via a pyrroloquinoline quinone (PQQ) linker (reviewed in Katz et al. [2003]) and attachment of apo-glucose dehydrogenase (lacking a PQQ cofactor) to gold nanoparticles via a PQQ-capped linker [Zayats et al., 2005]. These are described in further detail in Section 17.3.1. More recently, a substrate analog linker approach was applied to fungal laccase. A 2-aminoanthracene modifier on the electrode surface was found to stabilize laccase films, probably by interacting with the hydrophobic surface pocket where organic substrates are oxidized by the enzyme in its native environment (see Section 17.2.1 for further details) [Blanford et al., 2007].

In some cases, small biological redox partner proteins such as heme-containing cytochromes, ferredoxins comprising an iron–sulfur cluster, or azurin with a mononuclear Cu site have been used as natural mediators to facilitate fast electron exchange with enzymes. A specific surface site on the redox protein often complements a region on the enzyme surface, and enables selective docking with a short electron tunneling pathway. Examples of electrocatalysis of reactions relevant to fuel cells using this approach include H_2O_2 reduction at a gold electrode modified with cytochrome *c* followed by cytochrome *c* peroxidase [Heering et al., 2004] and O_2 reduction at gold modified with cytochrome *c* and cytochrome *c* oxidase [Katz et al., 1999].

Entrapment and multilayer adsorption of enzymes have been employed to increase current density, and, in these cases, mobile mediators are usually necessary to extend electron transfer beyond the electrode surface (represented in Fig. 17.4c). Heller and co-workers have tackled this problem by introducing mobile, tethered $Os^{III/II}$ redox groups into a hydrogel suitable for trapping a range of enzymes, including laccase, bilirubin oxidase, and glucose oxidase (see Sections 17.2.1 and 17.3.1) [Heller, 2006]. The Os centers cover a range of redox potentials depending on the ligands used, and with a large excess of the small, mobile mediator, enzyme active sites are engaged regardless of enzyme orientation with respect to the electrode. Alternative redox polymer approaches have also been reported. Morozov and co-workers have reported high electrocatalytic current densities for hydrogenases adsorbed on a carbon electrode modified with a polypyrrole polymer containing viologens as electron transfer mediators [Morozov et al., 2002]. Minteer and co-workers have made use of Nafion or chitosan to trap enzymes such as glucose oxidase and alcohol or aldehyde dehydrogenases in the presence of mediators such as $[Ru(2,2'\text{-bipyridyl})_3]^{2+}$ [Klotzbach et al., 2006]. For relatively thick layers of enzyme, the catalytic current is likely to become limited by substrate diffusion through the matrix or protein layer, and thin films on large surface area conducting supports are generally preferable.

17.1.3.2 The Electrode Surface: Materials and Modifications

It is widely believed that proteins denature on bare metal surfaces, although there are some reports of successful direct adsorption. Metals can be protected with self-assembling monolayers of organic molecules (usually long-chain alkanes or π-conducting systems) with thiol end groups [Chaki and Vijayamohanan, 2002; Fedurco, 2000]. The other end can carry methyl, carboxyl, or hydroxyl functionalities to create a hydrophobic, negatively charged, or hydrophilic surface, as required. A thin biocompatible environment can also be created by adsorption of peptides or immobilization of biological electron transfer partner proteins [Hasan et al., 2006; Heering et al., 2004; Katz et al., 1999]. Although such surfaces have proved useful for studying redox enzymes, in this chapter we focus on graphite as a cheap, readily available substrate for enzyme electrocatalysis, since this is more suitable for fuel cell electrodes.

At the microscopic level, the "edge" surface of pyrolytic graphite is rough, and nitrogen porosimetry measurements suggest that the "edge" surface area is 10^4 times greater than the geometric area [Blanford et al., 2007]. Current density can be increased by introducing further surface roughness, by adding nanoscopic structures such as carbon nanotubes (see Fig. 17.4d), or by using three-dimensional electrode lattices, carbon powders, or pastes. It may prove advantageous for protein immobilization to introduce structuring of the surface at a length scale comparable to that of the enzyme, i.e., 10–100 nm. Carbon electrodes are usually "naturally modified" with a multitude of C—O functionalities [Armstrong et al., 1987], the density of which depends on the chosen material and exposed crystal plane, and can also be controlled by pretreatments such as polishing, heating, and acid soaking. This yields a surface with apolar sites, polar groups, electroactive sites (quinol/quinone type), and, in particular, carboxylates, and these functional groups provide multiple docking sites for enzymes. The surface functionalities are also accessible to organic coupling chemistry, allowing tailoring of the surface properties and covalent immobilization of enzymes.

Owing to the presence of the many carboxylate groups, a pyrolytic graphite edge surface has an estimated pK of 5.6 [Armstrong et al., 1987]. The resulting overall negative charge at physiological pH is optimal for direct adsorption of positively charged proteins, and, in the presence of multivalent positive ions or positively charged co-adsorbates, PGE electrodes are also suitable for direct electrochemistry of negatively charged proteins [Hagen, 1989; Armstrong et al., 1987]. Although specific interactions (hydrogen bonding, salt bridging, hydrophobic, and coordinative) may play a role, these are not required, because multivalent counter-ions ($|Z| \geq 3$) tend to form a spatially correlated Stern layer that can overcompensate the original charge, resulting in charge inversion even at submillimolar concentrations [Shklovskii, 1999; Besteman et al., 2004].

17.2 ENZYMATIC CATHODES FOR O_2 REDUCTION

Ambient temperature catalysis of O_2 reduction at low overpotentials is a challenge in development of conventional proton exchange membrane fuel cells (polymer electrolyte membrane fuel cells, PEMFCs) [Ralph and Hogarth, 2002]. In this chapter, we discuss two classes of enzymes that catalyze the complete reduction of O_2 to H_2O: multi-copper oxidases and heme iron-containing quinol oxidases.

As discussed above, fuel oxidation in biology is coupled not only to respiratory O_2 reduction but also to reduction of a range of inorganic and organic oxidants, including fumarate, nitrate, nitrite, dimethylsulfoxide (DMSO), and peroxide. Direct electrocatalytic reduction of many of these substrates at enzyme-modified graphite electrodes has been reported (some examples are reviewed in Léger et al. [2003]). These electrodes would provide anaerobic alternatives to an O_2-reducing cathode system for an enzyme fuel cell, but the oxidants are significantly less convenient to supply than O_2 (air), and, with the exception of peroxide, give lower cell voltages. In this chapter, we therefore focus on enzymatic O_2 reduction as the cathode reaction for fuel cells.

17.2.1 Laccase and Bilirubin Oxidase

Laccase and bilirubin oxidase belong to a family of multi-copper oxidases that catalyze the four-electron reduction of O_2 directly to water (without releasing reactive oxygen intermediates) coupled to oxidation of an organic substrate [Solomon et al., 1996, 2001; Nakamura and Go, 2005]. Laccase and bilirubin oxidase both contain four Cu atoms (Fig. 17.5) (Plate 17.2). The mononuclear "blue" Cu center serves as the entry point for electrons donated by the organic substrate. In its oxidized state, this center is associated with a strong ligand-to-metal charge transfer (LMCT)

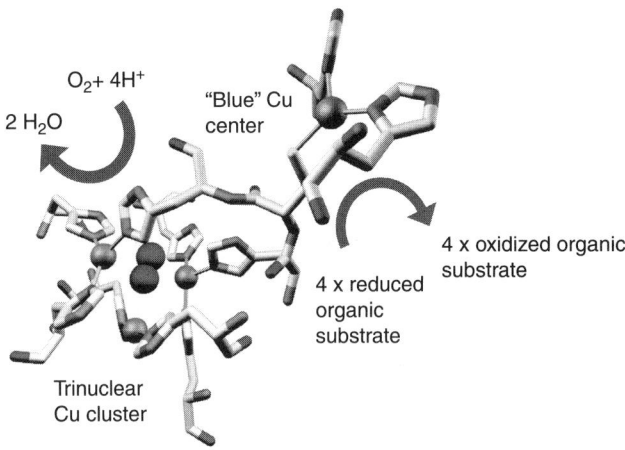

Figure 17.5 The protein environment around the Cu centers (gold spheres) of laccase from *Melanocarpus albomyces* (PDB file 1GW0) showing a substrate O_2 molecule bound in the trinuclear Cu site [Hakulinen et al., 2002]. The protein is depicted in stick representation with atoms in their conventional coloring. (Courtesy of Armand W. J. W. Tepper.) (See color insert.)

transition that gives rise to an electronic absorption ($\lambda_{max} \approx 600$ nm) and the strong blue color of the protein [Solomon et al., 1996]. Electrons move from this center through a His-Cys-His electron transfer pathway to a trinuclear Cu cluster located at a distance of about 1.3 nm. The substrate O_2 is bound at the trinuclear center in the crystallographic structure represented in Fig. 17.5. In the electrochemical approach, electrons for O_2 reduction are furnished by the electrode rather than by organic substrates, and the "blue" Cu center must be in close contact with the electrode surface.

Laccase, a well-studied multi-copper oxidase, couples O_2 reduction to the serial oxidation of four phenolic substrate molecules. In addition to applications in O_2 sensing and fuel cell catalysis, the ability of laccase to oxidize a wide range of organic molecules has led to widespread use of this enzyme in industrial catalysis, most notably in paper and textile bleaching and the bioremediation of wastewater [Couto and Herrera, 2006]. The usefulness of laccases as cathode catalysts for enzyme fuel cells depends on their catalytic efficiency k_{cat}/K_M and the midpoint potentials of the active sites. Laccases are categorized into two classes on the basis of the potentials of the "blue" Cu sites: for high potential laccases, the midpoint potential is about 680–790 mV, while for low-potential laccases, it is around 450–530 mV and limits the range of organic substrates that can be oxidized. The nature of the amino acid residue at the axial (fourth) position of the "blue" Cu center appears to strongly influence the reduction potential. Strong ligands favor Cu^{II} over Cu^{I}, whereas a weakly ligating residue (or no ligation at all) at this position destabilizes Cu^{II} with respect to Cu^{I} (consistent with the tendency of Cu^{I} to favor a three-coordination environment) and leads to an increase in the reduction potential [Solomon et al., 1996]. This reasoning is not sufficient, however, to explain the large difference in midpoint potentials between the high and low potential laccases. Replacing the axial nonligating phenylalanine with a methionine in laccase from *Trametes villosa*, for instance, results in a drop of the T1 midpoint potential of only 100 mV [Xu et al., 1999]. Clearly, other factors must come into play, such as solvent accessibility at the active site and charge distribution within the protein. This is an example of the role that subtle changes in the protein environment can play in exerting a strong influence over the catalytic properties of an enzyme active site. The reduction potentials of the other Cu sites in laccase are less well characterized.

Laccases usually have optimum activity under weakly acidic conditions (pH 4–5), whereas the optimum for bilirubin oxidase activity is close to neutral pH. Electrocatalytic O_2 reduction by laccases, as for Pt, is generally inhibited by chloride, and the enzymes typically retain only 40% activity at 100 mM NaCl [Barton et al., 2002]. In contrast, bilirubin oxidase retains substantial activity in the presence of chloride [Mano et al., 2003]. Bilirubin oxidase is therefore the more suitable enzyme for operation in most biological fluids. This enzyme is also tolerant to certain levels of alcohols, and has been used at the cathode of a membraneless ethanol/O_2 fuel cell operating at 1 mM ethanol in buffered aqueous solution [Topcagic and Minteer, 2006]. Laccase also has significant tolerance to alcohols: an RDE modified with laccase in an Os-containing redox hydrogel operating in 5 M methanol in aqueous buffer (pH 4) retained almost 80% of the activity recorded in methanol-free solution [Hudak and Barton, 2005]. A carbon paper cathode modified with laccase in the same redox hydrogel has been tested in a fuel cell in which either methanol or H_2 is oxidized at the Pt anode (separated from the cathode by a Nafion 1135 membrane). The cell polarization curves are almost identical for operation on 1 and 10 M methanol at the anode, indicating that the behavior of laccase is not significantly impaired by the increase in fuel crossover to the cathode compartment expected at the higher fuel concentration [Hudak and Barton, 2005].

17.2.1.1 Attachment Strategies It has been shown that a film of adsorbed laccase will exchange electrons directly with a PGE electrode, leading to electrocatalytic O_2 reduction, but the adsorbed film is very unstable [Blanford et al., 2007]. Several approaches have been employed to generate films of laccase that are stable for many days and show higher electrocatalytic current density.

A redox hydrogel approach for immobilizing and mediating electron transfer to glucose oxidase was suggested by Heller and co-workers in 1989, and this team has subsequently developed and elaborated upon the method for application to a number of redox enzymes, including laccase and bilirubin oxidase, and also ascorbate oxidase, horseradish peroxidase, and fructose dehydrogenase (see Section 17.3.1) [Pishko et al., 1990; Barton et al., 2004; Heller, 2006]. The technique involves attachment of an $Os^{III/II}$ complex to a hydrophilic polymer that can be crosslinked to form a redox-active hydrogel (Fig. 17.6a). An enzyme dissolved in the precursor solution with the redox polymer and crosslinker becomes incorporated into the hydrogel, and, upon crosslinking, is immobilized in the hydrogel film. Redox hydrogels are readily formed from soluble polymers such as polyvinylimidazole and polyvinyl alcohol, or from insoluble polymers such as polyvinylpyridine that can be made soluble by quaternization [Kenausis et al., 1996; Mano et al., 2003; de Lumley-Woodyear et al., 1995]. Copolymers of polyvinylimidazole and polyacrylamide have also been reported,

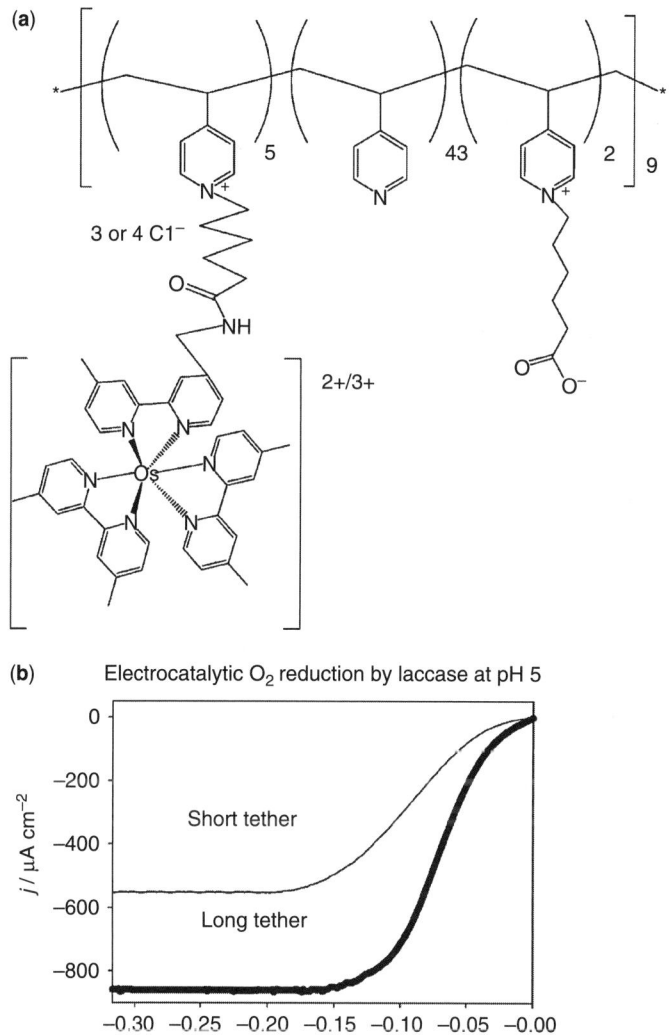

Figure 17.6 Redox hydrogel approach to immobilizing multiple layers of a redox enzyme on an electrode. (a) Structure of the polymer. (b) Voltammograms for electrocatalytic O_2 reduction by a carbon fiber electrode modified with laccase in the redox hydrogel shown in (a) (long tether) or a version with no spacer atoms in the tether between the backbone and the Os center (short tether). Reprinted with permission from Soukharev et al., 2004. Copyright (2004) American Chemical Society.

the acrylamide lending additional solubility to the polymer and lowering the net charge density.

A major advantage of redox polymers is their ability to form hydrated films with very high mediator concentration so that there is good electronic contact between the redox polymer and a large number of trapped enzyme molecules, regardless of orientation. Thus, the number of electrocatalytically active molecules is much higher than for a monolayer of enzyme directly attached to an electrode, and the current will be much greater, provided that there is efficient substrate transport in the hydrogel. The redox hydrogel also serves to trap and stabilize enzymes on an electrode, translating directly into technological applications. The enzyme may simply be sterically enmeshed within the hydrogel, but can also be immobilized by electrostatic interactions with the hydrogel or by covalent bonds. The net electrostatic charge on most enzymes at pH 7 is negative, and the charge on redox hydrogels, in which the backbone is generally neutral, is made positive by complexation with cationic complexes or by quaternization. Additional covalent bonds may be created in a number of ways. For instance, saccharide moieties on glycosylated enzymes such as laccase (see Fig. 17.3) can be oxidized to aldehydes, which can then be reacted with primary amines on the redox polymer to form Schiff bases.

In redox hydrogels, electron transport is accomplished by exchange between neighboring flexible redox centers that come into sufficiently close proximity for outer-sphere electron transfer through physical translation, and thus electron transport should obey Fick's laws of diffusion. Application of a potential gradient leads to a flux of electrons counter to the gradient, as the reduced form of the mediator is thermodynamically more stable at lower potential. The sum of these effects is that electron transport in redox mediators follows the well-known Nernst–Planck equation, except that the migration term is somewhat complicated by the requirement that the sum of the reduced mediator and oxidized mediator must be equal to the fixed total mediator concentration. Thus, electron transport (both diffusion and migration, assuming the Bose–Einstein relation) can be treated as proportional to mediator concentration and an apparent diffusion coefficient D_{app}.

Hydrogels typically have a low swelling ratio (≤ 2), which maintains a high concentration of the redox complex. A redox polymer hydrogel with an Os content of 10 wt% might contain up to 500 mM Os. A main disadvantage is that the redox species themselves are physically bound and immobilized to the hydrogel network leading to low values of the diffusion coefficient, typically in the range 10^{-8} to 10^{-9} cm^2 s^{-1} for Os complexes directly bound to a polyvinylimidazole backbone. There are several possible solutions to this problem. The first is to increase the mobility of the backbone by decreasing crosslinking. This approach has been demonstrated to effectively increase electron diffusion, but at the expense of lower stability in the presence of shear. An alternative is to elongate the side chain attaching the complex to the polymer backbone. This structural change enhances translational mobility of the redox complex independent of backbone mobility. Mao and co-workers have demonstrated that such a change can lead to diffusion coefficients in the 10^{-5} to 10^{-6} cm^2 s^{-1} range [Mao et al., 2003], and the effect of increasing the tether length on the limiting rate of electrocatalytic O_2 reduction by immobilized laccase is shown in Fig. 17.6b. Another approach is to increase the mediator concentration, possibly by increasing the fraction of backbone sites that are complexed, and D_{app} has been shown (both theoretically and experimentally) to increase linearly with mediator concentration [Mano et al., 2006]. Finally, the length scales over which electron transport takes place can be decreased by preparation of thin catalyst films, and this need not decrease the amount of catalyst if the film is prepared on a porous conducting support such as carbon nanotubes or carbon paper [Barton et al., 2007].

A rational strategy for direct attachment of laccase to graphite surfaces was reported recently [Blanford et al., 2007]. As described above, the crystallographic structure of *T. versicolor* laccase III reveals a wide, hydrophobic pocket at the surface of the protein close to the mononuclear type I Cu center at which organic substrates bind [Bertrand et al., 2002]. The proximity of the pocket to the "blue" Cu center provides a route for electrons to be transferred through the protein to the trinuclear active site where O_2 is reduced to water. A pyrolytic graphite electrode modified with 2-aminoanthracene (by electrochemical reduction of the anthracene-2-diazonium ion) forms a surface for direct adsorption of laccase. The laccase-modified electrode shows a well-defined electrocatalytic wave in the presence of O_2, with O_2 reduction commencing above 850 mV with respect to a standard hydrogen electrode (SHE) at pH 4 (Fig. 17.7) (Plate 17.3). The magnitude and long-term stability of the current response

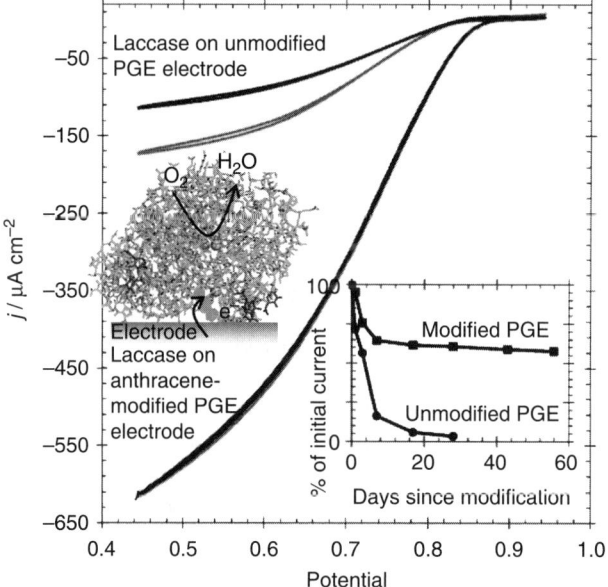

Figure 17.7 Electrocatalysis of O_2 reduction by *Pycnoporus cinnabarinus* laccase on a 2-aminoanthracene-modified pyrolytic graphite edge (PGE) electrode and an unmodified PGE electrode at 25 °C in sodium citrate buffer (200 mM, pH 4). Red curves were recorded immediately after spotting laccase solution onto the electrode, while black curves were recorded after exchanging the electrochemical cell solution for enzyme-free buffer solution. Insets show the long-term percentage change in limiting current (at 0.44 V vs. SHE) for electrocatalytic O_2 reduction by laccase on an unmodified PGE electrode (•) or a 2-aminoanthracene modified electrode (■) after storage at 4 °C, and a cartoon representation of the probable route for electron transfer through the anthracene (shown in blue) to the "blue" Cu center of laccase. Reproduced by permission of The Royal Society of Chemistry from Blanford et al., 2007. (See color insert.)

are far superior to that observed for laccase adsorbed at an unmodified electrode, indicating that the modified electrode both stabilizes and orientates laccase molecules for effective electron transfer (Fig. 17.7 inset). This approach would also be suitable for extension to high surface area forms of carbon.

17.2.2 Heme–Copper Oxidases

Reduction of O_2 directly to H_2O in biology is also catalyzed by heme–copper oxidases that carry out the last step in an aerobic respiratory chain, transferring electrons to O_2 (see Fig. 17.2). Electrons are supplied sequentially by either cytochrome c or the quinol pool, and, depending on the organism, these terminal oxidases are generally large membrane-bound enzyme complexes [Richter and Ludwig, 2003; Ferguson-Miller and Babcock, 1996; Schäfer et al., 1999]. The overall turnover rate of heme–copper oxidases is in the order of 1000 electrons s^{-1}, but for fuel cell applications their disadvantage relative to multi-copper oxidases is the fairly large overpotential for O_2 reduction. This is associated with their physiological role in coupling electron transfer to proton pumping across the membrane to establish a proton gradient that drives the formation of ATP (see Fig. 17.2) [Wikström and Verkhovsky, 2006; Moser et al., 2006]. Thus, in a single-compartment glucose/O_2 enzyme fuel cell with a cathode comprising cytochrome c oxidase interacting with cytochrome c tethered on a gold electrode, the open circuit voltage was less than 120 mV [Katz et al., 1999].

17.3 ENZYME-MODIFIED ANODES

17.3.1 Oxidation of Sugars and Other Alcohols

17.3.1.1 Enzymes for Alcohol Oxidation Several classes of enzymes catalyze the oxidation of alcohols to aldehydes or ketones, and their favored substrates range from simple primary alcohols to complex secondary alcohols, including sugars [Kroutil et al., 2004; Ameyama, 1982]. Further oxidation to carboxylic acids can be achieved by incorporating additional enzymes such as aldehyde dehydrogenase. Enzymes that utilize O_2 as the electron acceptor are termed oxidases, while those utilizing alternative oxidants are termed dehydrogenases.

Zinc-containing alcohol dehydrogenases take up two electrons and a proton from alcohols in the form of a hydride. The hydride acceptor is usually NAD(P)$^+$ (the oxidized form of nicotinamide adenine dinucleotide (NADH) or its phosphorylated derivative, NADPH). Several liver alcohol dehydrogenases have been structurally characterized, and Fig. 17.8 shows the environment around the catalytic Zn center and the bound NADH cofactor.

Alcohol dehydrogenases found in certain microorganisms utilize a pyrroloquinoline quinone (PQQ) or flavin cofactor to pass electrons released upon oxidation of alcohols to the heme electron-acceptor protein, cytochrome c. These membrane-associated alcohol dehydrogenases form part of a respiratory chain, and the energy from fuel oxidation therefore contributes to generation of a proton gradient across

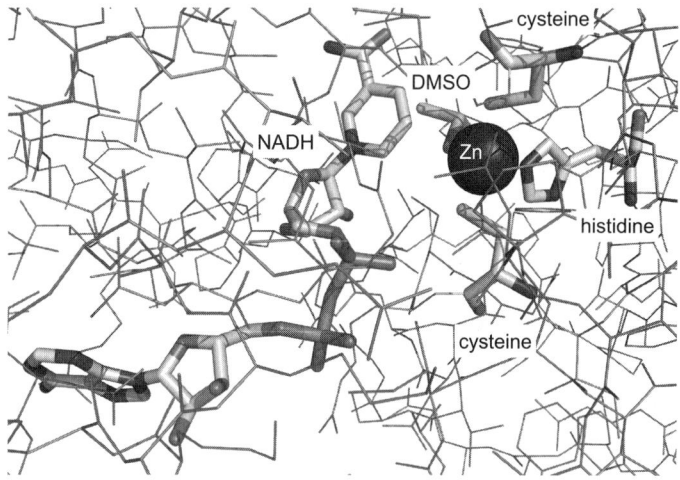

Figure 17.8 Catalytic zinc center of horse liver alcohol dehydrogenase revealed from an X-ray crystallographic structure (PDB file 2OHX) [Al-Karadaghi et al., 1994]. The bound NADH cofactor, a molecule of the inhibitor dimethylsulfoxide (DMSO), and the amino acid residues that coordinate the Zn are shown as sticks shaded according to the elements, and the Zn center is shown as a gray sphere, while the protein is shown in thin gray lines.

a biological membrane. A number of enzymes from this class have been isolated. Favored substrates range from ethanol and other primary aliphatic alcohols to sugars including D-glucose and D-fructose, and also include aldehydes [Ameyama, 1982]. The catalytic centers are buried in a substrate-accessible cleft in the protein, and mediators have been used in most reports of electrocatalysis (see, e.g., Zayats et al. [2005]). However, there are examples of flavin- and PQQ-dependent alcohol dehydrogenases that co-purify with a tightly bound cytochrome or incorporate heme electron transfer centers in addition to the catalytic site, and are suited for direct electron transfer [Ikeda et al., 1993; Kakehi et al., 2007; Kamitaka et al., 2007]. The structure of a PQQ-dependent alcohol dehydrogenase that incorporates a cytochrome electron transfer domain is shown in Fig. 17.9a. An artificial glucose dehydrogenase has also been genetically engineered, combining the PQQ catalytic domain of glucose dehydrogenase with the heme electron transfer domain of *Comamonas testosteroni* alcohol dehydrogenase [Okuda et al., 2007].

Yeasts and fungi express flavin-containing enzymes, including methanol oxidase and glucose oxidase, that use reduction of O_2 to H_2O_2 as the pathway for removal of electrons released during substrate oxidation [Fitzpatrick, 2001]. The most extensively studied and exploited enzyme in this group is glucose oxidase from the mold, *Aspergillus niger* (Fig. 17.9b), which is now commercially available. The enzyme oxidizes D-glucose to D-glucono-1,5-lactone and has been used in blood glucose sensing as well as in fuel cell electrocatalysis. The cofactor is buried in a cleft in the protein, and strategies for mediating electron transfer or providing a molecular wire to the flavin have been employed.

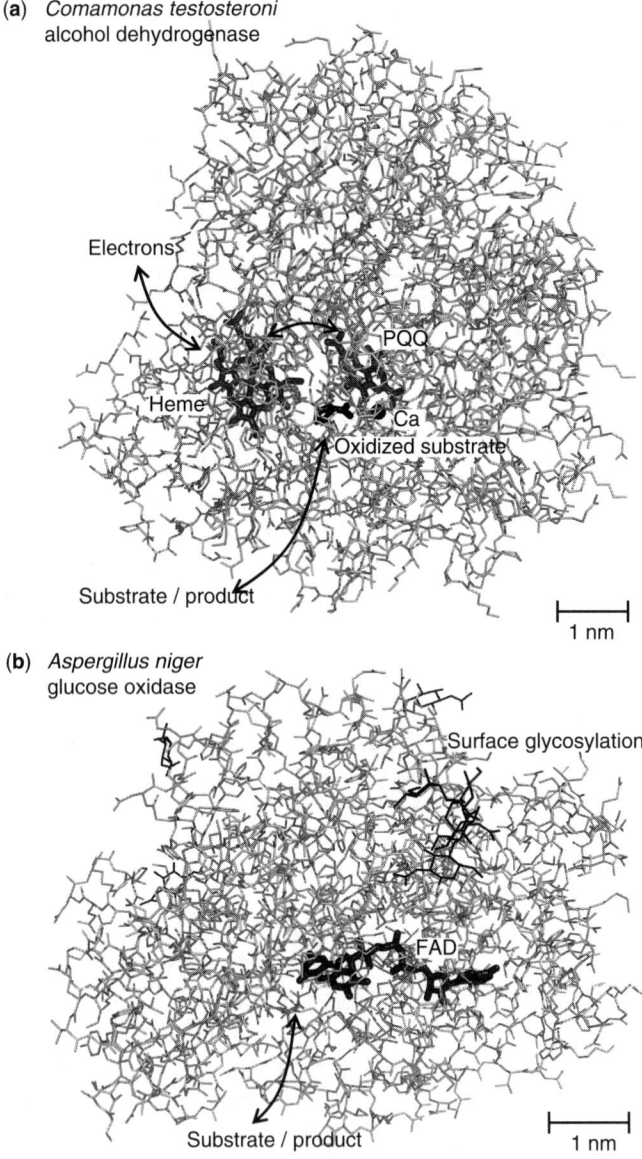

Figure 17.9 Structures of two alcohol-oxidizing enzymes. (a) Heme-containing pyrroloquinoline quinone (PQQ)-dependent alcohol dehydrogenase from *Comamonas testosteroni* (PDB file 1KB0) with a molecule of oxidized tetrahydrofuran bound at the active site [Oubrie et al., 2002]. (b) The environment around the flavin adenine dinucleotide (FAD) cofactor of glucose oxidase from the fungus *Aspergillus niger* (PDB file 1CF3) [Wohlfahrt et al., 1999]. In each case, the cofactors are shown as dark gray sticks, the protein is shown in thin gray sticks, and a Ca atom in (a) is shown as a gray sphere.

17.3.1.2 Electrocatalysis of Alcohol Oxidation by Enzymes The most significant challenge for electrocatalytic applications of Zn alcohol dehydrogenases is supply of the oxidized NAD(P)$^+$ cofactor. The cofactor is expensive, so a regeneration system is necessary. Nonenzymatic oxidation of NAD(P)H is difficult to achieve electrochemically at mild overpotentials [Blaedel and Jenkins, 1975] and is slow with O_2 as oxidant. Several approaches have made use of enzymes for regeneration of the oxidized cofactor. For example, Palmore and co-workers assembled an anode system for stepwise catalytic oxidation of ethanol to CO_2 by a series of dehydrogenases using a bacterial NADH oxidase with benzyl viologen as electron transfer mediator to regenerate the cofactor at an electrode [Palmore et al., 1998]. Such systems are complicated, because of the number of steps and components involved. Recently, it has been shown that a subcomplex comprising two subunits from the mitochrondrial respiratory Complex I exhibits direct NAD$^+$/NADH electrocatalysis at a graphite electrode, and this provides a promising alternative for cofactor regeneration [Barker et al., 2007].

There have been a number of reports of electrocatalysis of alcohol oxidation using immobilized PQQ-dependent alcohol dehydrogenases or flavin-containing alcohol dehydrogenases or oxidases with dissolved mediators in solution. Co-immobilizing the mediator with the enzyme is advantageous, as set out in Section 17.1, and several such strategies have been employed for electrocatalytic alcohol oxidation.

Heller and co-workers utilized Os-containing redox hydrogels, similar to those described in Section 17.2, to entrap glucose oxidase or lactate oxidase and facilitate electron transfer to the flavin [Kenausis et al., 1996]. The redox potential of the mediator groups can be tuned to a value close to that of the respective enzyme by modification of the ligands to the Os to minimize the overpotential for alcohol oxidation [Mano et al., 2003]. Supporting these polymers on high surface area multiscale carbon supports comprising multiwalled carbon nanotubes grown on carbon fiber paper results in high electrocatalytic currents for glucose oxidation: >20 mA cm^{-2} geometric area (Fig. 17.10) [Barton et al., 2007].

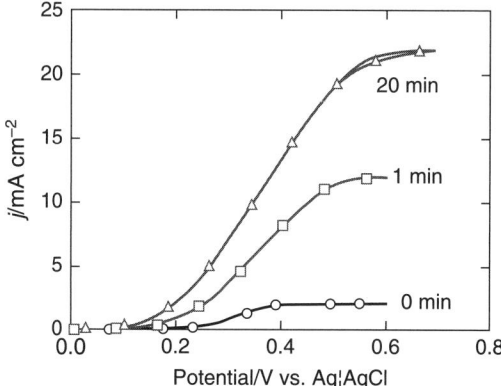

Figure 17.10 Electrocatalytic current (per geometric area) versus potential for glucose oxidation by glucose oxidase in an Os-containing redox polymer supported on carbon nanotubes grown for various periods (times indicated) on carbon paper. Reproduced by permission of ECS—The Electrochemical Society, from Barton et al., 2007.

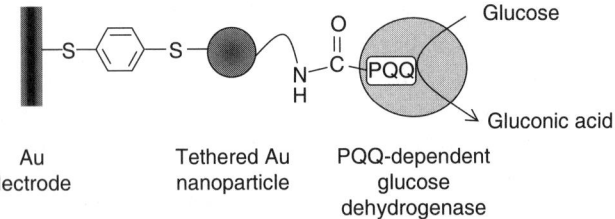

Figure 17.11 Schematic representation of an approach for achieving efficient electrocatalysis of glucose oxidation by glucose dehydrogenase on Au nanoparticles tethered on an Au electrode. The nanoparticles are modified with a PQQ-capped linker that interacts with the unoccupied PQQ site of cofactor-deficient glucose dehydrogenase [Zayats et al., 2005].

Willner and co-workers have developed an immobilization and mediation strategy that involves "wiring" a flavin cofactor to an electrode surface via a conducting linker that may incorporate a PQQ or ferrocene derivative. The immobilized flavin is then used to reconstitute cofactor-free glucose oxidase so that electrons can be channeled directly to or from the active site. Using variations on this method, glucose oxidase and glucose dehydrogenase (Fig. 17.11) have been attached to a range of conducting surfaces, including Au nanoparticles and carbon nanotubes, and very high rates of electron transfer through the wire (up to $5000\,s^{-1}$ at high overpotential) have been reported [Xiao et al., 2003; Zayats et al., 2005; Willner et al., 2007]. Enzyme electrodes assembled in this way have been used in a fuel cell [Katz et al., 1999].

There have been several recent reports of direct electrocatalytic oxidation of sugars by dehydrogenases that possess electron relay centers in addition to the PQQ or flavin active sites. Ikeda and co-workers studied a PQQ-dependent alcohol dehydrogenase from *Gluconobacter suboxydans* that has several heme-containing subunits and allows direct electrocatalysis of ethanol oxidation on a range of metal and carbon electrodes [Ikeda et al., 1993]. Related enzymes were used more recently by Minteer and co-workers for glycerol oxidation in a fuel cell [Arechederra et al., 2007]. Sode and co-workers have reported direct electrocatalysis of sugar oxidation at a carbon paste electrode modified with a thermostable flavin glucose dehydrogenase from *Burkholderia cepacia* that naturally incorporates a heme-containing subunit [Kakehi et al., 2007] or PQQ–glucose dehydrogenase genetically fused to a heme domain [Okuda et al., 2007]. In both of these reports, the catalytic currents are very low, indicating that few enzyme molecules are engaged in sugar oxidation. Impressive electrocatalytic current densities were achieved by Kano and co-workers using a bacterial D-fructose dehydrogenase that possesses a flavin catalytic site and a heme electron relay center at a carbon powder electrode in the presence of 200 mM D-fructose (Fig. 17.12) [Kamitaka et al., 2007].

Heme-containing dehydrogenase enzymes are promising for technological applications, and as more enzymes are isolated and studied, it is likely that systems with further attractive electrocatalytic properties will be uncovered.

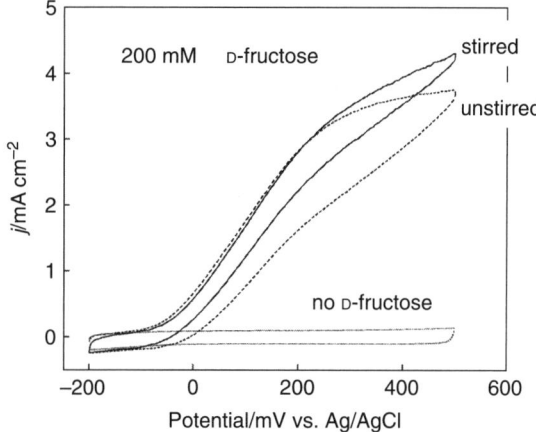

Figure 17.12 Direct electrocatalytic oxidation of D-fructose at a glassy carbon electrode painted with a paste of Ketjen black particles modified with D-fructose dehydrogenase from a *Gluconobacter* species. The enzyme incorporates an additional heme center allowing direct electron transfer from the electrode to the flavin active site. Cyclic voltammograms were recorded at a scan rate of 20 mV s^{-1} and at 25 ± 2 °C and pH 5.0. Reproduced by permission of the PCCP Owner Societies, from Kamitaka et al., 2007.

17.3.2 Oxidation of H_2

17.3.2.1 Electrocatalysis of H_2 Oxidation by Hydrogenases

Hydrogenase enzymes oxidize H_2 or reduce H^+ at bimetallic iron (FeFe), or nickel and iron (NiFe) catalytic centers that are deeply buried within the protein. While most transition-metal electrodes catalyze H_2 oxidation via homolytic cleavage, the bimetallic active sites catalyze heterolytic cleavage, stabilizing the resulting hydride intermediate [Cammack et al., 2001]. The crystal structure of an [NiFe]-hydrogenase is represented in Fig. 17.13, with the two protein subunits shown in thin gray lines and the metalloclusters shown as spheres [Fontecilla-Camps et al., 2007]. The NiFe active site is coordinated to the protein by bridging and terminal cysteine residues, and it is not possible to extract the intact catalytic center. As shown in Fig. 17.1, the iron is additionally coordinated by CO and CN^- ligands (assigned through infrared spectroscopic studies), which are unusual in biology. These ligands presumably play a key role in tuning the electronic and spin states of the metals. Fast electron tunneling through the enzyme is facilitated by a chain of iron–sulfur clusters spaced at approximately 1 nm intervals and connecting the surface of the protein to the buried catalytic site. Proton transport is accommodated by a network of charged amino acid side chains and ordered water molecules, and hydrophobic channels, identified by the trapping of xenon atoms within the protein when crystals are held under a pressure of xenon, may be important in regulating gas access to the catalytic site (for a review, see Fontecilla-Camps et al. [2007]).

Similar structural features are present in the [FeFe]-hydrogenases (FeS electron relay centers, CO and CN-ligands at the active site, and channels for gas access),

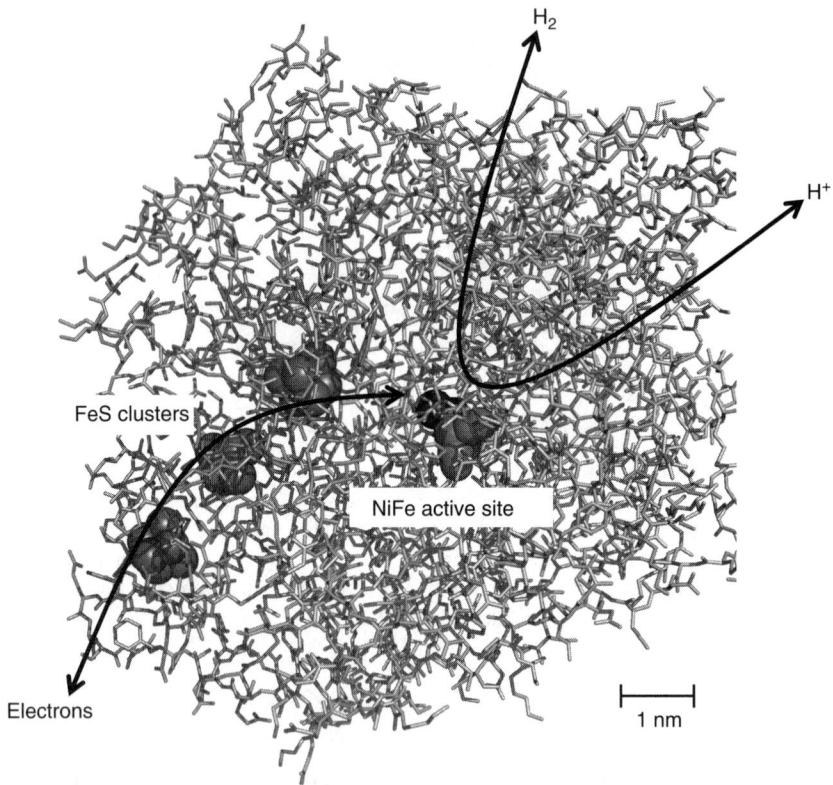

Figure 17.13 (a) Representation of the crystallographic structure of the [NiFe]-hydrogenase from the sulfate-reducing bacterium *Desulfovibrio fructosovorans* (PDB file 1YRQ). A number of hydrogenases have been shown to be active catalysts for both H^+ reduction and H_2 oxidation when adsorbed on a graphite electrode (for a review, see Vincent et al. [2007]). The small protein subunit incorporates three FeS clusters shown in gray spheres; the large protein subunit incorporates the NiFe active site, also shown in gray spheres. The protein is represented as thin gray sticks.

although they arose as a result of separate evolutionary pathways [Fontecilla-Camps et al., 2007]. The [FeFe]-hydrogenases can also be very active electrocatalysts (for a review, see Vincent et al. [2007]), but we focus in this chapter on catalysis by the [NiFe]-enzymes, which tend to be more robust in air.

In biology, the "electrical wiring" by a relay chain of FeS clusters between the active site and the surface of the protein allows electrons from H_2 oxidation to be taken up by cellular electron acceptors (often heme-containing oxidases; see Fig. 17.2), but this feature of hydrogenases also makes them well equipped for direct coupling to an electrode surface. The following sections describe electrochemistry at graphite electrodes modified with hydrogenase, and provide examples of the use of these electrodes in fuel cell catalysis.

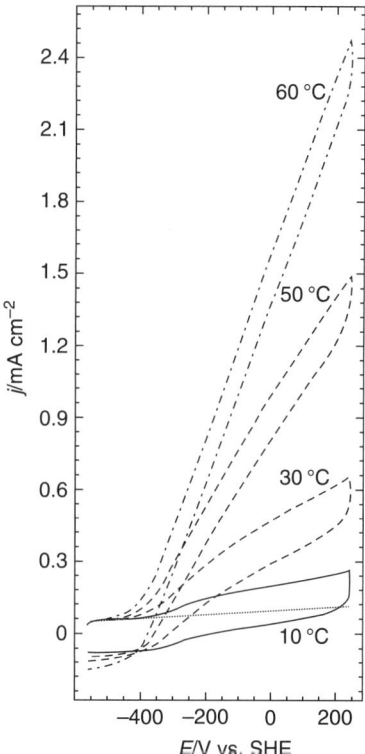

Figure 17.14 Cyclic voltammograms recorded at 1 V s^{-1} at a PGE RDE rotating at 2500 rev min^{-1}) modified by adsorption of a submonolayer film of [NiFe]-hydrogenase from the purple photosynthetic sulfur bacterium *Allochromatium vinosum* in buffered aqueous solution at pH 7.0 under an atmosphere of H$_2$ (1 bar). Reprinted with permission from Léger et al., 2002. Copyright (2002) American Chemical Society.

Electrocatalysis by a single layer of [NiFe]-hydrogenase on a PGE RDE is shown in Fig. 17.14 [Léger et al., 2002]. The film of enzyme was prepared by adsorption from a dilute solution of *Allochromatium vinosum* hydrogenase and a polyamine co-adsorbate, polymyxin, over a period of about 30 minutes, during which the electrode potential was cycled between −0.58 and 0.24 V versus SHE. Figure 17.14 shows catalytic cyclic voltammograms that result when the electrode is placed in enzyme-free, neutral aqueous solution in contact with 1 bar H$_2$. A positive current above about −0.4 V corresponds to H$_2$ oxidation by the enzyme and direct transfer of electrons from the enzyme to the electrode, i.e., with no external mediation. As the temperature is raised, the catalytic current increases, reaching over 2 mA cm^{-2} at 60 °C [Léger et al., 2002].

The shape of the hydrogenase catalytic voltammograms shown in Fig. 17.14 also changes as the temperature is raised. At 10 °C, the current tends towards a plateau at high overpotential as catalysis becomes limited by the inherent turnover frequency of the enzyme, but at higher temperatures, the current continues to increase linearly with electrochemical driving force. This has been attributed to a range of different orientations of the enzyme on the electrode that lead to a distribution of interfacial electron transfer rates such that some molecules require a high driving force to undergo catalysis [Léger et al., 2002]. At lower temperatures, electron transfer keeps up with catalysis, but as the enzyme turns over faster at higher temperatures, interfacial electron transfer becomes the rate-limiting step. This effect is also evident in two reported comparisons of H_2 oxidation catalyzed by hydrogenase-modified and platinized electrodes operating in aqueous solution: the precious metal catalyst is more active at potentials close to the value of $E(H^+/H_2)$, but, at higher overpotentials, both the platinized and hydrogenase-modified electrodes show a similar and diffusion-controlled level of H_2 oxidation [Jones et al., 2002; Karyakin et al., 2005]. This effect suggests that improved coupling of hydrogenase to the electrode would enhance electrocatalysis at lower applied potentials.

Non-turnover signals arising from electron transfer to the FeS relay clusters in the enzyme are just visible in voltammetry for high-coverage films of *A. vinosum* hydrogenase, suggesting that an upper limit for the electroactive coverage is about 3 pmol cm^{-2} (for a review, see Vincent et al. [2007]). The response in Fig. 17.14 therefore results from a low density of extremely active catalytic sites, and there is considerable scope for optimizing the current by increasing the effective electrode area.

A method was recently reported for covalently attaching hydrogenase to graphite or multiwalled carbon nanotubes via a carbodiimide/N-hydroxysuccinamide coupling reaction to aminophenyl functionalities introduced onto the nanotubes [Alonso-Lomillo et al., 2007]. This strategy relies upon a surface patch that is rich in glutamate residues close to the surface FeS cluster of *Desulfovibrio gigas* [NiFe]-hydrogenase to favor attachment close to the electron entry point, but probably still results in a distribution of orientations. The resulting hydrogenase film exhibits diffusion-controlled H_2 oxidation currents that persist for over 30 days. The authors also show that hydrogenase can be adsorbed onto unmodified carbon nanotubes, but this procedure results in a less active and less stable protein film. The negatively charged surface patch probably also explains the stabilization of directly adsorbed hydrogenase films in the presence of positively charged co-adsorbates such as polymyxin and polylysine (for a review, see Vincent et al. [2007]). Further advances are still needed for obtaining stable, high coverage films of hydrogenase that are orientated correctly for efficient interfacial electron transfer.

17.3.2.2 Tolerance of Hydrogenases to O_2, CO, and Sulfide

Figure 17.15 summarizes the reactions of [NiFe]-hydrogenases with small molecules that can poison Pt sites (for a review, see Vincent et al. [2007]). *A. vinosum*, the source of the [NiFe]-hydrogenase addressed in Fig. 17.14, lives in semi-aerobic soils or waterways, and therefore infrequently encounters high levels of O_2. The H_2 oxidation activity of this enzyme is fully inhibited by O_2, with oxygenic bridging ligands becoming trapped in the active site (Fig. 17.15). Under reducing potentials, it is possible to recover the activity almost fully, but the reaction, presumably requiring reductive removal of the bridging ligand as H_2O, is very slow (the half life is approximately 1 hour at 20 °C for re-activation of the hydroperoxide form) [Vincent et al., 2007]. For the purposes of operation of the enzyme within a fuel cell, or indeed probably a biological cell, the activity of *A. vinosum* hydrogenase is essentially destroyed by O_2.

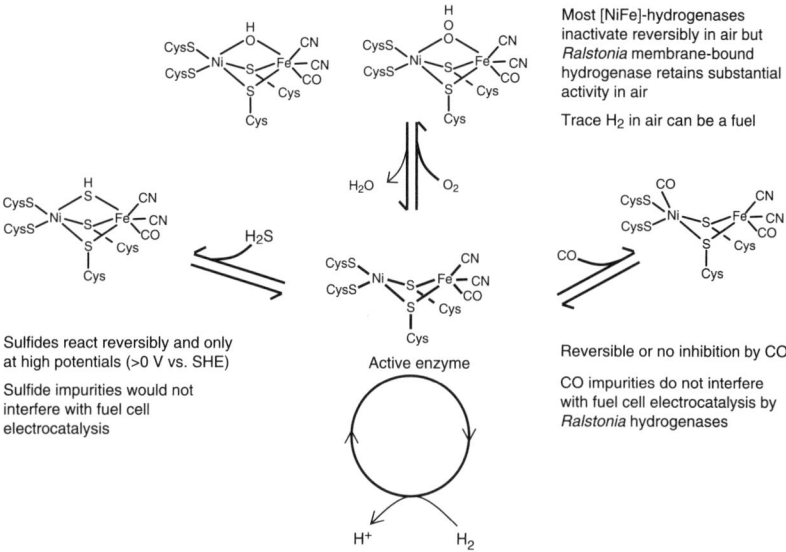

Figure 17.15 Reactions of the active site of [NiFe]-hydrogenases with small molecules that can poison Pt active sites, showing likely structures for the product species. Inhibition of many [NiFe]-hydrogenases by O_2, CO, and sulfides is reversible, while hydrogenases from *Ralstonia* even oxidize H_2 in the presence of these molecules (for a review, see Vincent et al. [2007]).

It has become apparent that some hydrogenases are not inhibited in this way by O_2. Certain microorganisms, such as the Knallgas bacteria *Ralstonia*, are suited to utilizing trace H_2 as an energy source in aerobic environments (see Fig. 17.2), and their hydrogenases must therefore remain catalytically active in air [Burgdorf et al., 2005]. Oxidation of H_2 in air presents interesting challenges: synthetic FeS-containing clusters that are able to bind H_2 tend to be sensitive to both O_2 and moisture, while conventional precious metal H_2 fuel cell catalysts are reactive to both H_2 and O_2.

Electrochemical methods have been used to address these questions for the NiFe membrane-bound hydrogenases from *Ralstonia* species that also adsorb readily on a graphite electrode [Vincent et al., 2005] (for a review, see Vincent et al. [2007]). Figure 17.16 shows the effect of increasing levels of O_2 on the activity of a membrane-bound [NiFe]-hydrogenase from the heavy-metal-resistant strain *R. metallidurans* CH34. Activity is measured as the electrocatalytic current response for a film of the enzyme adsorbed directly on a PGE RDE at close to 1 bar H_2. Introduction of 5 mbar O_2 has very little effect on the catalytic current, and even at 230 mbar O_2, above the level in air, substantial activity (about 50%) remains. The extremely high selectivity of this enzyme for H_2 has been utilized in a fuel cell operating on low levels of H_2 in air (see Section 17.4.3) [Vincent et al., 2006]. The origin of the O_2 tolerance remains unclear: the Fourier transform infrared (FTIR) spectrum for the *R. metallidurans* CH34 hydrogenase shows the same pattern of CO and CN^- ligands at the active site as that of O_2-sensitive [NiFe]-hydrogenases, suggesting that O_2-tolerance must arise from subtle changes further away from the active site [Vincent et al., 2005].

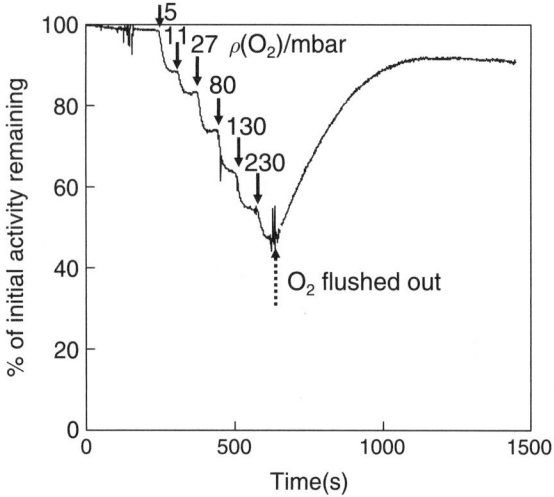

Figure 17.16 Electrocatalytic H_2 oxidation by *Ralstonia metallidurans* CH34 membrane-bound hydrogenase on a PGE RDE in the presence of O_2. The electrode is rotated at 2000 rev min^{-1} and polarized at $+0.142$ V vs. SHE in buffered aqueous solution at pH 5.6 and 30 °C, close to 1 bar H_2. Reprinted with permission from Vincent et al. [2007]. Copyright (2007) American Chemical Society.

Thus it may prove possible to improve O_2 tolerance further by directed mutation of residues within the protein.

Hydrogenases are typically inhibited by CO, but recover rapidly when the inhibitor is removed. However the membrane-bound hydrogenases from *Ralstonia* are almost completely insensitive to a large excess of CO [Vincent et al., 2005, 2006]. Certain hydrogenases thus offer opportunities for oxidation of H_2 in heavily contaminated gas streams. Sulfides have been shown to react with hydrogenases only at high potential, inhibiting H_2 oxidation above about 0 V vs. SHE, and thus would not interfere in fuel cell catalysis under typical operating conditions.

17.4 WORKING EXAMPLES OF ENZYMES AS FUEL CELL CATALYSTS: POSSIBILITIES FOR NOVEL DESIGNS AND APPLICATIONS

It is important to keep in mind that exploitation of metalloenzymes in catalysis is an emerging area, and the innovative concepts demonstrated by biofuel cells are far more significant than the magnitude of the power outputs recorded. Demonstrations of enzyme catalysis in fuel cells have generally been on a small scale, with power outputs typically in the microwatt to milliwatt range. Although these levels compare poorly with those of conventional fuel cells, advances in the attachment of enzymes to electrodes and the use of porous electrodes or electrodes modified with nanotubes or carbon powders are providing improvements in current suggesting that it should be possible to close that gap. Furthermore, niches exist where enzyme fuel cells are at an advantage owing to the unique properties of biological catalysts. In particular, the impressive substrate selectivity of enzymes leads to novel fuel cell designs and applications, and provides a benchmark for the development of catalysts for future technologies. While there are now a few examples of methanol/O_2 fuel cells that use modified precious metal catalysts (e.g., metal chalcogenides) for O_2-tolerant methanol oxidation [Alonso-Vante, 2003], there are no well-understood examples of O_2-tolerant H_2 oxidation at precious metals. In this section, we set out examples of membraneless fuel cells operating on fuels such as sugars, ethanol, glycerol, and H_2.

17.4.1 Sugars as Fuels

The most well-established use of enzymes in fuel cells is in generating electricity from sugar oxidation, using enzymes such as glucose oxidase, glucose dehydrogenase, or fructose dehydrogenase at the anode in combination with an O_2-reducing cathode modified with laccase or bilirubin oxidase. The availability of sugars and O_2 in blood and other biological fluids means that implantable devices should be possible, and thus a showcase application of biological fuel cells is likely to be in the area of power for biomedical devices.

Heller and co-workers, in a series of papers, describe the development of a low power sugar/O_2 biofuel cell device suitable for biomedical applications (for a review, see Heller [2004]). The electrodes consist of carbon fibers, approximately 7 μm in diameter, on which a hydrogel is deposited that incorporates enzymes and $Os^{III/II}$ redox mediator centers (as described in Sections 17.2.1 and 17.3.1) (Fig. 17.17). The enzyme at the anode is generally glucose oxidase, and laccase and bilirubin oxidase have been utilized at the cathode. Bilirubin oxidase is a more suitable complement for glucose oxidase, having maximum activity close to pH 7 and remaining active in the presence of chloride. Tuning the redox potential of the mediator in the anode and cathode polymer by modifying the ligands to Os meant that the electrodes can operate close to the onset potential for catalysis by glucose oxidase and bilirubin oxidase respectively, and the open circuit potential is around 0.8 V (Fig. 17.17 inset). This compares favorably with the open circuit potential of 120 mV reported for an earlier glucose/O_2 cell that utilized an enzyme with a high overpotential for O_2 reduction at the cathode, cytochrome c oxidase [Katz et al., 1999].

Recently, there has been interest in enzyme fuel cells that utilize high surface area forms of carbon modified with enzyme molecules that are able to undergo direct electron transfer. Kano and co-workers achieved current densities in the range 2–4 mA cm^{-2} by immobilizing laccase onto a mesoporous carbon aerogel and D-fructose dehydrogenase onto nanodimensional Ketjen black particles and then coating these materials onto a carbon paper support [Kamitaka, 2007]. A membraneless fuel cell incorporating these electrodes and operating in O_2-saturated aqueous solution at pH 5 containing 0.2 M fructose gave an open circuit voltage of 790 mV, and produced a current density greater than 1 mA cm^{-2} at a cell voltage of 410 mV in an unstirred solution.

17.4.1.1 Possibilities for Enzymes in Implantable Fuel Cells
There is significant and increasing demand for power supplies for implantable medical devices, including continuous glucose monitors for diabetic patients, thermal sensors for

monitoring tissue inflammation, implants that replace and enhance sight and hearing, pacemakers and defibrillators, and muscle actuation systems to assist paralysis victims. Enzyme fuel cells may be applicable in any of these applications, but significant challenges must be overcome in developing reliable devices.

A major challenge in providing electrical power for implantable devices is the isolation of toxic or bio-incompatible materials. As the size of the device decreases to a centimeter or millimeter scale, the parts responsible for isolation, such as canisters and seals, begin to determine the size of the device. An alternative is to design an electrochemical system that is compatible with the physiological environment and can take advantage of chemical species available in that environment, specifically the

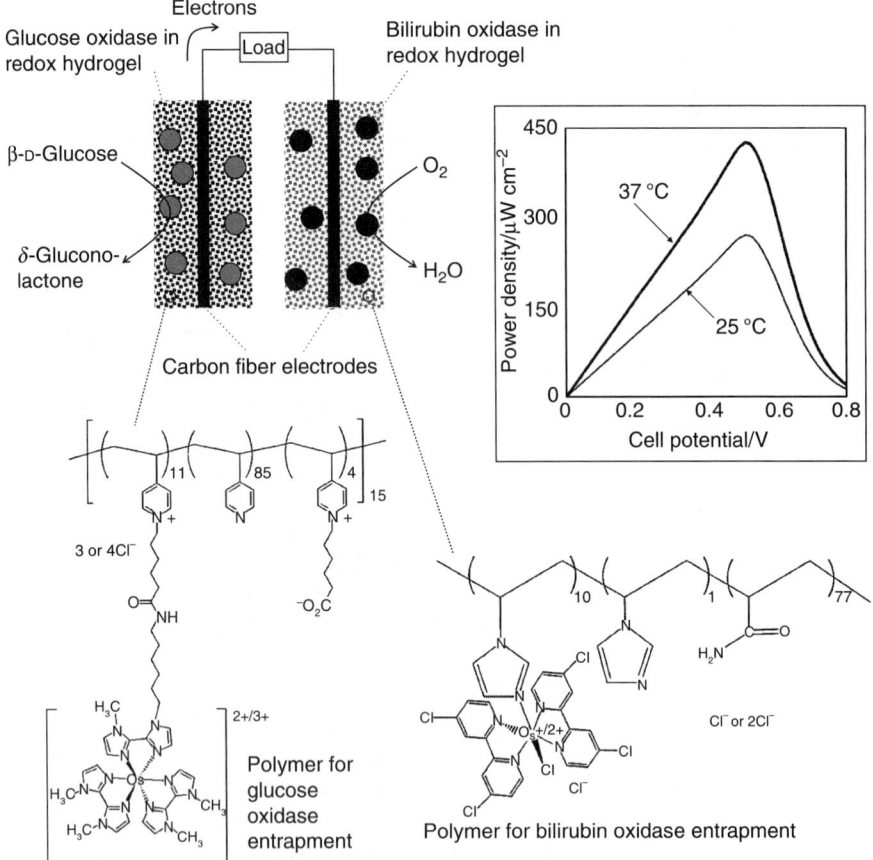

Figure 17.17 Schematic representation of a single-compartment glucose/O_2 enzyme fuel cell built from carbon fiber electrodes modified with $Os^{III/II}$-containing polymers that incorporate glucose oxidase at the anode and bilirubin oxidase at the cathode. The inset shows power density versus cell potential curves for this fuel cell operating in a quiescent solution in air at pH 7.2, 0.14 M NaCl, 20 mM phosphate, and 15 mM glucose. Parts of this figure are reprinted with permission from Mano et al. [2003]. Copyright (2003) American Chemical Society.

17.4 WORKING EXAMPLES OF ENZYMES AS FUEL CELL CATALYSTS

fuels (e.g., glucose and other sugars), O_2, and electrolytes (primarily NaCl). In theory, the device lifetime may be enhanced by replacing a battery with a fuel cell that scavenges ambient fuel and oxidant.

Heller and co-workers have demonstrated a glucose/O_2 fuel cell utilizing redox hydrogel glucose oxidase and bilirubin oxidase electrodes, operating in a grape in which the glucose concentration exceeded 30 mM and the pH was 5.4 (Fig. 17.18) [Mano et al., 2003]. Peak operation was obtained when the electrodes were implanted near the skin of the grape, presumably owing to the maximized O_2 flux. The maximum power, 240 $\mu W\ cm^{-2}$, was obtained with the cell operating at about 0.5 V, and 78% of the initial power was retained after 24 hours of continuous operation.

Fuel cells based on unmediated electrocatalysis by heme-containing sugar dehydrogenases have not yet been tested in biological fluids, but may be useful for implantable applications, as they avoid the need for toxic or expensive mediators and have minimal design constraints. Realistically, the lifetime of biofuel cells is still insufficient for biomedical applications requiring surgical installation.

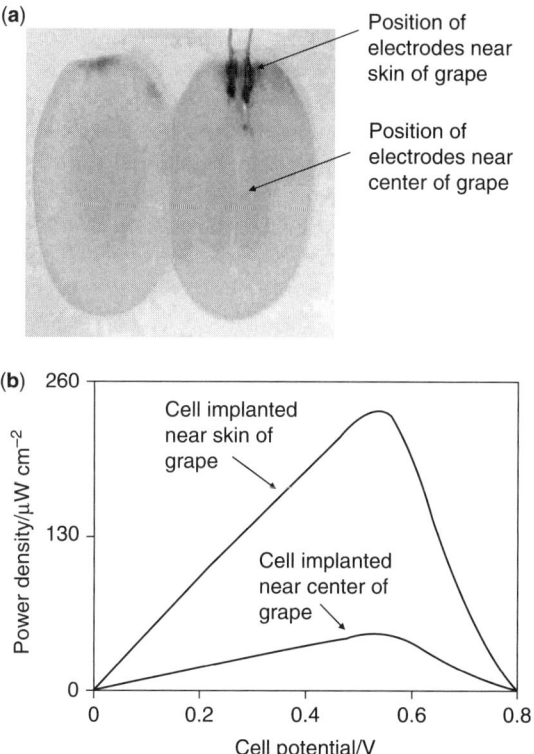

Figure 17.18 (a) Photograph of a whole and sliced grape showing the positions of implanted carbon fiber electrodes modified with enzymes embedded in redox hydrogels (as shown in Fig. 17.17). (b) Power output versus cell potential for the implanted fuel cell. Reprinted with permission from Mano et al. [2003]. Copyright (2003) American Chemical Society.

Applications feasible in the near term will most likely involve transcutaneous access to tissue-based glucose and reliance on O_2 from tissue or ambient air, involving little or no surgical intervention. These are likely to be disposable, low power devices with lifetimes of one week or less.

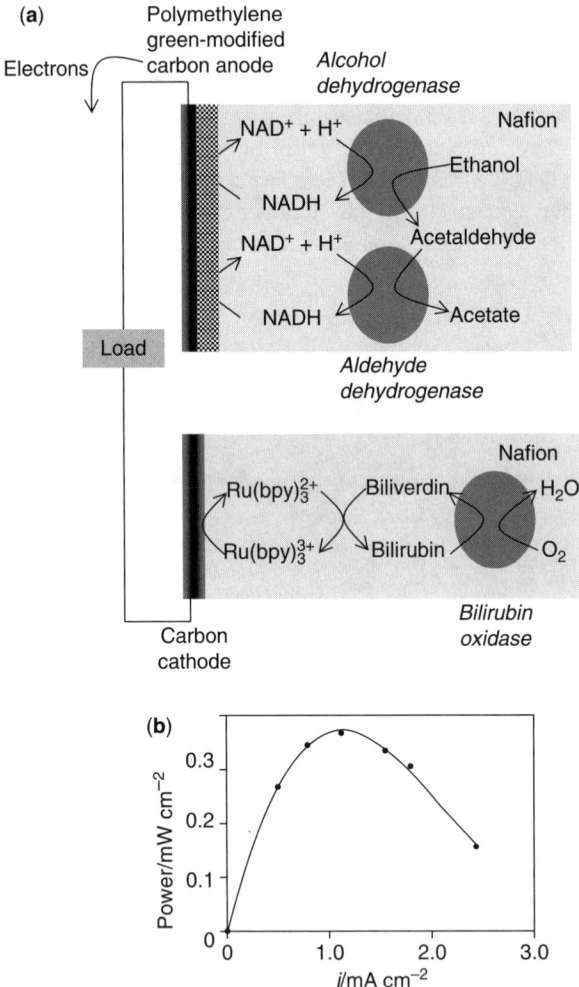

Figure 17.19 A membraneless ethanol/O_2 enzyme fuel cell. Alcohol dehydrogenase and aldehyde dehydrogenase catalyze a stepwise oxidation of ethanol to acetaldehyde and then to acetate, passing electrons to the anode via the mediator NAD^+/NADH. At the carbon cathode, electrons are passed via the $[Ru(2,2'\text{-bipyridyl})_3]^{3+/2+}$ and biliverdin/bilirubin couples to bilirubin oxidase, which catalyzes O_2 reduction to H_2O. (a) Schematic representation of the reactions occurring. (b) Power/current response for the cell operating in buffered solution at pH 7.15, containing 1 mM ethanol and 1 mM NAD^+. Panel (b) reprinted from Topcagic and Minteer [2006]. Copyright Elsevier, 2006.

17.4.2 Ethanol, Methanol, or Glycerol as Fuel

The alcohol tolerance of O_2 reduction by bilirubin oxidase means that membraneless designs should be possible provided that the enzymes and mediators (if required) are immoblized at the electrodes. Minteer and co-workers have made use of NAD^+-dependent alcohol dehydrogenase enzymes trapped within a tetraalkylammonium ion-exchanged Nafion film incorporating $NAD^+/NADH$ for oxidation of methanol or ethanol [Akers et al., 2005; Topcagic and Minteer, 2006]. The polymer is coated onto an electrode modified with polymethylene green, which acts as an electrocatalyst

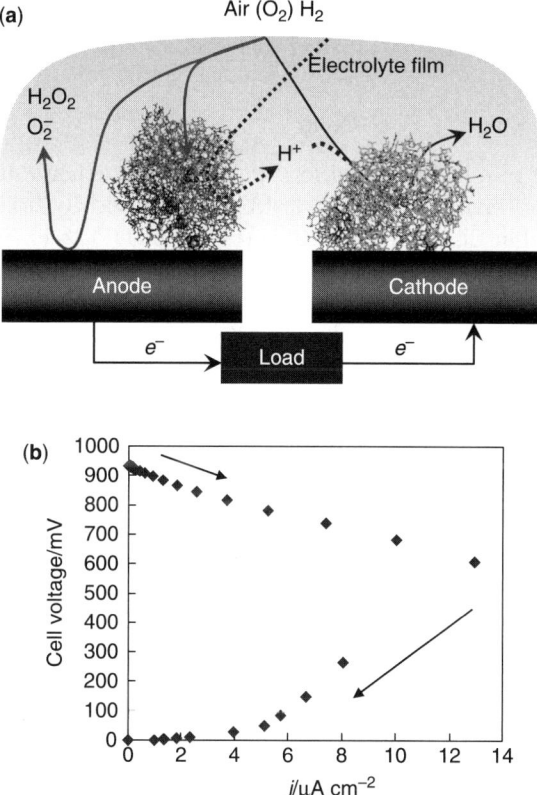

Figure 17.20 Use of enzyme catalysts in a membraneless fuel cell that operates on low levels of H_2 in air [Vincent et al., 2006]. (a) Schematic representation of the fuel cell. PGE electrodes modified with hydrogenase and laccase are inserted into a shallow tray of aqueous electrolyte (pH 5 citrate solution) in contact with an atmosphere of 3% H_2 in air. Blue lines indicate productive reactions: electrocatalytic H_2 oxidation at the anode and O_2 reduction at the cathode. Red lines indicate unproductive reactions: consumption of electrons at the anode by direct reduction of O_2 at bare graphite, generating radical species that may damage the enzymes, or reversible inhibition of hydrogenase by access of O_2 to the active site. (b) Typical cell voltage/current plot for this fuel cell obtained by applying a variable load, showing the rapid drop in current observed at high cell voltages, consistent with high potential inactivation of the hydrogenase at the anode. (See color insert.)

for NAD^+ regeneration. An anode for the oxidation of ethanol to acetate uses two enzymes that operate in series trapped together in the Nafion: alcohol dehydrogenase and aldehyde dehydrogenase. Combination of this anode with a mediated bilirubin oxidase O_2-reducing cathode, as shown in Fig. 17.19, gives a membraneless fuel cell that produces a small power output from oxidation of ethanol in air [Topcagic and Minteer, 2006].

Minteer and co-workers have also exploited the broad substrate specificity of PQQ-dependent alcohol dehydrogenase and aldehyde dehydrogenase from *Gluconobacter* species trapped within Nafion to oxidize either ethanol or glycerol at a fuel cell anode [Arechederra et al., 2007]. Although the alcohol dehydrogenase incorporates a series of heme electron transfer centers, it is unlikely that many enzyme molecules trapped within the mediator-free Nafion polymer are electronically engaged at the electrode.

17.4.3 H_2/O_2 Enzyme Fuel Cells

The O_2 tolerance of a hydrogenase from the aerobic bacterium *Ralstonia* has been exploited in a membraneless fuel cell that produces electricity from just 3% H_2 in air, a level of fuel that is too dilute to burn (Fig. 17.20) (Plate 17.4) [Vincent et al., 2006]. The design is far from optimized, and the power output is low owing to partial inactivation of the hydrogenase by O_2 at the low level of H_2 and to short-circuiting at the anode caused by O_2 reduction at bare regions of the graphite. However, the significant aspects of this demonstration are the selective catalysis of H_2 oxidation in the presence of an excess of O_2 and the simplicity of design. Both enzymes were directly adsorbed onto PGE strip electrodes immersed in a shallow tray of electrolyte in contact with still air containing a low level of H_2. It should therefore be possible to scale down such devices to provide localized micro- or nanoscale power sources. With selective enzyme catalysts, trace H_2 in air becomes a viable fuel source, even when contaminated with sulfides or CO [Vincent et al., 2007].

17.5 POSSIBILITIES FOR SCALING DOWN ENZYME ELECTRODES (NANOTECHNOLOGIES)

A number of reports of enzyme electrocatalysis on carbon nanotubes are now emerging. This approach may prove useful both for scaling up the current response in enzyme fuel cells (by increasing the effective electrode surface area) and for scaling down enzyme fuel cells for nanotechnology applications. The coupling between proteins and nanoscopic structures has been studied extensively, and is the subject of several reviews [Chen et al., 2007; Willner et al., 2006, 2007; Gooding, 2005; Katz et al., 2003]. There are examples in which either Au nanoparticles or carbon nanotubes provide efficient electronic coupling between an electrode and an enzyme. Willner and co-workers demonstrated that glucose oxidase shows unprecedented activity, albeit at high overpotentials, when the FAD cofactor is directly wired to an Au_{55} nanoparticle, which in turn is wired via an aromatic diol to an Au electrode [Xiao et al., 2003]. In a similar experiment, carbon nanotubes were used as a wire, with one end covalently attached to a Au electrode via cysteamine and the other to the flavin group of glucose oxidase [Patolsky et al., 2004]. Such a carbon nanotube "forest" electrode has also been applied to peroxidases [Gooding et al., 2003; Yu et al., 2003].

Of direct interest for biofuel cell applications are the reported reduction of O_2 by multi-copper oxidases on carbon nanotube electrodes [Yan et al., 2006; Zheng et al., 2006] and the oxidation of H_2 by hydrogenase covalently bound to carbon nanotubes [Alonso-Lomillo et al., 2007]. The hydrogenase/nanotube anode is extremely stable (>1 month), and shows 33-fold enhanced enzyme coverage compared with similarly treated graphite of the corresponding geometric surface area. *A. vinosum*

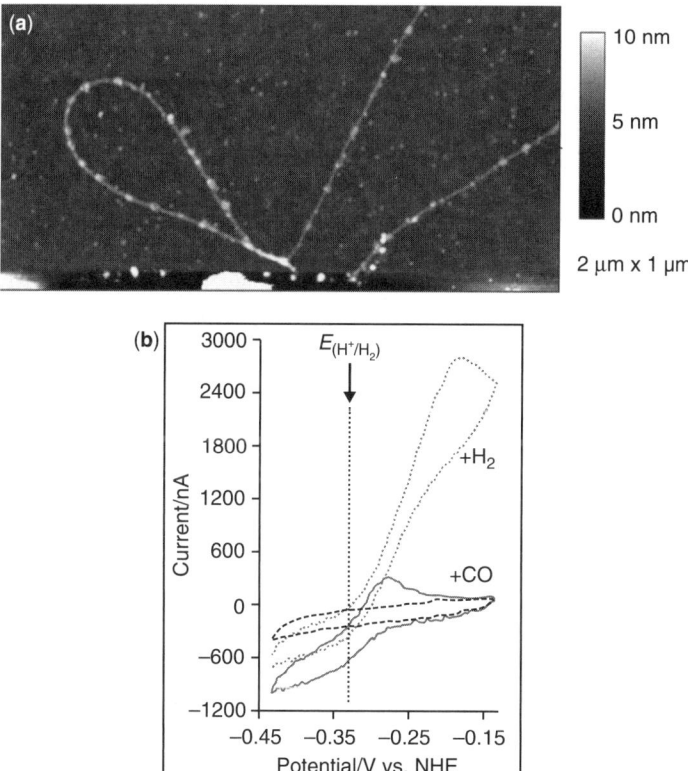

Figure 17.21 *Allochromatium vinosum* [NiFe]-hydrogenase adsorbed onto single-walled carbon nanotubes pretreated with polymyxin. (a) Tapping-mode atomic force microscopy image of carbon nanotubes grown on Si/SiO_2 by chemical vapor deposition (according to the method of [Kong et al., 1998]). Nanotubes were first incubated with 0.2 mg mL^{-1} polymyxin and then with 20 nm hydrogenase. (b) Cyclic voltammograms recorded at 5 mV s^{-1} showing electrocatalysis by a dense layer of carbon nanotubes [Heering et al., 2006], treated with 50 mg mL^{-1} polymyxin and then with 0.5 μM hydrogenase. The solid trace was recorded under argon, the dash–dotted trace with 0.4 mM H_2 in solution, and the dashed trace with 0.2 mM of the inhibitor CO. The buffer was 15 mM MES containing 100 mM NaCl, pH 5.7, and the macroscopic electrode area was 10 mm^2.

[NiFe]-hydrogenase will adsorb onto nanotubes pretreated with polymyxin as shown in Fig. 17.21a, and undergoes direct electrocatalysis of H_2 oxidation and production as shown in Fig. 17.21b [Hoeben et al., 2008]. An enzyme fuel cell has been constructed with glucose dehydrogenase at the anode, relying upon NAD^+ regeneration via methylene blue-modified nanotubes, combined with a cathode based on laccase adsorbed on carbon nanotubes [Yan et al., 2006]. These and other examples demonstrate that the graphite-like properties, large aspect ratio (a few nanometers in diameter, micrometers long), high surface area, and high conductivity of carbon nanotubes makes them an excellent material for wiring enzymes [Gooding, 2005; Heller et al., 2005; Tasis et al., 2006].

Microscopic or nanoscopic surface patterning can be applied using photolithography or electron-beam lithography (top-down chip technology), but can also be achieved, possibly more cheaply, by bottom-up assembly of nanoparticles or nanotubes on surfaces. The use of microscopic biofuel cells as localized power sources for autonomous "lab-on-a-chip" applications can be envisioned, for example powering microscopic sensors on an analytical microchip. Nanoscopic patterning can also enhance mass transport, reducing the need to stir or to create forced flow. An assembly of nanoscopic features may function as an array of nanoelectrodes, which promotes the even distribution of the adsorbed enzyme. The diffusion profile of the fuel to each of these small electrodes is radial. Because the flux is inversely proportional to the electrode radius and is time-independent, a nanoscopic electrode will yield an extremely high mass transport rate [Bard and Faulkner, 2001; Armstrong et al., 1993].

17.6 FUTURE PERSPECTIVES

Enzymes are efficient catalysts for cathodic and anodic reactions relevant to fuel cell electrocatalysis in terms of overpotential, active site activity, and substrate/reaction specificity. This means that design constraints (e.g., fuel containment and anode–cathode separation) are relaxed, and very simple devices that may take up ambient fuel or oxidant from their environment are possible. While operation is generally confined to conditions close to ambient temperature, pressure, and pH, and power densities over about 10 mW cm^{-2} are rarely achieved, enzyme fuel cells may be particularly useful in niche environments, for example scavenging trace H_2 released into air, or sugar and O_2 from blood. Thus, trace or unusual fuels become viable for energy production.

Vast numbers of different redox enzymes are expressed by microorganisms, and new variants on known enzymes are still being discovered. Organisms have adapted to life in a range of environments, and have therefore evolved to express enzymes with specific properties. Isolation of new enzymes is likely to lead to biological catalysts with novel properties, and the range can be extended further by targeted genetic modifications. Of particular interest will be enzymes with improved electron transfer properties (efficient electron relay chains) and even greater substrate and reaction specificity.

Enzymes are not as unstable as is commonly believed. This is demonstrated by a number of examples discussed in this Chapter: electrocatalysis of H_2 oxidation up to 60 °C by the hydrogenase from a bacterium that does not live at extreme temperatures [Léger et al., 2002], electrocatalytic currents that are stable over a month or more by laccase on a 2-aminoanthracene electrode [Blanford et al., 2007] or a covalently linked hydrogenase [Alonso-Lomillo et al., 2007], and electrocatalysis in demanding environments such as within a plant [Mano et al., 2003]. Isolation of novel enzymes may also lead to biological catalysts with enhanced stability.

Improved attachment and stabilization of enzymes on surfaces is one key area that requires significant attention in the development of enzyme electrocatalysis for fuel cells. Advances may come from a mixture of genetics (introduction of useful surface residues for chemical linkage) and electrode modification and coupling chemistry. It will also be important to increase electroactive coverage, and this may involve high suface area electrode supports or multilayer adsorption of enzyme, or a combination of these approaches. Design possibilities for enzyme fuel cells are at an early stage, and dialogue between enzyme electrochemists and fuel cell scientists may be important in advancing design of devices that exploit enzyme electrocatalysts.

The large molecular size and ambient operation of enzymes means that they are likely to be more suited to niche applications rather than to high-power devices, but there are important lessons to be learnt from biological catalysis that occurs in conditions under which conventional metal catalysts would fail. Development of synthetic catalysts inspired by the chemistry (although not necessarily the structures) of enzyme active sites may lead to future catalysts with new and improved properties.

ACKNOWLEDGMENTS

K. A. Vincent is a Royal Society University Research Fellow. H. A. Heering is financially supported by a VIDI grant from The Netherlands Organization for Scientific Research (NWO). Fraser A. Armstrong is thanked for helpful discussion and comments on the manuscript.

REFERENCES

Akers NL, Moore CM, Minteer SD. 2005. Development of alcohol/O_2 biofuel cells using salt-extracted tetrabutylammonium bromide/Nafion membranes to immobilize dehydrogenase enzymes. Electrochim Acta 50: 2521–2525.

Al-Karadaghi S, Cedergren-Zeppezauert ES, Hovmoller S. 1994. Refined crystal structure of liver alcohol dehydrogenase–NADH complex at 1.8 Å resolution. Acta Crystallogr DS0: 793–807.

Alonso-Lomillo MA, Rüdiger O, Maroto-Valiente A, Velez M, Rodríguez-Ramos I, Muñoz FJ, Fernández VM, de Lacey AL. 2007. Hydrogenase-coated carbon nanotubes for efficient H_2 oxidation. Nano Lett 7: 1603–1608.

Alonso-Vante N. 2003. Chevrel phases and chalcogenides. In: Vielstich W, Gasteiger HA, Lamm A, editors. Handbook of Fuel Cells—Fundamentals, Technology and Applications. Chichester: Wiley.

Ameyama M. 1982. Enzymatic microdetermination of D-glucose, D-fructose, D-gluconate, 2-keto-D-gluconate, aldehyde, and alcohol with membrane-bound dehydrogenases. Meth Enzymol 89: 20–29.

Arechederra RL, Treu BL, Minteer SD. 2007. Development of glycerol/O_2 biofuel cell. J Power Sources 173: 156–161.

Armstrong FA, Cox PA, Hill HAO, Lowe VJ, Oliver BN. 1987. Metal-ions and complexes as modulators of protein interfacial electron-transport at graphite-electrodes. J Electroanal Chem 217: 331–366.

Armstrong FA, Bond AM, Buchi FN, Hamnett A, Hill HAO, Lannon AM, Lettington OC, Zoski CG. 1993. Electrocatalytic reduction of hydrogen-peroxide at a stationary pyrolytic-graphite electrode surface in the presence of cytochrome-c peroxidase—A description based on a microelectrode array model for adsorbed enzyme molecules. Analyst 118: 973–978.

Bard AJ, Faulkner LR. 2001. Electrochemical Methods: Fundamentals and Applications. 2nd ed. New York: Wiley.

Barker CD, Reda T, Hirst J. 2007. The flavoprotein subcomplex of complex I (NADH : ubiquinone oxidoreductase) from bovine heart mitochondria: Insights into the mechanisms of NADH oxidation and NAD^+ reduction from protein film voltammetry. Biochemistry 46: 3454–3464.

Barton SC, Pickard M, Vazquez-Duhalt R, Heller A. 2002: Electroreduction of O_2 to water at 0.6 V (SHE) at pH 7 on the "wired" *Pleurotus ostreatus* laccase cathode. Biosens Bioelectron 17: 1071–1074.

Barton SC, Gallaway J, Atanassov P. 2004. Enzymatic biofuel cells for implantable and microscale devices. Chem Rev 104: 4867–4886.

Barton SC, Sun Y, Chandra B, White S, Hone J. 2007. Mediated enzyme electrodes which combine micro- and nanoscale supports. Electrochem Solid State Lett 10: B96–B100.

Bertrand T, Jolivalt C, Briozzo P, Caminade E, Joly N, Madzak C, Mougin C. 2002. Crystal structure of a four-copper laccase complexed with an arylamine: Insights into substrate recognition and correlation with kinetics. Biochemistry 41: 7325–7333.

Besteman K, Zevenbergen MAG, Heering HA, Lemay SG. 2004. Direct observation of charge inversion by multivalent ions as a universal electrostatic phenomenon. Phys Rev Lett 93: 170802/1–170802/4.

Binyamin G, Heller A. 1999. Stabilization of wired glucose oxidase anodes rotating at 1000 rpm at 37°C. J Electrochem Soc 146: 2965–2967.

Blaedel WJ, Jenkins RA. 1975. Study of the electrochemical oxidation of reduced nicotinamide adenine dinucleotide. Anal Chem 47: 1337–1343.

Blanford CF, Heath RS, Armstrong FA. 2007. A stable electrode for high-potential, electrocatalytic O_2 reduction based on rational attachment of a blue copper oxidase to a graphite surface. Chem Commun 1710–1712.

Burgdorf T, Lenz O, Buhrke T, van der Linden E, Jones AK, Albracht SPJ, Friedrich B. 2005. [NiFe]-hydrogenases of *Ralstonia eutropha* H16: Modular enzymes for oxygen-tolerant biological hydrogen oxidation. J Mol Microbiol Biotechnol 10: 181–196.

Cammack R, Frey M, Robson R. 2001. Hydrogen as a Fuel: Learning from Nature. London: Taylor & Francis.

Cardosi MF. 1994. Hydrogen peroxide-sensitive electrode based on horseradish peroxidase-modified platinized carbon. Electroanalysis 6: 89–96.

Chaki NK, Vijayamohanan K. 2002. Self-assembled monolayers as a tunable platform for biosensor applications. Biosens Bioelectron 17: 1–12.

Chen D, Wang G, Li JH. 2007. Interfacial bioelectrochemistry: Fabrication, properties and applications of functional nanostructured biointerfaces. J Phys Chem C 111: 2351–2367.

Couto SR, Herrera JTT. 2006. Industrial and biotechnological applications of laccases: A review. Biotechnol Adv 24: 500–513.

Davis G, Hill HAO, Aston WJ, Higgins IJ, Turner APF. 1983. Bioelectrochemical fuel cell and sensor based on a quinoprotein, alcohol dehydrogenase. Enzyme Microb Technol 5: 383–388.

de Jongh TE, van Roon A-MM, Prudêncio M, Ubbink M, Canters GW. 2006. Click chemistry with an active site variant of azurin. Eur J Inorg Chem 3861–3868.

de Lumley-Woodyear T, Rocca P, Lindsay J, Dror Y, Freeman A, Heller A. 1995. Polyacrylamide-based redox polymer for connecting redox centers of enzymes to electrodes. Anal Chem 67: 1332–1338.

Fedurco M. 2000. Redox reactions of heme-containing metalloproteins: Dynamic effects of self-assembled monolayers on thermodynamics and kinetics of cytochrome c electron-transfer reactions. Coord Chem Rev 209: 263–331.

Ferguson-Miller S, Babcock GT. 1996. Heme/copper terminal oxidases. Chem Rev 7: 2889–2907.

Fitzpatrick PF. 2001. Substrate dehydrogenation by flavoproteins. Acc Chem Res 34: 299–307.

Fontecilla-Camps JC, Volbeda A, Cavazza C, Nicolet Y. 2007. Structure/function relationships of [NiFe]- and [FeFe]-hydrogenases. Chem Rev 107: 4273–4303.

Gooding JJ. 2005. Nanostructuring electrodes with carbon nanotubes: A review on electrochemistry and applications for sensing. Electrochim Acta 50: 3049–3060.

Gooding JJ, Wibowo R, Liu JQ, Yang WR, Losic D, Orbons S, Mearns FJ, Shapter JG, Hibbert DB. 2003. Protein electrochemistry using aligned carbon nanotube arrays. J Am Chem Soc 125: 9006–9007.

Hagen WR. 1989. Direct electron-transfer of redox proteins at the bare glassy-carbon electrode. Eur J Biochem 182: 523–530.

Hakulinen N, Kiiskinen LL, Kruus K, Saloheimo M, Paananen A, Koivula A, Rouvinen J. 2002. Crystal structure of a laccase from *Melanocarpus albomyces* with an intact trinuclear copper site. Nature Struct Biol 9: 601–605.

Hasan MN, Kwakernaak C, Sloof WG, Hagen WR, Heering HA. 2006. *Pyrococcus furiosus* 4Fe-ferredoxin, chemisorbed on gold, exhibits gated reduction and ionic strength dependent dimerization. J Biol Inorg Chem 11: 651–662.

Heering HA, Wiertz FGM, Dekker C, de Vries S. 2004. Direct immobilization of native yeast Iso-1 cytochrome *c* on bare gold: Fast electron relay to redox enzymes and zeptomole protein-film voltammetry. J Am Chem Soc 126: 11103–11112.

Heering HA, Williams KA, de Vries S, Dekker C. 2006. Specific vectorial immobilization of oligonucleotide-modified yeast cytochrome *c* on carbon nanotubes. Chem Phys Chem 7: 1705–1709.

Heller A. 2004. Miniature biofuel cells. Phys Chem Chem Phys 6: 209–216.

Heller A. 2006. Electron-conducting redox hydrogels: design, characteristics and synthesis. Curr Opin Chem Biol 10: 664–672.

Heller I, Kong J, Heering HA, Williams KA, Lemay SG, Dekker C. 2005. Individual single-walled carbon nanotubes as nanoelectrodes for electrochemistry. Nano Lett 5: 137–142.

Hoeben FJM, Heller I, Albracht SPJ, Dekker C, Lemay SG, Heering HA. 2008. Polymyxin-coated Au and carbon nanotube electrodes for stable [NiFe]-hydrogenase film voltammetry. Langmuir 24: 5925–5931.

Hudak NS, Barton SC. 2005. Mediated biocatalytic cathode for direct methanol membrane–electrode assemblies. J Electrochem Soc 152: A876–A881.

Ikeda T, Kobayashi D, Matsushita F, Sagara T, Niki K. 1993. Bioelectrocatalysis at electrodes coated with alcohol dehydrogenase, a quinohemoprotein with heme c serving as a built-in mediator. J Electroanal Chem 361: 221–228.

Jones AK, Sillery E, Albracht SPJ, Armstrong FA. 2002. Direct comparison of the electrocatalytic oxidation of hydrogen by an enzyme and a platinum catalyst. Chem Commun 866–867.

Kakehi N, Yamazaki T, Tsugawa W, Sode K. 2007. A novel wireless glucose sensor employing direct electron transfer principle based enzyme fuel cell. Biosens Bioelectron 22: 2250–2255.

Kamitaka Y, Tsujimura S, Setoyama N, Kajino T, Kano K. 2007. Fructose/dioxygen biofuel cell based on direct electron transfer-type bioelectrocatalysis. Phys Chem Chem Phys 9: 1793–1801.

Karyakin AA, Morozov SV, Karyakina EE, Zorin NA, Perelygin VV, Cosnier S. 2005. Hydrogenase electrodes for fuel cells. Biochem Soc Trans 33: 73–75.

Katz E, Willner I, Kotlyar AB. 1999. A non-compartmentalized glucose |O_2 biofuel cell by bioengineered electrode surfaces. J Electroanal Chem 479: 64–68.

Katz E, Shipway AN, Willner I. 2003. Biochemical fuel cells. In: Vielstich W, Gasteiger HA, Lamm A, editors. Handbook of Fuel Cells—Fundamentals, Technology and Applications. Volume 4. Chichester: Wiley. pp. 354–381.

Kenausis G, Taylor C, Katakis I, Heller A. 1996. "Wiring" of glucose oxidase and lactate oxidase within a hydrogel made with poly(vinyl pyridine) complexed with [Os(4,4'-dimethoxy-2,2'-bipyridine)$_2$Cl]$^{+/2+}$. J Chem Soc Faraday Trans 92: 4131–4136.

Klotzbach T, Watt M, Ansari Y, Minteer SD. 2006. Effects of hydrophobic modification of chitosan and Nafion on transport properties, ion-exchange capacities, and enzyme immobilization. J Membrane Sci 282: 276–283.

Kong J, Soh HT, Cassell AM, Quate CF, Dai HJ. 1998. Synthesis of individual single-walled carbon nanotubes on patterned silicon wafers. Nature 395: 878–881.

Kroutil W, Mang H, Edegger K, Faber K. 2004. Biocatalytic oxidation of primary and secondary alcohols. Adv Synth Catal 346: 125–142.

Langry KC, Ratto TV, Rudd RE, McElfresh MW. 2005. The AFM measured force required to rupture the dithiolate linkage of thioctic acid to gold is less than the rupture force of a simple gold–alkyl thiolate bond. Langmuir 21: 12064–12067.

Lee CK, Wang YM, Huang LS, Lin SM. 2007. Atomic force microscopy: Determination of unbinding force, off rate and energy barrier for protein–ligand interaction. Micron 38: 446–461.

Léger C, Jones AK, Albracht SPJ, Armstrong FAA. 2002. Effect of a dispersion of interfacial electron transfer rates on steady state catalytic electron transport in [NiFe]-hydrogenase and other enzymes. J Phys Chem B 106: 13058–13063.

Léger C, Elliott SJ, Hoke KR, Jeuken LJC, Jones AK, Armstrong FA. 2003. Enzyme electrokinetics: Using protein film voltammetry to investigate redox enzymes and their mechanisms. Biochemistry 42: 8653–8662.

Lévy R, Maaloum M. 2005. Specific molecular interactions by force spectroscopy: From single bonds to collective properties. Biophys Chem 117: 233–237.

Liu X, Ibrahim SK, Tard C, Pickett CJ. 2005. Iron-only hydrogenase: Synthetic, structural and reactivity studies of model compounds. Coord Chem Rev 249: 1641–1652.

Mahadevan V, Gebbink RJMK, Stack TDP. 2000. Biomimetic modeling of copper oxidase reactivity. Curr Opin Chem Biol 4: 228–234.

Mano N, Mao F, Heller A. 2003. Characteristics of a miniature compartment-less glucose–O_2 biofuel cell and its operation in a living plant. J Am Chem Soc 125: 6588–6594.

Mano N, Soukharev V, Heller A. 2006. A laccase-wiring redox hydrogel for efficient catalysis of O_2 electroreduction. J Phys Chem B 110: 11180–11187.

Mao F, Mano N, Heller A. 2003. Long tethers binding redox centers to polymer backbones enhance electron transport in enzyme "wiring" hydrogels. J Am Chem Soc 125: 4951–4957.

Mayer D, Ataka K, Heberle J, Offenhaeusser A. 2005. Scanning probe microscopic studies of the oriented attachment and membrane reconstitution of cytochrome c oxidase to a gold electrode. Langmuir 21: 8580–8583.

Morozov SV, Vignais PM, Cournac L, Zorin NA, Karyakina EE, Karyakin AA, Cosnier S. 2002. Bioelectrocatalytic hydrogen production by hydrogenase electrodes. Int J Hydrogen Energy 27: 1501–1505.

Moser CC, Farid TA, Chobot SE, Dutton PL. 2006. Electron tunneling chains of mitochondria. Biochim Biophys Acta 1757: 1096–1109.

Nakamura K, Go N. 2005. Function and molecular evolution of multicopper blue proteins. Cellular Mol Life Sci 62: 2050–2066.

Nicholls DG, Ferguson SJ. 2002. Bioenergetics 3. London: Elsevier.

Okuda J, Yamazaki T, Fukasawa M, Kakehi N, Sode K. 2007. The application of engineered glucose dehydrogenase to a direct electron–transfer-type continuous glucose monitoring system and a compartmentless biofuel cell. Anal Lett 40: 431–440.

Oubrie A, Rozeboom HJ, Kalk KH, Huizinga EG, Dijkstra BW. 2002. Crystal structure of quinohemoprotein alcohol dehydrogenase from *Comamonas testosteroni*: Structural basis for substrate oxidation and electron transfer. J Biol Chem 277: 3727–3732.

Page CC, Moser CC, Chen X, Dutton PL. 1999. Natural engineering principles of electron tunneling in biological oxidation–reduction. Nature 402: 47–52.

Palmore GTR, Kim H-H. 1999. Electro-enzymic reduction of dioxygen to water in the cathode compartment of a biofuel cell. J Electroanal Chem 464: 110–117.

Palmore GTR, Bertschy H, Bergens SH, Whitesides GM. 1998. A methanol/dioxygen biofuel cell that uses NAD^+-dependent dehydrogenases as catalysts: Application of an electroenzymatic method to regenerate nicotinamide adenine dinucleotide at low overpotentials. J Electroanal Chem 443: 155–161.

Parkin A, Seravalli J, Vincent KA, Ragsdale SW, Armstrong FA. 2007. Rapid and efficient electrocatalytic CO_2/CO interconversions by *Carboxydothermus hydrogenoformans* CO dehydrogenase I on an electrode. J Am Chem Soc 129: 10328–10329.

Patolsky F, Weizmann Y, Willner, I. 2004. Long-range electrical contacting of redox enzymes by SWCNT connectors. Angew Chem Int Ed 43: 2113–2117.

Piontek K, Antorini M, Choinowski T. 2002. Crystal structure of a laccase from the fungus *Trametes versicolor* at 1.90 Å resolution containing a full complement of coppers. J Biol Chem 277: 37663–37669.

Pishko MV, Katakis I, Lindquist SE, Ye L, Gregg BA, Heller A. 1990. Direct electron exchange between graphite electrodes and an adsorbed complex of glucose oxidase and an osmium-containing redox polymer. Angew Chem 102: 109–111.

Ralph TR, Hogarth MP. 2002. Catalysis for low temperature fuel cells. Platinum Metals Rev 46: 3–14.

Richter OMH, Ludwig B. 2003. Cytochrome c oxidase—Structure, function, and physiology of a redox-driven molecular machine. Rev Physiol Biochem Pharmacol 147: 47–74.

Schäfer G, Engelhard M, Müller V. 1999. Bioenergetics of the archaea. Microbiol Mol Biol Rev 63: 570–620.

Shklovskii BI. 1999. Screening of a macroion by multivalent ions: Correlation-induced inversion of charge. Phys Rev E 60: 5802–5811.

Solomon EI, Sundaram UM, Machonkin TE. 1996. Multicopper oxidases and oxygenases. Chem Rev 96: 2563–2605.

Solomon EI, Chen P, Metz M, Lee SK, Palmer AE. 2001. Oxygen binding, activation, and reduction to water by copper proteins. Angew Chem Int Ed 40: 4570–4590.

Soukharev V, Mano N, Heller A. 2004. A four-electron O_2-electroreduction biocatalyst superior to platinum and a biofuel cell operating at 0.88 V. J Am Chem Soc 126: 8368–8369.

Tasis D, Tagmatarchis N, Bianco A, Prato M. 2006. Chemistry of carbon nanotubes. Chem Rev 106: 1105–1136.

Topcagic S, Minteer SD. 2006. Development of a membraneless ethanol/oxygen biofuel cell. Electrochim Acta 51: 2168–2172.

Vincent KA, Cracknell JA, Lenz O, Zebger I, Friedrich B, Armstrong FA. 2005. Electrocatalytic hydrogen oxidation by an enzyme at high carbon monoxide or oxygen levels. Proc Natl Acad Sci USA 102: 16951–16954.

Vincent KA, Cracknell JA, Clark JR, Ludwig M, Lenz O, Friedrich B, Armstrong FA. 2006. Electricity from low-level H_2 in still air—An ultimate test for an oxygen tolerant hydrogenase. Chem Commun 5033–5035.

Vincent KA, Parkin A, Armstrong FA. 2007. Investigating and exploiting the electrocatalytic properties of hydrogenases. Chem Rev 107: 4366–4413.

Weisel JW, Shuman H, Litvinov RI. 2003. Protein–protein unbinding induced by force: Single-molecule studies. Curr Opin Struc Biol 13: 227–235.

Wikstrom M, Verkhovsky MI. 2006. Towards the mechanism of proton pumping by the haem–copper oxidases. Biochim Biophys Acta 1757: 1047–1051.

Willner B, Katz E, Willner I. 2006. Electrical contacting of redox proteins by nanotechnological means. Curr Opin Biotech 17: 589–596.

Willner I, Baron R, Willner B. 2007. Integrated nanoparticle–biomolecule systems for biosensing and bioelectronics. Biosens Bioelectron 22: 1841–1852.

Wohlfahrt G, Witt S, Hendle J, Schomburg D, Kalisz HM, Hecht HJ. 1999. 1.8 and 1.9 Å resolution structures of the *Penicillium amagasakiense* and *Aspergillus niger* glucose oxidases as a basis for modelling substrate complexes. Acta Crystallogr D 55: 969–977.

Xiao Y, Patolsky F, Katz E, Hainfeld JF, Willner I. 2003. "Plugging into enzymes": Nanowiring of redox enzymes by a gold nanoparticle. Science 299: 1877–1881.

Xu F, Palmer AE, Yaver DS, Berka RM, Gambetta GA, Brown SH, Solomon EI. 1999. Targeted mutations in a *Trametes villosa* laccase. J Biol Chem 274: 12372–12375.

Yahiro AT, Lee SM, Kimble DO. 1964. Bioelectrochemistry. I. Enzyme utilizing bio-fuel cell studies. Biochim Biophys Acta 88: 375–383.

Yan YM, Zheng W, Su L, Mao LQ. 2006. Carbon-nanotube-based glucose/O_2 biofuel cells. Adv Mater 18: 2639–2643.

Yu X, Chattopadhyay D, Galeska I, Papadimitrakopoulos F, Rusling JF. 2003. Peroxidase activity of enzymes bound to the ends of single-wall carbon nanotube forest electrodes. Electrochem Commun 5: 408–411.

Yue PL, Lowther K. 1986. Enzymatic oxidation of C_1 compounds in a biochemical fuel cell. Chem Eng J 33: B69–B77.

Zayats M, Katz E, Baron R, Willner I. 2005. Reconstitution of apo-glucose dehydrogenase on pyrroloquinoline quinone-functionalized Au nanoparticles yields an electrically contacted biocatalyst. J Am Chem Soc 127: 12400–12406.

Zheng W, Li QF, Su L, Yan YM, Zhang J, Mao LQ. 2006. Direct electrochemistry of multicopper oxidases at carbon nanotubes noncovalently functionalized with cellulose derivatives. Electroanalysis 18: 587–594.

CHAPTER 18

Metalloporphyrin Catalysts of Oxygen Reduction

ROMAN BOULATOV

Department of Chemistry, University of Illinois, 600 South Mathews Avenue, Urbana, IL 61801, USA

18.1 INTRODUCTION

Sometime probably two billion years before humans became interested in efficient catalysts for four-electron, four-proton reduction of O_2 to H_2O, the so-called oxygen reduction reaction (ORR),

$$O_2 + 4H^+ + 4e^- \rightleftharpoons 2H_2O \qquad E^o = 1.23 \text{ V} \qquad (18.1)$$

primitive organisms on Earth faced a similar problem [Knoll, 2003]. The appearance of photosynthesis and the concomitant generation of O_2 produced evolutionary pressure to deal with this new chemical, first as a detoxification strategy for anaerobes and later as a new form of energy metabolism. The solution that life came up with is quite different from that dominating current research into low temperature fuel cell ORR catalysts: instead of nanoparticulate Pt or related bulk metals, biological O_2 reduction is catalyzed predominantly by hemes, Fe complexes of a planar macrocyclic ligand, porphyrin (Fig. 18.1). Biological ORR catalysts have to meet a rather different set of requirements than a fuel cell catalyst, yet the predominance of heme in aerobic metabolism over the past two billion years justifies a serious look at metalloporphyrins as potential ORR catalysts for low temperature fuel cells.

In addition to their proven capacity to catalyze a highly efficient and rapid reduction of O_2 under ambient conditions (e.g., cytochrome c oxidase, the enzyme that catalyzes the reduction of $>90\%$ of O_2 consumed by a mammal, captures $>80\%$ of the free energy of ORR at a turnover frequency of >50 O_2 molecules per second per site), metalloporphyrins are attractive candidates for Pt-free cathodes. Probably the major impetus for a search for Pt-free cathodic catalysts for low temperature fuel cells is

Fuel Cell Catalysis. Edited by Marc T. M. Koper
Copyright © 2009 John Wiley & Sons, Inc.

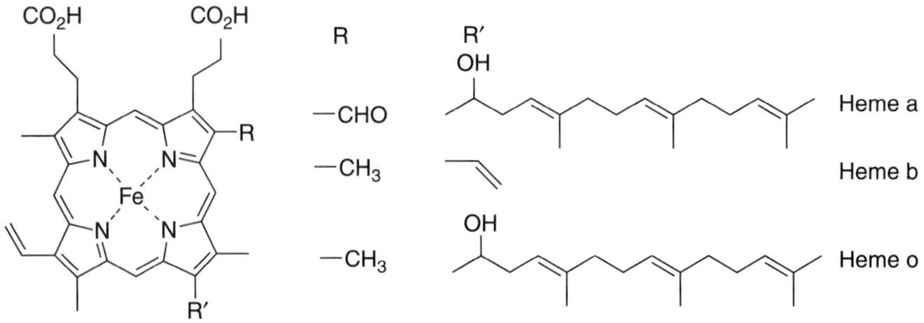

Figure 18.1 Chemical structures of some common naturally occurring hemes.

supply security, with the high cost or limited reserves of Pt being secondary motivators. First, over 90% of Pt in the early 2000s was mined in only two countries: South Africa and Russia [Dodelet, 2006]. In contrast, in April 2007, the 10 largest importing countries of crude oil into the US have accounted for only 75% of all imports, with over 20% coming from Canada and the UK [EIA, 2007]. Second, at a mid-2007 price of US$1300–1400 per troy ounce of Pt, and a Pt loading of 0.2 g_{Pt}/kW, Pt would contribute US$675 to the cost of a 75 kW fuel cell, which is not insignificant. Third, it is expected that the production of 10 million vehicles per year each incorporating a 75 kW fuel cell would require about 150 metric tons of Pt [Dodelet, 2006]; the annual demand for Pt in the early 2000s was about 200 metric tons, approximately 10% above the production in the same year. Although proven reserves of Pt exceeded 30,000 metric tons worldwide in the mid 2000s, wide use of Pt-utilizing fuel cells would demand that recovery of Pt from fuel cell stacks be much more complete than from autocatalysts.

The search for Pt-free ORR catalysts is being pursued along multiple avenues: this chapter summarizes the work reported by mid 2007 on ORR catalysis by metalloporphyrins, particularly Fe or Co porphyrin complexes, in four contexts: (i) respiratory ORR catalysis; (ii) ORR catalysis by simple metalloporphyrins; (iii) ORR catalysis by cofacial metalloporphyrins and porphyrinoids; and (iv) ORR catalysis by biomimetic metalloporphyrins. Metal complexes of porphyrinoids, such as corroles or phthalocyanines, are mentioned only when related directly to metalloporphyrin catalysis. Non-porphyrinoid molecular catalysts, such as Cu complexes and metal complexes of Schiff-type ligands, are omitted, as are catalysts obtained by pyrolysis of various precurosor complexes, including metalloporphyrin precursors. This last topic has been reviewed recently [Zagal et al., 2006]. The present review is not intended to be exhaustive, but rather representative of certain directions in studies of metalloporphyrins for ORR catalysis with the emphasis on molecular systems with as well-defined stereoelectronic properties as possible and on the molecular mechanism of O_2 reduction rather than catalyst performance parameters under realistic operating conditions of a fuel cell. The selection of the material in this review is necessarily subjective, and important work in the field omitted here is covered elsewhere [Zagal, 1992, 2003; Zagal et al., 2006; Yeager, 1984; Vasudevan et al., 1990; Adzic, 1998].

18.2 RESPIRATORY O$_2$ REDUCTION CATALYSIS

18.2.1 Basic Aspects of Energy Metabolism

Life requires constant dissipation of energy. This energy comes from the environment, in the form either of sunlight or of food (reduced organic matter). Respiration is the most efficient way to extract energy from food [Alberts et al., 2002]. In respiration, food is ultimately oxidized by an environmental oxidant, such as O$_2$ in aerobic respiration or a host of other bioavailable oxidants, from nitrate and sulfate to ferric ion in anaerobic respirations [Moodie and Ingledew, 1990] (Fig. 18.2). Energy metabolism is a multistep redox process, starting with food being converted to reduced electron carriers, such as the reduced forms of nicotinamide adenine dinucleotide and flavin adenine dinucleotide (NADH and FADH$_2$, respectively), which enter the respiratory cycle. This cycle proceeds by enzyme-catalyzed electron transfers between electron carriers of increasing redox potential, from strong reductants such as NADH

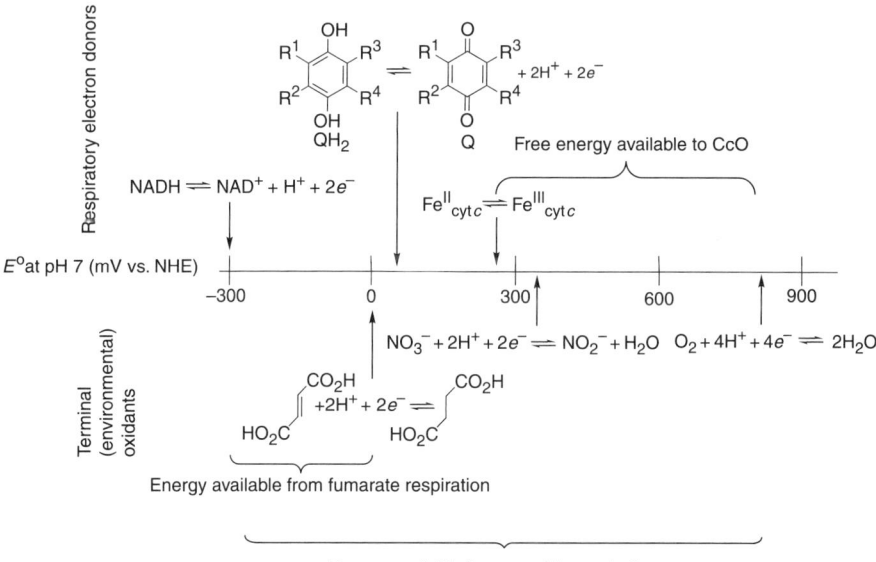

Figure 18.2 Summary of respiratory energy flows. Foods are converted into the reduced form of nicotinamide adenine dinucleotide (NADH), a strong reductant, which is the most reducing of the respiratory electron carriers (donors). Respiration can be based on a variety of terminal oxidants, such as O$_2$, nitrate, or fumarate. Of those, O$_2$ is the strongest, so that aerobic respiration extracts the largest amount of free energy from a given amount of food. In aerobic respiration, NADH is not oxidized directly by O$_2$; rather, the reaction proceeds through intermediate electron carriers, such as the quinone/quinol couple and cytochrome c. The most efficient respiratory pathway is based on oxidation of ferrocytochrome c (Fe$^{II}_{cytc}$) with O$_2$ catalyzed by cytochrome c oxidase (CcO). Of the 550 mV difference between the standard potentials of cytochrome c and O$_2$, CcO converts 450 mV into proton-motive force (see the text for further details).

to weak reductants, quinols, and ferrocytochrome c. Oxidation of these weakest biological reductants requires a terminal (environmental) oxidant, such as O_2. The stronger this terminal oxidant, the larger the amount of energy an organism would extract from a given amount of food (Fig. 18.2). Molecular oxygen is by far the strongest bioavailable oxidant on Earth; it has high permeation rates across biological membranes [Ligeza et al., 1998; Wittenberg and Wittenberg, 2003] and its fully reduced form, H_2O, is nontoxic. (However, partially reduced oxygen species, such as superoxide ($O_2^{\cdot-}$), hydroperoxyl radical ($HO_2^{\cdot}$), hydrogen peroxide (H_2O_2), and hydroxyl radical (HO˙) are highly toxic.) Hence, O_2 is an attractive terminal oxidant, and indeed all multicellular organisms are obligatory aerobes, whereas anaerobic respiration is limited to unicellular organisms.

Although respiration is a sequence of electron transfer steps, not unlike reduction of O_2 in a fuel cell, living organisms employ fundamentally different schemes to couple the free energy of O_2 reduction to their energy-dissipative processes. Living organisms cannot utilize electron currents; instead, a portion of the free energy of each electron transfer step in the respiratory chain is captured in the form of a transmembrane electrochemical proton (H^+) gradient [Nicholls, 1982]. This proton gradient powers the operation of ATP synthase that converts ADP (adenosine diphosphate) and phosphate ion into ATP (adenosine triphosphate) [Dimroth et al., 2003; Senios et al., 2002; Yoshida, 2001]. Exergonic hydrolysis of ATP powers most energy-consuming biological processes, from biosynthesis to motility [Alberts et al., 2002]. To effect the conversion of the free energy of redox processes into an electrochemical gradient during respiration, the suite of respiratory enzymes (e.g., NADH dehydrogenase, cytochmore bc_1, and cytochrome c oxidase; Fig. 18.3) is embedded in a relatively H^+-impermeable membrane [Schults and Chan, 2001; Hosler et al., 2006]. The respiratory enzymes maintain the membrane-encapsulated space at a pH different from that on the outside by two mechanisms. One involves active translocation of H^+ from the less acidic to the more acidic environment (proton pumps). The other relies on carrying out proton-consuming and proton-releasing redox processes at the opposite site of the membrane thanks to the proper location of catalytic moieties. As a result of relying on the proton-motive force (PMF) rather than electron-motive force (EMF) as do batteries and fuel cells, organisms use fairly reducing reductants for O_2 without losing much free energy of the ORR. For example, at pH 7, the standard redox potential of cytochrome c, which is the natural electron donor for cytochrome c oxidase, is >500 mV more reducing that that of the O_2/H_2O couple; however, only 100 mV of this difference is dissipated (see below).

Although it is highly exergonic, four-electron, four-proton reduction of O_2 to H_2O is slow without catalysis. The inertness of O_2 arises from it being a very weak base, having a low affinity for H atoms and an unfavorable one-electron reduction potential in neutral aqueous media, and being a ground-state triplet [Taube, 1986]. This electronic configuration disfavors direct reactions between O_2 and most organic compounds, which are ground-state singlets. This inertness of O_2 allows the existence of reduced organic matter in the oxidizing atmosphere of modern Earth. However, it also means that reduction of O_2 at rates and potentials that make the ORR useful for energy metabolism requires catalysis. This catalysis is affected by heme-containing enzymes generically called terminal oxidases, which are divided into two large

18.2 RESPIRATORY O_2 REDUCTION CATALYSIS

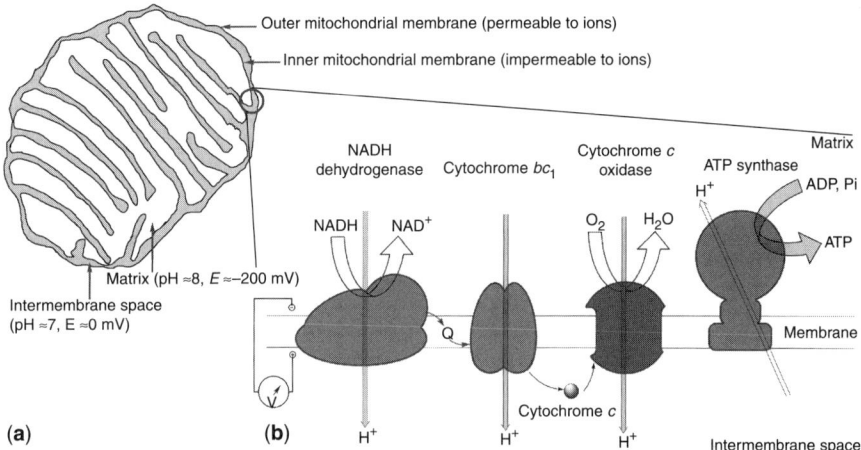

Figure 18.3 (a) Schematic representation of a mitochondrion, which has two distinct compartments (matrix and intermembrane space), separated by the inner mitochondrial membrane. (b) Schematic representation of a mammalian respiratory chain, which is embedded in the inner mitochondrial membrane. The chain comprises three sets of redox reactions, mediated by at least three enzymes. Electron transfer from the most reduced electron carrier (NADH) to lipid-soluble quinone is mediated by NADH dehydrogenase, and generates NAD^+ and quinol (see Fig. 18.2 for the chemical structure of the quinol). Quinol is oxidized back to quinone by ferricytochrome c at the enzyme cytochrome bc_1. The resultant ferrocytochrome c is the weakest biological reductant of the chain, and can only be oxidized by the terminal oxidant, O_2. The reaction is catalyzed by cytochrome c oxidase. Although thermodynamically favorable, direct reactions between various electron donors of the chain are slow. The respiratory enzymes capture a part of the free energy of each redox reaction by translocating protons from the matrix, which is deficient in H^+, to the intermembrane space (gray block arrows). The thermodynamically favorable backflow of H^+ can only occur through ATP synthase, the enzyme that converts the free energy of this flow into ATP. White and gray block arrows depict exergonic and endergonic processes, respectively.

classes: the well-studied superfamily of heme/Cu oxidases [Hosler et al., 2006; Brzezinski and Adelroth, 2006; Branden et al., 2006; Wikstrom, 2004; Pitcher and Watmough, 2004; Pereira et al., 2001; Mills, 2000; Michel, 1999; Rich and Moody, 1997] and a less well-understood group of cytochrome bd oxidases [Junemann, 1997]. Heme/Cu oxidases are widely distributed throughout the biosphere, whereas cytochromes bd are limited to certain prokaryotes (unicellular organisms lacking organells or nucleus). Heme/Cu oxidases contain a binuclear heme/Cu site, whereas the catalytic site in cytochromes bd is composed of two hemes. Cytochromes bd have been reviewed in the context of fuel cells elsewhere [Boulatov, 2006], and are not discussed here.

18.2.2 ORR Catalysis by Heme/Cu Terminal Oxidases

Three distinct types of heme/Cu oxidases are currently recognized: cytochrome c oxidases (CcOs or COX), quinol oxidases, and cytochrome cbb_3 oxidases [Pitcher and Watmough, 2004], with CcOs being limited to eukaryotic organisms. The

642 METALLOPORPHYRIN CATALYSTS OF OXYGEN REDUCTION

common features of all these enzymes are (i) a binuclear heme/Cu site at which O_2 is reduced and (ii) proton pumping. All these enzymes are embedded in a membrane. CcOs and cytochrome cbb_3 oxidases catalyze oxidation of reduced cytochrome c (ferrocytochrome c) by O_2. Certain CcOs can also utilize specific high-potential iron–sulfur proteins (HiPIPs) as electron donors [Pereira et al., 1999]. Quinol oxidases catalyze two-electron, two-proton oxidation of a stronger reductant, quinol, to quinone by O_2 [Abramson et al., 2000].

CcOs from various organisms, as well as CcOs and quinol oxidases, are fairly homologous structurally and functionally, and both are distinct from cytochromes cbb_3. Most, if not all, CcOs and quinol oxidases require two subunits for catalytic activity (subunits I and II; Fig. 18.4), although some, such as mammalian CcO, may contain as many as 11 more subunits of unknown functions [Abramson et al.,

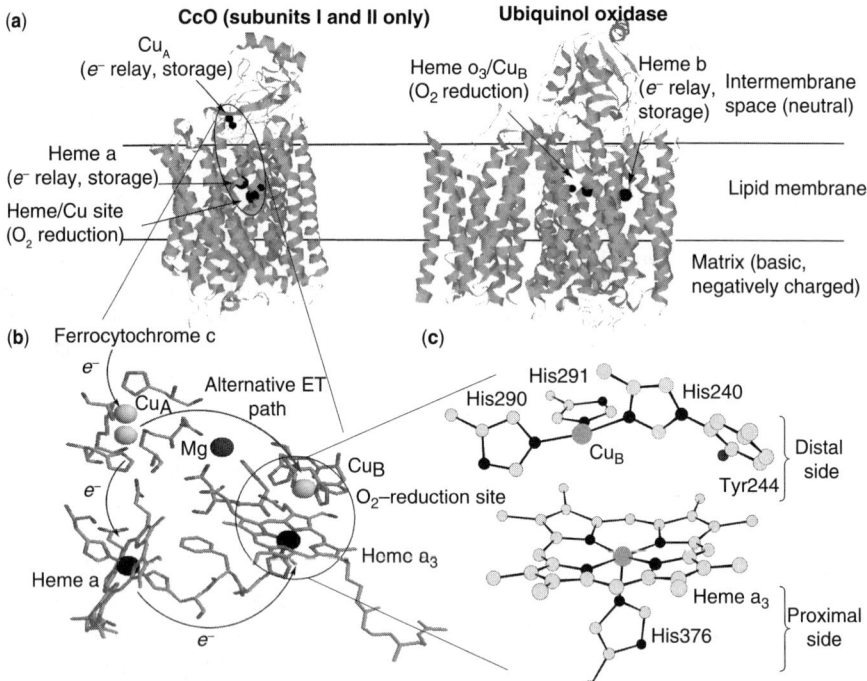

Figure 18.4 Structures of heme/Cu oxidases at different levels of detail. (a) Position of the redox-active cofactors relative to the membrane of CcO (left, only two obligatory subunits are shown) and quinol oxidase (right). (b) Electron transfer paths in mammalian CcO. Note that the imidazoles that ligate six-coordinate heme a and the five-coordinate heme a_3 are linked by a single amino acid, which can serve as a "wire" for electron transfer from ferroheme a to ferriheme a_3. (c) The O_2 reduction site of mammalian CcO; the numbering of the residues corresponds to that in the crystal structure of bovine heart CcO. The subscript 3 in heme a_3 and heme o_3 signifies the heme that binds O_2. The structures were generated using coordinates deposited in the Protein Data Bank, 1ar1 [Ostermeier et al., 1997]; 1fft [Abramson et al., 2000] (a) and 1occ [Tsukihara et al., 1996] (b, c).

2001]. Crystal structures of several CcOs and of a quinol oxidase have been reported [Abramson et al., 2000; Ostermeier et al., 1997; Tsukihara et al., 1996]. The O_2 reduction site is buried deep inside subunit I, approximately in the middle of the membrane that hosts the enzyme (Fig. 18.4). The site contains a heme coordinated by a single axial imidazole moiety at its proximal site (see Fig. 18.4c for the definition of the proximal and distal sites), and is reminiscent of the O_2-binding site of myoglobin, the simplest hemoprotein [Frauenfelder et al., 2003]. At the distal site of the heme, a Cu ion coordinated to three imidazoles is located. The Fe–Cu distance appears to vary around 5 Å, depending on the exogenous ligands at the two ions. The 5 Å distance is suitable for the formation of a bridging peroxide, although the intermediacy of such a species in the catalytic cycle remains uncertain. Finally, in most CcOs and quinol oxidases, one of the Cu-ligating imidazoles is linked covalently to a phenol residue of a tyrosine. This link is thought to be formed post-translationally, and is routinely taken as an indication that a phenoxyl radical is formed during the initial turnover of the enzyme, i.e., reduction of the very first molecule of O_2 after the enzyme is assembled [Rogers and Dooley, 2001].

In addition to the catalytic heme/Cu site, all heme/Cu oxidases contain a six-coordinate heme within the catalytic subunit, and all CcOs also contain a binuclear Cu site (so called Cu_A to distinguish it from the Cu ion of the O_2-reduction site, which is called Cu_B) in subunit II. Some CcOs have one more redox-active prosthetic group, a six-coordinate heme, usually located in the same subunit II. On the other hand, subunit II of quinol oxidases appears to lack redox cofactors. The physiological functions of these sites are (i) to conduct electrons from the docking site of the external electron donor (ferrocytochrome c or quinol) on the periphery of the enzyme to the catalytic site deep within the enzyme; (ii) to control the redox potentials of the catalytic site (redox cooperativity); (iii) possibly to store reducing equivalents during the turnover; and (iv) potentially in the proton pump [Brzezinski, 2006]. The presence of the additional electron relay sites in CcOs compared with quinol oxidases may be due to the different physicochemical properties of the electron donor that these two types of terminal oxidases utilize. Both quinol and quinone are lipophilic, water-insoluble molecules that are confined to the membrane within which quinol oxidase is embedded. Quinol binds the enzyme fairly close to the O_2 reducing site. In contrast, cytochrome c of CcO is a water soluble protein that binds CcO at a site exposed to the aqueous intermembrane space. Its docking site is at least several ångströms farther from the catalytic site, making direct electron tunneling from the ferrocytochrome c to the six-coordinate heme site too slow and requiring an intermediate electron relay site in the form of Cu_A.

In the thermodynamically redox-stable "resting state," CcOs all Cu ions are in the Cu^{II} state and all hemes are Fe^{III}. From this state, CcOs can be reduced by one to four electrons. One-electron reduced CcOs are aerobically stable with the electron delocalized over the Cu_A and heme a sites. The more reduced forms—mixed-valence (two-electron reduced), three-electron reduced, and fully (four-electron) reduced—bind O_2 rapidly and reduce it to the redox level of oxide (−2 oxidation state) within <200 μs [Wikstrom, 2004; Michel, 1999]. This rate is up to 100 times faster than the average rate of electron transfer through the mammalian respiratory chain under normal physiological conditions, which is one electron transfer per 5–20 ms [Alberts et al., 2002]. As a consequence of this mismatch in rates, (i) the predominant redox state of CcOs that bind O_2 under normal physiological conditions must be mixed-valence, whereas three-electron and fully reduced forms may be important under hypoxic conditions and (ii) in vivo oxidation of four molecules of ferrocytochrome c and reduction of one molecule of O_2 are kinetically uncoupled, i.e., once bound, O_2 is reduced all the way to H_2O *regardless* of the rate at which electrons flow in the rest of the respiratory chain. Conceivably, this mechanism minimizes the lifetime of CcO bound to any peroxo-level intermediates; thereby minimizing the probability that such an intermediate would dissociate from the enzyme into the solution as the enzyme waits for an electron from ferrocytochrome c.

CcOs accomplish this kinetic decoupling by at least two mechanisms. First, the redox potential of the catalytic heme/Cu site depends strongly on the redox state of the electron transfer site(s), the six-coordinate heme a, and/or Cu_A cofactors (redox cooperativity [Rich and Moody, 1997]). The atomic mechanism of this redox cooperativity is not known, but it manifests itself in the fact that in a singly reduced CcO, the electron is delocalized over the six-coordinate heme and Cu_A sites, whereas the O_2 reducing site remains in the aerobically stable Fe^{III}/Cu^{II} state. However, the arrival of the second electron (i.e., the formation of mixed-valence CcO) results in redistribution of *both* external electrons almost exclusively onto the catalytic site. The catalytic site is now in the Fe^{II}/Cu^{I} redox state, capable of binding O_2 and reducing it by four electrons. Hence, a major question in the mechanism of O_2 reduction by CcOs and quinol oxidases, which are generally assumed to follow similar mechanisms, is the origin of the additional two electrons required for the four-electron reduction of O_2.

The current consensus mechanism of O_2 reduction by mixed-valence CcO is shown schematically in Fig. 18.5. It is derived mostly from single-turnover spectroscopic experiments. Fully oxidized (compound H) and one-electron reduced (compound E) states of CcO contain an oxidized catalytic heme/Cu site (Fe^{III}/Cu^{II}), which has no affinity for O_2. As a result of the redox cooperativity between the Cu_A and heme a cofactors and the heme/Cu site, reduction of compound E by another molecule of ferrocytochrome c results in a two-electron reduction of the heme/Cu site (generating the Fe^{II}/Cu^{I} state), which rapidly binds O_2. The resulting adduct, called compound A, contains an oxyheme moiety that is very similar to that found in oxymyoglobin. It is regarded formally as a superoxo ($O_2^{\cdot -}$) complex of ferriheme, i.e., binding of O_2 to a ferroheme results in formal one-electron reduction of O_2. It is the inability of Fe^{III} to reduce O_2 by one electron that explains the lack of affinity of ferriheme to O_2. In compound A, this bound O_2 interacts little if at all with Cu_B or the phenol residue, as surmised from the fact the O–O vibrational frequency is very similar to that of isolated $O_2^{\cdot -}$. The next spectroscopically observable intermediate is so-called compound P_M. By this stage of the catalytic cycle, both atoms of the O_2 molecule are reduced to the "oxide" level (O^{2-}), with one atom having been released as H_2O (or bound to Cu^{II} as hydroxide, OH^-, according to calculations [Blombers, 2006]) and the other atom remaining bound to Fe as a terminal oxide. Two electrons for this four-electron reduction come from the oxidation of Fe from Fe^{II} in compound R to Fe^{IV} (ferryl) in compound P_M; oxidation of Cu_B yields one electron and oxidation of the

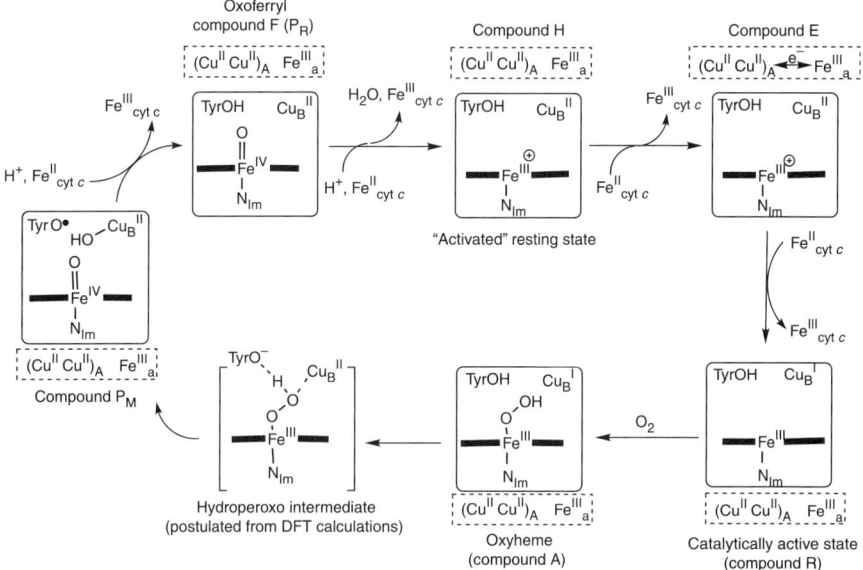

Figure 18.5 Plausible sequence of steps responsible for rapid and selective reduction of O_2 to H_2O by mixed-valence CcO. The square frames signify the catalytic site (Fig. 18.4c); imidazole ligation of Cu_B is omitted for clarity; in some or all intermediates, Cu_B may additionally be ligated by an exogenous ligand, such as H_2O (in Cu^I) or OH^- (in Cu^{II}); such ligation is not established, and hence is omitted in all but compound P_M and the putative hydroperoxo intermediate. The dashed frames signify the noncatalytic redox cofactors. Typically used phenomenological names of the spectroscopically observed intermediates (compounds A, E, H, etc.) are also indicated.

phenol of Tyr244 provides the fourth electron and a proton. Note that reduction of O_2 required no electron transfer between a ferrocytochrome c and CcO, because only those redox forms of CcO that contain all necessary reducing equivalents at the heme/Cu site have affinity to O_2.

The absence of any spectroscopically observable intermediates between compound A and compound P_M requires electron structure computations to understand the atomistic basis for rapid activation of the O–O bond in CcO. The most recently published computational studies [Blomberg and Siegbahn, 2006], which were carried out at the density functional theory (DFT) level (using the unrestricted B3LYP functional with double-ζ basis sets that included core potentials for Cu and Fe), suggest that this activation proceeds through a hydroperoxo intermediate, which is generated by oxidizing Cu_B and deprotonating the phenyl residue of Tyr244. These computations predict that a surprisingly large barrier (13 kcal/mol) separates this hydroperoxo intermediate from compound P_M. Depending on the computational procedure, the hydroperoxo intermediate has a very small to moderate (<10 kcal/mol) barrier for reversion to compound A, making it difficult, if not impossible, to observe spectroscopically, which is consistent with the experiment. No energy minima were found for a species in which Cu_B and Fe are bridged by a peroxo (O_2^{2-}) moiety, although this nuclear configuration lies on the reaction path to the hydroperoxo intermediate. Decay of the hydroperoxo intermediate to compound P_M is calculated to require oxidation of Tyr244 to the neutral phenoxyl radical and transfer of the terminal O of the hydroperoxyl group onto Cu_B as hydroxide. The A → P_M conversion is calculated to be close to thermoneutral (potential energy change between -1 and $+6$ kcal/mol).

This and related computational work suggests the importance of an interaction between the terminal O atom of the heme-bound O_2 ligand and the distal Cu_B and phenol of Tyr244 (Fig. 18.4c), which is reminiscent of the mechanism of the O–O bond activation in peroxidases. Just a single H^+ is needed to affect the four-electron reduction, and its primary role appears to be stabilizing the oxygen ligands at the peroxo level. The P_M intermediate is slowly (milliseconds) reduced to the resting state of the enzyme as electrons are delivered by ferrocytochrome c. This reduction is coupled to proton pumping.

The major conclusion from the studies of the mechanism of O_2 reduction by CcO is that *formation of a peroxo-level intermediate bridging two metal ions is not a prerequisite for four-electron reduction, at least in molecular complexes.*

As mentioned earlier, all terminal oxidases perform two bioenergetic roles. One is to clear the respiratory chain of low potential electrons by oxidizing the least-reducing respiratory electron carrier and enabling continuous flow of electrons from food to NAD^+ to quinones. The other is to convert part of the free energy of the reduction of the terminal oxidant into the proton-motive force. The redox potential of cytochrome c (about 250 mV at pH 7; Fig. 18.2) determines the overall potential drop (about 550 mV) available for the NADH dehydrogenase and cytochrome bc components of the respiratory chain (Fig. 18.3). Approximately 450 mV of the 550 mV difference between the standard redox potentials of the ferri-/ferrocytochrome c couple and the O_2/H_2O couple is captured by CcO as the electrochemical proton gradient. Two mechanisms are involved (Fig. 18.6). First, the enzyme draws four protons required for the reduction of O_2 from the basic, negatively charged site of the membrane (matrix; Fig. 18.3a). The four electrons, on the other hand, come from the opposite site of the membrane (the intermembrane space, Fig. 18.3a). The annihilation of these opposite charges at the O_2 reduction site—Reactions (18.3) and (18.4) in Fig. 18.6—is equivalent to the translocation of four charges across the membrane against the electrostatic potential. This consumes about 200 mV (the value of the transmembrane gradient) of the 550 mV difference between the standard potentials of the ORR and the oxidation half-reaction—Reaction (18.2) in Fig. 18.6 [Wikstrom, 2004]. CcO expends about another 220 mV by physically moving four protons from the matrix to the intermembrane space (IMS) of the mitochondrion (Fig. 18.3a), which comprises the proton pump. The mechanism of proton pumping is unknown [Brzezinski, 2006; Mills, 2000; Michel, 1999]. As a result, out of about 550 mV of the potential difference between the electron donor (ferrocytochrome c) and the electron acceptor (O_2), CcO captures about 450 mV (>80%) in a form that can be directly utilized by the cell to satisfy its energy-dissipating requirements. The energy-transducing efficiency of quinol oxidases is about 45%. These enzymes utilize a stronger reductant (e.g., ubiquinol, about 50 mV) [Ferguson-Miller, 1996] and operate against a lower electrochemical gradient (about180 mV) [Alberts et al., 2002].

18.3 EXPERIMENTAL APPROACHES TO STUDYING ORR CATALYSIS

Oxidation half-reaction	$(Fe^{II}_{cyt\,c})_{IMS} \rightleftharpoons (Fe^{III}_{cyt\,c})_{IMS} + e^-_{IMS}$	$\Delta E = -250$ mV	(18.2)
	$H^+_{matrix} \rightleftharpoons H^+_{heme/Cu}$	$\Delta E = -60$ mV	(18.3)
	$e^-_{IMS} \rightleftharpoons e^-_{heme/Cu}$	$\Delta E = -140$ mV	(18.4)
Proton pump	$H^+_{matrix} \rightleftharpoons H^+_{IMS}$	$\Delta E = -220$ mV	(18.5)
ORR	$(O_2 + 4e^- + 4H^+)_{heme/Cu} \rightleftharpoons 2H_2O$	$\Delta E = 800$ mV	(18.6)
Overall reaction	$O_2 + 4(Fe^{II}_{cyt\,c})_{IMS} + 8H^+_{matrix} \rightleftharpoons 2H_2O + 4(Fe^{III}_{cyt\,c})_{IMS} + 4H^+_{IMS}$	$\Delta E = 100$ mV	(18.7)

Figure 18.6 Energetics of the ORR at the heme/Cu site of CcO; the enzyme couples oxidation of ferrocytochrome c (standard potential about -250 mV; all potentials are listed with respect to a normal hydrogen electrode) to reduction of O_2 (standard potential at pH 7 800 mV). Of the 550 mV difference, only 100 mV is dissipated to drive the reaction; 220 mV is expanded to translocate four protons from the basic matrix compartment to the acidic IMS (intermembrane space). In addition 200 mV is converted into transmembrane electrostatic potential as ferrocytochrome is oxidized in the IMS, but the charge-compensating protons are taken from the matrix. The potentials are approximate.

18.3 EXPERIMENTAL APPROACHES TO STUDYING ORR CATALYSIS BY SYNTHETIC METALLOPORPHYRINS

Studies of ORR catalysis—Reaction (18.8) below—by metalloporphyrins typically aim at quantifying how selective the catalyst is towards the complete four-electron reduction of O_2 versus partial reduction, typically two-electron reduction to H_2O_2. Reduction of O_2 to $O_2^{\cdot-}$ (superoxide, or its conjugate acid, hydroperoxyl radical $HO_2^{\cdot}$) may also occur, particularly with Fe porphyrins, which are susceptible to autoxidation [Shikama, 1998] (see Section 18.6.2), but this pathway is rarely considered. This selectivity is often quantified as n_{av}, the average number of electrons by which one molecule of O_2 is reduced. It is related to α, the fraction of the four-electron pathway, by $n_{av} = 2(1 + \alpha)$ if reduction to H_2O_2 is the only competing process. A fairly small subset of reported studies have aimed at quantifying the kinetics of the reduction in terms of the bimolecular rate constant k. By examining how n_{av} and k depend on experimental conditions (potential of the electrons, H^+ and O_2 concentrations, and concentration or surface coverage of the catalyst) mechanistic inferences about Reaction (18.8) can sometimes be made, but the required careful mechanistic studies have only rarely been performed.

$$O_2 + 2(1 + \alpha)H^+ + 2(1 + \alpha)e^- + \text{catalyst} \xrightarrow{k} 2\alpha H_2O + (1 - \alpha)H_2O_2 + \text{catalyst} \quad (18.8)$$

Most studies of ORR catalysis by metalloporphyrins have been carried out using water-insoluble catalysts absorbed on a graphite electrode in contact with aqueous solution. In a limited number of cases, four other approaches have been used: catalysts imbedded in an inert film (i.e., Nafion or lipid) on the electrode surface; self-assembled monolayers of catalysts; catalysts in aqueous or mixed organic/aqueous solutions in contact with an electrode; and catalysis in mixed aqueous/organic medium using chemical reductants, such as ferrocenes, cobaltocene, or an organic macrocycle. Each method has its own strengths and limitations, and studying the same catalyst by more than one method, though rarely done, can yield mechanistic information that is not attainable from any one method.

In the most common approach, a water-insoluble metalloporphyrin is deposited on the surface of a rotating disk electrode (RDE) or on the disk of a rotating ring–disk electrode (RRDE; Fig. 18.7a) as a film of poorly defined morphology, either by spontaneous adsorption from a solution of the catalyst in an organic solvent or by evaporation of an aliquot of such a solution onto the electrode. It is impossible to know the

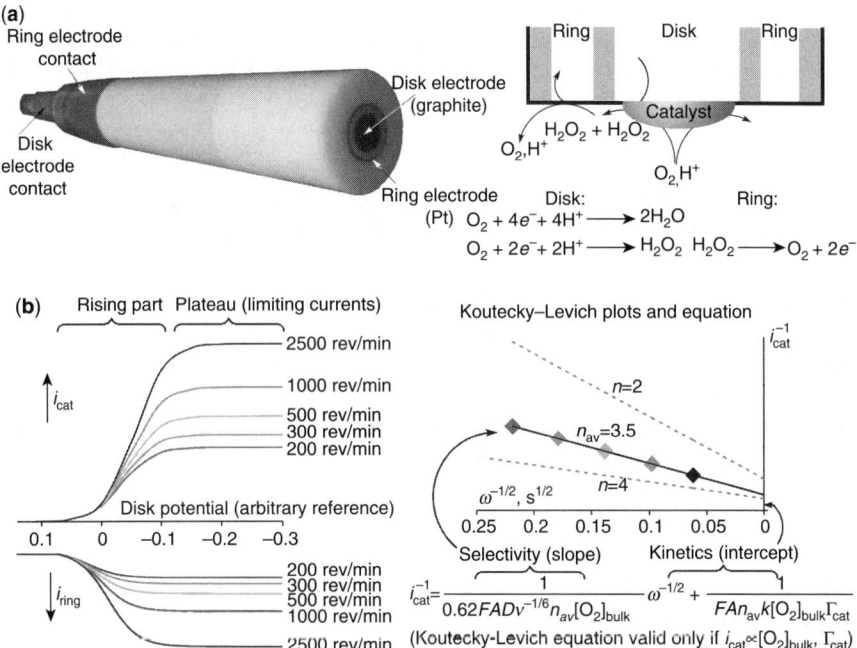

Figure 18.7 (a) Rotating ring–disk electrode (RRDE; left) and schematics of chemical processes at an RRDE in studies of metalloporphyrin-catalyzed ORR: the water-insoluble catalyst is deposited on the disk, and the electrode is immersed in aqueous electrolyte. Rotation of the electrode generates flow of the electrolyte to the electrode surface, which delivers O_2 and H^+ to the catalyst. The products of the reduction (H_2O and/or H_2O_2) diffuse and also are transported hydrodynamically from the disk to the ring electrode, which is set at a potential at which H_2O_2 is rapidly oxidized to O_2. The ratio of the ring and disk currents is proportional to the fraction of O_2 reduced to H_2O_2, i.e., selectivity of the catalyst to four-electron reduction. (b) Linear sweep voltammograms at different frequencies of electrode rotation. When the catalytic currents are proportional to bulk O_2 concentration and are limited by the rate of the catalytic reaction rather than by mass transport, the dependence of the catalytic currents on the rotational frequency of the electrode is described by the Koutecky–Levich equation. The slopes of the i_{cat}^{-1} versus $\omega^{-1/2}$ graphs (Koutecky–Levich graphs) are proportional to n_{av} and the intercept is proportional to k.

amount of catalyst immobilized on the electrode by spontaneous adsorption (although the amount of *electroactive* catalyst can be determined coulometrically). Nor is it generally possible to vary the amount of the deposited catalyst. The alternative (spin casting) is free of such limitations, but it may produce more morphologically heterogeneous catalytic films owing to nonuniform precipitation of the catalyst during the rapid evaporation of the solvent. The modified electrode is immersed in a buffered air- (or O_2-) saturated aqueous solution and its potential is scanned.

Rotating electrode voltammetry [Bard and Faulkner, 2001; Opekar and Beran, 1976], which is a forced convection/hydrodynamic method, allows one to quantify k and n_{av} as functions of the electrochemical potential by using an appropriate mathematical model to relate the observed catalytic currents to the rates of chemical reactions and mass transport processes. A general description of electrochemical kinetics at a chemically modified electrode was developed by Saveant and co-workers [Andrieux and Saveant, 1992]. Their model takes into account the three processes that can limit a catalytic current generated by a multilayer film of a catalyst on the surface of an electrode: (i) the turnover frequency of the catalytic cycle; (ii) the rate of reactant(s) transfer from the film–solution interface to the catalytic site(s); and (iii) the rate of electron transfer from the electrode–film interface (or charge-compensatory movement of counterions) to the catalytic site(s). In order to use rotating disk voltammetry to quantify the kinetics of the catalytic cycle, one must ensure that the last two processes are not rate-determining, which can usually be achieved by controlling the amount of catalyst deposited on the electrode (see below).

If the charge and reactant transfer within the film are much faster than the catalytic turnover, and the catalytic current is proportional to the concentration of O_2 in the catalytic film (i.e., the rate is first-order in O_2 and the catalyst is not even partially saturated with O_2), the electrode kinetics is adequately described by the so-called Koutecky–Levich equation (Fig. 18.7b) [Opekar and Beran, 1976]. This equation is a solution of the convection–diffusion equation in the case of the current being limited by (i) mass transport of the reactant from the bulk solution to the electrode surface (or, in the case of chemically modified electrodes, to the film–solution interface, since reactant transport within the film is assumed not to be kinetically limiting) and (ii) electrode reaction. To determine n_{av} and k of the catalytic film by the Koutecky–Levich equation, a series of linear sweep voltammograms is collected at different rotational frequencies ω of the modified electrode (Fig. 18.7b). Increasing the rotational frequency increases the flux of O_2 and H^+ to the electrode by decreasing the thickness of a layer of the aqueous electrolyte that rotates with the electrode (the stagnant layer [Opekar and Beran, 1976]). If the Koutecky–Levich equation is applicable and the data quality is adequate, inverse catalytic currents i_{cat}^{-1} are directly proportional to the inverse square root of the rotational frequency $\omega^{-1/2}$, (Koutecky–Levich plot; Fig. 18.7b). The slope of the corresponding linear regression line is $0.62FAD^{2/3}\nu^{-1/6}n_{av}[O_2]_{bulk}$ where F is Faraday's constant, A is the *geometric* area of the disk electrode, D is the diffusion coefficient of O_2, ν is the kinematic viscosity of the electrolyte and $[O_2]_{bulk}$ is the concentration of O_2 in the bulk of electrolyte. Either these parameters are taken from the literature or the factor $0.62FAD^{2/3}\nu^{-1/6}[O_2]_{bulk}$ is measured experimentally by carrying out O_2 reduction under identical experimental conditions using a catalyst whose n value is reliably established, for example graphite ($n = 2$) or thoroughly cleaned Pt ($n = 4$, Fig. 18.7b). The latter method may provide more precise estimates of n_{av}, as the values of D, v, and $[O_2]_{bulk}$ under given experimental conditions are usually not known with sufficient precision (for example, the literature values of D vary by a factor of 2 [Opekar and Beran, 1976] and solutions oversaturated in O_2 are easily obtained if O_2 is continuously bubbled in).

Reliable estimates of k typically cannot be obtained from a single Koutecky–Levich intercept, often referred to as i_k^{-1}, because of uncertainty in the amount of catalytically active porphyrin at the electrode. To obtain reliable estimates of k, Koutecky–Levich intercepts acquired at electrodes with varying surface coverage of the catalyst, Γ_{cat}, have to be plotted against Γ_{cat}. It is typical to observe a direct proportionality between i_k^{-1} and Γ_{cat} at low surface coverages of the catalyst (provided that O_2 reduction at the electrode material does not interfere) that plateaus as the coverage increases and may even decrease with a further increase in coverage (Fig. 18.8), signifying a change in the rate-determining step from catalytic reaction to mass or charge transport in the catalytic film. Only at the catalyst coverages that yield intercepts directly proportional to Γ_{cat} can the Koutecky–Levich equation be applied.

One also needs to be careful when using the slope of the Koutecky–Levich plot to determine n_{av} of the catalytic film. Examples of metalloporphyrin-catalyzed ORR have been reported where, above a certain value of the electrode rotational frequency, the catalytic currents became independent of ω, indicative of a breakdown of the Koutecky–Levich model, either because the rate of charge or substrate transfer within the film became rate-limiting or the catalyst became partially saturated with O_2 [Boulatov et al., 2002; Song et al., 1998; Collman et al., 1980]. In other cases, the i_{cat}^{-1} versus $\omega^{-1/2}$ graphs may remain mostly linear within the experimental

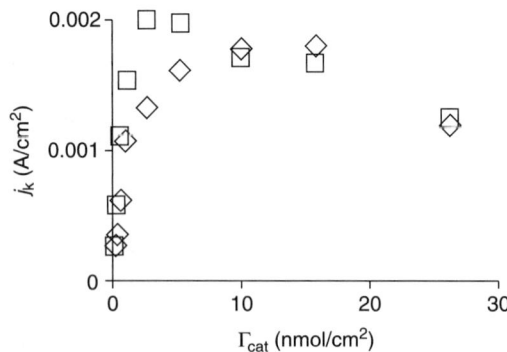

Figure 18.8 The inverse of the intercept of the Koutecky–Levich plots normalized to the geometric area of the electrode, j_k, versus surface coverage of two biomimetic Fe porphyrin catalysts deposited on an edge-plane graphite electrode [Boulatov et al., 2002]. At surface coverages $<1-2\,\text{nmol/cm}^2$ (depending on the catalyst), j_k is directly proportional to Γ_{cat}. Depositing more catalyst leads to a drop in j_k, suggesting a change in the turnover-determining step. The surface coverages are per geometric area; because of the high roughness of the edge-plane graphite, $1-2\,\text{nmol/cm}^2$ may correspond to only few monolayers of the catalyst.

uncertainty, but with less steep slopes than if the systems followed the Koutecky–Levich behavior. Application of the Koutecky–Levich equation in such situations leads to significantly overestimated catalytic selectivities. To test for this situation, one needs to establish that the Koutecky–Levich slopes are indeed directly proportional to bulk O_2 concentration, $[O_2]_{bulk}$ in as wide a range of O_2 concentrations as possible.

Although, in theory, the Koutecky–Levich equation can be applied to estimate n_{av} and k at any part of the voltammogram (provided that the conditions stated above are satisfied), for practical reasons only limiting (plateau) currents can be acquired with adequate reproducibility to yield suitable Koutecky–Levich plots.

The selectivity of the catalyst can also be assayed using an RRDE (Fig. 18.7a) [Bard and Faulkner, 2001; Opekar and Beran, 1976]. In this setup, the ring electrode serves as an electrochemical sensor of partially reduced oxygen species (H_2O_2 and O_2^-/HO_2) released by the catalytic film at the disk. While the disk potential is scanned, the (usually) Pt ring is maintained at a fixed oxidizing potential, so that a fraction of partially reduced oxygen that is hydrodynamically transported from the film–solution interface to the ring is oxidized to O_2. This oxidation generates a ring current. In order to obtain n_{av} from this ring current, one needs to know the fraction of the partially reduced oxygen species generated at the disk that are oxidized at the ring (the so-called "ring collection efficiency"). When the Pt ring can be set at a potential at which the rate of the analyte oxidation is much larger than the residence time of the analyte in the ring's vicinity (i.e., diffusion-limited oxidation), the collection efficiency can be calculated from the geometric parameters of the RRDE. However, because Pt is easily passivated toward H_2O_2 oxidation, sufficient overpotential to achieve diffusion-limited ring oxidation of H_2O_2 is probably not accessible under most experimental conditions. As a result, the ring collection efficiency of H_2O_2 is always lower than the theoretical value. Experimental methods to maintain collection efficiencies close to theoretical have been reported [Boulatov et al., 2002]. With such precautions, the ring current measurements are indispensable for estimating the catalytic selectivity at potentials where it cannot be obtained from Koutecky–Levich plots (e.g., the rising parts of the catalytic waves).

It is important to be mindful of the fact that catalytic properties of a film of a metalloporphyrin can be quite different from those of individual isolated molecules of the catalyst, thereby complicating mechanistic interpretation of the results. The major issue is the possibility of stepwise reduction of O_2 at multiple catalytic sites within a film. For example, the fraction of O_2 reduced to H_2O_2 by films of Fe tetraphenylporphyrin, Fe(TPP) (see Fig. 18.9 in the next section), decreases as the amount of the deposited catalyst increases [Collman et al., 2003a]. This dependence suggests that the putative ferric–hydroperoxo intermediate (see Fig. 18.11 in the next section) may undergo dissociative decomposition with release of H_2O_2 faster than reductive scission of the O–O bond. As a released molecule of H_2O_2 diffuses through the catalytic film, it could coordinate to other molecules of the catalyst (particularly if the concentration of O_2 in the film is low, as it is at the plateau of polarization curves). Once coordinated, H_2O_2 can undergo homolysis or heterolysis of the O–O bond, rather than dissociate from the porphyrin. The thicker the film, the longer the residence time of H_2O_2 in the film and the longer time H_2O_2 spends coordinated to a metalloporphyrin catalyst before escaping from the film. It was suggested that if two criteria are satisfied—(i) independence of the i_r/i_d ratios from the electrode rotation rate and (ii) the electrochemical reduction of H_2O_2 being significantly slower than that of O_2 by the same film—then the catalytic properties of the film reflect those of an individual isolated molecule.

Neither, however, rules out the intermediacy of free H_2O_2 by a film displaying $n_{av} > 2$. The first criterion does not take into account the fact that the H_2O_2 flux reaching the ring during O_2 reduction in a multilayer catalytic film may be determined mainly by the residence time of H_2O_2 in the film (which is independent of rotation rate) rather than the time an H_2O_2 molecule remains in the vicinity of the disk–electrolyte interface (which is inversely proportional to the electrode rotation rate). Under such conditions, the i_r/i_d ratio would be independent of rotation rate even if free H_2O_2 is generated and reduced further within the film. The second criterion neglects the fact that the catalytic rates of H_2O_2 reduction when H_2O_2 is added to an anaerobic electrolyte may be limited by the rate at which H_2O_2 penetrates the film–solution interface, rather than by the catalytic rate. In such a case, the magnitudes of catalytic currents of H_2O_2 reduction under anaerobic conditions do not reflect the rate at which the catalyst reduces/disproportionates H_2O_2 generated *within* the film as a result of incomplete O_2 reduction by individual molecules.

Three new criteria were proposed [Collman et al., 2003a] to establish that the four-electron reduction of O_2 by a catalytic film represents the selectivity of an individual molecule of the catalyst: (i) independence of the i_r/i_d ratio on the catalyst surface coverage; (ii) much higher stability of the catalyst in the ORR compared with reduction of H_2O_2 under conditions that reproduce the concentration of H_2O_2 in the film that would be generated if ORR proceeded by a step-wise mechanism; and (iii) the nature of the turnover-determining step.

ORR catalysis by Fe or Co porphyrins in Nafion [Shi and Anson, 1990; Anson et al., 1985; Buttry and Anson, 1984], polypyrrolidone [Wan et al., 1984], a surfactant [Shi et al., 1995] or lipid films [Collman and Boulatov, 2002] on electrode surfaces has been studied. The major advantages of diluting a metalloporphyrin in an inert film include the ability to study the catalytic properties of isolated molecules and the potentially higher surface loading of the catalyst without mass transport limitations. Stability of catalysts may also improve upon incorporating them into a polymer. However, this setup requires that the catalyst have a reasonable mobility in the matrix, and/or that a mobile electron carrier be incorporated in the film [Andrieux and Saveant, 1992]. The latter limits the accessible electrochemical potentials to that of the electron carrier.

Relatively little work has been done on ORR catalysis by self-assembled monolayers (SAMs) of metalloporphyrins. The advantages of this approach include a much better defined morphology, structure, and composition of the catalytic film, and the surface coverage, and the capacity to control the rate at which the electrons are transferred from the electrode to the catalysts [Collman et al., 2007b; Hutchison et al., 1993]. These attributes are important for deriving the catalytic mechanism. The use of optically transparent electrodes allows characterization of the chemical composition of the film by absorption spectroscopy by exploiting the very high extinction coefficients of metalloporphyrins around about 400 nm ($\varepsilon > 10^5\,M^{-1}\,cm^{-1}$). Potential disadvantages are the need for chemical moieties to allow chemisorption of the catalyst on the electrode surface (e.g., thiols for SAMs on Au), rapid degradation of the film, and reduction of O_2 on the bare electrode if the quality of the SAM is limited. Until recently, it was thought that SAMs were, in general, incompatible with RDEs or RRDEs [Collman et al., 2007b], but a decade-old strategy of attaching porphyrins to preformed SAMs of terminally functionalized alkyl thiols [Loetzbeyer et al., 1995] was recently extended to generate high quality SAMs compatible with RRDEs [Collman et al., 2007b].

In a now classical study by Murray and co-workers [Hutchison et al., 1993] a SAM of a modified Co tetraphenylporphyrin (see Fig. 18.12 in the next section) was formed. This SAM was found to catalyze two-electron reduction of O_2 in both acidic and basic media, similar to the behavior of graphite-adsorbed or dissolved Co(TPP). The selectivity of the SAM was quantified using interdigitated array electrodes. Remarkably, the catalyst retained its catalytic activity for over 10^5 turnovers, which probably exceeds the stability of a parent Co(TPP) catalyst absorbed on graphite, although estimates of stabilities of catalysts absorbed on graphite have substantial uncertainty because of the lack of data on the fraction of the electroactive catalyst that is capable of participating in the catalysis. Subsequent spectroelectrochemical measurements [Postlethwaite et al., 1995] demonstrated that (i) the onset of catalysis corresponds to the reduction of the SAM to the Co^{II} state; (ii) approximately a single monolayer of the metalloporphyrin is formed with coplanar orientation of the macrocycle relative to the surface; and (iii) the absorption spectra of the catalyst on the Au surface in both Co^{III} and Co^{II} states are quite similar to those in a CH_2Cl_2 solution. However, no information was obtained regarding the nature of the axial ligation of the Co ion, which is of course of primary importance in developing a mechanism of O_2 reduction (see the next section). This work was later extended to a cofacial Co porphyrin [Hutchison et al., 1997], a protoheme IX [Loetzbeyer et al., 1995] and a Ru porphyrin [Collman et al., 1996]. Very recently, O_2 reduction catalysis by highly elaborate Fe porphyrins that were designed to reproduce the heme/Cu site of cytochrome c oxidase attached to various SAMs has been reported [Collman et al., 2007b]. In this work, SAMs were used to control the rate of electron transfer between the electrode and the catalyst.

Several studies of electrocatalytic O_2 reduction by simple metalloporphyrins in aqueous or mixed organic/aqueous solutions have been reported [Shigehara and Anson, 1982; Forshey and Kuwana, 1983; LeMest et al., 1997; Su et al., 1990]. In this setup, the electrode serves as a source of electrons, but both the catalyst and the substrate are, ideally, in solution. A water-soluble catalyst precludes the use of an RRDE for detection of partially reduced oxygen, as the oxidation of the catalyst itself on the ring will generate unacceptably high background currents, and the chemisorption of the catalyst or its decomposition products on the Pt ring may passivate it toward H_2O_2 oxidation. As a result, stationary electrodes are typically employed in such studies. A cofacial bis-Co porphyrin (see Section 18.5) dissolved in 5.6 M triflic acid was reported to manifest very high stability and high selectivity toward four-electron O_2 reduction [LeMest et al., 1997]. Water-soluble Fe-tetrakis

(N-methyl-4-pyridyl)porphyrin cation (FeTMPyP; see Fig. 18.9 in the next section) was also found [Forshey and Kuwana, 1983] to reduce O_2 to H_2O over a wide range of pH. The proposed mechanism involves the intermediacy of the free H_2O_2, and was derived on the basis of mathematical simulation of the experimental cyclic voltammograms in the presence of O_2. However, because many simple metalloporphyrin catalysts adsorb strongly to graphite, which is commonly used as the working electrode in these studies, it is often difficult to rule out the possibility that the observed catalytic behavior arises from the electrode-adsorbed catalyst [Su et al., 1990].

Finally, studies of metalloporphyrin-catalyzed O_2 reduction by a soluble reductant, such as a ferrocene derivative, cobaltocene, or an organic macrocycle, have been reported [Fukuzumi et al., 1990, 2004; Collman et al., 1997; Anson, 1985]. Co porphyrins have been used in all but one study, since Fe^{II} porphyrins seem to degrade much faster under these conditions. The advantages of such a homogeneous setup include the capacity to monitor the reaction spectroscopically, which in favorable cases may allow identification of the dominant species during the catalysis. Likewise, the rates of catalytic reduction of H_2O_2 under anaerobic conditions are not influenced by the rate at which it traverses the solution/catalytic film interface, as is the case of surface-confined catalysts. Establishing that a catalyst reduces H_2O_2 much more slowly than O_2 under these circumstances can serve as convincing evidence against the stepwise four-electron reduction mechanism of O_2 through free H_2O_2. In this setup, determining the potential dependence of catalytic rates is more difficult. Also, reactions in which two molecules of the catalyst participate in reduction of a single O_2 molecule, such as the formation of μ-peroxo and μ-oxo dimers of simple Fe porphyrins, $(por)Fe-O_x-Fe(por)$, $x = 1, 2$ as kinetically important intermediates, are likely. Finally, whereas electron transfer from the electrode to a catalyst in physical contact with the electrode is probably always very rapid, reduction of catalytic intermediates by a dissolved reductant is limited at least by diffusion, potentially lowering the selectivity and/or lifetime of the catalyst.

Although all these methods allow one to determine n_{av}, they do not provide any direct information about the nature of the partially reduced oxygen byproducts that are responsible for $n_{av} < 4$. It is typical to assume that the byproduct is H_2O_2, but superoxide (O_2^- or its conjugate acid, hydroperoxyl radical, $HO_2^{\cdot}$, $pK(HO_2^{\cdot}) = 4.5$) and especially hydroxyl radical can also be generated. The nature of the partially reduced oxygen byproducts may affect the stability of the catalyst (see below) and determine stereoelectronic modifications of the catalyst that may lead to an increased n_{av}. $HO_2^{\cdot}$ is generated by autoxidation of Fe porphyrin/O_2 complexes; it undergoes rapid outer-sphere reduction to HO_2^-, whereas its conjugate base, $O_2^{\cdot -}$ is a fairly reactive and strong one-electron outer-sphere reductant, forming O_2. If the local concentration of $HO_2^{\cdot}/O_2^{\cdot -}$ is high, this species could undergo bimolecular disproportionation to O_2 and H_2O_2. At low overpotentials, thermodynamics limits the fraction of O_2 that can be reduced to $HO_2^{\cdot}/O_2^{\cdot -}$.

Hydroxyl radical is a strong indiscriminate outer-sphere oxidant (generating OH^-) and H-atom abstractor (generating H_2O) [Huie and Neta, 1999]. Simple Fe porphyrins are known to promote O–O bond *homolysis* in reaction with H_2O_2 [Watanabe, 2000]. Because of its high reactivity, once generated, ·OH probably reacts with the macrocycle or the peripheral subsituents of the same metalloporphyrin molecule. The resulting organic radical probably decomposes rapidly and the product(s) are reduced by the electrode or the chemical reductant that serves as the source of electrons for O_2 reduction. Therefore, generation of $^{\cdot}OH$ does not *directly* contribute to lowering of n_{av}; however, it does lead to rapid degradation of the catalyst. Low stability of a metalloporphyrin catalyst in ORR catalysis may be an indication of $^{\cdot}OH$ generation. This conclusion is strengthened if the incorporation of selective $^{\cdot}OH$ scavengers in the catalytic film increases the number of turnovers over which the film retains its catalytic selectivity. Scavengers of $O_2^{\cdot-}/HO_2$ and of $^{\cdot}OH$ incorporated in a catalytic film of a biomimetic Fe porphyrin (see Section 18.6) were used to determine that a very small amount of partially reduced oxygen species produced by this catalyst during O_2 reduction was mostly $O_2^{\cdot-}/HO_2$, which resulted from autoxidation of the catalyst/O_2 complex [Boulatov et al., 2002].

18.4 ORR CATALYSIS BY SIMPLE METALLOPORPHYRINS

The "simple" porphyrin category includes macrocycles that are accessible synthetically in one or few steps and are often available commercially. In such metalloporphyrins, one or both axial coordination sites of the metal are occupied by ligands whose identity is often unknown and cannot be controlled, which complicates mechanistic interpretation of the electrocatalytic results. Metal complexes of simple porphyrins and porphyrinoids (phthalocyanines, corroles, etc.) have been studied extensively as electrocatalysts for the ORR since the initial report by Jasinsky on catalysis of O_2 reduction in 25% KOH by Co phthalocyanine [Jasinsky, 1964]. Complexes of all first-row transition metals and many from the second and third rows have been examined for ORR catalysis. Of all simple metalloporphyrins, Ir(OEP) (OEP = octaethylporphyrin; Fig. 18.9) appears to be the best catalyst, but it has been little studied and its catalytic behavior appears to be quite distinct from that other metalloporphyrins [Collman et al., 1994]. Among the first-row transition metals, Fe and Co porphyrins appear to be most active, followed by Mn [Deronzier and Moutet, 2003] and Cr. Because of the importance of hemes in aerobic metabolism, the mechanism of ORR catalysis by Fe porphyrins is probably understood best among all metalloporphyrin catalysts.

18.4.1 Simple Fe Porphyrins

Simple Fe porphyrins whose catalytic behavior in the ORR has been studied fairly extensively are shown in Fig. 18.9. Literature reports disagree substantially in quantitative characterization of the catalytic behavior (n_{av}, overpotential, stability of the catalysts, pH dependence, etc.). It seems plausible that in different studies the same Fe porphyrin possesses different axial ligation, which depends on the electrolyte and possibly specific residues on the electrode surface; the thicknesses and morphologies of catalytic films may also differ among studies. All of these factors may contribute to the variability of quantitative characteristics. The effect of the supporting surface on

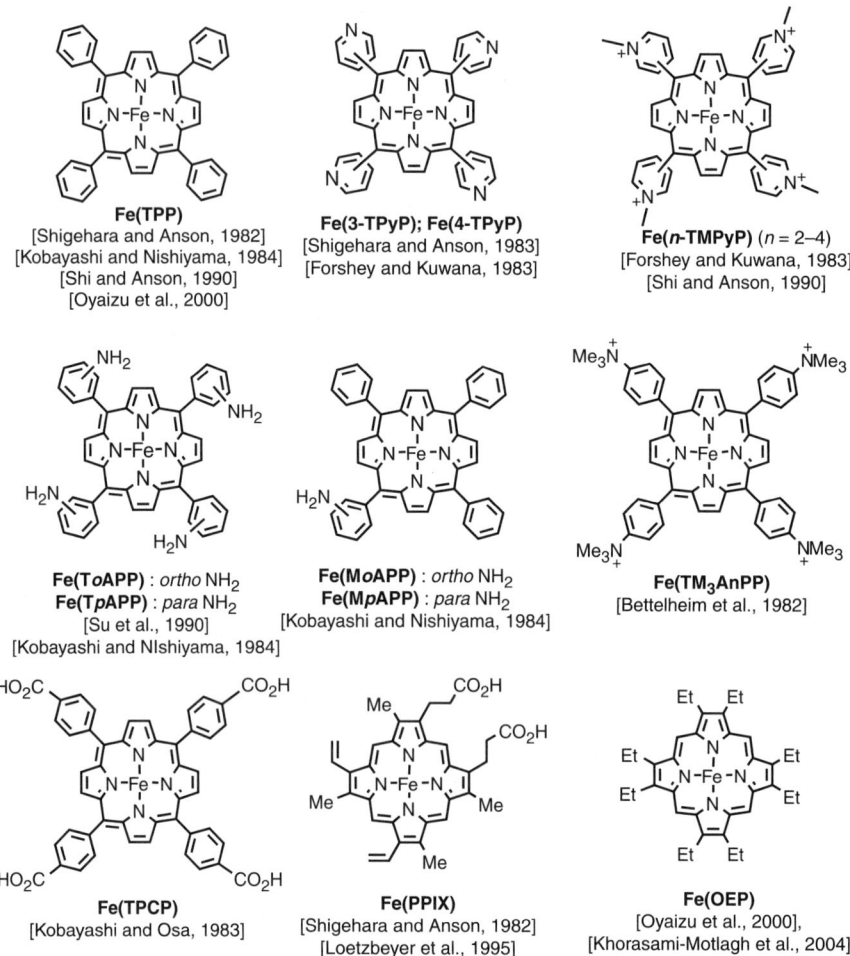

Figure 18.9 Chemical structures of simple Fe porphyrins whose catalytic properties in the ORR have been studied extensively. These properties are tabulated in Collman et al. [2004a].

the catalytic properties of Co porphyrins has been studied [Hutchison et al., 1997; Yamazaki et al., 2005]. Nonetheless, valuable generalizations can still be inferred from the literature data:

- Electrocatalytic reduction of both O_2 and H_2O_2 starts at potentials close to that of the $Fe^{III/II}$ couple in the absence of a substrate (which for most porphyrins is about 0.2–0 V with respect to a normal hydrogen electrode (NHE) at pH $\leq$ 6; the exception being Fe(TMPyP), $E \approx 0.5$ V). Catalytic reduction of H_2O_2 by simple *ferric* porphyrins is too slow to be detectable in typical electrocatalytic experiments; whereas ferrous porphyrins catalyze rapid reduction of H_2O_2,

18.4 ORR CATALYSIS BY SIMPLE METALLOPORPHYRINS

which may proceed through disproportionation of H_2O_2 and reduction of the resultant O_2.

- Depending on pH, increasing the acidity of the solution either makes the potential required to yield a fixed turnover frequency more oxidizing by 60 mV/pH or does not affect it. This pH dependence is in most cases the same as that of the $Fe^{III/II}$ couple in the absence of a substrate. These identical pH dependences suggest a pre-equilibrium between the ferric and ferrous forms of the catalyst followed, by a kinetically irreversible step that does not involve proton or electron transfer (e.g., O_2 binding).
- The apparent redox stoichiometry of O_2 reduction catalysis [n_{av}, Reaction (18.8)] is pH-independent, but for many catalysts depends strongly on the applied potential (Fig. 18.10). The apparent selectivity of Fe porphyrins deposited on the electrode surface typically increases with the amount of deposited catalyst.
- Usually, simple Fe porphyrins degrade rapidly during catalytic reduction of O_2 or of H_2O_2.

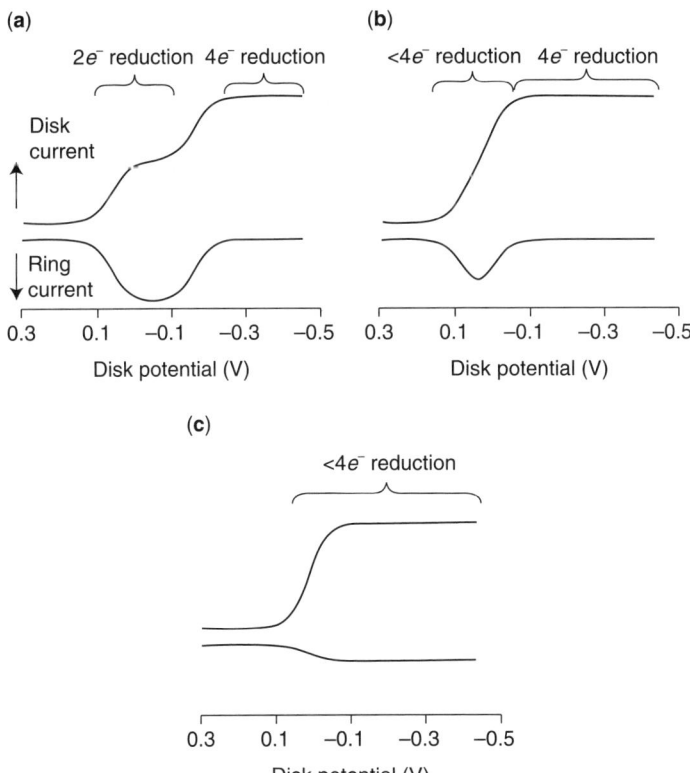

Figure 18.10 Three types of polarization curves typically manifested by simple Fe and Co porphyrins and cofacial metalloporphyrins (simulated voltammograms).

658   METALLOPORPHYRIN CATALYSTS OF OXYGEN REDUCTION

Most of these results can be understood within a generic mechanistic scheme (Fig. 18.11), which is based on numerous spectroscopic, computational, and electrochemical studies of reactions of O_2 or H_2O_2 with Fe porphyrins under stoichiometric conditions, and with heme enzymes under single-turnover conditions (see, e.g., Watanabe [2000]; Loew and Harris [2000]; Sono et al. [1996]). Indeed, such an O_2 reduction cycle at a heme site was characterized at atomic resolution by X-ray diffraction in two enzymes [Schlichting et al., 2000; Berglund et al., 2002]. Only ferrous porphyrins coordinated to an axial ligand (5-coordinate Fe^{II}) have any measurable affinity for O_2. Dioxygen is always bound end-on in such complexes. Without a nitrogenous base *trans* to O_2, dioxygen adducts of Fe porphyrins could only be prepared at 30 K in O_2 matrices [Proniewicz et al., 1991]. Under ambient conditions, 4- and 6-coordinate ferrous porphyrins do not bind O_2, but may reduce it to $O_2^{\cdot -}$ by an outer-sphere mechanism, depending on the $Fe^{III/II}$ potential [Shikama, 1998]. All ferric porphyrins are aerobically stable.

Anaerobic cyclic voltammetry suggests that simple Fe porphyrins deposited on an electrode in contact with an aqueous buffer contain two axial water molecules (or an

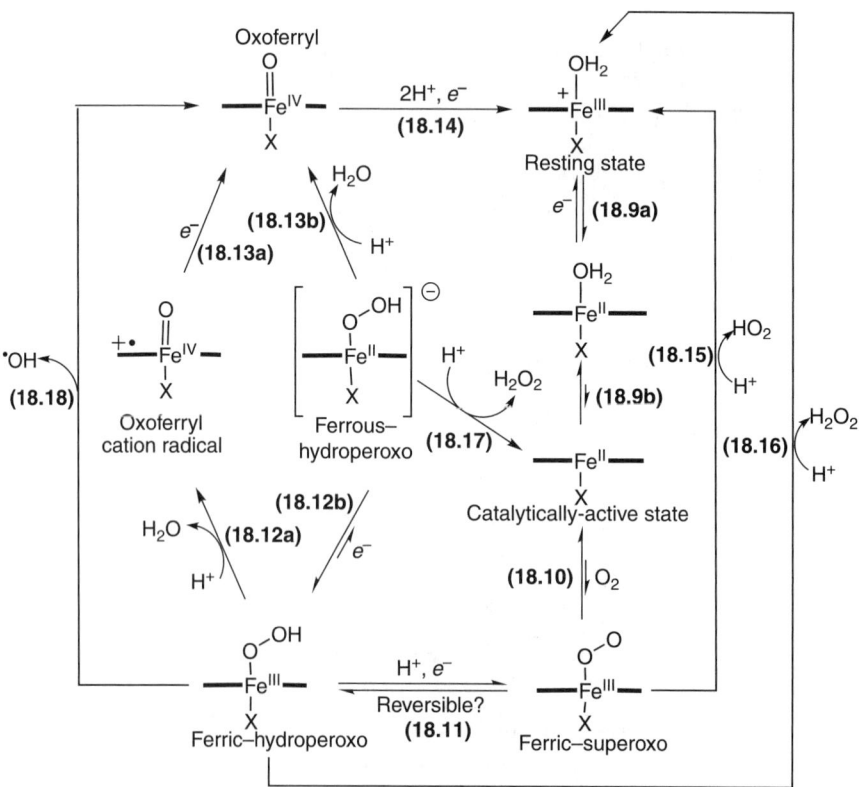

Figure 18.11 Plausible catalytic cycle for the ORR by simple Fe porphyrins adsorbed on the electrode surface and side Reactions (18.15)–(18.18). At pH < 3, the resting state of the catalyst is assumed to be ferric–aqua.

H_2O and HO^- ligands for ferric porphyrins at pH > 4) (see, e.g., Shigehara and Anson [1982]). This coordinatively saturated complex must lose one water molecule to bind O_2 [Reaction (18.9b) in Fig 18.11]. The equilibrium constant of this reaction was reported for one biomimetic Fe porphyrin on the graphite surface, the equilibrium being shifted strongly towards the 6-coordinate, aqua-ligated Fe^{II} porphyrin, ($K <$ 10^{-2} M) [Boulatov et al., 2002]. It is, however, also possible that a minority species, for example Fe^{II}(por) axially ligated by a carboxylate or quinol residue of the graphite electrode, is the catalytically active form.

Upon O_2 binding, the catalyst–O_2 adduct, which is formally a ferric–superoxo complex [Taube, 1986], is probably reduced to a ferric–hydroperoxo intermediate, whose subsequent reactions determine the product distribution of the catalysis. If both the distal and proximal environments around the Fe porphyrin are appropriate, as is in heme enzymes involved in O_2 metabolism, or in certain sophisticated biomimetic ORR catalysts (Section 18.6), then the O–O bond in the hydroperoxo intermediate invariably undergoes heterolysis. During O–O bond heterolysis, both oxygen atoms become reduced to the −2 ("oxide") redox state. If this heterolysis occurs in concert with an electron transfer from a nonheme group, as is postulated for cytochrome c oxidase (Section 18.2), then a ferryl intermediate is formed (Fig. 18.5). In other heme enzymes, both electrons come from the heme, yielding a high valence oxoferryl cation radical intermediate (Compound I) [Sono, 1996]. This reaction is facilitated both by the proximal ligand (e.g., imidazole, thiolate, or phenoxide) and the distal environment (see Fig. 18.4c for the definition of proximal and distal sides).

Since simple Fe porphyrins lack appropriate distal/proximal structures, it seems probable that O–O bond heterolysis in the ferric–hydroperoxo intermediate is inhibited relative to the competing reactions, which include O–O bond homolysis [Reaction (18.18) in Fig. 18.11], hydrolysis of intact H_2O_2 (18.16), and reduction to a ferrous–hydroperoxo intermediate (18.12b) under appropriately reducing potentials. The propensity of simple Fe^{III} porphyrins to induce O–O bond homolysis in H_2O_2 is well established from studies of stoichiometric reactions between H_2O_2 and such porphyrins [Watanabe, 2000]. Predominance of this homolytic reduction pathway (18.18) during catalytic O_2 reduction by electrode-confined Fe porphyrins would explain the common observation that such catalysts lose their catalytic capacity only after a few turnovers. (However, the uncertainty in the amount of catalytically active metalloporphyrin in the catalytic film makes any estimates of the turnover numbers only approximate.) In contrast, structurally analogous Co porphyrins, which catalyze reduction of O_2 only to H_2O_2, retain their activity over a substantially larger number of turnovers. This difference is consistent with the primary role of hydroxyl radicals, rather than H_2O_2 as sometimes suggested [Shigehara and Anson, 1982], in rapid degradation of Fe porphyrin catalysts.

An increase in the fraction of the four-electron reduction pathway at more reducing potentials (Fig. 18.10a, b) may be rationalized within at least two mechanisms. The first is based on the kinetic competition between the release of H_2O_2 from the ferric–hydroperoxo intermediate [Reaction (18.16) in Fig. 18.11] and its (reversible) reduction to a ferrous–hydroperoxo species, which undergoes rapid O–O bond heterolysis (18.13b). Because H_2O_2 and particularly HO_2^- are more basic ligands than H_2O, reduction of $[X(por)Fe^{III}(^-O_2H)]$ would require more reducing potentials than reduction of $[X(por)Fe^{III}(OH_2)]^+$ (the latter corresponds to the onset of catalytic O_2 reduction). Since O–O bond heterolysis in the *ferrous–hydroperoxo* intermediate does not require oxidation of the porphyrin, it is more facile than O–O bond heterolysis in the *ferric* analog. Hence, at potentials close to or more oxidizing than that of the $Fe^{III/II}$ couple, simple Fe porphyrins catalyze mainly two-electron reduction of O_2 to H_2O_2, with possibly an important side reaction of O–O bond homolysis (18.18). This reaction pathway dominates because the kinetics of O–O bond heterolysis (18.12a) is unfavorable in simple Fe porphyrins relative to the two competing pathways (18.16) and (18.18). However, at potentials more reducing than that of the $Fe^{III/II}$ couple under anaerobic conditions, one-electron reduction of the ferric–hydroperoxo intermediate becomes sufficiently favorable that the sequence of steps (18.12b)–(18.13b) becomes kinetically competitive with steps (18.16) and (18.18). The flux of H_2O_2 decreases, and the stability of the catalyst may increase.

The second mechanism often invoked to explain the increase in n_{av} of simple Fe porphyrins at potentials more reducing than that of the $Fe^{III/II}$ couple (under anaerobic conditions) is based on the fact that at such potentials the fraction of the catalyst in the 5-coordinate ferrous state is maximal because (i) the equilibrium (18.9) is shifted completely to the ferrous form and (ii) the concentration of O_2 in the catalytic film is low owing to mass transport limitations. The higher the concentration of the 5-coordinate ferrous porphyrin in the catalytic film, the greater the probability that any released H_2O_2 will re-enter the catalytic cycle by coordinating to a molecule of ferrous porphyrin and decay according to (18.13b) instead of (18.17).

Both mechanisms can also rationalize an increase in n_{av} due to the production of superoxide/HO_2 (18.16), which appears to dominate the flux of partially reduced oxygen species generated by certain biomimetic catalysts [Boulatov et al., 2002; Boulatov, 2004]. It remains to be established if either of these two mechanisms does indeed operate in simple Fe porphyrins, for example by carrying out single-turnover experiments similarly to the approach used to study ORR by cytochrome *c* oxidase.

Within the mechanism in Fig. 18.11, it seems implausible that simple Fe porphyrins can be effective ORR catalysts, since large overpotentials are required to access intermediates in which O–O bond heterolysis is facile. The only strategy discovered so far to facilitate this O–O bond heterolysis in the ferric–hydroperoxo intermediate is to control both the distal and the proximal environments of Fe porphyrins. In those cases, the overpotential of ORR reduction appears to be controlled by the potential of the $(por)Fe^{III/II}$ couple (see Section 18.6).

18.4.2 Simple Co Porphyrins and Anson Catalysts

Like their Fe analogs, Co^{III} porphyrins are aerobically stable. Co^{II}, being a d^7 ion, favors a square pyramidal coordination sphere, and simple Co^{II} porphyrins typically have fairly low affinity to axial ligands. As a result, many simple Co^{II} porphyrins are aerobically quite stable in the solid state, but oxidize slowly if dissolved in a coordinating solvent. Simple Co^{III} porphyrins are typically stronger oxidants than the Fe^{III}

analogs, which in theory should lead to lower overpotential of ORR catalysis. Unfortunately, the more oxidizing $Co^{III/II}$ potential also leads to lower affinity to O_2 (an exception being cofacial porphyrins; see Section 18.5). As a result, O_2 reduction catalysis by simple Co porphyrins often starts at potentials 0.2–0.4 V more reducing than that of the $Co^{III/II}$ couple [Song et al., 1998].

Simple Co porphyrins, whether adsorbed on an electrode, in a Nafion film, or in solution are generally thought to be largely two-electron catalysts (reduction of O_2 to H_2O_2) [Fukuzumi et al., 2004; Song et al., 1998; Bianchini and Zoellner, 1997; Anson et al., 1997; Postlethwaite et al., 1995; Hutchison et al., 1993]. A fair number of examples have been reported [Deronzier and Moutet, 2003; Yuasa et al., 1997, 2001; Song et al., 1998; Shi et al., 1997] of graphite-adsorbed Co porphyrins, including Co porphin (CoP; Fig. 18.12), manifesting voltammograms of the type shown in Fig. 18.10a. The regime with $n_{av} > 2$ is sometimes proposed to arise from catalysis by face-to-face dimers that are formed spontaneously upon absorption of the metalloporphyrin onto graphite. This conclusion is based on (i) the highly selective four-electron reduction of O_2 by certain cofacial porphyrins (Section 18.5) and (ii) the high propensity of simple neutral metalloporphyrins to π-stack. At present, however, there is little direct evidence supporting the presence of such dimers in catalytic films, and such dimers have never been shown to possess catalytic activity. An alternative explanation, invoking π-donation from the peripheral substituents of the porphyrin to Co, was proposed to explain some propensity of Co(TPHP) (Fig. 18.12) absorbed on graphite to reduce O_2 to H_2O [Anson et al., 1997]. The intermediacy of free H_2O_2 with this catalyst has been suggested. Compared with their Fe analogs, simple Co porphyrins that catalyze only two-electron reduction of O_2 to H_2O_2 usually retain their catalytic activity over a greater number of turnovers, which may be related to the fact that Co porphyrins do not typically promote homolysis of the O–O bond in free or metal-coordinated H_2O_2 [Marusak and Mears, 1995].

The π-donation was also invoked to explain a remarkable catalytic activity of Co porphyrins derivatized peripherally with Ru or Os ammine complexes (Fig. 18.12) [Anson et al., 1997]. Since their invention by Anson in the early 1990s, many Co porphyrins derivatized peripherally with metal complexes have been reported [Araki and Toma, 2006; Gadamsetti and Swavey, 2006]. Some of these catalysts manifest high selectivity for the four-electron pathway, albeit within a limited pH range [Araki and Toma, 2006]. In the original work, Co tetra-pyridylporphyrin was adsorbed on a graphite electrode and this modified electrode was exposed to a solution of $[Ru(NH_3)_5(OH_2)]^{2+}$, in which only the aqua ligand is labile and Ru^{II} has a higher affinity to pyridine than to H_2O. Based on an increase in the voltammetric response of the electrode after being exposed to a solution of $[Ru(NH_3)_5(OH_2)]^{2+}$, it was concluded that between 3 and 4 $[Ru(NH_3)_5]^{2+}$ moieties coordinated per molecule of Co(TPyP), presumably through the peripheral pyridyl groups. Using $[Ru(NH_3)_3(H_2O)_3]^{2+}$ instead of $[Ru(NH_3)_5(H_2O)]^{2+}$ was thought to yield a coordination polymer with one $[Ru(NH_3)_3(H_2O)]^{2+}$ moiety linking two Co(TPyP) molecules. Such electrodes catalyzed reduction of O_2 to a mixture of H_2O (70–90%) and H_2O_2. The catalysis started at potentials close to that of the $Ru^{III/II}$ couple (about 0.4 V vs. NHE). Similar results were observed with Co complexes of trispyridylmonophenylporphyrin,

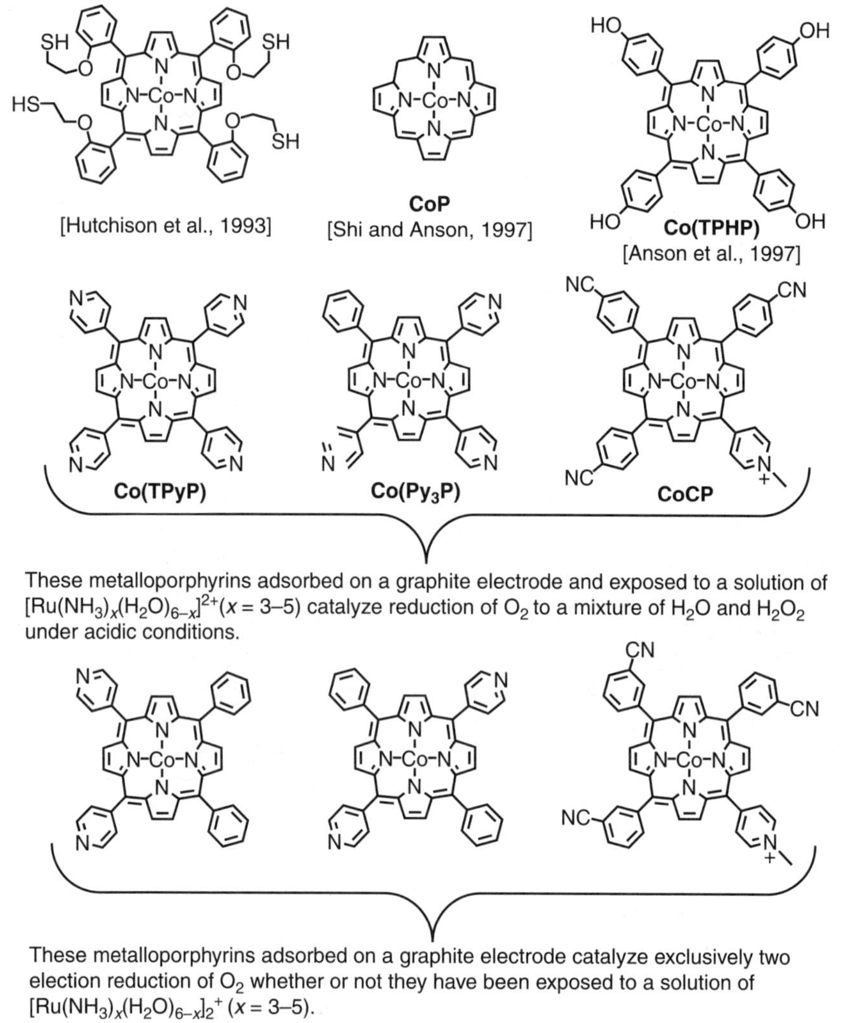

Figure 18.12 Chemical structures of selected simple Co porphyrins described in the text.

Co(Py$_3$P), and tris-(4-cyanophenyl)-(N-methyl-4-pyridyl)porphyrin CoCP (Fig. 18.12), in which the *para*-cyano groups were thought to coordinate to Ru^{II}. Using [Ru(edta)(OH$_2$)]$^-$ as a source of Ru or using other porphyrins with coordinating peripheral substituents (Fig. 18.12) did not produce catalysts superior to simple Co porphyrins. However, exposing graphite-adsorbed Co(TPyP) to a solution of [Os(NH$_3$)$_5$(OH$_2$)]$^{2+}$ appears to have yielded an improved catalyst, and it was recently reported that coordination of a single Pt complex to a Co porphyrin may also yield an ORR catalyst with $n_{av} > 2$ [Gadamsetti and Swavey, 2006]. The Co(TPyP) fragment remains the most widely used porphyrin for studying the effect of peripheral metal complexes on metalloporphyrin ORR catalysis [Araki and Toma, 2006].

Little is known about the stability of these porphyrins in O_2 reduction, how this peripheral substitution affects O_2 affinity of the metalloporphyrin, how the peripheral metal complexes perturb the energetics of various intermediates, and/or the kinetics of various steps or the mechanisms of O_2 reduction by these porphyrins. At present, it remains to be seen if the strategy of coordinating metal complexes on the periphery of a metalloporphyrin can be exploited in the *rational* design of new ORR catalysts.

18.5 COFACIAL PORPHYRINS AND RELATED COMPOUNDS

In cofacial porphyrins, two porphyrin moieties are constrained in an approximately stacked geometry by one to four covalent linkers (Fig. 18.13). The first cofacial porphyrins were reported in the literature contemporaneously by several groups in the late 1970s [Collman et al., 1994]. In their paper, Collman et al. first suggested that such porphyrins could be good ORR catalysts. In the next 30 years, this idea has been spectacularly demonstrated: certain Co and especially Ir derivatives appear to be the best molecular catalysts for the ORR yet found. Cofacial bismetalloporphyrins remain among the very few molecular catalysts definitely displaying bimetallic catalytic cooperativity. As such, they hold the potential to help rational design of efficient catalysts for the ORR and other multielectron processes. Recently, the concept of cofacial diporphyrins was extended to corroles [Kadish, 2005], and it has served as an inspiration for the design of new superstructured porphyrins [Rosenthal and Nocera, 2007].

Certain aspects of electrocatalysis by cofacial porphyrins have been reviewed previously [Collman et al., 1994, 2003a].

18.5.1 Molecular and Electronic Structure

Structurally, the cofacial porphyrins fall into two classes: in pillared (or Pacman) porphyrins the face-to-face arrangement of the two macrocycles is enforced by a single rigid linker, such as 1,2-phenylene, 1,8-biphenylene, or 1,8-anthracene, connected to each macrocycle at its *meso* position. In strapped cofacial porphyrins, at least two flexible chains connected at β-pyrrolic positions are used to hold the macrocycles together. Pillared cofacial porphyrins can typically be synthesized in fewer steps and better yields, although, even in the best cases, more than 15 steps are required, making these porphyrins prohibitively expensive for practical applications. Strapped cofacial porphyrins, on the other hand, allow a systematic variation in the length of the straps, potentially simplifying the acquisition and interpretation of structure/activity relationships.

Substantial effort has been invested in understanding how the linker(s) affect the metal–metal (M–M) distances, since the capacity of a cofacial porphyrin to accommodate a bridging diatomic molecule is often thought to be critical for achieving high catalytic activity. As of May 2007, 38 X-ray diffraction structures of cofacial porphyrins in Fig. 18.13 had been reported, with the DPA, DPB, DPD, and DPX

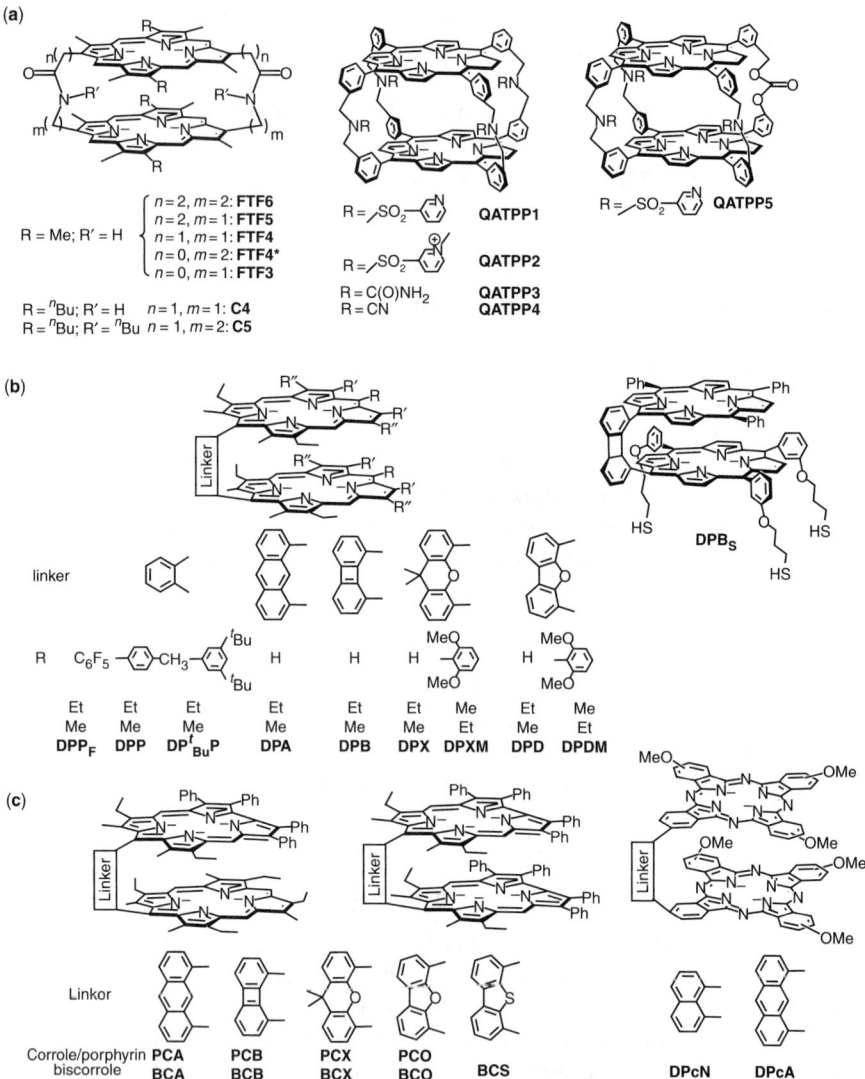

Figure 18.13 Chemical structures of selected cofacial strapped diporphyrins (a), pillared diporphyrins (b), and pillared porphyrin/corrole, dicorrole, and diphthalocyanine derivatives (c) whose metal complexes have been studied as ORR catalysts. Conventional notations for the structures are also listed (in **bold**). Other molecular architectures of cofacial porphyrins are known, but the corresponding complexes have not yet been studied as ORR catalysts.

series characterized most thoroughly. Several tentative inferences can be made from this data:

- Catalytic selectivity does not appear to correlate with the (average) M–M distance (in the absence of any ligands in the cavity) or, more broadly, the separation of the centers of the two porphyrins (Ct–Ct): the best ORR catalysts are bis-Co complexes of FTF4, DPD, and DPP, with the average Ct–Ct distance (in the absence of in-cavity ligands) of 3.42, 6.47, and 5.21 Å. Within this range, one also finds the M–M distances in two crystal structures of cofacial porphyrins whose bis-Co derivatives are essentially two-electron catalysts (C5 and FTF6).

- The DPD motif appears to be capable of accommodating the widest range of M–M distances (8.63–3.49 Å), followed by DPP (6.92–3.85 Å); in both cases, the flexibility is achieved by doming of the macrocycles. It is noteworthy that (DPD)Co_2 and (DPP)Co_2 are among the best ORR catalysts discovered so far (see Fig. 18.15 below).
- The two macrocycles appear to be largely parallel even in the DPD and DPP series, where the structure of the linker would seem to favor a conformer in which the planes of the macrocycles would be at an angle to each other. The parallel arrangement of the two macrocycles is thought to result from the drive to maximize π–π stacking.

In addition to the solid state structures, ultraviolet–visible (UV–vis), electron paramagnetic resonance (EPR), and redox properties of cofacial metalloporphyrins provide information about the separation of the two macrocycles. Historically, cofacial porphyrins have been separated into two groups [Collman et al., 1994]. In group 1 diporphyrins, the two macrocycles behave independently and similarly to monoporphyrins, whereas in group 2 diporphyrins, strong electronic interaction between the two porphyrins results in UV–vis, EPR, and redox properties that are significantly attenuated relative to monomeric porphyrins. For example, the Soret bands of group 2 diporphyrins are blueshifted relative to those of monoporphyrins or group 1 analogs. Whereas redox transformations of group 1 diporphyrin derivatives proceed in two-electron steps, those of group 2 are one-electron processes. Group 2 cofacial porphyrins have molecular orbitals (MOs) that are delocalized over both macrocycles, whereas MOs of group 1 are localized on each porphyrin. Thus, diamagnetism of doubly oxidized group 2 diporphyrins (e.g., [(FTF4)Zn_2]$^{2+}$ [LeMest et al., 1992]) is consistent with removal of an electron pair from a nondegenerate highest occupied MO (HOMO). In contrast, the magnetic properties of a group 1 metalloporphyrin, [(FTF6)Cu_2]$^{2+}$, are close to those of noninteracting (por$^{·+}$)Cu^{II} moieties.

In metal complexes of FTF5 and DPA, both single two-electron and double one-electron redox couples were observed. [(FTF5)Zn_2]$^{2+}$ is diamagnetic, whereas the EPR spectrum of [(DPA)Zn_2]$^{2+}$ was interpreted as that of a triplet that was complicated by aggregation. Why the ground electronic state of [(DPA)Zn_2]$^{2+}$ is a triplet remains to be established: a triplet porphyrin diradical was thought to be unique in porphyrin chemistry [LeMest et al., 1992].

As a result of strong electronic interactions between the two metalloporphyrin units, there is a substantial uncertainty in assigning oxidation states in mixed-valence group 2 complexes of redox-active metals, such as Co. Thus, although reduced neutral Co_2 derivatives can be reasonably well described as those of Co^{II}, the location (metal versus porphyrin) of the electron hole(s) in the singly and doubly oxidized derivatives is not known definitively, and may be very sensitive to the medium [LeMest et al., 1996, 1997]. For example, in benzonitrile, the UV–vis spectrum of [(FTF4)Co_2]$^{+}$

was interpreted as that of a π-cation radical [LeMest et al., 1997]. Frontier MOs in some cofacial diporphyrins appear to have substantial coefficients both on the metal ions and on the porphyrins [LeMest et al., 1997]. Therefore, oxidation state formalism may not always be useful in thinking about the electronic structure of the mixed-valence group 2 cofacial bismetalloporphyrins.

18.5.2 Chemistry of O_2 Adducts of Cofacial bis-Co Porphyrins

The dioxygen chemistry of group 2 biscobaltdiporphyrins (with the exception of (FTF3)Co$_2$; Fig. 18.13) provides a remarkable example of bimetallic cooperativity, and as a result appears to be strikingly different from that of metalloporphyrins containing other Groups 8 and 9 metals. Like their monomeric analogs, Co^{II} cofacial metalloporphyrins bind O$_2$ reversibly only in the presence of a nitrogenous heterocyclic base, such as imidazole. Most workers report that the resulting adducts are diamagnetic (their monomeric analogs have an $S = \frac{1}{2}$ ground state) and are typically interpreted as a complex of two d^6 Co^{III} ions bridged by a diamagnetic peroxo moiety, O_2^{2-} [Liu et al., 1985; Collman et al., 1980, 1983a]. Despite their diamagnetism, none of these adducts has ever been characterized by NMR. Controversially, a species obtained by exposing a solution of (FTF4)Co$_2$ in benzonitrile in the presence of N-methylimidazole to O$_2$ at low temperature was reported to give an EPR signal [LeMest et al., 1997].

Although oxidized monomeric Co porphyrins are inert toward O$_2$, singly and doubly oxidized group 2 Co$_2$ derivatives bind O$_2$ reversibly with unexpectedly high affinities. Even more surprising, a heterocyclic ligand *trans* to O$_2$ is not obligatory. For example, in a noncoordinating solvent (CH$_2$Cl$_2$), the half-saturation O$_2$ pressure $p_{1/2}(O_2)$ for [(DPB)Co$_2$]$^+$ is about 0.2 atm; it is about 0.01 atm in a weakly coordinating solvent, such as benzonitrile, whereas in the presence of 1,5-diphenylimidazole (Ph$_2$Im), the O$_2$ adduct could not be deoxygenated [Collman et al., 1992]. An excellent ORR catalyst, [(FTF4)Co$_2$]$^+$, in benzonitrile solution manifests very high O$_2$ affinity ($p_{1/2} = 0.4$ Torr, about 5×10^{-4} atm), which exceeds that of some myoglobins (0.4–10 Torr in aqueous buffers) [LeMest et al., 1997; Collman et al., 2004a]. Similarly unprecedented in metalloporphyrin chemistry is the fact that such adducts can be protonated *reversibly* in benzonitrile. Such protonation has been proposed [Rosenthal and Nocera, 2007; Fukuzumi et al., 2004] to be critical for the high catalytic activity of group 2 cofacial bis-Co porphyrins in the ORR.

Consensus appears to exist in the literature that O$_2$ adducts of singly oxidized bis-Co diporphyrins, [(dipor)Co$_2$O$_2$]$^+$, contain bridging superoxide. At room temperature, all of these adducts display isotropic or weakly anisotropic EPR signals centered at $g \approx 2$ with a 15-line superhyperfine splitting indicative of spin coupling to two equivalent ^{57}Co nuclei (nuclear spin 7/2). Collman and co-workers suggested that the anisotropy of the EPR signal correlated with poor four-electron selectivity [Collman et al., 1983a]: examples included pairs of structurally similar [(FTF4)Co$_2$(O$_2$)]$^+$ ($n_{av} \approx 4$) versus [(FTF4*)Co$_2$(O$_2$)]$^+$ ($n_{av} \approx 2$) and [(C5)Co$_2$(O$_2$)]$^+$ ($n_{av} \approx 2.4$) versus [(C4)Co$_2$(O$_2$)]$^+$ ($n_{av} > 3.2$; see Fig. 18.15 below). In contrast, Le Mest and co-workers reported that the EPR spectra of all dioxygenated group 2 Co$_2$ porphyrins are very similar: $g_{iso} = 2.02$ ($A_{iso,Co} = 12-14$ G), $g_{\parallel} = 2.09$, $g_{\perp} = 2.00$ ($A_{\parallel} = 15-17$ G) [LeMest et al., 1997]. Certainly, further work is required to use this EPR data to quantify the distribution of electron density in these adducts.

In benzonitrile, doubly oxidized Co_2 group 2 porphyrins also bind O_2, albeit more weakly than the singly oxidized analogs [LeMest et al., 1997]. The electronic configuration of the resulting adducts, $[(dipor)Co_2O_2]^{2+}$, is poorly understood. $[(FTF4)Co_2O_2]^{2+}$ was reported [LeMest et al., 1997] to give a single-line EPR spectrum indicative of a π-cation radical. The $[(dipor)Co_2O_2]^{2+}$ adduct is an even-electron system and must have two unpaired noncoupled electrons to give an EPR signal of an organic radical. The corresponding electronic state is unknown.

Stepwise oxidation of $[(FTF4)Co_2O_2]$ to the mono- and dications was reported to proceed reversibly in benzonitrile [LeMest et al., 1997]: such reversible redox chemistry has not been reported for any O_2 adducts of monomeric Co or Fe porphyrins.

There is a major controversy over which bridging mode (Fig. 18.14) the O_2 moiety adopts in the $[(dipor)Co_2O_2]^+$ adducts. Among the three conceivable modes, the μ-1,2 O_2 ligand is by far the most common in transition metal/O_2 chemistry, and is quite flexible with respect to the M–M separation. For example, the Co···Co distance in the Co–O–O–Co moiety in crystallographically characterized complexes varies between 4.9 and 2.8 Å. The wide range of M–M distances seen in crystal structures of derivatives of pillared cofacial porphyrins suggests that the diporphyrin core may also accommodate a range of Co···Co distances, including those optimal for the O_2 ligand coordinating in the μ-1,2 mode. On the other hand, the μ, $\eta^2:\eta^2$ binding mode is much less common, and is less flexible, imposing restrictions on the extent of lateral displacement of the two porphyrin rings that are exceeded by those seen in the crystal structures of most pillared porphyrins.

Resonance Raman studies of dioxygenated cofacial porphyrins similar to (FTF4)Co_2 and of $[(DPA)Co_2O_2]$ [Proniewicz et al., 1989] failed to differentiate the binding modes. The two adducts manifested similar O–O stretching frequencies $\nu(O_2)$, 1081 and 1085 cm^{-1}, despite the different solid state stereochemistry of the parent (deoxygenated) complexes. The $\nu(O_2)$ values are comparable to those in doubly bridged $[(NH_3)_8Co_2(\mu-O_2)(\mu-NH_2)]^{4+}$ adducts containing a $trans$-$\mu,\eta^1:\eta^1$-O_2^-

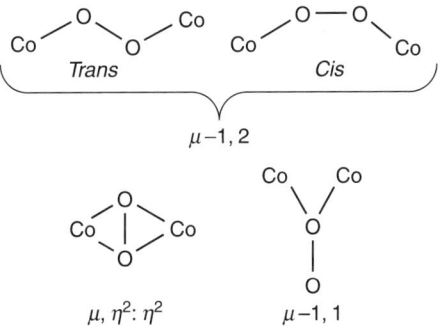

Figure 18.14 Three possible modes of O_2 binding in $[(dipor)Co_2]^{x+}$ ($x = 0, 1, 2$).

moiety [Barraclough et al., 1978], and lower than those in singly bridged [L$_8$Co$_2$(μ-O$_2$)]$^{5+}$ complexes (about 1100–1125 cm^{-1}; L = various neutral ligands) [Strekas and Spiro, 1975]. Although no unambiguous examples of CoO$_2$ units containing a side-on O$_2$ moiety are known, the ν(O$_2$) value of side-on O$_2$ in (TPP)FeO$_2$ is 89 cm^{-1} lower than that in the end-on isomer (1195 vs. 1106 cm^{-1}) [Proniewicz et al., 1991]. Finally, ν(Co–O) in [(DPA)Co$_2$O$_2$] (about 628 cm^{-1} in CH$_2$Cl$_2$) is also comparable to those in [(NH$_3$)$_8$Co$_2$(μ-O$_2$)(μ-NH$_2$)]$^{4+}$.

As mentioned early, the O$_2$ chemistry of (FTF3)Co$_2$ is distinct from that of other group 2 cofacial porphyrins, which may result from the (presumed) short separation of the two porphyrin macrocycles in FTF3. Exposure of solutions of (FTF3)Co$_2$ containing excess N-methylimidazole to O$_2$, followed by I$_2$, did not generate an EPR-active species [Collman et al., 1983a], in contrast to other group 2 porphyrins. It was proposed that (FTF3)Co$_2$ might have no affinity to O$_2$ under the experimental conditions and could not be oxidized with I$_2$, thereby remaining fully reduced and antiferromagnetically coupled throughout the manipulations. Alternatively, O$_2$ binding outside the cavity followed by formation of an intermolecular μ-peroxo complex and its decomposition (possibly assisted by the excess imidazole and moisture) can eventually yield a doubly oxidized diamagnetic [(FTF3)Co$_2$]$^{2+}$. Since FTF3Co$_2$ is a poor ORR catalyst, its O$_2$ chemistry has been little studied.

In summary, the cooperative behavior of the two metal ions determines the dioxygen chemistry of group 2 bis-Co porphyrins and certain flexible group 1 analogs (e.g., a mixture of (FTF6)Co$_2$ and [(FTF6)Co$_2$]$^{2+}$ in benzonitrile conproportionates in the presence of O$_2$ to yield a stable [(FTF6)Co$_2$O$_2$]$^+$ adduct containing a bridging superoxide ligand [LeMest et al., 1997]). Unlike monoporphyrin analogs, which bind O$_2$ only in the reduced (Co^{II}) form bearing an appropriate axial ligand, singly oxidized cofacial bis-Co porphyrins manifest unusually high O$_2$ affinities even in the absence of nitrogenous heterocycles. In all studied cases, O$_2$ binding to singly oxidized group 2 porphyrins (except FTF3) yields bridging μ-superoxo adducts, although at present the binding mode(s) of the μ-O$_2^-$ moiety is unknown. Whereas O$_2$ adducts of monometalloporphyrins rapidly autooxidize in the presence of protic sources, [(FTF4)Co$_2$(O$_2$)]$^+$ undergoes reversible protonation (pK_a of the conjugate acid in benzonitrile is 12 [Fukuzumi et al., 2004]) and reversible one-electron reduction and (possibly) one-electron oxidation [LeMest et al., 1998].

Little is known about the O$_2$ adducts of cofacial corrole/porphyrin, cofacial biscorrole, or cofacial phthalocyanine derivatives (Fig. 18.13).

18.5.3 ORR Catalysis by Cofacial Diporphyrins: Selectivities and Structure/Activity Relationships

The selectivities of metal complexes of cofacial porphyrinoids (porphyrins, corroles, and phthalocyanines) reported in the literature by mid-2007 are summarized in Fig. 18.15. The data are organized by the type of catalyst as well as in order of decreasing n_{av}. Whereas ORR catalysis by certain cofacial porphyrins, such as (FTF4)Co$_2$ and (DPY)Co$_2$ (Y = A, B) has been studied extensively by a number of groups, and the values of n_{av} are known with high degree of confidence, those for most other catalysts

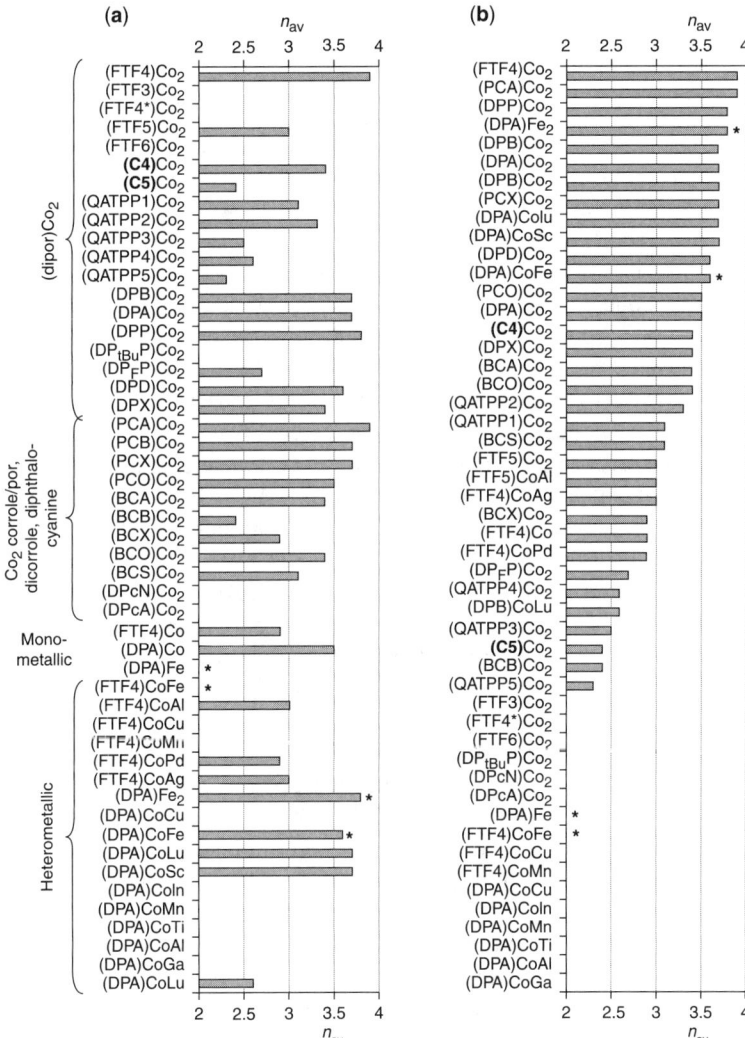

Figure 18.15 Summary of reported n_{av} at the plateau of the catalytic wave for cofacial metalloporphyrinoids adsorbed on a graphite electrode in contact with an aqueous acid (pH 0–1): (a) data organized by structural type; (b) data organized in order of decreasing selectivity toward the four-electron pathway. The asterisk identifies catalysts that manifest two catalytic waves (Fig. 18.10a); n_{av} of the first wave is listed; $n_{av} = 4$ at the plateau of the second wave. See Fig. 18.13 for the structures of the complexes. When contradictory values of n_{av} appear in the literature, the largest reported one is shown. Heterometallic complexes, (DPB)CoX (X = In, Mn, Ti, Al, Ga) were reported to manifest catalytic properties identical to those of the DPA analogs [Guilard et al., 1995]. The data are from Collman et al. [1994, 2003a], in addition to Rosenthal and Nocera [2007] for (DPD)Co$_2$, (DPDM)Co$_2$, (DPX)Co$_2$, and (DPXM)Co$_2$; Kadish et al. [2005] for the corrole-containing complexes; Kobayashi et al. [1990] for the phthalocyanines; and Guilard et al. [1995] for the bimetallic DPA and DPB series derivatives.

in Fig. 18.15 were reported in a single publication and may have a larger uncertainty. In other cases, the reports are contradictory: for example, reported values of n_{av} for a heterometallic complex, (DPA)CoFe, vary from 3.6 to 2, at the plateau of the first wave (the complex exhibits a voltammogram as in Fig. 18.10a) [Ni et al., 1987; Guilard et al., 1995]. As is the case for simple Fe and Co porphyrins, the catalytic performance of cofacial metalloporphyrins (and probably other porphyrinoids, although at present there are no data) appears to be quite sensitive to how the electrode surface was prepared and the catalyst deposited.

The data in Fig. 18.15 are for metalloporphyrins adsorbed on a graphite electrode in contact with an aqueous electrolyte with pH between 0 and 1, which are the conditions that have been most commonly used for ORR catalysis by cofacial metalloporphyrins. There is some controversy regarding the importance of pH and adsorption on graphite for achieving four-electron catalysis. The selectivity of (FTF4)Co$_2$ was reported to depend strongly on the pH of the aqueous electrolyte, with the predominantly four-electron pathway observed only at pH <3.5 [Collman et al., 1994]. In contract, the selectivity of (DPA)MM' (M = M' = Co, M = M' = Fe, or M = Co, M' = Fe) was found to be only weakly pH-dependent [Liu et al., 1985]. In solution, (FTF4)Co$_2$ was reported to be essentially a two-electron catalyst [Collman et al., 1994]. A self-assembled monolayer of a (DPB)Co$_2$ analog, (DPB$_S$)Co$_2$ (Fig. 18.13), was reported to catalyze O$_2$ reduction only to H$_2$O$_2$ [Hutchison et al., 1997]. In contrast, selectivities of (DPY)Co$_2$ (Y = A, B, D, X) dissolved in benzonitrile containing HClO$_4$ were reported to be identical to those of graphite-adsorbed catalysts [Fukuzumi et al., 2004].

At pH 0 the overpotential of ORR catalyzed by the best catalysts, (FTF4)Co$_2$ and (DPX)Co$_2$ (X = A, B, D, P, X) is about 0.55 V. Under identical conditions, most other bis-Co cofacial porphyrins manifest an overpotential of 0.6–1 V. In general, the overpotential and n_{av} do not correlate even for bis-Co derivatives. Replacing even one Co ion in a bis-Co porphyrin invariably increased the overpotential: for example, 0.6 V for (DPA)Co$_2$ versus 0.7 V for (DPA)Co (monometallic) and 0.9 V for (DPA)Fe$_2$. Likewise, replacing the porphyrin macrocycle with a porphyrinoid increased the overpotential: for example, 0.8 V for (PCA)Co$_2$ versus 0.6 V for (DPA)Co$_2$ (Fig. 18.13). As is the case for simple Fe and Co porphyrins, the stability of the mostly four-electron catalysts is low.

Among cofacial porphyrins, the most selective ORR catalysts are (FTF4)Co$_2$ and (DPY)Co$_2$ (Y = A, B, D, P, X), or the analogous porphyrin/corrole derivative, (PCA)Co$_2$, although the latter has not yet been studied extensively. It is not entirely clear if (DPA)Fe$_2$ is a more selective catalyst than a simple Fe porphyrin (Section 18.3). Overall, we lack adequate atomistic understanding of the structure/activity relationship in cofacial porphyrinoid catalysts, although many attempts have been made to rationalize experimental observations. Most hypotheses postulate that some interaction between the terminal O atom of the O$_2$ fragment bound to one metalloporphyrin fragment and the other porphyrin moiety (either with or without a metal) is required to achieve high n_{av}. Thus, the dramatic difference in the (FTFn)Co$_2$ series ($n = 3-6$) was attributed to the difficulty in forming intermediates containing bridging O_2 moieties for $n = 3$ or 6 because the required $Co\cdots Co$ distance could not be accommodated owing to restricted conformational flexibility of the peripheral straps [Collman et al., 1994]. (Note, however, that the two-electron catalyst (FTF6)Co_2 in benzonitrile solution forms a stable O_2 adduct, $[(FTF6)Co_2O_2]^+$, in which the superoxo moiety bridges the two Co^{III} ions [LeMest et al., 1997].) On the other hand, the DPY (Y = A, B, D, P, X) porphyrins are thought to be sufficiently flexible to accommodate a wide range of metal–metal distances [Rosenthal and Nocera, 2007].

Likewise, variations in selectivity within the pairs DPP versus $DP_{tBu}P$ (3.8 vs. 2); DPD versus DPDM (3.6 vs. 3), and DPX versus DPXM (3.4 vs. 2.8) were interpreted as a result of more substituted phenyl groups in the least selective member of the pair interfering with the formation of the O_2-derived bridges [Rosenthal and Nocera, 2007; Park et al., 1995]. The poor selectivity of the $(DP_FP)Co_2$ catalyst, whose perfluorophenyl substituents' steric requirements are unlikely to differ notably from those of the p-tolyl moieties in DPP, may illustrate the importance of electronic factors (cf. Anson's work mentioned in Section 18.4.2 regarding the enhanced four-electron selectivity of monomeric Co porphyrins bearing electron-donating peripheral subsituents, including π-basic metal complexes). The two-electron ORR catalysis by the cofacial phthalocyanines (DPcN)Co_2 and (DPcA)Co_2 (Fig. 18.13) was attributed to similar "electronic perturbations" [Kobayashi et al., 1990]. In contrast, a derivative of (FTF4)Co_2 bearing a single Cl atom at a *meso* position decreases the overpotential of O_2 reduction by about 40 mV and increases n_{av} at potentials <0.5 V (where the parent catalyst becomes increasingly less selective) [Collman et al., 1983b].

The stereoelectronic basis of the poor selectivity of (FTF4*)Co_2, (FTF5)Co_2, (C4)Co_2, (C5)Co_2 ($n_{av} < 3$; Fig. 18.15), which are structurally analogous to the most selective cofacial bis-Co porphyrin ORR catalyst, (FTF4)Co_2, remains unexplained. All five complexes bind O_2 in nonaqueous media exclusively in a bridging fashion (forming μ-peroxo or μ-superoxo derivatives, depending on the oxidation state; Section 18.5.2), and their electrochemical properties in the absence of a substrate suggest that the two π-systems of the two macrocycles interact strongly, and hence bimetallic cooperativity would be expected.

Much effort has been expanded in drawing mechanistic inferences from the observation that cofacial bismetalloporphyrins containing a non-redox-active metal ion are fairly selective catalysts (e.g., (DPA)CoM, where M = Lu, Sc, Al, Ag, Pd, 2H, i.e., monometallic porphyrins; Fig. 18.15). At least two hypotheses have been proposed: (i) polarization of the O–O bond in catalytic intermediates by the second ion (on an N–H moiety) acting as a Lewis acid [Collman et al., 1987, 1994] and (ii) spatial positioning of H^+ donors especially favorable for proton transfer to the terminal O atoms of coordinated O_2 [Ni et al., 1987; Rosenthal and Nocera, 2007]. To the best of my knowledge, neither hypothesis has yet been convincingly proven nor resulted in improved ORR catalysts. When seeking stereoelectronic rational of the observed n_{av} values, it is useful to be mindful that a fair number of simple Co porphyrins are also relatively selective ORR catalysts (Section 18.4.2).

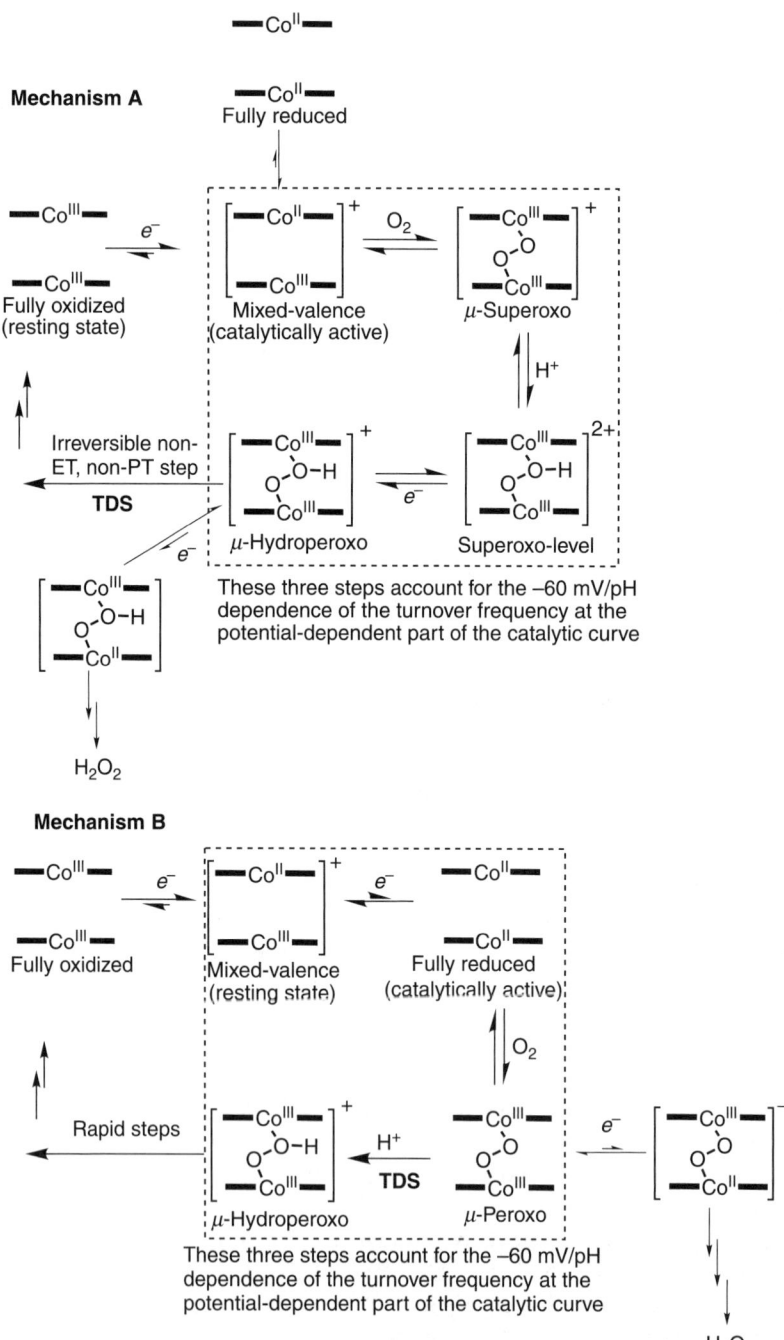

Figure 18.16 (*Continued*).

18.5.4 ORR Catalysis by Cofacial Diporphyrins: Plausible Catalytic Cycles

Little is known about the mechanism of O_2 reduction even in the case of the most well-studied catalysts, (FTF4)Co_2 and (DPY)Co_2 (Y = A, B). Catalysis starts at potentials between those of the $[(dipor)Co_2]^{2+/+}$ and $[(dipor)Co_2]^{+/0}$ pairs observed under anaerobic conditions [Collman et al., 1994]. In conjunction with the high O_2 affinity of $[(dipor)Co_2]^+$ observed in nonaqueous media, the onset of O_2 reduction is commonly interpreted to indicate that $[(dipor)Co_2]^+$ is the catalytically active redox state at potentials of the rising part of the O_2 reduction wave [Collman et al., 1994]. Certain polarographic studies of (FTF4)Co_2 were interpreted as indicative of O_2 binding primarily to the mixed-valence $Co^{III}Co^{II}$ redox state [Collman et al., 1988, 1994]. The change of the dominant redox form of the catalyst from $[(dipor)Co_2]^+$ to (dipor)Co_2 at about 0.5 V (vs. NHE at pH 0) was suggested to account for the decrease in the catalytic selectivity of graphite-adsorbed (FTF4)Co_2 at potentials more reducing than about 0.5 V (vs. NHE at pH 0), because the reduced catalyst was presumed to be more prone to binding O_2 outside the cavity and thus reducing it noncooperatively.

Unfortunately, little is known about the O_2 affinities of the various redox forms of (dipor)Co_2 in contact with an aqueous solution. For example, the presence of O_2 in the aqueous electrolyte does not appear to perturb the position of the $Co^{III}Co^{II}/Co_2^{II}$ redox wave in graphite-adsorbed (FTF4)Co_2 [Collman et al., 2003a], suggesting that the fraction of the oxygenated complexes, $[(FTF4)Co_2O_2]^{+/0}$ is quite low during the redox cycling as a result of either thermodynamic or kinetic limitations. In contrast, the positions of redox waves of (FTF4)Co_2 in anhydrous benzonitrile solutions are strikingly different under anaerobic and aerobic conditions [LeMest et al., 1995], indicative of rapidly attained highly favorable equilibrium: O_2 + (FTF4)$Co_2 \rightleftharpoons [(FTF4)Co_2O_2]^+ + e^-$. It seems plausible that competition between O_2 and H_2O for the Co sites may lower the O_2 affinity of graphite-adsorbed $[(FTF4)Co_2]^+$ in contact with an aqueous electrolyte.

Figure 18.16 Two plausible catalytic cycles of the ORR by cofacial bis-Co porphyrins that are consistent with the -60 mV/pH dependence of the turnover frequency at the rising part of the catalytic wave. Mechanisms A and B postulate that the mixed-valence $[(dipor)Co_2]^+$ and the fully reduced $[(dipor)Co_2]$ redox states, respectively, are the catalytically active form of the cofacial bis-Co porphyrin catalyst. The overall charges of, and the oxidation states of the Co ions in, the intermediates are used only for electron-bookkeeping purposes, and are not intended to imply a specific location of the hole or restrict the charge of any implicit ligand. The binding mode of O_2 is unknown; Co ions may have additional ligands in some or all redox states. ET and PT are electron and proton transfer, respectively. In mechanism B, both irreversible (turnover-determining step, TDS) protonation and reversible protonation, followed by an irreversible non-electron-transfer, non-proton-transfer (non-ET, non-PT) step would yield the -60 mV/pH dependence. The simplest sequence is shown following the principle of Occam's razor. Potential side reactions that account for the decrease in n_{av} with more reducing potentials are shown: they must include at least one electron transfer step to yield a potential-dependent selectivity

The half-wave potentials of (FTF4)Co_2-mediated O_2 reduction at pH 0–3 shifts by -60 mV/pH [Durand et al., 1983], which indicates that the turnover-determining part of the catalytic cycle contains a reversible electron transfer (ET) and a protonation, or two reversible ETs and two protonation steps. In contrast, if an irreversible ET step were present, the pH gradient would be $60/(n + \alpha)$ mV/pH, where n is the number of electrons transferred in redox equilibria prior to the irreversible ET and α is the transfer coefficient of the irreversible ET. The -60 mV/pH slope is identical to that manifested by simple Fe porphyrins (see Section 18.4.1). The turnover rate of ORR catalysis by (FTF4)Co_2 was reported to be proportional to the bulk O_2 concentration [Collman et al., 1994], suggesting that the catalyst is not saturated with O_2.

At least two catalytic cycles are consistent with these observations, depending on whether the catalytically active redox state of the cofacial bis-Co porphyrin is mixed-valence, $[(dipor)Co_2]^+$ (Fig. 18.16, mechanism A) or fully reduced, $[(dipor)Co_2]$ (Fig. 18.16, mechanism B). Since the catalysis occurs at potentials about 0.1 V more reducing than those of the $[(dipor)Co_2]^{2+/+}$ couple, the redox equilibrium $[(dipor)Co_2]^{2+} \rightleftharpoons [(dipor)Co^2]^+$ at the catalytic wave has a minimal impact on the molar fraction of the catalytically active species, and therefore cannot contribute to the observed pH dependence of the turnover rate. Hence, if the mixed-valence redox state, $[(dipor)Co_2]^+$, is indeed the catalytically active form, as is usually assumed [Collman et al., 1994], the catalytic cycle must contain another redox equilibrium before the turnover-determining step (TDS). The sequence of reversible protonation of the superoxo-level intermediate followed by its reversible one-electron reduction is one possibility. Superoxo adducts of a number of cofacial bis-Co porphyrins have been demonstrated to undergo reversible protonation in benzonitrile [Fukuzumi et al., 2004; LeMest et al., 1997], but the protonated form is not known to undergo reversible redox chemistry. Since only the solution chemistry of the protonated form was studied, if the peroxo-level intermediate undergoes rapid intramolecular rearrangement, very fast sweep rates may be required in solution voltammetry to observe redox equilibrium between $[(dipor)Co_2(O_2H)]^{2+}$ and $[(dipor)Co_2(O_2H)]^+$ species. For graphite-adsorbed species, the redox equilibria are assumed to establish very rapidly.

The alternative mechanism (Fig. 18.16, mechanism B) is based on the fully reduced $[(dipor)Co_2]$ state as the redox-active form of the catalyst. The redox equilibrium between the mixed-valence and fully reduced forms is shifted toward the catalytically inactive mixed-valence state, and hence controls the amount of catalytically active species in the catalytic cycle and contributes to the -60 mV/pH dependence. The fully reduced form is known to bind O_2 (probably reversibly) in organic solvents [LeMest et al., 1997; Fukuzumi et al., 2004], and the resulting diamagnetic adducts are typically viewed as a pair of Co^{III} ions bridged by a peroxide, which are of course quite common in the O_2 chemistry of nonporphyrin Co complexes. To obtain the -60 mV/pH dependence of the catalytic turnover rate, a protonation step is required either prior to the TDS or as the TDS. Mechanism B cannot be extended to monometallic cofacial porphyrins or heterometallic porphyrins with a redox-inert ion, but there is no reason to assume that the two classes of cofacial porphyrin catalysts, with rather different catalytic performance (Fig. 18.15), must follow the same mechanism.

Importantly, both mechanisms in Fig. 18.16 postulate a hydroperoxo complex to be the critical intermediate in the cycle, in parallel to the currently accepted mechanism of the ORR by cytochrome c oxidase (see Section 18.2) and cytochrome P450 enzymes. The major difference between the mechanisms in Fig. 18.16 and those in heme enzymes is the presumed importance of bridging binding modes of the O_2-derived ligands for cofacial bis-Co catalysts. The fascinating question is the fate of the μ-hydroperoxo intermediate (Fig. 18.16). In all heme enzymes involved in heterolysis of the O–O bond in a peroxo-level ligand, one electron always comes from oxidation of Fe^{III} to Fe^{IV}, whereas the other electron comes from an organic moiety, such as tyrosine (cytochrome c oxidase), porphyrin (catalase), probably an organic substrate (cytochrome P450 enzymes), etc. Little is known about Co analogs of the well-studied ferryl porphyrins, (por)Co=O and it is impossible at present to propose which decay path is most likely for the μ-hydroperoxo intermediate, $[(dipor)Co_2(O_2H)]^+$. Possibilities include spontaneous heterolysis or homolysis of O–O; and outer-sphere reduction, followed by O–O bond heterolysis to give a Co^{III}–OH and a Co=O fragment (this sequence can only be accommodated within mechanism B to preserve the -60 mV/pH dependence).

Both mechanisms in Fig. 18.16 accommodate the observed decrease in the catalytic selectivity at potentials more reducing than 0.5 V (vs. NHE at pH 0) [Collman et al., 1994]. For example, in mechanism A, the reversibly formed μ-hydroperoxo intermediate may undergo reversible one-electron reduction in competition with the TDS. If the potential of the $[(dipor)Co_2(O_2H)]^{+/0}$ pair is more reducing than that of the $[(dipor)Co_2(O_2H)]^{2+/+}$ pair, which seems plausible, this side equilibrium may be unfavorable at the rising part of the catalytic wave but may become shifted towards the reduced peroxo-level adduct, $[(dipor)Co_2(O_2H)]$, at potentials <0.5 V. The latter adduct, containing a formally Co^{II} ion, which is known to be labile, may be susceptible to hydrolysis, leading to the increasing production of H_2O_2 as the potentials become more reducing.

Likewise, within mechanism B, the μ-peroxo intermediate may be susceptible to reversible one-electron reduction to anionic $[(dipor)Co_2O_2]^-$, which may become important only at potentials <0.5 V. There is some indication that the formally peroxo adducts, $[(dipor)Co_2^{III}(O_2^{2-})]$, formed by addition of O_2 to fully reduced (dipor)Co_2, may undergo reversible reduction. Protonation of the anionic species may be followed by its hydrolysis, releasing H_2O_2.

Although impressive progress has been made in unraveling the mechanism of ORR catalysis by cofacial porphyrins, much remains to be learned before we can understand how this mechanism relates to those in heme enzymes and simple metalloporphyrins and use our mechanistic knowledge to rationally design improved metalloporphyrin catalysts for the ORR.

18.6 BIOMIMETIC ORR CATALYSTS

Biomimetic ORR catalysts are highly elaborate porphyrins designed to replicate the stereoelectronic properties of the O_2 reducing site of heme/Cu oxidases (see

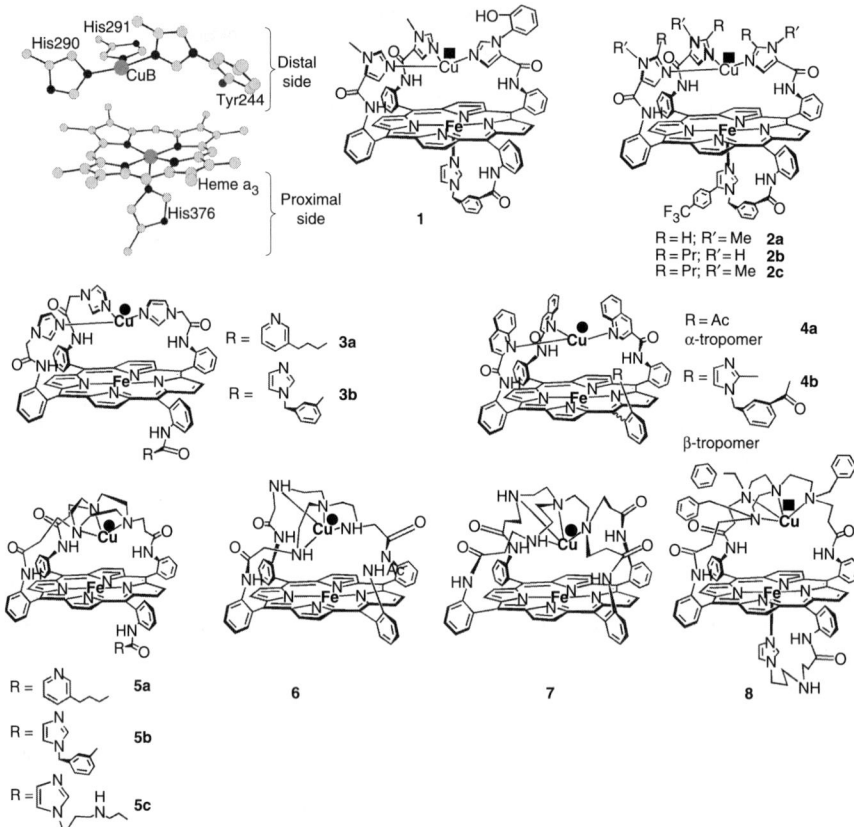

Figure 18.17 Biomimetic ORR catalysts.

Section 18.2). The latest generation of such catalysts (**1** in Fig. 18.17) reproduces the key features of the site: (i) the proximal imidazole ligation of the heme; (ii) the trisimidazole ligation of distal Cu; (iii) the Fe–Cu separation; and (iv) the distal phenol covalently attached to one of the imidazoles. As a result, binding of O_2 to compound **1** in its reduced (Fe^{II}/Cu^{I}) state appears to result in rapid reduction of O_2 to the level of oxides (−2 oxidation state) without the need for outer-sphere electron transfer steps [Collman et al., 2007b]. This reactivity is analogous to that of the heme/Cu site of cytochrome c oxidase (see Section 18.2).

Biomimetic studies typically have one or more of the following objectives: (i) to reproduce in a small synthetic molecule reactivity that was theretofore only observed in an enzyme; (ii) to understand the mechanisms of an enzymatic reaction and the relationship between the stereoelectronic attributes of the catalytic site and its reactivity; and (iii) to develop practical catalysts by exploiting and adopting solutions that evolved in Nature. Biomimetic studies of cytochrome c oxidase have been particularly impactfull in addressing aim (ii). On the other hand, this approach is yet to result in an understanding of the structure/activity relationships for metalloporphyrin-catalyzed ORR that would allow a rational design of metalloporphyrin-based catalysts for the operating conditions of practically useful fuel cells, rather than physiologically relevant ones. Finally, since even some simple Fe porphyrins catalyze four-electron reduction of O_2 under specific conditions (physical contact with an electrode, pH, potentials), achieving goal (i) means demonstrating highly selective ORR catalysis under conditions where simple Fe porphyrins are clearly inadequate catalysts.

So far, certain biomimetic catalysts (**1** and **2b** in Fig. 18.17) have been shown to reduce O_2 to H_2O under a slow electron flux at physiologically relevant conditions (pH 7, 0.2–0.05 V potential vs. NHE) and retain their catalytic activity for $>10^4$ turnovers. Probably, only the increased stability of the turning-over catalyst is of relevance to the development of practical ORR catalysts for fuel cells. In addition, biomimetic catalysts of series **1**, **2**, **3**, and **5**, and catalyst **4b** are the only metalloporphyrins studied in ORR catalysis with well-defined proximal and distal environments. For series **2**, which is by far the most thoroughly studied series of biomimetic ORR catalysts, these well-defined environments result in an effective catalysis that seems to be the least sensitive among all metalloporphyrins to the electrode material (whether the catalyst is adsorbed or in the film) and to chemicals present in the electrolyte or in the O_2 stream, including typical catalyst poisons (CO and CN^-).

18.6.1 Chemistry of O_2 Adducts of Biomimetic Catalysts

Stoichiometric reactions of O_2 with porphyrins **1**, **2a–c**, **5c**, and **8** in organic solvents were studied, in addition to an analog of **5c** with Co replacing Fe, **5cCo** (Fig. 18.18). Dioxygen adducts with **5c**, **5cCo**, and **8** were reported to be stable at room temperature in air and to contain a bridging peroxo moiety, based on infrared spectroscopy, in addition to mass spectroscopy (**5cCo**) or ^{1}H-NMR (**5c**, **8**) [Collman et al., 2003a]. Note that ^{1}H-NMR spectroscopy revealed that the adducts were diamagnetic, but did not provide information about the binding mode of the O_2 ligand. These adducts could not be deoxygenated under vacuum, but could be reduced by cobaltocene, $CoCp_2$. For example, titration of a solution of **5cCoO$_2$** with $CoCp_2$ resulted in spectral changes in the 400–600 nm region that were interpreted as indicative of the presence of only two metalloporphyrin species, the oxygenated porphyrin **5cCoO$_2$** and free, fully reduced porphyrin **5cCo** (Co^{II}/Cu^I), until 4 equivalents of $CoCp_2$ had been added. From these results, it was concluded that reduction of **5cCoO$_2$** to [**5cCoO$_2$**]$^-$ was rate-limiting. Similarly, 4 equivalents were required to reduce **8O$_2$**, and again, only two species (**8O$_2$** and **8**) could be detected by UV–vis when less than 4 equivalents of $CoCp_2$ were used. In contrast, **5c** was regenerated from **5cO$_2$** upon addition of only 2 equivalents of $CoCp_2$. Unlike experiments in which O_2 reduction by ferrocenes is catalyzed by cofacial porphyrins in an organic solvent containing an acid (Section 18.5), titrations of **5cO$_2$**, **5cCoO$_2$**, and **8O$_2$** with $CoCp_2$ were carried out under anhydrous conditions excluding protic sources, which poses two questions: (i) What chemical form are the O^{2-} atoms derived from reduction of **5cCoO$_2$** or **8O$_2$** with 4 equivalents of $CoCp_2$ (and what is the form of the O_2^{2-} species upon reduction of

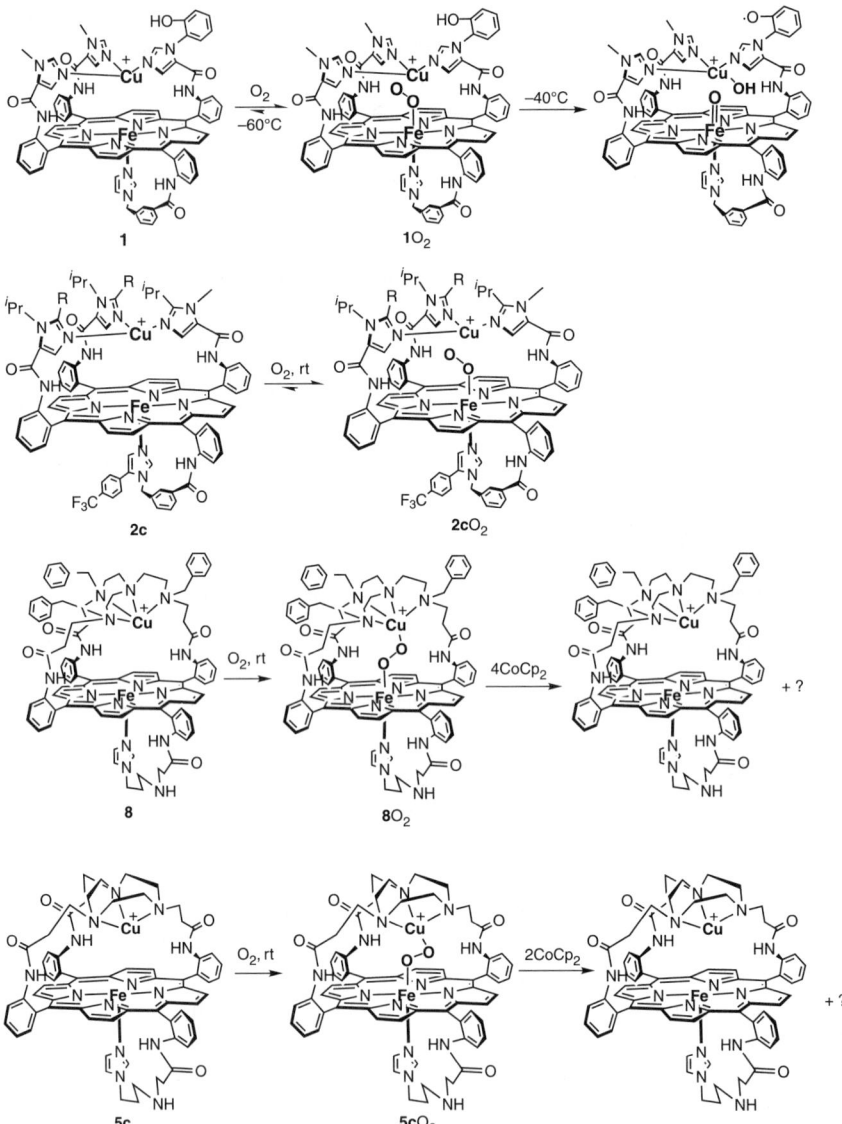

Figure 18.18 O$_2$ chemistry of biomimetic ORR catalysts.

5cO$_2$ with 2 equivalents of CoCp$_2$)? (ii) How can these results be translated to electrocatalytic reduction in aqueous electrolytes?

Reactivity of compounds **1** and **2c** with O$_2$ is remarkably different from that of **5c**, **5cCo**, and **8** (Fig. 18.18). Metalloporphyrin **2c** binds O$_2$ reversibly at room temperature under anhydrous conditions to produce a diamagnetic ferric–superoxo/Cu^I adduct as demonstrated by resonance Raman spectroscopy using ^{16}O$_2$ and ^{18}O$_2$

isotopes and by ^{1}H- and ^{19}F-NMR spectroscopy [Collman et al., 2003b]. Similar results were observed for Co analogues of **2b** and **2c**, in which Fe is replaced by Co, whose paramagnetism required characterization by EPR rather than by NMR in addition to resonance Raman spectroscopy [Collman et al., 2002a]. Although adduct **2c**O_2 is stable at room temperature it undergoes rapid autoxidation in the presence of H^+ sources (compare this with the reversible protonation of the μ-superoxo complexes of cofacial bis-Co porphyrins; Section 18.5). Although Fe and Cu ions in series **2** metalloporphyrins (Fig. 18.18) are positioned sufficiently close to be bridged by a peroxide (and indeed PhNO, which is isoelectronic with O_2, binds series **2** metalloporphyrins in a bridging mode), a bridged peroxide is not formed. It is not at present known if the preference for a ferric–superoxo/Cu^I isomer versus the Fe^{III}–O–O–Cu^{II} alternative is thermodynamic or kinetic. Notably, O_2 binding to the heme/Cu site in cytochrome c oxidase also generates a ferric–superoxo/Cu^I intermediate (compound A in Fig. 18.5). The distal Cu^I ion in **2c**, although not interacting directly with the bound superoxide, had a large effect on the O_2 affinity of metalloporphyrin **2c**. The Cu-free (Fe-only) form of **2c** manifested very low O_2 affinity such that the O_2 adduct could only be observed at $-60\,°C$, and even at this low temperature the Fe-only complex rapidly underwent irreversible oxidation.

Compound **1** (Fig. 18.18) reversibly forms an analogous ferric–superoxo/Cu^I adduct at $-60\,°C$, as demonstrated by resonance Raman spectroscopy. However, warming the sample to $-40\,°C$ results in a rapid four-electron reduction of the bound O_2 ligand, generating a ferryl/Cu^{II}/phenoxyl radical derivative (Fig. 18.18) [Collman et al., 2007a].

18.6.2 ORR Catalysis by Biomimetic Metalloporphyrins

ORR catalysis by biomimetic metalloporphyrins is typically studied using porphyrins adsorbed on a graphite electrode in contact with an aqueous electrolyte buffered at pH 7, to reproduce physiological conditions. For most catalysts, only selectivity at the plateau of polarization curves at a single pH was reported; in a few cases, the difference in catalytic selectivity by the bimetallic (FeCu) and monometallic (Fe-only) forms of porphyrins was also reported. When this difference exists (as in the case of series **4** catalysts; Fig. 18.17) it provides support that the bimetallic catalyst does not lose Cu when adsorbed on the electrode. However, when the behavior of the two forms was identical, or when only the bimetallic catalyst was studied, no evidence was provided that Cu was indeed present in the adsorbed porphyrin. Overall, the biomimetic catalysts appear to manifest higher n_{av} values and retain their catalytic activity for more turnovers than does Fe tetraphenylporphyrin; exceptions are the bimetallic forms of series **4** porphyrins [Ricard et al., 2001]. Contradictory results were reported regarding the effect of Cu on n_{av}: ORR catalysis by Cu-free (Fe-only) forms of series **3** and **5** and by porphyrins **6–8** have not been reported [Collman et al., 2003a]; the bimetallic, FeCu forms of series **4** porphyrins were reported to manifest *lower* n_{av} than (Cu-free) Fe-only forms [Ricard et al., 2001]. Distal Cu had relatively minor effect on ORR catalysis by **1** and series **2** porphyrins when the electron flux from the electrode was rapid [Boulatov et al., 2002; Boulatov, 2004;

Collman et al., 2007b]. However, when the electron delivery was slow, only the bimetallic forms manifested ORR catalysis (it was presumed that the monometallic Cu-free (Fe-only) catalysts degraded rapidly under these conditions) [Collman and Boulatov, 2002; Collman et al., 2007b].

ORR catalysis by series **2** metalloporphyrins has been studied most extensively [Boulatov et al., 2002; Collman et al., 2002a, 2003b; Shiryaeva et al., 2003], and the results can be summarized as follows:

- The Fe-only form of these metalloporphyrins is a highly selective ORR catalyst when adsorbed on a graphite or Au electrode. It operates at an overpotential of about 0.55 V at pH 7 and $n_{av} > 3.9$ (Fig. 18.19) and retains these characteristics for $>10^4$ turnovers; the catalytic selectivity is independent of the amount of deposited catalyst.
- Detectable amounts of partially reduced oxygen species are generated only at potentials more oxidizing than 0.1 V (vs. NHE at pH 7); the primary product was identified to be superoxide by incorporating into catalytic films selective scavengers of H_2O_2, $O_2^{\cdot -}$, and $\cdot OH$. These catalysts do not appear to generate significant amounts of $\cdot OH$ as do simple Fe porphyrins (see Section 18.4).
- In contrast to simple metalloporphyrins, or cofacial diporphyrins, the catalytic performance of these biomimetic catalysts improves at higher pH; as a result, the smallest overpotential was observed at pH 8 (0.5 V) and at pH > 8 no partially reduced oxygen species could be detected at any potential.
- These metalloporphyrins are unique among Fe and Co porphyrins in their high catalytic efficiency of electroreduction of H_2O_2 (at potentials <0.75 V vs. NHE at pH 7), as well as disproportionation and oxidation of H_2O_2 (at potentials >0.8 V).
- Despite the high catalytic activity toward H_2O_2 reduction and disproportionation, ORR catalysis does not appear to proceed via free H_2O_2 as inferred from

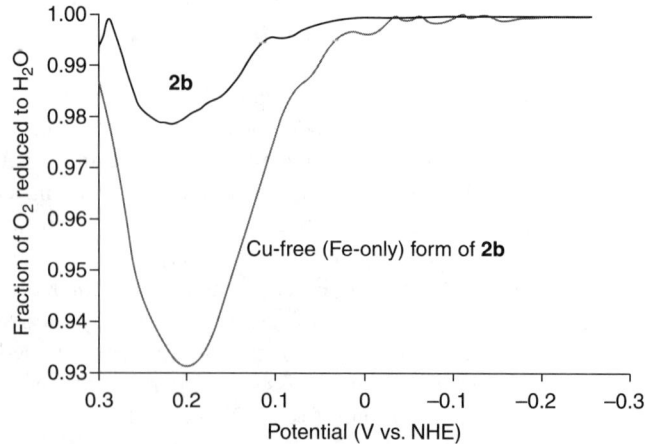

Figure 18.19 Selectivity toward four-electron oxygen reduction by graphite-adsorbed catalysts **2b** (Fig. 18.17) in the bimetallic (FeCu) and monometallic (Fe-only) forms at pH 7.

the fact that the average catalyst molecule reduces >1000-fold greater number of O_2 molecules compared with H_2O_2 in the absence of air before losing its catalytic capability.
- When adsorbed on graphite or Au, the catalytic characteristics of the FeCu form are very similar to those of the Fe-only analog, with about 50 mV smaller overpotential, a fourfold smaller fraction of O_2 reduced to O_2^-/HO_2^- ($n_{av} > 3.95$; Fig. 18.19) and three- to fivefold lower susceptibility to poisoning by CO, CN^-, and N_3^-.
- When dispersed in a lipid film on a surface of a graphite electrode, the FeCu catalysts are clearly superior to the Fe-only forms in catalytic selectivity, stability, and turnover frequency.

A detailed mechanism of ORR catalysis for electrode-adsorbed complexes was proposed (Fig. 18.20) [Boulatov et al., 2002; Boulatov, 2004], based on determination of the following:

- The pH dependence of (i) the turnover frequency (-30 mV/pH at pH < 8 and -60 mV/pH at pH > 8) and (ii) the $Fe^{III/II}$ and $Cu^{II/I}$ potentials E_1.
- The reduction potential of the ferryl species (por)Fe^{IV}=O/(por)Fe^{III}(OH), E_f, in a range of pH, and the pK_a of the H_2O ligand in ferric–aqua form [(por)Fe(OH$_2$)]$^+$, using standard cyclic voltammetry at different pH of the anaerobic aqueous electrolyte (por is the superstructured porphyrin in Fig. 18.17).
- The equilibrium constant K for (por)Fe(OH$_2$) $\rightleftharpoons$ (por)Fe, which determines the molar fraction of the 5-coordinate redox-active Fe^{II} catalyst. This constant was estimated from analysis of the catalytic turnover frequencies in the presence of varying concentrations of an inhibitor, CN^-, which competes with both O_2 and H_2O for the 5-coordinate Fe^{II} porphyrin.
- The low limit on the rate constant k_{hetero} of O–O bond heterolysis in the putative ferric–hydroperoxo intermediate by analyzing the turnover frequency of H_2O_2 reduction at potentials 0.6–0.4 V (vs. NHE at pH 7).

Only three steps of the proposed mechanism (Fig. 18.20) could not be carried out individually under stoichiometric conditions. At pH 7 and the potential-dependent part of the catalytic wave (>150 mV vs. NHE), the -30 mV/pH dependence of the turnover frequency was observed for both Fe/Cu and Cu-free (Fe-only) forms of catalysts **2**, and therefore it requires *two* reversible electron transfer steps prior to the turnover-determining step (TDS) and one proton transfer step either prior to the TDS or as the TDS. Under these conditions, the resting state of the catalyst was determined to be ferric–aqua/Cu^{II}, which was in a rapid equilibrium with the fully reduced ferrous–aqua/Cu^I form (the $Fe^{III/II}$ and $Cu^{II/I}$ potentials were measured to be within <20 mV of each other, as they are in cytochrome c oxidase, resulting in a two-electron redox equilibrium). This first redox equilibrium is biased toward the catalytically inactive fully oxidized state at potentials >0.1 V, and therefore it controls the molar fraction of the catalytically active metalloporphyrin. The fully reduced ferrous–aqua/Cu^I form is also in a rapid equilibrium with the catalytically active 5-coordinate ferrous porphyrin. As a result of these two equilibria, at 150 mV (vs. NHE), only <0.1%

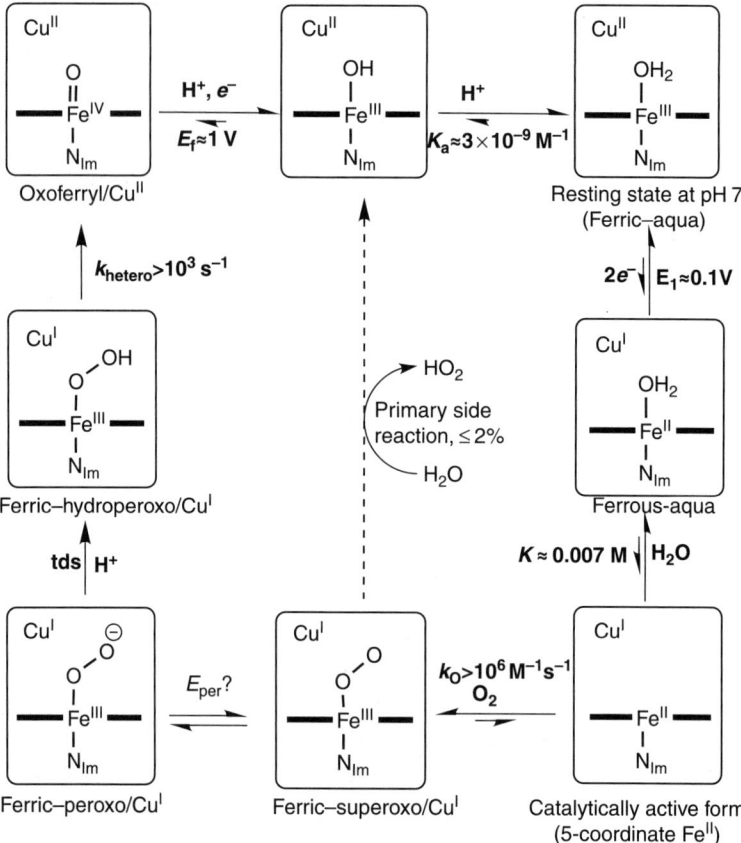

Figure 18.20 A plausible ORR catalytic cycle by biomimetic catalysts **2** (Fig. 18.17). Cu is ligated by three imidazoles (omitted for clarity) and potentially an exogenous ligand, whose nature is not known. All intermediates other than ferric–peroxo and ferric–hydroperoxo were prepared independently.

of a series **2** porphyrin is in the catalytically active form, 5-coordinate Fe^{II}. The dioxygen reactivity of 5-coordinate Fe^{II} forms of series **2** porphyrins was studied under stoichiometric conditions (see the previous section): the resulting O_2 adduct was formulated as ferric–superoxo/Cu^I; the same isomer was assumed to be present on the electrode during ORR catalysis. To give the observed -30 mV/pH dependence of the turnover frequency, another redox equilibrium was proposed, between the ferric–superoxo and the ferric–peroxo intermediates, followed by an irreversible protonation (the TDS), presumably to yield a ferric–hydroperoxo species. A slow protonation was the simplest way to account for both the -30 mV/pH dependence and the fact that all other steps were too fast to give the overall measured turnover frequency of 2 s^{-1} (at 150 mV). The structures of the peroxo-level intermediates were not determined, and in these intermediates the terminal O atom of the peroxo ligand may interact with distal Cu, forming a μ-peroxo complex.

The observed facile electroreduction and electro-oxidation of H_2O_2 at potentials >0.2 V very likely proceeded through the same ferric–hydroperoxo intermediate as did ORR catalysis. Therefore, the turnover frequency of H_2O_2 reduction placed the lower limit on the rate of O–O bond heterolysis in the hydroperoxo–ferric intermediate (note that O–O bond homolysis was ruled out by the use of ˙OH scavengers and the long (>10^4 turnovers) lifetime of the catalyst during the ORR). The resulting oxoferryl/Cu^{II} complex could be prepared independently by electrochemical oxidation of the resting state of the catalyst at about 1 V (vs. NHE at pH 7). At the rising part of the ORR catalytic wave, this oxoferryl/Cu^{II} intermediate undergoes rapid, essentially irreversible, proton-coupled reduction to a ferric–hydroxo/Cu^{II} complex. The ferric–hydroxo and ferric–aqua complexes were found to be in a rapid equilibrium (the pK_a of the coordinated water was estimated to be 8.5), thus regenerating the resting state of the catalyst.

The key feature of this mechanism is the generation of the ferric–hydroperoxo intermediate in the TDS, which means that the (pseudo)-first-order rate constant for the decay of this intermediate was >10-fold the pseudo-first-order rate constant of the preceding protonation. Hence, this intermediate has a very short lifetime (or, equivalently, is present in a very low molar fraction during steady-state ORR catalysis), which minimizes the probability that it will release H_2O_2. Probably because of distal Cu^I, no evidence for O–O bond homolysis in the ORR or H_2O_2 reduction catalysis was observed (the Cu-free analog did generate ˙OH radicals in H_2O_2 reduction, but not in the ORR). However, even the Fe-only (Cu-free) metalloporphyrins retained their catalytic activity over 10^2-10^3 times as many turnovers as simple Fe porphyrins, probably because the imidazole ligand *trans* to the peroxo moiety lowers the activation barrier of the heterolytic pathway relative to O–O bond homolysis [Watanabe, 2000; Sono et al., 1996].

The catalytic cycle in Fig. 18.20 also rationalizes the potential-dependent n_{av} of series **2** catalysts (Fig. 18.19). The primary partially reduced oxygen species was determined to be superoxide, $O_2^{\cdot-}$, by using $O_2^{\cdot-}$ scavengers incorporated in catalytic films. Superoxide is produced by autoxidation, i.e., heterolysis of the Fe–O bond in the ferric–superoxo intermediate [Shikama, 1998], probably induced by protonation of the terminal O atom in bound O_2. The hypothesis of protonation-assisted autoxidation was supported by the observation that n_{av} at the rising part of catalytic curves was smaller in acidic media (more superoxide was produced), whereas no partially reduced oxygen species were detected at any potentials in basic (pH > 8) electrolytes. The autoxidation rate constant at pH 7 was estimated to be 0.03 s^{-1} (for the Fe-only forms of series **2** catalysts) and <0.01 s^{-1} for the FeCu forms.

Release of superoxide during ORR catalysis indicates that the ferric–superoxo intermediate (Fig. 18.20) has a substantial residence time at 0.2 V (the potential of the maximum production of superoxide), suggesting that the potential of the ferric–superoxo/ferric–peroxo couple, E_{per} (Fig. 18.20), is more reducing than 0.2 V. The fraction of superoxide detected at potentials >0.2 V probably reflects the fact that $O_2^{\cdot-}$, which is a strong outer-sphere reductant [Huie and Neta, 1999], was oxidized by the mostly ferric catalytic film before it could escape the film. There are two plausible explanations for the decrease in the fraction of superoxide byproduct released at

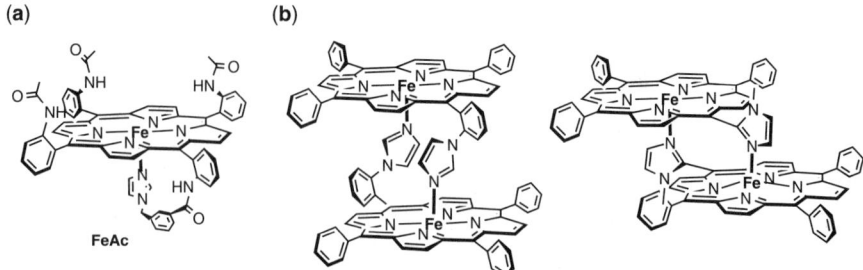

Figure 18.21 (a) FeAc is a simplified version of the biomimetic catalysts in series **1–4** (Fig. 18.17); when adsorbed on a graphite electrode, it manifests catalytic behavior comparable to that of series **2** metalloporphyrins. (b) The simplest structural motif that ensures the axial imidazole ligation of the Fe^{II} site [Khvostichenko et al., 2007; Yang et al., 2008].

potentials <0.2 V. One is that it reflects the decrease in the fraction of the ferric–superoxo intermediate, which would require E_{per} to lie somewhere between about 0.1 and 0.2 V for the ferric–superoxo/ferric–peroxo equilibrium to become shifted significantly toward the peroxo intermediate at potentials <0.2 V. The other possibility is that the fraction of the ferric–superoxo intermediate was not perturbed much at potentials >0.1 V, i.e., that E_{per} < 0.1 V, but the decrease was caused by recapture of O_2^- released from one molecule of catalyst by another molecule in the Fe^{II} state (yielding a ferric–peroxo intermediate that re-enters the cycle; Fig. 18.20). The probability of such recapture is increased at more reducing potentials by (i) a decrease in the concentration of O_2 in the catalytic film due to mass transport limitations (plateau part of polarization curves) and (ii) a shift in the $Fe^{III/II}$ equilibrium toward Fe^{II} at potentials <0.1 V.

The combination of selectivity (n_{av} > 3.9), overpotential (about 0.5 V at the turnover frequency of >2 s^{-1} mol O_2/mol) and lifetime (>10^4 turnovers) make series **2** metalloporphyrins, even in the monometallic (Fe-only) form, probably the best among the known metalloporphyrin ORR catalysts. These catalysts perform best in pH 8 electrolytes rather than in the more popular acidic (pH 0) media. Their unique performance characteristics appear to be largely due to the presence of the imidazole ligand *trans* to coordinated O_2. This assertion is supported by the fact that the metalloporphyrin FeAc (Fig. 18.21a) is only slightly inferior to series **2** as an ORR catalyst ($n_{av} \approx 3.85$; a turnover frequency > 2 s^{-1} was attained at an overpotential of about 0.6 V at pH 7, and the catalytic performance was retained for at least 10^4 turnovers). Series **2** metalloporphyrins have also been shown to be efficient electrocatalysts for reduction of oxohalides (e.g., ClO_2^-), and organic peroxides and peroxoacids [Collman et al., 2002c, 2004b].

Based on the well-studied catalytic mechanism in series **2** metalloporphyrins (Fig. 18.20), further developments of the heme/imidazole motif in the context of fuel cell catalysis have to be directed towards the following:

- Simplifying the synthesis of the catalysts (e.g., the highly optimized convergent synthesis of monometallic series **2** metalloporphyrins required between 14 and 17 steps [Collman et al., 2002d]—FeAc (Fig. 18.21a) is available in 7 steps).

- Making the Fe$^{III/II}$ potential more oxidizing—even at the expense of lowering the O$_2$ affinity.
- Decreasing the affinity of the catalyst in the Fe^{II} redox state for H$_2$O.

It appears that the heme/imidazole motif can be realized in metalloporphyrins that are available in just two steps (Fig. 18.21b). However, it is not yet known how to accomplish objectives 2 and 3. It is also important to understand the mechanism of catalyst degradation during the ORR and to identify alternative functional groups that may increase catalyst stability: to be useful in fuel cells, a metalloporphyrin catalyst would probably have to retain its catalytic properties over at least 4×10^6 turnovers (about 1000 hours of operation at a turnover frequency of 1 s^{-1}), i.e., more than a hundred times longer than the most stable metalloporphyrin catalysts reported to date.

18.7 SUMMARY AND CONCLUSIONS

Most of the O$_2$ consumed by aerobic organisms (including all mammals) is reduced to H$_2$O at a heme group of cytochrome c oxidase. Under physiological conditions, each molecule of this enzyme catalyzes reduction of up to 50 molecules of O$_2$ per second (turnover frequency of 50 s^{-1}), capturing over 80% of the free energy of oxygen reduction to satisfy the energy needs of the organism. Although the O$_2$-reducing site of cytochrome c oxidase is bimetallic (in addition to an Fe porphyrin, it has a Cu ion), four-electron oxygen reduction can be catalyzed efficiently by a 5-coordinate heme, provided that outer-sphere electron transfer steps do not limit the turnover.

The prevalence of the heme in O$_2$ metabolism and the discovery in the 1960s that metallophthalocyanines adsorbed on graphite catalyze four-electron reduction of O$_2$ have prompted intense interest in metalloporphyrins as molecular electrocatalysts for the ORR. The technological motivation behind this work is the desire for a Pt-free cathodic catalyst for low temperature fuel cells. To date, three types of metalloporphyrins have attracted most attention: (i) simple porphyrins that are accessible within one or two steps and are typically available commercially; (ii) cofacial porphyrins in which two porphyrin macrocycles are confined in an approximately stacked (face-to-face) geometry; and (iii) biomimetic catalysts, which are highly elaborate porphyrins designed to reproduce the stereoelectronic properties of the O$_2$-reducing site of cytochrome oxidase.

Although simple porphyrins are attractive in terms of cost, they are generally poor catalysts because they manifest one or more of (i) low selectivity for the four-electron reduction; (ii) low turnover numbers (the number of molecules of the substrate that one molecule of catalyst reduces before losing its catalytic activity); and (iii) high overpotential. Simple Fe and Co porphyrins appear to be fairly inefficient in catalyzing heterolysis of the O–O bond in peroxo-level intermediates. As a result, bond homolyses, generating highly destructive hydroxyl radicals and/or release of free H$_2$O$_2$, are important processes in ORR catalysis by these porphyrins.

A bis-Co cofacial porphyrin, reported in the early 1980s, is among the best molecular catalysts ever found for the ORR in acidic media in terms of overpotential (about

0.55 V) and selectivity ($n_{av} \approx 3.9$). It appears fairly well accepted that in the very best cofacial porphyrin catalysts, bimetallic cooperativity plays a critical role. The catalytic mechanism remains to be adequately elucidated, complicating rational improvement of these fascinating compounds. Cofacial porphyrins remain too unstable and prohibitively expensive for practical applications.

Biomimetic metalloporphyrins provide examples of some of the best molecular ORR catalysts in neutral (pH 7–8) media, with a 0.55 V overpotential and $n_{av} > 3.9$ being retained over $>10^4$ turnovers. Among all metalloporphyrin ORR catalysts, ORR catalysis by biomimetic metalloporphyrins is the best understood mechanistically. In the context of fuel cell catalysis, further development of biomimetic catalysts needs to focus on (i) simplifying the synthesis and identifying the stereoelectronic moieties that are most critical to the catalysis; (ii) increasing the $Fe^{III/II}$ potential, even at the expense of O_2 affinity; (iii) decreasing the affinity of the ferrous porphyrin to H_2O; and (iv) increasing the turnover number by at least 100-fold.

Reduction of O_2 at a metalloporphyrin site appears to proceed invariably through partially reduced oxygen intermediates, such as bound peroxide. This mechanism, however, does not necessitate large overpotentials or low selectivity, since coordination of peroxide to one or more metal ions can stabilize it both relative to O_2 (making the $O_2 \to O_2^{2-}$ conversion nearly thermoneutral) and relative to free H_2O_2 (making the two-electron reduction kinetically unfavorable). The efficiency of ORR catalysis in monomeric metalloporphyrins seems to be controlled in large part by a ligand *trans* to bound O_2. Hence, understanding how to control this axial ligation in a cost-effective manner may be a promising strategy to develop metalloporphyrins that would be of use in low temperature fuel cells.

REFERENCES

Abramson J, Riistama S, Larsson G, Jasaitis A, Svensson-Ek M, Laakkonen L, Puustinen A, Iwata S. 2000. The structure of the ubiquinol oxidase from *Escherichia coli* and its ubiquinone binding site. Nat Struct Biol 7: 910.

Abramson J, Svensson-Ek M, Byrne B, Iwata S. 2001. Structure of cytochrome *c* oxidase: A comparison of the bacterial and mitochondrial enzymes. Biochim Biophys Acta 1544: 1.

Adzic R. 1998. Recent advances in the kinetics of oxygen reduction. In: Lipkowski J, Ross PN, editors. Electrocatalysis. New York: Wiley. p. 197.

Alberts B, Johnson A, Lewis J, Raff M, Walter P. 2002. Molecular Biology of the Cell. 4th ed. New York: Garland Science.

Andrieux CP, Saveant J-M. 1992. Catalysis at redox polymer coated electrodes. In: Murray RW, editor. Molecular Design of Electrode Surfaces. New York: Wiley. p. 207.

Anson FC, Ni CL, Saveant JM. 1985. Electrocatalysis at redox polymer electrodes with separation of the catalytic and charge propagation roles. Reduction of dioxygen to hydrogen peroxide as catalyzed by cobalt(II) tetrakis(4-*N*-methylpyridyl)porphyrin. J Am Chem Soc 107: 3442.

Anson FC, Shi C, Steiger B. 1997. Novel multinuclear catalysts for the electroreduction of dioxygen directly to water. Acc Chem Res 30: 437.

Araki K, Toma HE. 2006. Supramolecular porphyrins as electrocatalysts. In: Zagal JH, Bedioui F, Dodelet J-P, editors. N_4-Macrocyclic Metal Complexes. New York: Springer. p. 255.

Bard AJ, Faulkner LR. 2001. Electrochemical Methods. New York: Wiley.

Barraclough CG, Lawrance GA, Lay PA. 1978. Characterization of binuclear μ-peroxo and μ-superoxo cobalt(III) amine complexes from Raman spectroscopy. Inorg Chem 17: 3317.

Berglund GI, Carlsson GH, Smith AT, Szöke H, Henriksen A, Hajdu J. 2002. The catalytic pathway of horseradish peroxidase at high resolution. Nature 417: 463.

Bettelheim A, Parash R, Ozer D. 1982. Catalysis of oxygen cathodic reduction by adsorbed iron(III)-tetra(N,N,N-trimethylanilinium)porphyrin on glassy carbon electrodes. J Electrochem Soc 129: 2247.

Bianchini C, Zoellner RW. 1997. Activation of dioxygen by cobalt group metal complexes. AdvInorg Chem 44: 263.

Blomberg MRA, Siegbahn PEM. 2006. Quantum chemistry applied to the mechanisms of transition metal containing enzymes—Cytochrome c oxidase, a particularly challenging case. J Comput Chem 27: 1373.

Boulatov R. 2004. Understanding the reaction that powers this world: Biomimetic studies of respiratory O_2 reduction by cytochrome oxidase. Pure Appl Chem 76: 303.

Boulatov R. 2006. Billion-years old oxygen cathode that actually works: Respiratory oxygen reduction and its biomimetic analogs. In: Zagal JH, Bedioui F, Dodelet J-P, editors. N_4-Macrocyclic Metal Complexes. New York: Springer. p. 1.

Boulatov R, Collman JP, Shiryaeva IM, Sunderland CJ. 2002. Functional analogs of the O_2 reduction site of cytochrome oxidase: Mechanistic aspects and possible effects of Cu_B. J Am Chem Soc 124: 11923

Branden G, Gennis RB, Brzezinski P. 2006. Transmembrane proton translocation by cytochrome c oxidase. Biochim Biophys Acta 1757: 1052.

Brzezinski P, Adelroth P. 2006. Design principles of proton-pumping haem–copper oxidases. Curr Opin Struct Biol 16: 465.

Buttry DA, Anson FC. 1984. New strategies for electrocatalysis at polymer-coated electrodes. Reduction of dioxygen by cobalt porphyrins immobilized in Nafion coatings on graphite electrodes. J Am Chem Soc 106: 59.

Collman JP, Boulatov R. 2002. Electrocatalytic O_2 reduction by synthetic analogs of the heme/ Cu site of cytochrome oxidase incorporated in a lipid film. Angew Chem Int Ed 41: 3487.

Collman JP, Denisevich P, Konai Y, Marrocco M, Koval C, Anson FC. 1980. Electrode catalysis of the four-electron reduction of oxygen to water by dicobalt face-to-face porphyrins. J Am Chem Soc 102: 6027.

Collman JP, Anson FC, Barnes CE, Bencosme CS, Geiger T, Evitt ER, Kreh RP, Meier K, Pettman RB. 1983a. Further studies of the dimeric β-linked "face-to-face four" porphyrin: FTF4. J Am Chem Soc 105: 2694.

Collman JP, Bencosme CS, Barnes CE, Miller BD. 1983b. Two new members of the dimeric β-linked face-to-face porphyrin family: FTF4* and FTF3. J Am Chem Soc 105: 2704.

Collman JP, Hendricks NH, Kim K, Bencosme CS. 1987. The role of Lewis acids in promoting the electrocatalytic four-electron reduction of dioxygen. J Chem Soc Chem Commun 1537.

Collman JP, Hendricks NH, Leidner CR, Ngameni E, L'Her M. 1988. Multilayer activity and implications of hydrogen peroxide in the catalytic reduction of dioxygen by a dicobalt cofacial bis(porphyrin) (Co_2FTF4). Inorg Chem 27: 387.

Collman JP, Hutchison JE, Lopez MA, Tabard A, Guilard R, Seok WK, Ibers JA, L'Her M. 1992. Synthesis and characterization of a superoxo complex of the dicobalt cofacial diporphyrin [(μ-O$_2$)Co$_2$(DPB)(1,5-diphenylimidazole)$_2$][PF$_6$], the structure of the parent dicobalt diporphyrin Co$_2$(DPB), and a new synthesis of the free-base cofacial diporphyrin H$_4$(DPB). J Am Chem Soc 114: 9869.

Collman JP, Wagenknecht PS, Hutchison JE. 1994. Molecular catalysts for multielectron redox reactions of small molecules: The "Cofacial metallodiporphyrin" approach. Angew Chem Int Ed 33: 1537.

Collman JP, Ennis MS, Offord DA, Chng LL, Griffin JH. 1996. Electrocatalytic reduction of dioxygen by diruthenium cofacial diporphyrins axially-bound to a gold-supported, self-assembled monolayer. Inorg Chem 35: 1751.

Collman JP, Fu L, Herrmann PC, Zhang X. 1997. A functional model related to cytochrome c oxidase and its electrocatalytic four-electron reduction of O$_2$. Science 275: 949.

Collman JP, Berg KE, Sunderland CJ, Aukauloo A, Vance MA, Solomon EI. 2002a. Distal metal effects in cobalt porphyrins related to CcO. Inorg Chem 16: 6583.

Collman JP, Boulatov R, Shiryaeva IM, Sunderland CJ. 2002b. The distal Cu ion protects functional heme/Cu analogs of cyctochrome oxidase from inhibition by cyanide and CO. Angew Chem Int Ed 41: 4139.

Collman JP, Sunderland CJ, Boulatov R. 2002c. Biomimetic studies of terminal oxidases: Trisimidazole picket metalloporphyrins. Inorg Chem 41: 2282.

Collman JP, Boulatov R, Sunderland CJ, Shiryaeva IM, Berg KE. 2002d. Electrochemical metalloporphyrin-catalyzed reduction of chlorite. J Am Chem Soc 124: 10670.

Collman JP, Boulatov R, Sunderland CJ. 2003a. Functional and structural analogs of the dioxygen reduction site in terminal oxidases. In: Kadish KM, Smith KM, Guilard R, editors. The Porphyrin Handbook. Boston: Academic Press. p. 1.

Collman JP, Shiryaeva IM, Boulatov R. 2003b. Effect of electron availability on selectivity of O$_2$ reduction by synthetic monometallic Fe porphyrins. Inorg Chem 42: 4807.

Collman JP, Sunderland CJ, Berg KE, Vance MA, Solomon EI. 2003c. Spectroscopic evidence for a heme-superoxide/Cu(I) intermediate in a functional model of cytochrome c oxidase. J Am Chem Soc 125: 6648.

Collman JP, Boulatov R, Sunderland CJ. 2004a. Functional analogs of cytochrome c oxidase, myoglobin and hemoglobin. Chem Rev 104: 561.

Collman JP, Kaplun M, Sunderland CJ, Boulatov R. 2004b. Electrocatalytic reduction of ROOH. J Am Chem Soc 126: 11166.

Collman JP, Decreau RA, Yan Y, Yoon J, Solomon EI. 2007a. Intramolecular single-turnover reaction in a cytochrome c oxidase model bearing a Tyr244 mimic. J Am Chem Soc 129: 5794.

Collman JP, Devaraj NK, Decréau RA, Yang Y, Yan Y-L, Ebina W, Eberspacher TA, Chidsey CED. 2007b. A cytochrome c oxidase model catalyzes oxygen to water reduction under rate-limiting electron flux. Science 315: 1565.

Deronzier A, Moutet J-C. 2003. Electrochemical reactions catalyzed by transition metal complexes. In: Ward MD, editor. Comprehensive Coordination Chemistry, II. Volume 9. Amsterdam: Elsevier. p. 471.

Dimroth P, Ballmoos Cv, Meier T, Kaim G. 2003. Electrical power fuels rotary ATP synthase. Structure 11: 1469.

Dodelet J-P. 2006. Oxygen reduction in pem fuel cell conditions: Heat-treated non-precious metal-N_4 macrocycles and beyond. In: Zagal JH, Bedioui F, Dodelet J-P, editors. N_4-Macrocyclic Metal Complexes. New York: Springer. p. 83.

Durand RR Jr, Bencosme CS, Collman JP, Anson FC. 1983. Mechanistic aspects of the catalytic reduction of dioxygen by cofacial metalloporphyrins. J Am Chem Soc 105: 2710.

EIA 2007. http://tonto.eia.doe.gov/dnav/pet/pet_move_impcus_a2_nus_ep00_im0_mbbl_m.htm.

Ferguson-Miller S, Babcock GT. 1996. Heme/copper terminal oxidases. Chem Rev 96, 2889–2907.

Forshey PA, Kuwana T. 1983. Electrochemistry of oxygen reduction. 4. Oxygen to water conversion by iron(II)(tetrakis(N-methyl-4-pyridyl)porphyrin) via hydrogen peroxide. Inorg Chem 22: 699.

Frauenfelder H, McMahon BH, Fenimore PW. 2003. Myoglobin: The hydrogen atom of biology and a paradigm of complexity. Proc Natl Acad Sci USA 100: 8615.

Fukuzumi S, Mochizuki S, Tanaka T. 1990. Efficient catalytic systems for electron transfer from an NADH model compound to dioxygen. Inorg Chem 29: 653.

Fukuzumi S, Okamoto K, Gros CP, Guilard R. 2004. Mechanism of four-electron reduction of dioxygen to water by ferrocene derivatives in the presence of perchloric acid in benzonitrile, catalyzed by cofacial dicobalt porphyrins. J Am Chem Soc 126: 10441.

Gadamsetti K, Swavey S. 2006. Electrocatalytic reduction of oxygen at electrodes coated with a bimetallic cobalt(II)/platinum(II) porphyrin. J Chem Soc Dalton Trans 5530.

Guilard R, Brandes S, Tardieux C, Tabard A, L'Her M, Miry C, Gouerec P, Knop Y, Collman JP. 1995. Synthesis and characterization of cofacial metallodiporphyrins involving cobalt and lewis acid metals: New dinuclear multielectron redox catalysts of dioxygen reduction. J Am Chem Soc 117: 11721.

Hosler JP, Ferguson-Miller S, Mills DA. 2006. Energy transduction: Proton transfer through the respiratory complexes. Annu Rev Biochem 75: 165.

Huie RE, Neta P. 1999. Chemistry of reactive oxygen species. In: Gilbert DL, Colton CA, editors. Reactive oxygen species in biological systems. New York: Kluwer. p. 33.

Hutchison JE, Postlethweite TA, Murray RW. 1993. Molecular films of thiol-derivatized tetraphenylporphyrins on gold: Film formation and electrocatalytic dioxygen reduction. Langmuir 9: 3277.

Hutchison JE, Postlethwaite TA, Chen C-h, Hathcock KW, Ingram RS, Ou W, Linton RW, Murray RW, Tyvoll DA, Chng LL, Collman JP. 1997. Electrocatalytic activity of an immobilized cofacial diporphyrin depends on the electrode material. Langmuir 13: 2143.

Jasinski R. 1964. A new fuel cell cathode catalyst. Nature 201: 1212.

Junemann S. 1997. Cytochrome *bd* terminal oxidase. Biochim Biophys Acta 1321: 107.

Kadish KM, Shao J, Ou Z, Zhan R, Burdet F, Barbe J-M, Gros CP, Guilard R. 2005. Electrochemistry and spectroelectrochemistry of heterobimetallic porphyrin-corrole dyads. Influence of the spacer, metal ion, and oxidation state on the pyridine binding ability. Inorg Chem 44, 9023–9038.

Khorasani-Motlagh M, Noroozifar M, Ghaemi A, Safari N. 2004. Iron(III) octaethylporphyrin chloride supported on glassy carbon as an electrocatalyst for oxygen reduction. J Electroanal Chem 565: 115.

Khvostichenko D, Yang Q, Boulatov R. 2007. Simple heme dimers with strongly cooperative ligand binding. Angew Chem Int Ed 46: 8368.

Knoll AH. 2003. Life on a Young Planet. Princeton: Princeton University Press.

Kobayashi N, Nishiyama Y. 1984. Catalytic electroreduction of molecular oxygen at glassy carbon electrodes with immobilized iron porphyrins containing zero, one, or four amino groups. J Electroanal Chem 181: 107.

Kobayashi N, Osa T. 1983. Catalytic electroreduction of molecular oxygen using meso-tetrakis(p-carboxyphenyl)porphinatoiron in water. J Electroanal Chem 157: 269.

Kobayashi N, Lam H, Nevin WA, Janda P, Leznoff CC, Lever ABP. 1990. Electrochemistry and spectroelectrochemistry of 1,8-naphthalene- and 1,8-anthracene-linked cofacial binuclear metallophthalocyanines. New mixed-valence metallophthalocyanines. Inorg Chem 29: 3415.

Le Mest Y, L'Her M. 1995. Electrochemical generation of a new type of dioxygen carrier complex. Reversible fixation of dioxygen by the highly electron-deficient two-electron oxidized derivative of a dicobalt face-to-face diporphyrin. J Chem Soc Chem Commun 1441.

Le Mest Y, L'Her M, Hendricks NH, Kim K, Collman JP. 1992. Electrochemical and spectroscopic properties of dimeric cofacial porphyrins with nonelectroactive metal centers. Delocalization processes in the porphyrin π-cation-radical systems. Inorg Chem 31: 835

Le Mest Y, L'Her M, Saillard JY. 1996. Electrochemical and spectroscopic behavior of dicobalt cofacial diporphyrins. The redox sites revisited. Inorg Chim Acta 248: 181.

Le Mest Y, Inisan C, Laouenan A, L'Her M, Talarmin J, El Khalifa M, Saillard J-Y. 1997. Reactivity toward dioxygen of dicobalt face-to-face diporphyrins in aprotic media. Experimental and theoretical aspects. Possible mechanistic implication in the reduction of dioxygen. J Am Chem Soc 119: 6095.

Ligeza A, Tikhonov AN, Hyde JS, Subczynski WK. 1998. Oxygen permeability of thylakoid membranes: Electron paramagnetic resonance spin labeling study. Biochim Biophys Acta 1365: 453.

Liu HY, Abdalmuhdi I, Chang CK, Anson FC. 1985. Catalysis of the electroreduction of dioxygen and hydrogen peroxide by an anthracene-linked dimeric cobalt porphyrin. J Phys Chem 89: 665.

Loetzbeyer T, Schuhmann W, Schmidt H-L. 1995. Direct electrocatalytic H_2O_2 reduction with hemin covalently immobilized at a monolayer-modified gold electrode. J Electroanal Chem 395: 341.

Loew GH, Harris DL. 2000. Role of the heme active site and protein environment in structure, spectra and function of the cytochrome P450s. Chem Rev 100: 407.

Marusak RA, Mears CF. 1995. Exploration of selected pathways for metabolic oxidative ring opening of benzene based on estimates of molecular energetics In: Valentine JS, Foote CS, Greenberg A, Liebman JE, editors. Active Oxygen in Biochemistry. New York: Chapman and Hall. p. 336.

Michel H. 1999. Cytochrome c oxidase: Catalytic cycle and mechanisms of proton pumping— A discussion. Biochemistry 38: 15129.

Mills DA, Florens L, Hiser C, Qian J, Ferguson-Miller S. 2000. Where is 'outside' in cytochrome c oxidase and how and when do protons get there? Biochimica et Biophysica Acta, Bioenergetics 1458, 180–187.

Moodie AD, Ingledew WJ. 1990. Microbial anaerobic respiration. Adv Microb Physiol 31: 225.

Ni CL, Abdalmuhdi I, Chang CK, Anson FC. 1987. Behavior of four anthracene-linked dimeric metalloporphyrins as electrocatalysts for the reduction of dioxygen. J Phys Chem 91: 1158.

Nicholls DG. 1982. Bioenergetics: An Introduction to the Chemiosmotic Theory. London: Academic Press.

Opekar F, Beran P. 1976. Rotating disk electrodes. J Electroanal Chem 69: 1.

Ostermeier C, Harrenga A, Ermler U, Michel H. 1997. Structure at 2.7 Å resolution of the *Paracoccus denitrificans* two-subunit cytochrome c oxidase complexed with an antibody Fv fragment. Proc Natl Acad Sci USA 94: 10547.

Oyaizu K, Haryono A, Natori J, Shinoda H, Tsuchida E. 2000. Electroreduction of μ-oxo iron(III) porphyrins adsorbed on an electrode leading to a cofacial geometry for the iron(ii) complex: Unexpected active site for the catalytic reduction of O_2 to H_2O. Bull Chem Soc Jpn 73: 1153.

Park GJ, Nakajima S, Osuka A, Kim K. 1995. Electrocatalytic four-electron reduction of dioxygen by 1,2-phenylene-bridged dicobalt diporphyrins. Chem Lett 255.

Pereira MM, Carita JN, Teixeira M. 1999. Membrane-bound electron transfer chain of the thermohalophilic bacterium *Rhodothermus marinus*: Characterization of the iron–sulfur centers from the dehydrogenases and investigation of the high-potential iron–sulfur protein function by in vitro reconstitution of the respiratory chain. Biochemistry 38: 1276.

Pereira MM, Santana M, Teixeira M. 2001. A novel scenario for the evolution of haem–copper oxygen reductases. Biochim Biophys Acta 1505: 185.

Pitcher RS, Watmough NJ. 2004. The bacterial cytochrome cbb_3 oxidases. Biochim Biophys Acta 1655: 388.

Postlethwaite TA, Hutchison JE, Hathcock KW, Murray RW. 1995. Optical, electrochemical and electrocatalytic properties of self-assembled thiol-derivatized porphyrins on transparent gold-films. Langmuir 11: 4109.

Proniewicz LM, Odo J, Goral J, Chang CK, Nakamoto K. 1989. Resonance Raman spectra of dioxygen adducts of pillared cobalt cofacial diporphyrins. J Am Chem Soc 111: 2105.

Proniewicz LM, Paeng IR, Nakamoto K. 1991. Resonance raman spectra of two isomeric dioxygen adducts of iron(II) porphyrins and π-cation radical and nonradical oxoferryl porphyrins produced in dioxygen matrixes: Simultaneous observation of more than seven oxygen isotope sensitive bands J Am Chem Soc 113: 3294.

Ricard D, Didier A, L'Her M, Boitrel B. 2001. Application of 3-quinolinoyl picket porphyrins to the electroreduction of dioxygen to water: Mimicking the active site of cytochrome c oxidase. ChemBioChem 2: 144.

Rich PR, Moody AJ. 1997. Cytochrome c oxidase. In: Graber P, Milazzo G, editors. Bioenergetics. Basel: Birkhauser. p. 418.

Rogers MS, Dooley DM. 2001. Posttranslationally modified tyrosines from galactose oxidase and cytochrome c oxidase. Adv Protein Chem 58: 387.

Rosenthal J, Nocera DG. 2007. Role of proton-coupled electron transfer in O–O bond activation. Acc Chem Res 40: 543.

Schlichting I, Berendzen J, Chu K, Stock AM, Maves SA, Benson DE, Sweet RM, Ringe D, Petsko GA, Sligar SG. 2000. The catalytic pathway of cytochrome P450cam at atomic resolution. Science 287: 1615.

Schults BE, Chan SI. 2001. Structures and proton-pumping strategies of mitochondrial respiratory enzymes. Annu Rev Biophys Biomol Struct 30: 23.

Senios AE, Nadanaciva S, Weber J. 2002. The molecular mechanism of ATP synthesis by F_1F_0-ATP synthase. Biochim Biophys Acta 1553: 188.

Shi C, Anson FC. 1990. Catalytic pathways for the electroreduction of oxygen by iron tetrakis(4-N-methylpyridyl)porphyrin or iron tetraphenylporphyrin adsorbed on edge plane pyrolytic graphite electrodes. Inorg Chem 29: 4298.

Shi C, Mak KW, Chan K-S, Anson FC. 1995. Enhancement by surfactants of the activity and stability of iridium octaethyl porphyrin as an electrocatalyst for the four-electron reduction of dioxygen. J Electroanal Chem 397: 321.

Shi C, Steiger B, Yuasa M, Anson FC. 1997. Electroreduction of O_2 to H_2O at unusually positive potentials catalyzed by the simplest of the cobalt porphyrins. Inorg Chem 36: 4294.

Shigehara K, Anson FC. 1982. Electrocatalytic activity of three iron porphyrins in the reduction of dioxygen and hydrogen peroxide at graphite cathodes. J Phys Chem 86: 2776.

Shikama K. 1998. The molecular mechanism of autoxidation for myoglobin and hemoglobin: A venerable puzzle. Chem Rev 98: 1357.

Shiryaeva IM, Collman JP, Boulatov R, Sunderland CJ. 2003. Nonideal electrochemical behavior of biomimetic iron porphyrins: Interfacial potential distribution across multilayer films. Anal Chem 79: 494.

Song E, Shi C, Anson FC. 1998. Comparison of the behavior of several cobalt porphyrins as electrocatalysts for the reduction of O_2 at graphite electrodes. Langmuir 14: 4315.

Sono M, Roach MP, Coulter ED, Dawson JH. 1996. Heme-containing oxygenases. Chem Rev 96: 2841.

Strekas TC, Spiro TG. 1975. Resonance Raman spectra of superoxide-bridged binuclear complexes. μ-Superoxo-decacyanodicobaltate(5 −) and μ-superoxo-decaamminedicobalt (5 +). Inorg Chem 14: 1421.

Su YO, Kuwana T, Chen SM. 1990. Electrocatalysis of oxygen reduction by water-soluble iron porphyrins. Thermodynamic and kinetic advantage studies. J Electroanal Chem 288: 177.

Taube H. 1986. Interaction of dioxygen species and metal ions—Equilibrium aspects. Prog Inorg Chem 34: 607.

Tsukihara T, Aoyama H, Yamashita E, Tomizaki T, Yamaguchi H, Shinzawa-Itoh K, Nakashima R, Yaono R, Yoshikawa S. 1996. The whole structure of the 13-subunit oxidized cytochrome c oxidase at 2.8 Å. Science 272: 1136.

Vasudevan P, Santosh, Mann N, Tyagi S. 1990. Transition metal complexes of porphyrins and phthalocyanines as electrocatalysts for dioxygen reduction. Transition Metal Chemistry, 15, 81–90.

Wan G-X, Shigehara K, Tsuchida E, Anson FC. 1984. Virtues of a copolymer containing pyrrolidone and iron porphyrin groups in the catalysis of the reduction of dioxygen at graphite electrodes. J Electroanal Chem 179: 239.

Watanabe Y. 2000. High-valent intermediates. In: Kadish KM, Smith KM, Guilard R, editors. The Porphyrin Handbook. San Diego: Academic Press. p. 97.

Wikstrom M. 2004. Cytochrome c oxidase. Biochim Biophys Acta 1655: 241.

Wittenberg JB, Wittenberg BA. 2003. Myoglobin function reassessed. J Exp Biol 206: 2011.

Yamazaki S, Yamada Y, Ioroi T, Fujiwara N, Siroma Z, Yasuda K, Miyazaki Y. 2005. Estimation of specific interaction between several co porphyrins and carbon black: Its influence on the electrocatalytic O_2 reduction by the porphyrins. J Electroanal Chem 576: 253.

Yang Q, Khvostichenko D, Atkinson J, Boulatov R. 2008. Simple dimer containing dissociatively stable mono-imidazole ligated ferrohemes. Chem Commun (8): 963.

Yeager E. 1984. Electrocatalysis for O_2 reduction. Electrochim Acta 29: 1527.

Yoshida M, Muneynki E, Hisabori T. 2001. ATP synthase—A marvelous rotary engine of the cell. Nature Reviews Molecular Cell Biology 2, 669–677.

Yuasa M, Nishihara R, Shi C, Anson FC. 2001. A comparison of several meso-tetraalkyl cobalt porphyrins as catalysts for the electroreduction of dioxygen. Polym Adv Technol 12: 266.

Zagal JH. 1992. Metallophthalocyanines as catalysts in electrochemical reactions. Coord Chem Rev 119, 86–136.

Zagal JH. 2003. Macrocycles. In: Vielstich W, Lamm A, Gasteiger HA, editors. Handbook of Fuel Cells—Fundamentals, Technology and Applications. Chichester: Wiley. p. 544.

Zagal JH, Bedioui F, Dodelet J-P, editors. 2006. N_4-Macrocyclic Metal Complexes. New York: Springer.

INDEX

Ab initio thermodynamics, 129–155
Acetaldehyde oxidation, 196–197, 624
Acetic acid, 192–198, 394–395
Active sites in electrocatalysis, 93–124, 159–198, 237, 250, 253
Adiabatic and non-adiabatic electron transfer, 34
Alcohol oxidation by enzymes, 610–613
Alloy/bimetallic catalysts, 6–7, 70–71, 245–266, 317–337
Anderson–Newns Hamiltonian, 33–34
Anion adsorption effects, 143, 174–175, 208–239, 254, 281–283, 336, 525, 535–536
Au(100) surface reconstruction, 141–148, 248
Auger Electron Spectroscopy, 255, 259, 468

Bilirubin oxidase, 603–606, 621–626
Biomimetic catalysts, 679–686
Bond-breaking electron transfer reactions, 43–44

Carbon corrosion, 300
Carbon monoxide adsorption, 248, 250, 255, 325–327, 347, 386–391, 528–532
Carbon monoxide oxidation
  By adatom-modified platinum surfaces, 232–235
  By gold, 175–176, 571–572, 580–583
  By platinum, 161–173, 176–177, 382–386, 395–396, 417–425, 539–544, 570–571
  By platinum–ruthenium, 325–327, 396–399, 484–496
  By platinum-tin, 255–257

By rhodium, 173–175
DFT calculations, 118–120
Carbon monoxide tolerance, 318–320, 330, 346, 465, 618–620
Catalyst degradation, 301, 302
Catalyst stability, 7, 300–303, 305, 330
Catalyst utilization, 3–5
Cofacial porphyrins, 363, 653, 657, 663–675
Computational electrocatalyst screening, 77–87
Core-shell nanoparticles, 7
Cytochrome c, 603, 610, 637–647
Cytochrome c oxidase, 603, 621, 637–647

d band, 45, 49–52, 259–261, 272–273, 323, 327, 524, 537, 550
Density functional theory (DFT) calculations, 57–87, 93–124, 141–155, 166, 176, 184, 190, 191, 275, 284, 290, 292, 298, 299, 470, 483, 493
Density of states, 46, 70, 261, 272, 515
Descriptor, 12, 79–83, 107, 111, 291, 293
Differential Electrochemical Mass Spectrometry (DEMS), 179–180, 186, 188, 191, 411–453

Electric field effects in DFT, 76–77, 98–100, 145–146
Electrochemical Quartz Crystal Microbalance (EQCM), 320–321
Enzyme catalysts, 593–629, 637–647
Enzyme immobilization, 600–604, 614
Ethanol oxidation, 192–196, 334–336, 355–359, 606, 613–614, 624–626

Fuel Cell Catalysis. Edited by Marc T. M. Koper
Copyright © 2009 John Wiley & Sons, Inc.

695

Experimental electrocatalyst screening, 77–87
Extended Hückel model, 48

Formaldehyde oxidation, 185, 192, 425–432, 436, 438–440, 441–451
Formic acid oxidation, 177–184, 235–239, 264–265, 392–393, 425–429, 435–438, 441–451, 544–545
Fourier Transform Infrared FTIR Spectroscopy, 179, 187, 194, 196, 233, 251–253, 255, 323, 356–358, 378, 530, 544, 548

Glucose oxidation, 596–628
Gold electrocatalyst, 175–176, 184, 192, 195, 567–586

Hammer–Nørskov d-band model, 70, 272–273, 327
Heme-copper oxidase, 610
High Throughput Synthesis of Nanoparticles, 572–574
Hydrogen (underpotential) adsorption, 60–63, 254, 471–484, 526
Hydrogen evolution reaction (HER), 31, 79–87
Hydrogen oxidation reaction (HOR), 52, 327–330, 531–533, 596–600, 615–620
Hydrogen peroxide, 273–275, 331–332, 366, 534, 570, 578
Hydrogenase, 615–620

Impedance Spectroscopy, 21–22
Inner sphere electron transfer, 47–48
Ion transfer reaction, 39–40

Laccase, 598–610
Low Energy Electron Diffraction (LEED), 246–247, 471–472
Low Energy Ion Scattering (LEIS), 250–251

Koutecky–Levich plot, 288, 319, 534, 648–651

Marcus theory, 31, 37, 38
Mean-field approximation, 61, 163
Metalloporphyrin catalysts, 362–366, 637–686

Methanol oxidation reaction (MOR), 184–192, 237, 346–359, 391–392, 433–434, 437, 440–451, 545–549, 595, 625
  DFT calculations, 114–118
  Reaction mechanism, 181, 412, 448
Methanol tolerance, 298, 300, 362, 549, 606, 615
Mixed-metal monolayer catalysts, 292–296
Molecular catalysts, 362–366, 637–686
Monte Carlo simulations, 61, 174

Nanoparticle size effects in electrocatalysis, 3, 186, 198, 330–331, 507–552, 567–586
Nanoparticle structure, 263, 512–516
Nanoparticle thermodynamics, 509–511
Nucleation and growth model for CO oxidation, 163

Ostwald ripening, 304, 511
Oxygen reduction reaction (ORR), 1–28, 66–73, 259–263, 271–306, 330–336, 359–366, 533–538, 568–570, 578–580, 637–686
  By enzymes, 598–610
  By metal-macrocycle catalysts, 25–26, 362–366, 637–686
  DFT calculations, 66–73, 112–113, 120–123
  Influence of surface oxidation, 12–28
  Potential energy surface, 66–73
  Reaction order, 21–22
  Tafel slope, 18–20, 276–277, 297
Oxygen tolerance, 618–620
Outer sphere electron transfer, 33–38

Palladium electrocatalysts, 183
Palladium-alloy electrocatalysts, 298–300
Pareto-optimal plot, 85
Platinum-alloy electrocatalysts, 6, 70–71, 284–288, 317–337
  Platinum–bismuth, 86–87, 224
  Platinum–chromium, 361–362
  Platinum–cobalt, 71, 257–260, 319, 321–330, 334–335
  Platinum–iron, 319, 321, 334–335
  Platinum–molybdenum, 253, 319–320

Platinum–nickel, 71–73, 260–263, 334–335
Platinum–ruthenium, 118, 253, 257, 319–330, 346–354, 396–402, 465–498
Platinum–tin, 254, 256–257, 320
Platinum-monolayer electrocatalysts, 288–292, 488–494
Platinum nanoparticle agglomeration, 300, 541
Platinum (nanoparticle) dissolution, 301–306
Platinum–palladium layers, 264–265
Platinum-skin catalysts, 70–71, 77, 257–263, 321–323, 335–336
Platinum surface modification by adatoms, 208–239
Platinum surface oxidation, 16, 71–73, 109, 148–155, 276–280, 528, 530
Potential of maximum entropy, 229–230
Potential of zero charge, 98, 142, 227, 524, 529
Pourbaix diagram, 28, 63–65, 301
Proton transfer reactions, 41–42

Rhodium electrocatalyst, 173–175, 195–196
Rotating (ring-)disk electrode, 256, 260, 265, 318–320, 521, 648
Ruthenium electrocatalyst, 465–498

Sabatier principle, 69, 77, 79, 273, 291, 496
Scanning Tunneling Microscopy (STM), 249–250, 321–322, 475, 476, 478, 482

Scanning Electron Microscopy (SEM), 249–250, 397
Stepped surfaces, 163–164, 172–174, 176, 188, 190, 195–197, 536
Sum Frequency Generation (SFG), 375–403
Support effects in electrocatalysis, 567–586
Surface diffusion, 163, 173–177
Surface Enhanced Infrared Reflection Adsorption Spectroscopy (SEIRAS), 183
Surface Enhanced Raman Spectroscopy (SERS), 176, 184, 194
Surface X-Ray Scattering (SXS), 247–248

Tafel plots and Tafel slopes
  For carbon monoxide oxidation, 164–166, 175
  For formic acid oxidation, 182
  For methanol oxidation, 189, 350–351
  For oxygen reduction, 18–20, 276–277, 297, 364–365, 522, 534, 538
Theoretical Standard Hydrogen Electrode, 58–60, 101
Transmission Electron Microscopy (TEM), 263, 514

Ultra-High Vacuum (UHV) Techniques, 246–247, 465–498

Volcano plot, 9, 27, 71, 81, 259–260, 264, 273, 289, 291

Water in the double layer, 74–75, 103–105, 148